International Association
of Geodesy Symposia

*Fernando Sansò, Series Editor*

# International Association of Geodesy Symposia

*Fernando Sansò, Series Editor*

---


*Symposium 101:* Global and Regional Geodynamics
*Symposium 102:* Global Positioning System: An Overview
*Symposium 103:* Gravity, Gradiometry, and Gravimetry
*Symposium 104:* Sea SurfaceTopography and the Geoid
*Symposium 105:* Earth Rotation and Coordinate Reference Frames
*Symposium 106:* Determination of the Geoid: Present and Future
*Symposium 107:* Kinematic Systems in Geodesy, Surveying, and Remote Sensing
*Symposium 108:* Application of Geodesy to Engineering
*Symposium 109:* Permanent Satellite Tracking Networks for Geodesy and Geodynamics
*Symposium 110:* From Mars to Greenland: Charting Gravity with Space and Airborne Instruments
*Symposium 111:* Recent Geodetic and Gravimetric Research in Latin America
*Symposium 112:* Geodesy and Physics of the Earth: Geodetic Contributions to Geodynamics
*Symposium 113:* Gravity and Geoid
*Symposium 114:* Geodetic Theory Today
*Symposium 115:* GPS Trends in Precise Terrestrial, Airborne, and Spaceborne Applications
*Symposium 116:* Global Gravity Field and Ist Temporal Variations
*Symposium 117:* Gravity, Geoid and Marine Geodesy
*Symposium 118:* Advances in Positioning and Reference Frames
*Symposium 119:* Geodesy on the Move
*Symposium 120:* Towards an Integrated Global Geodetic Observation System (IGGOS)
*Symposium 121:* Geodesy Beyond 2000: The Challenges of the First Decade
*Symposium 122:* IV Hotine-Marussi Symposium on Mathematical Geodesy
*Symposium 123:* Gravity, Geoid and Geodynamics 2000
*Symposium 124:* Vertical Reference Systems
*Symposium 125:* Vistas for Geodesy in the New Millennium
*Symposium 126:* Satellite Altimetry for Geodesy, Geophysics and Oceanography
*Symposium 127:* V Hotine Marussi Symposium on Mathematical Geodesy
*Symposium 128:* A Window on the Future of Geodesy
*Symposium 129:* Gravity, Geoid and Space Missions

# Gravity, Geoid and Space Missions

GGSM 2004
IAG International Symposium
Porto, Portugal
August 30 - September 3, 2004

Edited by
Christopher Jekeli
Luisa Bastos
Joana Fernandes

Springer

Volume Editors

Professor Christopher Jekeli
Geodetic Science
Ohio State University
470 Hitchcock Hall
2070 Neil Ave.
Columbus, OH 43210
USA

Dr. Luisa Bastos
Faculdade de Ciências da Universidade Porto
Observatório Astronómico
Alameda do Monte da Virgem
4430-146 V.N. GAIA
Portugal

Professor Joana Fernandes
Faculdade de Ciências da Universidade Porto
Departamento de Matemática Aplicada
Rua do Campo Alegre 687
4169-007 Porto
Portugal

Series Editor

Professor Fernando Sansò
Polytechnic of Milan
D.I.I.A.R. - Surveying Section
Piazza Leonardo da Vinci, 32
20133 Milan
Italy

Library of Congress Control Number: 2005927793

**ISSN 0939-9585**
ISBN 3-540-26930-4 Springer Berlin Heidelberg New York



Springer is a part of Springer Science+Business Media
springeronline.com

Printed in Germany



Cover design: design & production GmbH, Heidelberg
Production: Almas Schimmel
Typesetting: Camera ready by the authors
Printing and Binding: Mercedes-Druck, Berlin

Printed on acid-free paper 32/3141/as 5 4 3 2 1 0

# Preface

The IAG International Symposium on Gravity, Geoid, and Space Missions 2004 (GGSM2004) was held in the beautiful city of Porto, Portugal, from 30 August to 3 September 2004. This symposium encompassed the themes of Commission 2 (Gravity Field) of the newly structured IAG, as well as interdisciplinary topics related to geoid and gravity modeling, with special attention given to the current and planned gravity-dedicated satellite missions. The symposium also followed in the tradition of mid-term meetings that were held between the quadrennial joint meetings of the International Geoid and Gravity Commissions. The previous mid-term meetings were the International Symposia on Gravity, Geoid, and Marine Geodesy (Tokyo, 1996), and Gravity, Geoid, and Geodynamics (Banff, 2000).

GGSM2004 aimed to bring together scientists from different areas in the geosciences, working with gravity and geoid related problems, both from the theoretical and practical points of view. Topics of interest included the integration of heterogeneous data and contributions from satellite and airborne techniques to the study of the spatial and temporal variations of the gravity field. In addition to the special focus on the CHAMP, GRACE, and GOCE satellite missions, attention was also directed toward projects addressing topographic and ice field mapping using SAR, LIDAR, and laser altimetry, as well as missions and studies related to planetary geodesy.

The Science Committee for the Symposium comprised Christopher Jekeli (President), Ilias N. Tziavos, Roger Haagmans, René Forsberg, Luisa Bastos, and Joana Fernandes, while its local organization was under the direction of Luisa Bastos, Joana Fernandes and Machiel Bos of the Faculty of Science, University of Porto. In addition, many colleagues associated with Commission 2 and the IAG organized the nine sessions of the Symposium as follows:

1. Gravity field modeling from satellite missions
   Pieter Visser (Delft University of Technology, The Netherlands)
   Roger Haagmans (ESA, The Netherlands)

2. Airborne and satellite gravimetry instrumentation
   René Forsberg (Geodynamics Department, Danish National Space Center)
   Luisa Bastos (University of Porto, Portugal)

3. Regional geoid modeling
   Urs Marti (Swiss Federal Office of Topography, Switzerland)
   Ilias Tziavos (Aristotle University of Thessaloniki, Greece)

4. Radar and laser surface mapping from satellites
   Philippa Berry (De Montfort University, U.K.)
   Bill Carter (University of Florida, U.S.A.)

5. Topographic data bases and gravity modeling
   Steve Kenyon (NGA, U.S.A.)
   Per Knudsen (KMS, Denmark)

6. Satellite altimetry, oceanography, and the geoid
   Dave Sandwell (University of California, San Diego, U.S.)
   Joana Fernandes (University of Porto, Portugal)

7. Terrestrial gravity instrumentation, networks, and geodynamics
   Shuhei Okubo (Earthquake Research Institute, University of Tokyo, Japan)
   Tonie van Dam (European Centre for Geodynamics and Seismology, Luxemburg)

8. Temporal gravity variations: modeling and measurements
   C.K. Shum (Ohio State University, U.S.A.)
   Martin Vermeer (Helsinki University of Technology, Switzerland)

9. Planetary gravity fields and models
   Dave Smith (NASA, U.S.A.)
   Georges Balmino (CNES, France)

The Symposium attracted 258 papers, of which 108 were presented orally and 150 as posters. It was truly an international scientific event as all six continents and 39 countries were represented by a total of 234 participants. A Proceedings of the Symposium was published in the form of a CD with most of the oral and poster presentations, as well as many corresponding journal-style papers. Of the latter, a portion were selected and reviewed for inclusion in this volume and they represent the high level of activity and advanced research in gravity and geoid modeling that was displayed by all contributions to this symposium. Although Session 9 papers were not submitted to this volume it is anticipated that the relationship between gravity field modeling and planetary geodesy particularly encouraged in this symposium will be strengthened in future symposia and other similar events.

Special recognition and gratitude go to the Session co-conveners whose hard work in organizing their sessions and guiding the reviews of the submitted papers resulted in a very successful Symposium and a high quality scientific volume. The organization of a Symposium of this magnitude is never easy, but it unfolded flawlessly due to the expert preparation and continual attentiveness of the organizing committee, headed by Luisa Bastos, Joana Fernandes and Machiel Bos. The collection of material for the CD Proceedings and the assembly of the papers for this volume was also done professionally and efficiently and special thanks are due the Faculty of Science, University of Porto, for helping to support these publications.

The success of the Symposium also depends to a great extent on the financial sponsorship of interested and supporting organizations and institutions in the form of cash and travel re-imbursements, especially for students and colleagues from developing countries. We gratefully acknowledge financial contributions from IAG, IUGG, NASA, ESA, GRICES, FCT and the University of Porto.

Finally, I wish to extend my personal appreciation and congratulations to Luisa Bastos, Joana Fernandes and Machiel Bos for their devotion to this symposium and their perseverance in completing the publications. It is not easy, even in these modern times, to coordinate such affairs over intercontinental distances, but they maintained a schedule and succeeded splendidly in every respect.

Christopher Jekeli                                        Ohio State University, March 2005

# Contents

## Session 1:

**Session 4:**

**Session 5:**



**Session 6:**



**Session 7:**

**Session 8:**

# Initial results from retracking and reprocessing the ERS-1 geodetic mission altimetry for gravity field purposes.

Ole B. Andersen, P. Knudsen,
Danish Space Center, Juliane Mariesvej 30, DK-2100 Copenhagen, Denmark. Email: oa@spacecenter.dk,

P. A. M. Berry, E. L. Mathers. EAPRS Laboratory, De Montfort University, The Gateway, Leicester. LE1 9BH, UK.

R. Trimmer, S. Kenyon
National Geospatial-Intelligence Agency, GIMG, St Louis. MS, USA

**Abstract.** All present global marine gravity fields are based on the 1Hz ERS-1 Geodetic Mission (GM) altimeter data combined with other altimetric datasets. Close to the coast (<25 km) this investigation shows that the altimetric gravity field determination degrades due to a combination of several factors, where the main reason is the degradation of the quality of the altimeter data. By starting out from the original waveform data and retracking the entire ERS-1 GM mission using a highly advanced expert based system of multiple retrackers, the return time from both open ocean and coastal sea surface as well as from all ice-covered regions within the coverage of the ERS-1 can be derived with higher accuracy than presently available.

Initial results of the combined effort to improve the ERS-1 GM altimetric dataset through retracking and regression to 2Hz (3km) are presented, and its effort on gravity field modeling close to the coast and in Polar Regions is discussed. Close to the coast and in particular in Polar Regions the use of multiple retrackers leads to considerably more and better data than in the normal 1 Hz data delivered by the European Space Agency (ESA).

Extensive comparison with marine gravity field data by the National Geospatial-Intelligence Agency is also presented to document the improvement on gravity field determination

**Keywords. Satellite altimetry, retracking, gravity field determination, coastal, polar regions.**

## 1 Introduction

One of the outstanding problems in global marine gravity field determination is the degradation of the gravity field close to the coast as well as in Polar Regions.

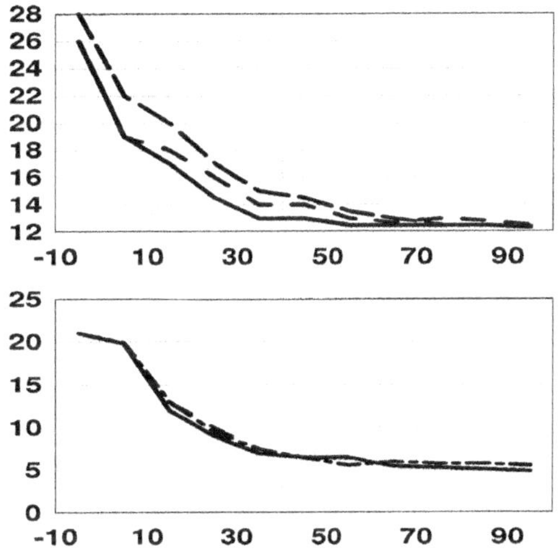


**Figure 1.** Standard deviation (mGal) of gravity comparison with marine data as a function of depth in meters (upper figure) and distance in kilometers to the coast (lower figure). The fields are KMS99 (Andersen and Knudsen, 1998) as solid line, SS 11.2 (Sandwell and Smith, 1997) long dash, and GSFC00.1 (Wang, 2000) short dash

Figure 1 shows the standard deviation of different global gravity fields as a function of depth (upper figure) and distance to the shore (lower figure) for coastal regions. The comparison with global marine gravity fields has been carried out with more than

1000 marine gravity observations in each 10 m depth or 10 km distance interval on the x-axis. In both figures the coast is to the left. For both the comparison with depth and distance, there is clear degradation in the accuracy from roughly 25 km to the coast in terms of distance, or for depths shallower than 20 meters. The degradation can be directly related to the degradation of the coverage and accuracy of the altimetric observations. Overall, the largest contributors are as follows:

1. Lack of data due to editing and automatic criteria applied to data.
2. Errors due to bad retracking of the data.
3. Change in sea state /ocean variability.
4. Errors in applied models for geophysical and range corrections, particularly the tidal model

Besides these altimetric related problems, there will be errors in the gravity field computation due to the presence of the coast, combined with the use of approximate methods which are required because the vast number of datapoints (mainly FFT methods).

The papers by Andersen et al., (1999, 2004) deal with the effect on gravity field modelling from improving the corrections applied to the altimetric observations.

In this paper we will present a way to improve the coverage of data in shallow water and Polar Regions, and the errors due to bad retracking. Several other scientific groups are currently investigating the use of retracking to enhance the quality of the altimetric observation from the ERS-1 geodetic mission altimetry. The current Sandwell and Smith (1997) global marine gravity field version 11.2 is produced from retracked ERS-1 altimetry. Laxon and McAdoo (1998) have retracked altimetry in the Arctic Ocean, Hwang (2003) has retracked altimetry in the China Sea, and Fairhead et al. (2004) have retracked data in several coastal regions.

Our methods differs from all the approaches above, as it includes a full retracking of the 20 Hz waveform data for the whole ERS-1 Geodetic Mission data (GM) using a suite of different retrackers for different conditions, with the use of a new method to regress these data onto 1 or 2 Hz values. We will show initial results to demonstrate the improvement in gravity field determination in a test region around Hawaii. Finally we will demonstrate the importance of retracking on data coverage in Polar Regions.

## 2 Retracking the ERS-1 GM data.

Whilst individual open ocean echoes generally correspond well to a mathematical model, which enables precise range to surface information to be retrieved, echoes in coastal zones are frequently distorted and therefore fail the standard automated processing procedures, and are rejected.

A variety of different echo shapes appear in the coastal zones caused by a variety of surface effects: these include land contamination of the echo, off-ranging to inland water, and the presence of unusually calm water in sheltered areas. However, although coastal zone echoes are complex and rapidly changing, almost all these waveforms can be successfully retracked. Within 5 km of the coastline, the majority of waveforms do not conform to the normal ocean model echo shape and are hence not retracked in the original data provided by ESA.

The EAPRS rule-based expert system (Berry et al., 1997) was designed to retrack the complex waveform shapes collected over land. Recently, the SOC ocean retracker (Challenor and Srokosz, 1989) has been added to the system, which now contains 11 retracking algorithms making the system capable of retracking waveforms everywhere.

For the implemented SOC ocean retracker, the 5-point Maximum Likelihood Estimate fit was changed to a 3-point least-squares fit because thermal noise was suppressed on ERS-1, and the bottom part of the leading edge is missing (so sea surface skewness cannot be calculated).

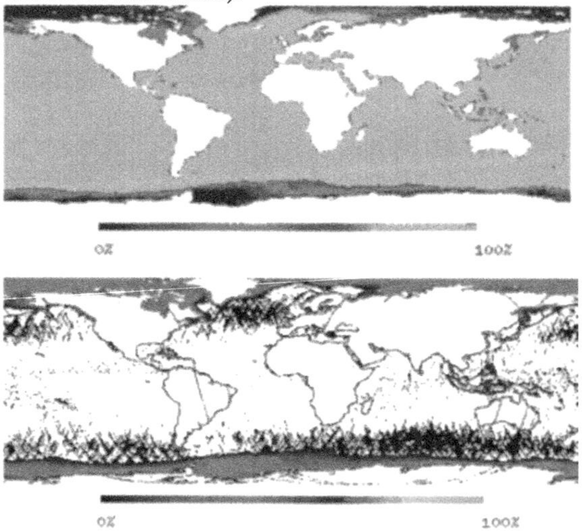


**Figure 2** Percentage of the total number of points retracked using two different retrackers in the ESPRS system. The upper figure is the number of data retrieved by the SOC ocean retracker, and the lower figure is the sea-ice retracker. The data in the lower figure are normally rejected from the altimeter data delivered by ESA.

Figure 2 shows the additional amount of data retrieved by the EAPRS system compared with the normal ocean retracking performed by ESA. The upper figure shows the percentage of data retrieved by the normal ocean retracking which is very close to 100 percent in the open ocean, and the lower figure shows the additional data obtained from one of the 10 additional retrackers in the EAPRS system. In this case it is the sea-ice retracker. Note that the sea-ice retracker turns on in small numbers in coastal regions and in the presence of strong currents. This is because the echoes are similar in shape to one category of sea-ice echoes.

## 3 Regressing altimetry onto 2Hz

With both the GEOSAT and the ERS-1 GM, the average track spacing between the combined set of tracks is around 4 km at the Equator. The distance between the 1 Hz individual altimetric observations along the track is 7 km. In order to increase the spatial resolution along track we decided to investigate the possibility of regressing the data onto 2 Hz values, which would give a 3.5 km distance between the observations. The 2Hz data would be computed from 10 observations versus 20 observations for the 1Hz average.

Normally, the 1 Hz average values are computed in fixed 1-second bins by removing the two observations furthest away from the mean and then by averaging the remaining observations. We propose to use an alternative technique for the computation of 2 Hz data, which is a local regression smoothing called "Loess" (Cleveland, 1979) which stands for local weighted linear regression to smooth data. The regression approach has the advantage over averaging, that it is possible to interpolate data closer to the coast (it only requires one 20 Hz observation closer to the coast). It furthermore uses a robust weight function that is resistant to outliers.

Initially gross-errors and data over land are removed by comparing with the KMS04 Mean sea surface height. The Loess scheme works by computing tricubic weigh functions for each 20 Hz observation depending on the along-track distance to the position to be computed (Cleveland, 1988). Subsequently, a linear least squares regression is performed using a first-degree polynomial. The smoothed or regressed point is then given as the weighed regression at the point of interest. By comparing all the residuals to the regressed point and computing the median of the absolute deviations, outliers can be detected and removed, and the regression can be repeated iteratively until all outliers have been removed.

## 4 Gravity field test around Hawaii.

Hawaii was chosen as the test region for the retracking and regression analysis as this region has very good marine gravity observations, and extreme gravity field changes, varying by more than 700 mGal across the islands.

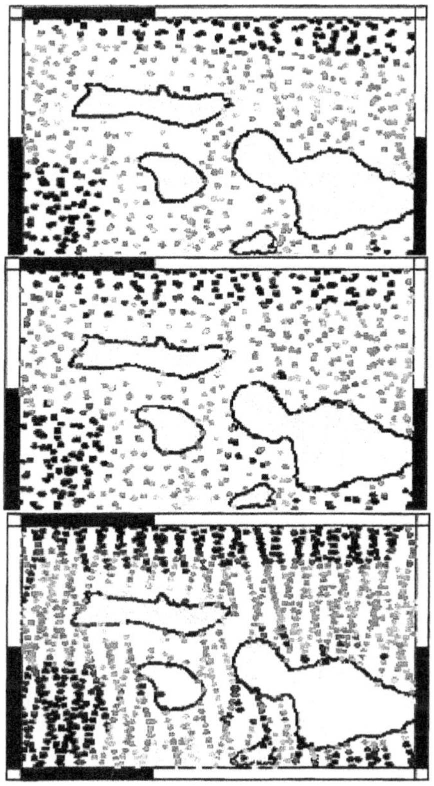


**Figure 3** Residual geoid heights (to EGM96) of 1 Hz altimetric observations retracked by the ESA retracker (upper figure) and 1 Hz observations retracked by the EAPRS. (middle figure) and 2 Hz observations (lower figure).

Figure 3 shows crossover adjusted residual geoid height observations compared to EGM96 of the normal ESA retracked observations and the observations retracked by the EAPRS expert retracker system. The expert system clearly retracks more point close to the coast. There are however some visible differences, particularly at the on-set of new observations after passage of the islands. This is due to the fact that the ocean retrackers in the expert system has not yet been adjusted properly, and hence shows an offset compared with the other retrackers. This is primarily expressed as a bias; the expert system is currently being tuned using a global ocean cross-calibration of retracker heights to remove this offset.

**Table 1** Comparison with 76789 marine gravity observations within the Hawaii region. RADS indicate ESA retracked data. All values are in mGal.

| | Mean | Std Dev | Max dev. |
|---|---|---|---|
| RADS | -1.76 | 12.82 | 144.53 |
| EAPRS 1Hz | -2.11 | 12.59 | 156.21 |
| EAPRS 2 Hz | -2.10 | 12.01 | 152.78 |

Table 1 shows a comparison with marine gravity observations within the Hawaii region bound by (15°<lat<29°, 188°<long<208°) in a gravity field computation following the methods by Andersen and Knudsen, (1998). Only ERS-1 GM data have been used for the gravity field computation.

Even with the known offset in one of the retrackers in the EAPRS system these data perform slightly better in a comparison with marine gravity observations that the ESA retracked data (12.82 to 12.59 mGal respectively).

With the additional enhancement of using 2 Hz data rather than 1 Hz data, further improvements are obtained (12.82 to 12.01 mGal respectively) which shows the importance of the increased data coverage in regions of large gravity field variations.

Further investigations in coastal regions is currently ongoing to substantiate and validitate these preliminary results as well as to fine-tune the dataediting and gravity predictions softward.

## 5 Polar Regions.

Retracking will provide the most new valuable observations in the Polar Regions where most observations fail to be retracked by the standard ESA ocean-model retracker when sea-ice is present. This was also demonstrated in Figure 2, which showed the number of additional data that could be retrieved with the sea-ice retracker.

Figure 4 shows the vast improvement in spatial coverage of good altimetric observations in complex coastal Polar Regions. The region east of Greenland is notorious for the presence of sea ice. It is roughly 600 km by 400 km in extent, and the number of ERS-1 data points that can be retrieved from retracking is increased from 350 datapoints to 11.200 datapoints when including, in particular, the sea-ice retracker. The figure also demonstrates apparent inaccuracies in the coastline model used for the plot, and illustrates the possibility of retracking data even in deep fjords, where altimetry has not previously been available.

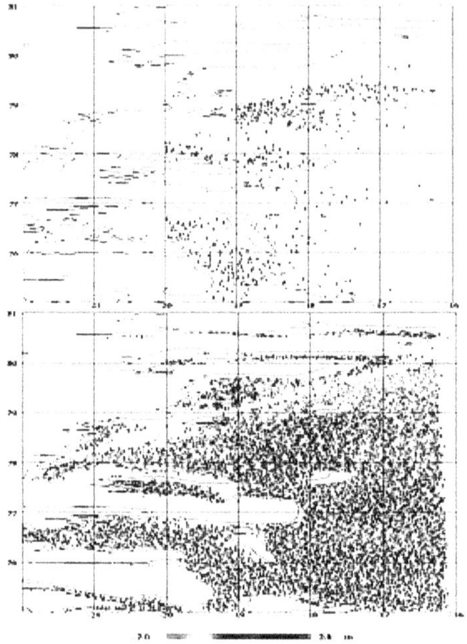

**Figure 4** Unadjusted altimetric geoid height observations in the ice-covered regions east of Greenland. The upper figure shows the 1Hz data in the Radar Altimeter Database System (RADS), which uses the data delivered by the ESA retracking system, and the lower figure shows the 1 Hz retracked altimetric data.

A comparison with 1288 marine in the permanently ice-covered region east of Greenland is shown in Table 2. The Laxon and McAdoo (1998) polar gravity field and the KMS02 gravity field were included. In this region the great impact of retracking is clearly demonstrated. For KMS02 the lack of ocean retracked altimetry shows that this field is not an improvement over EGM96, whereas the retracked gravity field data by Laxon and McAdoo performs substantially better.

**Table 2** Comparison with 1288 marine observations within the square bounded by 77°N to 79°N, and 5°W to 15°W from the Nordic gravity data base. All values are in mGal.

| | Mean | Std Dev | Max dev. |
|---|---|---|---|
| EGM96 | -0.76 | 17.63 | 122.91 |
| Laxon&McAdoo (97) | 1.10 | 6.59 | 56.21 |
| KMS02 | 0.98 | 18.58 | 139.28 |
| EAPRS 2 Hz | -2.10 | 5.78 | 43.93 |

## 6 Outlook

Within the coastal zone a variety of echo shapes can be returned to the altimeter. Some are caused by calmer patches of water, especially prevalent in sheltered areas and within coves and inlets. Others have been contaminated by the land, or show off-ranging to still inland water, especially close to river deltas or harbours.

However, even these complex shapes can be successfully retracked using the EAPRS expert system, vastly improving the data retrieval in coastal zones.

By using a new method to regress the 20 Hz ERS-1 geodetic mission altimetric data onto 2 Hz values corresponding to roughly 3.5km along track, we can obtain a spatial distribution of the altimetric data that has roughly the same along-track and across track spacing. On average it is 4 km across track spacing and 3.5 km along-track spacing.

Initial comparison with marine gravity field data already reveals that even at this preliminary stage (where residual offsets are known to exist between the different retrackers in the EAPRS system) we can obtain better comparison with the marine gravity observations than that obtained when using the data retracked by ESA.

**Acknowledgment.** The authors would like to thank R. Scharroo and colleagues for the 1 Hz (RADS) and ESA for providing waveform data for the analysis.

**Information.** The KMS2002 gravity field and the described gravity field (upon completion) will be available from the authors on CD-ROM (oa@spacecenter.dk) or from anonymous ftp at: ftp://ftp.kms.dk/pub/GRAVITY, www.research.dk/GRAVITY.

## References

Andersen, O. B., and P. Knudsen, Global Marine Gravity Field from the ERS-1 and GEOSAT Geodetic Mission Altimetry, J. Geophys. Res., 103, 8129-8137, 1998.

Andersen, O. B., and Knudsen, P. The Role of Satellite Altimetry in Gravity Field Modelling in Coastal Areas, *Physics and Chemistry of the Earth (A)*, Vol. 25, No. 1, pp. 17-24., 2000

Andersen, O. B., P. Knudsen, S. Kenyon and R. Trimmer, Recent improvement in the KMS global marine gravity field, Boll Geofis. teor. ed Applic. Vol, 40, 3-4, pp 369-377, 1999.

Andersen O. B. and P. Knudsen, Global Marine Gravity Field from the ERS-1 and GEOSAT Geodetic Mission Altimetry, *J. Geophys. Res., 103*(C4), 8129-8137, 1998.

Andersen, O. B., P. Knudsen and R. Trimmer, Improved high resolution gravity field mapping (the KMS02 Global Marine gravity field). In press, IAG symposium, 126, Sapporo, 2004

Berry, P.A.M., H. Bracke, and A. Jasper, 1997. Retracking ERS-1 altimeter waveforms over land for topographic height determination: an expert systems approach, ESA Pub. SP414 Vol. 1, 403-408.

Berry, P.A.M., J.D. Garlick, and E.L. Mathers, 2004. Global scale monitoring of land surface water using multi-mission satellite radar altimetry. EGU 1st General Assembly, Nice.

Challenor, P.G., and M.A. Srokosz, 1989. The extraction of geophysical parameters from radar altimeter return from a nonlinear ocean surface, in Brooks, S.R. (ed.) Mathematics in Remote Sensing, Institute of Mathematics and its Applications, pp 257-268.

Cleveland, W.S. (1979) "Robust Locally Weighted Regression and Smoothing Scatterplots," Journal of the American Statistical Association, Vol. 74, pp. 829-836.

Cleveland, W.S. and Devlin, S.J. (1988) "Locally Weighted Regression: An Approach to Regression Analysis by Local Fitting," Journal of the American Statistical Association, Vol. 83, pp. 596-610.

Fairhead , J. D., C. M. Green, K. M. U. Fletcher, Global mapping deep water hydrocarbon plays of the continental margins, ASEG 17th Geophysical conference and exhibition, Sydney, extended abstract, 2004,

Hwang, G. et. al., Satellite radar waveform retracking for improved coastal marine gravity anomaly accuracy in the Taiwan Strait., proceedings of the International Workshop on Satellite Altimetry for Geodesy, Geophysics and Oceanography: Summer Lecture Series and IAG symporium 126, 2003.

Laxon and McAdoo, 1998, Satellites Provide New Insights into Polar Geophysics, EOS, Transactions AGU, 79 (6), 69-72.

Sandwell, D. T., and Smith, W.H.F. Marine Gravity Anomaly from Geosat and ERS-1 Satellite Altimetry, Journal of Geophysical Research, Vol. 102, pp. 10039-10054, 1997.

Wang, Y. M. GSFC00 mean sea surface, gravity anomaly, and vertical gravity gradient from satellite altimeter data., J. Geophys res., 106, C12, 31167-31174, 2001

# Space-borne gravimetry: determination of the time variable gravity field

P.N.A.M. Visser and E.J.O. Schrama
Department of Earth Observation and Space Systems (DEOS), Faculty of Aerospace Engineering, Delft University of Technology, Kluyverweg 1, 2629 HS, Delft, The Netherlands, Fax: +31 15 2785322, email: Pieter.Visser@lr.tudelft.nl, E.J.O.Schrama@lr.tudelft.nl

**Abstract.** The gravity field of the Earth can be divided into a dominant quasi-static part and several relatively small but significant temporal constituents. Important examples of temporal sources are ocean tides, atmospheric pressure variations, and geophysical signals like those of continental hydrology and ocean bottom pressure variations predicted by the ECCO ocean model. Space-borne gravimetry, such as by the GRACE system, aims at observing temporal changes of the Earth's gravity field, including those induced by continental hydrology. A case study, based on a simulated gravity field retrieval for a 1-year GRACE-type mission, has been conducted to analyze the separability of continental hydrology from other temporal gravity sources.

It has been investigated how typical differences between recent ocean tide models and between global atmospheric pressure variation maps affect the observations (low-low satellite-to-satellite range-rate tracking (SST) and orbital positions from GPS high-low SST) and retrieved gravity field spherical harmonic expansions. In addition, the aliasing of signals predicted by the ECCO model and the effect of low-low SST observation noise and uncertainties in the recovered orbital positions has been analyzed.

It is concluded that large scale features of continental hydrology can be observed by a GRACE-type mission, provided that the low-low SST observations have a precision at the level of 1 $\mu$m/s at 1 Hz, and when great care is taken with the gravity field recovery approach.

**Key words.** low-low satellite-to-satellite tracking, temporal gravity, hydrology, atmosphere, tides, ocean models

---

## 1 Introduction

Satellite Laser ranging (SLR) to satellites such as LAGEOS-1 and -2 has proved the possibility for observing temporal gravity in the very long wavelength domain leading for example to intriguing results for the time evolution of the Earth's oblateness or predominantly the $J_2$ term (Cox and Chao 2002; Cazenave and Nerem 2002). The CHAMP (launch July 15, 2000, (Reigber et al. 1999)) and GRACE (launch March 17, 2002, (Tapley and Reigber 1999)) satellite missions have opened the possibility for observing temporal gravity from space on a much more detailed scale than ever before and impressive results have already been obtained (Tapley et al. 2004). These results indicated the possibility for observing changes in continental hydrology in very large basins, such as in the Amazone area in South-America. However, results also indicated that great care needs to be taken when modeling other temporal gravity field sources such as ocean tides and changes in the atmospheric mass distribution. In addition, it was found that the gravity field reduction process is very sensitive to the parameterization of the gravity field estimation problem (arc length, empirical accelerations, accelerometer biases and scale factors). It was concluded that SLR tracking remains to be an important asset when analyzing geocenter variations (spherical harmonic degree 1 terms) and changes in $J_2$, and is a prerequisite for diagnosing possible problems in the processing of CHAMP and GRACE observations taken by the on-board science instruments.

The problem of temporal aliasing of different gravity field sources has been studied extensively,*cf.* (Han et al. 2004) and (Velicogna et al. 2001). We have built a simulation tool around the GEODYN software package (Rowlands et al. 1995) that allows to study the observability and separability of different gravity field sources (static and temporal) for several gravity field mission concepts and scenarios, including CHAMP- and GRACE-type missions, and possible future missions such as GOCE (ESA 1999) and GRACE/GOCE follow-ons. This tool allows a rigorous parameterization of the gravity field estimation problem and long data periods. It has been used for a case study where a one-year GRACE-type mission is defined for observing mass changes due to continental hydrology. The observability is studied in the presence of typical low-low SST observation noise levels and coupling with (errors in the modeling of) other temporal gravity field sources, including atmospheric mass redistributions, ocean tides and mass changes inflicted by ocean bottom pres-

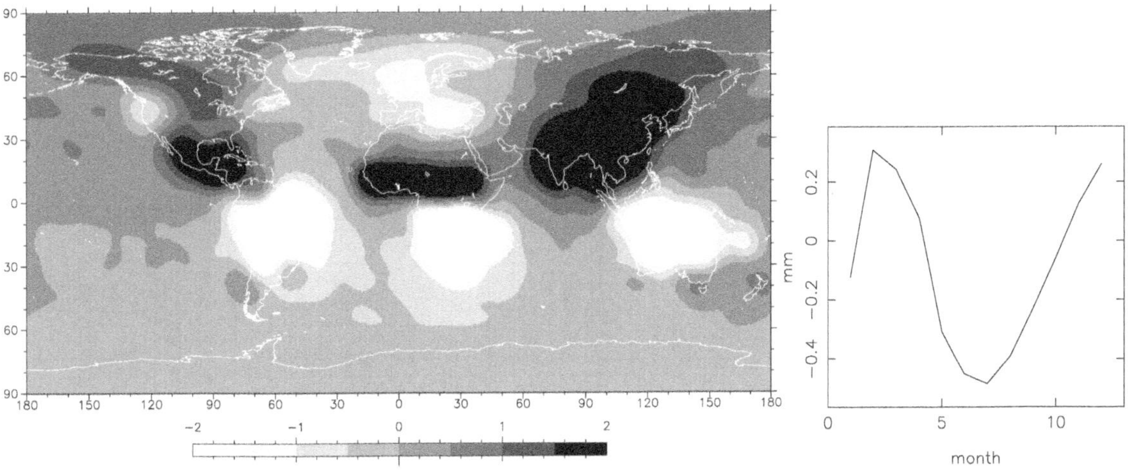


**Fig. 1.** Dominant mode (EOF) of mass changes due to continental hydrology for 2000 (Fan and van den Dool 2004). Please note the scale of the time pattern is in mm.

sure variations making use of ECCO ocean models (ECCO homepage 2004).

The simulation tool makes use of numerical integration of the equations of motion and variational equations for the estimated gravity field and orbit parameters, and offers the possibility to describe the Earth's gravity field as a sum of a baseline static part and (different) combinations of temporal sources. Currently all sources are represented by spherical harmonic expansions, although also space localized functions such as gravity anomalies and density layers are possible. For the case study to be outlined in the next section, the following models were used:

- Static gravity field model: GGM01S;
- Temporal gravity:
  - Continental hydrology (Fan and van den Dool 2004);
  - Ocean tides: FES99 and GOT99.2b;
  - Atmospheric mass variations: ECMWF and NCEP;
  - Ocean bottom pressure: ECCO.

The real world was modeled by the GGM01S static field (GRACE CSR home page 2004) in combination with GOT99.2b (Ray 1999) ocean tides, ECMWF based atmospheric mass variations, oceanic mass redistribution according to ECCO models, and continental hydrology (Fan and van den Dool 2004). The GGM01S model is a GRACE-based satellite only solution to degree and order 120, but was truncated at degree and order 50 in the case study. The two ocean tide models were derived using different methodologies. GOT99.2b is an empirical model, whereas FES99 makes use of hydrodynamical equations. The temporal background gravity field models were developed complete to degree and order 20 (making in certain cases for example use of Love numbers for converting equivalent water heights to Stokes coefficients, see (Schrama 2003)). Figure 1 displays the dominant mode (first EOF or Empirical Orthogonal Function) of the continental hydrology model in terms of geoid variations for a one year period. This dominant mode represents about 80% of the amplitude, or about 60% of the energy of the total signal. Clearly visible are relatively large fluctuations in the area covering part of the Southern states of the U.S, Mexico and Latin America, the area from the Sahel to South-Africa, the Amazone and Zambesi basins and areas in East-Asia. Also clearly visible is the dominant annual signature (right part of the Figure). The objective of the case study is to investigate whether this signature can be recovered by a GRACE-type mission in the presence of observation errors (low-low SST and GPS-based orbit reconstruction errors), mismodeling of ocean tides (using FES99 as reference model, (Lefèvre et al. 2002)) and atmospheric mass variations (using NCEP reanalysis surface pressure data as reference), and ignoring ECCO predicted gravity changes (see also Table 1). The static gravity field model is assumed to be a long-period averaged solution with negligible errors.

## 2 Temporal gravity

The signal and/or model uncertainty size derived from the spherical harmonic expansions of the different gravity field sources is displayed in Figure 2. In fact, for the atmospheric mass variations the signal size is the average of 366 daily spherical harmonic expansions complete to degree and order 20 (in the following referred to as 20x20) using daily pressure fields from the year 1992, for ECCO from 2000, and for continental hydrology the average of 12 monthly 20x20 fields for 2000. It can be seen that the error

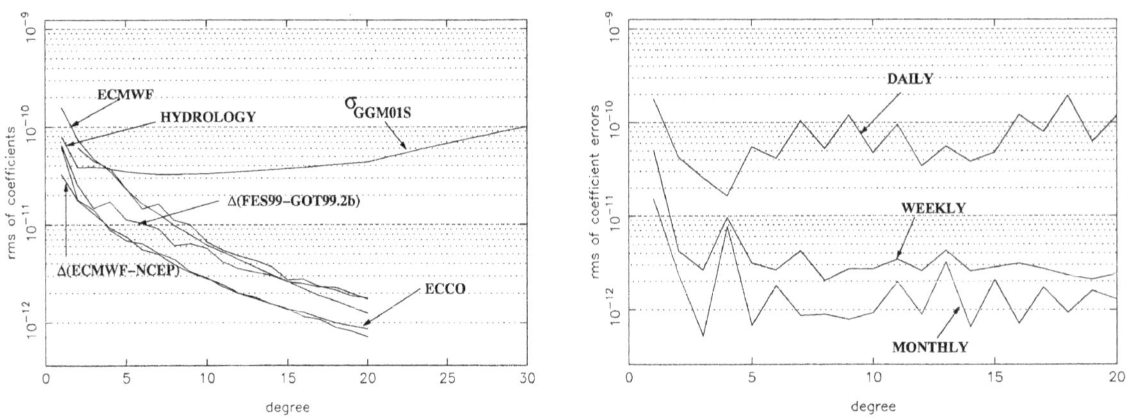


**Fig. 2.** Magnitude of temporal gravity field sources and model differences, and quality of GGM01S (left), and effect of observation noise on gravity field recovery accuracy for different observation period durations (right). The temporal gravity field sources include ocean tide model differences between FES99 and GOT99.2b, gravity field changes due to atmospheric mass redistributions (total signal according to ECMWF, and differences between ECMWF and NCEP), and gravity field variations due to continental hydrology and ECCO ocean models.

level of GGM01S, which is based on 111 days of GRACE observations, is above the signal size of continental hydrology up to degree 5. However, it is fair to assume that significant improvements will be made as time progresses resulting in more observations and a better understanding of the behavior of the GRACE system. Moreover, the objective is to study temporal gravity, although it is realized that errors in the static gravity field model might affect the recovery of temporal gravity, which is an interesting topic for future research. The signal size of the atmospheric mass variations is of the same order of magnitude as those inflicted by continental hydrology. Assuming that the differences between ECMWF and NCEP atmospheric pressure fields are representative for the accuracy with which atmospheric mass variations can be modeled, $\Delta$(ECMWF-NCEP), continental hydrology can still be observed. The signal predicted by the ECCO model has a size comparable to the differences between the two atmospheric pressure field models and is in fact much below the continental hydrology signal. The uncertainty in ocean tide modeling, reflected by $\Delta$(FES99-GOT99.2b), intersects the continental hydrology signal around spherical harmonic degree 15 (see Figure 2). Based on these results, it may be concluded that an effort is required to further improve ocean tide modeling.

## 3 Case study

Gravity field recovery simulation experiments have been conducted for a one year period, or 366 days for 2004 (leap year). It has to be noted that some temporal gravity sources that were used in the simulations are for 1992 and 2000 (Section 1). It is fair to assume that these data sets realistically reflect the signal magnitudes and time signatures that can be expected. A GRACE-type mission is selected, consisting of two satellites flying en echelon in 440 km altitude orbits with an inclination of 89° and separation of 200 km. Gravity field models are estimated from low-low SST observations and orbit positions (inertial Cartesian $x, y, z$ coordinates) resolved from the GPS high-low SST observations. The low-low SST observations are sampled at 30-s intervals and Gaussian noise is added with a standard deviation of 0.2 $\mu$m/s (equivalent to 1 $\mu$m/s at 1 Hz). The orbit coordinates of the two satellites are assumed to have an accuracy of 1 cm (Gaussian) and are sampled at 2-min intervals (Table 1).

**Table 1.** Definition of case study: truth, reference and error models.

| | |
|---|---|
| Truth model | |
| static gravity field: | GGM01S |
| continental hydrology: | Fan & Dool, 2004 |
| ocean tides: | GOT99.2b |
| ocean bottom pressure: | ECCO |
| atmospheric pressure: | ECMWF |
| Reference model | |
| static gravity field: | GGM01S |
| continental hydrology: | none |
| ocean tides: | FES99 |
| ocean bottom pressure: | none |
| atmospheric pressure: | NCEP |
| Observation errors (Gaussian) | |
| low-low SST: | $\sigma = 0.2$ $\mu$m/s @ 30-s |
| orbit coordinates: | $\sigma(x, y, z) = 1$ cm @ 2-min |

The one-year simulated observation data set is divided into daily periods and for each day normal equations are computed for a 20x20 spherical harmonic gravity field model (including degree 1 terms) and for epoch state vectors for the two satellites ($2 \times 6 = 12$ unknowns per day, each state vector consisting of 3 position and 3 velocity coordinates). It is fair to assume that above degree 20, the temporal gravity field signals have a very low signal magnitude (Figure 2) and it is also assumed that a high-accuracy, higher resolution background model is available for the static gravity field (in this case the 50x50 truncated GGM01S model). However, for future more advanced and precise gravity field missions, the simulations can be extended to (much) higher degrees, requiring extensive (but feasible) computer resources. The daily normal equations can be combined to obtain gravity field solutions for different period lengths. For example, a weekly solution is obtained by combining 7 daily normal equations solving for 84 ($7 \times 12$) epoch state vector unknowns and one 20x20 gravity field spherical harmonic expansion.

## 4 Results

Before conducting the gravity field recovery in the presence of all error sources according to Table 1, the effect of different temporal gravity field sources on the low-low SST range-rate observations was assessed. The signal Root-Mean-Square (RMS) of the low-low SST range-rate observations is typically around 20 cm/s, and is dominated by the $J_2$ term. For all 366 days, the RMS is computed by estimating only the 12 ($2 \times 6$) epoch state vector parameters. Finally, the RMS of the 366 daily RMS values is displayed in Table 2. Continental hydrology causes an RMS signal of about 0.18 $\mu$m/s, compared to 0.14 $\mu$m/s for the FES99/GOT99.2b ocean tide model differences, 0.43 $\mu$m/s for atmospheric mass variations predicted by ECMWF pressure fields, 0.22 $\mu$m/s for the differences between ECMWF and NCEP, and 0.10 $\mu$m/s for mass variations induced by the ECCO model. These numbers indicate that atmospheric mass variations need to be accurately modeled, that ocean tide model uncertainties compete with the continental hydrology signal and that the ECCO model results in relatively small low-low SST range-rate perturbations.

In a second step, the separate effect of observation errors on the achievable gravity field recovery error was assessed by generating 366 daily, 52 weekly and 12 monthly solutions. The annual averages of the degree RMS values of spherical harmonic coefficient errors is displayed in Figure 2. It can be seen that for weekly and monthly solutions the errors are below the continental hydrology signal, but that this is not the case for daily solutions (which can be anticipated considering that the satellites complete 16 orbital revolutions per day, but that a 20x20 model is solved for; in all cases no regularization was applied). The dominant error mode (EOF analysis) of the daily gravity field solutions in term of geoid is displayed in Figure 3 displaying a pattern commensurate with the daily ground tracks of the satellite pair. A similar pattern was predicted by an EOF analysis of the dominant eigenvectors of the daily inverses of the normal equations.

**Table 2.** RMS of low-low SST observation residuals (30-s sampling) due to different temporal gravity field sources (366 daily arcs)

| Source | RMS ($\mu$m/s) |
|---|---|
| Continental hydrology: | 0.179 |
| Tide model differences: | 0.142 |
| Atmosphere: ECMWF | 0.432 |
| Atmosphere: ECMWF-NCEP | 0.227 |
| ECCO | 0.101 |

Based on the previous results, it was decided to generate a time series of 52 weekly (or 52 7-day) gravity field solutions in the presence of all error sources listed in Table 1. Again, an EOF analysis was conducted. Figure 4 clearly reveals the dominant mode caused by errors in the recovery of degree 1 terms, which are heavily correlated with the $2 \times 6$ epoch state vector. It is obvious that the continental hydrology signal (Figure 1) is completely obscured by this mode. This error mode can again be predicted by error propagation, and thus seems to indicate an inherent weakness in the gravity field recovery approach (which might be solved by adding certain types of tracking data to other satellites, such as SLR tracking of LAGEOS-1/2). It was also found that the FES99 and GOT99.2b ocean tide model differences cause relatively large perturbations in the degree one gravity field terms indicating the need for co-estimation of tide model terms and/or independent tide model improvement. However, generating a second series of weekly gravity field solutions without solving for the degree one terms results in a dominant mode as displayed in Figure 5, clearly revealing the most important features of the continental hydrology signal (Figure 1). Striking differences can be observed in the Antarctic region, which can be attributed to large differences between the ECMWF and NCEP atmospheric pressure fields. Although the time signature is rather noisy, it displays a clear annual period comparable to the true annual pattern. It was found that this noisy behavior is reduced significantly when making monthly gravity field solutions.

It is interesting to compare as well the fourth EOF of the case where the gravity degree 1 terms (3 coefficients leading to possibly 3 dominant EOFs) are

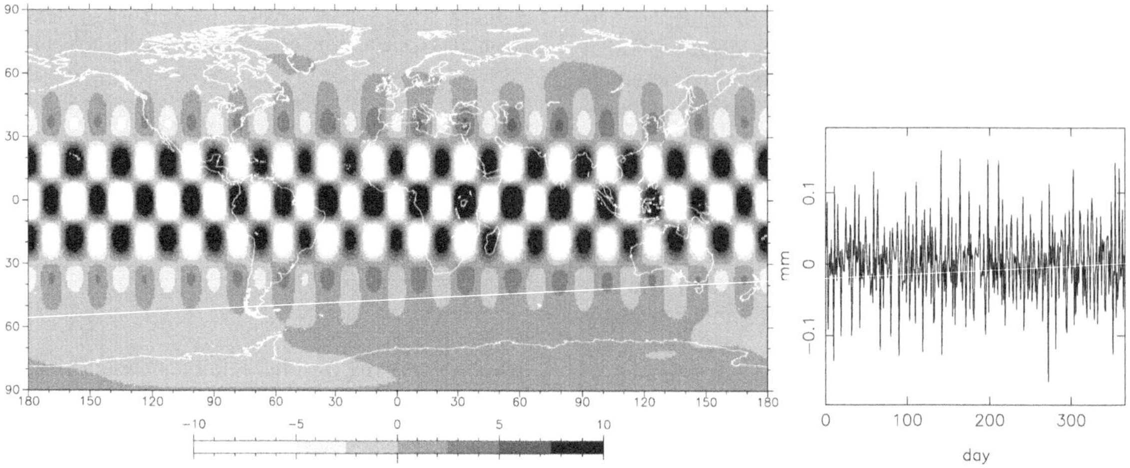


**Fig. 3.** Dominant mode (EOF) of recovered daily gravity field solutions with only observation noise switched on.

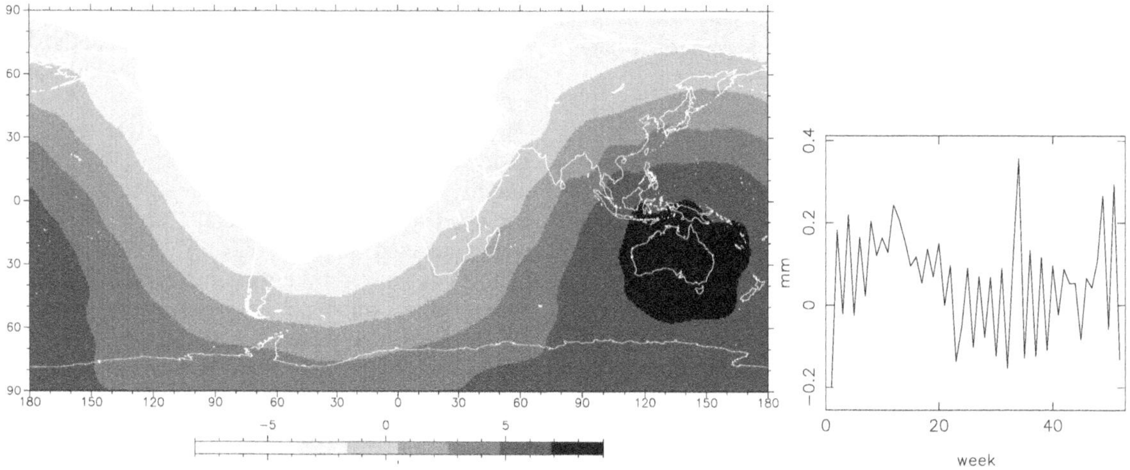


**Fig. 4.** Dominant mode (EOF) of recovered weekly gravity field solutions with all error sources switched on (degree one terms included).

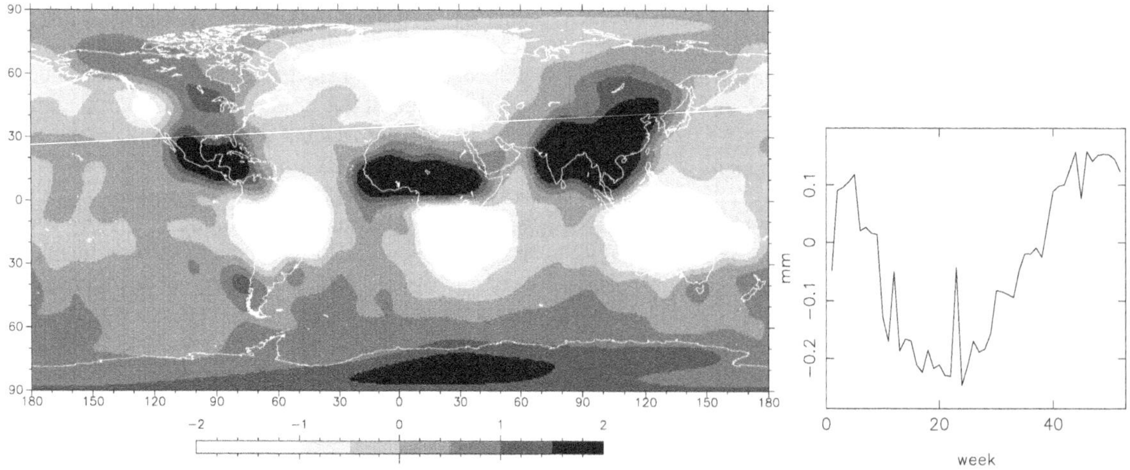


**Fig. 5.** Dominant mode (EOF) of recovered weekly gravity field solutions with all error sources switched on (degree one terms ignored).

estimated with the first EOF as displayed in Figure 4. It was found that these EOFs compare very well: the correlation is bigger than 90%.

## 5 Conclusions

Temporal gravity field variations such as induced by continental hydrology is observable with GRACE-type missions.

Great care needs to be taken with the parameterization of the gravity field recovery. First of all, the arc length and the combined estimation of non-gravitational (for example satellite epoch state vectors) and gravitational parameters needs to be carefully defined and investigated. Second, the period for which gravity field solutions are to be generated needs to be balanced with the required precision, temporal and spatial resolution levels.

It was found that the degree one terms can be seriously affected by uncertainties in ocean tide models. Also, there are indications that degree one terms are weakly observable by the investigated mission concept in combination with the adopted gravity field recovery approach. For overcoming this weakness, continued high-quality SLR tracking is instrumental and will in combination with GRACE-type observations guarantee high-precision temporal gravity modeling from the very long to the medium wavelength domain (degree 1 - 20).

The low-low SST observations need to have a high precision level, of the order of 1 $\mu$m/s at 1 Hz. In order to be able to observe continental hydrology, mass variations due to ocean tides and atmosphere need to be modeled with great precision.

Finally, it can be concluded that a tool has been implemented that can be used for gravity field mission analysis, opening the possibility to assess in a closed-loop the effect of observation noise, satellite configuration, mismodeling of (combinations of) gravity field sources and gravity field recovery reduction approach and parameterization.

*Acknowledgments.* The Center for Space Research, University of Texas at Austin, Texas, is acknowledged for providing the GGM01S model. The GEODYN software was kindly provided by the Goddard Space Flight Center, Greenbelt, Maryland. The ECCO models are a a contribution of the Consortium for Estimating the Circulation and Climate of the Ocean (ECCO) funded by the National Oceanographic Partnership Program. The ECCO models were converted to spherical harmonic expansions of equivalent water height and provided by J. Kusche, DEOS, Delft, The Netherlands.

## References

Cazenave, A., and R.S. Nerem (2002), Redistributing Earth's Mass, *Science, 297*, 783–784, August 2002.

Cox, C.M., and B.F. Chao (2002), Detection of Large-Scale Mass Redistribution in the Terrestrial System Since 1998, *Science, 297*, 831–833, August 2002.

ECCO homepage, *http://www.ecco-group.org/*, last accessed: July 2004

ESA (1999), Gravity Field and Steady-State Ocean Circulation Mission, *Reports for Mission Selection, The Four Candidate Earth Explorer Core Missions, SP-1233(1)*, European Space Agency, July 1999.

Fan, Y., and H. van den Dool (2004), Climate Prediction Center global monthly soil moisture data set at 0.5° resolution for 1948 to present, *J. Geophys. Res., 109*(D10), 8 pp., May 2004.

GRACE CSR homepage, *http://www.csr.utexas.edu/grace/gravity*, last accessed: July 2004

Han, S.-C., C. Jekeli, and C.K. Shum (2004), Time-variable aliasing effects of ocean tides, atmosphere, and continental water mass on monthly mean GRACE gravity field, *J. Geophys. Res., 109*(B4), doi:10.1029/2003JB002501.

Lefèvre, F., F.H. Lyard, C. LeProvost, and E.J.O. Schrama (2002), FES99: A Global Tide Finite Element Solution Assimilating Tide Gauge and Altimetric Information, *Journal of Atmospheric and Oceanic Technology, 19*(9), 1345–1356.

Ray, R D (1999), A global ocean tide model from Topex/Poseidon altimetry: GOT99.2, *NASA Tech. Memo. 209478*, 58 pp., Goddard Space Flight Center.

Reigber, Ch., P. Schwintzer, and H. Lühr (1999), The CHAMP geopotential mission, in *Bollettino di Geofisica Teorica ed Applicata, Vol. 40, No. 3-4, Sep.-Dec. 1999, Proceedings of the 2nd Joint Meeting of the International Gravity and the International Geoid Commission, Trieste 7-12 Sept. 1998, ISSN 0006-6729*, edited by I. Marson and H. Sünkel, pp. 285–289.

hspace*-3mm Rowlands, D., J.A. Marshall, J. McCarthy, D. Moore, D. Pavlis, S. Rowton, S. Luthcke, and L. Tsaoussi (1995), GEODYN II system description, Vols. 1-5, *Contractor report*, Hughes STX Corp., Greenbelt, MD.

Schrama, E.J.O. (2003), Error characteristics estimated from CHAMP, GRACE and GOCE derived geoids and from satellite altimetry derived mean dynamic topography, *Space Science Reviews, 108*, 179–193.

Tapley, B.D., and C. Reigber (1999), GRACE: a satellite-to-satellite tracking geopotential mapping mission, in *Bollettino di Geofisica Teorica ed Applicata, Vol. 40, No. 3-4, Sep.-Dec. 1999, Proceedings of the 2nd Joint Meeting of the International Gravity and the International Geoid Commission, Trieste 7-12 Sept. 1998, ISSN 0006-6729*, edited by I. Marson and H. Sünkel, p. 291.

Tapley, B.D., S. Bettadpur, J.C. Ries, P.F. Thompson, and M.M. Watkins (2004), GRACE Measurements of Mass Variability in the Earth System, *Science, 305*, 1503–505.

Velicogna, I., J. Wahr, and H. Van den Dool (2001), Can surface pressure be used to remove atmospheric contributions from GRACE data with sufficient accuracy to recover hydrological signals?, *J. Geophys. Res., 106*(B8), doi: 10.1029/2001JB900228.

# Satellite clusters for future gravity field missions

Nico Sneeuw
Department of Geomatics Engineering, University of Calgary, sneeuw@ucalgary.ca,
currently at: Universität Stuttgart, Geodätisches Institut

Hanspeter Schaub
Aerospace and Ocean Engineering Department, Virginia Tech, schaub@vt.edu

**Abstract.** The current missions CHAMP and GRACE have already contributed drastically to our knowledge of the Earth's gravity field in terms of accuracy, homogeneity and time- and space-resolution. The future mission GOCE will further add to that in terms of spatial resolution. Nevertheless, each of these missions has its own limitations. At the same time several geoscience disciplines push for ever higher requirements on spatial resolution, time resolution and accuracy. Future gravity field missions will need to address these requirements.

A number of new technologies may enable these future missions. They include laser tracking and atomic interference. Most likely, a mission that implements such technologies, will make use of the concept of formation flying. This paper will discuss the feasibility of low-Earth satellite clusters. It focuses in particular on the stability of satellite formations under the influence of perturbations by the Earth's flattening. Depending on initial conditions several types of relative $J_2$ orbits can be attained.

By interpreting the low-low satellite-to-satellite tracking observable as gradiometry this paper furthermore indicates how satellite clusters may be employed in satellite gravimetry.

## 1 Limitations of current and planned gravity missions

**CHAMP, GRACE and GOCE.** The satellite mission CHAMP currently provides static gravity field solutions at dm-level geoid accuracy up to an effective maximum spherical harmonic degree of around 60, cf figure 1. Recovery of the time-variable field seems to be at—or rather below—the edge of feasibility. Although CHAMP is still in orbit, delivering quality science data, this combination of resolution and accuracy is the natural limitation of the mission.

The accuracy of GRACE-derived geoids, on the other hand, is at mm-level around these degrees. It achieves its resolution around degree 120 at which the geoid accuracy is at the dm-level again. Future data and modeling improvements will likely push the limit towards a maximum degree of 150. Moreover, GRACE provides monthly solutions that clearly reveal time-variable gravity (Tapley et al., 2004).

The gradiometer mission GOCE, due for launch in 2006, aims at cm-accuracy and a spatial resolution corresponding to maximum degree 300. This high resolution, combined with the relatively short mission duration, does not allow time-variable gravity recovery, although time variations will alias into the static solution.

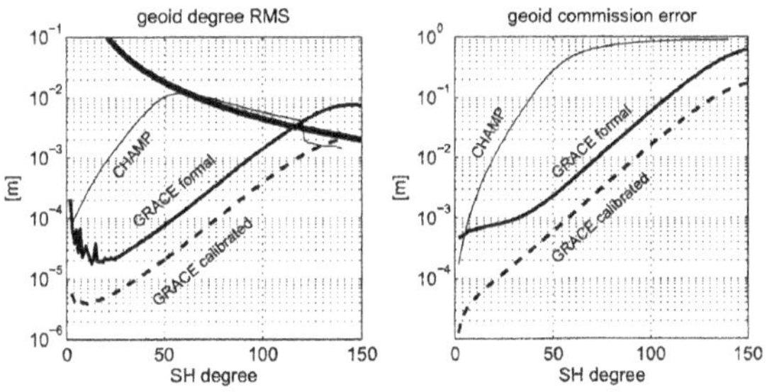


**Fig. 1.** CHAMP and GRACE gravity recovery performance: geoid RMS (left) and cumulative geoid error (right) as function of spherical harmonic degree. The CHAMP curves represent the model EIGEN2. Those of GRACE refer to EIGEN-GRACE02S.

Despite the wealth of new gravity field information and despite the many new scientific issues that can be addressed, these missions are limited in spatial resolution, temporal behaviour (resolution and mission duration) and accuracy of the resulting gravity field recovery. The key limitations, at least from a gravity recovery viewpoint, are:

- Sampling and resolution: missions are designed for either spatial or spectral resolution. A simultaneous high spatial and spectral resolution is fundamentally impossible with a single mission.
- Aliasing: unmodeled phenomena with sub-monthly period will alias into the monthly GRACE solutions. Time-variable signal will also map into the static GOCE field.
- Monitoring: limited mission durations of 5 year

(CHAMP, GRACE) or 1 year (GOCE).

- Gravity signal: when interpreted as an along-track gravity gradient, the GRACE-observable is seen to be a relatively weak component, see below.

The limitations of GRACE and GOCE are analyzed from a more technological viewpoint by Aguirre-Martinez and Sneeuw (2003).

**Future low-low SST.** At the same time Earth scientist are driving the requirements for ever higher accuracies and resolutions. Moreover, similar to satellite radar altimetry, there is a growing demand for a monitoring facility rather than a few individual satellite missions. Studies into next-generation gravity field missions tend to focus on low-low satellite-to-satellite tracking (SST). Indeed, the accuracy gain that is potentially achieved by laser SST over a GRACE-type radio link is far larger than the expected future improvements in gradiometry technologies.

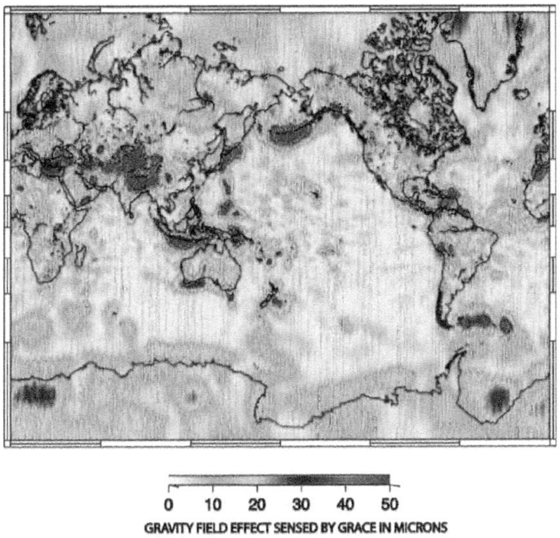


**Fig. 2.** GRACE *first light*: map of gravity field effect on inter-satellite baseline.

The key GRACE-type SST observable is the inter-satellite distance and relative velocity in a leader-follower configuration at near-polar inclination. This type of observable inherently suffers from the weakness that it is mainly sensitive along the line-of-sight, i.e. in North-South direction. This was demonstrated by the very first release of a GRACE map, cf. Figure 2, which clearly demonstrated a sensitivity towards East-West features in the Earth's gravity field. Note, for instance, the weak presence of Andes or Rocky Mountains in the map.

The observable approximates the along-track in-line gravity gradient term $V_{xx}$. In terms of gravity gradiometry this is known to be a relatively weak term. Its spectral content is approximately one half of the radial gravity gradient term $V_{zz}$. More importantly, the directional sensitivity of the observable also translates into a non-isotropic error behaviour.

**Formation flying.** Formation flying, which is currently receiving much attention internationally, may solve some of the aforementioned issues. A satellite formation may consist of any number of satellites that are performing a relative motion around a common center. A GRACE-type leader-follower formation can be seen as a simple example of formation flight. In general, satellites may perform more complicated elliptical or circular relative motion. Obviously, when the distances between these satellites would be measured, the gravitational signal would include radial information. Moreover, a relative inclination might be achieved that would lead to cross-track information going into the observable. Such observables could address several of the aforementioned weaknesses, most notably the spectral content and the non-isotropy of the low-low SST observable. Including cross-track information may also reduce the aliasing problem.

For these reasons this paper will mainly investigate the feasibility of formation flying in a realistic gravity field. It will then be discussed how to use formation flying in a gravity field mission.

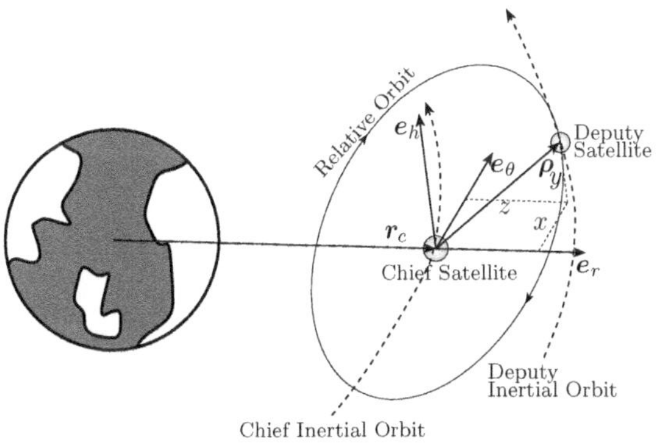


**Fig. 3.** Illustration of a general spacecraft formation with out-of-plane relative motion.

## 2 Feasibility of formation flying in a $J_2$ gravity field

### 2.1 Equations of relative motion

Let us adopt the following formation flying notation. A set of deputy satellites are to fly about a chief location as shown in Figure 3. This location could be an actual spacecraft, or simply a reference point. The inertial chief position vector is $\boldsymbol{r}_c$, while $\boldsymbol{\rho}$ is the deputy relative position vector. The rotating Hill frame $\mathcal{O} = \{\boldsymbol{e}_\theta, \boldsymbol{e}_h, \boldsymbol{e}_r\}$ is defined with $\boldsymbol{e}_r$ being along the chief orbit radial, $\boldsymbol{e}_h$ being along the chief orbit plane normal, and $\boldsymbol{e}_\theta$ completing the right handed coordinate system. The angular rate of the Hill frame (chief motion) is $\dot{\theta}$. The deputy position vector $\boldsymbol{\rho}$ is then expressed in the Hill frame through

$$ {}^{\mathcal{O}}\boldsymbol{\rho} = (x, y, z)^T \tag{1} $$

The general equations of motion of a deputy satellite with respect to a chief is given by Schaub and Junkins (2003):

$$ \ddot{x} + z\ddot{\theta} + 2\dot{z}\dot{\theta} - x\left(\dot{\theta}^2 - \frac{\mu}{r_c^3}\right) = a_x \tag{2a} $$

$$ \ddot{y} + \frac{\mu}{r_c^3}y = a_y \tag{2b} $$

$$ \ddot{z} - z\left(\dot{\theta}^2 + 2\frac{\mu}{r_c^3}\right) - x\ddot{\theta} - 2\dot{x}\dot{\theta} = a_z \tag{2c} $$

where $\mu$ is the gravitational constant and $(a_x, a_y, a_z)$ are non-Keplerian forces acting on the deputy satellite. They could be due to atmospheric drag, $J_2$ gravitational oblateness effects, or control thrusters.

Many missions consider formations where the chief motion is essentially circular with a near zero eccentricity $e$. In this case the chief rate $\dot{\theta}$ is constant and equal to the mean orbit rate $n = \sqrt{\mu/r_c^3}$. The equations of motion simplify to the well-known linearized Hill equations (Hill, 1878), see also (Clohessy and Wiltshire, 1960):

$$ \ddot{x} + 2n\dot{z} = a_x \tag{3a} $$

$$ \ddot{y} + n^2 y = a_y \tag{3b} $$

$$ \ddot{z} - 2n\dot{x} - 3n^2 z = a_z \tag{3c} $$

We will refer to them as HE in the sequel. Eq. (3) has been used extensively in spacecraft formation flying mission analysis and control research. They are reasonable as long as $(x, y, z)$ are small compared to the chief orbit radius $r_c$.

Often it is convenient to work in non-dimensional states. Let $(u, v, w) = (x, y, z)/r_c$ be non-dimensional deputy relative position coordinates. If the true anomaly $f$ is used as the independent angle instead of time, then the general first order equations of motion are given through:

$$ u'' + 2w' = \alpha_u \tag{4a} $$

$$ v'' + v = \alpha_v \tag{4b} $$

$$ w'' - 2vv' - \frac{3w}{1 + e\cos f} = \alpha_w \tag{4c} $$

Many further forms of the relative motion have been developed. Schweighart and Sedwick (2002) developed an extension to the HE which includes linear $J_2$ oblateness perturbations in the equations of motion. Humi and Carter (2003) have shown solutions with special forms of quadratic drag. An excellent survey of relative motion state transition matrices is found in (Carter, 1998).

## 2.2 First order analytical solutions

If the chief motion can be modeled as circular, then the HE can be solved analytically. Assuming no perturbations or thrusting is present ($a_x = a_y = a_z = 0$), all possible deputy relative motions can be expressed in closed form (Schaub and Junkins, 2003):

$$ x(t) = -2A_0 \sin(nt + \alpha) - \frac{3}{2}ntz_{\text{off}} + x_{\text{off}} \tag{5a} $$

$$ y(t) = B_0 \cos(nt + \beta) \tag{5b} $$

$$ z(t) = A_0 \cos(nt + \alpha) + z_{\text{off}} \tag{5c} $$

Note that the out-of-plane motion is decoupled from the in-plane motion. The integration constants can be expressed in terms of initial conditions through:

$$ A_0 = \frac{1}{n}\sqrt{\dot{z}_0^2 + (2\dot{x}_0 + 3nz_0)^2} \tag{6a} $$

$$ B_0 = \frac{1}{n}\sqrt{\dot{y}_0^2 + (ny_0)^2} \tag{6b} $$

$$ \alpha = \arctan\left(\frac{\dot{z}_0}{3nz_0 + 2\dot{x}_0}\right) \tag{6c} $$

$$ \beta = \arctan\left(\frac{-\dot{y}_0}{ny_0}\right) \tag{6d} $$

$$ z_{\text{off}} = \frac{2}{n}(\dot{x}_0 + 2nz_0) \tag{6e} $$

$$ x_{\text{off}} = x_0 - \frac{2\dot{z}_0}{n} \tag{6f} $$

These equations are very convenient to explore what possible *natural* and unforced formation shapes are feasible. For example:

- If $B_0 = 0$, a purely in-plane relative motion is achieved which is always a 2:1 ellipse (the CartWheel-mode);
- If $B_0 = \sqrt{3}A_0$, the relative motion is circular with radius $2A_0$ (the LISA-mode);

- If $B_0 = 2A_0$, one achieves elliptical motion with a circular cross-section in the local horizon plane $e_\theta$-$e_h$ (the TechSat21-mode).

The latter two configurations also require either $\alpha = \beta$ or $\alpha = \beta + \pi$.

If the chief motion is not circular, then the solution in Eq. (5) is no longer valid. Even small amounts of eccentricity can produce modeling errors comparable to those produced by $J_2$ gravitational perturbations or atmospheric drag. Carter (1998) presents an analytical solution to the linearized relative motion where the true anomaly is used as the independent variable. However, this solution does not provide that elegant geometrical insight the classical HE solution provides. In (Schaub, 2004) the first order $(u, v, w)$ non-dimensional relative motion is expressed in terms of orbit element differences. This analytical relative motion solution is valid for chief motions of any eccentricity, but also uses true anomaly as the independent variable. The orbit element difference based solution is written in terms of static offsets and sinusoidal components, and has a similar geometric structure as the analytical HE solution. Even for highly eccentric chief motions, the first order out-of-plane relative motion is still decoupled from the in-plane motion.

### 2.3 Bounded relative motion constraints

To avoid having the formation drift apart, bounded relative motion solutions are sought. If no perturbations are present, then the nonlinear bounded relative motion constraint is simply that all orbit periods must be equal. This is equivalent to requiring that the semi-major axis differences $\delta a$ be zero. This bounded motion constraint is valid for both circular and eccentric orbits, as well as small and large relative orbit dimensions.

The orbit element constraint $\delta a = 0$ can be approximated using Hill frame Cartesian coordinates by taking a first order expansion. An equivalent approach is to look at the analytical HE solution in Eq. (5). The only secular growth occurs in the along track direction through the $-3/2ntz_{\text{off}}$ term. For this secular growth term to be zero, we find that the initial Cartesian coordinate conditions must satisfy

$$\dot{x}_0 + 2nz_0 = 0 \tag{7}$$

This first order approximation of $\delta a = 0$ assumes that the chief is circular and that the relative orbit radius $\rho$ is small compared to the chief orbit radius $r_c$. However, this condition can be applied at any point within the orbit. If the chief motion has a small but non-zero eccentricity, then the first order bounded relative motion constraint is written as (Inalhan et al., 2002; Schaub and Junkins, 2003):

$$\dot{x}_0 + (2 + 3e)nz_0 = 0 \tag{8}$$

if the initial time is set at perigee.

If the gravitational $J_2$ perturbation is present, then all orbits experience short and long period perturbations. Only the ascending node $\Omega$, argument of periapsis $\omega$ and initial mean anomaly $M_0$ will experience secular drift. Their mean rates are given by (Schaub, 2004; Schaub and Alfriend, 2001):

$$\frac{\mathrm{d}\Omega}{\mathrm{d}t} = -\frac{\epsilon(a,e)}{2} n \cos i \tag{9a}$$

$$\frac{\mathrm{d}\omega}{\mathrm{d}t} = \frac{\epsilon(a,e)}{4} n \left(5\cos^2 i - 1\right) \tag{9b}$$

$$\frac{\mathrm{d}M_0}{\mathrm{d}t} = \frac{\epsilon(a,e)}{4} n\,\eta \left(3\cos^2 i - 1\right) \tag{9c}$$

with $\epsilon(a,e) = 3J_2\big(r_{\text{eq}}/a(1-e^2)\big)^2$ and where $\eta = \sqrt{1-e^2}$ is an eccentricity measure. The distance $r_{\text{eq}}$ is Earth's equatorial radius. Note that only $a$, $e$ and $i$ control the secular drift rate of the remaining three orbit elements. This drift could be compensated for by thrusting. However, this will quickly consume a lot of fuel. Schaub and Alfriend (2001) introduce the concept of $J_2$-invariant relative orbits. Here the relative orbit geometry is designed such that while all orbits are still drifting, on average, they will drift at equal rates. To achieve this, the following mean relative drift rates are set to zero:

$$\delta\dot{\theta} = \delta\dot{\omega} + \delta\dot{M}_0 = 0 \tag{10a}$$

$$\delta\dot{\Omega} = 0 \tag{10b}$$

The first condition guarantees no in-plane drift and leads to the orbit element constraint equation

$$\frac{\delta a}{a} = \frac{J_2}{2}\frac{a^2}{r_{\text{eq}}^2}\frac{1}{\eta}(4 + 3\eta)(1 + 5\cos^2 i)\delta\eta \tag{11a}$$

The second conditions controls the out-of-plane drift. It yields the orbit element constraint

$$\delta\eta = -\frac{\eta}{4}\tan i\,\delta i \tag{11b}$$

By choosing either a difference in eccentricity, inclination, or semi-major axis, the other two orbit element differences are then dictated through the constraints in Eqs. (11a) and (11b). Note that in order to have either a difference in eccentricity or inclination, a non-zero difference in semi-major axis is required.

This is a departure from the Keplerian bounded relative motion results. For near-polar chief motions with inclination differences, the $J_2$-invariance constraints may result in very large along-track relative orbit dimensions. To avoid this, the 2nd constraint (11b) is typically dropped and any out-of-plane secular drift will have to be compensated for through thrusting (Schaub and Junkins, 2003). When designing $J_2$-invariant relative orbits, the motion is typically described in mean element space, cf. (Schaub and Alfriend, 2001; Brouwer, 1959). To map between the osculating (instantaneous) orbit elements and the mean orbit elements (long period and secular terms removed), the Brouwer-Lyddane theory can be used (Lyddane, 1963; Brouwer, 1959).

## 3 Gravity mapping from satellite formations

Future low-low sst missions, whether formation flying or not, will most likely employ laser technology for the intersatellite link. Bender et al. (2003) discuss heterodyne laser interferometry, whereas McGuirk et al. (2002) discuss atomic interference. Differential accelerometry seems feasible at a level of $10^{-12}\,\mathrm{m\,s^{-2}}/\sqrt{\mathrm{Hz}}$. Over a baseline of 1 km this would translate already into gradiometry at the $10^{-6}\,\mathrm{E}/\sqrt{\mathrm{Hz}}$ level. The baseline length immediately scales into the error level.

The range rate $\dot\rho$ between two satellites is the projection of the relative vectorial velocity $\dot{\boldsymbol\rho}$ on the line-of-sight unit vector $\boldsymbol e$, e.g. (Rummel et al., 1978):

$$\dot\rho = \dot{\boldsymbol\rho}\cdot\boldsymbol e \qquad (12a)$$

$$\Rightarrow \ddot\rho = \ddot{\boldsymbol\rho}\cdot\boldsymbol e + \frac{1}{\rho}\left(\dot{\boldsymbol\rho}\cdot\dot{\boldsymbol\rho} - \dot\rho^2\right) \qquad (12b)$$

Using Newton's equations, the vectorial acceleration difference $\ddot{\boldsymbol\rho}$ equals the difference in gravitational attraction $\nabla V$ between the forces. The scalar range acceleration $\ddot\rho$ can be obtained from the observed range rate by numerical differentiation. To extract the gravitational information, one should further correct for the relative velocity terms at the right of (12b).

**Gradiometry from satellite formations.** Alternatively, when dividing $\ddot{\boldsymbol\rho}\cdot\boldsymbol e$ by the baseline, one obtains the in-line gravity gradient in the baseline direction $\boldsymbol e^{\mathrm T}\boldsymbol V\boldsymbol e$, with $\boldsymbol V$ the gravity gradient tensor. With the baseline close to along-track direction, this observable is mainly $V_{xx}$. Again, one should correct for the relative velocity terms at the right hand side of (12b). Moreover, one has to account for a linearization error in the approximation $V_{xx} \approx (V_{x,2} - V_{x,1})/(x_2 - x_1)$.

In a satellite formation, the baseline performs a full revolution in the Hill frame $\mathcal{O}$, i.e. the direction $\boldsymbol e$ rotates once every orbital revolution. Thus the observed gravity gradient $\boldsymbol e^{\mathrm T}\boldsymbol V\boldsymbol e$ contains projections of several tensor components $V_{ij}$, $i,j \in \{x,y,z\}$. The gravity gradient tensor $\boldsymbol V$ transforms under a rotation of the coordinate frame as $R\boldsymbol V R^{\mathrm T}$, in which $R$ denotes the rotation matrix. Let us consider one satellite pair only in the simplest formation, namely the 2:1-ellipse in the orbital plane. Now assume a time-variable rotation $\alpha$ about the $y$-axis, such that the two satellites are always on the new $x'$-axis. The coplanar gradients $V_{xx}, V_{xz}, V_{zz}$ project onto the observable as follows:

$$V_{x'x'} = \cos^2\alpha V_{xx} + 2\cos\alpha\sin\alpha V_{xz} + \sin^2\alpha V_{zz}$$

The observable $V_{x'x'}$ ($= \boldsymbol e^{\mathrm T}\boldsymbol V\boldsymbol e$) contains the required gravity observable already. However, if one wants to disentangle the 3 contributing tensor components in the Hill frame, 3 independent intersatellite distances need to be tracked. With 3 different angles $\alpha$ one would have 3 simultaneous equations of the above kind, leading to an instantaneous determination of $V_{xx}$, $V_{xz}$ and $V_{zz}$. This can either be realized by a CartWheel of 3 satellites, measuring in a triangle, or by 6 satellites, measuring along the spokes of the wheel, cf. figure 4. The spokes configuration may be easier to realize at the cost of more satellites. The intersatellite links in the triangular formation are dependent. Technologically that may be more demanding, but it has the added benefit that the required orientations are better constrained.

**Fig. 4.** Potential coplanar configurations for measuring the in-plane $V_{xx}$, $V_{xz}$ and $V_{zz}$ simultaneously: triangle edges (left) or spokes (right).

Gradiometry of out-of-plane components $(V_{xy}, V_{yy}, V_{yz})$ can only be achieved through non-coplanar satellite configurations. A relative inclination of the formation w.r.t. the orbit plane can be represented by a rotation $R_x(\beta)$. Along the same lines of arguing as above it can be demonstrated that all gravity gradient tensor components will generally project onto a particular $V_{x'x'}$. To disentangle this projection, 6 instantaneous intersatellite distances should be measured. Thus formation flying offers a way of full-tensor gravity gradiometry.

## 4 Conclusion

When designing future gravity field missions, formation flying is a viable alternative to leader-follower low-low SST configurations. Despite the presence of perturbations—the strongest being the Earth's oblateness—stable configurations exist. The homogeneous Hill equations demonstrate which natural formation shapes are possible. In its simplest form, a 2:1 relative ellipse, the radial gravity gradient $V_{zz}$ is projected onto the SST observable. Thus, the inherent weakness and the non-isotropic behaviour of the conventional low-low SST observable can be solved by formation flying.

Gravity field recovery can be based on observed range rates $\dot{\rho}$. Alternatively they may be differentiated numerically into $\ddot{\rho}$, which can be interpreted as differences in the gravitational attraction between the satellites. Moreover, $\ddot{\rho}$ can be recast into a gravity gradient observable $\boldsymbol{e}^{\mathrm{T}}\boldsymbol{V}\boldsymbol{e}$. With sufficiently many satellites linked together in a strategic way, one can even achieve full-tensor gravity gradiometry.

If the relative orbits comprise a cross-track motion, the corresponding observables gain sensitivity in East-West direction. Although this may be helpful in dealiasing signals, the fundamental temporal-spatial sampling problem of a gravity field satellite mission is not addressed. To overcome aliasing multiple-formation configurations must be considered, such as the planned geomagnetic field mission SWARM: one satellite pair at the same altitude but with different right-ascension, plus a single higher satellite.

These results are only the first step towards establishing formation flying technology as a powerful tool for future gravity field satellite missions. In future closed-loop simulations the gravity recovery performance of such missions must be carefully assessed under a variety of mission parameters and formation designs. Moreover, it is obvious that such missions will prompt further technological questions, e.g. into orbit-, attitude- and drag-control. These and other issues must be addressed in future studies.

Finally, if the individual satellites can be designed and launched in a cost-effective way, a formation flying mission would be suitable as a long-term monitoring mission.

**Acknowledgments.** Nico Sneeuw gratefully acknowledges the support of the Alexander von Humboldt Foundation and of the GEOIDE Network of Centres of Excellence.

## References

Aguirre-Martinez, M. and Sneeuw, N. (2003). Needs and tools for future gravity measuring missions. *Space Science Reviews*, 108(1-2):409–416.

Bender, P. L., Hall, J. L., Ye, J., and Klipstein, W. M. (2003). Satellite-satellite laser links for future gravity missions. *Space Science Reviews*, 108(1-2):377–384.

Brouwer, D. (1959). Solution of the problem of artificial satellite theory without drag. *The Astronomical Journal*, 64(1274):378–397.

Carter, T. E. (1998). State transition matrix for terminal rendezvous studies: Brief survey and new example. *Journal of Guidance, Control and Dynamics*, 31(1):148–155.

Clohessy, W. H. and Wiltshire, R. S. (1960). Terminal guidance system for satellite rendezvous. *Journal of the Aerospace Sciences*, 27(9):653–658.

Hill, G. W. (1878). Researches in the lunar theory. *Am. Journal of Math.*, I:5–26,129–147,245–260.

Humi, M. and Carter, T. (2003). The Clohessy-Wiltshire equations can be modified to include quadratic drag. In *Proc. AAS/AIAA Space Flight Mechanics Meeting*, number AAS 03–240, Ponce, Puerto Rico. AAS/AIAA.

Inalhan, G., Tillerson, M., and How, J. P. (2002). Relative dynamics & control of spacecraft formations in eccentric orbits. *Journal of Guidance, Control and Dynamics*, 25(1):48–59.

Lyddane, R. H. (1963). Small eccentricities or inclinations in the Brouwer theory of the artificial satellite. *The Astronomical Journal*, 68(8):555–558.

McGuirk, J. M., Foster, G. T., Fixler, J. B., Snadden, M. J., and Kasevich, M. A. (2002). Sensitive absolute-gravity gradiometry using atom interferometry. *Physical Review A*, 65:doi 10.1103/PhysRevA.65.033608.

Rummel, R., Reigber, Ch., and Ilk, K.-H. (1978). The use of satellite-to-satellite tracking for gravity parameter recovery. In *Proc. European Workshop On Space Oceanography, Navigation And Geodynamics (SONG)*, volume SP-137, pages 153–161, Schloss Elmau. ESA.

Schaub, H. (2004). Relative orbit geometry through classical orbit element differences. *Journal of Guidance, Control and Dynamics*, 27(5):839–848.

Schaub, H. and Alfriend, K. T. (2001). $J_2$ invariant relative orbits for spacecraft formations. *Celestial Mechanics and Dynamical Astronomy*, 79(2):77–95.

Schaub, H. and Junkins, J. L. (2003). *Analytical Mechanics of Space Systems*. AIAA Education Series, Reston, VA.

Schweighart, S. A. and Sedwick, R. J. (2002). A high fidelity linearized $J_2$ model for satellite formation flight. *Journal of Guidance, Control, and Dynamics*, 6(25):1073–1080.

Tapley, B. D., Bettadpur, S., Ries, J. C., Thompson, P. F., and Watkins, M. M. (2004). GRACE measurement of mass variability in the Earth system. *Science*, 305:503–505.

# A Preliminary Gravitational Model to Degree 2160

N.K. Pavlis, S.A. Holmes
Raytheon ITSS Corporation, 1616 McCormick Drive, Upper Marlboro, Maryland 20774, USA
Nikolaos_Pavlis@raytheon.com Fax: +301-883-4140

S.C. Kenyon, D. Schmidt, and R. Trimmer
National Geospatial-Intelligence Agency, Arnold, Missouri 63010, USA

**Abstract.** The National Geospatial-Intelligence Agency (NGA) of the USA has embarked upon the development of a new Earth Gravitational Model (EGM), to support future realizations of NGA's World Geodetic System. Current plans call for the development of the new EGM (EGM05) by the end of 2005. The new model will be complete to degree and order 2160, and aims at a ±15 cm global Root Mean Square (RMS) geoid undulation error requirement. The new model will combine optimally the gravitational information that is extracted from dedicated geopotential mapping satellite missions (CHAMP, GRACE), with the information contained within a global gravity anomaly database of 5′x5′ resolution. This paper describes the development of a Preliminary Gravitational Model (PGM2004A). We developed PGM2004A by combining the GRACE-only model GGM02S, with a 5′x5′ global gravity anomaly database compiled by NGA. PGM2004A is complete to degree and order 2160, and is accompanied by propagated error maps at 5′x5′ resolution, accounting for the *entire* bandwidth of the model (from degree 2 to degree 2160), for various model-derived gravimetric quantities ($\Delta g, N, \xi, \eta$). We have evaluated PGM2004A through comparisons with *independent* data including GPS/Leveling data, astronomic deflections of the vertical over the conterminous US (CONUS), and altimeter data from TOPEX. The results of these comparisons indicate that the goal set for EGM05 is well within reach. We summarize in this paper our current status and technical accomplishments, and discuss briefly our next steps towards the development of EGM05.

**Keywords.** Earth gravitational model, spherical and ellipsoidal harmonics, high-degree expansion.

## 1 Introduction

The new satellite missions GRACE (JPL, 1998) and GOCE (ESA, 1999) (to be launched in 2006) promise quantum leaps in the accuracy and resolution of *satellite-only* gravitational models. Nevertheless, there is still a need to combine that information with terrestrial gravity and satellite altimetry data in an optimal fashion. Such combination could permit the "seamless" extension of the gravitational spectrum (beyond the resolution recoverable from space techniques), taking advantage of the rich high frequency content in surface gravimetric and altimetric data. Recognizing that need, NGA has embarked upon the development of a new Earth Gravitational Model (EGM05) that will be complete to degree and order 2160. EGM05 is expected to be a *composite* model like EGM96 (Lemoine et al., 1998). A comprehensive combination solution will define the model up to the maximum degree that will be recoverable from GRACE data (~140 or so). Within this comprehensive solution, the GRACE-derived information will be combined with surface gravity data and a Mean Sea Surface, to estimate *simultaneously* gravitational potential coefficients, Dynamic Ocean Topography (DOT) coefficients, and coefficients representing biases within the surface gravity data. Pavlis and Kenyon (2003) discuss some of the design considerations and numerical aspects pertinent to such a combination solution. A complete error covariance matrix will accompany this comprehensive combination solution. Beyond the maximum degree recoverable from GRACE data, and up to degree 2160, EGM05 will be determined from the analysis of a complete, global set of 5′x5′ gravity anomaly data, using block-diagonal and numerical quadrature harmonic analysis techniques.

## 2 The Development of PGM2004A

While the development of EGM05 is in progress, the currently available data permitted the development of a Preliminary Gravitational Model designated PGM2004A. This model was developed primarily in order to gauge our progress and, more importantly, to identify areas requiring improvement. In terms of data compliment and estimation approach, PGM2004A is very similar to the block-diagonal combination solutions described by Pavlis in Lemoine et al. (1998, Chap. 8). The GRACE-derived satellite-only model GGM02S (Ries, personal communication, 2004) and a 5′x5′ global gravity anomaly database are combined within a block-diagonal least squares adjustment, to yield the coefficients of PGM2004A up to degree and order 2159, as well as their error estimates. This adjustment does not make use of any *a priori* constraints. Spectral weights are used for the surface gravity information, thus affecting a degree-dependent weighting of the 5′ gravity anomaly data, designed to compensate for the neglected error correlations among these data. This approach circumvents many of the shortcomings of the weighting schemes used for corresponding data in previous models. The PGM2004A coefficients of degree 2160 were estimated separately using numerical quadrature harmonic analysis. The main steps of the development of PGM2004A are:

(1) A block-diagonal least squares fit as described in detail by Pavlis in Lemoine et al. (1998, Chap. 8) creates the normal equations implied by the 5′ gravity anomaly data, in terms of ellipsoidal harmonics. This fit also produces a model (designated PGM2004SG) representing *only* the surface gravity information.

(2) The GGM02S coefficients and their standard deviations are converted from spherical to ellipsoidal harmonic representation using Jekeli's (1988) algorithm.

(3) The (strictly diagonal) normal equations of GGM02S (*only* the standard deviations of the model coefficients were available to us) are combined within a least squares adjustment with the surface gravity block-diagonal normal equations. We will use the complete error covariance matrix accompanying the GRACE-derived coefficients, when such a matrix becomes available. This adjustment produces the ellipsoidal harmonic spectra (signal and error) of the combined solution (PGM2004A).

(4) The ellipsoidal harmonic signal and error spectra of PGM2004A are converted to the corresponding spherical ones (Jekeli, 1988).

In the development of PGM2004A, the analytical continuation necessary to reduce the surface free-air $\Delta g$ to the ellipsoid (a surface of revolution whose symmetries give rise to the block-diagonal normal equation patterns) was omitted. This affects mostly coefficients beyond degree ~90, since lower degree coefficients are dominated by the GRACE information. Analytical continuation will be one of the focal points of our future analysis work.

## 3 Gravity Anomaly Data Used

A complete global file of 5′x5′ area-mean surface free-air gravity anomalies was compiled by NGA, and became available for this study in May 2004. The gravity anomalies within this "merged" file originate from terrestrial, airborne, and altimetry-derived sources. Within this file, approximately 17% of the Earth's area is occupied by "fill-in" values. These were computed from a composite gravitational model comprised of GGM02S up to degree 100 and EGM96 from degree 101 to 360, plus the gravity anomaly contribution implied by the Residual Terrain Model (RTM) effect (Forsberg, 1984). Figure 1 shows the distribution by source of these 5′ $\Delta g$. Table 1 summarizes their essential statistics.

**Table 1.** Statistics of the 5′ gravity anomaly data (mGal).

| Source | % Area | Min. | Max | RMS | RMS $\sigma$ |
|---|---|---|---|---|---|
| ArcGP | 4.0 | -186.0 | 235.6 | 31.3 | 8.1 |
| Altimetry | 67.0 | -360.4 | 377.9 | 29.2 | 2.5 |
| Terrestrial | 12.3 | -223.1 | 425.3 | 34.8 | 6.5 |
| Fill-in | 16.7 | -318.7 | 582.5 | 44.0 | 18.6 |
| Non Fill-in | 83.3 | -360.4 | 425.3 | 30.2 | 3.8 |
| All | 100.0 | -360.4 | 582.5 | 32.9 | 8.4 |

Notice that the area void of high quality 5′ data (i.e., that occupied by the "fill-in" values) is also the "roughest" area of the gravity anomaly field. This affects the anomaly degree variances recovered from PGM2004A, as we discuss next.

## 4 The Estimated Spectra

Following the procedure outlined in Section 2, we used the 5′ gravity anomaly data discussed in Section 3, to estimate first (through a block-diagonal least squares fit) the ellipsoidal spectrum representing *only* the surface gravity information (PGM2004SG). Statistics of the residual 5′ gravity anomalies from this fit are shown in Table 2.

**Table 2.** Statistics of the residual 5′ gravity anomalies produced by PGM2004SG using a block-diagonal least squares fit to degree 2159 and numerical quadrature for degree 2160 (units are mGal).

| Source | % Area | Min. | Max | RMS |
|---|---|---|---|---|
| ArcGP | 4.0 | -51.1 | 75.1 | 1.87 |
| Altimetry | 67.0 | -117.5 | 53.3 | 0.43 |
| Terrestrial | 12.3 | -57.8 | 55.1 | 1.75 |
| Fill-in | 16.7 | -57.5 | 95.4 | 0.93 |
| All | 100.0 | -117.5 | 95.4 | 0.88 |

These residuals represent primarily signal and/or noise present in the 5′ $\Delta g$ *beyond* degree and order 2160. Wenzel (1998) reported corresponding RMS residual misfits of ±5.3 mGal, ±5.1 mGal, and ±7.9 mGal for GPM98A, B, and C, respectively (considering in all cases expansions to degree 1799). These values are to be compared with the ±0.88 mGal RMS misfit of our analysis.

The PGM2004SG normal equations were then combined with the GGM02S information to produce the PGM2004A model. Figure 2 shows the gravity anomaly degree variances, for the signal and error spectra involved in the combination solution. Figure 2a shows the low degree portion of these degree variances. Our weighting of the 5′ $\Delta g$ allows the GGM02S model to dominate the combination solution up to degree and order 90 or so. In this study we considered the error spectrum of GGM02S to be well "calibrated" and did not alter this spectrum in any fashion. The surface gravity information dominates above degree 130. The "transition" from GRACE to surface gravity is effected within a relatively narrow degree range $(90 < n < 130)$, primarily because the error spectra of PGM2004SG and GGM02S intersect at a rather steep angle. Figure 2b shows the spectra for the entire bandwidth of PGM2004A. The signal spectrum of PGM2004A suffers a small "jump" discontinuity at degree 360. This is most likely due to the lack of high quality 5′ data over some of the "roughest" areas of the anomaly field of the Earth, as we discussed in Section 3. PGM2004A's signal spectrum dips below its error spectrum around degree 2010.

In addition to an error spectrum, PGM2004A is accompanied by global 5′x5′ geographic grids of its propagated (commission) error on geoid undulations, gravity anomalies, deflections of the vertical, and gravity disturbances. These grids were computed using the technique of Pavlis and Saleh (this issue), and, *for the first time*, correspond to the entire bandwidth of the model, up to degree and order 2160. Figure 3 shows the geoid undulation propagated error of PGM2004A to degree 2160.

**Table 3.** Propagated error of various gravimetric quantities implied by PGM2004A to degree 2160.

| Region | RMS $\sigma(N)$ (cm) | RMS $\sigma(\Delta g)$ (mGal) | RMS $\sigma(\xi)$ (arcsec) | RMS $\sigma(\eta)$ (arcsec) |
|---|---|---|---|---|
| "T/P" | 15.8 | 5.3 | 0.78 | 0.80 |
| CONUS | 15.7 | 5.4 | 0.81 | 0.81 |
| Land | 27.1 | 9.8 | 1.46 | 1.46 |
| Ocean | 16.3 | 5.5 | 0.82 | 0.83 |
| Globe | 20.1 | 7.0 | 1.05 | 1.06 |

The availability of these geographic grids of the propagated error estimates on various gravimetric functionals, for the *entire* model bandwidth, allows error estimation over any geographic area of interest. These area-specific error estimates can then be compared to the *observed* performance of the model in the particular geographic area, as this performance may be deduced from comparisons with independent data (provided of course that independent test data are available in the particular area). This, in turn, permits optimal weighting and "calibration" of the errors assigned to the gravity data over a particular area to be performed with much higher fidelity than it was possible in previous high-degree gravitational model developments. Table 3 summarizes the propagated errors of various gravimetric quantities implied by PGM2004A to degree 2160, over some geographic areas of interest ("T/P" denotes the ocean area within latitude ±66°). In the next section, we discuss some comparisons with independent data, whose results in certain cases can be compared to the propagated errors shown in Table 3. We emphasize here that the weighting technique used in the development of PGM2004A has not been "calibrated" – it represents our *first* attempt at estimating appropriate weights for the 5′ gravity anomaly data. Our final solution will be considered well "calibrated" when its propagated errors will match its observed performance in comparisons with independent data (after consideration of omission errors and errors associated with the independent data).

## 5 Model Evaluation

In the following comparisons we consider 4 models:

(1) EGM96 (Lemoine et al., 1998) to $n = 360$.
(2) The composite model "G02S/EGM96" consisting of: GGM02S $(2 \le n \le 100)$ and EGM96 $(101 \le n \le 360)$.
(3) PGM2004A truncated to $n = 360$.
(4) PGM2004A complete to $n = 2160$.

### 5.1 Comparisons with TOPEX Altimetry

We have used the Sea Surface Heights (SSH) of a 6-year mean track of TOPEX/Poseidon, sampled at the 1 Hz rate. A 1000 m depth "mask" was used in all the comparisons, to exclude shallow water SSH data that may be less reliable (e.g., due to tidal model errors). In addition, we developed a Dynamic Ocean Topography (DOT) model to degree and order 60, based on the GSFC00.1 Mean Sea Surface (Wang, 2001) and the GGM02S-implied geoid.

This DOT model is subtracted from the 1 Hz TOPEX SSH, to yield an estimate of the geoid undulation, that is compared to the corresponding model-derived value. Apart from the residual SSH (499,234 values in total), we also use residual along-track slopes (476,653 values in total), which provide a good measure of the accuracy of model-derived deflections of the vertical over ocean areas. Table 4 summarizes the results of our comparisons.

**Table 4.** Comparisons with residual SSH and along-track slopes from TOPEX altimeter data.

| Model (Nmax) | Residual SSH (cm) | | Along-Trk. Slope (″) |
|---|---|---|---|
| | Max ‖•‖ | Std. Dev. | Std. Dev. |
| EGM96 (360) | 331 | 20.7 | 1.92 |
| G02S/EGM96 (360) | 301 | 19.4 | 1.92 |
| PGM2004A (360) | 297 | 17.8 | 1.86 |
| PGM2004A (2160) | 177 | 10.4 | 0.44 |

The results of Table 4 show that our combination solution up to degree 360 outperforms both EGM96 and the composite model G02S/EGM96 in this test. The extension of PGM2004A to degree 2160 reduces the standard deviation of the residual SSH by about 60%, and of the residual along-track slopes by about a factor of 4.2. The observed performance of PGM2004A in this test compared to its propagated errors (Table 3, "T/P") indicates that our weighting of the oceanic (mostly altimetric) gravity anomalies may be pessimistic.

## 5.2 Comparisons with Astronomic Deflections of the Vertical over CONUS

We have used 3561 astronomic deflections of the vertical over the conterminous US (CONUS). Jekeli (1999) describes this dataset, and discusses in detail the appropriate systematic corrections that should be applied to its values before comparing them to model-derived estimates. Table 5 summarizes the results of our comparisons. In this test, we also used the 1′x1′ gravimetric deflections computed at NGS (see www.ngs.noaa.gov/GEOID/DEFLEC99 for details), which we obtained in October 2002.

**Table 5.** Comparisons with 3561 astronomic deflections of the vertical over CONUS (units are arc seconds).

| Model (Nmax) | RMS $\Delta\xi$ | RMS $\Delta\eta$ |
|---|---|---|
| EGM96 (360) | 2.80 | 3.22 |
| G02S/EGM96 (360) | 2.79 | 3.22 |
| PGM2004A (360) | 2.75 | 3.17 |
| PGM2004A (2160) | 1.22 | 1.27 |
| DEFLEC99 (1′→10800) | 0.91 | 0.92 |

As expected, there is little (but noticeable) distinction in this comparison between the three models extending to degree 360. A major reduction of the RMS differences occurs when PGM2004A is extended to degree and order 2160, and its performance begins to approximate that of the detailed gravimetric deflections computed at NGS. Although the NGS file outperforms PGM2004A by ~30%, we should note here that PGM2004A's performance is somewhat hindered in this test by the omission of the analytical continuation corrections in its development. We will revisit this in future models, which will include the analytical continuation corrections.

## 5.3 Comparisons with GPS/Leveling Data

Over several years we maintain a global database of GPS/Leveling (GPS/L) data, generously contributed by various colleagues. Currently, our database contains a total of 12684 points, distributed over 46 countries. 6169 of these points are located within CONUS, and 1930 points are in Canada. The majority of our GPS/L data holdings are located over North America, Europe, and Australia, but points over South America, Asia, and Africa are also present in our database. The geographic distribution of these GPS/L stations is uneven, and in many occasions clusters of stations located extremely close to each other are present in the data. This affects significantly the statistics of our comparisons, and may produce misleading results. After careful inspection of the geographic distribution of the GPS/L data within each source, we created a "thinned" version of our database containing 10767 points in total, 4649 of which are in CONUS. Within our database, some of the GPS/L sources provide geoid undulations $(N)$, while others provide height anomalies $(\zeta)$. We account for this using Rapp's (1997) formulation. Details of our testing procedure include:

(1) We computed the $\zeta$ to $N$ conversion using the same coefficient set as in EGM96, which extends only to degree 360 (Rapp in Lemoine et al., 1998, section 5.2.1). In the future, we will replace this set with one computed from gravitational and elevation models complete to degree 2160.

(2) The GPS/L data over CONUS are segmented by state (see also Smith and Roman, 2001).

(3) We apply a ±2 meter edit to the differences between GPS/L undulations (or height anomalies) and model-derived values.

(4) We compute statistics of GPS/L undulations minus model-derived values after removing a bias, as well as after removing a linear trend.

**Table 6.** GPS/Leveling Comparisons over CONUS.

| Model (Nmax) | Bias Removed | | Linear Trend Removed | |
|---|---|---|---|---|
| | Number Passed Edit | Weighted Std. Dev. (cm) | Number Passed Edit | Weighted Std. Dev. (cm) |
| EGM96 (360) | 4547 | 21.4 | 4544 | 18.1 |
| G02S/EGM96 (360) | 4614 | 19.1 | 4611 | 17.7 |
| PGM2004A (360) | 4627 | 18.3 | 4624 | 16.9 |
| PGM2004A (2160) | 4648 | 9.7 | 4645 | 7.3 |
| G99SSS (1′→10800) | 4649 | 9.1 | 4646 | 5.7 |

**Table 7.** GPS/Leveling Comparisons Globally.

| Model (Nmax) | Bias Removed | | Linear Trend Removed | |
|---|---|---|---|---|
| | Number Passed Edit | Weighted Std. Dev. (cm) | Number Passed Edit | Weighted Std. Dev. (cm) |
| EGM96 (360) | 10571 | 29.3 | 10528 | 25.8 |
| G02S/EGM96 (360) | 10645 | 24.4 | 10602 | 22.0 |
| PGM2004A (360) | 10657 | 22.8 | 10614 | 20.5 |
| PGM2004A (2160) | 10680 | 15.5 | 10637 | 12.6 |

Tables 6 and 7 summarize our GPS/L comparison results. Table 6 includes the statistics of the differences with the detailed (1′x1′) gravimetric geoid G99SSS (Smith and Roman, 2001). The results of Tables 6 and 7 are quite reassuring. Moving from EGM96 to newer models (expected to be more accurate), the number of points passing the ±2 m editing criterion is monotonically increasing, while the standard deviation of the differences is monotonically decreasing. There is noticeable difference in the performance of the model G02S/EGM96, and that of our combination solution PGM2004A, when both are considered up to degree 360. To degree 2160, PGM2004A performs tantalizing close to the detailed geoid G99SSS. The latter however was developed based on EGM96, and therefore does not benefit the long wavelength improvements brought about by GGM02S. Comparing the *observed* performance of PGM2004A, to the propagated errors of Table 3, we conclude that our weighting of the surface gravity data (at least over the areas where GPS/L data are available) may be pessimistic.

## 6 Summary and Future Work

This paper described the development and evaluation of a global gravitational model (PGM2004A), complete to degree and order 2160, from the combination of 5′ gravity anomaly data with the satellite-only model GGM02S. PGM2004A is a preliminary solution developed in preparation for EGM05. Its performance indicates that the goals set by NGA for EGM05 are well within reach. Future work will focus on improving the 5′x5′ gravity anomaly database, the error estimation associated with the 5′ data, and their analytical continuation.

***Acknowledgements.*** We thank Byron Tapley and the UT/CSR GRACE team for providing the model GGM02S, and Jarir Saleh (Raytheon) for his careful editing of the GPS/Leveling data and for creating numerous graphics used in this study.

## References

Forsberg, R. (1984). A study of terrain reductions, density anomalies and geophysical inversion methods in gravity field modelling. *Rep. 355*, Dep. of Geod. Sci. and Surv., Ohio State Univ., Columbus, OH.

Jekeli, C. (1988). The exact transformation between ellipsoidal and spherical harmonic expansions. *manusc. geod.*, ***13** (2)*, 106-113.

Jekeli, C. (1999). An analysis of vertical deflections derived from high-degree spherical harmonic models. *J. Geod.*, *73*, 10-22.

Lemoine, F.G., et al. (1998). The development of the joint NASA GSFC and the National Imagery and Mapping Agency (NIMA) geopotential model EGM96. *NASA Tech. Publ. TP-1998-206861*, NASA GSFC.

Pavlis, N.K., S.C. Kenyon (2003). Analysis of surface gravity and satellite altimetry data for their combination with CHAMP and GRACE information. In: *Gravity and Geoid 2002*, I.N. Tziavos (Ed.), Thessaloniki, Greece.

Rapp, R.H. (1997). Use of potential coefficient models for geoid undulation determinations using a spherical harmonic representation of the height anomaly/geoid undulation difference. *J. Geod.*, *71*, 282-289.

Smith, D.A., D.R. Roman (2001). GEOID99 and G99SSS: 1-arc-minute geoid models for the United States. *J. Geod.*, *75*, 469-490.

Wang, Y.M. (2001). GSFC00 mean sea surface, gravity anomaly, and vertical gravity gradient from satellite altimeter data, *J. Geophys. Res.*, ***106** (C12)*, 31167-31175

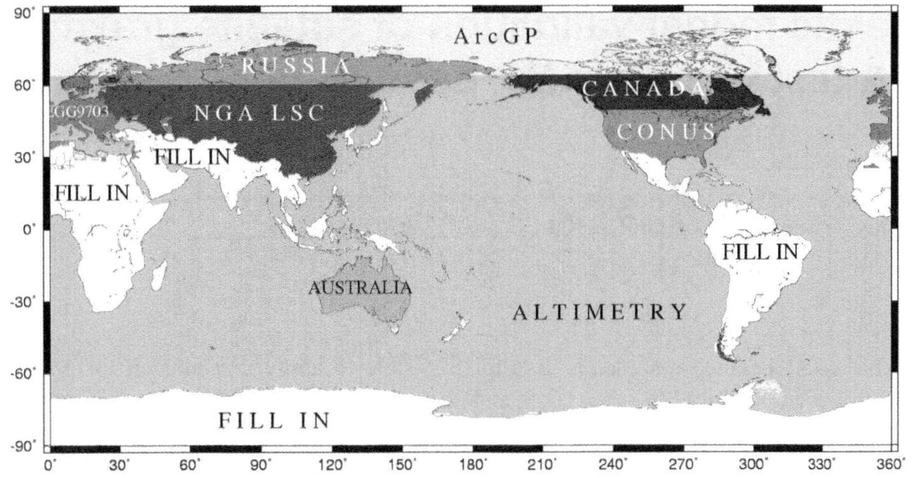


**Fig. 1** Geographic Distribution of 5′ Gravity Anomaly Data Sources.

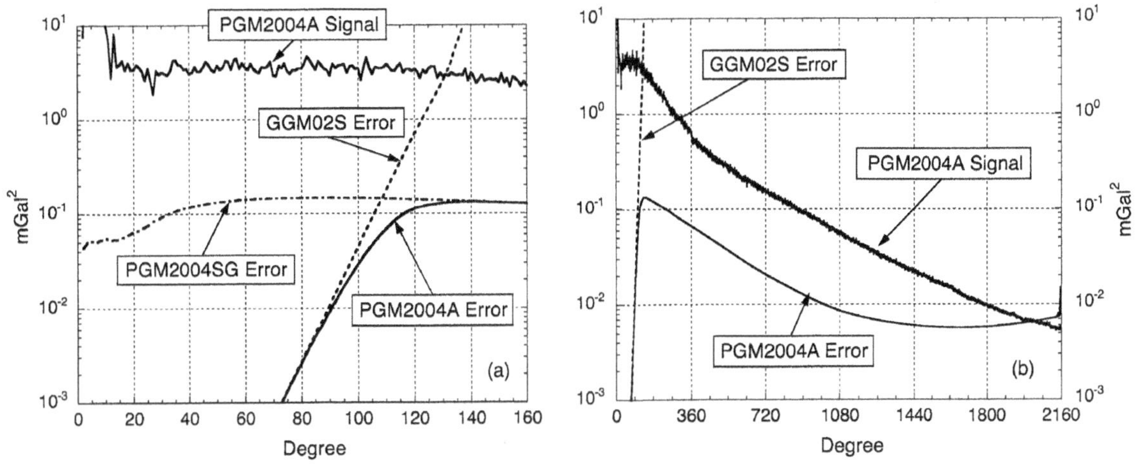


**Fig. 2** Gravity Anomaly Degree Variances from Ellipsoidal Harmonic Coefficients: (a) to degree 160, (b) to degree 2160.

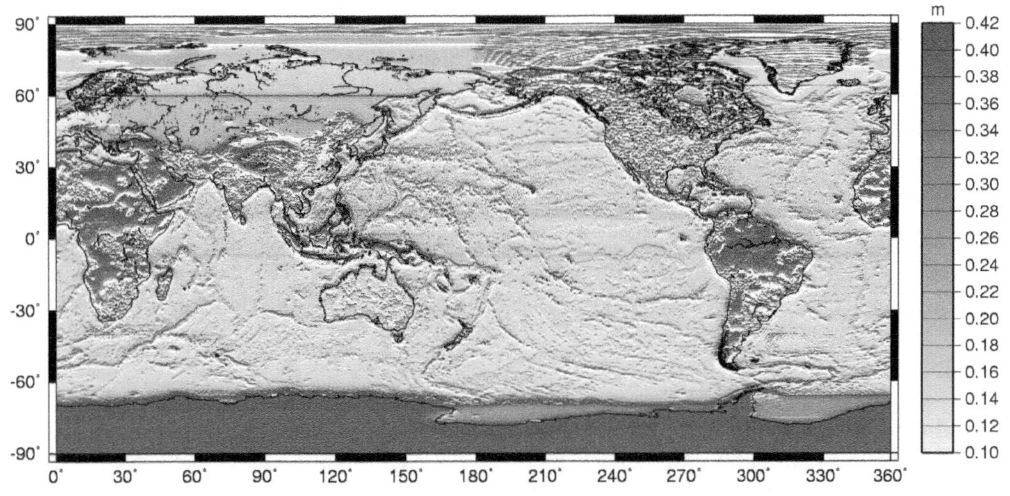


**Fig. 3** 5′x5′ Geoid Undulation Propagated (Commission) Error from PGM2004A to Degree 2160.

# Stochastic model validation of satellite gravity data: A test with CHAMP pseudo-observations

J.P. van Loon, J. Kusche
DEOS,
TU Delft, Kluyverweg 1, PO box 5058, 2600 GB Delft, The Netherlands

**Abstract.** The energy balance approach is used for a statistical assessment of CHAMP orbits, data and gravity models. It is known that the quality of GPS–derived orbits varies and that CHAMP accelerometer errors are difficult to model. The stochastic model of the in–situ potential values from the energy balance is therefore heterogeneous and it is unclear if it can be described accurately using a priori information. We have estimated parameters of this stochastic model in an iterative variance–component estimation procedure, combined with an outlier rejection method. This means we solve simultaneously for a spherical harmonic model, for polynomial coefficients absorbing accelerometer drift, for sub–daily noise variance components, and for a variance parameter that controls the influence of an a–priori gravity model (EGM96). We develop here, for the first time, a fast Monte Carlo variant of the Minimum Norm Quadratic Unbiased Estimator (MINQUE) as an alternative for the fast Monte Carlo Maximum Likelihood VCE that we have introduced earlier. In this way, spurious data sets could be indicated and downweighted in the least-squares estimation of the unknown parameters. Using only 299 days of CHAMP kinematic orbit data, the quality of the estimated global gravity model was found comparable to the EIGEN-3P model. Monte Carlo Variance Components Estimation appears to be a valid method to estimate the stochastic model of satellite gravity data and thus improves the least squares solution considerably.

**Keywords.** CHAMP, energy balance approach, statistical assessment, variance components

---

## 1 Energy Balance Approach

The theory of the energy balance approach and its potential application to LEO satellite experiments goes back to the 60's, and has been considerably revived recently, see Jekeli (1999) and Visser et al. (2003). The launch of the CHAMP satellite has made it possible to apply this method for gravity field recovery due to the high quality GPS receivers and an on board accelerometer to measure the effect of non-conservative forces (e.g. air-drag and solar radiation pressure). Various groups have demonstrated that the energy balance approach is a valid method to compute the Earth's gravity field from CHAMP reduced-dynamic or kinematic orbits, see Gerlach et al. (2003), Howe et al. (2003), Löcher and Ilk (2004) and Kusche and van Loon (2004).

The principle of the energy balance approach is to find a balance between the kinetic energy of the satellite and its potential energy. We use the following formulation, which is based on expressing all quantities of interest in an inertial coordinate system:

$$T(t) - V^{\mathrm{ref}}(t) = \delta V(t) + R(t) + \\ + E_0 + \int_{t_0}^{t} \boldsymbol{f} \cdot \dot{\boldsymbol{x}}\, d\tau + \int_{t_0}^{t} \nabla V^{\mathrm{tides}} \cdot \dot{\boldsymbol{x}}\, d\tau \qquad (1)$$

Here $T = \frac{1}{2}|\dot{\boldsymbol{x}}|^2$ is the kinetic potential, $V^{\mathrm{ref}}$ is a static reference potential appearing time–dependent in inertial coordinates, $\delta V$ is a residual geopotential that we parameterize by spherical harmonics whose coefficients $\delta c_{lm}, \delta s_{lm}$ are to be solved for, $R$ is the potential rotation term which approximates the potential contribution $\int_{t_0}^{t} \frac{\partial V}{\partial t}\, d\tau \approx -\omega_e(x_1\dot{x}_2 - x_2\dot{x}_1)$ (up to a constant, see Jekeli 1999) of the rotating earth in inertial space, and $E_0$ is a constant. Furthermore, $\boldsymbol{f}$ are corrected measurements from CHAMP's STAR accelerometer to account for non–conservative forces, and the last term on the right–hand side accommodates for tidal effects by evaluating the corresponding work integral. We model the direct attraction by sun and moon from JPL DE ephemeris, the solid earth tides following the IERS conventions, plus ocean tides (GOT 99.2, see Ray (1999)).

We have estimated simultaneously corrections to the spherical harmonic coefficients, sub–daily polynomial coefficients describing residual (after applying bias and scale factors from the accelerometer files) drift of the accelerometer, and sub–daily variance components of the in–situ potential values. A known obstacle for this type of analysis is the selection of 'good' orbits. In our approach, arcs showing spurious behavior are effectively down-weighted within an iterative variance component estimation (VCE) process, which improves our gravity field solution significantly. As a by–product, a variance (or regularization) factor for controlling the influence of an a–priori gravity model is determined in the same way.

## 2 Statistical Assessment and Estimation Procedure

The energy balance approach uses eq (1) for combining orbit, accelerometry data, reference geopotential model, tidal corrections and auxiliary information (e.g. earth rotation) into a preprocessed stream of pseudo–observations, $\delta V(t)$, which can be used without further linearization to estimate the $\delta c_{lm}, \delta s_{lm}$. Consequently, position errors $\boldsymbol{\epsilon}_{\boldsymbol{x}}$, velocity errors $\boldsymbol{\epsilon}_{\dot{\boldsymbol{x}}}$, accelerometry errors $\boldsymbol{\epsilon}_{\boldsymbol{f}}$ and tide model errors affect the in–situ potential differences (see Visser et al., 2003) approximately as

$$\epsilon_V = \dot{\boldsymbol{x}} \cdot \boldsymbol{\epsilon}_{\dot{\boldsymbol{x}}} - \nabla V^{\text{ref}} \cdot \boldsymbol{\epsilon}_{\boldsymbol{x}} - \omega_e(\epsilon_{x_1}\dot{x}_2 - \epsilon_{x_2}\dot{x}_1)$$
$$-\omega_e(x_1\epsilon_{\dot{x}_2} - x_2\epsilon_{\dot{x}_1}) + \int_{t_0}^{t} \boldsymbol{\epsilon}_{\boldsymbol{f}} \cdot \dot{\boldsymbol{x}}\, d\tau + \delta V^{\text{tides}} \tag{2}$$

It is clear that accelerometer biases and scaling errors, predominantly in the in–flight axis, cause in first approximation a linear drift in the $\delta V$ measurement. Knowing the variance of the potential difference error, $\sigma^2(\delta V) = E\{\epsilon_V^2\}$, we can set up a weighted least square adjustment which would suppress spurious arcs by down-weighting. This noise variance, however, is non-stationary and difficult to assess a priori. The orbits, on the other hand, are given in batches of 1.5–20 hours length dependent on the POD analysis strategy. Here we assign an unknown variance component $\sigma_k^2(\delta V)$ to each batch and estimate it jointly with the residual gravity field and with parameters that account for accelerometer drift. Written as a Gauss–Markov model, this is

$$\boldsymbol{A}_k \begin{pmatrix} \boldsymbol{x}_{\text{SH}} \\ \boldsymbol{x}_{\text{ACC}_1} \\ \cdots \\ \boldsymbol{x}_{\text{ACC}_p} \end{pmatrix} - \boldsymbol{y}_k = \boldsymbol{\epsilon}_k \qquad \begin{aligned} D(\boldsymbol{\epsilon}_k) &= \sigma_{(k)}^2 \boldsymbol{I} \\ D(\boldsymbol{x}_{\text{SH}}) &= \sigma_{(0)}^2 \boldsymbol{R} \end{aligned} \tag{3}$$

for $k = 1 \ldots p$, where $\boldsymbol{x}_{\text{SH}}$ contains spherical harmonic coefficients, $\boldsymbol{x}_{\text{ACC}_k}$ and $\sigma_{(k)}^2$ are drift parameters and variance components for the $k$–th data set, and $\sigma_{(0)}^2$ is a regularization parameter if needed. This requires an iterative VCE strategy involving repeated re–weightings of the contributions, synthesis of potential residuals, and repeated solutions of the overall least squares problem. It will be discussed in what follows.

## 3 Variance Component Estimation

Two different Variance Component Estimators have been used in this test set up: iterated MINQUE and iterated Maximum Likelihood (ML) VCE. At convergence, the results of the iterated ML VCE should equal those of the iterated MINQUE technique (when we assume Gaussian distributions as usual), but in practice one performs only a couple of iteration steps. Iterated MINQUE is known to converge faster towards unbiased estimates. More information on both estimators can be found in Rao and Kleffe (1988).

MCVCE, a fast Monte Carlo implementation of iterative maximum–likelihood VCE along theses lines has been developed in Koch and Kusche (2002), Kusche (2003), and tested in Kusche and van Loon (2004) on real CHAMP data. Mayer-Gürr et al (in press) used it for deriving the ITG–CHAMP01 gravity model series.

We develop here for the first time a Monte Carlo variant of the MINQUE technique. In MINQUE, at each iteration an equation system of the type $\boldsymbol{u} = \boldsymbol{S\sigma}$ has to be solved. The $k = 0 \ldots p$ entries of $\boldsymbol{u}$ hold the weighted residual square sum for each batch of data (including $\boldsymbol{R}$–weighted residual SH coefficients w.r.t. to the prior model for $k = 0$ in our case), $\boldsymbol{\sigma}$ holds the $p + 1$ variance components we are interested in, and $\boldsymbol{S}$ is basically composed from $tr(\boldsymbol{N}^{-1}\boldsymbol{N}_k)$ and $tr(\boldsymbol{N}^{-1}\boldsymbol{N}_k\boldsymbol{N}^{-1}\boldsymbol{N}_l)$ expressions (see e.g. Rao and Kleffe (1988), Grafarend and Schaffrin (1993)). $\boldsymbol{N}^{-1}$ is the inverse accumulated normal matrix, and the $\boldsymbol{N}_k = \boldsymbol{A}_k^T\boldsymbol{A}_k$ are data batch normal matrix contributions (including $\boldsymbol{R} = \boldsymbol{N}_0$). We have suggested in earlier publications to replace $tr(\boldsymbol{N}^{-1}\boldsymbol{N}_k)$ by $\boldsymbol{z}_k^T\boldsymbol{A}_k\boldsymbol{q}_k$ where $\boldsymbol{q}_k$ follows from solving $\boldsymbol{N}\boldsymbol{q}_k = \boldsymbol{A}_k^T\boldsymbol{z}_k$ and $\boldsymbol{z}_k$ is a random vector of the same dimension as $\boldsymbol{y}_k$ but zero expectation and unit covariance matrix. It is now straightforward to devise a Monte Carlo variant of MINQUE (we will call it MCMINQUE) by setting

$$tr(\boldsymbol{N}^{-1}\boldsymbol{N}_k\boldsymbol{N}^{-1}\boldsymbol{N}_l) \simeq \boldsymbol{q}_k^T\boldsymbol{N}_l\boldsymbol{q}_k = ||\boldsymbol{A}_l\boldsymbol{q}_k||^2 \tag{4}$$

or, equivalent, using $\boldsymbol{q}_l^T\boldsymbol{N}_k\boldsymbol{q}_l = ||\boldsymbol{A}_k\boldsymbol{q}_l||^2$. Taking the expectation (with respect to the distribution of the random vector) proves that this is an unbiased estimate, which of course can be improved by averaging multiple realizations of the $z_k$. These expressions can be further refined for separating local ($x_{\mathrm{ACC}_k}$) and global ($x_{\mathrm{SH}}$) normal equation contributions. In the MCMINQUE approximation, the entries of the MINQUE matrix $S$ are given by the $l_2$–norm of the $(p+1)(p+2)/2$ synthetic data vectors that are generated if one passes standard random noise for each data batch in turn and zeros for the others through the estimation, and predicts one of the other batches respectively. The inverse of the matrix $S$ is often used as a simple estimate for the variance-covariance matrix of the estimated variance components, see e.g. Sjöberg (1983): $Q_{\hat{\sigma}} = 2S^{-1}$.

## 4 Outlier rejection

VCE techniques are suitable for down-weighting a degraded batch of data and for weighting a prior gravity model. But in case that just one data point contains an outlier this should not affect the whole batch. It is therefore necessary to fit an outlier rejection method within the VCE scheme. Due to the size of the problem, we cannot base outlier rejection on the classical data snooping or Pope's test in their original setting.

We have therefore chosen to use an approximation of the Pope's test, i.e. to refer the observation residual to the estimated standard deviation of the observation group (from VCE). This is a valid approximation as this standard deviation only differs from the standard deviation of the residual by only a few percent, as we found in explicit computations. Since the redundancy of the problem is very high, the teststatistic has a distribution close to the normal distribution. The critical value is set to 3.0, equivalent to a level of significance above 99.5 percent. The least-square solution is then recomputed without the detected outliers. It is then possible to repeat the outlier detection, however we found that 1 iteration is sufficient.

## 5 Test setup

We have investigated about 2 years of kinematic CHAMP orbits, which were kindly provided by D. Svehla, IAPG, TU Munich. These orbits are processed following the zero–differencing strategy, see Svehla and Rothacher (2003), and were provided with full 3D variance–covariance information per data point. The 2–year data set was first divided into four groups, each with a time span of six months. A group was then split into several (2000-3000) smaller data batches. Each time when the GPS a priori sigma exceeds 10 cm, these data were removed and the data batch split into separate batches. The 500 largest data batches within a group were selected and used in further computations. The length of such a data batch differed from 1.5 hours to 20 hours, with a total of 299 days (2000 batches). Pseudo-observations were computed using the kinematic orbit data, accelerometer data and quaternion data, according to eq. (1). A 7–point Lagrange interpolator was used to compute the kinematic velocities. A normal matrix has been computed for each data batch used. Each normal matrix is weighted within our least-squares procedure according to the inverse of the estimated variance component. The gravity model was parameterized complete to degree 75.

## 6 Results

The first group of data batches was used to compare the MCMINQUE and the MCVCE methods. The square root of the variance components, computed with the MCMINQUE method, can be found in figure 1, together with the a priori standard deviation of the kinematic orbit, a by-product of the GPS processing. Note the different units.

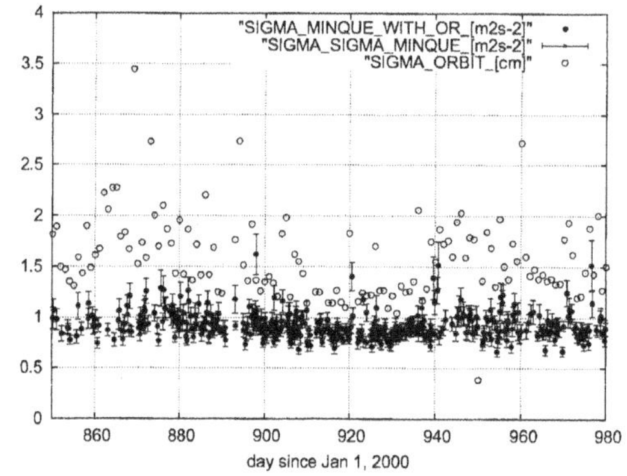


**Fig. 1** Estimated standard deviations (MCMINQUE) of the pseudo-observations of group 1 (with error bar from $S$), together with estimated standard deviations of the kinematic orbit.

The graph suggests that smooth and noisy periods are visible in our estimates as well, but the variations are less pronounced. This can be explained because noise in the GPS position time series (passed through the differentiation to velocities) is not the only contributing factor to noise in the residual potential pseudo-observations and a part of the noise is of course absorbed in the gravity model. On average, we estimate $0.9m^2s^{-2}$ for the noise level in the pseudo-observations. This would correspond to a velocity error of $0.1mm/s$ and a position error of a few mm. As one would expect from literature (e.g. Lucas, 1985), the MCMINQUE method converged faster (i.e. less iteration steps) than the MCVCE method. However, as the computations of MCMINQUE are more time-consuming, the MCVCE proved to be a more efficient method. The computed weights ($w_i = \hat{\sigma}_{(i)}^{-2}$) are almost identical to each other for both methods, as can been seen in figures (2) and (3).

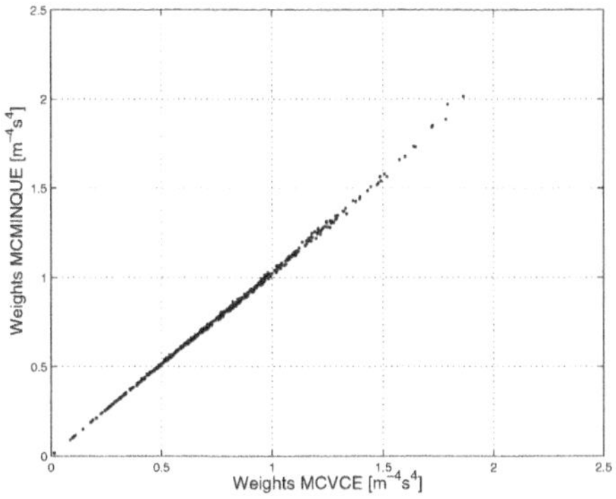


**Fig. 2** Comparison between the weights computed with MCVCE and with MCMINQUE with all outliers present.

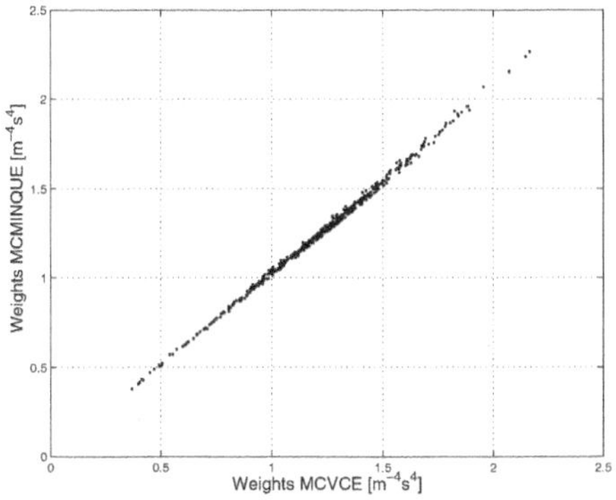


**Fig. 3** Comparison between the weights computed with MCVCE and with MCMINQUE with the largest outliers removed.

The two figures clearly show a down-weighting of the data sets in the presence of outliers. The removal of a single outlier in one particular data set decreases the variance component of that data set and increases the weight of the data set within the least-squares procedure. Looking at the figures more thoroughly, one can see a slight scaling factor ($\approx 1.03$) between the estimated weights of both methods. However, such a scaling factor has no influence on the final solution, as the accumulated group normal matrix will be rescaled when combining the data with an existing global gravity field model. We therefore advise to use the faster MCVCE method instead of the MCMINQUE method.

To test the effect of MCVCE weighting and the outlier algorithm, we will compare four different solutions:

- Sol. 1: outliers present, equal weighting.
- Sol. 2: outliers present, MCVCE weighting.
- Sol. 3: outliers removed, equal weighting.
- Sol. 4: outliers removed, MCVCE weighting.

In all computations, the four accumulated group normal matrices and right-hand-side vectors, are combined with the EGM96 model (Lemoine et al., 1998). In the computations of solutions 1 and 3, this was achieved using equal weights for each normal matrix. In the computations of solutions 2 and 4, use was made of MCVCE to estimate the weights of the normal matrices. The absolute weights, at convergence, computed for solution 4, were 0.573, 0.539, 0.189 and 0.439 for data groups 1,2,3 and 4 resp. The EGM96 model was down-weighted by the factor 0.216. This is equivalent to an increase of the standard deviations by a factor $\approx 2$. Similar weights were found in the computation of solution 2.

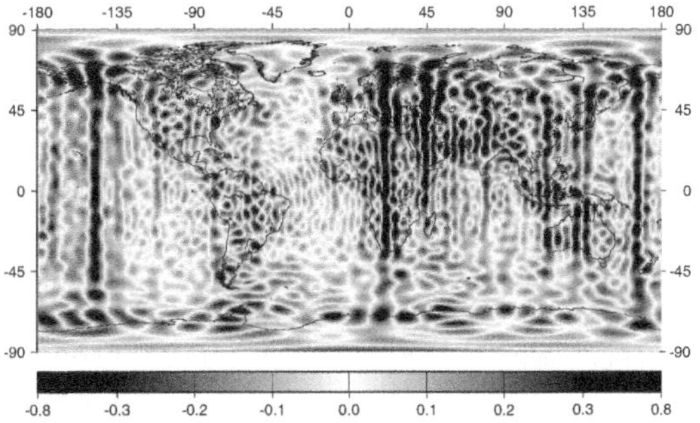

**Fig. 4** Geoid differences (L=50) between Solution 1 and the EIGEN-GRACE02S model [m].

The computation of solution 1 does not account for any outlier detection or variance component estimation. It just accumulates all normal matrices and computes the solution in an unweighted least-squares approach. Comparing solution 1 to the EIGEN-GRACE02S model (Reigber et al., submitted) clearly shows some bad data batches within the solution (fig. 4). The rms of the geoid differences, weighted proportional to the grid size they represent, was 14.8 cm for the truncated model up to degree and order 50.

In solution 2, the MCVCE weighting clearly detected and consequently downweighted the spurious data batches present in the data, as can be seen in figure 5. These particular data batches all lie within the time-span of day 131 to 134 of the year 2003. Monte Carlo VCE down-weighted the data sets up to a factor 1000, so basically removed the spurious data. Apart from these data, MCVCE was also usefull to distinquish between the quality of the good data sets, as the estimated $\hat{\sigma}_{(i)}$ varied between 0.7 and 1.8 $m^2s^{-2}$, being on average 0.9 $m^2s^{-2}$. The rms geoid difference with EIGEN-GRACE02S improved from 14.8 centimeters to 7.6 centimeters (L=50).

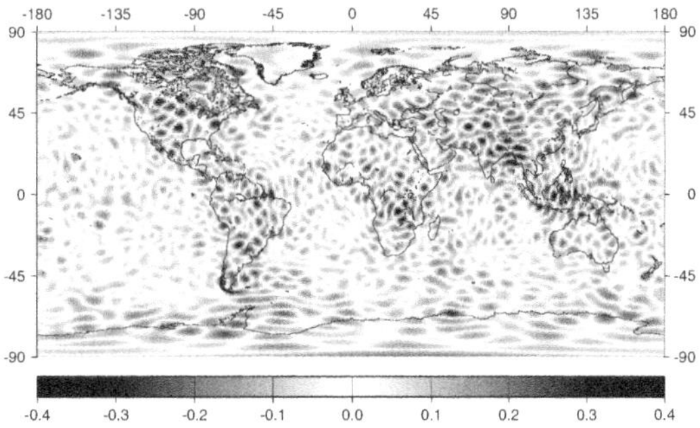


**Fig. 5** Geoid differences (L=50) between Solution 2 and the EIGEN-GRACE02S model [m].

In the third solution, most of the outliers were removed with the approximation of the Pope's test, as was mentioned earlier. The reduced normal matrices and right-hand-side vectors were weighted equally, as was done with solution 1. Again, geoid differences with EIGEN-GRACE02S clearly show trackiness. This shows, that the outlier detection method was unable to identify these bad data sets, since the estimated variance components (which were not used this time for weighting the normal matrices) had high values. The test statistic was therefore well below the critical value and the observations were not detected as outliers. Only outliers present in relatively good data batches could be estimated as outliers. The geoid differences to EIGEN-GRACE02S were very close to those of solution 1, reducing the rms to 14.6 cm.

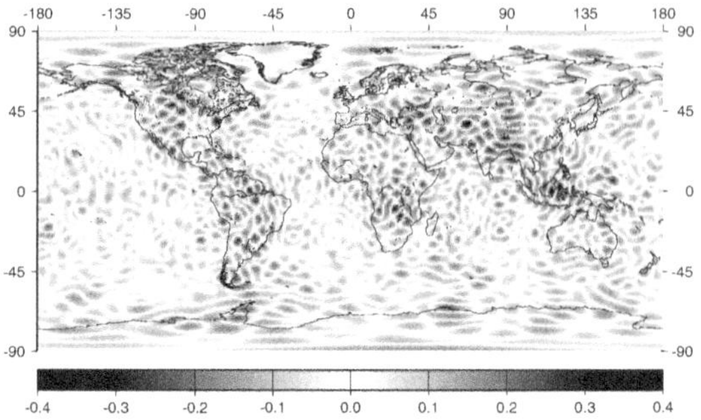


**Fig. 6** Geoid differences (L=50) between Solution 4 and the EIGEN-GRACE02S model [m].

Solution 4 is, as expected, closest to the EIGEN-GRACE02S model. Outliers are detected and removed, and MCVCE is used to weight the normal matrices and the prior (model) information. Spurious data batches were again down-weighted considerably, removing the stripe pattern from the plot of geoid differences (fig. 6). The geoid differs by 7.0 cm rms from EIGEN-GRACE02S at degree 50, and 19.2 cm at degree 75. This is comparable to the EIGEN-3P model (Reigber et al., 2004) which differs from the EIGEN-GRACE02S model by a rms of 5.9 cm (L=50) and 39.3 cm (L=75). Our model appears better at high degrees because it involves EGM96.

## 7 Conclusions

We have discussed a statistical assessment of CHAMP orbits and data within the energy balance method. Non–stationary noise has been modeled with piecewise constant variance. In addition to our previously suggested Monte Carlo implementation of maximum likelihood VCE, we have developed and tested a Monte Carlo implementation of the MINQUE strategy. We have proven that we can efficiently estimate individual noise levels for data batches and a variance factor for combining CHAMP data with EGM96 using these estimation techniques in combination with outlier rejection, and that gravity solutions using an optimally weighted least squares procedure are superior to heuristic weighting. Ongoing research includes using more data, and accounting for time–wise correlations.

## Acknowledgements

We are grateful to GFZ Potsdam for providing

CHAMP ACC data. Thanks go also to IAPG, TU Munich, for providing CHAMP kinematic orbits. J.v.L. acknowledges financial support by the GO-2 program (SRON EO-03/057).

## References

Gerlach C, Földvary L, Švehla D, Gruber Th, Wermuth M, Sneeuw N, Frommknecht B, Oberndorfer H, Peters Th, Rothacher M, Rummel R, Steigenberger P (2003). *A CHAMP–only gravity field model from kinematic orbits using the energy integral.* GRL 30(20), 2037, doi:10.1029/2003GL018025

Grafarend EG, Schaffrin B (1993) *Ausgleichungsrechnung in linearen Modellen.* BI Verlag, Mannheim

Howe E, Stenseng L, Tscherning CC (2003) *Analysis of one month of state vector and accelerometer data for the recovery of the gravity potential.* Adv Geosciences 1:1–4.

Jekeli C (1999) *The determination of gravitational potential differences from satellite–to–satellite tracking.* Cel Mech Dyn Astr 75: 85–100.

Koch K–R, Kusche J (2002) *Regularization of geopotential determination from satellite data by variance components.* J Geodesy 76: 259–268.

Kusche J (2003) *A Monte–Carlo technique for weight estimation in satellite geodesy.* J Geodesy 76: 641–652.

Kusche J, van Loon JP (2004) *Statistical assessment of CHAMP data and models using the energy balance approach.*, in Reigber et al., *Earth observation with CHAMP, results from three years in orbit*, Springer.

Lemoine FG, Kenyon SC, Factor JK, Trimmer RG, Pavlis NK, Chinn DS, Cox CM, Klosko SM, Luthcke SB, Torrence MH, Wang YM, Williamson RG, Pavlis EC, Rapp RH, Olsen TR (1998) *The development of the joint NASA GSFC and the National Imagery and Mapping Agency (NIMA) geopotential model EGM96.* NASA/TP–1998–206861, NASA–GSFC, Greenbelt MD

Löcher A, Ilk KH (2004) *Energy balance relations for validation of gravity field models and orbit determination*, in Reigber et al., *Earth observation with CHAMP, results from three years in orbit*, Springer.

Lucas JR (1985) *A variance component estimation method for sparse matrix applications* NOAA Technical Report NOS 111 NGS 33, National Geodetic Survey, Rockville

Mayer–Gürr T, Ilk KH, Eicker A, Feuchtinger M (in press) *ITG-CHAMP01: A CHAMP gravitiy field model from short kinematical arcs of a one-year observation period*, accepted for J Geodesy

Rao CR, Kleffe J (1988) *Estimation of variance components and applications.* North–Holland, Amsterdam

Ray RD (1999) *A global ocean tide model from TOPEX / POSEIDON altimetry: GOT99.2*, NASA/TM-1999-209478, NASA–GSFC, Greenbelt, MD

Reigber C, Schmidt R, Flechtner F, König R, Meyer U, Neumayer KH, Schwintzer P, Zhu SY *EIGEN gravity field model to degree and order 150 from GRACE Mission data only*, submitted to Journal of Geodynamics.

Reigber C, Jochman H, Wünsch J, Petrovic S, Schwintzer P, Barthelmes F, Neumayer K-H, König R, Förste C, Balmino G, Biancale R, Lemoine J-M, Loyer S, Perosanz F (2004) *Earth gravity field and seasonal variability from CHAMP*, in Reigber et al., *Earth Observation with CHAMP, results from three years in orbit*, Springer.

Švehla D, Rothacher M (2003) *Kinematic and reduced–dynamic precise orbit determination of low earth orbiters.* Adv Geosciences 1: 47–56.

Sjöberg LE (1983) *Unbiased estimation of variance-covariance components in condition adjustment with unknowns - A MINQUE approach.* Zeitschrift für Vermessungswesen 108:9, p. 382–387

Visser P, Sneeuw N, Gerlach C (2003) *Energy integral method for gravity field determination from satellite orbit coordinates.* J Geodesy 77: 207–216.

# Analysis of $J_2$-Perturbed Relative Orbits for Satellite Formation Flying

C. Xu, R. Tsoi, N. Sneeuw
Department of Geomatics Engineering, University of Calgary
2500 University Drive, NW, Calgary, AB, Canada, T2N 1N4
Tel: +1(403)220-4984 Fax: +1(403)284-1980 Email:{xuc, rtsoi, sneeuw}@ucalgary.ca

**Abstract.** We study the concept of satellite formation flying in a geodetic context, namely as a viable alternative for future gravity field satellite missions. The feasibility of formation flight is demonstrated. In particular the stability of such a formation in a $J_2$ gravity field is investigated. To this end three orbit computation approaches are compared: 1) numerical integration of Newton's equations (NN) of motion, 2) numerical integration of Hill equations (HE), and 3) a new set of nontrivial, non-homogeneous analytical solutions of HE.

Hill equations provide an elementary description of relative orbital motion. In order to accommodate $J_2$ gravitational perturbations we modify the HE in several steps: evaluating the $J_2$ disturbing force function on the nominal orbit; changing the orbital rotation rate (frequency matching), due to in-orbit $J_2$ precession; as well as evaluating the time-averaged $J_2$ gravity gradient tensor. The resulting HE are solved analytically.

The orbit simulations show that the analytical solution of the modified HE are consistent with their numerically integrated counterpart. Differences with respect to the reference NN method remain, which means that not all $J_2$ effects have been captured yet in the modified HE. The usefulness of HE as a formation design tool are demonstrated by simulations of circular relative motion.

**Keywords.** formation flying, Hill equations (HE), $J_2$-perturbed orbits, numerical integration

## 1 Introduction

The current gravity field satellite missions CHAMP, GRACE and GOCE show certain limitations in terms of spatial and temporal resolution and accuracy (Aguirre-Martinez & Sneeuw, 2003). Future gravity field satellite missions are being discussed already with satellite-to-satellite tracking using laser technology as the key geodetic observable. However, a GRACE-type observable is inherently non-isotropic, due to the mostly north-south orientation of the laser link and to its scalar character. In order to enhance the spectral content, future geodetic satellite missions will most likely make use of the concepts of satellite formation flying, e.g. LISA (NASA, 2004) or Cartwheel concepts, cf. (Sneeuw & Schaub, 2004).

Hill equations (HE), cf. (Hill, 1878), are an elementary and often used tool to describe the relative motion between a chief and a deputy satellite in a reference frame (Hill frame), which co-rotates on a circular orbit, cf. (Schaub & Junkins, 2003, ch. 14). They are also referred to as Clohessy-Wiltshire (CW) equations. The HE perform well in a central force field. In a realistic orbit scenario, i.e. with the Earth's dynamic flattening giving rise to major orbit perturbations, the standard HE perform poorly. Much attention was paid recently in the literature to accommodate $J_2$-effects in the HE, e.g. (Schaub, 2002; Schweighart & Sedwick, 2002).

The objective of the present paper is to modify the HE systematically. The following steps are implemented: 1) The $J_2$ disturbing force is evaluated (analytically) on the nominal circular orbit; 2) the orbital speed is modified to accommodate in-orbit $J_2$ precession (referred to as frequency matching; and 3) the time-averaged $J_2$ gravity gradient tensor is evaluated to accommodate the main part of the separation between the actual and nominal satellite location.

The HE are a set of constant coefficient, linearized differential equations (2) defined in a central force field. The orientation of the Hill frame is defined by: $x$ along-track, $y$ cross-track and $z$ radial direction, cf. Fig. 1. The orbital frequency vector $\boldsymbol{\omega}$ of the rotating Hill frame satisfies Kepler's third law:

$$\boldsymbol{\omega} = [0 \;\; n \;\; 0]^{\mathrm{T}} \text{ , with } \quad n = \sqrt{\frac{GM}{a^3}} \qquad (1)$$

where $GM = 3.986005 \cdot 10^{14}\,\mathrm{m}^3/\mathrm{s}^2$ is the Earth gravitational constant, and $a$ is the semi-major axis of the satellite orbit. The HE can be written as:

$$\begin{array}{lllll} \ddot{x} & + 2n\dot{z} & & = 0 \\ \ddot{y} & & + n^2 y & = 0 \\ \ddot{z} & - 2n\dot{x} & - 3n^2 z & = 0 \end{array} \quad (2)$$

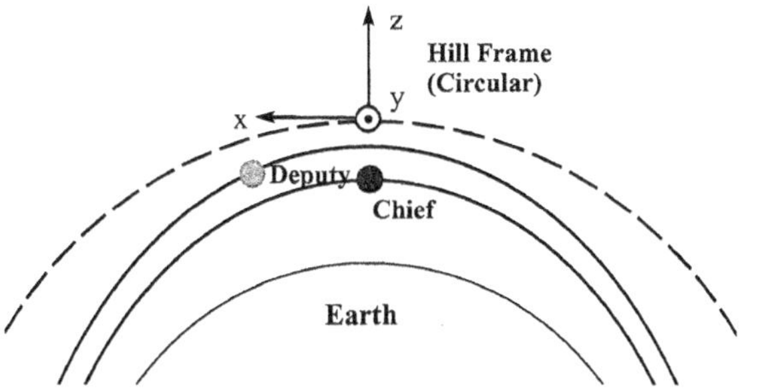


Figure 1: Orientation of the reference Hill frame

Without any perturbations or thrusting, the analytical homogeneous solutions of HE describe relative orbit motion very well, e.g. (Kaplan, 1976):

$$\begin{aligned} x(t) &= \frac{2}{n}\dot{z}_0 \cos nt + \left(\frac{4}{n}\dot{x}_0 + 6z_0\right) \sin nt \\ &\quad -(3\dot{x}_0 + 6nz_0)t + x_0 - \frac{2}{n}\dot{z}_0 \\ y(t) &= y_0 \cos nt + \frac{\dot{y}_0}{n} \sin nt \\ z(t) &= \left(-\frac{2}{n}\dot{x}_0 - 3z_0\right) \cos nt + \frac{\dot{z}_0}{n} \sin nt \\ &\quad + \frac{2}{n}\dot{x}_0 + 4z_0 \end{aligned} \quad (3)$$

where $(x_0,\, y_0,\, z_0)$ and $(\dot{x}_0,\, \dot{y}_0,\, \dot{z}_0)$ are the initial states of positions and velocities in the reference Hill frame respectively.

Apparently, in the coupled along-track and radial directions, the bias and drift terms cancel under a proper choice of initial conditions, while the cross-track direction is decoupled from the in-plane motion, cf. (Schaub, 2002). A certain relative motion, e.g. a 2 by 1 ellipse in the orbital plane, is achieved by setting $x_0 = \frac{2}{n}\dot{z}_0$, $z_0 = 0$, and $\dot{x}_0 = 0$, cf. (Sneeuw, 2002; Schweighart & Sedwick, 2002).

## 2 Analysis of $J_2$-Perturbed Relative Orbits

The purely central force field—one of the underlying assumptions of the HE—makes the homogeneous solution of HE unsuitable to simulate more realistic scenarios. The second degree zonal spherical harmonic of the Earth, $J_2 = 1082 \cdot 10^{-6}$, produces the primary disturbing potential due to the Earth oblateness, see (Seeber, 1993):

$$R_{2,0} = -\frac{1}{2} GM\, J_2 \frac{{a_e}^2}{r^3} \left(3 \sin^2 \varphi - 1\right), \quad (4)$$

where $a_e = 6378\,137$ m is the semi-major axis of the Earth, $r$ is the radial distance and $\varphi$ is the geocentric latitude of the satellite in the Earth fixed frame.

The $J_2$ generated perturbations include secular, long-periodic, and short-periodic components, which consequently will disturb the satellite orbits, for instance, precession of ascending node $\dot{\Omega}$, and drifts in argument of perigee $\dot{\omega}$ and the mean anomaly $\delta\dot{M}$. The $J_2$ potential component can be also expressed in the Hill frame based on $\sin\varphi = \sin I \sin u$, where $I$ is the inclination of the orbit, and $u$ is the argument of latitude:

$$R_{2,0} = -\frac{1}{2} GM\, J_2 \frac{{a_e}^2}{r^3} \left(3 \sin^2 I \sin^2 u - 1\right). \quad (5)$$

The corresponding force functions, which will be applied to the right hand side of the HE (2), are:

$$\begin{aligned} \boldsymbol{f} &= \nabla R_{2,0} = [f_x \;\; f_y \;\; f_z]^{\mathrm{T}} \quad (6) \\ f_x &= C_1 \sin^2 I \sin 2u \\ f_y &= C_1 \sin 2I \sin u \\ f_z &= C_1 \left(1 - \frac{3}{2}\sin^2 I + \frac{3}{2}\sin^2 I \cos 2u\right) \end{aligned}$$

with the coefficient $C_1 = -\frac{3}{2} GM\, J_2 \frac{{a_e}^2}{r^4}$.

To be more precise and realistic, the second order derivative (tensor components) of $J_2$ disturbing potential term must also be included to capture the variations in field strength of $R_{2,0}$ due to orbit perturbations, which in turn are also caused by $R_{2,0}$. Since this term is a function of time, which may cause trouble for determining the analytical solutions, the time-average of the second order derivative tensor vector $\overline{\boldsymbol{R}}_{2,0}$ will be used, cf. (Schweighart & Sedwick, 2002):

$$\begin{aligned} \overline{\boldsymbol{R}}_{2,0} &= \frac{1}{2\pi} \int_0^{2\pi} \boldsymbol{R}_{2,0}(u) \mathrm{d}u \\ &= n^2 \begin{bmatrix} -3C_2 & 0 & 0 \\ 0 & -C_2 & 0 \\ 0 & 0 & 4C_2 \end{bmatrix} \quad (7) \\ C_2 &= \frac{3 J_2 a_e^2}{8 r^2} \left(1 + 3\cos 2I\right). \end{aligned}$$

Note that these time-averaged tensor components will come back in the frequency matching section 3, but are not included in the calculations in this section.

Three different methods will be used to analyze the relative orbits in the rotating reference Hill frame:

1) numerical integration of Newton's Equations of motion (NN);
2) numerical integration of HE (HN);
3) nontrivial, non-homogeneous analytical solutions of HE (HA).

### 2.1 Numerical Integration of Newton's Equation

In a two-body problem, the absolute satellite motion in the inertial frame is described by Newton's equation of motion:

$$\ddot{\boldsymbol{r}} = -\frac{GM}{r^3}\boldsymbol{r} + \boldsymbol{k}\,, \tag{8}$$

where $\boldsymbol{r}$ is the geocentric position vector of the satellite, $\ddot{\boldsymbol{r}}$ is the corresponding acceleration, and $\boldsymbol{k}$ is the perturbing force vector. In our case, the disturbing force is due to the $J_2$ perturbation potential (4). Hence, by taking first order derivatives of the potential, the differential equations of absolute motion in the inertial frame can be decomposed into three acceleration vector components as $\boldsymbol{r} = (X, Y, Z)$:

$$\begin{aligned} \ddot{X} &= -GM\frac{X}{r^3}\left[1 - J_2\frac{3}{2}\left(\frac{a_e}{r}\right)^2\left(5\frac{Z^2}{r^2} - 1\right)\right] \\ \ddot{Y} &= -GM\frac{Y}{r^3}\left[1 - J_2\frac{3}{2}\left(\frac{a_e}{r}\right)^2\left(5\frac{Z^2}{r^2} - 1\right)\right] \\ \ddot{Z} &= -GM\frac{Z}{r^3}\left[1 - J_2\frac{3}{2}\left(\frac{a_e}{r}\right)^2\left(5\frac{Z^2}{r^2} - 3\right)\right]. \end{aligned} \tag{9}$$

The orbits of the chief $\boldsymbol{r}_c$ and deputy $\boldsymbol{r}_d$ satellites are obtained respectively by numerical integration of (9). Their difference $\boldsymbol{r}_i$ describes the relative motion in the inertial frame:

$$\boldsymbol{r}_i = \boldsymbol{r}_d - \boldsymbol{r}_c\,. \tag{10}$$

Three rotations and a permutation are required to obtain the relative motion $\boldsymbol{r}_s$ in the Hill frame:

$$\boldsymbol{r}_s = \begin{bmatrix} 0 & 1 & 0 \\ 0 & 0 & 1 \\ 1 & 0 & 0 \end{bmatrix} R_3(u)R_1(I)R_3(\Omega)\boldsymbol{r}_i\,, \tag{11}$$

with $\boldsymbol{r}_s=(x, y, z)$ the position vector in the Hill frame and $R_3(u)$, for instance, the $3 \times 3$ rotation matrix:

$$R_3(u) = \begin{bmatrix} \cos u & \sin u & 0 \\ -\sin u & \cos u & 0 \\ 0 & 0 & 1 \end{bmatrix}. \tag{12}$$

Since there are no assumptions and approximations involved in equations of motion (9), they describe the real motion of the satellite in the presence of $J_2$ perturbations. The NN method is considered the benchmark or reference in the following analyses.

### 2.2 Numerical Integration of the HE

Note that no disturbing forces are involved in the HE (2), so they are homogeneous differential HE. Apparently, as mentioned earlier, if we change the right hand side of the homogeneous HE with the corresponding $J_2$ disturbing force components (6), the modified, non-homogeneous, second order differential HE can be written as follows:

$$\begin{aligned} \ddot{x} + 2n\dot{z} &= f_x \\ \ddot{y} + n^2 y &= f_y \\ \ddot{z} - 2n\dot{x} - 3n^2 z &= f_z \end{aligned} \tag{13}$$

The relative motion $\boldsymbol{r}_s = (x, y, z)$ in the Hill frame can be obtained directly by numerical integration.

### 2.3 Analytical Solution of the HE

Studying the $J_2$ disturbing force in the Hill frame (6), both the along-track and radial direction forces have a two cycle per revolution (CPR) frequency component (a sinusoid function of $2u$), whereas, the force in the cross-track direction is at the orbital frequency. We decompose the different force components and get the following non-homogeneous HE:

$$\begin{aligned} \ddot{x} + 2n\dot{z} &= A_x \cos 2nt + B_x \sin 2nt + C_x \\ \ddot{y} + n^2 y &= A_y \cos nt + B_y \sin nt + C_y \\ \ddot{z} - 2n\dot{x} - 3n^2 z &= A_z \cos 2nt + B_z \sin 2nt + C_z \end{aligned} \tag{14}$$

The corresponding disturbing force components in the presence of $J_2$ effect are expressed by:

$$f_x = C_1 \sin^2 I \sin 2u \quad \Rightarrow \begin{cases} A_x = C_1 \sin^2 I \sin 2u_0 \\ B_x = C_1 \sin^2 I \cos 2u_0 \\ C_x = 0 \end{cases} \tag{15a}$$

$$f_y = C_1 \sin 2I \sin u \quad \Rightarrow \begin{cases} A_y = C_1 \sin 2I \sin u_0 \\ B_y = C_1 \sin 2I \cos u_0 \\ C_y = 0 \end{cases} \tag{15b}$$

$$f_z = C_1\left(1 - \frac{3}{2}\sin^2 I + \frac{3}{2}\sin^2 I \cos 2u\right) \quad \Rightarrow \begin{cases} A_z = C_1 \frac{3}{2} \sin^2 I \cos 2u_0 \\ B_z = -C_1 \frac{3}{2} \sin^2 I \sin 2u_0 \\ C_z = C_1(1 - \frac{3}{2}\sin^2 I) \end{cases} \tag{15c}$$

where the coefficient $C_1$ is defined in (6), and $u_0$ is the initial state of the argument of latitude. Although the expression of the HE becomes lengthier, the analytical solution of the above linear differential equations is straightforward.

## 3 Frequency Matching Approach — Modified Numerical & Analytical Solutions of HE

The $J_2$ perturbation force will disturb the Kepler elements of the orbit, and it will consequently change the orbital frequency. Under this influence, the relative distance between the chief and deputy satellites will become larger and larger, and eventually the linearized equations break down. To accommodate this problem in the differential HE, the reference Hill frame frequency should be adjusted to match the perturbed orbital frequency using a frequency matching approach, in other words, the frequency of the Hill frame will be speeded up. In order to capture the effects of the $J_2$ perturbation force, the time-averaged value of the $J_2$ disturbing tensor term (7) should also be calculated in this step.

If we neglect the effect of nodal precession on in-plane orbital velocity, i.e. on the rotation rate of the Hill frame, the new frequency vector can be defined as, cf. (Schweighart & Sedwick, 2001, 2002):

$$\begin{aligned} \boldsymbol{\omega_p} &= [0 \ n_p \ 0]^{\mathrm{T}} \\ n_p &= n\sqrt{1+\frac{3J_2 a_e^2}{8r^2}(1+3\cos 2I)} \end{aligned} \quad (16)$$

The original HE are formulated in a frame that rotates at the orbital frequency $n$. Many terms cancel or are simplified because the (kinematic) rotation rates relate to the (dynamic) force field through Kepler's third law. In the presence of the $J_2$ perturbations, a new rotation frequency $n_p$ is required that is decoupled from Kepler's law. As a consequence the modified HE will become far more elaborate. The HE need to be modified to take into account:

1) $J_2$ perturbation forces on the nominal orbit, as done already in section 2,

2) a change in the orbital frequency due to $J_2$ precession,

3) and the time-averaged $J_2$ gravity gradient tensor in order to accommodate the fact that the $J_2$ disturbing forces are not sensed on the nominal orbit but on the perturbed one instead.

Consequently, the HE (13) are modified as:

$$\begin{aligned} \ddot{x} + 2n_p\dot{z} & &&= f_x \\ \ddot{y} & + (3n_p^2 - 2n^2)y &&= f_y \\ \ddot{z} - 2n_p\dot{x} & - (5n_p^2 - 2n^2)z &&= f_z + C_3 \end{aligned} \quad (17)$$

The disturbing force function is defined in (6). $C_3 = (n_p^2 - n^2)a$ in the radial direction is a correction term due to frequency matching.

Correspondingly, the force function can be decomposed into combinations of sinusoid functions with the new matching frequency.

$$\begin{aligned} f_x &= A_x \cos 2n_pt + B_x \sin 2n_pt + C_x \\ f_y &= A_y \cos n_pt + B_y \sin n_pt + C_y \\ f_z &= A_z \cos 2n_pt + B_z \sin 2n_pt + C_z \end{aligned} \quad (18)$$

The modified HE (17) are constant coefficient linear differential equations. Again, the solutions with the force functions (18) are straightforward, although the expressions become very lengthy, cf. (Schweighart & Sedwick, 2001, 2002).

Studying the set of analytical solutions, two new frequencies are involved only in the decoupled cross-track direction, which are $n_p$ and $\sqrt{(3n_p^2 - 2n^2)}$. While two other new frequencies are involved in both $x$ and $z$ directions since they are coupled, which are $2n_p$ and $\sqrt{(2n^2 - n_p^2)}$.

## 4 Simulation Results and Discussion

To illustrate the accuracy performance of the aforementioned methods including NN, HN, HA, with and without frequency matching approach, orbit simulation of the chief and deputy satellites motion is carried out. Table 4 shows the Kepler elements of the chief satellite orbit and the initial relative state between the chief and deputy satellites in the reference Hill frame. This scenario describes the deputy satellite is 1 km ahead of the chief satellite in the along-track direction at the initial state. The eccentricity of the reference orbit (Hill frame) is chosen to be zero to satisfy one of the assumptions. The initial state has been specially chosen to satisfy the following conditions $\dot{y}_0 = \frac{n}{2}x_0$ and $\dot{z}_0 = \sqrt{3}\dot{y}_0$. Under a zero perturbation environment, these specific values constrain the relative formation flying as a purely circular motion, see (Sneeuw, 2002). The duration of the simulation extends for 16 revolutions with 10 seconds sampling increment.

Figure 2 and Figure 3 illustrate the relative motion between the chief and deputy satellites using the methodology described in section 2 without frequency matching approach. Since there are no assumptions and approximations involved in Newton's

Table 1: Parameters of chief satellite and initial states in Hill frame

| Kepler element | value | relative state | value |
|---|---|---|---|
| $a$ | 7100 km | $x$ | 1000 m |
| $e$ | 0.005 | $y$ | 0 m |
| $I$ | 70° | $z$ | 0 m |
| $\Omega$ | 30° | $\dot{x}$ | 0 m/s |
| $\omega$ | 0° | $\dot{y}$ | 0.914 m/s |
| $M$ | 10° | $\dot{z}$ | 0.528 m/s |

equations of motion (9), we choose NN as the benchmark results. Figure 2 plots the nearly circular relative motion and shows different projections of the NN approach onto different planes. Apparently, there are some drift terms in all three directions in the presence of $J_2$ perturbations. In addition, we can see that an approximate 2 by 1 ellipse in the orbital plane (along-track vs. radial) and also a line with a 30° relative inclination in the cross-track vs. radial directions can be achieved. Compared with the NN method, both HN and HA methods describe the relative motion well. However, the relative distance drift away with the time.

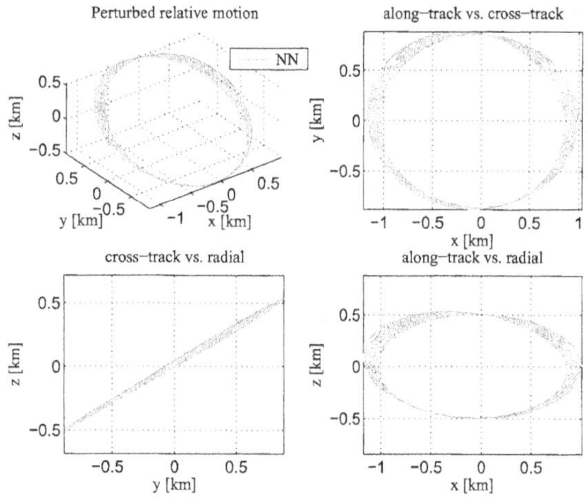


Figure 2: Projection of relative motion without frequency matching onto different planes

Figure 3 shows the differences between the numerical and analytical solutions of HE (HN and HA) with the benchmark NN. Apparently, there are also some big differences between HN and HA. HN results are much closer to the reference NN. The reason is that when we derive the analytical solutions of HE (HA) based on (14), the perturbation force components are considered as constant, e.g. the force coefficient $A_x$, $B_x$, however, when we use HN to numerically integrate the orbit, these force components are variant with time. This situation should be improved by the frequency matching approach as mentioned in section 3.

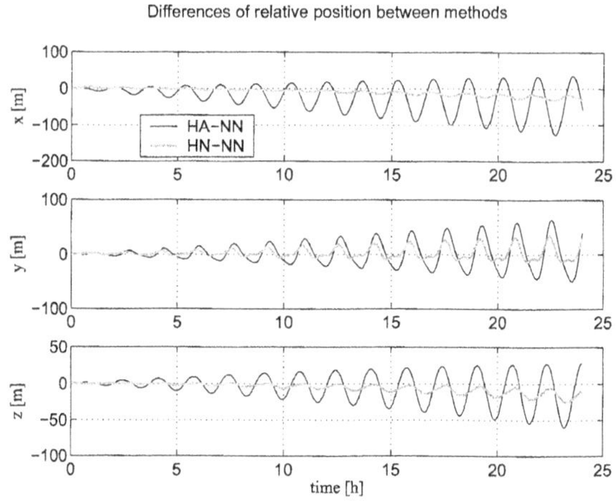


Figure 3: Compared to benchmark (NN) without frequency matching

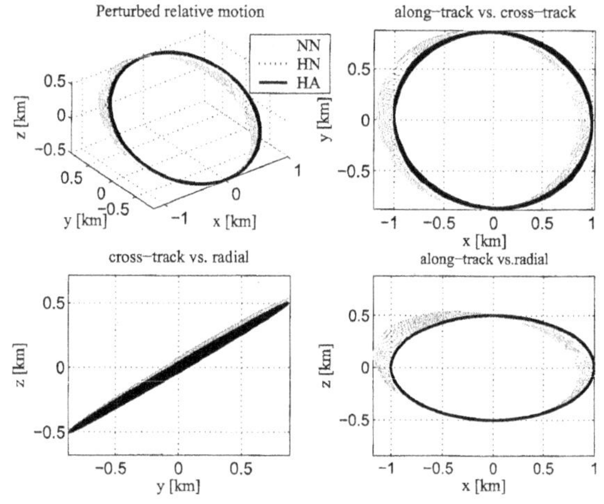


Figure 4: Projection of relative motion with frequency matching onto different planes

Figure 4 shows the relative motion between the chief and deputy satellites using the frequency matching approach of NN, HN and HA. Both of HN and HA show a consistent relative motion in the presence of $J_2$ perturbations because both drift and amplifying terms have almost been cancelled due to frequency matching. There are still some deviations of these solutions from NN at the level of several meters. This deviation may originate from a change in orbital frequency due to nodal precession. However, HN and HA can capture the characteristics of the relative motion under the influence of $J_2$ disturbing forces.

Figure 5 show the differences between HN and HA. They are consistent at the mm-level.

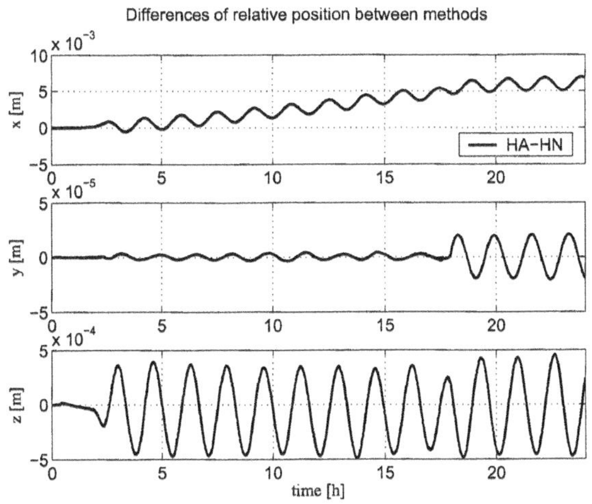


Figure 5: Difference between HN and HA with frequency matching

## 5 Conclusion and Future Work

The satellite formation flying concept is a viable alternative for future geodetic satellite missions, addressing current limitations in spectral content of the signal, spatial resolution and temporal sampling characteristics. When analyzing the relative motion between the chief and deputy satellites, the typical homogeneous solutions of HE cannot capture the effect of $J_2$ perturbations due to certain assumptions and limitations.

Orbit simulations have been carried out. The results show that a modification of the Hill equations (HE), that only takes into account the $J_2$ perturbation forces on the nominal orbit, cannot properly describe relative motion. To achieve a consistent analytical solution, the orbit frequency needs to be tuned and the $J_2$ disturbing forces must be evaluated at the real orbit location through the use of a time-averaged $J_2$ gravity gradient tensor. The analytical solution HA is then shown to be consistent with numerical solution HN at the mm-level.

However the deviations of these solutions from straightforward integration of the Newton equations is still at the level of several meters. This deviation, which may originate from a change in orbital frequency due to nodal precession, is not yet understood and is subject for future research.

The numerical examples demonstrate that relative motion can be designed in a realistic gravity field. Figures 2 and 4 in particular show the feasibility of 2:1 elliptical relative motion (in the orbital plane) and circular relative motion.

## References


Aguirre-Martinez M, Sneeuw N (2003) Needs and tools for future gravity measuring missions. *Space Science Reviews* **108**(1-2):409–416.

Hill G (1878) Research in the lunar theory. *American Journal of Mathematics I* pp. 5–26,129–147,245–260.

Kaplan M (1976) *Modern Spacecraft Dynamics & Control*. John Wiley and Sons, New York.

NASA (2004) LISA website. Retrieved August 20, 2004, from http://lisa.jpl.nasa.gov/.

Schaub H (2002) Spacecraft relative orbit geometry description through orbit element differences. *14th US National Congress of Theoretical and Applied Mechanics, VA* .

Schaub H, Junkins JL (2003) *Analytical Mechanics of Space Systems*. AIAA Education Series, Reston,VA.

Schweighart S, Sedwick R (2001) A perturbative analysis of geopotential disturbances for satellite formation flying. In: *Proceeding of the IEEE Aerospace Conference*, vol. 21, pp. 1001–1019.

Schweighart S, Sedwick R (2002) High-fidelity linearized $J_2$ model for satellite formation flight. *Journal of Guidance, Control, and Dynamics* **25**(6):1073–1080.

Seeber G (1993) *Satellite Geodesy*. Walter de Gruyter.

Sneeuw N (2002) LISA/Cartwheel orbit type for future gravity field satellite missions. In: *Proc. W.A. Heiskanen Symposium in Geodesy, Columbus, OH.*

Sneeuw N, Schaub H (2004) Satellite clusters for next generation gravity field missions. In: *IAG Proceedings for Gravity, Geoid and Satellite Missions 2004*, IAG.

# GOCE Gravity Field Processing

R. Pail
Institute of Navigation and Satellite Geodesy, Graz University of Technology
Steyrergasse 30, 8010 Graz, Austria, Tel: ++43 316 873 6359, E-mail: pail@geomatics.tu-graz.ac.at

W.-D. Schuh
Institute of Theoretical Geodesy, University of Bonn
Nussallee 17, 53115 Bonn, Germany, Tel: ++49 228 73 26 26, E-mail: schuh@mail.geod.uni-bonn.de

M. Wermuth
Institute of Astronomical and Physical Geodesy, Technical University Munich
Arcisstrasse 21, 8033 Munich, Germany, Tel: ++49 89 289 23181, E-mail: wermuth@bv.tu-muenchen.de

**Abstract.** A concept for an operable software system for the processing of a high-accuracy, high-resolution spherical harmonic model of the Earth's gravity field from GOCE observables (satellite gravity gradiometry (SGG), satellite-to-satellite tracking in high-low mode (hl-SST)) is presented. The software architecture and the data flow are briefly described, and the main software components and recent developments are presented. Selected numerical simulations are performed to demonstrate the functionality of the software. They are based on a realistic mission scenario. Special emphasis is placed on the impact of the new GOCE mission design, i.e. the gravity gradients defined in the Gradiometer Reference Frame (GRF), which deviates from the actual flight direction (Local Orbit Reference Frame; LORF) by a few degrees, and the resulting modified error budget of the GOCE gradiometer. Additionally, the benefits of a combination of the SGG and hl-SST components are presented and discussed.

**Keywords.** GOCE · Gravity field modeling · Gravity gradients · Gradiometer Reference Frame · Energy integral · Combined solution

## 1 Introduction

The dedicated satellite gravity mission GOCE (Gravity field and steady-state Ocean Circulation Explorer; ESA (1999)), the first Earth Explorer Core Mission, in the context of ESA's Living Planet programme, strives for a high-accuracy, high-resolution global model of the Earth's static gravity field.

GOCE is based on a sensor fusion concept: satellite-to-satellite tracking in the high-low mode (hl-SST) using GPS, and satellite gravity gradiometry (SGG). The GOCE mission, when successfully completed, will provide a huge data set consisting of several hundred million orbit data plus very precise gravity gradiometry data. This data contains abundant information about the gravity field of the Earth on a near-global scale, from very low (derived mostly from hl-SST) to high (derived mostly from SGG) frequencies.

The mathematical model for the parameterization of the Earth's gravity field is based on a series expansion into spherical harmonics. The goal is to resolve the model complete to degree and order 250, yielding approximately 63 000 unknown spherical harmonic coefficients. The determination of these coefficients from the complementary hl-SST and SGG data sets is a demanding numerical and computational task, and therefore efficient solution strategies are required to solve the corresponding large normal equation systems. During the last decade, several approaches have been developed to solve this large system of equations (e.g., Rummel et al. (1993), Schuh (1996), Klees et al. (2000), Pail and Plank (2002), Migliaccio (2003)). In Pail and Plank (2002, 2004), the rigorous solution of the large normal equation matrix by means of a parallel processing strategy implemented on a Beowulf cluster, the Distributed Non-approximative Adjustment (DNA) approach, was proposed. While direct methods perform an epoch-wise processing of the gravity field observations, the semianalytic approach considers the observations along a satellite track as a time-series (Rummel et al. (1993), Sneeuw (2002), Pail et al. (2003)).

The present paper outlines the concept for the installation of an operational software system, the processing strategy and the software architecture for the computation of a high-accuracy, high-resolution spherical harmonic model of the static Earth's gravity field (including a quality description in terms of a full variance/covariance matrix). This software system is part of the "GOCE High-Level Processing Facility" (HPF; Rummel et al. (2004)). The HPF is an assignment by the European Space Agency (ESA) for the installation of a de-centralized operable software system for the scientific processing of GOCE Level 1b data into the gravity field model. This work is done by the European GOCE Gravity Consortium (Rummel et al. (2004)), a co-operation of 10 European research institutions.

The operational software sub-system (HPF WP 6000) for the gravity field processing presented in this paper will be implemented and integrated, with a target date of March 2006, by the partner institutions

- Institute of Navigation and Satellite Geodesy, Graz University of Technology (task leader)
- Space Research Institute, Austrian Academy of Sciences
- Institute of Theoretical Geodesy, Univ. of Bonn
- Institute of Astronomical and Physical Geodesy, Technical University Munich

## 2 Software architecture and data flow

The software system is composed of two main components:

1. Quick-Look Gravity Field Analysis

   Computation of fast approximate gravity field solutions based on SGG and hl-SST data, for the purpose to derive a fast diagnosis of the GOCE system performance and of the level 1B input data in parallel to the mission.

2. Core Solver

   Rigorous high-precision solution of the very large normal equation systems applying parallel processing strategies. The Core Solver is composed of the Final Solver, taking the full normal equation matrix into account, and the Tuning Machine, being based on the method of preconditioned conjugate gradients, which will verify and tune the involved software components of the Final Solver in many respects. Analogously to the quick-look analysis, for the hl-SST processing the energy integral approach is applied.

The software system for time-wise gravity field processing to be developed, implemented and integrated will be conceived in a highly modular manner that allows the investigation of specific aspects of gravity field modelling such as filtering, numerical stability and optimum regularization of the normal equation systems, complementarity of SST and SGG, and their optimum weighting. This flexibility will support an adaptation of the software to unforeseen changes in the mission. Figure 1 shows the architectural layout of the software sub-system.

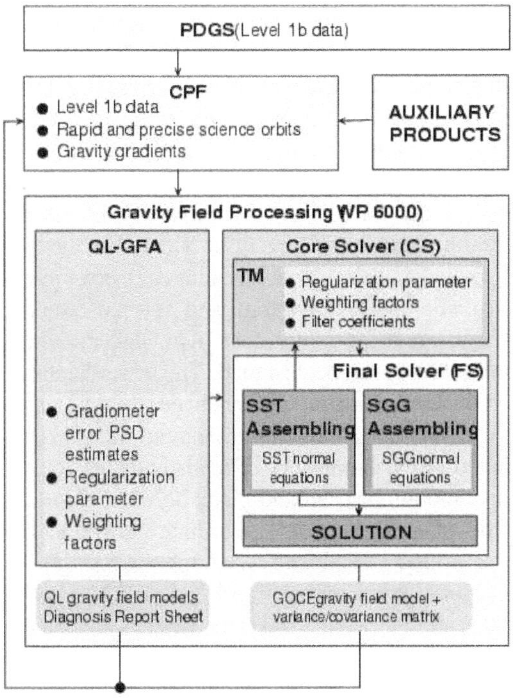


**Fig. 1** Architecture of the GOCE gravity field processing.

The Level 1b data will be processed in the framework of the Payload Data Ground Segment (PDGS). The data distribution among the 10 groups contributing to the HPF will be performed by the rolling archive of the Central Processing Facility (CPF) operated by SRON (National Institute of Space Research, Utrecht, NL). There, in addition to the Level 1b data also several products produced in the framework of the HPF are stored and distributed. These products include the rapid science and precise science orbit solutions, the externally calibrated gravity gradients, temporal correction time series for the gravity gradients, and auxiliary products.

### 2.1 Quick-Look Gravity Field Analysis (QL-GFA)

As the first module of gravity field processing, the Quick-Look Gravity Field Analysis (QL-GFA) will perform a fast check of the SGG and hl-SST input data in parallel to the mission. Quick-look gravity field models (SGG only, SST only, combined SST+SGG) are computed for the purpose of a fast analysis of the information content of the input data on the level of the gravity field solution, and statistical hypothesis test strategies in time and frequency domain are applied to evaluate the quality of the input data. Additionally, quick-look gravity solutions are compared with reference gravity field models, and are statistically tested using the confidence level of the QL-GFA solution and the covariance information of the reference gravity field model. The QL-GFA will also provide rough estimates for the regularization parameter and the optimum weighting factors of the individual measurement components. Additionally, the gradiometer error PSD (power spectral density) will be estimated from the residuals of a SGG-only gravity field analysis. Therefore, the question whether the a priori gradiometer error model is realistic can be answered, and optimal filters for an high-precision adjustment (e.g., Pail and Plank (2002, 2004)) can be designed. The target latency of the QL-GFA products varies between 1 and 3 days.

The QL-GFA method was already successfully applied in the framework of realistic GOCE closed-loop simulations (Pail and Plank (2002, 2004)), also in the case of short data sets, data gaps, and non-closing orbits (Pail and Wermuth (2003), Pail et al. (2003), Preimesberger and Pail (2003)). In Pail (2002) and Pail et al. (2003) the strategy for the estimation of the gradiometer error PSD is outlined, and in the latter paper several hypothesis test strategies for statistically testing both the residuals of the adjustment, and the quick-look gravity field solutions in terms of spherical harmonic coefficients, are presented and discussed.

QL-GFA solutions complete up to degree/order 250 can be processed within one hour on a standard PC. The efficiency and speed of QL-GFA is founded mainly on the application of FFT techniques, the assumption of block-diagonality of the normal equation matrix, and also on a simplified filter strategy in the spectral domain to cope with the colored noise characteristics of the gradiometer (Cesare (2002)). Deviations from this assumption are incorporated by means of an iterative procedure. A detailed discussion of the theory and the mathematical models of the QL-GFA software can be found in Pail et al. (2003).

Output products of the QL-GFA will be quick-look gravity field models, a Diagnosis Report Sheet describing the results of the quality check of the input data, and an estimate of the GOCE error PSD.

## 2.2 Core Solver (CS)

The objective is to compute a high-accuracy, high-resolution spherical harmonic model including a quality description of the static Earth's gravity field from GOCE SGG and SST observations. The parameterization of the model will be complete at least to degree and order 200. However, a resolution up to degree and order 250 is envisaged, depending on the actual accuracy of the SGG observations. Additionally, a quality description in terms of a full variance-covariance matrix will be provided.

The Core Solver is based on an epoch-wise (time-wise) processing of SST and SGG observations following the least-squares principle, applying parallel processing strategies implemented on a Beowulf cluster, by distributing the computational load to the CPUs of standard PCs. It performs a rigorous solution of the very large normal equation systems.

The Core Solver is composed of the Final Solver (FS) and the Tuning Machine (TM). The TM consists of two main modules:

- pcgma (pre-conditioned conjugate gradient adjustment; Schuh (1996, 2002)): It acts as a stand-alone solution strategy, and is used to verify and tune the involved software components of the Core Solver in many respects, e.g., to derive optimum regularization and weighting parameters.
- Data analysis tool: The data inspection and filter design application is used to verify diverse data and products involved in the gravity field processing, and to derive the filter coefficients which will be used in the Final Solver.

The FS consists of the following main modules:

- SST assembling: The information content of the SST data is exploited by making use of the precise kinematic GOCE orbits. The principle of energy conservation is applied. The feasibility of the approach was demonstrated by, e.g., Földvary et al. (2003), Badura et al. (2004). Since for this component the involved matrices are small, this application shall run on a serial hardware platform with a runtime of a few hours.
- SGG assembling: Given the precise GOCE orbit, the calibrated gravity gradients defined in the GRF are directly related to the unknown potential coefficients resulting in the linear observation model for all relevant tensor components, allowing to exploit the high degree of precision and resolution of the data. The complications arising from the colored noise of the gradiometer are managed by a recursive filter procedure in time domain (Schuh (1996), Pail and Plank (2002)).

- Solution: The mathematical models for SGG and SST data are combined to the overall mathematical model by means of superposition of the normal equations, applying an optimum weighting of the individual data types. The ill-posedness of the normal equations due to the polar gaps and the downward continuation are managed by regularization techniques (Metzler (2004)). The solution is computed applying a parallelized Cholesky algorithm.

Output products of the Core Solver will be a GOCE gravity field model in terms of spherical harmonic coefficients including a full variance-covariance matrix.

## 3 Simulations and Results

In the following, two case studies will be presented. In the first one the accuracy of the gravity field solution in the case of the new GOCE mission design, i.e. the gravity gradients given in the GRF, and with a lower accuracy compared to the original gradiometer specification (Cesare (2002)), will be evaluated, applying the QL-GFA software. In the second case study the potential of a high-precision combined gravity field solution, taking SGG and hl-SST observations into account, will be investigated.

### 3.1 Case study 1: GRF problem

#### Test data description

A realistic sun-synchronous 59-days repeat orbit with an inclination of i = 96.6° and a mean altitude of approx. 250 km, based on the global gravity model OSU91A (Rapp et al. (1991)), complete to degree/order $l_{max}$ = 80, was used. These orbit data were generated applying numerical orbit integration. The orientation of the satellite with respect to the LORF will have two major components: It is expected to be yaw steered, i.e. piloting w.r.t. the LORF, experiencing the maximum drag force, and it will perform a roll motion due to the latitude-dependence of the magnetic torquers' accuracy. These orientation angles of the GRF w.r.t. the LORF were modelled to be of sinusoidal shape with a 1/rev. characteristics, and with an amplitude of the yaw angle of ±3.5°, and the roll angle of ±2.2°. In practice, this orientation of the GRF w.r.t. the LORF will not be perfectly known. Therefore, in the following case studies these rotation angles were optionally superposed by a white noise time series with an amplitude of $\sigma$ = 20 arcsec. Considering the actual performance of the star tracker, this is a quite pessimistic assumption.

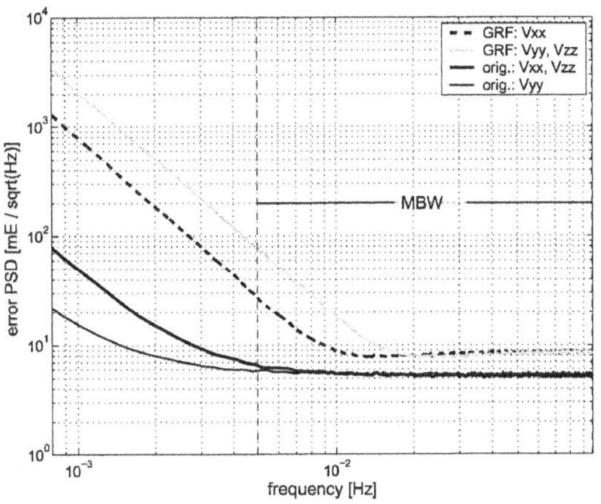


**Fig. 2** Gradiometer error PSD of the main diagonal components $V_{XX}$, $V_{YY}$, and $V_{ZZ}$; original performance vs. new GRF specifications.

Gradiometer measurement time series representing the three main diagonal elements of the gradient tensor $V_{XX}$, $V_{YY}$ and $V_{ZZ}$, defined in the GRF (according to the rotation defined above), were simulated along the orbit with a sampling interval of $\Delta$t = 5 s, leading to a data volume of more than 3 million observations. The SGG data contain gravity field information based on the OSU91A model complete up to degree/order $l_{max}$ = 180. The instrument noise characteristics were simulated based on figures of the new GOCE gradiometer error PSD. Figure 2 shows the spectral properties of the (smoothed) noise time series in terms of the square root of the PSD. An identical spectral noise behaviour of the $V_{YY}$ and $V_{ZZ}$ components is assumed, while the error PSD of the $V_{XX}$ component is considerably smaller. As a reference, also the error curves related to the original gradiometer specification (Cesare (2002)) are given. Obviously, the main problem of the new specification is the decreased accuracy in the lower frequency range of the original measurement bandwidth (MBW).

#### QL-GFA: Gravity Field Solutions

In the first simulation, the SGG observations ($V_{XX}$, $V_{YY}$ and $V_{ZZ}$) were superposed by a colored noise time series according to the new error specifications. In this simulation it is assumed that the rotation between LORF and GRF is perfectly known. The gray curve in Figure 3 shows the deviations of the estimated coefficients, parameterized complete to degree and order $l_{max}$ = 180, from the initial "true" OSU91A model represented by the degree median,

$$\sigma_l = \text{median}_m \left\{ |\bar{R}_{lm}^{(est)} - \bar{R}_{lm}^{(OSU)}| \right\} \qquad (1)$$

where $\bar{R}_{lm} = \{\bar{C}_{lm}, \bar{S}_{lm}\}$ are the harmonic coefficients, $(est)$ denotes the estimated quantities, and $(OSU)$ refers to the reference model OSU91A.

As a reference, the dash-dotted black curve shows the same configuration, but applying the original gradiometer error as specified by Cesare (2002) to the SGG data, demonstrating the degradation of the gravity field solution due to the degraded gradiometer performance.

Applying white noise of $\sigma$ = 20 arcsec to the rotation matrix results in the dotted black curve. Evidently, small errors in the orientation information hardly affect the accuracy of the gravity field solution. In order to demonstrate the effect of larger uncertainties of these rotation angles, the last simulation was recomputed, but now applying an unrealistically high rotation error of $\sigma$ = 5 arcmin. The solid black curve shows the corresponding results. Clearly, errors in the orientation information mainly affect the higher harmonic degrees $l$, while the effect in the low degree range of the spectrum is negligible.

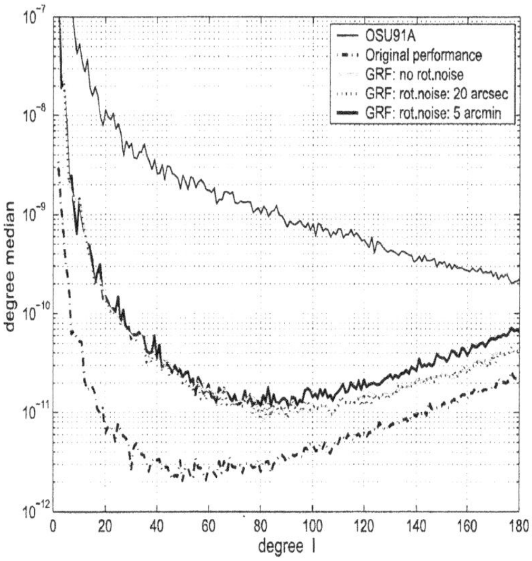


**Fig. 3** Degree median of diverse simulation scenarios: SGG-only solutions, resolved complete up to degree/order 180.

The errors of the estimated coefficients can also be expressed in terms of global gravity anomaly differences with respect to OSU91A. These cumulative gravity field errors were analyzed in the latitudinal range of 80°N to 80°S, excluding the polar gap regions where no GOCE observations are performed. Table 1 summarizes the main statistical parameters of the simulations described above.

Another valuable feature of the QL-GFA, the estimation of the gradiometer error PSD from the residuals of the adjustment, is discussed, e.g., in Pail (2004), but will not be further elaborated here.

**Table 1** Statistical comparison of several simulation results in terms of the minimum, maximum and RMS of cumulative gravity anomaly deviations $[mGal]$ ($1\ mGal = 10^{-5}\ m\,s^{-2}$) from the 'true' OSU91A reference model.

| Data sets | min. [mGal] | max. [mGal] | RMS [mGal] |
|---|---|---|---|
| Original performance | -2.34 | 2.15 | 0.45 |
| GRF, no rot. error | -7.71 | 8.36 | 1.71 |
| GRF, rot. error $\sigma$ = 20 arcsec | -7.72 | 8.38 | 1.71 |
| GRF, rot. error $\sigma$ = 5 arcmin | -9.20 | 8.18 | 1.97 |

### 3.2 Combined Gravity Field Solution

In this second case study a combined gravity field solution applying the DNA is performed. This solution strategy uses the full normal equation matrices for both GOCE observation types (SST and SGG), and assembles and solves the system in a rigorous sense. Very large matrices are involved for the SGG component, and thus also for the final solution. Since the storage requirements for the normal equations of system complete to $l_{max}$ = 250 requires 15 GBytes of RAM, parallel capacities have to be used to assemble the normal equations, and to solve and invert the system (Plank and Badura (2004)).

The case study is again based on the 59 days repeat orbit, with a simulated SGG signal complete to $l_{max}$ = 250 (3 main diagonal elements of the gradient tensor: $V_{XX}$, $V_{YY}$, $V_{ZZ}$), leading to 3 million observations and approximately 63 000 unknowns in the least squares adjustment. The observations were superposed by a measurement noise time-series according to the original noise behavior of the gradiometer (Cesare (2002)). For the combined solution we included SST normal equations parameterized up to a maximum degree of $l_{max}$(SST) = 80, with the orbit information of the GOCE satellite superposed by a white noise time series with a standard deviation of 0.3 mm/s (Badura et al. (2004)). In this simulation the combination was done by a simple superposition of the two normal equation systems, although in real GOCE data processing more sophisticated strategies (optimum weighting, ...) will have to be applied.

For the processing 32 PCs (64 CPUs) were applied for the DNA approach, with each CPU holding $\sim$ 240 MB of the normal equations. The DNA needed 50 hours for the assembling of the normal equations. For the solution and the inversion of the full system, 40 PCs (80 CPUs) were used. The solution was performed in 5 hours and 20 minutes, while the processing time for the inversion was 16 hours, including the time required to read the normal equations from disk.

Fig. 4 shows the cumulative geoid height errors (absolute values) up to degree/order 200, in the latitudinal band of ±80°, related to this solution. The main statistical parameters are: min = -0.12 m, max = 0.12 m, $\sigma$ = 0.023 m. The corresponding cumulative errors at degree/order 250 are: min = -0.57 m, max = 0.59 m, and $\sigma$ = 0.123 m.

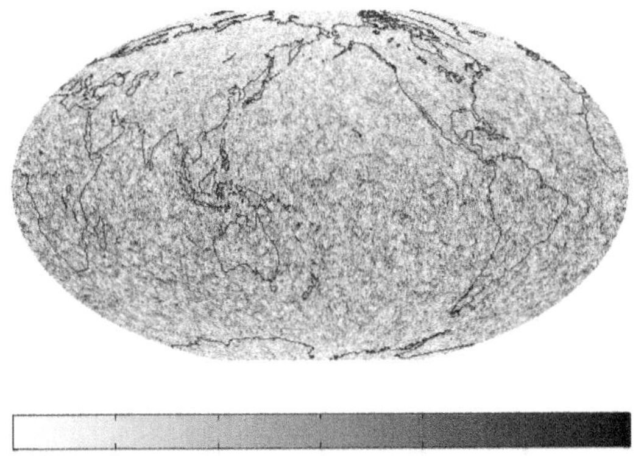


**Fig. 4** Deviations of the rigorous combined solution from the initial Earth model OSU91A in terms of geoid heights [m] up to degree/order 200.

## Acknowledgments


Financial support for this study came from the ASA (Austrian Space Agency) contract ASAP-CO-008/03. This study was performed in the framework of the ESA-project GOCE High-Level Processing Facility (Main Contract No. 18308/04/NL/MM).


## References


Badura, T., Klostius, R., Gruber, C., Sakulin, C. (2004). Derivation of a CHAMP only global gravity field model applying the Energy Integral Approach. Submitted to: Stud. Geophy. Geod. (in review).

Cesare, S. (2002). Performance requirements and budgets for the gradiometric mission. Technical Note, GOC-TN-AI-0027, Alenia Spazio, Turin, Italy.

ESA (1999). Gravity Field and Steady-State Ocean Circulation Mission. Reports for mission selection, The four candidate Earth explorer core missions, SP-1233(1), European Space Agency, Noordwijk.

Földvary, L., Svehla, D., Gerlach, Ch., Wermuth, M., Gruber, T., Rummel, R., Rothacher, M., Frommknecht, B., Peters, T., Steigenberger, P. (2003). Gravity Model TUM-2Sp Based on the Energy Balance Approach and Kinematic CHAMP Orbits. Proc. $2^{nd}$ CHAMP Science Meeting, GFZ Potsdam, Sprt. 1-4, 2003 (in print).

Klees, R., Koop, R., Visser, P.N.A.M., and van den IJssel, J. (2000). Efficient gravity field recovery from GOCE gravity gradient observations. J. Geod., 74, pp. 561-571.

Metzler, B. (2004). Core Solver: Combined SST+SGG solution. ASAP-Project "'GOCE DAPC Graz", Bridging Phase, Final Report, Contract ASAP-CO-008/03, WP Ib-4.4, pp. 167-180, Graz.

Migliaccio, R., Reguzzoni, M, and Sansó, F. (2003). Space-wise approach to satellite gravity field determinations in the presence of coloured noise. Submitted to J. Geod.

Pail, R. (2002). In-orbit calibration and local gravity field continuation problem. ESA-Project 'From Eötvös to mGal+", Final Report, ESA/ESTEC Contract 14287/00/NL/DC, WP 1, pp. 9 - 112, European Space Agency, Noordwijk.

Pail, R. (2004). GOCE Quick-Look Gravity Field Analysis: Treatment of gravity gradients defined in the Gradiometer Reference Frame. Proceedings of the $2^{nd}$ Internat. GOCE User Workshop, Frascati, March 2004.

Pail, R., Lackner, B., and Preimesberger, T. (2003). Quick-Look Gravity Field Analysis (QL-GFA). DAPC Graz, Phase Ia, Final Report, WP Ia-4.1, pp. 107 - 161, Graz.

Pail, R., and Plank, G. (2002). Assessment of three numerical solution strategies for gravity field recovery from GOCE satellite gravity gradiometry implemented on a parallel platform. J. Geod., 76, pp. 462 - 474.

Pail, R., and Plank, G. (2004). GOCE Gravity Field Processing Strategy. Stud. Geophys. Geod., 48, pp. 289-308.

Pail, R., and Wermuth, M. (2003). GOCE SGG and SST quick-look gravity field analysis. Advances in Geosciences, 1, pp. 5 - 9.

Plank, G., and Badura, T. (2004). Combined SST and SGG GOCE Gravity Field. Proc. GGSM 2004, Porto.

Preimesberger, T., and Pail, R. (2003). GOCE quick-look gravity solution: application of the semianalytic approach in the case of data gaps and non-repeat orbits. Studia geoph. et geod., 47, pp. 435 - 453.

Rapp, R., Wang, Y., and Pavlis, N. (1991). The Ohio state 1991 geopotential and sea surface topography harmonic coefficient models. OSU Report, 410, Department of Geodetic Science and Surveying, The Ohio State University, Columbus.

Rummel, R., van Gelderen, M., Koop, R., Schrama, E., Sansó, F., Brovelli, M., Miggliaccio, F., and Sacerdote, F. (1993). Spherical harmonic analysis of satellite gradiometry. Neth. Geod. Comm., Publications on Geodesy, 39, Delft, The Netherlands.

Rummel R, Gruber T, Koop R (2004) High Level Processing Facility for GOCE: Products and Processing Strategy. Proc. of the $2^{nd}$ Internat. GOCE User Workshop, Frascati, March 2004

Schuh, W.-D. (1996). Tailored Numerical Solution Strategies for the Global Determination of the Earth's Gravity Field, Mitteilungen geod. Inst. TU Graz, 81, Graz Univ. of Technology, Graz.

Schuh, W.-D. (2002). Improved modelling of SGG-data sets by advanced filter strategies. ESA-Project 'From Eötvös to mGal+", Final Report, ESA/ESTEC Contract 14287/00/NL/DC, WP 2, pp. 113 - 181, ESA, Noordwijk.

Sneeuw, N. (2002). A semi-analytical approach to gravity field analysis from satellite observations. Dissertation, DGK, Reihe C, Munich, 527, Bayerische Akademie d. Wissenschaften, Munich.

# GRACE Gradiometer

M.A. Sharifi, W. Keller
Department of Geodesy and GeoInformatics, University of Stuttgart, Geschwister-Scholl-Str. 24D, D-70174 Stuttgart, Germany

**Abstract.** Improving the accuracy of the spherical harmonic coefficients of the Earth's gravity field and its temporal variations at long and medium spatial-scales with unprecedented accuracy is the primary science objective of the GRACE mission. The line of sight (LOS) acceleration difference between the satellite pair is the most frequently utilized form of the observable. It is the simplest form of the observable which can be easily employed. Nevertheless, the observable is a two-point function and has no direct relationship with the field geometry at the evaluation point.

In this paper, as the alternative, *gradiometry approach* is proposed. Being a one-point function and having a direct relation with the field geometry (curvature of the field at the point) are two noteworthy achievements of the alternative formulation. Besides, using an observation quantity that is related to the second instead of the first-order derivatives of the gravitational potential amplifies the high-frequency part of the signal.

Complexity of the derived mathematical model and its proper treatment is the severe problem for the gradiometry approach. Herein, mixed gravitational acceleration-gradient model and also use of the available Earth' gravity model as a priori information on the low-degree harmonics are addressed.

The first recently released EIGEN2 CHAMP-only Earth's gravity model was employed for numerical analysis. Error analysis showed that the residuals of the estimated degree variances were of about $10^{-4}$ for $n \leq 90$. Also, the gravity anomaly residuals were less than 5 mGal for most points on the Earth.

**Keywords.** GRACE Gradiometer, Gradiometry, Sequential solution

## 1 Introduction

The line of sight (LOS) acceleration difference between a satellite pair has been frequently used for mapping the field globally (e.g. Hajela , 1974; Rummel , 1980; Garcia , 2002; Han et al. , 2003). The idea can also be applied to the GRACE observable as the first realization of the LL-SST mode.

Moreover, the GRACE configuration can be viewed as a huge one-component gradiometer with an arm length of 250 km. Rummel (2003) showed that the accuracy of this virtual one-dimensional gradiometer is about $10^{-6}\,\mathrm{E}/\sqrt{\mathrm{Hz}}$. Consequently, the GRACE configuration can be considered as potentially a precise one-dimensional virtual gradiometer. The advantage of looking GRACE observations as gradiometer data over looking at them as satellite-to satellite tracking data is that in the gradiometry mode the gravity field recovery can be done in a space-wise approach. The space-wise approach leads to an inversion-free recovery algorithm, which makes all the measures obsolete, which have to be taken to stabilize the linear system of equations in the satellite to satellite mode.

The paper starts with the mathematical description of gravitational field recovery by satellite-to-satellite tracking in the low-low mode as it has already been worked out in the contribution Rummel et al. (1978) and Rummel (1980) . This traditional approach relates the observation to a two-point function of the potential namely the gravitational acceleration difference of the satellites, projected onto the inter-satellite unit vector. The gravitational acceleration is proportional to the first-order derivative of the potential. For a high-resolution determination of the gravitational potential, an observation would be desirable which is a one-point function and relates to higher order derivatives of the potential, since for a one-point functional of the gravity field recovery in the space-wise model becomes possible.

Hence, we consider the GRACE mission as a one-axis gradiometer and formulate the problem in terms of the gravitational acceleration tensor components. Besides some remainder terms which can be modelled with sufficient accuracy, the gravitational acceleration gradient is related to the second order derivatives of the gravitational potential at the mid-point of the satellites configuration. In this way, the two-point first order problem is replaced by a one-point second order problem, promising a higher resolution of gravitational potential recovery.

Due to the complexity, the developed mathematical model can not be fully coded. Therefore, for ease of computation, we have to approximate the model with numerically applicable forms. Including the higher order terms beyond the linear term of the Taylor expansion makes the numerical computation very complicated. From the numerical point of view, the only possible form would be the linear approximation of the equation. On the other hand, excluding the higher order terms of the expansion results in truncation error whose contribution to the observation equation is considerable. Hence, using either a higher order approximation of the equation or modified linear approximation is inevitable.

Herein, we introduce two modified gradiometry algorithms which result in simple practical mathematical models. Mixed gravitational acceleration-gradient model and use of the available Earth' gravity model as a priori information on the low-degree harmonics are formulated. In both cases, increment to the low-degree harmonics and the higher degree coefficients are simultaneously estimated.

The final part of the paper is dedicated to recovery of the gravitational field and the analysis of the results. As a priori information, we will employ the first recently released EIGEN2 CHAMP-only Earth's gravity model to show the practical performance of the modified formulation. The article will end in some conclusions and recommendations.

## 2 Mathematical Formulation

The key observables of the GRACE mission are the inter-satellite distance $\rho$, and its first and second time derivatives $\dot{\rho}$ and $\ddot{\rho}$. These principal scalar quantities measured by the K-band Ranging system (KBR) are considered as the LL-SST information. They can be related to the Earth's gravitational field if the absolute positions of the spacecrafts, i.e. $\mathbf{r}_1$ and $\mathbf{r}_2$ are known. Therefore, the two satellites have been equipped with dual-frequency Blackjack GPS receivers to provide the HL-SST information. Eq. (1) connects the LL-SST observations to those of the HL-SST at each evaluation point (Rummel et al. , 1978):

$$\Delta\ddot{\mathbf{r}} \cdot \mathbf{e} = \ddot{\rho} + \rho^{-1}(\dot{\rho}^2 - \|\Delta\dot{\mathbf{r}}\|^2) \qquad (1)$$

where, $\mathbf{e} = \rho^{-1}(\mathbf{r}_2 - \mathbf{r}_1)$ is the unit vector along the LOS. $\Delta\dot{\mathbf{r}} = \dot{\mathbf{r}}_2 - \dot{\mathbf{r}}_1$ and $\Delta\ddot{\mathbf{r}} = \ddot{\mathbf{r}}_2 - \ddot{\mathbf{r}}_1$ are the difference of the velocities and accelerations of the two satellites here expressed in an inertial frame. We consider Eq. (1) as the basic equation of the $HL{+}LL$ combination observable and modify it in each case accordingly.

### 2.1 Gravitational Acceleration Difference

In the absence of non-gravitational forces, the left-hand-side of Eq. (1) can be considered as the Earth's gravitational acceleration difference along LOS, $\Delta\Gamma^{\mathrm{LOS}}$.

$$\Delta\Gamma^{\mathrm{LOS}} = \ddot{\rho} + \frac{\dot{\rho}^2}{\rho} - \frac{\|\Delta\dot{\mathbf{r}}\|^2}{\rho} \qquad (2)$$

The left-hand-side of Eq. (2) is a function of the gravitational potential partial derivatives and the inter-satellite unit vector $\mathbf{e}$, whereas the right-hand-side is the observational quantity. Using a sequence with an adequate number of observations, we set up the linear system of observation equations and recover the spherical harmonic coefficients. More details can be found in Keller and Sharifi (2004) and the references cited therein.

### 2.2 Gravitational Acceleration Gradient

As mentioned earlier, combining the two SST concepts, as shown in Fig. 1, makes the twin satellites to appear as a very accurate one-component gradiometer. Rummel (2003) showed that the accuracy of this virtual one-dimensional gradiometer is about $10^{-6}$ E/$\sqrt{\mathrm{Hz}}$. This unique characteristic of GRACE is a motivation to switch from the first derivatives of the gravitational potential to the second derivatives of the field. In other words, we write the observation equation (Eq. 2) as a function of the gravitational acceleration gradient components instead of the gravitational potential gradient.

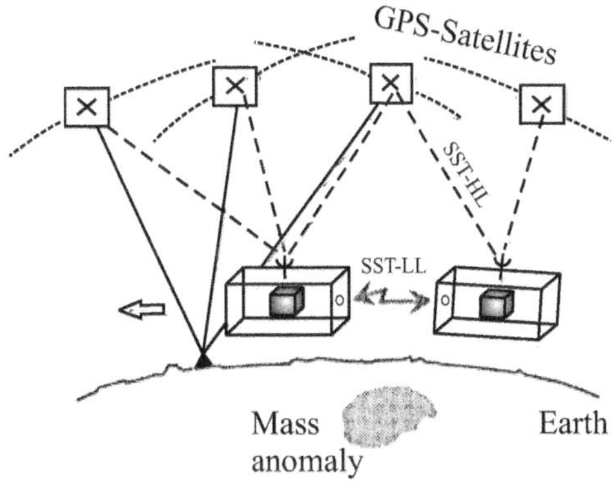


**Fig. 1.** Gradiometry with the GRACE twin satellites (from Rummel et al. (2002))

To derive the respective mathematical formulae, we expand the gravitational acceleration at the two satellites' respective positions around the mid-point using Taylor expansion. Subtracting the resultant expression yields (Keller and Sharifi , 2004):

$$\Delta\Gamma = \sum_{j=1:2:\infty} \frac{2^{1-j}}{j!}\,(\nabla^j \otimes \Gamma^T_{\mathrm{mid}}) \cdot \Delta\mathbf{r}^j, \qquad (3)$$

where, $\otimes$ is Kronecker product symbol and $\Gamma_{\text{mid}}$ is the gravitational acceleration at the mid-point of the satellites configuration. The left-hand-side of Eq. (3) is a two-point first order quantity, whereas the right-hand-side is a one-point higher order (at least second order) one. Consequently, inserting Eq. (3) into Eq. (2) results in the sought-after formulation.

Obviously, the expansion (Eq. 3) contains partial derivatives of the Earth's gravitational potential higher than the second order. Including the partial derivatives beyond the linear term makes the mathematical model rather complicated. Therefore, we will consider the linear term of the expansion and modify the equations to minimize the linearization error. Herein, we also assume the Earth's gravitational force as the only governing force field.

### 2.2.1 Linear Approximation

Assuming $j = 1$ in Eq. (3) dismisses the summation out and makes the equation as simple as possible:

$$\Delta\Gamma \doteq (\nabla \otimes \Gamma^T) \cdot \Delta\mathbf{r} = \mathbf{G}\Delta\mathbf{r}, \tag{4}$$

where, $\mathbf{G}$ is the Earth's gravitational gradient tensor. Inserting Eq. (4) into Eq. (2) and dividing both sides of the equation by $\rho$ results in Eq. (5), which is called *linear gradiometry equation*:

$$\mathbf{e}^T\mathbf{G}\mathbf{e} \doteq \frac{\ddot{\rho}}{\rho} + \frac{\dot{\rho}^2}{\rho^2} - \frac{\|\Delta\dot{\mathbf{r}}\|^2}{\rho^2} \tag{5}$$

Right-hand-side of Eq. (5) is a linear function of the Earth's gravitational gradient tensor elements. Comparing Eqs. (2) and (5), leads to a criteria for evaluation of the linearization error:

$$|\mathbf{e}^T\mathbf{G}\mathbf{e} - \frac{1}{\rho}\Delta\Gamma^{\text{LOS}}| \leq \epsilon \tag{6}$$

The linear approximation is valid as long as the linearization error is negligible. Otherwise, the error degrades the model and the linear approximation of the grtadiometry equation will collapse.

Keller and Sharifi (2004) investigated the linearization error for the GRACE configuration and showed that the error is at the level of $0.55\,\mathrm{E}(1\,\mathrm{E} = 1\,\textit{Eötvös Unit} = 10^{-9}\,\mathrm{s}^{-2})$, which can not be neglected. Thus, we should either include at least the cubic term or modify the model to lower the linearization error. Due to the complexity of the cubic approximation, we prefer to retain the linear approximation and apply the remove-restore technique to reduce influence of the cubic term.

### 2.2.2 Mixed Mathematical Model

The largest orbit perturbation for all satellite orbits is the so-called $J_2$-effect caused by the flattening of the Earth; and the effect of the next three zonal harmonics in the expansion of the Earth's gravitational field is about two orders of magnitude smaller than the perturbation from $J_2$.

Keller and Heß (1998) and Keller and Sharifi (2004) showed that the linearization error reduced to few ten mE by introducing an ellipsoidal reference field. Nevertheless, the estimation process leads to unacceptable solution at the presence of the linearization error residual corresponding to the gravitational disturbing potential. Thus, instead of introducing an ellipsoidal reference field, we split the gravitational potential ($V$) into a low-degree spheroidal reference field ($V_l$) and an incremental one ($V^l$). Accordingly, we consider observation equation of Eqs. (2) and (5) types for the first $l$ and the higher degree ($> l$) terms of the gravitational potential harmonic expansion respectively. Therefore, Eq. (5) can be recast into:

$$\frac{1}{\rho}\Delta\Gamma_l^{\text{LOS}} + \mathbf{e}^T \cdot \mathbf{G}^l \cdot \mathbf{e} \doteq \frac{\ddot{\rho}}{\rho} + \frac{\dot{\rho}^2}{\rho^2} - \frac{\|\Delta\dot{\mathbf{r}}\|^2}{\rho^2} \tag{7}$$

where, $\mathbf{G}^l$ stands for the gradient tensor corresponding to the higher degree harmonics of the gravitational potential expansion. Analogously, we redefine the linearization error criteria:

$$|\mathbf{e}^T\mathbf{G}^l\mathbf{e} - \frac{1}{\rho}\Delta\Gamma^{l\,\text{LOS}}| \leq \epsilon. \tag{8}$$

An appropriate choice of $l$ leads to some negligible linearization error residual. For instance, for $l = 10$, it is at the level of few mE (Keller and Sharifi , 2004). Consequently, all the spherical harmonic coefficients are estimated all together in a linear system of equations with reasonable accuracy.

Compared with Eq. (5), Eq. (7) contains less systematic error. In contrast, it is partially a two-point first order problem. In the following subsection, we improve this deficiency by introducing the sequential solution.

### 2.2.3 Sequential Estimation

Up to now, different global gravity models of the Earth have been released to public and many more may be developed later on. Combining the existing models with any new set of observations, carried out on the Earth's gravity field, is of particular interest to geoscientists. In other words, hybrid solution would be without doubt one of the most interesting challenges of the coming years.

As already discussed, the linearization error is the greatest single obstacle to the linear gradiometry equation. On the other hand, the low-degree harmonics' contribution is the most dominant one. Therefore, we consider one of the available Earth's gravity models and utilize the low-degree coefficients of the model as a priori information. Accordingly, any quantity corresponding to the low-degree harmonics can be split into the approximate value plus the respective correction. The first term on the left-hand-side of Eq. (7), for instance, can be written as:

$$\frac{1}{\rho}\Delta\Gamma_l^{\mathrm{LOS}} = \underbrace{\frac{1}{\rho}\Delta\Gamma_{0l}^{\mathrm{LOS}}}_{\text{approximate value}} + \underbrace{\frac{1}{\rho}\delta\Delta\Gamma_l^{\mathrm{LOS}}}_{\text{correction}}. \quad (9)$$

Replacing the correction term by the corresponding expression of the gradiometry type yields:

$$\frac{1}{\rho}\Delta\Gamma_l^{\mathrm{LOS}} = \mathbf{e}^T\delta\mathbf{G}_l\mathbf{e} + \frac{1}{\rho}\Delta\Gamma_{0l}^{\mathrm{LOS}}. \quad (10)$$

Inserting Eq. (10) into Eq. (7) results in Eq. (11), which is called *sequential gradiometry observation equation*:

$$\mathbf{e}^T(\delta\mathbf{G}_l + \mathbf{G}^l)\mathbf{e} \doteq \frac{\ddot{\rho}}{\rho} + \frac{\dot{\rho}^2}{\rho^2} - \frac{\|\Delta\dot{\mathbf{r}}\|^2}{\rho^2} - \frac{1}{\rho}\Delta\Gamma_{0l}^{\mathrm{LOS}} \quad (11)$$

Eq. (11) results from application of standard sequential adjustment to Eq. (7). The linearization error criteria is modified as:

$$|\mathbf{e}^T(\delta\mathbf{G}_l+\mathbf{G}^l)\mathbf{e}-\frac{1}{\rho}(\Delta\Gamma^{\mathrm{LOS}}-\Delta\Gamma_{0l}^{\mathrm{LOS}})| \leq \epsilon. \quad (12)$$

We implemented the idea based on some simulated data and the achieved results will be presented in section. (3). So, in brief:

- the two-point first order problem is replaced by a one-point second order problem,
- the linearization error reduces to an acceptable level,
- both the low- and the high-degree harmonic coefficients are estimated.

Eventually, It should be noted that the first two terms of the right-hand-side of Eqs. (2), (5), (7) and (11) are computed by means of the GRACE ranging data. For the third term, GPS and Doppler observations have to be used. The modelling of this term from GPS and Doppler observations was investigated by Keller and Heß (1998). They showed that this term can be modelled with an accuracy of about $10^{-13}\,\mathrm{s}^{-2}$ under realistic assumptions, which is sufficient for the purpose of the presented study.

## 3 Numerical Analysis

Numerical studies are based on the IAG simulated data of the Earth's gravity field dedicated satellite missions (Ilk et al. , 2003). As the pseudo-real gravity field of the Earth, *EGM96* (Lemoine et al. , 1998) complete to degree 300, has been considered. Moreover, we utilize EIGEN2 *CHAMP*-only (Reigber et al. , 2003) as a priori information.
First, the sequential gradiometry equation's truncation error and random error of the approximate value are evaluated. Finally, we will recover some low-degree coefficients of the gravitational potential based on the sequential gradiometry approach. In both cases, a one-month span of the GRACE observations is considered.

### 3.1 Evaluation of the Truncation Error and Random error of the Approximate Value

Using Eq. (12), we can determine the truncation error of the sequential observation equation. The evaluation was done for a few low degrees of the spherical harmonics and the results of a one-day span of the mission were shown in Fig. (2). As shown in Fig. (2),

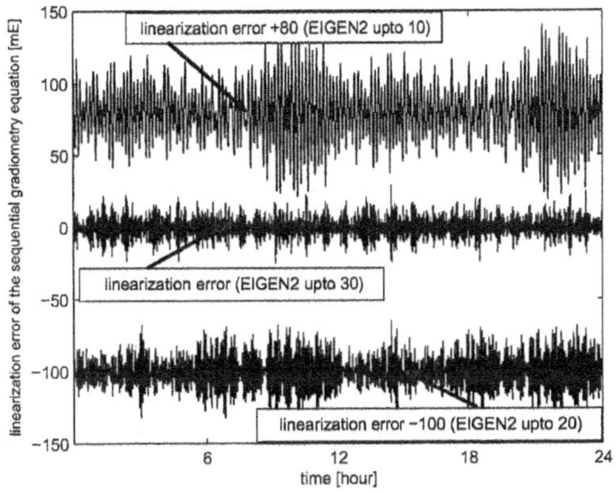


**Fig. 2.** Linearization error of the sequential gradiometry equation, Eq. 12, (EGM96 upto 90 as the pseudo-real field)

increasing the degree $l$ decreases the linearization error. On the other hand, stepping the degree up increases random error of the approximate value. As a representative example, using the variance components provided in EIGEN2 data file, the error was estimated and the results were depicted for one revolution of the mission for $l = 30$ and $l = 50$ in Figs. (3) and (4) respectively. The variance-covariance matrix has a dominant block diagonal structure in both cases. However, the diagonal elements correspond-

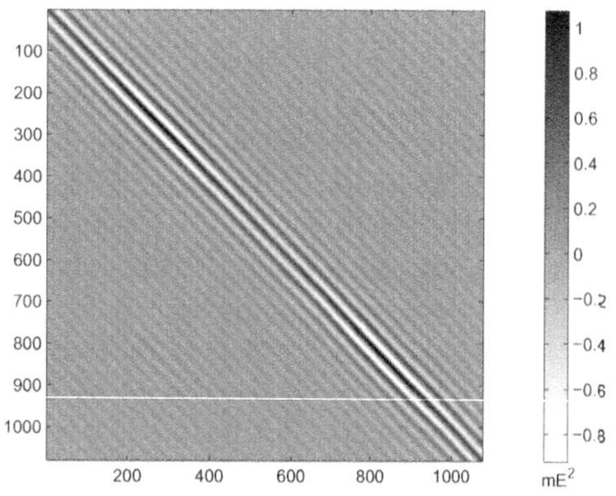


**Fig. 3.** Variance-covariance matrix structure of the approximate value corresponding to one revolution of the GRACE mission (pseudo-real field: EIGEN2 upto $l = 30$).

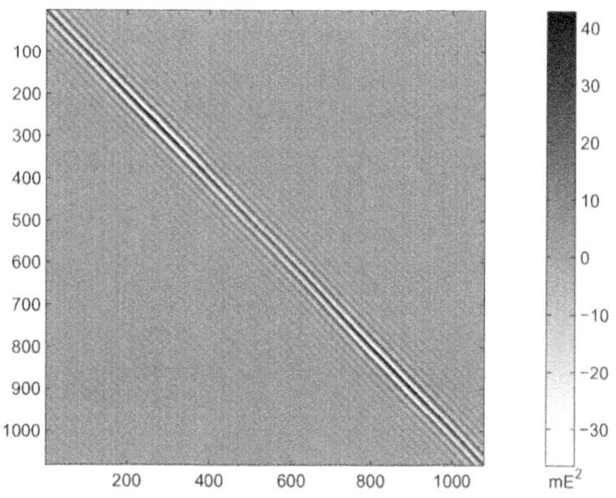


**Fig. 4.** Variance-covariance matrix structure of the approximate value corresponding to one revolution of the GRACE mission (pseudo-real field: EIGEN2 upto $l = 50$).

ing to $l = 30$ are about $1\,\mathrm{mE}^2$, whereas they exceed $40\,\mathrm{mE}^2$ for $l = 50$. Compared with the linearization error of the sequential gradiometry equation (Eq. 12), random error of the approximate value is negligible for $l = 30$. Nevertheless, subtracting the approximate value corresponding to $l = 50$ will double the diagonal elements of the variance-covariance matrix of the reduced observations. Hence, we employ the approximate value corresponding to $l = 30$ whose respective uncertainties are really negligible.

## 3.2 Recovery of the Spherical Harmonic Coefficients

In this subsection, we analyze the sequential gradiometry approach performance. We considered $n = 90$ as the maximum degree of the sought-after spherical harmonics. Therefore, to avoid the *omission* errors, the simulated observations only contain the respective signals ($n, m$ upto 90). We plugged the sequence of the simulated observations in Eq. (11) to estimate the spherical harmonic coefficients. The achieved results, as well as the original coefficients (EGM96), were plotted in Fig. (5) in terms of degree variances. Besides the degree variances, the figure shows estimation error of the coefficient . As

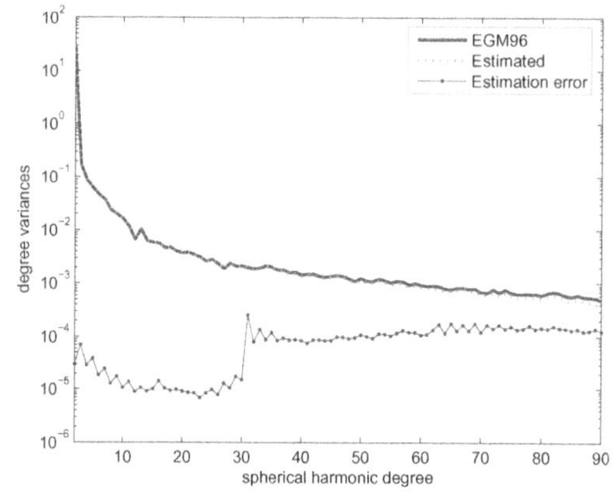


**Fig. 5.** Estimated degree variances upto 90 based on EIGEN2 upto 30 as a priori information.

seen in Fig. (5), estimation error is lower than $10^{-4}$ for $l \leq 30$. The error steps up to $10^{-4}$ for $l > 30$. This jump is in accordance with the split point of the Earth's gravitational potential in the sequential formulation. Therefore, it indicates that the higher value of $l$ leads to a better accuracy, at least for $n \leq l$. However, as mentioned earlier, the high-degree of $l$ will dramatically increase uncertainty of the reduced observations. Then, a medium degree of $l$ would be an optimal choice.

Moveover, the gravity anomaly is computed on a regular $2° \times 2°$ grid on the mean sphere using both the estimated coefficients and EGM96's. As we see in Fig. (6), the gravity anomaly errors are less than 5 mGal for most points on the Earth. The error does not exceed 15 mGal.

# 4 Conclusion

The GRACE mission is the first mission that has realized satellite-to-satellite tracking concept in LL-mode. Despite the mission realization, the idea has been investigated theoretically since 1970 (e.g. Wolff , 1969). Consequently, different approaches have been introduced by many authors and researchers.

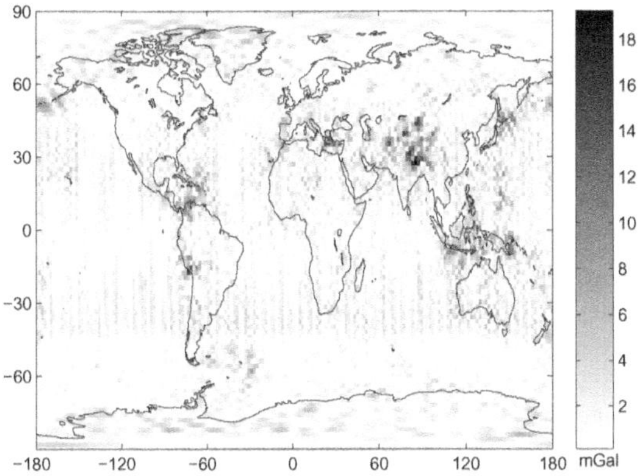


**Fig. 6.** Gravity anomaly residuals (absolute values).

Among them, LOS acceleration difference between the two satellites has been of particular interest. It is a two-point first-order problem. However, a one-point second-order formulation is far preferable to a two-point first-order formulation. For instance, having direct relationship with the gravity field geometry and promising improved gravitational field resolution are two noteworthy achievements of the sought-after formulation. Moreover, a relatively long inter-satellite range is a motivation to consider the spacecrafts as one-dimensional virtual gradiometer.

In this regards, we derived the desired formulation simply by expanding the LOS acceleration differences around the mid-point of the satellite configuration. We utilized linear term of the expansion called linear gradiometry equation, because of its simplicity. However, because of considerable linearization error, we introduced the sequential gradiometry equation as an intermediate solution to lower the linearization error.

EGM96 upto 90 and EIGEN2 CHAMP-only model upto 30 were respectively employed as the pseudo-real field and a priori information on the low-degree harmonics of the gravitational potential. The spherical harmonic degree variances estimated with an accuracy of about $10^{-4}$. Also, Gravity anomaly residuals were less than 5 mGal for most points on the Earth's surface. To sum up, the estimated results indicate the high level performance of the proposed method.

**Acknowledgements** M.A. Sharifi expresses his deep thanks for the financial support provided by IAG for participating in Gravity, Geoid and satellite missions symposium (GGSM2004).

The authors appreciate invaluable remarks of the anonymous reviewers.

## References

Garcia, R. V. (2002). Local Geoid Determination from GRACE Mission. Report No. 460, Dept. of Geod. Sci., Ohio State University, Columbus.

Hajela, D. P. (1974). Improved Procedures for the recovery of 5° mean gravity anomalies from ATS-6/GEOS-3 satellite-to-satellite range-rate observation. Report No. 276, Dept. of Geod. Sci., Ohio State University, Columbus.

Han, S. C., Jekeli, C., Shum, C. K., (2003). Static and temporal gravity field recovery using grace potential difference observables. *Advances in Geosciences*, 1: 19-26.

Ilk, K. H., Visser, P., Kusche, J. (2003). Satellite Gravity Field Missions. Final Report Special Commission 7, Travaux IAG, Vol. 32, general and technical reports 1999-2003, Sapporo.

Keller, W., Heß, D. (1998). Gradiometrie mit GRACE. *ZfV* 124: 137-144.

Keller, W. Sharifi, M. A. (2004). Satellite Gradiometry Using a Satellite Pair. *J. Geodesy*, under review.

Lemoine, F. G., Kenyon, S. C., Factor, J. K., Trimmer, R. G., Pavlis, N. K., Chinn, D. S., Cox, C. M., Klosko, S. M., Luthcke, S. B., Torrence, M. H., Wang, Y. M., Williamson, R. G., Pavlis, E. C., Rapp, R. H., Olson, T. R. (1998). The development of the joint NASA GSFC and the National Imagery and Mapping Agency (NIMA) geopotential model EGM96. NASA/TP-1998-206861, National Aeronautics and Space Administration, Washington, DC.

Reigber, C., Schwintzer, P., Neumayer, K.-H., Barthelmes, F., König, R., Förste, C., Balmino, G., Biancle, R., Lemoine, J.-M., Loyer, S., Bruinsma, S., Perosanze, F., and Fayard, T. (2003). The CHAMP-only Earth gravity field model EIGEN2. *Adv. Space Res.* accepted.

Rummel R, Reigber, C., Ilk, K. H. (1978). The Use of Satellite-to-satellite Tracking for Gravity Parameter Recovery. *Proc. of the European workshop on Space Oceanography, Navigation and Geodynamics*, ESA SP-137, pp. 153-161.

Rummel, R. (1980). Geoid height, Geoid height differences, and mean gravity anomalies from "low-low" satellite-to-satellite tracking- an error analysis. Report No. 306, Dept. of Geod. Sci., Ohio State University, Columbus.

Rummel, R., Balmino, G., Johannessen, J., Visser, P., Woodworth, P. (2002). Dedicated Gravity Field Missions - Principles and Aims, *J. Geodynamics* 33: 3-20.

Rummel, R. (2003). How to Climb the Gravity Wall. *Space Science Reviews*, vol. 108: 1-14.

Wolff M (1969). Direct Measurements of the Earth's Gravitational Potential Using a Satellite Pair. *J. Geophys Res.*, 74: 5295-5300.

# Modelling the Earth's gravity field using wavelet frames

I . Panet [1,2], O. Jamet [1]
[1] Laboratoire de Recherche en Géodésie, Institut Géographique National, France

M. Diament [2]
[2] Laboratoire de Gravimétrie et Géodynamique, Département de Géophysique Spatiale et Planétaire, Institut de Physique du Globe de Paris, France

A. Chambodut [3]
[3] Laboratoire de Géomagnétisme et Paléomagnétisme, Institut de Physique du Globe de Paris, France

**Abstract** Present and forthcoming satellite gravity missions provide us with new and unique datasets in order to model the Earth's gravity field at 100 km resolution. These new models will bring significant advances in our understanding of the Earth's structure and dynamics. However, it will be necessary to combine satellite data with surface and airborne measurements in order to improve the short wavelength components of the gravity field. The derived regional models with an increased spatial resolution will be used to carry out geodynamic studies at lithospheric or crustal scale. Whereas the classical spherical harmonics decomposition leads to strong numerical difficulties when dealing with local features, wavelet-based representations can handle the local scales as well as the global ones; they should thus be extremely useful to derive local models taking into account data of different origins. Here we describe the construction of wavelet frames on the sphere based on the Poisson multipole wavelet. Those wavelets are of special interest for field modelling since their shape is linked to the potential of multipole sources, and their scaling parameter to the multipole depth (Holschneider et al., 2003). We also compute a local wavelet decomposition of the gravity field at high resolution from evenly and unevenly distributed data using least squares collocation. Our first results show the efficiency of such a representation.

**Keywords** Spherical wavelets, multipoles, frame, covariances.

## 1 Introduction

Gravity field observation from space provides promising advances for improving measurement of the geoid at 100 km resolution, and of its time variations at 400 km resolution. The derived models of the static and time-varying gravity field will lead to a better understanding of the internal geodynamic processes and of the superficial fluid envelops. They will also contribute to improve global and national height references. However, it will be necessary to densify them locally with surface and airborne measurements in order to improve their resolution for local studies. If the spherical harmonic analysis is well suited for global representations, it is very demanding for such high resolution models since the number of functions involved is quite large. Moreover, the systems to solve are badly conditioned when the data do not cover the whole sphere. That is the reason why other methods have been investigated last years: orthonormalization of the spherical harmonics over a bounded domain (Albertella et al., 1999; Hwang, 1993), spherical cap harmonic analysis (de Santis et al.,1997), and a promising approach: wavelet frames (Freeden et al., 1998; Holschneider et al., 2003).

Discrete wavelet frames are based on the discretization of the continuous wavelet transform on the sphere. Many mother wavelets can be used, and among them, axisymmetric wavelets are of special interest. Indeed, they can be expressed in a simple way as series of Legendre polynomials. In the present paper, we focus on a particular family of axisymmetric wavelets, well-suited for modelling of potential fields: the Poisson multipole wavelets. Let us notice that wavelets based on the Poisson kernel were already used in one or two dimensions to analyse and interprete gravity and magnetic anomalies (Sailhac et al., 2000, Martelet et al., 2001).

Contrary to the spherical harmonics basis, wavelet frames are not made of orthonormal functions. Thus, computing the dual frame is required in order to derive the covariances between wavelet coefficients involved in collocation. We finally derive local representations underlining the interest of wavelets for regional, high resolution modelling.

## 2 Frames of wavelets

### 2.1 Frames

A collection of functions $\{g_n\}_{n=0,1,\ldots}$ in a Hilbert space $H$ is a frame of $H$ if we have for all $s \in H$:

$$A\|s\|^2 \leq \sum_n |\langle g_n, s\rangle|^2 \leq B\|s\|^2 \tag{1}$$

The constants ($0 < A \leq B < \infty$) are called frame-bounds, and $\langle g_n, s\rangle$ denotes the scalar product of $g_n$ with $s$. Such a family provides a complete and stable representation of the space $H$, which may be redundant. $s \in H$ can be written as:

$$s = \sum_n \langle s, \tilde{g}_n\rangle \, g_n \tag{2}$$

where $\{\tilde{g}_n\}_{n=0,1,\ldots}$ is the dual frame (see section 3.2). When $A = B$, the frame is tight. No direction of space $H$ is privileged. In the opposite case, $A$ and $B$ reflect the extremal sensitivities of the frame.

In the following, the space $H$ is the space of square integrable functions on the unit sphere (admitting an harmonic continuation outside this sphere). The considered frames are spherical wavelet frames. The wavelets are characterized by a scale parameter, defining their spectral content, and a position parameter, defining the point around which they concentrate their energy in space domain. We discretize the continuous wavelet transform via these two parameters. Scales are discretized in order to cover the whole spectrum. For each chosen scale, we discretize the positions in order to cover the whole sphere homogeneously. This defines successive generations $j$ of wavelets.

### 2.2 The Poisson multipoles wavelets

The Poisson multipole wavelets, introduced by Holschneider et al. (2003), are of special interest for potential field modelling. Indeed, their shape may be identified with the potential of a multipolar source. The wavelet $\psi_i^{a,n,\vec{e}}$ at point $\vec{x}$ on the unit-sphere is expressed as:

$$\psi_i^{a,n,\vec{e}}(\vec{x}) = N_i \sum_\ell (a\ell)^n e^{-a\ell} Q_\ell \left(\frac{\vec{e}}{\|\vec{e}\|} \cdot \vec{x}\right) \tag{3}$$

with :

$$N_i = \left(\langle \psi_i^{a,n,\vec{e}}(\vec{x}), \psi_i^{a,n,\vec{e}}(\vec{x})\rangle\right)^{-\frac{1}{2}} \tag{4}$$

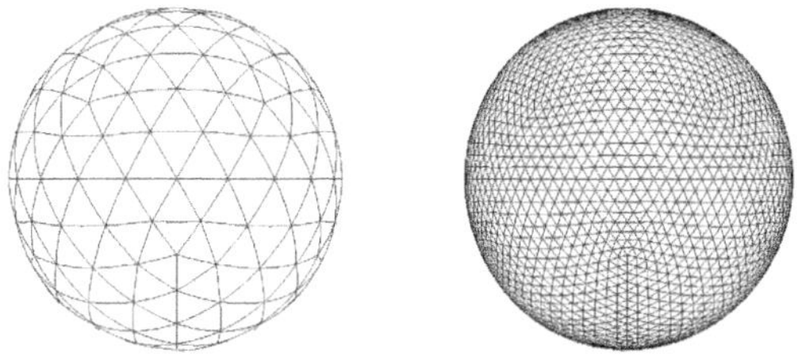

Figure 1: Meshes at generations 3 (*left*) and 5 (*right*).

Its harmonic continuation at point $\vec{x}$ outside the sphere $\Psi_i^{a,n,\vec{e}}$ is given by:

$$\Psi_i^{a,n,\vec{e}}(\vec{x}) = \tag{5}$$
$$N_i \sum_\ell \frac{1}{\|\vec{x}\|^{\ell+1}} (a\ell)^n e^{-a\ell} Q_\ell \left(\frac{\vec{e}}{\|\vec{e}\|} \cdot \frac{\vec{x}}{\|\vec{x}\|}\right)$$

In the above equations, $N_i$ is a $L_2$-normalisation factor, $\ell$ is the degree of the Legendre polynomial $P_\ell$ and $Q_\ell$ is the related reproducing kernel: $Q_\ell = \frac{2\ell+1}{4\pi} P_\ell$. $a$ is the scale parameter, $\vec{e}$ is the position of the multipole, and $n$ its order. Increasing $n$ improves the frequency localization but deteriorates the space localization. The depth of the multipole in the case of the unit sphere is linked to the scale parameter by $\|\vec{e}\| = e^{-a}$.

### 2.3 Discretization of scales and positions

We chose a simple sampling of the sphere, based on the successive subdivisions of the facets of a regular convex polyhedron centred with respect to the sphere. The directions of the vertexes are then projected onto the sphere. The more we subdivide the polyhedron's facets, the finer the mesh: thus, we create easily a set of hierarchical meshes associated with wavelets at different scales. The polyhedron chosen is the icosahedron, since it leads to very regular meshes, as shown by Figure 1 (the dispersion of distances between points comes to about 10% of the mean value). The construction of an icosahedric mesh was implemented based on $C^{++}$ code from Richard J. Bono freely distributed on *http://www.applied-synergetics.com*, following Kenner (1976).

To provide a regular coverage of the spectrum, we discretize the scale parameter according to a geometric progression: the scale $a_j$ at level $j$ is given by: $a_j = C.\gamma^{j-1}$. We chose $C = 3$ and $\gamma = \frac{1}{2}$. Thus, the spectrum is covered (see Figure 2); and the multipoles regularly sample the Earth's interior.

The number of wavelets for each scale is large enough so that the wavelet set shows a good sensitivity

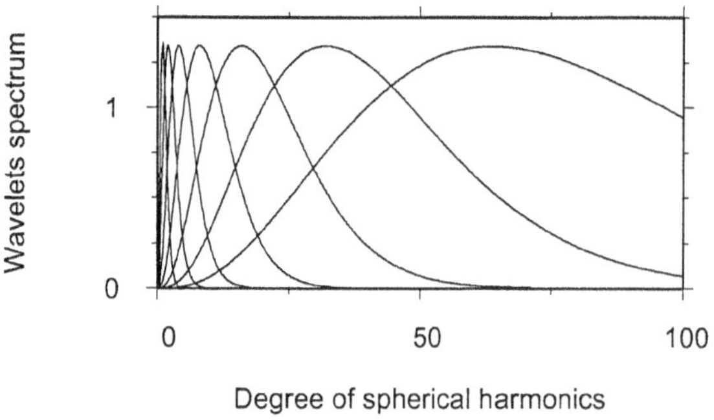


Figure 2: Spectrum of the wavelets $(a_j\ell)^n e^{-a_j\ell}$ as a function of spherical harmonic degree $\ell$ for generations $j = 1$ to $j = 7$ of the wavelet frame (order $n = 3$ multipoles).

to all degrees and orders of the spherical harmonics. Figure 3, presented in Sect. 3, can be interpreted as a mean number of wavelets per degree/order taking into account a spectral ponderation: the observed values are clearly larger than 1 (case of the spherical harmonics basis) and never tend to 0.

### 2.4 Frames of spherical wavelets

We did not prove that the above family of wavelets is actually a frame of $H$. In particular, results describing appropriate discretizations of positions via fundamental systems for band-limited wavelets (Freeden, 1998) can not be applied to wavelets with an infinite spectrum. However, numerical tests seem to support the hypothesis that the wavelet set is a frame (see for instance Sect. 3.2). In any case, the wavelet family provides an approximation of the gravity field. Its quality depends on the generating properties of the wavelets. Thus, a bad representation of the gravity field can be symptomatic of ungenerated harmonic spaces whereas a good one guaranties that the wavelets at least numerically generate those spaces. Numerical tests (Chambodut et al., submitted) at a global scale show that the first generations of wavelets numerically generate spherical harmonics. Given the process of construction of the wavelets set, one can reasonably assume the next generations to show the same approximating properties.

## 3 Collocation in the non-orthogonal case

Our aim is to derive a high resolution wavelet model of the disturbing potential based on gravity disturbances data. In this section, we derive the collocation formulas in the general frame settings.

### 3.1 Deriving a model from observations

Taking into account the multipolar nature of the wavelets, we express the disturbing potential $T$ as a linear combination of wavelets:

$$T = \sum_i \alpha_i \Psi_i \qquad (6)$$

Thus, wavelet coefficients can be interpreted as a ponderation of multipolar sources. Gravity disturbances $\delta g$ are hereafter deduced in the spherical approximation:

$$\delta g = \sum_i \alpha_i \frac{\partial \Psi_i}{\partial r} \qquad (7)$$

or, in matricial notation: $\delta g = M \cdot \alpha$. $\alpha$ is the vector of unknown coefficients, $M$ the observation matrix. We denote $b$ the vector of measurements (of gravity disturbances). The solution vector is given after Moritz (1989):

$$\alpha = K \cdot M^t (M \cdot K \cdot M^t + W)^{-1} \cdot b \qquad (8)$$

where $W$ is the covariance matrix of the noise of the measurements and $K$ the covariance matrix of the coefficients, detailed in the following.

### 3.2 Dual frame

If the wavelets set is a frame, the coefficients are defined as the scalar products between restriction of the disturbing potential to the unit sphere $T_\Sigma$ and the dual frame: $\alpha_i = \langle T_\Sigma, \tilde{\psi}_i \rangle$. The dual frame is given by (Mallat, 1999):

$$\tilde{\psi}_i = (U^* U)^{-1} \psi_i \qquad (9)$$

where $U$ is the operator defined by:

$$\forall i, \ \forall s \in H, \ \ Us[i] = \langle s, \psi_i \rangle \qquad (10)$$

and $U^*$ its adjoint. This definition is equivalent to: $\langle \psi_j, \tilde{\psi}_i \rangle = \delta_{ij}$ when the wavelets set is independant.

However, the computation of the dual frame based on the above relationship is computationally demanding, unless the frame is tight. Indeed, a tight frame with frame-bounds equal to $A$ verifies:

$$\tilde{\psi}_i = \frac{1}{A} \psi_i \qquad (11)$$

We provide here a qualitative argument supporting the hypothesis that the described wavelet family is close to be tight. We already discussed the generating properties of the wavelet set. In this section, we focus on its isotropy in the space $H$. Let us define $E_m^\ell$ as:

$$E_m^\ell = \sum_i |\langle \psi_i, Y_m^\ell \rangle|^2 \qquad (12)$$

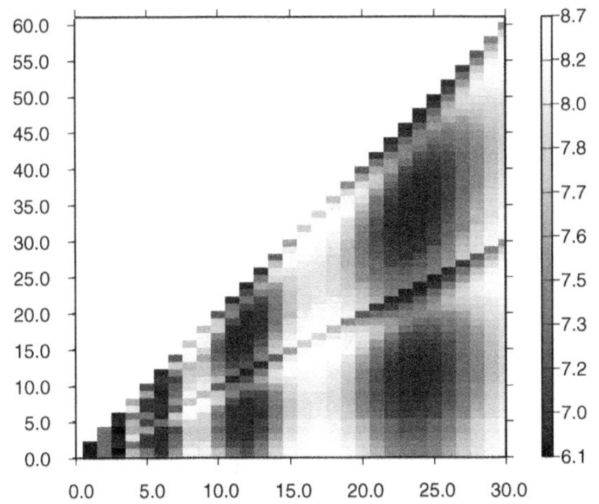


Figure 3: Value of $E_m^\ell$ for the first 7 generations of wavelets, for all degrees and orders lower than 30. *Abscissa* : degree of spherical harmonics, *ordinate* : order of spherical harmonics.

$E_m^\ell$ increases when the redundancy of the wavelet family $(\psi_i)_i$ in the direction of the spherical harmonic $Y_m^\ell$ increases. A necessary condition for the frame to be tight is that $E_m^\ell$ be constant. This condition is not sufficient. Though, since the directions $Y_m^\ell$ sample regularly the directions of $H$, we consider that it supports the tightness hypothesis. Figure 3 plots the values of $E_m^\ell$ taking into account the first 7 generations of wavelets: they are all coming around 7.5, with relative variations of 18%. Adding further scales would not change this result significantly since the next generations of wavelets have almost no power at degrees lower than 30. Thus, the wavelet set is rather isotropic in the directions $Y_m^\ell$, and we assume the framebounds to be close. Further experimental results will confort this hypothesis.

### 3.3 Covariance matrix $K$

Thus, we computed the covariances between wavelet coefficients applying the tight frame approximation. We denote $K(\vec{r}, \vec{r_1})$ the covariance function of gravity disturbances at points $\vec{r}$ and $\vec{r_1}$ on the sphere. We make the assumption that $K(\vec{r}, \vec{r_1})$ only depends on the spherical distance between $\vec{r}$ and $\vec{r_1}$. In this case, it can be written as a series of Legendre polynomials after Moritz (1989):

$$K(\vec{r}, \vec{r_1}) = \sum_\ell c_\ell P_\ell(\vec{r} \cdot \vec{r_1}) \qquad (13)$$

The coefficients $c_\ell$ are equal to the variance of gravity anomalies for degree $\ell$. We assume that the power spectrum of the gravity potential follows Kaula's rule of quadratic decrease (Kaula, 1966). Thus, the power spectrum of the gravity disturbance at degree $\ell$ decreases as:

$$c_\ell = (\ell+1)^2(2\ell+1)\frac{\alpha}{\ell^4} \qquad (14)$$

Let us now denote $K_{\vec{r},\vec{r_1}}$ the operator associating to each square integrable function on the sphere $f$ its scalar product with $K(\vec{r}, \vec{r_1})$:

$$K_{\vec{r},\vec{r_1}} f(\vec{r_1}) = \langle K(\vec{r}, \vec{r_1}), f(\vec{r}) \rangle \qquad (15)$$

We derive the covariance between two coefficients $\alpha_i$ and $\alpha_{i'}$ as a scalar product between the corresponding wavelets. This comes from the formulas of covariance propagation:

$$K_{i,i'} \approx \frac{1}{A^2} \langle K_{\vec{r},\vec{r_1}} \psi_i(\vec{r_1}), \psi_{i'}(\vec{r_1}) \rangle \qquad (16)$$

## 4 Results

### 4.1 Data

In this section, we derive wavelet models of the gravity field applying our approach. The studied area is in the northern part of the Andes. Synthetic datasets are obtained by sampling the EGM96 gravity disturbances model from Lemoine et al. (1998). We applied a gaussian filtering to the coefficients up to degree 360 (with an attenuation of 0.6 at degree 250) in order to avoid artificial oscillations. We considered 2 datasets, the one made of regularly distributed data, the other made of randomly distributed data. The regular distribution of data counts 1369 samples, with one data per bin of $0.25°$ in the area of $-5°/-14°$ lat. N, and 278°/287° long. E. The irregular one counts 576 samples between $0°/-18°$ lat. N, and 275°/293° long. E. The concentration of data is higher in the northern half (394 data) whereas the southern half is more sparsely covered (182 data).

### 4.2 Parameters

We inverted these datasets on generations 2 to 8 of the frame for the case of irregular data distribution (1920 wavelets), and 2 to 10 for the case of regular data distribution (7189 wavelets). Indeed, since gravity disturbances have no component on degree 1, we do not expect the first generation of wavelets (mainly centred on degree 1) to be significant in the representation. We applied a spatial selection of the wavelets : only wavelets whose influence radius intersects the area under study are selected. The influence radius of a wavelet is based on its spatial variance as defined by Freeden (1998).

Matrix $W$ is diagonal, assuming an uncorrelated noise. As the data are perfect, the choice of $W$ is arbitrary. We considered a noise of $10^{-2}$ mGals for the regular distribution and $2.5\ 10^{-3}$ mGals for the irregular one. The covariance matrix $K$ is evaluated assuming that $\tilde{\psi}_i = \frac{1}{7.5}\psi_i\ \forall i$. Lastly, the filter applied to the synthetic data is taken into account within the observation equations.

## 4.3 Results

Results for the regular case are presented in Figure 4. Wavelets succeed in representing the local gravity disturbances. Wavelet model shows visually no difference with the EGM 96 model. Residuals between synthetic data and wavelet model amount to a few microGals. Residuals between the EGM 96 model and the wavelet model are of same order that measurements residuals in the central part of the area. We could not avoid small edge effects, around 0.1 to 0.2 mGals, due to the spatial selection of the wavelets. A possible explanation is that the vertices of the meshes show a slight obliquity with respect to meridians and parallels, thus, two edges out of four are privileged. However, this issue still has to be investigated in more details.

Results for the irregular case are presented in Figure 5. The wavelets handle the gaps without oscillating, and restitute the main features of the gravity disturbances. Residuals between synthetic data and wavelet model mainly amount to a few tens of microGals. Residuals between the EGM 96 model and the wavelet model are smaller in the norther part of the area, reflecting the higher density of data. They increase in the areas of strong gravity variations since the available data do not constrain sufficiently the model.

## 5 Conclusion

These tests proved the ability of a subset of the wavelet frame to represent the gravity field at a rather high resolution in a delimited area, and to cope with an irregular distribution of data. They validate the approximations made in the estimation of the dual frame. The frame used here may be too redundant: good results were also obtained in the regular case with less wavelets (4607 wavelets only). Lastly, let us notice that the wavelet coefficients can be interpreted for geophysical purposes, the generations 1 to 10 corresponding to multipoles located at varying depths from the core up to the Earth's crust. Applying this method, we intent to derive local refinements of the global gravity model from current and planned space missions by jointly modelling two

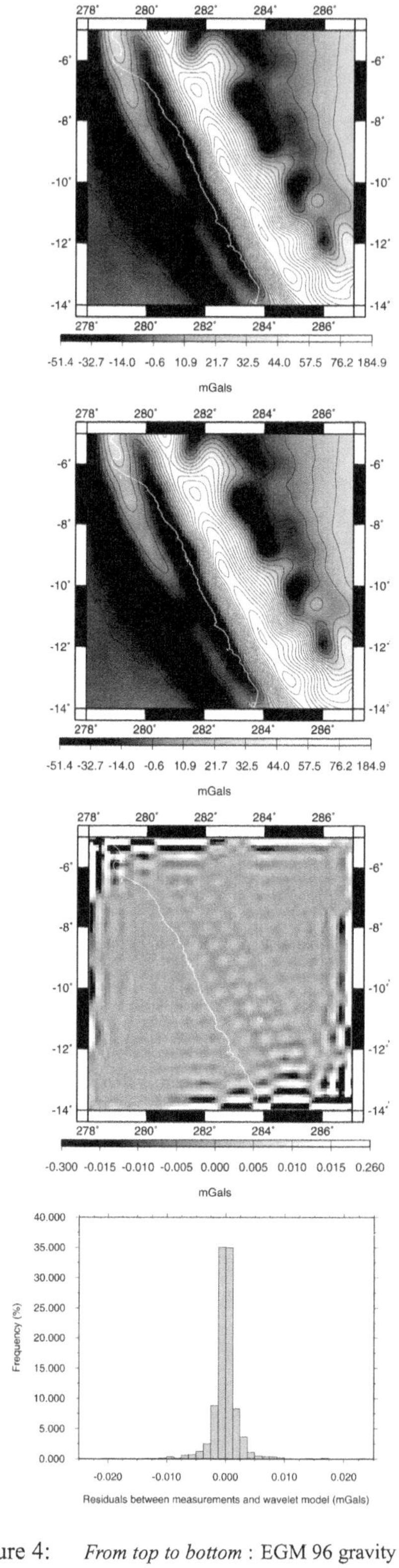


Figure 4: *From top to bottom* : EGM 96 gravity disturbances model; wavelet gravity disturbances model computed from the regular distribution of data; residuals between above wavelet model and EGM 96; residuals between wavelet model and synthetic data.

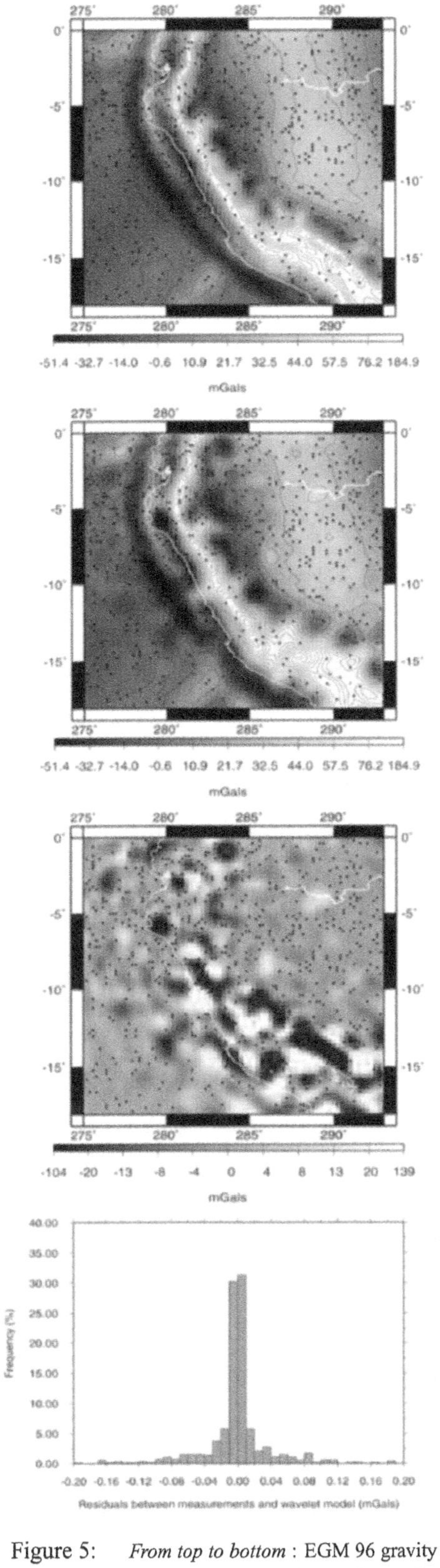


Figure 5: *From top to bottom* : EGM 96 gravity disturbances model (black dots represent the data); wavelet gravity disturbances model computed from the irregular distribution of data; residuals between above wavelet model and EGM 96; residuals between wavelet model and synthetic data.

datasets: the first one based on a satellite-derived gravity model, and the other one made of ground measurements, bringing the high frequency content.

## References

Albertella, A., Sanso, F. and Sneeuw, N., 1999. Band-limited functions on a bounded spherical domain: the Slepian problem on the sphere. *J. of Geod.*, **73**, 436-447.

Chambodut, A., Panet, I., Mandea, M., Diament, M., Jamet, O., Holschneider, M.. Wavelet frames: an alternative to the spherical harmonics representation of potential fields. *Geophys. Journ. Int.*, submitted.

De Santis, A. and J.M., Torta, 1997. Spherical cap harmonic analysis : a comment on its proper use for local gravity field representation, *J. of Geod.*, **71**, 526-532.

Freeden, W., T., Gervens and M., Schreiner, M., 1998. Constructive Approximation on the Sphere (With Applications to Geomathematics), *Oxford Science Publication*, Clarendon Press, Oxford.

Holschneider, M., A., Chambodut and M., Mandea, 2003. From global to regional analysis of the magnetic field on the sphere using wavelet frames, *Phys. Earth Planet. Inter.*, **135**, 107-124.

Hwang, Ch., 1993. Spectral analysis using orthonormal functions with a case study on the sea surface topography. *Geoph. Journ. Int.*, **115**, 1148-1160.

Kaula, W.M., 1966. Theory of satellite geodesy, Waltham, Blaisdell.

Kenner, H., 1976. Geodesic math and how to use it, Berleley CA: University of California Press.

Lemoine, F.G. et al. 1998. The development of the joint NASA GSFC and the National Imagery and Mapping Agency (NIMA) geopotential model EGM96. NASA/TP - 1998 - 206861, Greenbelt, Maryland.

Mallat, S., 1999. A wavelet tour of signal processing. Academic Press, 2nd edition.

Martelet, G., Sailhac, P., Moreau, F., Diament, M., 2001. Characterization of geological boundaries using 1D wavelet transform on gravity data: Theory and application to the Himalayas. *Geophysics*, **66**, 4, 1116-1129.

Moritz, H., 1989. Advanced physical geodesy, Karlsruhe:Wichmann, 2. ed.

Sailhac, P., Galdeano, A., Gibert, D., Moreau, F., Delor, C.., 2000. Identification of sources of potential fields with the continuous wavelet transform: Complex wavelets and application to aeromagnetic profiles in French Guiana. *J. Geophys. Res.*, **105**, B8, 19455-19475.

# Numerical Velocity Determination and Calibration Methods for CHAMP Using the Energy Balance Approach

M. Weigelt, N. Sneeuw
University of Calgary, Department of Geomatics Engineering, mlbweige@ucalgary.ca

**Abstract.** More than two years of data of the CHAMP satellite mission is available and the usage of the energy balance approach for global gravity field recovery has been successfully implemented by several groups around the world. This paper addresses two important aspects of the data processing. First, high-quality gravity recovery requires numerical differentiation of kinematic positions. Two methods are investigated using simulated and real dynamic data. It is shown that a third order Taylor differentiator is sufficient to reach good results. Second, drift due to the accelerometer bias has to be corrected. Two possible approaches are discussed: cross-over calibration on the one hand, calibration w.r.t. a reference model on the other hand. Currently the crossover calibration fails due to the insufficient accuracy of the crossover determination whereas the calibration w.r.t. a reference model gives good results.

**Keywords.** CHAMP, energy balance, FFT, Taylor differentiator, high-pass filter, crossover calibration

## 1 Introduction

Several groups showed successfully the feasibility of the energy balance approach for gravity field recovery (*Gerlach et al.*, 2003a; *Han et al.*, 2002). The basic characteristic of this approach is the use of GPS derived position and velocity data and the correction for non-gravitational forces derived from accelerometer data.

The derivation of the energy integral starts from the equation of motion in the rotating frame (*Schneider*, 1992, §5.4). Unit mass ($m = 1$) is assumed.

$$\ddot{\boldsymbol{x}} = \boldsymbol{f} + \boldsymbol{g} + \sum_k \boldsymbol{g}_k - \boldsymbol{\omega} \times (\boldsymbol{\omega} \times \boldsymbol{x}) - 2\boldsymbol{\omega} \times \dot{\boldsymbol{x}} - \dot{\boldsymbol{\omega}} \times \boldsymbol{x} \quad (1)$$

with the dissipative forces $\boldsymbol{f}$, the force $\boldsymbol{g}$ of the static gravitational field, the sum of all time variable gravitational forces $\sum_k \boldsymbol{g}_k$, the centrifugal acceleration $\boldsymbol{\omega} \times (\boldsymbol{\omega} \times \boldsymbol{x})$, the Coriolis acceleration $2\boldsymbol{\omega} \times \dot{\boldsymbol{x}}$ and the Euler acceleration $\dot{\boldsymbol{\omega}} \times \boldsymbol{x}$ on the right hand side. Multiplication with velocity, integration over time, introduction of a normal part $U$ of the gravitational potential and reordering yields on the left hand side the disturbing potential $T$ and the integration constant $c$ (*Sneeuw et al.*, 2003):

$$T + c = E_{\text{kin}} - U - Z - \int \left( \boldsymbol{f} + \sum_k \boldsymbol{g}_k \right) d\boldsymbol{x} \quad (2)$$

On the right side we have the kinetic energy $E_{\text{kin}}$ which can be calculated from the velocity $\dot{\boldsymbol{x}}$ of the satellite. The normal gravitational potential $U$ as well as the centrifugal potential $Z$ can be determined from the position of the satellite $\boldsymbol{x}$ and additional parameters (e.g. WGS84). $\int \boldsymbol{f}\, d\boldsymbol{x}$ is the dissipated energy, which is an integral of CHAMP's accelerometer data $\boldsymbol{f}$ along the orbit. Finally $\int \sum_k \boldsymbol{g}_k\, d\boldsymbol{x}$ contains both known sources (e.g. tides) that can be corrected, and unknown sources of gravity field changes. If we assume a geoid height accuracy target of $0.1\,\text{m}$, which is equivalent to an accuracy in the disturbing potential of $1\,\text{m}^2/\text{s}^2$, the requirement on the velocity determination would be $0.14\,\text{mm}/\text{s}$ (*Sneeuw et al.*, 2003).

The gained knowledge about calibration and numerical differentiation is valuable not only for the CHAMP mission but also for other satellite missions and is moreover independent from the usage of the energy balance approach.

## 2 Velocity Determination

In the calculations purely kinematic CHAMP orbits are used which avoid the introduction of a priori gravity field information. The disadvantage is that kinematic orbit determination yields only position and the velocity has to be derived by numerical differentiation. The ideal differentiator can be described by the spectral transfer, e.g. (*Antoniou*, 1993):

$$H\left(e^{i\omega T}\right) = i\omega \quad \text{for} \quad 0 \leq |\omega| < \frac{\omega_s}{2} \quad (3)$$

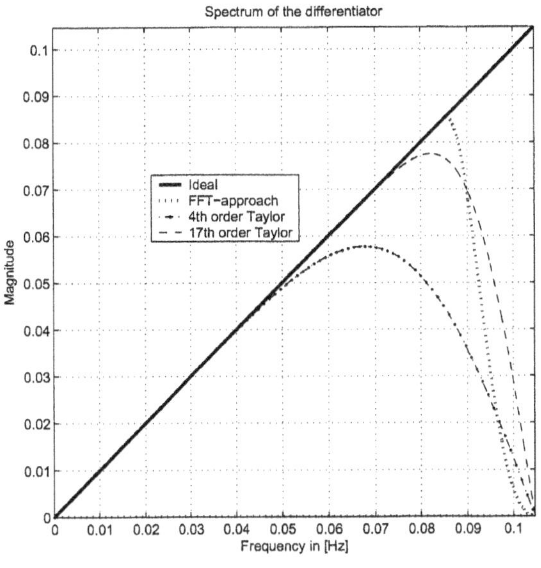


**Fig. 1.** Magnitude response of several differentiators

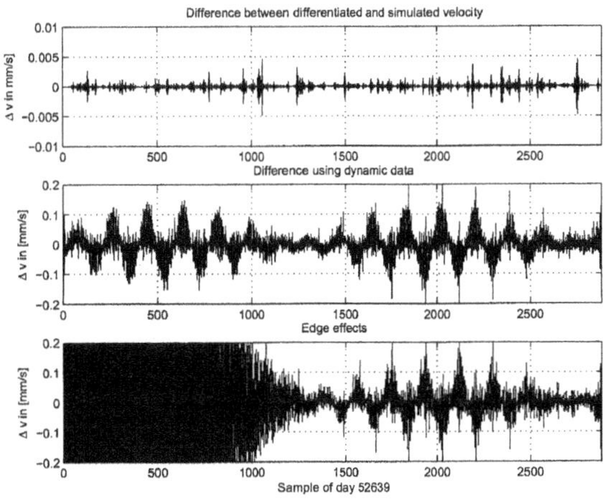


**Fig. 2.** Test results with simulated data (top), with real dynamic data (middle), edge effect (bottom)

Therefore, the ideal differentiator is represented by the angle bisection of the first and third quadrant as shown in figure 1 (solid line). Note that the frequency axis is already adapted to the case of CHAMP. In the following two methods are presented which are both approximations to the ideal differentiator. This approximation is desired since differentiators act as a high-pass filter and therefore tend to amplify noise. By adjusting the properties of each method the amplification of noise in the data can be minimized.

## 2.1 FFT-Method

The first method presented here utilizes a Fourier (FFT) approach. The idea is to determine the spectral content of the signal, i.e. of the position, and multiply the result with $i\omega$. After back-transformation to the time domain the signal represents the velocities. The advantage of the approach is the easy application and modification of the differentiator itself.

It is obvious that problems with leakage and aliasing arise. Also data gaps cause problems in the transformation. In order to overcome the latter problem the kinematic position data gaps are filled with dynamic position data. This is a major drawback of the approach since it can introduce high frequencies due to jumps between the kinematic and dynamic position. In order to minimize leakage windowing techniques are used. Applying a window results in major differences at the edges of the data. In order to minimize errors due to this approach data before and after the desired area of interest must be added. In addition to the applying of the windows in the time domain, a window must also be applied to the differentiator in the frequency domain in order to minimize the sidelobes of the corresponding time domain function of the differentiator. This window will also act as a low-pass filter and therefore gives control over the amplification of the noise.

In practice the calculation was performed on a daily basis. For each day one complete day of data was added before and afterwards. Half of these were used to apply the window. The differentiator is used complete up to about 80% of the frequency content and a Blackman window is used for windowing (*Lyons*, 2001). The top panel in figure 2 shows results with noiseless simulated data for CHAMP which were developed by the IAG Special Commission SC7 (*Ilk*, 2001). The data is created using numerical integration with the EGM96 model up to degree and order 300. For this plot the position of the satellite is differentiated and compared to the simulated velocity. Note that the difference is at the level of $10^{-3}$ mm/s for the simulated data. Computations with real dynamic data provided by the IAPG, TU Munich (*Švehla and Rothacher*, 2004) are shown in the middle panel of figure 2. It is obvious that the data is more noisy and therefore the simulated results are overoptimistic. One loses about two orders of magnitude in accuracy due to the noise. However, the results are still at the level of 0.14 mm/s. In this scenario the data availability was ideal, which means there was enough data before and after the area of interest available.

The lower panel shows the edge effect which originate from the usage of the windowing technique. If a major data gap exists (e.g. data for one day is not available) and/or filling with dynamic position data is not possible a lot of valuable data is lost. In case of CHAMP around 1500 samples are necessary to reach the level of $0.14\,\mathrm{mm/s}$. With its 30 second data sampling this means a loss of data of half a day or 8 revolutions.

The second problem is the global behavior of the approach. Changes due to orbit maneuvers or sudden changes in the environment are aliased over the whole interval of calculation. In practice this means that the differentiator causes a up to one day delay in the reaction to these events and contaminates data before these events.

## 2.2 FIR-Method

The FFT-method uses multiplication in the frequency domain which is identical to a convolution in the time domain. Therefore, the differentiator can be implemented as finite impulse response (FIR) filter. The design of the filter is based on a central finite difference Taylor approximation of the first derivative (*Bruton et al.*, 1999). The coefficients of $n$-th order can be calculated in a fast and effective way using closed-form expressions, cf. (*Khan and Ohba*, 1999).

The usage of a FIR-filter overcomes the two obstacles of the FFT-method. For the calculation of one value only data in the size of $2n + 1$ elements for a $n$-th order differentiator is necessary. Therefore, the differentiator responds quickly to changes due to orbit maneuvers and changes in the environment. The delay is maximum in the order of the differentiator. The edge effect still exists but is much smaller than in case of FFT. For a $n$-th order differentiator $n$ elements at the beginning and at the end of an area of interest is lost. Therefore, the smaller the order of the differentiator the closer one gets to data gaps. The filling with dynamic position data becomes unnecessary if a reasonable small order for the differentiator is used. This leads to a more rigorous data treatment.

However, it is necessary to use a reasonable high order to achieve results on the level of $0.14\,\mathrm{mm/s}$. The property of the Taylor differentiator is that the higher the order the better the approximation to the ideal differentiator, cf. figure 1. Therefore, the task is to choose the minimum possible and maximum necessary order which will mainly depend on the frequency content of the signal.

Several differentiators with different order size have been tested. Figure 3 shows results with the simulated data and clearly demonstrates that the order of the differentiator influences the approximation of the ideal differentiator and the accuracy of the solution. The higher the order the better the result. However, this counts for the noiseless case only. Figure 4 shows test results performed with dynamic data. It is obvious that there is virtually no difference between the three solutions. This is supported by the RMS-values of the three solutions which are all almost identical in the area of $0.05\,\mathrm{mm/s}$. Therefore, the conclusion is that in case of CHAMP an order of 3 is sufficient. For completeness it should be mentioned that an order less than three gives unacceptable results (e.g. RMS $= 2\,\mathrm{cm/s}$ for order 2).

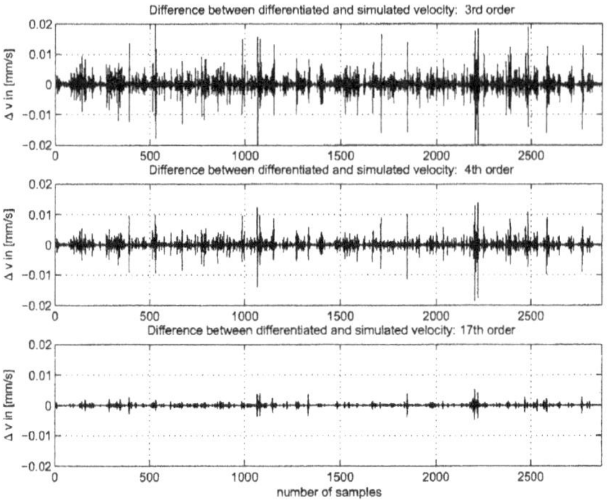


**Fig. 3.** Test results with simulated data: 3rd order (top), 4th order (middle), 17th order (bottom)

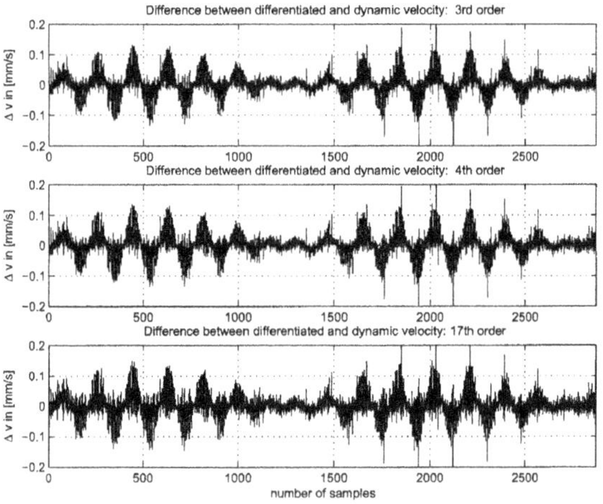


**Fig. 4.** Test results with dynamic data: 3rd order (top), 4th order (middle), 17th order (bottom)

# 3 Calibration Methods

In the data processing a calibration is necessary to account for the deficiencies of the accelerometer. The calibration does not aim at an explicit determination of scale and bias parameters of the three accelerometer components. Instead, the potential signal itself is calibrated. Two methods have been tested for this purpose. The first one is crossover analysis which consists of an adjustment of crossover points under the assumption that the potential in the Earth-fixed frame is constant. The second method is calibration by fitting polynomials to residuals of the uncalibrated potential w.r.t. a priori models.

### 3.1 Crossover Calibration

The basic procedure for crossover calibration is already described in (*Sneeuw et al.*, 2003). *Gerlach et al.* (2003b) use a 9th order polynomial to fit a period of 11 days. Here different strategies for the proper choice of a polynomial fit are studied. Beside the fit over a time period of 27 days also daily fits and the interpretation of a crossover as a first order Taylor differentiator is investigated.

In the first approach which is similar to (*Gerlach et al.*, 2003b), the potential difference at the crossover points are collected for one continuous arc and a higher order polynomial is fitted through the data points. The higher order polynomial is necessary due to the changing behavior, e.g. the temperature dependence of the accelerometer, but is restricted to a maximum order of 15 in order to avoid the removal of time variable signals of the gravity field. Figure 5 shows the difference between the calibrated disturbing potential from the calculation and the potential along the orbit determined using the TUM2S model (*Wermuth et al.*, 2004) after fitting a

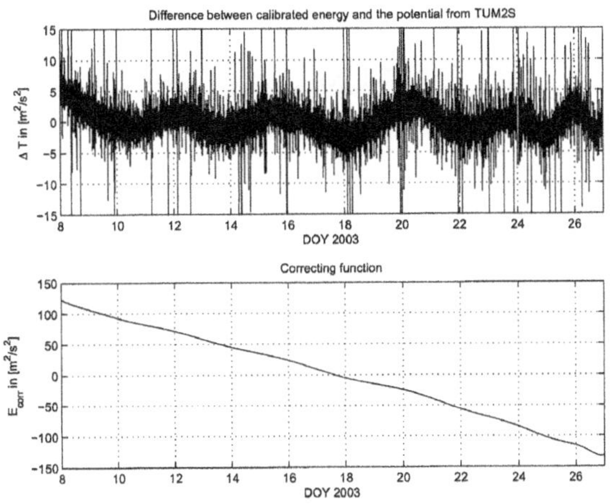


**Fig. 5.** Modelling with high-order polynomials: difference between calibrated energy and the disturbing potential from TUM2S(top); the correcting function (bottom)

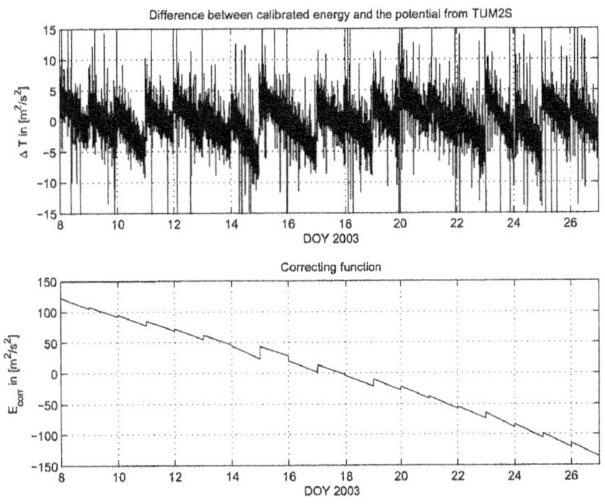


**Fig. 6.** Modelling with first order polynomials on a daily basis: difference between calibrated energy and the disturbing potential from TUM2S (top); the correcting function (bottom)

13th-order polynomial. It is obvious that an oscillation with an amplitude of about $5\,m^2/s^2$ is remaining which is related to the shape and thus the order of the polynomial. In the current tests no order of polynomial was found that would solve this problem.

In a second attempt to refine this procedure a first order polynomial is fitted to the data on a daily basis which is similar to the determination of a bias parameter once per day. A problem which arises immediately is that the correcting function contains discontinuities between days, cf. figure 6 (bottom). Besides this the calibration method is not able to correct for all the drift in the data. The reason is the noisy data used for the polynomial fit which is explained in more detail at the end of this section.

Till now the crossover difference was attributed to the time point of the second pass of the satellite in the crossover point. However, this is not rigorous. In fact, we have a drift in the data due to the accelerometer bias. This means that the longer the time between the two passes of the satellite at the crossover location the bigger the difference in the energy. The top panel in figure 7 visualizes this by showing the difference in energy of an ascending and a descending arc versus the time difference. Since the crossover with the longer time difference tend to be in the end of the day an oscillation is introduced which leaded to the failed calibration in the first attempt. A more rigorous approach is to divide the energy difference by the time difference and interpret the result as a first order Taylor differentiator, thus determine a time series of the first derivative. A high order polynomial can be

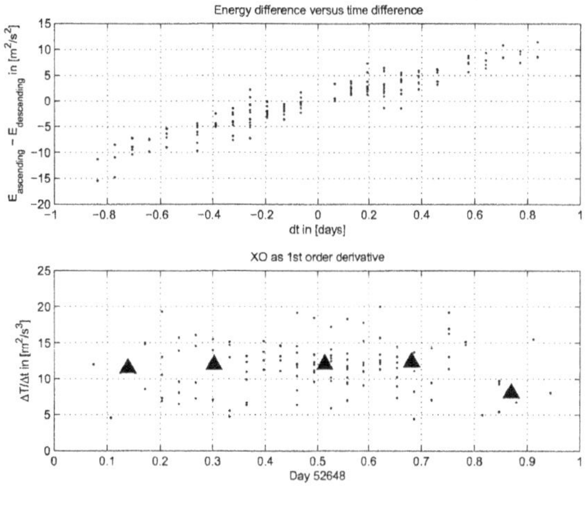


**Fig. 7.** Energy difference versus time difference (top); crossover interpreted as 1st order Taylor differentiator (bottom) and result of the calibration (bottom)

fitted and integrated analytically. The crossover data turned out to be very noisy as can be seen in the bottom panel of figure 7. In order to smooth the data the mean value is determined several times per day (▲). The top panel of figure 8 shows again the difference between the calibrated disturbing potential from the calculation and the potential along the orbit determined using the TUM2S model. It is obvious that also this method failed since some drift is still remaining.

No calibration method using crossovers seems to work but actually the results are inconclusive. During the investigation it turned out that the determination of the crossover location by crossing third order polynomials is not precise enough. The error for the crossover location lies in the range of several km. Determining the RMS-value of the gradient in north-south and east-west direction of the disturbing potential yields that an accuracy of $\approx 200\,\mathrm{m}$ in the determination of the crossover location is necessary in order to achieve an accuracy of $1\,\mathrm{m}^2/\mathrm{s}^2$ in the disturbing potential.

A further more general drawback of this method is that not more than 170 points per day can be used for calibration and the accuracy for these points depends heavily on their location. Moreover, the calculation is performed arc-wise depending on the length of coherent accelerometer data. Therefore, the crossover calibration easily fails for very short arcs which results in a loss of valuable data.

### 3.2 Calibration w.r.t. a priori models

A simple way to correct the drift is to compare the uncalibrated energy with the potential from a priori models and fit a first order polynomial to the residuals. This means the correction of the slope can also be seen as removing the very low frequencies from the spectrum of the data.

The calculation was first done on a daily basis resulting again in discontinuities in the correction function. In order to overcome this problem the approach can be refined using a time series for the correcting function. The time series is created by determining polynomial coefficients e.g. on a daily basis and interpolate the coefficients to the data resulting

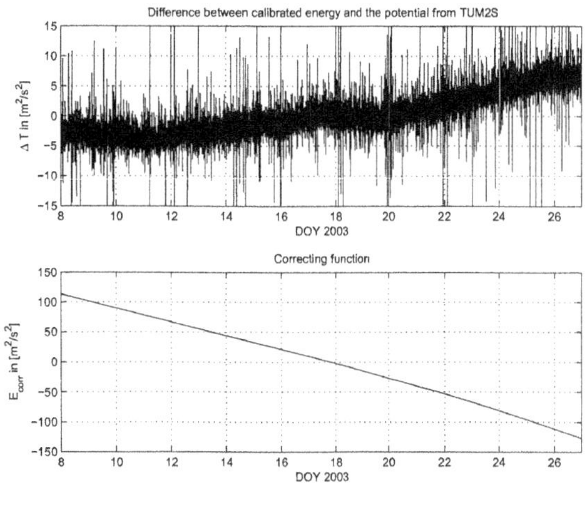


**Fig. 8.** Modelling interpreting crossover as first order Taylor differentiator: difference between calibrated energy and the disturbing potential from TUM2S (top); the correcting function (bottom)

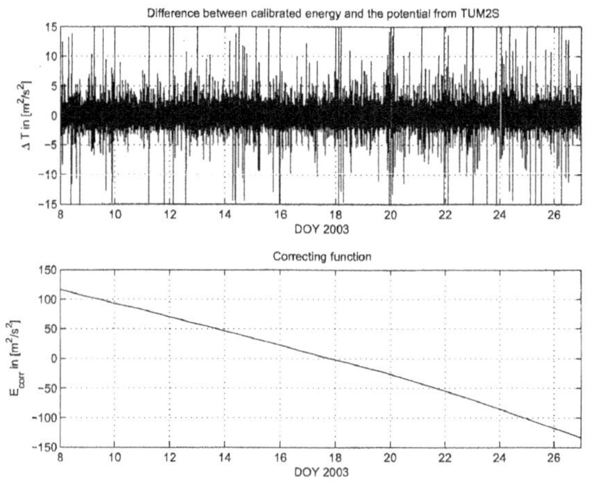


**Fig. 9.** Modelling by fitting first order polynomials to residual: difference between calibrated energy and the disturbing potential from TUM2S (top); the correction function (bottom).

in a smooth transition between the determined coefficients. As can be seen in figure 9 this approach is capable of removing the drift from the data .

However, it is obvious that a dependence on the a priori gravity field used for the polynomial fitting might exist and must be investigated. Current results show that the difference between solutions are on a level below 1 $m^2/s^2$ and might be negligible. Yet, the current method used for the spherical harmonic analysis only gives a rough solution and further investigation is still necessary.

## 4 Conclusions

In the investigation of the two methods for the numerical differentiation it is shown that the FFT-method has some major drawbacks which are overcome by the usage of the FIR-method. Moreover, it has been shown that the order of the FIR-filter must be higher than 2. A 4th order FIR-filter is sufficient, though. A further investigation by using e.g. IIR-filters seems not necessary since the order is reasonably small and, therefore, data gaps can be treated in a rigorous way without the loss of many data points.

Two calibration methods have been investigated and advantages and disadvantages have been discussed. The crossover approach only mildly depends on the use of a priori information but easily fails for short arcs and demands a highly accurate crossover location determination. Missing the latter one caused at the current stage the failure of the calibration procedure. The big advantage of the usage of a priori knowledge is that every point is contributing to the calibration. Although the approach is successfully calibrating the data it is not yet clarified if the solution depends on the a priori gravity field used for the polynomial fitting. This will be investigated in future work as well as a refined calculation for the crossover location. If this dependency exists a combined method seems possible. In the first step crossover calibration is used to achieve preliminary results of the gravity field and virtually use no a priori information. However this will lead to the loss of data from short arcs where crossover calibration is not possible. In order to make use of this information the preliminary solution is then used for the second calibration method as a priori model yielding an independent solution.

The future work will also contain the transition to the GRACE satellite mission. The calibration can be applied directly since the technology for the two satellites is similar to CHAMP. The differentiation methods will not only be needed for the determination of the velocity from kinematic orbits but also for the high-precision determination of the inter-satellite range rate $\dot{\rho}$, range acceleration $\ddot{\rho}$ and the change of the direction vector $\dot{e}_{12}$.

## References


Antoniou, A. (1993), *Digital filters: Analysis, design and applications*, New York: McGraw-Hill.

Bruton, A., C. Glennie, and K. P. Schwarz (1999), Differentiation for high-precision GPS velocity and acceleration determination, *GPS Solutions*, *2*(4), 7–21.

Gerlach, C., N. Sneeuw, P. Visser, and D. Švehla (2003a), CHAMP gravity field recovery with the energy balance approach: first results, In: C. Reigber, H. Lühr, and P. Schwintzer (eds), *First* CHAMP *Mission Results for Gravity, Magnetic and Atmospheric Studies*, Springer, pp. 134–139.

Gerlach, C., N. Sneeuw, P. Visser, and D. Švehla (2003b), CHAMP gravity field recovery using the energy balance approach, *Adv. Geosciences*, *1*, 73–80.

Han, S., C. Jekeli, and C. Shum (2002), Efficient gravity field recovery using in situ disturbing potential observabels from CHAMP, *Geophys. Res. Lett.*, *29*(16), 1789, doi:10.1029/2002GL015180.

Ilk, K. H. (2001), Special commission SC7, gravity field determination by satellite gravity gradiometry, *http://www.geod.uni-bonn.de/SC7/sc7.html*.

Khan, I., and R. Ohba (1999), Closed-form expressions for the finite difference approximations of first and higher derivatives based on taylor series, *Journal of Computational and Applied Mathematics*, *46*, doi:10.1049/ip-vis:19990380.

Lyons, R. (2001), *Understanding Digital Signal Processing*, Prentice Hall PTR.

Schneider, M. (1992), *Grundlagen und Determinierung*, vol. I of *Himmelsmechanik*, BI-Wissenschaftsverlag, (In German).

Sneeuw, N., C. Gerlach, D. Švehla, and C. Gruber (2003), A first attempt at time-variable gravity recovery from CHAMP using the energy balance approach, In: I. Tziavos (ed), *Gravity and Geoid 2002*, pp. 237–242.

Švehla, D., and M. Rothacher (2004), Two years of CHAMP kinematic orbits for geosciences, *Geophysical Research Abstracts*, *6*, 06645, SRef-ID:1607-7962/gra/EGU04-A-06645.

Wermuth, M., D. Švehla, L. Földvary, C. Gerlach, T. Gruber, B. Frommknecht, T. Peters, M. Rothacher, R. Rummel, and P. Steigenberger (2004), A gravity field model from two years of CHAMP kinematic orbits using the energy balance approach, *Geophysical Research Abstracts*, *6*, 03843, SRef-ID:1607-7962/gra/EGU04-A-03843.

# Upward Continuation of Ground Data for GOCE Calibration/Validation Purposes

K.I. Wolf and H. Denker
Institut für Erdmessung, University of Hannover
Schneiderberg 50, D-30167 Hannover, Germany
Email: wolf@ife.uni-hannover.de, Fax: +49 511 762 4006

**Abstract.** With the upcoming ESA satellite mission GOCE, gravitational gradients (2nd derivatives of the Earth's gravitational potential) will be measured globally, except for the polar gaps. An accuracy of a few mE (1 mE = $10^{-3}$ Eötvös, 1 E = $10^{-9}s^{-2}$) is required to derive, in combination with satellite-to-satellite tracking (SST) measurements, a global geopotential model up to about spherical harmonic degree 200 with an accuracy of 1 . . . 2 cm in terms of geoid undulations and 1 mgal for gravity anomalies, respectively.

To meet these requirements, the gradiometer will be calibrated and validated internally as well as externally. One strategy for an external calibration or validation includes the use of ground data upward continued to satellite altitude. This strategy can only be applied regionally, because sufficiently accurate ground data are only available for selected areas.

In this study, gravity anomalies over Europe are upward continued to gravitational gradients at GOCE altitude. The computations are done with synthetic data in a closed-loop simulation. Two upward continuation methods are considered, namely least-squares collocation and integral formulas based on the spectral combination technique. Both methods are described and the results are compared numerically with the ground-truth data.

**Keywords.** gradiometry, upward continuation, least-squares collocation, spectral combination, GOCE, calibration, validation

## 1 Introduction

The ESA mission GOCE, which is in preparation for launch in 2006/07, will for the first time apply satellite gradiometry, i.e. the measurement of differential accelerations over short baselines in one satellite. Satellite gradiometry, in combination with GPS-SST, is used to determine the Earth's gravity field, aiming at an accuracy of 1 cm for the geoid and 1 mgal for gravity anomalies at 100 km resolution. For this purpose, several calibration steps (in orbit and post-processing) are required to guarantee an accuracy of the gravitational gradients at the mE level.

In this paper, the major part of the external calibration process is discussed, being the upward continuation of terrestrial gravity anomalies for the computation of calibration gradients. A similar calibration approach was studied in Arabelos and Tscherning (1998), where parametric least-squares collocation (LSC) was used simultaneously for the upward continuation of ground gravity data and the estimation of the calibration parameters for GOCE gradients. A comparison between different upward continuation methods, like LSC, integral formulas (IF), mass modelling techniques or the least-squares adjustment of harmonic expansions, is done in Pail (2002) for the radial gravitational gradients, using a synthetic Earth model for the generation of the ground-truth data.

In this study, synthetic data sets based on existing geopotential models are used for a closed-loop computation. All six components of the gradient tensor $T_{ij}$, with $i, j = x, y, z$ are computed at GOCE altitude ($R_E + 250$ km) with LSC and IF based on the spectral combination technique. The methodology and numerical experiments are described. The results show that the terrain heights of the ground gravity points should be considered to improve the predictions. Consequently, a two step concept for the upward continuation is proposed at the end of this paper.

## 2 Synthetic Data

Synthetic data sets are produced for a closed-loop computation. For this purpose, a blended geopotential model ($GPM^{true}$) is created by combining the coefficients from an actual GRACE GPM ($l$ = 0...89, JPL (2003)), EGM96 ($l$ = 90...360, Lemoine et al. (1998)), and GPM98C ($l$ = 361...1300, Wenzel (1999)). From this ground-truth model, the following input data sets are derived, see also Fig. 1:

A. Two sets of gravity anomalies are derived from $GPM^{true}$. The first data set $\Delta g^{true}$ (without noise) consists of the $GPM^{true}$ values to degree $l_{max} = 1300$. The second data set $\Delta g$ (with noise) is created from the first one by adding 1 mgal white noise. The anomalies are computed in a 5′ geographical grid on top of the topography, and for testing purposes also directly on the ellipsoid.
B. Two geopotential models to degree $l_{max} = 360$, serving as reference models in the remove-restore procedure, are derived from $GPM^{true}$. The first model, $GPM_0^{true}$ (without noise), simply consists of the $GPM^{true}$ coefficients up to degree $l_{max} = 360$, while for the second (clone) model, $GPM_0$, noise is added according to the standard deviations of the coefficients.
C. Gradients $T_{ij}^{true}$ are computed from $GPM^{true}$ at GOCE altitude, serving as ground-truth data for the upward continuation results, where $i, j = x, y, z$, with $x$ pointing North, $y$ East and $z$ Radial.

For the following computations, the remove-restore procedure is applied based on a reference geopotential model. For this purpose, residual gravity anomalies $\Delta g'$ are generated by subtracting the long-wavelength effects of a reference geopotential model ($\Delta g_0$) from the observations $\Delta g$:

$$\Delta g' = \Delta g - \Delta g_0 \ . \tag{1}$$

The residual anomalies are then upward continued. The predicted residual $T'_{ij}$ have to be complemented by the $T_{ij,0}$ values from the geopotential model to get finally the complete signal $T_{ij}$. In the numerical tests, the above described terrestrial $\Delta g$ and $GPM$ data sets are employed.

In order to check the noise generation for the reference geopotential model, the gravity anomalies from the (clone) $GPM_0$ model are subtracted from the corresponding $GPM_0^{true}$ values. The differences are depicted in Fig. 2. The standard deviation of the differences is 13.0 mgal, agreeing well with the accumulated error degree variances from the $GPM_0$ model of 10.9 mgal. In practice, the errors over the Atlantic Ocean are probably smaller because of the high quality and homogeneity of the altimeter data used in the geopotential model development. However, the clone $GPM_0$ including noise should allow more realistic results than an error-free model. The error-free case was considered already in Wolf and Müller (2004).

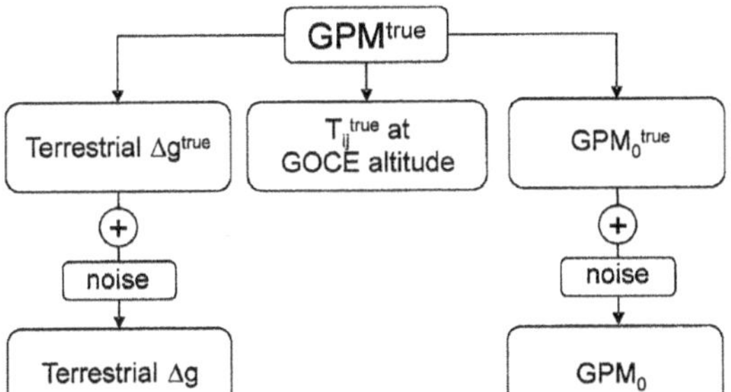


Fig. 1. Scheme for the generation of the synthetic data sets based on $GPM^{true}$.

## 3 Method I: Least-squares Collocation (LSC)

The theoretical background of LSC is described in detail in Moritz (1980). The basic formulas of LSC are

$$\hat{T}'_{ij} = C_{T'_{ij}\Delta g'} \cdot (C_{\Delta g'\Delta g'} + E_{\Delta g\Delta g})^{-1} \cdot \Delta g' \tag{2}$$

and

$$E_{T_{ij}T_{ij}} = C_{T'_{ij}T'_{ij}} - C_{T'_{ij}\Delta g'} \cdot (C_{\Delta g'\Delta g'} + E_{\Delta g\Delta g})^{-1} \cdot C^T_{T'_{ij}\Delta g'} , \tag{3}$$

where $C_{T'_{ij}\Delta g'}$, $C_{\Delta g'\Delta g'}$ and $C_{T'_{ij}T'_{ij}}$ are the signal covariance matrices, $E_{\Delta g\Delta g}$ and $E_{T_{ij}T_{ij}}$ are the error covariance matrices, and $\hat{T}'_{ij}$ and $\Delta g'$ are the residual signals (gradients, gravity anomalies).

The complete formalism for the derivation of the signal covariances is described in Tscherning (1976b). All covariances are based on the covariance function $C_{TT}$ of the disturbing potential $T$

$$C_{TT} = \sum_{l=2}^{\infty} \sigma_l^2(T) \left(\frac{R_B^2}{r_P r_Q}\right)^{l+1} P_l(cos(\psi)) \ , \tag{4}$$

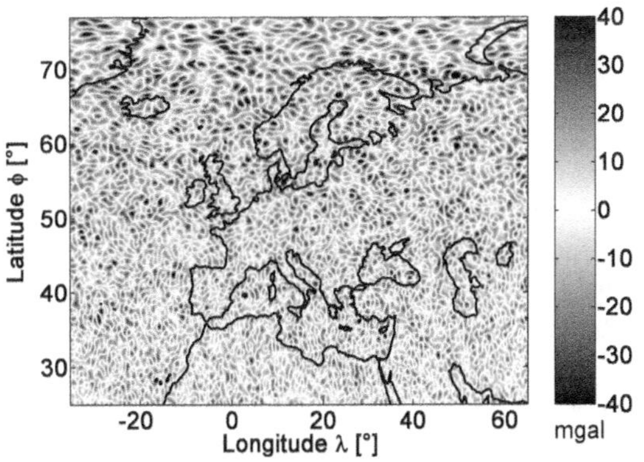


Fig. 2. Gravity anomalies differences between clone $GPM_0$ and $GPM_0^{true}$.

with the degree variances $\sigma_l^2(T)$ of $T$, the radius $R_B$ of the Bjerhammer sphere, the radii $r_P$, $r_Q$ of the points P and Q, and the Legendre polynomials $P_l(cos(\psi))$ with respect to the spherical distance $\psi$.

The covariance functions depend on the error degree variances of the geopotential model up to degree $l_{max}$, and above degree $l_{max}$ the Tscherning and Rapp (1974) degree variance model is applied. For the computation of the signal covariance values, subroutines implemented by C. C. Tscherning (Tscherning and Rapp (1974), Tscherning (1976b)) are used in an updated version.

## 4 Method II: Integral Formulas (IF)

The basic integration formula for the disturbing potential $T$, using the spectral combination technique according to Wenzel (1982), is

$$T(r_P,\theta_P,\lambda_P) = \frac{R}{4\pi}\iint_\sigma K(\psi_{PQ},r_P,R)\cdot\Delta g'(R,\theta_Q,\lambda_Q)d\sigma\ , \quad (5)$$

with the kernel function

$$K(\psi_{PQ},r_P,R)=\sum_{l=2}^{k_{max}}\left(\frac{R}{r_P}\right)^{l+1}\frac{2l+1}{l-1}w_lP_l(\cos(\psi)), \quad (6)$$

where $r_P$, $\theta_P$, $\lambda_P$, $R$, $\theta_Q$ and $\lambda_Q$ are the spherical coordinates of points P and Q, and $w_l$ are the spectral weights. In addition, $k_{max}$ is the maximum degree considered for the terrestrial gravity data, depending on the data resolution.

The spectral weights are derived by a least-squares adjustment:

$$w_l = \frac{\sigma_l^2(\epsilon_{\Delta g_0})}{\sigma_l^2(\epsilon_{\Delta g_0})+\sigma_l^2(\epsilon_{\Delta g})}\ . \quad (7)$$

The $\sigma_l^2(\epsilon_{\Delta g_0})$ and $\sigma_l^2(\epsilon_{\Delta g})$ are the error degree variances of the geopotential model and of the terrestrial gravity data, respectively. The $\sigma_l^2(\epsilon_{\Delta g})$ are computed on the basis of the white noise assumption introduced in section 2. The resulting spectral weights are shown in Fig. 3. Due to the high long wavelength accuracy of the geopotential model (based on GRACE), the spectral weights of the terrestrial data are below 0.5 up to degree 52.

For the derivation of the integral formulas for the gradients $T_{ij}$ from Eq. (5), the following partial derivatives are needed (for details, see e.g. Moritz (1971), Tscherning (1976a), Thalhammer (1994) or Ditmar and Klees (2002)):

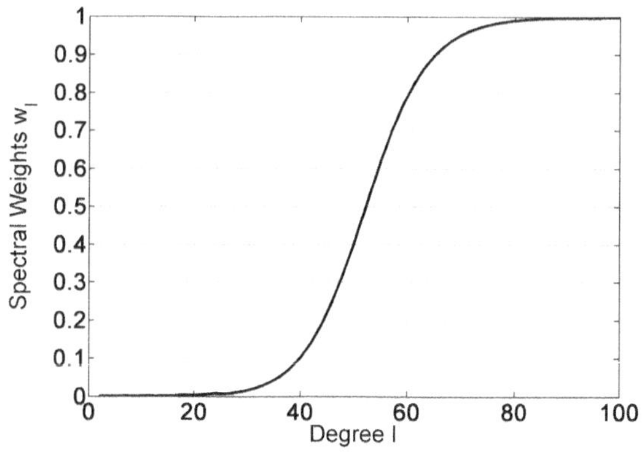


**Fig. 3.** Spectral Weights based on uncorrelated 5′ gravity anomalies and geopotential model errors.

$$\frac{\partial^2}{\partial i\partial j} = \begin{bmatrix} \frac{\partial^2}{\partial x^2} & \frac{\partial^2}{\partial x\partial y} & \frac{\partial^2}{\partial x\partial z} \\ \frac{\partial^2}{\partial y\partial x} & \frac{\partial^2}{\partial y^2} & \frac{\partial^2}{\partial y\partial z} \\ \frac{\partial^2}{\partial z\partial x} & \frac{\partial^2}{\partial z\partial y} & \frac{\partial^2}{\partial z^2} \end{bmatrix} \quad (8)$$

with

$$\frac{\partial^2}{\partial x^2} = \frac{1}{r}\frac{\partial}{\partial r} + \frac{1}{r^2}\frac{\partial^2}{\partial\theta^2}\ ,$$

$$\frac{\partial^2}{\partial x\partial y} = \frac{\partial^2}{\partial y\partial x} = \frac{1}{r^2\sin(\theta)}\left(\cot(\theta)\frac{\partial}{\partial\lambda} - \frac{\partial^2}{\partial\lambda\partial\theta}\right)\ ,$$

$$\frac{\partial^2}{\partial x\partial z} = \frac{\partial^2}{\partial z\partial x} = \frac{1}{r^2}\frac{\partial}{\partial\theta} - \frac{1}{r}\frac{\partial}{\partial r\partial\theta}\ ,$$

$$\frac{\partial^2}{\partial y^2} = \frac{1}{r}\frac{\partial}{\partial r} + \frac{\cot(\theta)}{r^2}\frac{\partial}{\partial\theta} + \frac{1}{r^2\sin^2(\theta)}\frac{\partial}{\partial\lambda^2}\ ,$$

$$\frac{\partial^2}{\partial y\partial z} = \frac{\partial^2}{\partial z\partial y} = \frac{1}{r\sin(\theta)}\left(-\frac{1}{r}\frac{\partial}{\partial\lambda} + \frac{\partial^2}{\partial r\partial\lambda}\right)\ , \text{and}$$

$$\frac{\partial^2}{\partial z^2} = \frac{\partial^2}{\partial r^2}\ .$$

Furthermore, the following derivatives of the kernel function $K$ in Eq. (6) are introduced:

$$K_\psi = \frac{\partial K(\psi,r,R)}{\partial\psi},\quad K_r = \frac{\partial K(\psi,r,R)}{\partial r}\ ,$$

$$K_{\psi\psi} = \frac{\partial^2K(\psi,r,R)}{\partial\psi^2},\ K_{rr} = \frac{\partial^2K(\psi,r,R)}{\partial r^2}\ , \quad (9)$$

$$K_{\psi r} = \frac{\partial^2K(\psi,r,R)}{\partial\psi\partial r} = \frac{\partial K_\psi}{\partial r} = \frac{\partial K_r}{\partial\psi}.$$

Combining Eqs. (8) and (5) and considering Eq. (9) yields:

$$T_{xx} = \frac{R}{4\pi} \iint_\sigma \Big( \frac{1}{r} K_r + \frac{1}{2r^2} \Big( (K_{\psi\psi} + K_\psi \cot(\psi)) + (K_{\psi\psi} - K_\psi \cot(\psi)) \cos(2\alpha) \Big) \Big) \Delta g'(R, \theta_Q, \lambda_Q) d\sigma \,,$$

$$T_{xy} = \frac{R}{4\pi} \iint_\sigma \frac{1}{2r^2} (K_{\psi\psi} - K_\psi cot(\psi)) \sin(2\alpha) \Delta g'(R, \theta_Q, \lambda_Q) d\sigma \,,$$

$$T_{xz} = \frac{R}{4\pi} \iint_\sigma \frac{1}{r} \left( \frac{1}{r} K_\psi - K_{\psi r} \right) \cos(\alpha) \Delta g'(R, \theta_Q, \lambda_Q) d\sigma \,,$$

$$T_{yy} = \frac{R}{4\pi} \iint_\sigma \Big( \frac{1}{r} K_r + \frac{1}{2r^2} \Big( (K_{\psi\psi} + K_\psi cot(\psi)) - (K_{\psi\psi} - K_\psi cot(\psi)) \cos(2\alpha) \Big) \Big) \Delta g'(R, \theta_Q, \lambda_Q) d\sigma \,,$$

$$T_{yz} = \frac{R}{4\pi} \iint_\sigma \frac{1}{r} \left( \frac{1}{r} K_\psi - K_{\psi r} \right) \sin(\alpha) \Delta g'(R, \theta_Q, \lambda_Q) d\sigma \quad \text{and}$$

$$T_{zz} = \frac{R}{4\pi} \iint_\sigma K_{rr} \Delta g'(R, \theta_Q, \lambda_Q) d\sigma \,. \qquad (10)$$

In the above equations, the derivatives of the Legendre polynomials with respect to $\cos(\psi)$ have to be computed; recursion formulas for this purpose are given in Wenzel (1985). The practical evaluation of the equations is done by 1D FFT according to Haagmans et al. (1993), as used already by Denker (2003) for the radial component $T_{zz}$.

## 5 Numerical Experiments

The LSC and IF are applied to data with a resolution of 5′, covering an area from 35°...59° N and 0°...24° E. The predictions are compared in the central 8° by 8° area in a closed-loop computation with the ground-truth gradients $T_{ij}^{true}$. The total observation set comprises 82,944 points. The handling of this amount of data does not cause any computational problems for the IF and 1D FFT approach. Regarding LSC (Eqs. (2) and (3)), a "52 GB normal equation system" has to be solved, which is only possible on a supercomputer. However, for the case that the input data are gridded regularly and located at the same height, the matrix $C_{\Delta g' \Delta g'}$ has a block Toeplitz structure. This is assumed in the numerical computations and leads to a significant reduction of the memory requirements (183 MB). Furthermore, with the special routines of the SLICOT-package, see Brenner et al. (1999), the system $[(C_{\Delta g' \Delta g'} + E_{\Delta g \Delta g})^{-1} \Delta g']$ can be solved even on a one-processor system.

In a first test, the upward continuation is done with IF and error-free data (i.e. gravity anomalies $\Delta g^{true}$ and $GPM_0^{true}$) on the ellipsoid. The maximum differences between the predictions and the ground-truth gradients $T_{ij}^{true}$ are below 0.1 mE in the central 8° by 8° area, thus validating the formalism and the software.

A second closed-loop computation is carried out with noisy $\Delta g$, in combination with the reference model $GPM_0$, including errors as described in section 2. In this analysis, the data points are located on the topography of the Earth using heights from a digital terrain model (DTM). Both, LSC and IF are used for the upward continuation and show nearly the same results (Table 1). The implicit extrapolation behaviour of LSC, described in de Min (1995), may cause the small differences between the LSC and IF results. In contrast to LSC, the IF approach strictly uses data in the given 24° by 24° area. A more detailed discussion and theoretical comparison of the IF and LSC approach is given in Moritz (1976).

The largest differences in the central 8° by 8° computation area are located over the Alps (Fig. 4). Apart from the strong gravity signal there, the DTM heights of the points differ significantly (up to

**Table 1.** Statistics of $T_{ij}^{true}$ and differences to IF and LSC results (mE), $\Delta g$ on DTM.

| Dataset | Min | Max | Mean | Std |
|---|---|---|---|---|
| $T_{xx}^{true}$ | -361.8 | 192.1 | -148.2 | 129.6 |
| $T_{xx}^{true}$ - IF | -2.9 | 1.8 | 0.2 | 1.1 |
| $T_{xx}^{true}$ - LSC | -2.6 | 1.8 | 0.1 | 1.0 |
| IF - LSC | -0.3 | 0.4 | 0.0 | 0.2 |
| $T_{xy}^{true}$ | -158.2 | 169.6 | -30.2 | 62.2 |
| $T_{xy}^{true}$ - IF | -0.8 | 0.8 | 0.0 | 0.4 |
| $T_{xy}^{true}$ - LSC | -0.8 | 0.7 | 0.0 | 0.3 |
| IF - LSC | -0.3 | 0.4 | 0.0 | 0.1 |
| $T_{xz}^{true}$ | -343.2 | 333.4 | 15.7 | 150.7 |
| $T_{xz}^{true}$ - IF | -2.7 | 2.7 | 0.0 | 1.2 |
| $T_{xz}^{true}$ - LSC | -2.5 | 2.5 | 0.0 | 1.1 |
| IF - LSC | -0.6 | 0.4 | 0.0 | 0.2 |
| $T_{yy}^{true}$ | -171.9 | 44.9 | -77.8 | 57.6 |
| $T_{yy}^{true}$ - IF | -2.1 | 1.2 | 0.0 | 0.6 |
| $T_{yy}^{true}$ - LSC | -1.7 | 1.0 | 0.0 | 0.5 |
| IF - LSC | -1.0 | 0.8 | 0.0 | 0.3 |
| $T_{yz}^{true}$ | -174.6 | 275.7 | 0.2 | 74.0 |
| $T_{yz}^{true}$ - IF | -1.3 | 1.9 | 0.2 | 0.7 |
| $T_{yz}^{true}$ - LSC | -1.1 | 1.4 | 0.2 | 0.5 |
| IF - LSC | -0.7 | 0.6 | 0.0 | 0.3 |
| $T_{zz}^{true}$ | -213.9 | 429.2 | 225.9 | 173.6 |
| $T_{zz}^{true}$ - IF | -2.9 | 4.6 | -0.2 | 1.4 |
| $T_{zz}^{true}$ - LSC | -2.7 | 4.0 | -0.1 | 1.2 |
| IF - LSC | -0.9 | 1.3 | -0.1 | 0.4 |

3000 m). When repeating the computations with observations located directly on the ellipsoid, the differences over the Alps decrease (see Table 2 and Fig. 5). The remaining differences reflect the noise introduced in the synthetic data.

The main reason for the discrepancies between the two cases with $\Delta g$ on the terrain and on the ellipsoid is the constant radius approximation, which is used in the implementation of the LSC and IF approach. In LSC, the actual point heights can be introduced in principle, but then one loses the block Toeplitz structure of the matrix $C_{\Delta g' \Delta g'}$. In the IF approach, varying terrain heights cannot be considered directly. Therefore, a two step upward continuation procedure should be applied by harmonically continuing the terrestrial data to a reference sphere in a first step, followed by the IF and LSC approach as described above. This will be investigated in the future in more detail.

Using Eq. (3) of LSC, prediction errors are estimated for grid points in the central 8° by 8° area with a resolution of 30′. The errors are homogeneous over this area and vary less than 0.1 mE. The average standard deviations are: $\sigma_{T_{xx}} = 0.9$ mE, $\sigma_{T_{xy}} = 0.5$ mE, $\sigma_{T_{xz}} = 1.0$ mE, $\sigma_{T_{yy}} = 0.9$ mE, $\sigma_{T_{yz}} = 1.0$ mE, and $\sigma_{T_{zz}} = 1.4$ mE. These accuracy estimates agree well with the differences shown in Table 2.

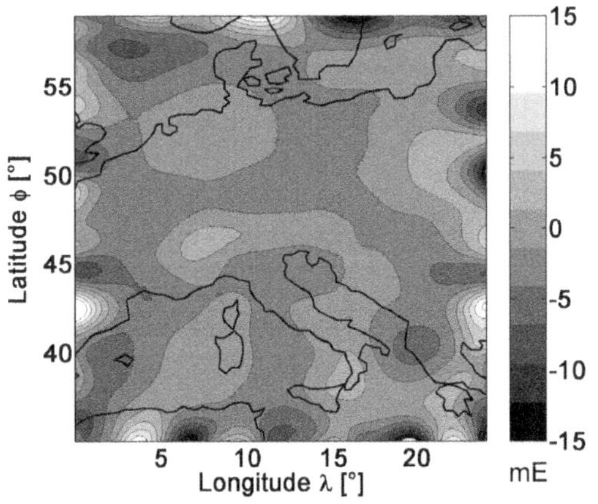


**Fig. 4.** Differences between ground-truth and LSC predictions for $T_{zz}$ in the 24° by 24° area, $\Delta g$ located on DTM.

**Table 2.** Statistics of the differences between $T_{ij}^{true}$ and IF results (mE), $\Delta g$ on ellipsoid.

| Dataset | Min | Max | Mean | Std |
|---|---|---|---|---|
| $T_{xx}^{true}$ - IF | -0.8 | 0.9 | 0.0 | 0.3 |
| $T_{xy}^{true}$ - IF | -0.5 | 0.4 | 0.0 | 0.1 |
| $T_{xz}^{true}$ - IF | -0.8 | 0.9 | 0.1 | 0.4 |
| $T_{yy}^{true}$ - IF | -1.1 | 0.9 | 0.1 | 0.4 |
| $T_{yz}^{true}$ - IF | -1.0 | 0.8 | 0.1 | 0.4 |
| $T_{zz}^{true}$ - IF | -1.3 | 1.2 | -0.1 | 0.6 |

In addition, the effect of the maximum degree of the reference geopotential model was investigated by truncating the model at spherical harmonic degree $l_{max} = 90$. The tests with the gravity data set located on the ellipsoid show that the residual anomalies increase to $\pm 28.31$ mgal, in contrast to $\pm 19.68$ mgal for the reference model to degree 360. Consequently, larger residual signals are to be handled in the computation formulas, which leads to slightly increased standard deviations of the differences (0.7 mE, 0.3 mE, 0.7 mE, 0.4 mE, 0.7 mE, 0.9 mE, for $T_{xx}$, $T_{xy}$, $T_{xz}$, $T_{yy}$, $T_{yz}$, $T_{zz}$, respectively) in the closed-loop computations, as compared to Table 2. Therefore, a high-degree reference geopotential model should be considered in order to diminish the effect of spherical approximations, which are used in the computation formulas.

## 6 Conclusions

Gravitational gradients have been computed from terrestrial gravity anomalies and a geopotential model. Two methods, namely least-squares collocation (LSC) and integral formulas (IF) based on the spectral combination technique, are investigated using synthetic data in a closed-loop computation. The synthetic gravity anomalies are generated on the ter-

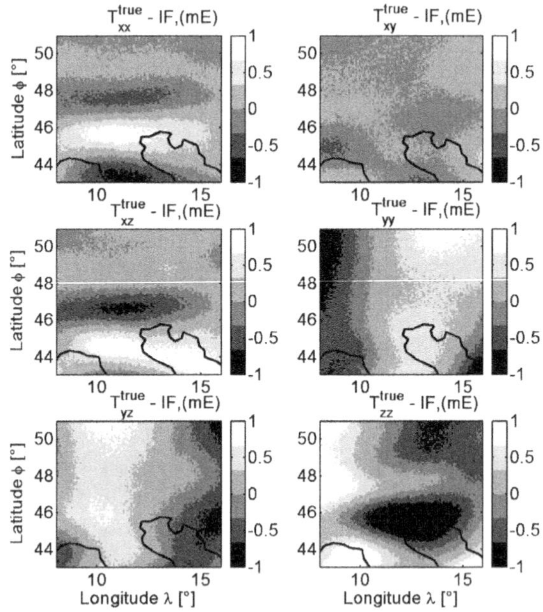


**Fig. 5.** Differences between ground-truth and IF predictions for all $T_{ij}$ in the central 8° by 8° area, $\Delta g$ located on ellipsoid.

rain and on the ellipsoid. The standard deviations of the differences from the closed-loop computations are varying from 0.3 to 1.4 mE for $\Delta g$ on the terrain and from 0.1 to 0.6 mE for $\Delta g$ on the ellipsoid. The differences between the two cases ($\Delta g$ on terrain and ellipsoid) are mainly caused by the constant radius approximation used in the implementation of the LSC and IF approach. Therefore, the actual terrain heights should be considered in the modelling techniques. A practical solution may be a two step upward continuation procedure, where the terrestrial data are first harmonically continued to a reference sphere, followed by the gradient computation at satellite altitude. This will be investigated in the future. Nevertheless, the closed-loop differences comply with the accuracy requirements for the GOCE calibration/validation.

Regarding the numerical experiments, the assumption of white noise is probably not valid for real data sets. Therefore, further investigations will be done with coloured noise.

To summarise, both methods used in this study are suitable for the upward continuation, needed for the GOCE calibration/validation. Larger data sets can be handled with the IF approach, while LSC provides location-dependent prediction errors.

**Acknowledgments.** We like to thank C.C. Tscherning for providing his covariance computation subroutines. This research was financially supported by the Geotechnologien-Projekt of the German Federal Ministry of Education and Research (BMBF) and the German Research Foundation (DFG). This is publication no. GEOTECH-89 of this project.

## References

Arabelos, D. and Tscherning, C. C. (1998). Calibration of Satellite Gradiometer Data Aided by Ground Gravity Data. *Journal of Geodesy*, (72):617–625.

Brenner, P., Mehrmann, V., Sima, V., Van Huffel, S., and Varga, A. (1999). SLICOT – A Subroutine Library in Systems and Control Theory. Report 97-3, NICONET.

de Min, E. (1995). A Comparison of Stokes' Numerical Integration and Collocation, and a new Combination Technique. *Bulletin Géodésique*, (69):223–232.

Denker, H. (2003). Computation of Gravity Gradients Over Europe For Calibration/Validation of GOCE Data. In *Proceedings of the Gravity and Geoid Meeting of the IAG, Thessaloniki, August 26-30, 2002*, pages 287–292.

Ditmar, P. and Klees, R. (2002). A Method to Compute the Earth's Gravity Field from SGG/STT Data to be Acquired by the GOCE Satellite. Delft university press, Delft.

Haagmans, R., de Min, E., and van Gelderen, M. (1993). Fast Evaluation of Convolution Integrals on the Sphere Using 1D FFT, and a Comparison With Existing Methods for Stokes' Integral. *manuscripta geodaetica*, 18(5):227–241.

JPL (2003). *JPL – GRACE Homepage*. Jet Propulsion Laboratory. `http://podaac.jpl.nasa.gov/grace/`.

Lemoine, F. G., Kenyon, S. C., Factor, J. K., Trimmer, R. G., Pavlis, N. K., Chinn, D. S., Cox, C. M., Klosko, S. M., Luthcke, S. B., Torrence, M. H., Wang, Y. M., Williamson, R. G., Pavlis, E. C., Rapp, R. H., and Olson, T. R. (1998). The Development of the Joint NASA GSFC and NIMA Geopotential Model EGM96. Technical Paper NASA/TP-1998-206861, NASA.

Moritz, H. (1971). Kinematical Geodesy II. Department of Geodetic Science, Ohio State University, Report 165.

Moritz, H. (1976). Integral Formulas and Collocation. *manuscripta geodaetica*, 1(1):1–40.

Moritz, H. (1980). *Advanced Physical Geodesy*. Herbert Wichmann Verlag, Karlsruhe.

Pail, R. (2002). In-orbit Calibration and Local Gravity Field Continuation Problem. In *ESA From Eötvös to Milligal+ Final Report, Contract 14287/00/NL/GD*, pages 9–112. ESA/ESTEC.

Thalhammer, M. (1994). The Geographical Truncation Error in Satellite Gravity Gradiometer Measurements. *manuscripta geodaetica*, (19):45–54.

Tscherning, C. C. (1976a). Computation of the Second-Order Derivatives of the Normal Potential Based on der Representation by Legendre Series. *manuscripta geodaetica*, 1(2):71–92.

Tscherning, C. C. (1976b). Covariance Expressions for Second and Lower Order Derivatives of the Anomalous Potential. Department of Geodetic Science, Ohio State University, Report 225.

Tscherning, C. C. and Rapp, R. H. (1974). Closed Covariance Expressions for Gravity Anomalies, Geoid Undulations, and Deflections of the Vertical Implied by Anomaly Degree Variance Models. Department of Geodetic Science, Ohio State University, Report 208.

Wenzel, H.-G. (1982). Geoid Computation by Least-Squares Spectral Combination Using Integral Kernels. In *Proceedings of the General Meeting of the IAG, Tokyo, May 7–15, 1982, IAG Symposia*, pages 438–453. Springer Verlag.

Wenzel, H.-G. (1985). Hochauflösende Kugelfunktionsmodelle für das Gravitationspotential der Erde. Wiss. Arb. der Fachr. Verm.wesen der Univ. Hannover 137.

Wenzel, H.-G. (1999). Schwerefeldmodellierung durch ultra hochauflösende Kugelfunktionsmodelle. *Zeitschrift für Vermessungswesen*, 124(5):144–154.

Wolf, K. I. and Müller, J. (2004). Prediction of Gravitational Gradients Using Simulated Terrestrial Data for GOCE Calibration. In Meurers, B. and Pail, R., editors, *Österreichische Beiträge zu Meteorologie und Geophysik, Proceedings of the 1st Workshop on International Gravity Field Research, Graz, May 8-9 2003*, volume 31, pages 31–36. Zentralanstalt für Meteorologie und Geodynamik.

# Global Gravity Field Solutions Based on a Simulation Scenario of GRACE SST Data and Regional Refinements by GOCE SGG Observations

A. Eicker, T. Mayer-Guerr, K.H. Ilk
Institute of Theoretical Geodesy,
University of Bonn, Nussallee 17, D-53115 Bonn, Germany
annette@geod.uni-bonn.de

**Abstract.** GOCE (Gravity Field and Steady-State Ocean Circulation Explorer) has the potential of deriving the global gravity field with unprecedented accuracy in the high resolution spectral part. The usual way is to model the gravity field by spherical harmonics up to a degree limited by the numerical stability of the recovery procedure. A disadvantage of this kind of gravity field representation is the lack of flexibility in modeling the inhomogeneous gravity field of regions with variable rough gravity field features. An alternative approach is to determine a global gravity field solution with high long and medium wavelength accuracy, e.g. based on GRACE SST observations up to a moderate degree, and improve this global solution in regions with characteristic gravity field features by an adapted regional recovery procedure. The individual gravity field features in these regions can be modeled by space localizing base functions like spherical spline functions. The advantage of this method is the possibility of adjusting the spline representation and the recovery procedure according to the regional gravity field structures and the specific data distribution. As a first indicator of a rough gravity field the structure of the topography or geophysical a-priori information can be used as a criterium. The resolution of the regional gravity field can be further improved by a subsequent iteration step. If neccessary, several regional solutions with global coverage can be merged by means of quadrature methods to obtain a global solution. Simulation results are presented to demonstrate this approach. Due to the regionally adapted recovery strategies this method provides better results than calculating a spherical harmonics solution by recovering the potential coefficients directly.

**Keywords.** GOCE, GRACE, gravity field recovery, regional solutions, space localizing base functions, Gauss-Legendre-Quadrature

## 1 Introduction

The approach presented here is based on the combination of an accurate long to medium wavelength global solution, in this case derived from the satellite mission GRACE (Tapley et al. (2004)), with high resolution regional refinements calculated from GOCE (ESA (1999)) observations. Due to the lack of real GOCE data the calculations are performed according to a simulated GRACE-GOCE scenario. The global solution is parameterized by spherical harmonic coefficients and the regional solutions are represented by spherical splines as space localizing base functions. This procedure provides several advantages. The regional approach allows to exploit the individual signal content in the observations and a tailored regularization for regions with different gravity field characteristics. The advantage compared to a uniform global regularization is that the regularization factor is selected for each region individually. A global regularization causes an overall filtering of the observations leading to a mean dampening of the global gravity field features. By an individually adapted regularization it is possible to extract more information out of the given data than would be possible with a global gravity field determination. Regions with a smooth gravity field signal for example can be regularized stronger without dampening the signal. In addition, the resolution of the gravity field determination can be chosen for each region individually according to the spectral behavior of the signal in the specific region.

Furthermore the regional approach has the advantage of dealing with regions with different data coverages more easily. If no data at all is available (e.g. the polar gap) the regional refinement can be skipped. For regions with sparse data coverage a coarser parameterization can be selected. If spherical harmonics are used, the resolution can only be defined globally and data gaps are reflected in the accuracy of all spherical harmonic coefficients. Furthermore, because of stability problems the spherical harmonic series has to be limited to an upper degree despite the fact that gravity signal is still available in regions with rough gravity field features.

Another aspect especially relevant for the GOCE mission with its potential to recover the high resolution gravity field is the fact that regional solutions contain less unknown parameters and therefore the computation procedure is simplified. This enables to reduce the computation costs significantly.

## 2 Mathematical Model

The observations of the GOCE gradiometer are the second derivatives of the gravitational potential $\nabla\nabla V(P)$. They are observed along the satellite orbit at a regular sampling rate. Every observation constitutes an observation equation if the gravity field is parameterized adapted to the specific task. The parameterization of the potential is performed in terms of spherical harmonics for the global solution, in case of a regional gravity field recovery the potential has to be parameterized by space localizing base functions. It can be modeled as a sum of base functions as follows

$$V(\mathbf{r}_P) = \sum_{i=1}^{I} a_i \, \Phi(\mathbf{r}_P, \mathbf{r}_{Q_i}). \tag{1}$$

$a_i$ are the unknown field parameters arranged in a column matrix, $\mathbf{r}_P$ denominates the field point and the $I$ nodal points $\mathbf{r}_{Q_i}$ indicate where the base functions $\Phi(\mathbf{r}_P, \mathbf{r}_{Q_i})$ are located. These base functions can be described in the following way:

$$\Phi(\mathbf{r}_P, \mathbf{r}_{Q_i}) = \sum_{n=0}^{N_{max}} k_n \left(\frac{R_E}{r}\right)^{n+1} P_n(\mathbf{r}_P, \mathbf{r}_{Q_i}). \tag{2}$$

As the maximum degree used for the regional recovery corresponds to a spherical harmonic expansion up to a maximum degree $N_{max} = 300$, the spline kernel has been constructed so that the expected unknown gravity field features will be represented as well as possible. Therefore the expansion has been truncated at degree $N_{max} = 300$. The coefficients $k_n$ are the difference degree variances of the gravity field spectrum to be determined,

$$k_n = \sum_{m=0}^{n} (\Delta c_{nm}^2 + \Delta s_{nm}^2), \tag{3}$$

$R_E$ is the mean equator radius of the Earth, $r$ the distance of a field point from the geo-center and $P_n(\mathbf{r}_P, \mathbf{r}_{Q_i})$ are the Legendre polynomials depending on the spherical distance between a field point $P$ and the nodal points $Q_i$ of the set of base functions. With this definition the base functions can be interpreted as isotropic and homogeneous harmonic spline functions (Freeden et al. (1998)). The nodal points are generated on a grid by a uniform subdivision of an icosahedron of twenty equal-area spherical triangles (Fig. 1). This refined grid is illustrated in Fig. 2. In this way the global pattern of spline nodal points $Q_i$ shows approximately uniform nodal point distribution.

The observation equation for gradiometer measurements is obtained by differentiating the potential twice:

$$\nabla\nabla V(P) = \sum_{i=1}^{N} a_i \nabla\nabla\Phi(\mathbf{r}_P, \mathbf{r}_{Q_i}). \tag{4}$$

The observation equations are established for short arcs over the selected regional recovery area, while the coverage with short arcs should be slightly larger than the recovery region itself to prevent the solution from geographical truncation effects. Every short arc builds a partial normal equation. To consider different accuracies of the short arcs these normal equations are combined by estimating a variance factor for every arc by means of variance component estimation as described by Koch and Kusche (2001). In case of a regularization the regularization factor can be determined by this same procedure of variance component estimation as well.

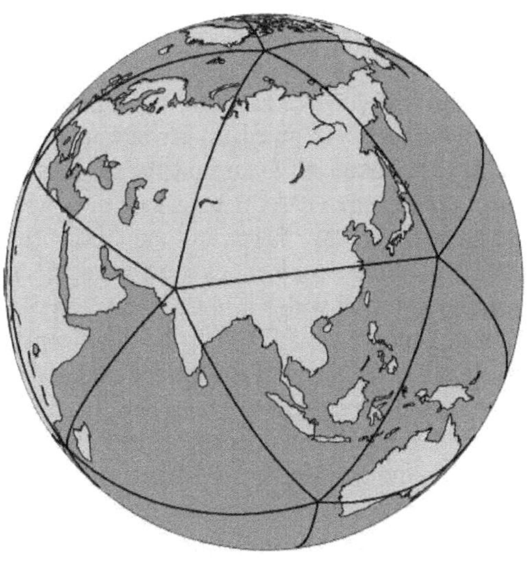

**Fig. 1** Icosahedron Grid

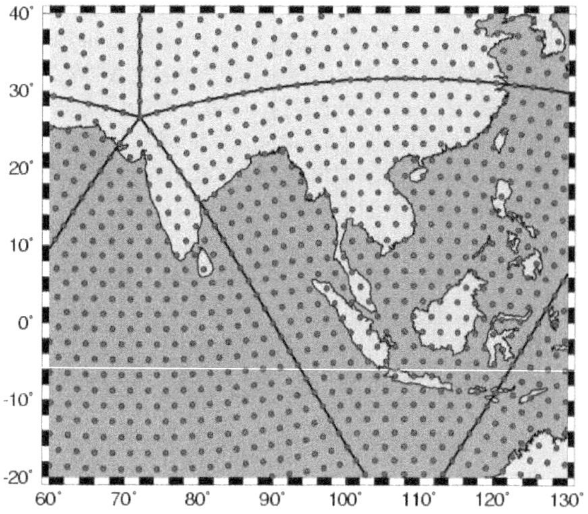


**Fig. 2** Refined Grid

## 3 Simulation Scenario

In our approach a precise global gravity field solution is refined by high resolution regional solutions. A simulated GRACE-like solution as outlined, e.g. by Ilk et al. (1998) serves as the global solution as this mission will provide excellent results especially in the long and medium wavelength part of the gravity field spectrum. The regional refinements of this solution are then calculated on the basis of GOCE-like observations (Mayer-Guerr et al. (2003)), as they cover the short periodic part of the gravity field spectrum with superior accuracy. In order to achieve a consistent data set both the GRACE and the GOCE solution were calculated from simulated observations on the basis of the EGM96 (Lemoine et al. (1998)) up to degree $n = 300$. The observations for GRACE as well as for GOCE were simulated for a period of 30 days with a sampling rate of 5 sec.

In case of the GRACE satellites the simulated orbit positions were corrupted by a white noise with a standard deviation of 3 cm and the intersatellite ranges between the GRACE twin satellites with a white noise of 10 $\mu$m. Regarding the GOCE satellite the gradiometer observations were corrupted by a colored noise model with a standard deviation of 1.2 mE. In this model the PSD is assumed as constant in the measurement band of the gradiometer and features increasing energy in the long wavelength part of the spectrum as the gradiometer is not able to recover these low frequencies. For the satellite positions again an accuracy of 3 cm was assumed.

From this simulated data set first the GRACE solution up to a spherical harmonic degree of $n = 150$ was derived. The regional refinements were then calculated as residual fields to this global solution. Using a parameterization by splines as space localizing base functions, the spline kernels are to represent the gravity field features to be determined. Therefore up to degree $n = 150$, we used the error degree variances of the GRACE solution as coefficients $k_n$ according to equation (3). These error degree variances represent the signal which is still in the data in addition to the GRACE solution. Above degree $n = 150$ the degree variances of the EGM96 were used for the coefficients $k_n$. The degree variances are shown in Fig. 3. Alternatively, the spline kernel could have been calculated according to Kaula's rule of thumb above degree 150.

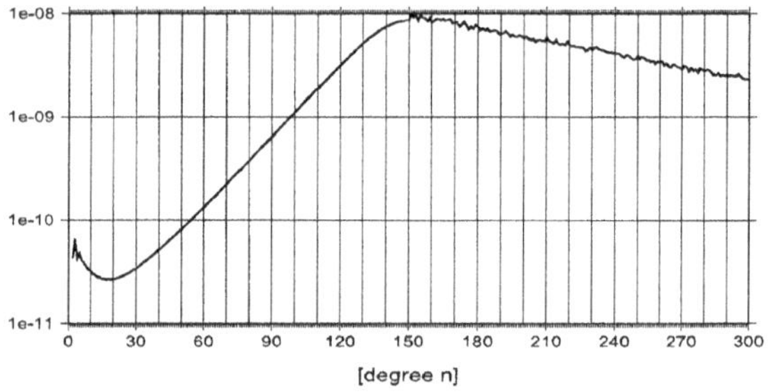


**Fig. 3** Degree variances for the spherical harmonic expansion

As there is no signal content in the GRACE field higher than degree 150 there is no jump in the degree variance graph. Even if in case of real data a jump occurred, this would have only minor influence on the shape of the base function. The normalized kernel is displayed in Fig. 4. The peak is located at the nodal points $Q_i$ and the x- and y-axis indicate the spherical distance from this nodal point.

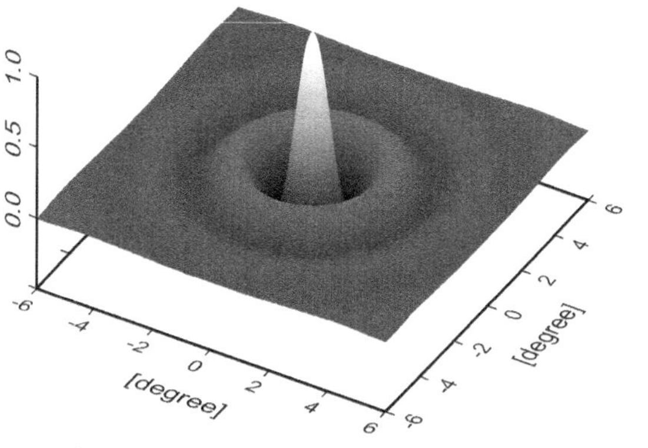


**Fig. 4** Normalized spline kernel

## 4 Regional Gravity Field Recovery

To combine the advantages of the satellite mission GOCE to cover almost the complete Earth with the advantages of regional focussing techniques, as pointed out in chapter 1, the surface of the Earth has been divided into patches and for each patch the gravity field has been determined separately. The disturbing potential is represented by splines located at nodal points with a mean distance of approximately 67 km. Concerning the number of unknown parameters this corresponds to a resolution of a spherical harmonic degree 300. This leads to a number of about 5000 to 9000 spline parameters to be determined for each region, the size of the patches being limited by storage restrictions.

Due to the noise especially in the high frequencies a Tikhonov regularization has been applied. The regional recovery approach offers the possibility of calculating an individually adapted regularization parameter for each regional patch by means of the variance component estimation procedure according to Koch and Kusche (2001).

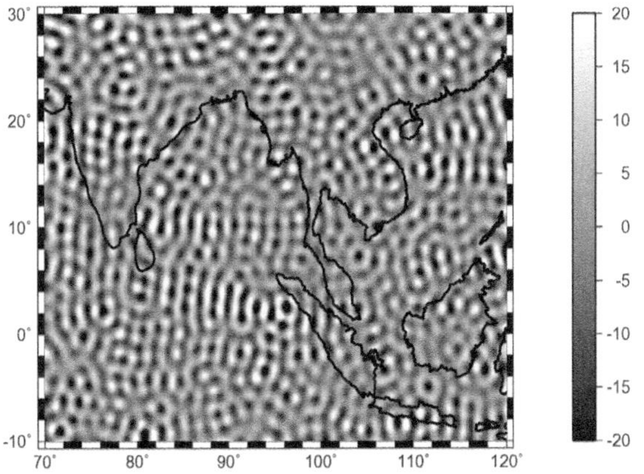


**Fig. 5** Differences in geoid undulations between a regional solution and the global gravity field model EGM96 [cm], RMS: 10.1 cm

Fig. 5 shows the differences of the regional solution for the Himalayan region, a region with extremely rough gravity field signal, compared to the pseudo-real field EGM96. The comparison has only been performed up to degree $n = 240$ because the higher degrees are too strongly corrupted by noise.

The regional recovery procedure offers a chance to deal with the polar gap problem in a tailored way, as in regions without any data the regional refinement can either be skipped or the regularization parameter can be adjusted accordingly.

Regional refinements have been computed covering the complete Earth. To avoid truncation effects at the boundaries an overlapping border of 10° has been taken into account. Fig. 6 shows some examples of the regional solutions and exemplary borders.

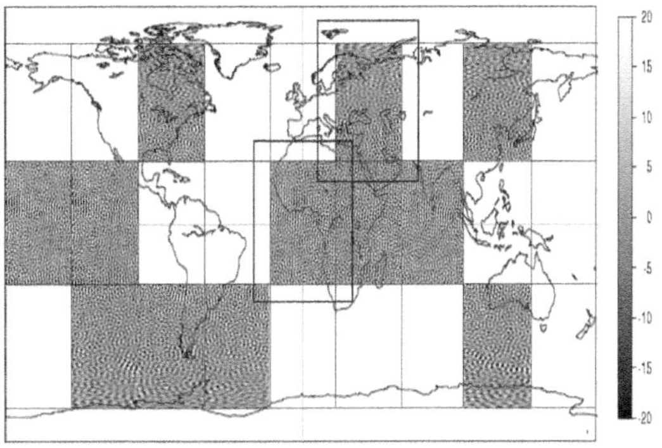


**Fig. 6** Regional solutions covering the complete Earth, differences in geoid undulations to the EGM96, [cm]

Taking a closer look at two adjacent regional solution patches (Fig. 7) reveals how well their residual patterns match together despite the fact that the recovery parameters such as variance factors and the regularization parameter are selected independently for each geographical region.

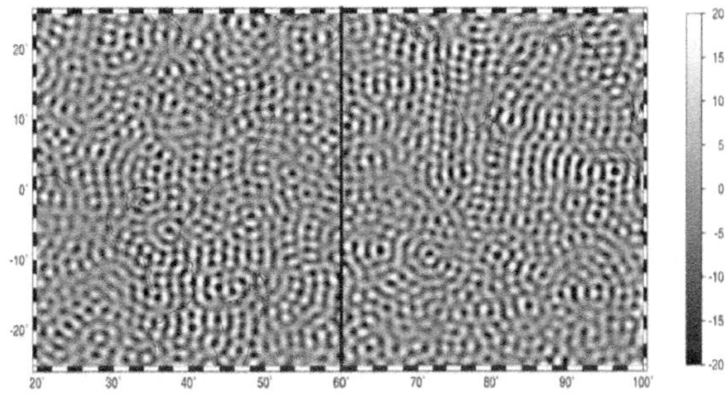


**Fig. 7** Two adjacent regional solutions, differences in geoid undulations to the EGM96, [cm]

## 5 Combination of regional solutions

For many applications it seems to be useful to derive a global gravity field model by spherical harmonics without losing the details of a regional zoom-in. This can be performed by a direct stable computation step. One possibility to merge the regional solutions to a global one is to compute gravity functionals in the specific regions, in principle with arbitrary resolution and to compute the spherical harmonic coefficients by numerical quadrature. An alternative is to derive these coefficients analytically, but an adequate technique is not available yet. In our approach the coefficients of the spherical harmonic expansion are calculated by means of the Gauss-Legendre-Quadrature (see for example Stroud and Secrest (1966)). This method is also referred to as Neumann's method, as described in Sneeuw (1994) among different other quadrature methods.

$$\left\{ \begin{array}{c} c_{nm} \\ s_{nm} \end{array} \right\} = \frac{R_E}{GM \cdot 4\pi} \times \sum_{k=1}^{K} V_k P_{nm}(\cos\theta_k) \left\{ \begin{array}{c} \cos(m\lambda_k) \\ \sin(m\lambda_k) \end{array} \right\} w_k, \quad (5)$$

with the area weights

$$w_k = \frac{2}{(1-t_k^2)(P'_{N+1}(\cos(\theta_k)))^2}. \quad (6)$$

With $V_k$ being the gravitational potential at the $K$ nodes of the quadrature, $P_{nm}$ being the associated Legendre functions and $P'_{N+1}$ being the first derivative of the Legendre polynomial of degree $N+1$, when $N$ is the maximum degree to be determined. This method requires the data points to be located at a specific grid, called Gauss-Legendre-Grid. From the regional spline solutions the gravitational potential can be calculated at the nodes of this gridwithout loss of accuracy. It has equi-angular spacing along circles of latitude, along the meridians the nodes are located at the zeros of the Legendre Polynomials of degree $N+1$. This quadrature method has the advantage of maintaining the orthogonality relations of the Legendre functions despite the discretization procedure, which allows an exact calculation of the potential coefficients. The grid used for the calculations has a spacing of $\Delta\lambda = 0.5°$ which corresponds to a number of 360 circles of latitude. It shall be pointed out that the direct computation of the spherical harmonic coefficients by solving the improperly posed downward continuation cannot provide a stable solution up to an (arbitrarily) high degree as it can be achieved here.

## 6 Results

The regional spline solutions have been merged to a global gravity field by predicting the gravity field functionals to the nodal points of a Gauss-Legendre-Grid. From this global field a spherical harmonics expansion has been calculated using the Gauss-Legendre-Quadrature. The differences of this combined global solution compared to the pseudo-real field EGM96 are displayed in Fig. 8. Again, the comparison has only been performed up to a spherical harmonics degree of $n = 240$.

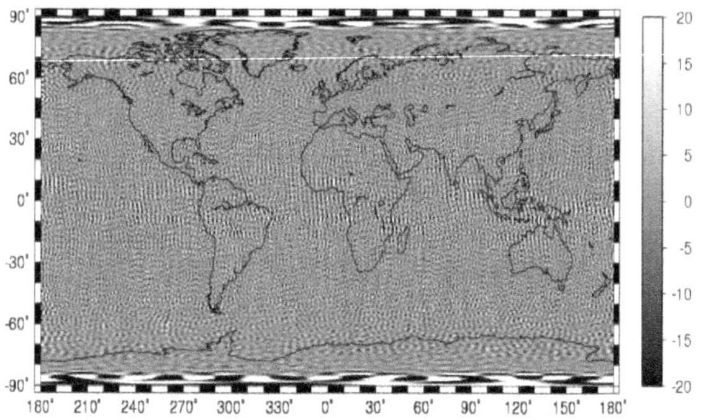


**Fig. 8** Differences in geoid undulations between regional solution and EGM96 [cm], RMS: 8.6 cm

The results show that an exact transformation of the individual regional patches to one global spherical harmonic solution is possible by this quadrature method. The differences between our solution and the EGM96 have been calculated on a 0.5° × 0.5° grid of point geoid undulations, revealing a weighted global RMS of 8.6 cm, which has been calculated including the poles. This accuracy corresponds to the noise model described in chapter 3 with the gradiometer observations being corrupted with a colored noise of 1.2 mE. These results reveal the practicability of the presented procedure.

## 7 Conclusions and Outlook

Combining regional solutions to a global gravity field solution seems to be a reasonable alternative to deriving a global gravity field solution directly. The tailored calculation of a regularization parameter for each region allows a tailored filtering according to the individual gravity field features. The method is modest in terms of computation costs, as the complete adjustment problem is split up into much smaller problems. This procedure enables the computation of a global GOCE solution up to an arbitrary resolution on a single PC. This approach allows the calculation of a spherical harmonic expansion as well, with the advantage that the quadrature procedure does not limit the resolution to an upper degree but allows us to extract all possible information out of the given data. Adjacent regional patches fit remarkably well, even though the recovery parameters are selected individually for each region. Remaining differences are due to the different selection of the regularization parameter and to truncation errors. Further improvements are expected for example by refining the regularization strategy and tailoring it more accurately to the demands of the gravity field features in the specific regions. So far, the regions are selected according to a geographical grid, but it is planned for the future to select the region boundaries such that the global gravity field is divided into regions with homogeneous gravity field features. Furthermore, a more precise selection of the base functions and the nodal point distribution is intended in order to achieve an even better adjustment to the signal content in the regional areas. In this context multiresolution strategies might be successful as well. Other aspects that have to be taken into consideration for further investigations are for example the aliasing effects originating from the patching of several regional solutions and the problem that certain global conditions (such as the conservation of the center of mass) might be violated when merging the regional solutions. Another problem that is to be solved in the future is the fact that it is necessary to take overlapping boundaries into account for the determination of the regional refinement patches in order to prevent the solution from geographical truncation errors and to ensure a smooth matching. But in this way the same data is used for the calculation of two adjacent patches and therefore the patches are not independent. This might be especially important for the derivation of a global solution by merging the patches.

## Acknowledgments

The support by BMBF (Bundesministerium fuer Bildung und Forschung) and DFG (Deutsche Forschungsgemeinschaft) within the frame of the Geotechnologien-Program is gratefully acknowledged.

## References

European Space Agency (1999) Gravity Field and Steady-State Ocean Circulation Explorer Mission (GOCE). Report for mission selection, in The four candidate Earth explorer core missions, SP-1233 (1), Nordwijk, The Netherlands.

Freeden W, Gervens T, Schreiner M (1998) Constructive Approximation on the Sphere. Oxford University Press, Oxford.

Ilk KH, Feuchtinger M, Mayer-Guerr T (2003) Gravity Field Recovery and Validation by Analysis of Short Arcs of a Satellite-to-Satellite Tracking Experiment as CHAMP and GRACE. accepted for publication in the IAG proceedings of the General Assembly of the IUGG 2003, Sapporo, Japan.

Koch KR, Kusche J (2001) Regularization of geopotential determination from satellite data by variance components. Journal of Geodesy 76 (5):259-268.

Lemoine FG, Kenyon SC, Factor JK, Trimmer RG, Pavlis NK, Chinn DS, Cox CM, Klosko SM, Luthcke SB, Torrence MH, Wang YM, Williamson RG, Pavlis EC, Rapp RH, Olson TR (1998) The development of the joint NASA GSFC and the National Imagery and Mapping Agency (NIMA) geopotential model EGM96. NASA/TP-1998-206861, Goddard Space Flight Center, Grenbelt, MD.

Mayer-Guerr T, Ilk KH, Eicker A (2003) Regional Gravity Field Recovery from GOCE Gradiometer Measurements and SST-high-low Observations – a Simulation Study. Proceedings of the 1st Workshop on International Gravity Field Research 2003, Graz, Austria.

Sneeuw N (1994) Global Spherical Harmonic Analysis by Least Squares and Numerical Quadrature Methods in Historical Perspective. Geophys. J. Int., 118: 707-716.

Stroud AH, Secrest D (1966) Gaussian Quadrature Formulas. Prentice-Hall, Englewood Cliffs, N.J.

Tapley BD, Bettadpur S, Watkins M, Reigber C (2004) The gravity recovery and climate experimant: mission overview and early results. Geophys Res Lett 31, L09607: doi10.1029/2004GL019920.

# Effect of geopotential model errors on the projection of GOCE gradiometer observables

Gy. Tóth[1], L. Földváry [1, 2]

[1] *Budapest University of Technology and Economics, Department of Geodesy and Surveying and Physical Geodesy and Geodynamics Research Group of BUTE-HAS, H-1521 Budapest, Hungary, Műegyetem rkp. 3, Fax: +36 14633192,* gtoth@sci.fgt.bme.hu.

[2] *Institute of Astronomical and Physical Geodesy, Technical University of Munich, Arcisstraße 21., D-80333 München, Germany* , foeldvary@bv.tu-muencen.de

**Abstract** The forthcoming GOCE mission will provide gravity gradient observations along its orbit at varying altitude. It is necessary for certain data processing strategies to project the GOCE gravity gradients to a mean reference sphere. In the present simulation study the radial distance of the projection is in the order of 10 km, and can be done using the Taylor expansion of the gravity gradients.

In this paper we present an error analysis of such a projection. The omission of higher-order terms of the Taylor expansion and commission errors of the geopotential model are discussed.

The paper presents an error analysis study based on simulated GOCE gradiometry. The results are validated with stringent accuracy requirements of the GOCE mission.

**Keywords.** gravity gradient tensor, geopotential model, space gradiometry, projection error

## 1 Introduction

Observables of Satellite Gravity Gradiometry (SGG) are the elements of the Eötvös tensor. Four components, three diagonal and one off-diagonal, of the SGG tensor will be observed with high accuracy in the measurement bandwidth (MBW) of 5 to 100 mHz in the Gradiometer Reference Frame (GRF). In this reference frame the $x$-axis on average is in the velocity direction, the $z$-axis approximately radially outward and the $y$-axis complements the right-handed frame. Simulations of SGG measurement errors range from 6 mE/√Hz (1E = $10^{-9}$ $s^{-2}$) in the high frequency part of the MBW to 15÷79 mE/√Hz in the low frequency part for the diagonal gravity gradients ($V_{xx}$, $V_{yy}$, $V_{zz}$). See for reference Cesare (2002) and Floberhagen et al. (2004).

There are different data processing strategies for GOCE gradiometry. The time-wise approach describes the gravity gradients with Keplerian elements, showing the explicit time dependence of gravity gradient measurements via the time coordinate (mean anomaly) along the quasi-fixed orbit defined by the other 5 elements (cf. Colombo 1986; Koop 1993). In case of the space-wise approach the time dependency is implicit via the time dependence of the orbit. Certain data processing strategies would require a pre-processing stage, e.g. rotation into another frame, interpolation, reduction to a sphere, etc. For example, combination with ground data may require integration over a certain region on a sphere (Haagmans et al., 2003). The application of such approximations introduces errors into the solution. The question is whether these errors are negligible compared to measurement errors.

Several authors analyzed the accuracies of GOCE gradients in various reference frames and error assessment using along track interpolation (Müller, 2003; Bouman and Koop, 2003). The present paper deals with the error analysis of the projection of gravity gradient observations along an orbit to a mean sphere, using the Taylor expansion of the gravity gradients. In the first part of the paper we overview the theoretical background. Subsequently, geopotential model induced omission and commission errors of the projection in the natural geographical frame (the Earth-fixed reference frame (ERF) defined by $x$-axis pointing towards North, $y$-axis to East and the $z$ upwards) are discussed.

A GOCE-like orbit has been simulated using the orbit integrator at IAPG Munich. The simulated orbit spans 90 days. The orbit data sampled at 1 Hz.

## 2 Radial projection of GOCE gravity gradients in the geographical frame

The projection of gravity gradients along an orbit (i.e. GOCE) onto a sphere means a change along the radial direction, $r$. Therefore it is done by the Taylor expansion of the gravity gradients by $r$.

*Correspondence to*: Gy. Tóth

$$V_{ij}^{\text{sphere}}(\varphi,\lambda,r^{\text{sphere}}) = V_{ij}^{\text{obs}}(\varphi,\lambda,r) + \frac{\partial V_{ij}^{\text{obs}}(\varphi,\lambda,r)}{\partial r}dr + \frac{1}{2}\frac{\partial^2 V_{ij}^{\text{obs}}(\varphi,\lambda,r)}{\partial r^2}dr^2 + \ldots \qquad (1)$$

where $i,j=x,y,z$ in the ERF. If the projection distance $dr$ is small, i.e. of order 10 km, then higher order terms of the Taylor expansion can be neglected. The magnitude of the second order term is the most important in this respect, but it can be neglected, as we will see later on. This is true for a nearly circular orbit where the deviations of the satellite positions from a mean sphere remain relatively small. The contribution of the second order term will be investigated in this study as well.

We make use of the spherical harmonic representation of the gravity gradients and of its radial derivatives in eq. (1) of the gravity gradient observations are expressed by spherical harmonics. We derive all the third derivatives of the gravitational potential, since no complete derivatives in the literature was found. However, due to the space limit of the present paper, we do not show the derivations here. For the radial derivatives (cf. eq. 2a-j), on the other hand, we can refer e.g. to Rummel (1997), where all six second derivatives of the geopotential can be found in the ERF. Since the $z$-axis is equivalent to the radial direction, we need six third derivatives of the geopotential $V$ with respect to $z$ in this recent study. It is straightforward to compute from the above mentioned six second derivatives all the required radial derivatives, therefore they can readily be checked. Since every component of degree $\ell$ of any element of the Eötvös tensor, i.e. $V_{ij}$ is a homogeneous function of degree $-(\ell+3)$, its first and second radial derivatives can be computed by multiplying all degree $\ell$ terms by the factors $-(\ell+3)/r$ and $-(\ell+3)(\ell+4)/r^2$, respectively (Rummel et. al, 1993). We consider especially important the third radial derivative $V_{zzz}$, since this is the largest term in the context of GOCE Any gravity field functional computed from a geopotential model has two main sources of error:

- propagation of errors of geopotential coefficients (*commission error*)
- errors introduced by neglecting coefficients higher than the maximum degree of the model (*omission error*)

These errors propagate into errors of the radial projection of gravity gradients using eq. (1).

The computation of the commission error of the radial derivatives in eq. (1) is straightforward if the variance-covariance matrix of geopotential coefficients is available. Strict error propagation using the full variance-covariance matrix is a computationally demanding task, however. Therefore we have used only the error standard deviations of the geopotential model and the errors of the coefficients are assumed to be uncorrelated.

$$\begin{aligned}
V_{xxx} &= \frac{1}{r^3\cos^3\varphi}V_{\lambda\lambda\lambda} - \frac{3\tan\varphi}{r^3\cos\varphi}V_{\varphi\lambda} + \frac{3}{r^2\cos\varphi}V_{r\lambda} - \frac{2}{r^3\cos^3\varphi}V_{\lambda}\\
V_{xxy} &= \frac{1}{r^3\cos^2\varphi}V_{\varphi\lambda\lambda} + \frac{2\tan\varphi}{r^3\cos^2\varphi}V_{\lambda\lambda} - \frac{\tan\varphi}{r^3}V_{\varphi\varphi} + \frac{1}{r^2}V_{r\varphi} - \frac{1}{r^3\cos^2\varphi}V_{\varphi}\\
V_{xxz} &= \frac{1}{r^2\cos^2\varphi}V_{r\lambda\lambda} - \frac{2}{r^3\cos^2\varphi}V_{\lambda\lambda} - \frac{\tan\varphi}{r^3}V_{r\varphi} + \frac{1}{r}V_{rr} + \frac{2\tan\varphi}{r^3}V_{\varphi} - \frac{1}{r^2}V_r\\
V_{xyy} &= \frac{1}{r^3\cos\varphi}V_{\varphi\varphi\lambda} + \frac{2\tan\varphi}{r^3\cos\varphi}V_{\varphi\lambda} + \frac{1}{r^2\cos\varphi}V_{r\lambda} + \frac{2\tan^2\varphi}{r^3\cos\varphi}V_{\lambda}\\
V_{xyz} &= \frac{1}{r^2\cos\varphi}V_{r\varphi\lambda} - \frac{2}{r^3\cos\varphi}V_{\varphi\lambda} + \frac{\tan\varphi}{r^2\cos\varphi}V_{r\lambda} - \frac{2\tan\varphi}{r^3\cos\varphi}V_{\lambda}\\
V_{yyy} &= \frac{1}{r^3}V_{\varphi\varphi\varphi} + \frac{3}{r^2}V_{r\varphi} - \frac{2}{r^3}V_{\varphi}\\
V_{yyz} &= \frac{1}{r^2}V_{r\varphi\varphi} - \frac{2}{r^3}V_{\varphi\varphi} + \frac{1}{r}V_{rr} - \frac{1}{r^2}V_r\\
V_{yzz} &= \frac{1}{r}V_{rr\varphi} - \frac{2}{r^2}V_{r\varphi} + \frac{2}{r^3}V_{\varphi}\\
V_{zzx} &= \frac{1}{r\cos\varphi}V_{rr\lambda} - \frac{2}{r^2\cos\varphi}V_{r\lambda} + \frac{2}{r^3\cos\varphi}V_{\lambda}\\
V_{zzz} &= V_{rrr}
\end{aligned} \qquad \text{(2a-j)}$$

Since we are interested in the error PSD, which is the error variance at various frequencies, it is unclear, at least to us, how to derive easily this error PSD from the process error variances. These (spatial) error variances of course can be derived easily through error propagation from error standard deviations of the geopotential coefficients, but we need the PSD, not the (spatial) error variances of the signal. Therefore, we followed a more pragmatic approach here, i.e. to simulate the actual errors and then it is straightforward to estimate the error PSD.

First we generated a non-correlated normally distributed random noise model of geopotential coefficients within the error variances of the geopotential model. Only one realization of this model was considered for each spherical harmonic expansion degree. Then this error model served the purpose of commission error computations, so all subsequent commission error tests were performed by subsequent realizations of these pseudo-random geopotential models.

The omission error in principle can be derived from a suitably high degree reference model, i.e. higher than the geopotential model used for projection, for evaluating the error effect of the unused degrees. The practical solution to the problem of finding a suitable degree of the reference model depends on many factors. These are the required gravity field functional, computation altitude and error tolerance for the omission error. The chosen maximal degree of the omission error reference model was 720, since the signal degree variance of the third radial derivative, $V_{zzz}$, falls off rapidly at the altitude of GOCE (250 km), and reaches the level of about $10^{-10}$ mE/km at this degree (cf. Fig. 3)

## 3 Computations and results

The mean reference sphere was defined by the mean height of the GOCE orbit. The distribution of

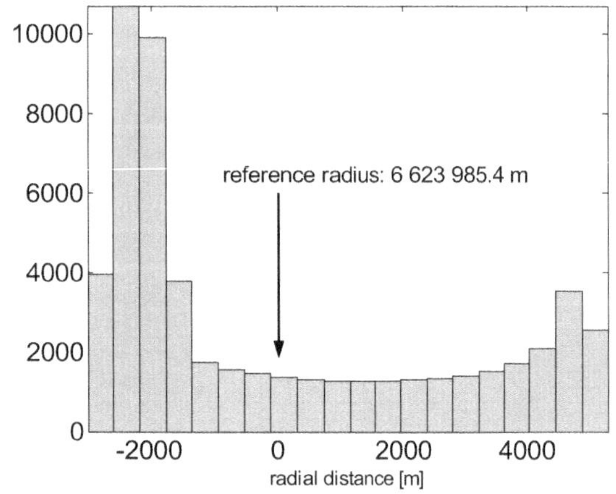


**Fig. 1** Histogram of the radial distances of 55 000 simulated GOCE orbit points with respect to a mean reference sphere of radius 6 623 985.4 m.

radial positions with respect to this mean reference sphere is shown in Fig. 1. The distribution is obviously not normal, but condensed around two peaks. This is due to the elliptical orbit geometry.

In all our tests we have examined extensively the most critical term, the radial projection of the $V_{zz}$ gravity gradient, but also tested other components. Our computations were performed in the ERF geographical reference frame and not in the GRF. Though we feel that results in GRF may give useful hints on the projection of GOCE observables, especially for the tested $V_{zz}$ component.

The computations of the commission error of the radial projection were done with a combined model. It is defined by a linear transition of the coefficients from GRACE GGM01C to EGM96 models between spherical harmonic degrees 81-90. This way strong contribution of a state-of-the-art satellite-only gravity model on long-wavelength has been combined with the more reasonable short-wavelength information of a combined model containing terrestrial data as well. Though a combined model was used in this study, we should keep in mind that it is not an inconsistent geopotential model, created with the purpose of simulating a better geopotential model as recently available. The error propagation of the projection using this model was determined and power spectra were computed at different spherical harmonic degrees (Fig. 2). In the low frequency part of the spectrum the models above degree 120 gave almost a constant commission error level of about 2 mE/√Hz. It means that commission errors of the radial projection mostly contributed by coefficients up to degree 120, and the higher degree coefficients altering the characteristics of the falloff of the error in the 0.01 – 0.03 Hz frequency band (harmonic degrees 50-160).

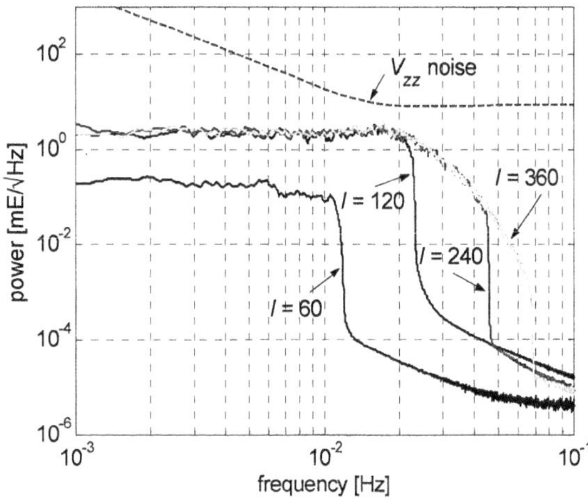


**Fig. 2** Commission error PSD of the radial projection of the $V_{zz}$ component from the combined GRACE GGM01C – EGM96 geopotential model at different spherical harmonic degrees. In total 55000 consecutive points of the 1 sec simulation (i.e. ~10 orbit revolutions) were projected to a mean sphere. The line labeled '$V_{zz}$ noise' shows the estimated gradiometer noise level.

Degrees close to degree 100 are the weakest part of the current geopotential models with respect to the projection procedure. We mean that in the vicinity of this wavelength the error spectrum approaches the gradiometer noise level most closely (cf. figure 2).

Next, omission errors were computed at 55000 points by comparison to the degree 720 solution (Fig. 5). According to the expectations, the omission error PSD from the projection is dependent on the maximum degree of the geopotential model used. The omission error within the MBW (i.e. between 0.005 mHz and 0.1 mHz) is well below the gradiometer noise level with a maximum degree of 240 or higher. For degree 240 expansion or higher the power spectral density of omission errors of the radial projection is below 0.1 mE/√Hz across the whole spectrum (cf. Fig. 5).

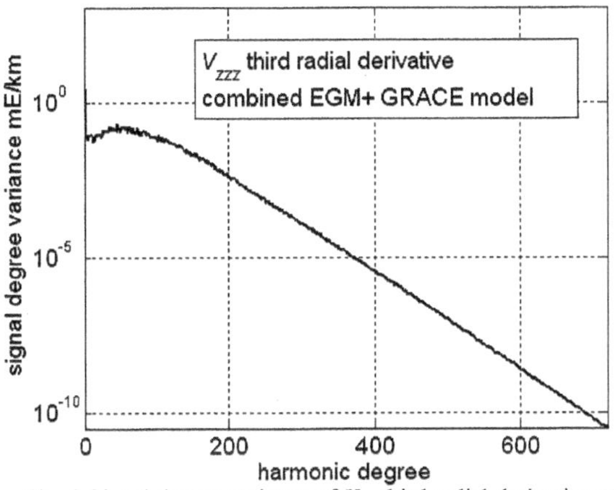


**Fig. 3** Signal degree variance of $V_{zzz}$ third radial derivative at the altitude of GOCE. The combined GRACE GGM01C+EGM96 geopotential model was used to maximum degree 720.

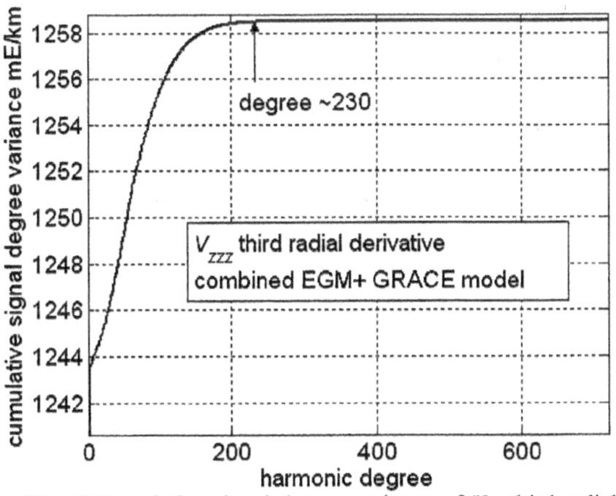


**Fig. 4** Cumulative signal degree variance of $V_{zzz}$ third radial derivative at the altitude of GOCE. The combined GRACE GGM01C+EGM96 geopotential model was used to maximum spherical harmonic expansion degree 720.

The cumulative signal degree variances of the third radial derivative (Fig. 4) confirm that most of the power in the third radial derivative is contained below the spherical harmonic degree of about 230 at the altitude of GOCE. Therefore it is permitted to use a high degree reference model (in this study the maximum degree of the reference model was 720) to evaluate the omission error spectrum of the projection.

From figures 2 and 5 it is obvious, that projection errors with a use of a geopotential model up to degree and order 240 introduces almost an order of magnitude smaller error than that of observations.

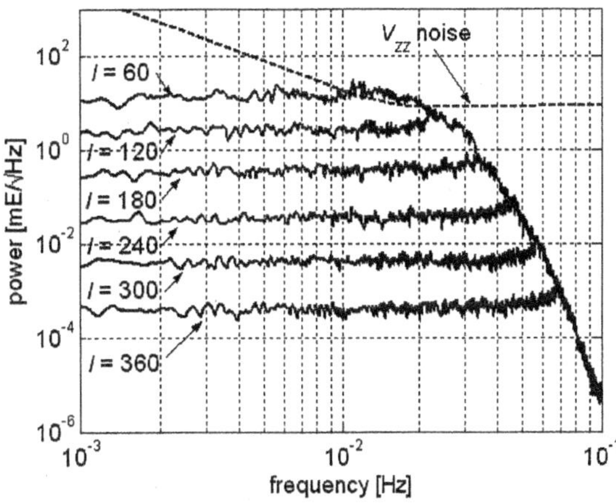


**Fig. 5** Omission error PSD of projection of $V_{zz}$ gravity gradients to a mean sphere in GRF. The combined GRACE GGM01C+EGM96 geopotential model was used to maximum degree 720 as reference (zero omission error).

The actual commission error of the projection depends somehow on the accuracy of the chosen geopotential model, at least below harmonic degree of about 90 (cf. Fig . 6). Therefore it is recommended to use a more accurate gravity model than recent geopotential solutions.

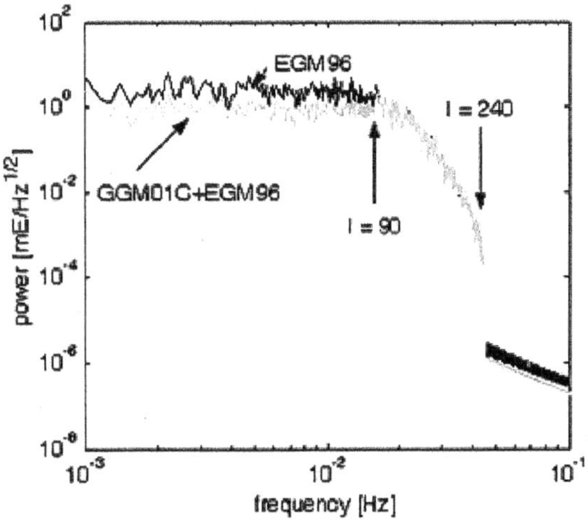


**Fig. 6** Comparison of commission error PSD of projection of $V_{zz}$ gravity gradients to a mean sphere in GRF. The combined GRACE GGM01C+EGM96 geopotential model was compared to the EGM96 model up to maximum spherical harmonic expansion degree 240. Below harmonic degree $l$ = 90 (15 mHz), the combined GRACE model is clearly superior to EGM96.

Finally the effect of the quadratic term in eq. (1) was investigated. The results for the $V_{zz}$ component in the ERF can be seen in Fig. 7. The quadratic errors are much below the observation errors at every frequency, therefore the use of this term is unnecessary.

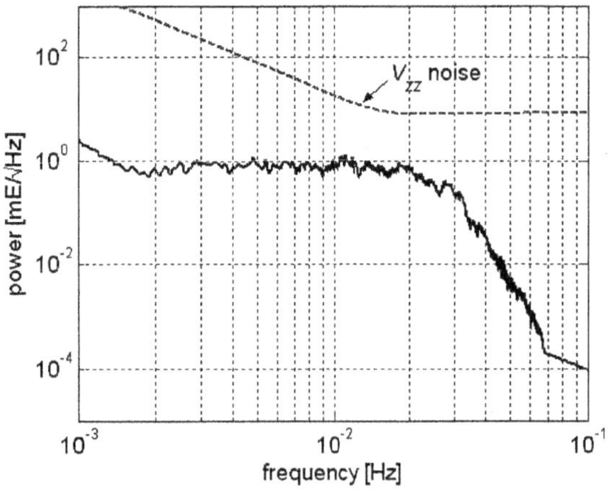


**Fig. 7** PSD of the quadratic term of $V_{zz}$ in Eq. (1) for radial projection to a mean sphere. The combined GRACE GGM01C and EGM96 geopotential model to degree and order 360 was used for the computation.

## 4 Conclusions and recommendations

Our main conclusion from the above study is that a radial projection procedure of GOCE observables distributed globally using a geopotential model is reasonable only if an error level of about 2-3 mE/√Hz is tolerable on the most critical 10-30 mHz part of the gradiometer spectrum. This error level is only by a factor of 2-4 smaller than the current predictions on the gradiometer performance. The error itself comes mainly from the errors of the geopotential coefficients above about degree 80 (cf. Fig. 2). In any case the expansion degree of the used model should be at least about 240 to reduce the omission error to be negligible compared to the gradiometer errors.

The analysis performed is relevant only to the case when gravity gradients are in the Earth-fixed reference frame. The real observations will be in the gradiometer reference frame, and therefore the proper orientation of the gradiometer axes should be taken into account. It depends mainly on the performance of the star sensors. The orbit errors, on the other hand, will play a much smaller role in gradiometry errors, since displacement of gravity gradients with some centimeters (typical accuracy of precise orbit determination) is much below the measurement accuracy.

It is sometimes desirable to project the observations to other locations both horizontally and vertically, e.g. to look at crossovers (Bouman and Koop, 2003). This is another interesting topic, which may be addressed in a forthcoming paper.


## Acknowledgements

The above investigations were funded by OTKA projects T037929, T046418 and Hungarian Space Office (project TP 205).


## References


Bouman, J, Koop, R (2003). Error assessment of GOCE SGG data using along track interpolation. Advances in Geosciences, No. 1, pp 27-32.

Cesare, S. (2002) Performance requirements and budgets for the gradiometric mission. Technical Note, GOC-TN-AI-0027, Alenia Spazio, Turin, Italy.

Colombo, O (1986). Notes on the mapping of the gravity field using satellite data. In: Sünkel H (ed) Mathematical and numerical techniques in physical geodesy. Lecture Notes in Earth Sciences, Vol 7. Springer, Berlin Heidelberg New York, pp 260-316.

Floberhagen, R, Demond, F-J, Emanuelli, P, Muzi, D, Popescu, A (2004). Development status of the GOCE programme. Paper presented at the 2nd GOCE User Workshop, ESA ESRIN, 8-10. March, 2004, Italy. Available at http://earth.esa.int/goce04.

Haagmans, R, Prijatna, K, Omang, O (2003). An Alternative Concept for Validation of GOCE Gradiometry Results Based on Regional Gravity. 3rd Meeting of the IGGC, Tziavos (ed.), Gravity and Geoid 2002, pp 281-286.

Koop, R (1993). Global gravity field modeling using satellite gravity gradiometry. Publ. Geodesy, New Series, No. 38. Netherlands Geodetic Commission, Delft.

Müller, J (2003). GOCE gradients in various reference frames and their accuracies. Advances in Geosciences, No. 1, pp 33-38.

Rummel, R, van Gelderen, M, Koop, R, Schrama, E, Sansò, F, Brovelli, M, Miggliaccio, F, Sacerdote, F (1993). Spherical harmonic analysis of satellite gradiometry. Publ. Geodesy, New Series, No. 39. Netherlands Geodetic Commission, Delft.

Rummel, R. (1997). Spherical Spectral Properties of the Earth's Gravitational Potential and ist First and Second Derivatives. In: Geodetic Boundary Value Problems in View of the One Centimeter Geoid. Eds.: F. Sansò, R. Rummel, Lecture Notes Earth Sciences, 65, 359 - 404, Springer.

# Comparison of some robust parameter estimation techniques for outlier analysis applied to simulated GOCE mission data

B. Kargoll
Institute of Theoretical Geodesy (ITG),
University of Bonn, Nussallee 17, D-53115 Bonn, Germany

**Abstract.** Until now, methods of gravity field determination using satellite data have virtually excluded robust estimators despite the potentially disastrous effect of outliers. This paper presents computationally-feasible algorithms for Huber's M-estimator (a classic robust estimator) as well as for the class of R-estimators which have not traditionally been considered for geodetic applications. It is shown that the computational time required for the proposed algorithms is comparable to the direct method of least squares. Furthermore, a study with simulated GOCE satellite gradiometry data demonstrates that the robust gravity field solution remains almost unaffected by additive outliers. In addition, using robustly-estimated residuals proves to be more efficient at detecting outliers than using residuals resulting from least squares estimation. Finally, the non-parametric R-estimators make less assumptions about the measurement errors and produce similar results to Huber's M-estimator, making that class a viable robust alternative.

**Keywords.** GOCE, satellite gradiometry, robust parameter estimation, rank norm, outlier diagnostics

## 1 Introduction

The GOCE satellite mission, currently under preparation for its launch in 2006, will provide tens of millions of satellite gradiometry (SGG) observations used to recover the detailed structures of the Earth's gravity field. The global gravity field will be resolved up to degree and order 250 resulting in more than 60,000 estimated spherical harmonic coefficents. To tackle this huge adjustment problem the method of least squares has been accepted as the traditional estimator to be used. In addition to its computational feasibility, the least squares estimator produces unbiased estimates with minimal variances under certain assumptions, and comprises a unified theory including consistent variance-covariance information of the estimates, tests of model adequacy, parameter tests, and outlier tests.

However, the usual assumptions underlying the least squares procedure such as normality and outlier-freeness of the observations cannot be accepted at face value (even though the term *outlier* usually implies the association with a visibly extreme observation, it is, in the context of the current paper, understood to be any observation stemming from some *contaminating distribution* different from the main distribution of the errors, and could thus comprise the case of a blunder). Outliers, even in low numbers, are known for distorting parameter and accuracy estimates, rendering them potentially useless. The common practice of detecting outliers from least squares residuals may be fruitless due to the tendency of the estimated trend to be drawn towards extreme data values and the masking effect of multiple outliers (see for instance Rousseeuw and Leroy, 2003, p. 226 and p. 234, respectively). The traditional approach to dealing with outliers was pioneered by the work of Baarda (1968) and is based on an iterative elimination of the most prominent outlier candidates ("data snooping"). With each iteration the partially cleaned data set is readjusted, which, however, may become computationally very expensive when a huge amount of observations and a large number of potential outliers are involved.

As a remedy to this problem, the current paper investigates in robust estimators, which potentially become less affected by extreme values and are able to highlight outliers in the residuals by unmasking them. Despite such benefits, these methods apparently have been ignored in the field of gravity field determination from satellite data, mainly because of the high computational effort usually associated with robust techniques. The main purpose of this paper is to show that in fact estimates with good robustness properties may be obtained in a computational time comparable to the least squares approach. Huber's classical $M$-estimator using metrically Winsorized residuals (Huber, 1981, p. 179ff.) and $R$-estimators based on rank statistics, also mentioned by Huber (1981, p. 163) and worked out for the linear model by Hettmansperger (1984) and Hettmansperger and McKean (1998), are considered, because they are found to be particularly suitable for the huge adjustment problem encountered in the GOCE mission. As it appears that $R$-estimators have not been used for geodetic applications, a more detailed review of the underlying theoretical ideas is given in Sect. 2.

Two essential factors, computation time and quality of estimates, were compared between a robust estimation approach and a *direct* approach to gravity field determination from SGG data (see for instance Pail and Plank, 2002; Schuh, 1996). The outlier study investigating these factors and the corresponding results are presented in Sect. 3.

It should be noted that the intention of this paper is not to discredit the least squares approach, but to promote the use of robust estimates in a complementary way as a reference to check least squares residuals for abnormal behaviour possibly caused by undetected, masked outliers.

## 2 Theory and implementation

The proposed robust estimators are derived in the context of a linear model of the form

$$\mathbf{y} = \mathbf{X}\boldsymbol{\beta} + \mathbf{e} \tag{1}$$

where $\mathbf{y}$ is an $n \times 1$ random vector of observations, $\mathbf{e}$ is an $n \times 1$ random vector of unobservable disturbances, $\mathbf{X}$ is an $n \times u$ matrix of fixed coefficients, and $\boldsymbol{\beta}$ is a $u \times 1$ vector of unknown parameters. For now let the disturbances $e_1, e_2, \ldots, e_n$ be independently, identically, but not necessarily normally distributed random variables. The parameters are estimated by minimizing some function of the residuals. In robust estimation "less rapidly increasing functions" (Huber, 1981, p. 162) are used instead of the quadratic function in least squares. Therefore, differences between $M$- and $R$-estimators are mainly determined by the choice of this function with all arising *technical* implications. However, it is seen in the following subsections that *in practice* the implementation of these two classes of estimators is very similar.

### 2.1 Huber's $M$-estimator

$M$-estimates are obtained by minimizing

$$Q(\mathbf{e}) = \sum_{i=1}^{n} \rho(e_i) \tag{2}$$

where $\rho$ is a symmetric function of the residuals $e_i = y_i - x_i^T\boldsymbol{\beta}$ with $x_i^T$ the $i-th$ row of $\mathbf{X}$. The gradient of $Q$ with respect to $\boldsymbol{\beta}$ is given by

$$\nabla Q(\mathbf{e}) = -\mathbf{X}^T\psi(\mathbf{e}). \tag{3}$$

$M$-estimates are not automatically scale invariant so that $\mathbf{e}$ must be divided by some scale factor $\sigma$. Consequently, (3) becomes

$$\nabla Q(\mathbf{e}, \sigma) = -\mathbf{X}^T\psi(\mathbf{e}/\sigma). \tag{4}$$

For Huber's $M$-estimator $\psi(\cdot)$ is defined as

$$\psi(e_i/\sigma) := \begin{cases} e_i/\sigma & : \text{ if } |e_i| < c\sigma \\ c \,\mathrm{sign}(e_i) & : \text{ if } |e_i| \geq c\sigma \end{cases} \tag{5}$$

where $c$ is a constant whose value depends on the percentage of outliers in the observations (see Huber, 1981, p. 87). The values for 1% and 5% are $c \approx 2.0$ and $c \approx 1.4$, respectively. Note that the least squares estimates are obtained by setting $\psi(e_i) := e_i$ (cf. for instance Koch, 1999, Chaps. 3 and 4, as a reference of the method of least squares in linear models). With (5) the *Winsorized residuals* are defined as

$$e_i^* := \psi(e_i/\sigma)\,\sigma. \tag{6}$$

The estimates $\tilde{\boldsymbol{\beta}}$ are obtained from the Newton step

$$\tilde{\boldsymbol{\beta}}^{(k+1)} = \tilde{\boldsymbol{\beta}}^{(k)} + (\mathbf{X}^T\mathbf{X})^{-1}\mathbf{X}^T\mathbf{e}^* \tag{7}$$

where the relaxation factor was set equal to 1. The scale factor may be computed from the residuals after each step by $\tilde{\sigma} = 1.483\,\mathrm{med}_i\{|e_i^*|\}$.

### 2.2 Construction of $R$-estimators

In this section only the most basic results from the work of Jaeckel (1972), Hettmansperger (1984), and Hettmansperger and McKean (1998) are stated in order to develop a practical and intuitive approach to the material. Attention is focussed on demonstrating the similarities and differences of $R$-estimators to Huber's $M$-estimator and the method of least squares.

The goal of the commonly-used least squares estimator is to minimize the variance of the residuals $\mathbf{y} - \mathbf{X}\boldsymbol{\beta}$. Since few extreme values may cause an unreasonable increase in variance, Jaeckel (1972) discussed an alternative measure of variability which is less sensitive to outliers. This measure of dispersion $D(\cdot)$ is defined as

$$D(\mathbf{z}) = \sum_{i=1}^{n} a(i) z_{(i)} \tag{8}$$

where $a(1) \leq \ldots \leq a(n)$ is a nonconstant set of *scores* satisfying $\sum_{i=1}^{n} a(i) = 0$, $\mathbf{z}$ is any realvalued $n \times 1$ vector, and $z_{(i)}$ are the ordered, non-decreasing elements of $\mathbf{z}$. Now, values $1, \ldots, n$, which are denoted as the *ranks* $R(\cdot)$ of the elements of $\mathbf{z}$ are assigned to the ordered $z_{(1)}, \ldots, z_{(n)}$. Then (8) is equivalent to

$$D(\mathbf{z}) = \sum_{i=1}^{n} a(R(z_{(i)}))z_{(i)}. \quad (9)$$

Substituting the arbitrary $\mathbf{z}$ by the residuals from (1) yields a rank estimate of $\boldsymbol{\beta}$ that minimizes

$$D(\mathbf{e}) = \sum_{i=1}^{n} a(R(e_i))e_i \quad (10)$$

where $e_i = y_i - x_i^T \boldsymbol{\beta}$ with $x_i^T$ being the $i - th$ row of $\mathbf{X}$. It is remarkable that (10) could have been defined in terms of the $R$ pseudo norm

$$||v||_R = \sum_{i=1}^{n} a(R(e_i))\, e_i \quad (11)$$

which substitutes 'one half' of the residuals from the $L_2$ norm

$$||v||_{L_2} = \sum_{i=1}^{n} e_i\, e_i \quad (12)$$

by rank-transformed and re-weighted residuals

$$e_i^{**} := a(R(e_i)). \quad (13)$$

Jaeckel (1972) shows that $D(\mathbf{e})$ is a nonnegative, continuous, and convex function of $\boldsymbol{\beta}$ which attains its minimum with bounded $\boldsymbol{\beta}$ if $\mathbf{X}$ has full rank, which are also familiar properties of the Gauss-Markov model of full rank. However, in contrast to the latter, the measure of dispersion $D(\mathbf{e})$ is not a quadratic function of the residuals, but rather linear, potentially reducing the effect of outliers on the estimates (Hettmansperger, 1984, p. 233).

The partial derivatives of $D(\mathbf{y} - \mathbf{X}\boldsymbol{\beta})$ with respect to $\boldsymbol{\beta}$ exist almost everywhere with gradient

$$\nabla D(\mathbf{y} - \mathbf{X}\boldsymbol{\beta}) = -\mathbf{S}(\mathbf{y} - \mathbf{X}\boldsymbol{\beta}) \quad (14)$$

where

$$\mathbf{S}(\mathbf{y} - \mathbf{X}\boldsymbol{\beta}) = \mathbf{X}^T \mathbf{a}(R(\mathbf{y} - \mathbf{X}\boldsymbol{\beta})). \quad (15)$$

Setting the gradient to approximately zero yields the $R$ normal equations

$$\mathbf{X}^T \mathbf{a}(R(\mathbf{y} - \mathbf{X}\boldsymbol{\beta})) \approx 0, \quad (16)$$

which are solved by $\tilde{\boldsymbol{\beta}}$. Note that the gradient need not necessarily attain exactly zero due to its non-continuous range. The normal equations (16) cannot be solved directly, and furthermore, the dispersion function, being essentially a decreasing step function, is not ideally suited for gradient methods. Therefore, Hettmansperger and McKean (1998, p. 184) suggest constructing a Newton-type algorithm in analogy to Huber's $M$-estimator, based on linearization of $\mathbf{S}(\mathbf{y} - \mathbf{X}\boldsymbol{\beta})$. Let $\boldsymbol{\beta}_0$ denote the true parameters and the scale factor $\tau_\varphi = 1/\int f^2(x)dx$ where $f$ is the density function of the disturbances. Then, according to Hettmansperger and McKean (1998, p. 162) the linearization is given by

$$\mathbf{S}(\mathbf{y} - \mathbf{X}\boldsymbol{\beta}) = \mathbf{S}(\mathbf{y} - \mathbf{X}\boldsymbol{\beta}_0) - \frac{1}{\tau_\varphi}\mathbf{X}^T\mathbf{X}(\boldsymbol{\beta} - \boldsymbol{\beta}_0) + o_p(1)$$

from which a quadratic function $Q(\cdot)$ approximating the dispersion function $D(\cdot)$ is constructed by integration as

$$\begin{aligned} Q(\mathbf{y} - \mathbf{X}\boldsymbol{\beta}) &= \frac{1}{2\tau_\varphi}(\boldsymbol{\beta} - \boldsymbol{\beta}_0)^T\mathbf{X}^T\mathbf{X}(\boldsymbol{\beta} - \boldsymbol{\beta}_0) \\ &\quad -(\boldsymbol{\beta} - \boldsymbol{\beta}_0)^T\mathbf{S}(\mathbf{y} - \mathbf{X}\boldsymbol{\beta}_0) \\ &\quad +D(\mathbf{y} - \mathbf{X}\boldsymbol{\beta}_0). \end{aligned}$$

Jaeckel (1972; Lemma 1) proved that $Q(\cdot)$ is indeed a good local approximation. The estimate

$$\tilde{\boldsymbol{\beta}} = \boldsymbol{\beta}_0 + \tau_\varphi(\mathbf{X}^T\mathbf{X})^{-1}\mathbf{X}^T\mathbf{a}(R(\mathbf{y} - \mathbf{X}\boldsymbol{\beta}_0)) \quad (17)$$

minimizes the quadratic approximation $Q(\cdot)$ and solves the linearization.

Turning attention to the practical implementation of this rank-based estimator, (17) would be computed as the first Newton step by substituting initial parameter values for the true parameters. Consequently, (17) becomes

$$\tilde{\boldsymbol{\beta}}^{(k+1)} = \tilde{\boldsymbol{\beta}}^{(k)} + \tilde{\tau}_\varphi^{(k)}(\mathbf{X}^T\mathbf{X})^{-1}\mathbf{X}^T\mathbf{e}^{**(k)}. \quad (18)$$

The scale factor $\tilde{\tau}_\varphi^{(k)}$ can be estimated from the residuals of each preceeding step. A computationally feasible estimator is derived in Hettmansperger and McKean (1998, pp. 181-184), which was used for the following simulations. The final estimate of $\tau_\varphi$ is used for the computation of the covariance matrix of the estimated parameters $\Sigma\{\tilde{\boldsymbol{\beta}}\} = \tilde{\tau}_\varphi^2(\mathbf{X}^T\mathbf{X})^{-1}$.

The optimal method of generating the scores $a(i) = \varphi(i/(n+1))$ through some score function $\varphi(u)$ depends on the distribution of the disturbances, which in the linear model (1) was not necessarily assumed to be Gaussian. In the current paper the score functions $\varphi(u) = \sqrt{12}(u - 1/2)$ and $\varphi(u) = \text{sign}(u - 1/2)$ generating the Wilcoxon and the sign pseudo-norm are considered for the following reasons. Hettmansperger and McKean (1998) show in theory that the Wilcoxon pseudo-norm exploits the information contained in the observations almost as efficiently as least squares when the errors are Gaussian and outlier-free, and have good robustness when outlying observations are present. The sign pseudo norm is used, because it is equivalent to the well-known $L_1$ norm. However, while the estimates generated by the sign pseudo-norm can be easily computed by means of a block algorithm (see Sect. 2.3), the evaluation of the $L_1$ norm, usually based on a simplex-type algorithm, would not be possible for the given problem as it requires that $\mathbf{X}$ be stored as one piece in the working memory.

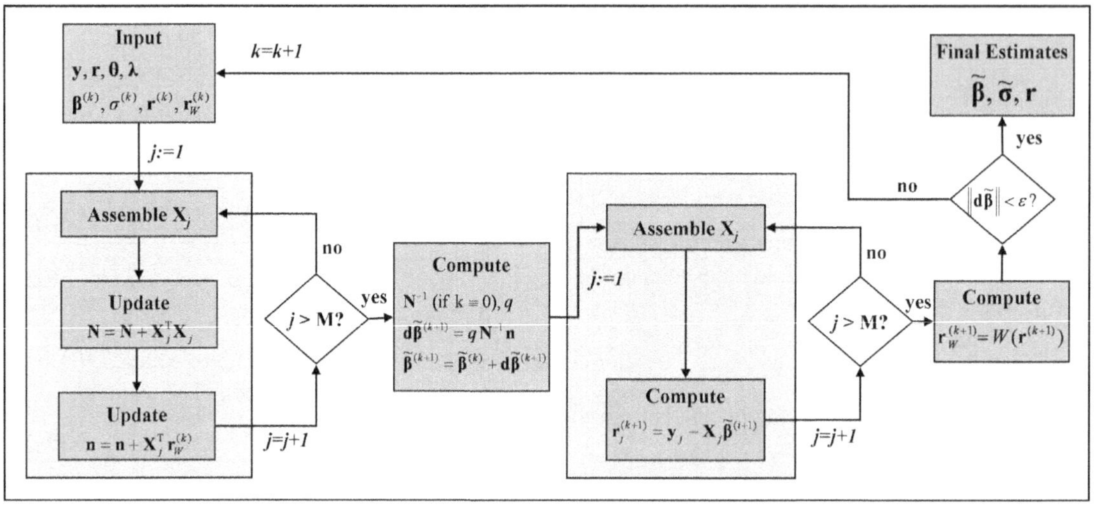


**Fig. 1** Flowchart of the proposed robust gravity field solver. Input are SGG observations $y_i$ of the three diagonal tensor components, spherical positions $(r_i, \theta_i, \lambda_i)$, and start values of the parameters $\boldsymbol{\beta}$, the scale factor $\sigma$, and the (modified) residuals generated by $\boldsymbol{\beta}_0$. $\mathbf{X}_j$ denotes the $j$-th block ($j = 1, \ldots, M$) of the design matrix, which is used to compute the normal equation matrix $\mathbf{N}$ and the right hand side $\mathbf{n}$ of the system $\mathbf{N}\boldsymbol{\beta} = \mathbf{n}$. $\mathbf{r}_W$ is the vector of residuals modified by a weight function $W(\cdot)$ according to (6) or (13). If the parameter update exceeds $\varepsilon$, the next Newton step $(k+1)$ is performed, otherwise the algorithm terminates with final estimates. The final residuals $\boldsymbol{r}$ are studentized and used for outlier detection.

## 2.3 Implementation of the robust gravity field solver

The functional model for the adjustment of GOCE SGG observations is obtained by taking the second derivatives of the mathematical representation of the Earth's gravitational potential

$$V(r,\theta,\lambda) = \frac{GM}{r}\left\{1 + \sum_{\ell=2}^{\ell_{max}} \sum_{m=0}^{\ell} \left(\frac{a}{r}\right)^{\ell} \times \bar{P}_{\ell m}(\cos\theta)\left(\bar{C}_{\ell m}\cos m\lambda + \bar{S}_{\ell m}\sin m\lambda\right)\right\}$$

where $G$ denotes the geocentric gravitational constant, $M$ the Earth's mass, and $a$ the semi-major axis. The triple $(r, \theta, \lambda)$ represents the spherical coordinates of a point, $\ell$ and $m$ the degree and order, $\ell_{max}$ the maximum degree of the expansion, and $\bar{P}_{\ell m}$ the fully normalized associated Legendre functions. The model as linear functions of the desired parameters $\bar{C}_{\ell m}$, $\bar{S}_{\ell m}$ (the fully normalized harmonic coefficients) can be expressed as the linear model (1).

As the design matrix $\mathbf{X}$ eventually contains millions of observations, it becomes far too large to be processed in one piece. Consequently, it is not possible to compute the Newton steps (see Fig. 1 for the processing flow chart) as in (7) or (17). Therefore, $\mathbf{X}$ is assembled in parts, with each part $\mathbf{X}_\mathrm{j}$ ($j = 1, \ldots, M$) containing 750 rows. The normal equation matrix is computed within the first Newton step by $\mathbf{N} := \mathbf{X}^T\mathbf{X} = \sum_{j=1}^{M} \mathbf{X}_\mathrm{j}{}^T\mathbf{X}_\mathrm{j}$, and after inversion, $\mathbf{N}^{-1}$ is stored for the following steps. This procedure works for, say $\ell_{max} = 90$, as shown in the performed simulation study (see Sect. 3). To reach the GOCE mission goal of a resolution of $\ell_{max} = 250$, the algorithm must be modified, as $\mathbf{N}$ would also exceed the working memory (see *Outlook*).

The residuals generated by the parameter start values $\boldsymbol{\beta}_0$ are modified to $\mathbf{r}_W$ according to (6) or (13). Using $\mathbf{n} := \mathbf{X}^T\mathbf{r}_W = \sum_{j=1}^{M} \mathbf{X}_\mathrm{j}{}^T\mathbf{r}_{W,j}$ and setting the relaxation parameter $q := 1$ for Huber's $M$-estimator, or $q := \tilde{\tau}_\varphi$, respectively, for one of the $R$-estimators, the parameter update $\boldsymbol{d\tilde{\beta}} = q\mathbf{N}^{-1}\mathbf{n}$ is computed. The new residuals are obtained piece by piece by assembling $\mathbf{X}$ block-wise again. In case the parameter update exceeds a prescribed $\varepsilon$ the next Newton step is performed with updated start values. Otherwise, the current estimates are saved as the final solution. The residuals are then used for subsequent outlier analysis.

## 3 Simulation Study

The goal of the current simulation study is, firstly, to investigate the convergence rate of the robust estimators, because the computation time of each Newton step corresponds approximately to the entire computation time of the least squares estimation (about 4 hours on a single 3.06 GHz processor with 1 GB RAM). Secondly, the quality of the robust gravity field solutions is compared to the least squares solution. Finally, the success of outlier detection is evaluated by analyzing studentized residuals.

### 3.1 The Test Data

The observation functionals were computed on a sun-synchronous orbit of 23 days with an initial altitude of 250 km and an inclination of 96.6°. They were sampled equally at a rate of 4 s yielding altogether 496,430 positions and 1,489,290 values of the three main diagonal elements of the gradient tensor. The trend, computed from EGM96 coefficients up to degree and order 90, was superimposed by white noise with standard deviation $\sigma = 1$ mE. From these observations two data sets containing additional outliers (generated as realizations of uniformly distributed random variables between 3 and 50 mE) were deduced. The first set contains 1% additive outliers and the second 5%, distributed randomly over the zz-component. The observations of the xx- and the yy-component were not altered. Since the true outlier distribution will be unknown, a rather pessimistic measurement scenario was simulated by selecting the outlier ratio and bandwidth as specified above.

### 3.2 Results

The absolute differences between the estimated and the reference solution (least squares parameters estimated from the observations containing no outliers) were computed. Fig. 2 shows that the mean and median values over all orders of the same degree are ten times larger for the least squares estimates (LSE) than for the three robust solutions. Huber's $M$-estimates (HME) are equal to the Wilcoxon norm estimates (WNE), while the sign norm estimates (SNE) performs slightly worse than the WME.

Table 1 summarizes the geoid height differences between the reference solution and the estimated solutions. The differences between the reference solution and the geoid heights computed from the true EGM96 model are also given. It is seen that the robust solutions are already as close as a few millimeters to the reference values after one Newton iteration, and they converge fully after the second iteration. The least squares geoid heights differ significantly from the reference heights, and they explode for the second data set containing 5% outliers (lower part of Table 1). Using the worsened least squares estimates as start values, the robust estimates converge only after five iterations. In comparison to LSE the robustly estimated geoid heights were considerably less affected by the outliers.

Table 2 gives a summary of the performance of outlier detection by means of internally studentized residuals, defined as $rs_i = r_i/(\tilde{\sigma}\sqrt{1-h_i})$ for the least squares residuals and $rs_i = r_i/(\tilde{\sigma}\sqrt{1-Kh_i})$ for the residuals of the $R$-estimates. The latter are modified by $K$, as $R$-estimators do not project $\mathbf{y}$ orthogonally into the column space of $\mathbf{X}$ (see Hettmansperger and McKean, 1998, p. 197ff.). $h_i$ denotes the $i$-th diagonal element of the orthogonal projector $\mathbf{X}(\mathbf{X}^T\mathbf{X})^{-1}\mathbf{X}^T$. All robust estimators detect 99.8% of the outliers (indicating a very high test power), especially 'unmasking' all outliers larger than 5 mE. The undetected, small outliers are located in the range of the measurement noise, which makes them hard to identify. By contrast, the least squares studentized residuals are much smaller (because the estimated standard deviation is inflated by the outliers), leaving even a high number of large outliers undetected.

Approximately 4% of the "good" observations were wrongly marked as outliers when using one of the robust estimators. This number could be improved only at cost of the test power, i.e. a larger number of outliers would remain undiscovered. For example, if one decreased the error number from 4% down to 0.2% by raising the threshold, one would diminish the performance rate by approximately 1%. However, for the robust estimators the choice of the actual test power is not a crucial point, because none of the observations are deleated, but their residuals downweighted.

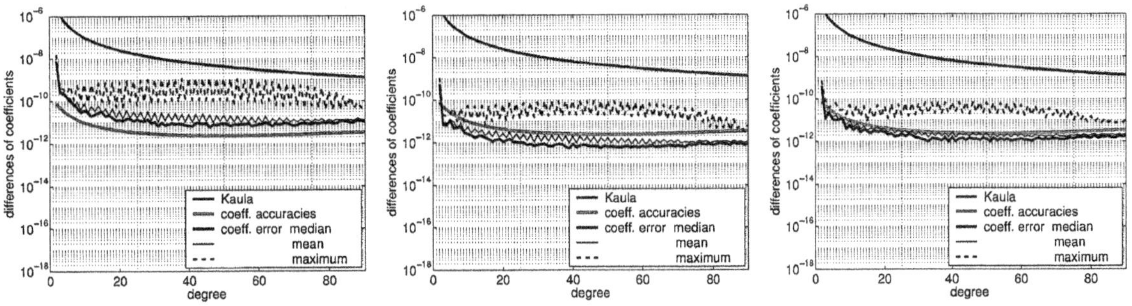


**Fig. 2** Median, mean and maximum values of absolute differences between estimated (with 5% outliers) and reference (without outliers) coefficients over all orders of the same degree. From left to right: Least squares (LSE), Wilcoxon norm (WNE), and sign norm estimates (SNE) (the figure for Huber's $M$-estimates is the same as for the WNE and was omitted).

**Table 1.** Reconstruction of second-level information on a $1^\circ \times 1^\circ$ grid: differences between the geoid heights in meters computed from the true model (EGM96) and estimated solutions ("Reference": Least squares solution without outliers; "Least Squares": Least squares with outliers); upper part: 1% outliers, lower part: 5% outliers.

| | global | | $-80^\circ \leq \phi \leq 80^\circ$ | | local | | | |
|---|---|---|---|---|---|---|---|---|
| **1% outliers** | min | max | min | max | min | max | mean | $\sigma$ |
| Reference | -0.012 | +0.013 | -0.012 | +0.013 | -0.008 | +0.008 | -0.000 | 0.003 |
| Least Squares | -0.055 | +0.082 | -0.055 | +0.077 | -0.031 | +0.033 | -0.000 | 0.009 |
| **1.iteration** | | | | | | | | |
| Wilcoxon | -0.015 | +0.015 | -0.015 | +0.015 | -0.009 | +0.011 | -0.000 | 0.003 |
| Sign | -0.017 | +0.017 | -0.017 | +0.017 | -0.014 | +0.013 | -0.000 | 0.004 |
| Huber | -0.016 | +0.015 | -0.016 | +0.015 | -0.010 | +0.011 | -0.000 | 0.003 |
| **2.iteration** | | | | | | | | |
| Wilcoxon | -0.015 | +0.015 | -0.015 | +0.015 | -0.010 | +0.010 | -0.000 | 0.003 |
| Sign | -0.017 | +0.018 | -0.017 | +0.018 | -0.014 | +0.015 | -0.000 | 0.004 |
| Huber | -0.014 | +0.014 | -0.014 | +0.014 | -0.009 | +0.010 | -0.000 | 0.003 |
| **5% outliers** | min | max | min | max | min | max | min | $\sigma$ |
| Reference | -0.012 | +0.013 | -0.012 | +0.013 | -0.008 | +0.008 | -0.000 | 0.003 |
| Least Squares | -0.236 | +0.333 | -0.236 | +0.333 | -0.111 | +0.120 | -0.002 | 0.035 |
| **1.iteration** | | | | | | | | |
| Wilcoxon | -0.027 | +0.022 | -0.027 | +0.022 | -0.009 | +0.011 | -0.000 | 0.003 |
| Sign | -0.038 | +0.041 | -0.038 | +0.041 | -0.027 | +0.024 | -0.000 | 0.007 |
| Huber | -0.060 | +0.080 | -0.060 | +0.080 | -0.032 | +0.034 | -0.001 | 0.010 |
| **5.iteration** | | | | | | | | |
| Wilcoxon | -0.026 | +0.028 | -0.026 | +0.028 | -0.015 | +0.013 | -0.000 | 0.004 |
| Sign | -0.023 | +0.022 | -0.023 | +0.021 | -0.018 | +0.015 | -0.000 | 0.005 |
| Huber | -0.026 | +0.027 | -0.026 | +0.027 | -0.015 | +0.013 | -0.000 | 0.004 |

**Table 2.** Outlier detection for the second data set containing 5% outliers. perf: percentage of correctly identified outliers (second column), error: percentage of observations wrongly marked as outliers (third column); columns 4-7: numbers of unidentified outliers of given sizes. The last row contains the distribution of the implemented outliers.

| | perf | error | 3-4 mE | 4-5 mE | 5-6 mE | >6mE |
|---|---|---|---|---|---|---|
| Least Squares | 89.3% | 0.0% | 519 | 523 | 493 | 1,123 |
| Huber | 99.8% | 4.1% | 51 | 8 | 0 | 0 |
| Wilcoxon | 99.8% | 4.3% | 50 | 5 | 0 | 0 |
| Sign | 99.8% | 4.3% | 52 | 6 | 0 | 0 |
| Total outliers | - | - | 519 | 525 | 499 | 23,279 |

## 4 Discussion and Outlook

It was seen that Huber's $M$-estimator and the $R$-estimators remain robust when a small percentage of the observations are contaminated by additive outliers. Robustly estimated spherical harmonics coefficients and derived second-level products such as geoid heights became far less affected than with the least squares approach. Consequently, the "unmasked" outliers were detected almost perfectly by comparing the robustly estimated studentized residuals with a threshold value. Huber's $M$-estimator and the Wilcoxon norm estimator produced very similar results, and were slightly superior to the less efficient sign norm estimates (which is equivalent to the $L_1$ norm). All robust estimates converged after a few iterations when heavily distorted least squares start values were used. When valid *a priori* information was used, they converged within one step, i.e. the computational effort was essentially the same as for computing the least squares solution, making robust procedures feasible.

For the future it is intended to robustly estimate models up to degree and order 250 (the planned resolution of the GOCE mission). Since the proposed algorithm allows the block-wise processing of the normal equations, this can be easily accomplished by implementation on a parallel computer system.

## Acknowledgments

The support by BMBF through the GOCE-GRAND project within the "Geotechnologien-Programm" is gratefully acknowledged.

## References

Baarda, W. (1968). *A testing procedure for use in geodetic networks.* Publications on Geodesy by the Netherlands Geodetic Commission 2(5).

Hettmansperger, T.P. (1984). *Statistical inference based on ranks.* John Wiley, New York.

Hettmansperger, T.P. and J.W. McKean (1998). *Robust nonparametric statistical methods.* Arnold, London.

Huber, P.J. (1981). *Robust Statistics.* John Wiley, New York.

Jaeckel, L.A. (1972). *Estimating regression coefficients by minimizing the dispersion of the residuals.* The Annals of Mathematical Statistics 43(5):1449-1458.

Koch, K.-R. (1999). *Parameter estimation and hypothesis testing in linear models.* Springer, Berlin/Heidelberg.

Pail, R. and G. Plank (2002). *Assessment of three numerical solution strategies for gravity field recovery from GOCE satellite gradiometry implemented on a parallel platform.* Journal of Geodesy 76:462-474.

Rousseeuw, P.J. and A.M. Leroy (2003). *Robust regression and outlier detection.* John Wiley, New York.

Schuh, W.D. (1996). *Tailored numerical solution strategies for the global determination of the Earth's gravity field.* Mitteilungen der geodaetischen Institute der TU Graz 81.

# Comparison of outlier detection algorithms for GOCE gravity gradients

J. Bouman (1), M. Kern (2), R. Koop (1), R. Pail (2), R. Haagmans (3), T. Preimesberger (4)
(1) SRON National Institute for Space Research, Sorbonnelaan 2, 3584 CA Utrecht, The Netherlands
(2) Institute of Navigation and Satellite Geodesy, TU Graz, Steyrergasse 30, 8010 Graz, Austria
(3) Science and Applications Department, ESA/ESTEC, Keplerlaan 1, 2200 AG Noordwijk, The Netherlands
(4) Austrian Academy of Sciences, Space Research Institute, Schmiedlstrasse 6, 8042 Graz, Austria

**Abstract.** GOCE will be the first satellite ever to measure the second derivatives of the Earth's gravitational potential in space. It will be possible to derive a high accuracy and high resolution model of the gravitational field if systematic errors and/or outliers have been removed from the data. It is necessary to detect outliers in the data pre-processing because undetected outliers may lead to erroneous results when the data are further processed, for example in the recovery of a gravity field model. Outliers in the GOCE gravity gradients will be searched for and detected in the gravity field analysis pre-processing step.

In this paper, a number of algorithms are discussed that detect outliers in the diagonal gravity gradients. One of them combines wavelets with either a statistical method or filtered gradients with an identification rate of about 90% or more. Another high performing algorithm is the combination of three methods, that is, the tracelessness condition (a physical property of the diagonal gradients), comparison with model or filtered gradients, and along-track interpolation of gradient anomalies. Using two sets of simulated gravity gradients, the algorithms are compared in terms of their identification rate and number of falsly detected outliers. In addition, it is shown that the quality of the gravity field solution is very much affected by outliers. Undetected outliers can degrade the gravity field solution by up to twenty times as compared with a solution without outliers.

**Keywords.** Outliers · GOCE mission · Gradiometry · Statistical tests

## 1 Introduction

The main goal of the GOCE mission (expected launch in August 2006) is to provide unique models of the Earth's gravity field and of its equipotential surface, as represented by the geoid, on a global scale with an accuracy of 1 cm at 100 km resolution (ESA 1999). To this end, GOCE will be equipped with a GPS receiver for high-low satellite-to-satellite tracking (SST-hl), and with a gradiometer for observation of the gravity gradients (GG). Only the latter will be considered in this paper.

Even after the in-flight calibration, the observations will be contaminated with stochastic and systematic errors. Systematic errors include GG scale factor errors and biases (Cesare 2002) which are corrected for in the external calibration step (see e.g. Arabelos and Tscherning 1998; Bouman et al. 2004). In addition, outliers in the GOCE gravity gradients are searched for and detected in the gravity field analysis (GFA) pre-processing step. If some remain undetected, they may seriously affect the accuracy of the final GOCE gravity field model (Kern et al. 2004).

A vast number of outlier detection methods exists and a selection is discussed in this paper. Outliers are searched for in simulated gravity gradient time series contaminated with noise and outliers. The performance of the methods is evaluated with respect to the detection rate and the type I error (rejecting correct data). Section 2 details several outlier detection methods and section 3 shows numerical examples. One alternative for the time-wise methods studied here, is presented by (Tscherning 1991).

## 2 Outlier detection methods

### 2.1 TSOFT outlier detection algorithm

The TSOFT outlier detection algorithm is based on the algorithm presented by Vauterin and Van Camp (2004). The idea is to low-pass filter the gravity gradient time series which tends to reduce the outliers. If for certain points the difference between the filtered and unfiltered time series is above a certain threshold, $thr_1$, then these points are likely to be outliers. The effect of the low-pass filter is not only a reduction of the size of the outliers, but also a redistribution of the power over neighbouring points. In addition, an outlier that is close to another outlier may mask that outlier. Therefore, iteration is necessary, replacing the detected outliers in the original time series by the filtered values. This updated time series is then again low-pass filtered and again outliers can be searched for, etc. The final low-pass filtered time series, with most outliers removed, is tested against the original time series with outliers. If the difference is above a threshold, $thr_2 \leq thr_1$, then an outlier is detected. The second threshold can be smaller than the first one since the final low-pass filtered time series is affected less by outliers than the filtered times series in the iteration. The thresholds themselves have to be determined using simulated data or by trial and error.

## 2.2 Wavelet outlier detection algorithm

Single and higher level Haar wavelet can be used to detect outliers. The wavelet outlier detection is explicitely explained in the paper by Kern et al. (2004). It searches for discontinuities in the signal. A single level outlier detection algorithm may be formulated as follows.

[1] Compute detailed and smoothed wavelet coefficients using the forward wavelet transformation ($k = 1, \ldots, m/2 - 1$)

$$\begin{aligned} s_{1,k} &= (x_{2k} + x_{2k+1})/\sqrt{2}, \\ d_{1,k} &= (x_{2k} - x_{2k+1})/\sqrt{2}, \end{aligned} \tag{1}$$

with data vector $x_i = \{x_1 \ldots x_m\}$.

[2] Threshold the detailed coefficients by setting a threshold $td$

$$d'_{1,k} = \begin{cases} d_{1,k} & \text{for } |d_{1,k}| < td, \\ 0 & \text{otherwise.} \end{cases} \tag{2}$$

[3] Reconstruct the signal

$$\begin{aligned} x^w_{2k} &= (s_{1,k} + d'_{1,k})/\sqrt{2}, \\ x^w_{2k+1} &= (s_{1,k} - d'_{1,k})/\sqrt{2}. \end{aligned} \tag{3}$$

[4] Compute the residual signal using the reconstructed signal $x^w$

$$r_i = x_i - x^w_i, i = 1, \ldots, m. \tag{4}$$

[5] Apply a pattern recognition program on the residuals to identify the position of the outliers.

## 2.3 Dixon test

The Dixon test is a hypothesis test that uses the ratio of differences between a possible outlier and its nearest or next-nearest neighbour (data excess) to the range. The data have to be normally distributed. Let the data vector $x_i = \{x_1 \ldots x_m\}$ be sorted in ascending order and let $t_{1-\alpha}(m-1)$ be the Student's t-distribution, which depends on the significance level $\alpha \in (0, 1)$ and the number of observations $m$. An $\alpha$-outlier region for upper outliers is defined as

$$\text{out}(\alpha, m) := \{r_j > t_{1-\alpha}(m-1); j = 1, 2\} \tag{5}$$

where the test functions are given as (Barnett and Lewis 1994)

$$r_1 = \frac{x_m - x_{m-1}}{x_m - x_2}, \; r_2 = \frac{x_m - x_{m-1}}{x_m - x_3}. \tag{6}$$

If one of the test functions $r_j$ exceeds the critical value $t_{1-\alpha}(m-1)$, the largest observation is an outlier or the distribution is not normal. The test statistic $r_1$ does not contain the smallest value $x_1$ to avoid masking effects (large denominator). Similarly, the test statistic $r_2$ can be used to avoid masking effects from the smallest two values ($x_1$ and $x_2$). Because the Dixon test is a very robust method, one may expect that it also works in the presence of data gaps. This was, however, not investigated in this paper.

## 2.4 Tracelessness condition

The sum of the diagonal gravity gradients, also called Laplace's equation or tracelessness condition, has to be zero, which is a physical property of the gravity gradients. The gravitational potential is a harmonic function outside the attracting masses (Heiskanen and Moritz 1967). However, before external calibration, the gradients suffer from systematic errors of which a bias and scale factor errors are the most important. The effect of a scale factor error is the largest at a frequency of 0 Hz or the mean value. Of course, also the bias is manifest at this frequency. Therefore, the following condition equation is considered

$$E\{V_{xx} + V_{yy} + V_{zz}\} - \text{ median } = 0 \tag{7}$$

where $E$ is the expectation operator and the median is the median of the point-wise Laplace's equation of the time series considered. Note that in the GOCE case the rotational terms, caused by rotation of the satellite, have been removed as good as possible from the gradients using the differential accelerations (Cesare 2002). The w-test is used, i.e., if the tracelessness condition is violated then an outlier is detected. The trace is weighted with the a priori error of the GOCE gravity gradients neglecting along-track error correlations (Bouman 2004). The major drawback of the tracelessness condition is that the outlier detection is ambiguous, i.e., one cannot discriminate between outliers on $V_{xx}, V_{yy}$ and $V_{zz}$. The advantage is that it is a sensitive method. The smaller the signal-to-noise-ratio (SNR), the easier it is to detect outliers. In fact, the SNR can not be smaller since the sum of the diagonal gradients should be zero.

### 2.5 Gravity gradient anomalies

The GOCE gravity gradients could be confronted with gravity gradients generated from a global Earth gravity field model. If the difference between the two, weighted with the sum of the respective errors, is above a certain threshold, then an outlier is detected (the median of differences is subtracted to account for the GOCE gravity gradient bias and scale factor error). More details on this w-test are given in Bouman (2004). The advantage of this method with gravity gradient anomalies is that all gradients can be tested separately and point-wise. A disadvantage is that the accuracy of the model gradients may be low compared to the GOCE gradients, which makes this a less sensitive method. In addition, the two sets of gradients have different measurement bandwiths, which may restrict the test.

Alternatively, one could consider the along-track interpolation of GG anomalies. An anomaly at time $t = t_i$ is compared with the predicted anomaly at time $t = t_i$, where the prediction is based on anomalies at $t = t_j, j \neq i$. Many interpolation methods could be used; splines are used here since they are simple and fast and the interpolation errors are small (Bouman 2004). The advantage of the along-track interpolation is that the gradients can be tested separately, but several points are combined which may lead to masking effects, that is, outliers close to each other can not be well separated.

### 2.6 Combination solutions

One possibility to improve the results is to combine two or more of the methods described above. The combination of the TSOFT algorithm and the wavelet method are considered, while the latter is also combined with the Dixon test. A data point is flagged as an outlier if it is detected in both methods.

Also considered is the combination of the tracelessness condition, gravity gradient differences and the interpolation of these differences. Since the tracelessness condition is a sensitive but ambiguous method, the other two methods are used to confirm a detected outlier by the tracelessness condition. In other words, if an outlier on $V_{xx}$, $V_{yy}$ and $V_{zz}$ is detected by the tracelessness condition and this outlier is confirmed by either the gradient differences or the interpolation, a data point is flagged as an outlier.

### 2.7 Other methods

Besides the above methods, the so-called m-estimator (Mayer 2003) and two thresholding methods (Kern et al. 2004) were studied. The m-estimator has a high outlier detection rate, but it also has a large type I error, that is, up to one out of five observations is erroneously detected as an outlier in the tests made. This method will therefore not be discussed. The threshold methods detect an outlier if the difference between the value at a given data point and the mean or median is above a certain threshold. The threshold may be linked to the standard deviation of the data or to some fixed value. These methods, however, suffer from a relatively large type I error while the number of detected outliers is relatively small. Therefore, these methods are not considered here.

## 3 Numerical results

Two data sets with different characteristics were studied. One is a small data set with a length of 1 day which contains various types of outliers. The second data set has a length of 59 days and contains single and bulk outliers. This set allows for gravity field analysis.

### 3.1 Small data set with various outliers

The first data set used in this study consists of the diagonal gravity gradients $V_{xx}$, $V_{yy}$ and $V_{zz}$ which were simulated using EGM96 (Lemoine et al. 1998) for a 1 day orbit with a sampling rate of 1 s. Simulated, correlated noise was added to the signals, the data statistics are given in Table 1. The model gradients which are required for some methods were generated using OSU91A (Rapp et al. 1991). A first test was done that used the noisy gradients without any outliers (case 1a). The type I error is (close to) zero as one would hope. However, this is not to be expected for the tracelessness condition, model gradients and spline interpolation. These all use the w-test with a critical value of $k = 2$, which would mean that approximately 4.6% of the observations is rejected although they are correct. For the tracelessness condition and spline interpolation, however, the type I error is 0%. This may be due to the error correlation between the simulated gradients which is neglected. The model gradients have a larger type I error but this is dominated by the model error, that is, the difference between EGM96 and OSU91A. The type I error is probably larger than expected because we have used a simple scale unit matrix as error covariance matrix.

**Table 1** Noise, outlier and anomaly properties, values in [mE].

| Small set | (86,351 pts) | $V_{xx}$ | $V_{yy}$ | $V_{zz}$ |
|---|---|---|---|---|
| noise | mean | 1443.7 | -805.2 | 2248.9 |
| | rms | 2.2 | 4.4 | 5.7 |
| outliers | mean | 0.5 | 0.3 | 0.0 |
| | rms | 58.9 | 27.9 | 52.7 |
| | number | 3,891 | 420 | 1,988 |
| anomalies | mean | 0.0 | 1.5 | -1.5 |
| | rms | 36.4 | 35.3 | 58.9 |
| **Large set** | **(5,097,835 pts)** | $V_{xx}$ | $V_{yy}$ | $V_{zz}$ |
| noise | mean | 0.0 | 0.0 | 0.0 |
| | rms | 10.1 | 2.7 | 10.0 |
| outliers | mean | 0.0 | 0.0 | 0.0 |
| | rms | 78.5 | 78.5 | 78.5 |
| | number | 83,153 | 83,153 | 83,153 |
| anomalies | mean | 0.0 | -0.4 | 0.4 |
| | rms | 37.2 | 35.3 | 60.0 |

**Table 2** Type I error for case 1a (no outliers, small data set); TS – TSOFT algorithm, W – wavelet detection, TSW – TSOFT + wavelet, TR – tracelessness condition, M – model gradients, SP – spline interpolation, TMS – TR + M + SP.

| Method | $V_{xx}$ | $V_{yy}$ | $V_{zz}$ |
|---|---|---|---|
| TS | 0.1% | 0.1% | 0.1% |
| W | 0.0% | 0% | 0.2% |
| TSW | 0.2% | 0.1% | 0.4% |
| TR | 0.0% | 0.0% | 0.0% |
| M | 6.2% | 6.0% | 5.9% |
| SP | 0% | 0% | 0% |
| TMS | 0% | 0% | 0% |

A second test was done with outliers on all three gradients with an absolute size varying between 0.07 E and 0.1 E (case 1b). The outliers on $V_{xx}$ are randomly distributed single outliers. The outliers on the $V_{yy}$ component are an offset of 0.5 E during one minute ($t = 20 - 79$ s) and a bulk of outliers during six minutes ($t = 5000 - 50359$ s). Finally, the outliers on the $V_{zz}$ component consist of randomly distributed 'twangs', i.e., outliers at $t = t$ that are followed by an other outlier of opposite sign and of the same size at $t = t + 1$. In total there are 3891, 420 and 1988 outliers on the $V_{xx}$, $V_{yy}$ and $V_{zz}$ component respectively, see also Table 1.

Outlier detection results are shown in Table 3. Rows 1-3 show the TSOFT algorithm (TS), wavelets (W) and their combination (TSW) respectively. Rows 4 and 5 show the tracelessness condition (TR) and the model gradients (M). Row 6, TS1, shows the results for filtered gradients, that is, the GG with outliers were filtered and these were used as model gradients to compute GG anomalies. A 2nd order low-pass Butterworth filter with a cut-off frequency of 0.2 Hz was used. The acronym TS1 is used because this method is similar to one step of the TSOFT algorithm, although the low-pass filter is different. The last three rows show respectively the results for spline interpolation (SP), the combination of the tracelessness condition, model gradients, splines (TMS) and the combination of the tracelessness condition, filtered gradients and splines (TFS).

Most of the single outliers $V_{xx}$ are detected by all methods (we consider percentages above 95% to be good). The type I error is large for filtered gradients and splines because outliers are spread out over several data points in both methods (we consider type I errors above 5% to be too large). The 'twangs' on $V_{zz}$ are also detected by most methods. The wavelet detection algorithm has problems because it takes the difference between two consecutive data points. The detection rate for model gradients is somewhat low due to the larger GOCE GG error and the larger difference between the 'true' GG and the model GG (OSU91A and EGM96). The offset on $V_{yy}$ causes problems for many methods. The filter methods, TS and TS1, as well as spline interpolation fail to detect the offset. An offset tends to cancel in these methods. The wavelet outlier detection rate is low as it not only fails to identify the offset, but it also has problems with the bulk outliers. In general, the combination of different outlier detection methods gives a higher detection rate and a low type I error. The detection of the offset remains a problem, also in the combination solutions, with the exception of the combination with model gradients which detects most of the $V_{yy}$ outliers.

## 3.2 Large data set and gravity field retrieval

The second data set used in this study also consists of the diagonal gravity gradients $V_{xx}$, $V_{yy}$ and $V_{zz}$ which were simulated using OSU91A for a 59 day orbit with a sampling rate of 1 s (over half a million data points). Simulated, correlated noise was added to the signals. (A test with no outliers gives roughly the same percentage of type I errors as for the small data set except for the tracelessness condition which has a type I error of 4.7%. The simulated GG errors for the large data set show no correlation between the different GG.) In addition, outliers were added to all three gradients with an absolute size varying between 0.05 E and 1.8051 E (case 2). The outliers were randomly distributed single outliers as well as bulk outliers, see Table 1 for data statistics.

Besides the detection methods discussed before, the combination of wavelets and the Dixon test was

**Table 3** Detected outliers for case 1b (outliers on all three diagonal gradients, small data set); TS – TSOFT algorithm, W – wavelet detection, TSW – TSOFT + wavelet, TR – tracelessness condition, M – model gradients, TS1 – filtered gradients, SP – spline interpolation, TMS – TR + M + SP, TFS – TR + TS1 + SP.

| | $V_{xx}$ | | $V_{yy}$ | | $V_{zz}$ | |
|---|---|---|---|---|---|---|
| Method | correct | type I | correct | type I | correct | type I |
| TS | 99.9% | 1.3% | 83.6% | 0.1% | 100% | 0.1% |
| W | 95.8% | 0.1% | 46.9% | 0.0% | 62.8% | 0.7% |
| TSW | 100% | 3.0% | 87.6% | 0.1% | 100% | 0.9% |
| TR | 99.9% | 2.6% | 99.8% | 6.7% | 99.9% | 4.8% |
| M | 93.6% | 5.9% | 92.6% | 6.0% | 76.9% | 5.8% |
| TS1 | 99.8% | 23.5% | 84.5% | 0.0% | 100% | 2.2% |
| SP | 98.9% | 11.6% | 77.6% | 0.0% | 99.5% | 2.0% |
| TMS | 99.8% | 0.5% | 98.6% | 0.4% | 99.7% | 0.4% |
| TFS | 99.9% | 0.7% | 87.1% | 0% | 99.9% | 0.1% |

**Table 4** Detected outliers for case 2 (outliers on all three diagonal gradients, large data set); TS – TSOFT algorithm, W – wavelet detection, TSW – TSOFT + wavelet, WD – wavelet + Dixon test, WDQL – wavelet + Dixon + QL-GFA, TR – tracelessness condition, M – model gradients, TS1 – filtered gradients, SP – spline interpolation, TMS – TR + M + SP, TFS – TR + TS1 + SP.

| | $V_{xx}$ | | $V_{yy}$ | | $V_{zz}$ | |
|---|---|---|---|---|---|---|
| Method | correct | type I | correct | type I | correct | type I |
| TS | 99.0% | 4.4% | 99.0% | 4.3% | 99.0% | 4.4% |
| W | 86.4% | 0.1% | 86.7% | 0.1% | 85.6% | 0.3% |
| TSW | 99.6% | 6.5% | 99.6% | 6.3% | 99.6% | 6.7% |
| WD | 97.0% | 0.0% | 98.4% | 0.0% | 96.2% | 0.1% |
| WDQL | 100.0% | 0.1% | 100.0% | 0.0% | 100.0% | 0.1% |
| TR | 99.8% | 7.7% | 99.9% | 7.7% | 99.9% | 7.7% |
| M | 96.8% | 5.9% | 97.5% | 6.2% | 92.3% | 5.9% |
| TS1 | 99.0% | 5.6% | 99.7% | 10.5% | 98.9% | 5.6% |
| SP | 86.8% | 2.2% | 98.6% | 4.7% | 86.8% | 2.2% |
| TMS | 97.7% | 0.6% | 99.8% | 0.8% | 94.6% | 0.6% |
| TFS | 99.6% | 0.4% | 99.9% | 0.8% | 99.6% | 0.4% |

added (WD). The cleaned GG from this method are used in Quick-Look Gravity Field Analysis (QL-GFA) to compute a global gravity field model (Pail and Preimesberger 2003). This gravity field model is used to compute GG along the orbit. Then, an additional search in the residuals between these GG and the observed GG is done in an iterative manner (WDQL).

As with the small data set, wavelets perform worse for bulk outliers (row W of Table 4). The detection rates for $V_{xx}$ and $V_{zz}$ are lower than for $V_{yy}$ using splines because of the higher noise level of the former two. The combination algorithms detect almost all outliers while the type I error is small. One exception is the combination of TSOFT and wavelets, which has a large type I error. The best results are obtained by the wavelet-Dixon method in QL-GFA. Almost all outliers are detected, while the type I error is very small. The major advantage of the GFA is that the data are not only 'compared' along track or point-wise, which is the drawback of the other pre-processing algorithms considered here, but that the least-squares adjustment combines all observations.

The effect of undetected outliers can be disastrous, see Table 5. Although only 1.6% of the observations contain outliers, the gravity field solution has a very low accuracy if the outliers are not removed. Shown are the gravity anomaly differences between OSU91A and a QL-GFA solution up to degree and order 250. The error standard deviation is twenty times as high compared to a solution where no outliers are present (126.0 mGal and 6.7 mGal respectively). The wavelet - Dixon combination gives a considerable improvement compared to no outlier detection, see Table 5 and Fig. 1. It does not, however, detect all bulk outliers, which cause a visible track (Fig. 1). The best combination solution that uses pre-processing only (TFS) gives a small gravity anomaly difference (9.8 mGal). Finally, the wavelet - Dixon combination in the GFA (WDQL) gives a gravity field anomaly error which is almost at the level of no outliers (7.0 mGal), see again Table 5.

**Table 5** Gravity anomaly error for case 2 (large data set); difference between OSU91A and QL-GFA up to degree and order 250, excluding polar caps of 10°.

| Method | error rms [mGal] |
|---|---|
| no outliers | 6.7 |
| all outliers | 126.0 |
| WD | 24.9 |
| TFS | 9.8 |
| WDQL | 7.0 |

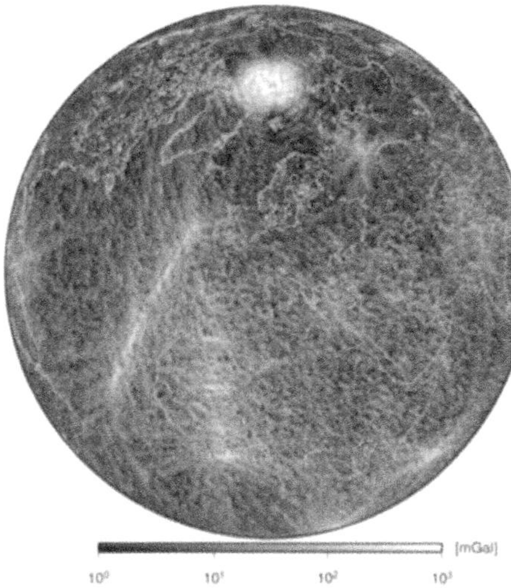


**Fig. 1** Gravity anomaly differences OSU91A – QL-GFA (log scale), pre-processing outlier detection using WD.

## 4 Conclusions and outlook

Several outlier detection algorithms have been compared. While single outliers and 'twangs' can be detected at high rates, bulk outliers and offsets cause problems in almost all methods. Generally, a combination of methods improves the detection results. In particular, the combinations WD and TFS yield highest detection rates while having a small type I error. After applying the pre-processing methods, the overall rms of the gravity field solution can be reduced by an additional search inside the gravity field solver.

The results for the model gradients may improve as more accurate gravity models become available. Especially at the time GOCE flies, preliminary GOCE gravity field models could be used. Future studies may include orbit errors, various GG error scenarios, uncertainties in the GG a priori error model, etc.

## 5 Acknowledgements

Financial support for the second author came from an external ESA fellowship. Financial support for the fourth author came from the ASA contract ASAP-CO-008/03. This study was performed in the framework of the ESA project GOCE High-level Processing Facility (No. 18308/04/NL/MM). All this is gratefully acknowledged. Also the remarks by two anonymous referees are acknowledged.

## References

Arabelos D, Tscherning CC (1998) Calibration of satellite gradiometer data aided by ground gravity data. Journal of Geodesy. 72: 617-625

Barnett V, Lewis T (1994) Outliers in statistical data. 3rd edition, John Wiley and Sons, Chichester, New York

Bouman J (2004) Quick-look outlier detection for GOCE gravity gradients. Paper presented at the IAG Porto meeting (GGSM 2004)

Bouman J, Koop R, Tscherning CC, Visser P (2004) Calibration of GOCE SGG data using high-low SST, terrestrial gravity data and global gravity field models. Journal of Geodesy, 78, DOI 10.1007/s00190-004-0382-5

Cesare S (2002) Performance requirements and budgets for the gradiometric mission. Issue 2 GO-TN-AI-0027, Preliminary Design Review, Alenia, Turin

ESA (1999) Gravity field and steady-state ocean circulation mission. Reports for mission selection. The four candidate Earth explorer core missions. ESA SP-1233(1). European Space Agency, Noordwijk

Heiskanen W, Moritz H (1967) Physical Geodesy. W.H. Freeman and Company, San Francisco

Kern M, Preimesberger T, Allesch M, Pail R, Bouman J, Koop R (2004) Outlier detection algorithms and their performance in GOCE gravity field processing. Accepted for publication in Journal of Geodesy

Lemoine F, Kenyon S, Factor J, Trimmer R, Pavlis N, Chinn D, Cox C, Klosko S, Luthcke S, Torrence M, Wang Y, Williamson R, Pavlis E, Rapp R, Olson T (1998) The development of the joint NASA GSFC and the National Imagery and Mapping Agency (NIMA) geopotential model EGM96. TP 1998-206861, NASA Goddard Space Flight Center, Greenbelt

Mayer C (2003) Wavelet modelling of ionospheric currents and induced magnetic fields from satellite data. PhD thesis. Geomathematics Group, Department of Mathematics, University of Kaiserslautern, Germany

Pail R, Preimesberger T (2003) GOCE quick-look gravity solution: application of the semianalytic approach in the case of data gaps and non-repeat orbits. Studia geophysica et geodaetica. 47:435-453

Rapp R, Wang Y, Pavlis N (1991) The Ohio State 1991 geopotential and sea surface topography harmonic coefficient models. Rep 410, Department of Geodetic Science and Surveying, The Ohio State University, Columbus

Tscherning CC (1991) The use of optimal estimation for gross-error detection in databases of spatially correlated data. Bulletin d'Information, no. 68, 79-89, BGI

Vauterin P, van Camp M (2004) TSoft Manual. Version 2.0.14, Royal Observatory of Belgium, Bruxelles, Belgium. Available at http://www.astro.oma.be/SEISMO/TSOFT/tsoft.html

# Using the EIGEN-GRACE02S Gravity Field to Investigate Defectiveness of Marine Gravity Data

Wolfgang Bosch
Deutsches Geodätisches Forschungsinstitut (DGFI), Marstallplatz 8, 80539 München, Germany

**Abstract.** The latest GRACE-only gravity field model, EIGEN-GRACE02S, is used to investigate if high resolution marine gravity data exhibit medium or long wavelength errors. The "trackiness" of the GRACE gravity field requires a rigorous smoothing down to about 4° block mean values or a harmonic degree at or below 40. The anomaly differences between GRACE and the marine gravity is found to be inconspicuous, transforms however to large scale pattern of geoid height differences with ± 0.4m amplitude, located differently for the marine data sets investigated.

**Keywords.** Marine Gravity, GRACE, gravity field models, satellite altimetry

## 1 Introduction

The GRACE gravity mission already led to dramatic improvements of the Earth gravity field (Tapley et al., 2004). In spite of these improvements the very high frequency information of the gravity field will be further based on surface gravity. The most recent data sets of marine gravity data provide a grid resolution of 2'x2'. Even a smoothed resolution of 6'x6' would correspond to a spherical harmonic degree of 1800 - far beyond the maximum degree considered for GRACE or GOCE gravity field models. Thus the high frequency information of the marine gravity field will be further based on satellite altimetry.

The recovery of marine gravity anomalies from altimetry is based on inversion of Stokes or Vening-Meinesz formulas (Heiskanen & Moritz, 1967) and as such rather sensitive to errors: the process implies differentiation enhancing high frequency errors. Moreover, the inversion is realized either by Least Squares Collocation (LSC) or Fast Fourier Transform (FFT), applied to a sequence of local areas in order to avoid - in case of LSC - the inversion of huge matrices or to ensure the validity of planar approximation in case of FFT. A remove-restore technique is finally applied to localize the inversion: residual quantities of a state-of-the-art gravity field are first subtracted. After inversion the gravity anomalies of the reference field are restored.

These two approaches justify to investigate if the marine gravity data has been recovered without systematic errors. The present investigation focuses on possible medium and long wavelength errors. The GRACE-only gravity fields are independent of the marine gravity data and allow for the first time to study these errors in detail.

Is it possible to localized and quantify medium and long-wavelength errors of the marine gravity data? First, some information about the latest GRACE-only gravity field is given. It is shown that looking for medium and long-wavelength errors it is necessary to rigorously smooth both, the altimetric gravity data and the GRACE-only gravity field. The differences between GRACE and marine gravity is then performed in the spatial domain (in terms of block mean values) and the spectral domain (in terms of spherical harmonics). The gravity anomaly differences appear to be inconspicuous, imply however, remarkable large scale pattern for the corresponding geoid height differences with ± 0.4m amplitude - located differently for the two latest marine data sets investigated.

## 2 GRACE-only gravity field

The EIGEN-GRACE02S is the latest published static gravity field model (Reigber et al., 2004) de-

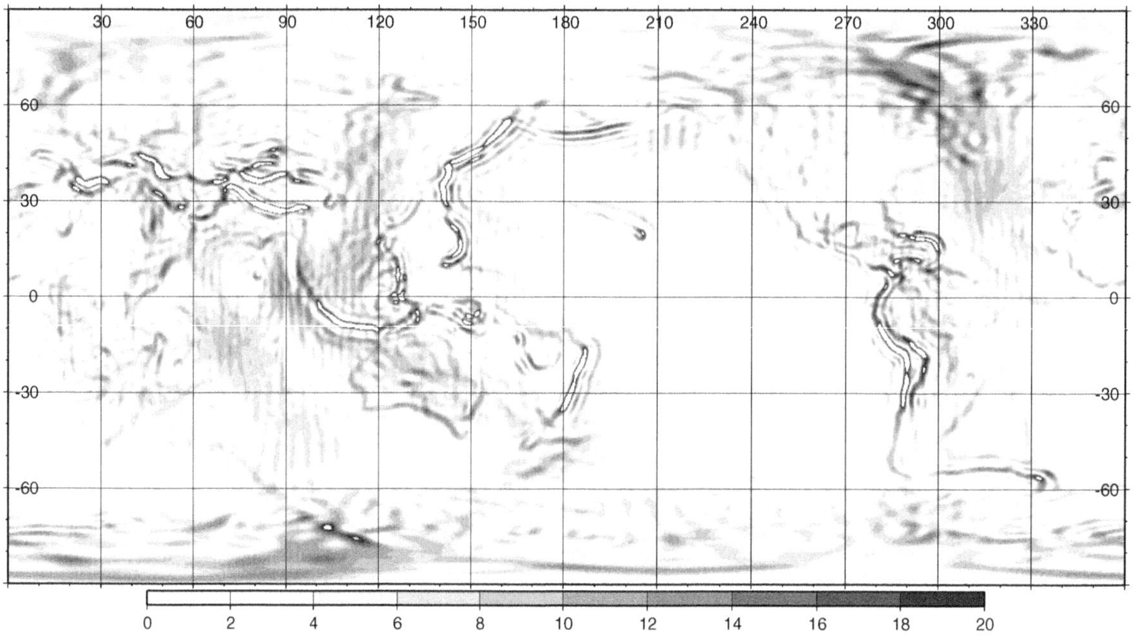


**Fig. 1** *The latest GRACE-only gravity field still exhibit a pronounced „trackiness". The figure shows artificially illuminated gradients of EIGEN-GRACE02S. Gradients of EIGEN-GRACE01S and GGM01S look very similar. The „trackiness" (most properly caused by inconsistent treatment of neighbouring tracks) implies that only smoothed versions of these gravity fields can be used to validate the long wavelength of the marine gravity data.*

rived solely from 110 days GRACE tracking data, in particular the very precise inter-satellite observations. This "satellite-only" gravity field is independent from any surface gravity data and therefore well suited for the objective of the present paper: investigate possible long wavelength errors of the marine gravity data. EIGEN-GRACE02S is represented by spherical harmonic coefficients complete up to degree and order 150.

Previous gravity field models from the CHAMP and GRACE missions exhibit a remarkable "trackiness", patterns significantly correlated with the ground tracks of the satellites. The reason for these patterns is not completely understood. The inter-satellite observation are primarily sensitive to the along track component of the gravity gradient, $V_{xx}$. It seems that $V_{xx}$ of neighbouring tracks is not consistently treated. Unfortunately, these meridional patterns are still existing in the EIGEN-GRACE02S gravity field as illustrated in Figure 1. This implies that a considerable smoothing has to be applied before any comparison with the marine surface data is performed.

## 3 Marine Gravity Data

We investigate three marine data sets, all of them derived from satellite altimetry. The two most recent data sets provide a spatial resolution of 2'x2' (there are even versions with a 1'x1' resolution) and are obtained by analysing the so called geodetic phases of the Geosat and ERS-1 altimeter missions, both with an extremely dense ground track spacing. These two data sets are:

- Version 11.2 of Sandwell & Smith marine gravity data (in the following abbreviated by SSv11.2), which includes retracked ERS-1 data, reducing the signal to noise ratio by 40% (Sandwell & Smith 1997).
- KMS2002 marine gravity provided by Andersen & Knudsen (2005) and documented in Andersen & Knudsen (1997)

KMS2002 was derived by inverting Stokes formula by means of least squares collocation (LSC). The SSv11.2 data set is obtained by inverting Vening-Meinesz formula by Fast Fourier Transform (FFT). Because, up to now, the EGM96 gravity field was taken as state-of-the-art and also used here as a refer-

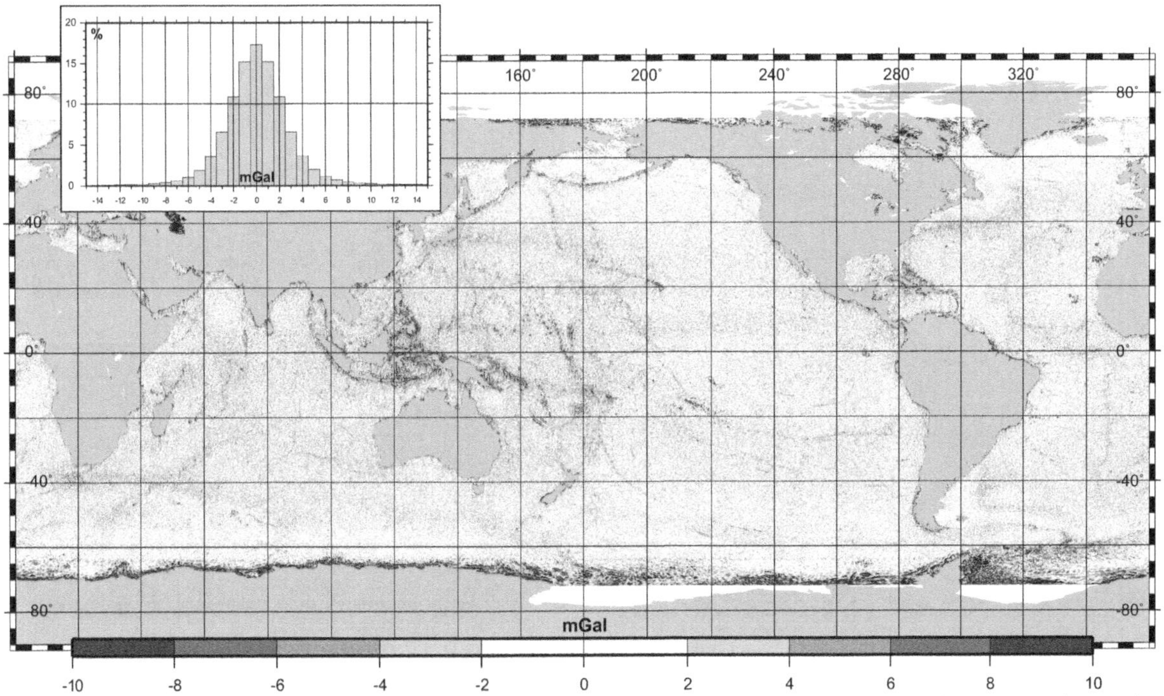


**Fig. 2** *Gravity anomaly differences [mGal] between SSv11.2 and KMS2002 on the basis of 6'x6' block mean values within the common latitude range ± 72°. There is a mean difference of 0.06 ± 4,58 mGal which reduces to 0.03 ± 2.67 mGal if only 3% of the anomaly differences exceeding a magnitude of 10 mGal are excluded. Note, the gray scale is choosen to visualize the location of pattern with larger differences - only a colour scale allows to identify neighbouring differences with opposing sign.*

ence, the present investigations were also applied to

- the 30'x30' marine gravity data of NIMA used nearly a decade ago for the development of the EGM96 gravity field (Lemoine et al., 1998).

The NIMA data set is essentially based on the geodetic mission of Geosat with very little ERS-1 data at higher latitudes.

## 4 Comparisons in the Space Domain

In a first step, the two high resolution data sets, SSv11.2 and KMS2002, were averaged to 6'x6' block mean values and both limited to the common latitude range ±72°. Because both data sets are augmented by non-marine gravity, a land ocean mask was then applied to both data sets in order to limit any further comparison to marine data only. The differences between both data sets are shown in Figure 2. On the basis of the 6'x6' block mean values a mean of 0.06 ± 4,58 mGal (0.03 ± 2.67 mGal after removal of outliers) was found - a rather excellent agreement. Figure 2 indicates that larger differences between SSv11.2 and KMS2002 are found in the Southern Ocean, below the latitude range, covered by TOPEX/Poseidon and/or in sea ice covered areas. The two data sets also differ at the coast, in areas of strong currents and rough bathymetry.

Gravity anomaly differences between EIGEN-GRACE02S and the two marine data sets were then performed by computing and subtracting 6'x6' block mean values from the complete GRACE model. These differences were then further smoothed to larger block mean values. Averaging to block sizes of 2.5° or less were unsatisfactory: with this block size the meridional pattern of the GRACE gravity field model was reproduced and dominated the geographical distribution of differences. Reasonable results were obtained only after smoothing to 3° or 4° block mean values (see Figure 3). Even a smoothing to 3° blocks still exhibits the trackiness artifact in the southern ocean. Only when smoothed to 4° block mean values (corresponding to a harmonic series up to degree and order 45) the meridional stripes disappeared.

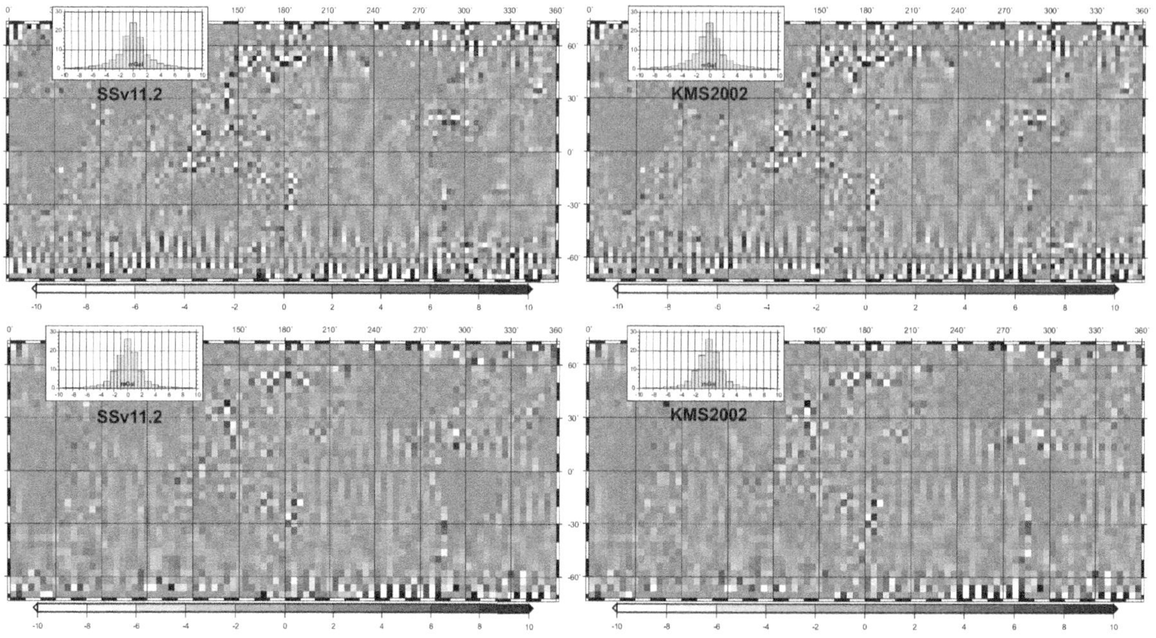


**Fig. 3** *Spatially smoothed gravity anomaly differences [mGal] between EIGEN-GRACE02S and (left column) SSv11.2 marine gravity and KMS2002 marine gravity (right column). The top panels show 3° block mean values which seem to reproduce the EIGEN-GRACE02S meridional pattern in the southern ocean - above all around 55°S. Only when smoothed to 4° block mean values most of these meridional pattern of the anomaly differences disappear.*

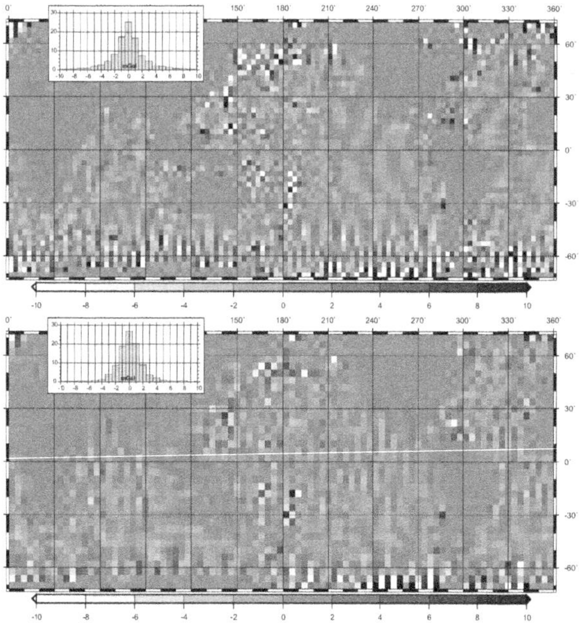

**Fig. 4** *Spatially smoothed anomaly differences [mGal] between EIGEN-GRACE02S and the NIMA marine anomaliy data set. The top panel again shows 3° block mean values, the bottom panel 4° block means values. The distribution of differences is similar as in Figure 3.*

In general, large anomaly differences between the GRACE gravity model and the marine data sets are found in the southern ocean - above all around 55°S, but also in the North Pacific, an area north-west of New Zealand, and in coastal areas of the North Atlantic and the Caribbean Sea. The distribution of differences is very similar for SSv11.2 and KMS2202 - a consequence of the good overall agreement between the two marine data sets.

The same type of comparison has also been performed with the NIMA data set. The results are shown in Figure 4 with similar geographical distribution of smoothed anomaly differences.

## 5 Comparison in the Spectral Domain

For the sphere it is more appropriate to perform comparisons in terms of spherical harmonics - because the block mean values at higher latitude do not account for the convergence of the meridians. To perform a comparison in the spectral domain the anomaly differences, now in terms of 30'x30' block mean values were expanded into spherical harmonic series up to degree and order 120. This expansion was performed by integration, taking advantage of the orthogonality relation of spherical harmonics. Non-ocean blocks

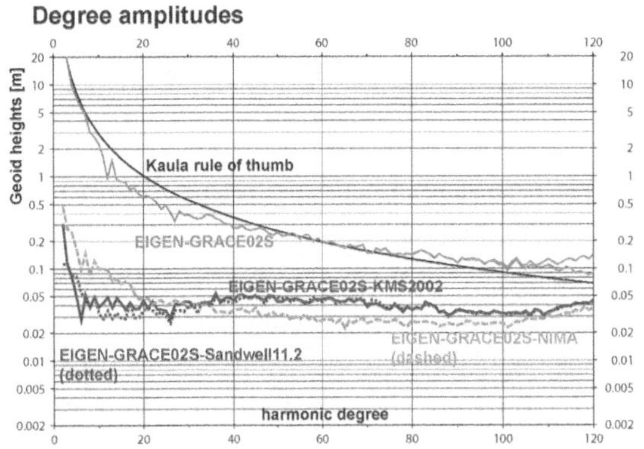


**Fig. 5** *Degree amplitudes, expressed as geoid heights [m] for EIGEN-GRACE02S and the differences to the marine data sets, SSv11.2 (dotted), KMS2002 (solid), and NIMA (dashed).*

and blocks at latitudes above ±72° were ignored (or – equivalent – taken with zero mean values). Degree 120 is far beyond the harmonic degree relevant for the investigation of medium and long wavelength errors. Gibbs effect and frequency folding may appear, but will not affect the low degree harmonics to be considered here.

Figure 5 shows the degree amplitudes (in terms of geoid height differences) of the spectra of the anomaly differences between EIGEN-GRACE02S and SSv11.2, KMS2002, and the NIMA data sets. It is remarkable that the difference spectra for the high resolution marine data sets show a nearly constant power between degree 5 and 120. It is therefore difficult to decide for the most appropriate truncation of the difference spectra. The comparison in the spatial domain have shown that above degree 40 the spectral power is already dominated by the trackiness of the EIGEN-GRACE02S.

The harmonic series of anomaly differences were truncated at degree 40 and transformed back to both, anomaly differences as well as geoid height differences. Figure 6 shows the results: limited to degree 40 the spectra of anomaly differences shows an inconspicuous distribution for both, the SSv11.2 and the KMS2002 data sets. Anomaly differences above 2 mGal are found below 60°S, in the Indonesian Sea, at the coast of South America, and at the center of the Gulf stream. The SSv11.2 data set exhibits slightly higher anomaly differences in the southern ocean.

However, if the difference spectra for both data sets is used to compute geoid height differences, large scale pattern appear with amplitudes up to about ± 0.4 meter – located differently for SSv11.2

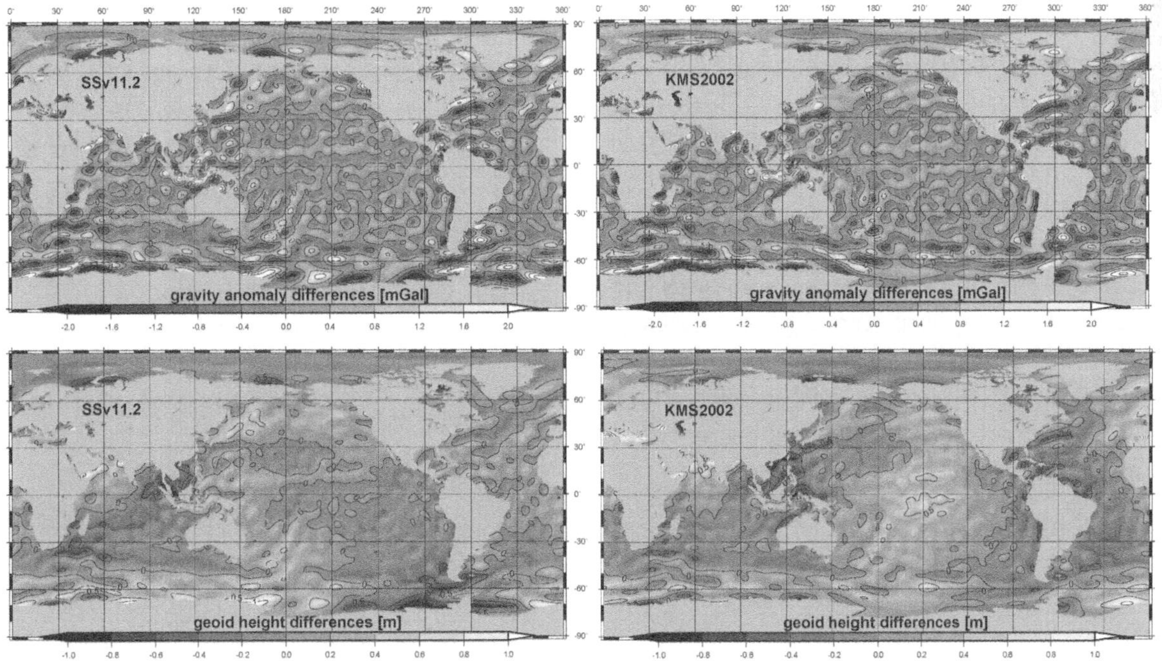


**Fig. 6** *Gravity anomaly differences [mGal], top panels, and geoid height differences [m], bottom panels, for the difference spectra EIGEN-GRACE02S minus SSv11.2 (left hand) and minus KMS2002 (right hand), truncated at degree and order 40. Note, the different large scale geoid height differences in the central Pacific and Atlantic.*

(lower left plot of Figure 6) and KMS2002 (lower right plot). While the geoid differences in the Indian Ocean look rather similar, significant differences between SSv11.2 and KMS2002 exist in the tropical and subtropical areas of the Pacific and the Atlantic Ocean.

Of course, both data sets have been treated in just the same way. Therefore the differences must be attributed to differences in the marine gravity data. However, the magnitude of the differences may depend on the harmonic degree used to truncate the spectral representation of the anomaly differences.

Up to now there is no explanation for the different large scale pattern that were identified.

## Conclusion

In summary the investigation showed

- The „trackiness" of EIGEN-GRACE02S requires a rigorous smoothing and limits any validation of marine gravity data to wavelength at or below degree 40 (500 km). This also limits the use of the GRACE gravity field models as reference for the remove-restore technique.
- A few areas with anomaly differences of up to ± 10 mGal have been identified. The differences for both, the SSv11.2 and the KMS2002 data sets look rather similar. However, these differences imply an effect on geoid heights with up to ± 0.4 m amplitudes and remarkable large scale pattern - located at different areas.

In general, the trackiness of EIGEN-GRACE02S makes it difficult to identify the right cut-off frequency for the spectral representation of anomaly differences. Improved and smoother GRACE gravity field models are required to obtain a more reliable estimate of possible long wavelength errors of the marine gravity data.

## References

Andersen, O. B. and P. Knudsen (1997): Global marine gravity field from the ERS-1 and Geosat geodetic mission altimetry. J. Geophys. Res., Vol. 103 , No. C4 , p. 8129

Andersen O. B., P. Knudsen and R. Trimmer (2005): Improving high resolution altimetric gravity field mapping (KMS2002 global marine gravity field). In: Proceedings of the 26th IAG general assembly 'A window on the future of geodesy', Sapporo, Japan, 2003, IAG symposia, Vol. 128, Springer, 326-331

Heiskanen W.A. and H. Moritz (1967): Physical Geodesy. W.H. Freeman and Company, San Francisco.

Lemoine F.G., et al. (1998): The Developement of the Joint NASA GSFC and the National Imagery and Mapping Agency (NIMA Geopotential Model EGM96. NASA/TP-1998-206861, NASA Goddard Space Flight Center, Greenbelt, Maryland

Reigber Ch., R. Schmidt, F. Flechtner, R. König, U. Meyer, K.-H. Neumayer, P. Schwintzer, and S.Y. Zhu (2004): An Earth gravity field model complete to degree and order 150 from GRACE: EIGEN-GRACE02S. Journal of Geodynamics

Sandwell D.T., and W. Smith (1997): Marine gravity from Geosat and ERS-1 satellite altimetry. J. Geophys. Res., Vol. 102, No. B5, pp.10039-10054

Tapley B. D. , S. Bettadpur, M. M. Watkins and Ch. Reigber (2004): The Gravity Recovery and Climate Experiment: Mission Overview and Early Results. Geophys. Res. Lett., 31, L09607, doi:10.1029/2004GL019920, 2004.

# Determination of gravity gradients from terrestrial gravity data for calibration and validation of gradiometric GOCE data

M. Kern
Institute of Navigation and Satellite Geodesy, TU Graz, Steyrergasse 30, 8010 Graz, Austria
R. Haagmans
Science and Applications Department, ESA/ESTEC, Keplerlaan 1, 2200 AG Noordwijk ZH, The Netherlands

**Abstract.** The satellite mission GOCE (Gravity field and steady-state ocean explorer) is the first gravity field mission of ESA's Living Planet Programme. Measurement principles are satellite-to-satellite tracking (SST) and, for the first time, satellite gravity gradiometry (SGG). To meet the mission goal of a 1-2 cm geoid at a spatial resolution of about 100 km, the satellite instruments will be calibrated in pre-flight mode and prior to the measurement phases (in-flight mode). Moreover, external calibration and validation of the measurements is performed using gravity information over well-surveyed areas.

In this paper, all components of the gravity tensor are determined from terrestrial gravity data. Integral formulas based on the extended Stokes and Hotine formulas are used. It is shown that the entire tensor can be computed with an accuracy of 1.5-2.5 mE in the local North-East-Up coordinate system. In addition, the effect of white noise and a bias in the terrestrial data is studied.

**Keywords.** Calibration/Validation · GOCE mission · Gradiometry · Upward continuation

## 1 Introduction

The GOCE mission is the first satellite mission with a gradiometer on board. The gradiometer consists of six three-axis accelerometers mounted in pairs along three orthogonal arms. The accelerometer readings allow for the determination of the common-mode and differential-mode signals, which are used to derive the gravity gradients. The highest precision is achieved in the measurement bandwidth (MBW) between 5 and 100 mHz (Cesare 2002).

The observations will be burdened with (systematic and stochastic) errors and to meet the mission goal, the measurements have to be calibrated and validated. Besides internal calibrations (pre-flight and in-orbit), a number of external calibration and validation concepts for SGG measurements have been suggested: Existing gravity field models are used in Visser et al (2000), Bouman et al (2004). Terrestrial gravity data over well-surveyed areas are proposed in Arabelos and Tscherning (1998), Haagmans et al. (2002), Pail (2002), Müller et al (2004). Finally, cross-over techniques for (internal) calibration/validation are presented in Müller et al. (2004).

The upward continuation of terrestrial gravity data using least-squares collocation has been investigated in Arabelos and Tscherning (1998) and Bouman et al. (2004). Denker (2002) uses integral formulas for the upward continuation and transformation and finds results for the radial component at the level of a few mE ($1\,\mathrm{E} = 10^{-9}\,\mathrm{s}^{-2}$). First results for all components of the gravity gradients based on Stokes's formula are shown in Reed (1973). In this paper, Reed's results are extended by deriving the kernel functions in the spectral domain. Also, the derivations are done for Hotine's formula. Note that covariance expressions for least-squares collocation are given in Tscherning (1993).

## 2 Extended Stokes and Hotine formula

The following derivations are based on the extended Stokes and Hotine formulas. The extended Stokes formula is given as

$$T(r,\Omega) = \frac{R}{4\pi}\iint\limits_{\Omega} S(\psi,r)\Delta g(r,\Omega')\,d\Omega' \qquad (1)$$

where $T$ is the disturbing potential, $\Delta g$ are gravity anomalies and $R$ is the radius of the reference sphere. The extended (Pizzetti-) Stokes function $S = S(\psi, r)$ is (Heiskanen and Moritz 1967)

$$\begin{aligned} S &= t\left[\frac{2}{D}+1-3D-tx\left(5+3\ln\frac{1-tx+D}{2}\right)\right] \\ &= \sum_{l=2}^{\infty}\frac{2l+1}{l-1}t^{l+1}P_l(x) \end{aligned} \qquad (2)$$

with

$$t = R/r \quad \text{and} \quad D = \sqrt{1-2tx+t^2} \qquad (3)$$

$l$ is the degree and $x = \cos\psi$ is the cosine of the spherical distance between the running point and the computation point (Heiskanen and Moritz 1967)

$$x = \sin\varphi \sin\varphi' + \cos\varphi \cos\varphi' \cos(\lambda' - \lambda) \quad (4)$$

$r, \varphi, \lambda$ are the geocentric radius, spherical latitude and longitude and $\Omega$ stands for the pair of angular spherical coordinates $(\varphi, \lambda)$. $P_l$ is the $l^{\text{th}}$-degree Legendre polynomial. The extended Hotine formula for gravity disturbances $\delta g$ can be written as

$$T(r, \Omega) = \frac{R}{4\pi} \iint_{\Omega} H(\psi, r) \delta g(r, \Omega')\, d\Omega' \quad (5)$$

where the extended Hotine function $H = H(\psi, r)$ is (Pick et al. 1973, eq. 1572)

$$\begin{aligned} H &= \frac{2t}{D} - \ln\left(\frac{v}{1-x}\right) \\ &= \sum_{l=0}^{\infty} \frac{2l+1}{l+1} t^{l+1} P_l(x) \end{aligned} \quad (6)$$

and $v = D + t - x$. Note the difference in the denominator $l + 1$ and $l - 1$ in Eq. (2) and (6) and the summation start $l = 2$ and $l = 0$, respectively. Band-limited kernel functions can easily be derived by truncating the spectral kernel functions. Also, appropriate weighting functions may then be introduced (see e.g. Haagmans et al. 2002).

## 3 Derivatives of the extended Stokes and Hotine formula

Gravity gradients at satellite altitude can be derived using the derivatives of the extended Stokes and Hotine formulas. The first derivatives of the disturbing potential with respect to the spherical coordinates can be transformed to the local cartesian system NEU ($x, y, z$; $x$ pointing north, $y$ pointing east and $z$ pointing radially outwards) as follows (e.g. Tscherning 1976)

$$\begin{aligned} T_x &= \frac{1}{r} T_\varphi \\ T_y &= \frac{1}{r\cos\varphi} T_\lambda \\ T_z &= T_r \end{aligned} \quad (7)$$

Note that $T_x$ stands for $\partial T / \partial x$ and so on. Similarly, the second derivatives are given as (Koop and Stelpstra 1989)

$$\begin{aligned} T_{xx} &= \frac{1}{r^2} T_{\varphi\varphi} + \frac{1}{r} T_r \\ T_{xy} &= \frac{1}{r^2\cos\varphi} T_{\varphi\lambda} + \frac{\sin\varphi}{r^2\cos^2\varphi} T_\lambda \\ T_{xz} &= \frac{1}{r} T_{r\varphi} - \frac{1}{r^2} T_\varphi \\ T_{yy} &= \frac{1}{r^2\cos^2\varphi} T_{\lambda\lambda} - \frac{\tan\varphi}{r^2} T_\varphi + \frac{1}{r} T_r \\ T_{yz} &= \frac{1}{r\cos\varphi} T_{\lambda r} - \frac{1}{r^2\cos\varphi} T_\lambda \\ T_{zz} &= T_{rr} \end{aligned} \quad (8)$$

Eq. (8) assembles the tensor $\boldsymbol{T}^{\text{NEU}}$. The computation of $T_{xx}$ requires the determination of $T_{\varphi\varphi}$ and $T_r$. For the other components, other derivatives are needed. Hence, partial derivatives of the extended Stokes and Hotine integral with respect to $r$, $\varphi$ and $\lambda$ have to be determined. The first partial derivatives of the extended Stokes function are given as (Ecker 1969)

$$\begin{aligned} T_r &= \frac{R}{4\pi} \iint_{\Omega} S_r \Delta g(r, \Omega')\, d\Omega' \\ T_\varphi &= \frac{R}{4\pi} \iint_{\Omega} S_\psi \frac{\partial\psi}{\partial\varphi} \Delta g(r, \Omega')\, d\Omega' \\ T_\lambda &= \frac{R}{4\pi} \iint_{\Omega} S_\psi \frac{\partial\psi}{\partial\lambda} \Delta g(r, \Omega')\, d\Omega' \end{aligned} \quad (9)$$

Analogously, the second derivatives are given as (Reed 1973)

$$\begin{aligned} T_{rr} &= \frac{R}{4\pi} \iint_{\Omega} S_{rr} \Delta g' d\Omega' \\ T_{\varphi\varphi} &= \frac{R}{4\pi} \iint_{\Omega} \left[ S_{\psi\psi} \left(\frac{\partial\psi}{\partial\varphi}\right)^2 + S_\psi \frac{\partial^2\psi}{\partial\varphi^2} \right] \Delta g' d\Omega' \\ T_{\lambda\lambda} &= \frac{R}{4\pi} \iint_{\Omega} \left[ S_{\psi\psi} \left(\frac{\partial\psi}{\partial\lambda}\right)^2 + S_\psi \frac{\partial^2\psi}{\partial\lambda^2} \right] \Delta g' d\Omega' \\ T_{\varphi\lambda} &= \frac{R}{4\pi} \iint_{\Omega} \left[ S_{\psi\psi} \frac{\partial\psi}{\partial\varphi} \frac{\partial\psi}{\partial\lambda} + S_\psi \frac{\partial^2\psi}{\partial\varphi\partial\lambda} \right] \Delta g' d\Omega' \\ T_{r\varphi} &= \frac{R}{4\pi} \iint_{\Omega} \left[ S_{r\psi} \frac{\partial\psi}{\partial\varphi} \right] \Delta g' d\Omega' \\ T_{r\lambda} &= \frac{R}{4\pi} \iint_{\Omega} \left[ S_{r\psi} \frac{\partial\psi}{\partial\lambda} \right] \Delta g' d\Omega' \end{aligned} \quad (10)$$

where $\Delta g(r, \Omega')$ is abbreviated with $\Delta g'$. For Hotine's function, Eqs. (9) and (10) contain gravity disturbances $\delta g$. Partial derivatives of the spherical distance with respect to the geocentric latitude and longitude can be derived using Eq. (4) and formulas from spherical trigonometry, see e.g Heiskanen and Moritz (1967, pg. 113):

$$\frac{\partial \psi}{\partial \varphi} = -\cos\alpha', \quad \frac{\partial \psi}{\partial \lambda} = -\cos\varphi\sin\alpha'$$

$$\frac{\partial^2 \psi}{\partial \varphi^2} = \sin^2\alpha' \cot\psi$$

$$\frac{\partial^2 \psi}{\partial \lambda^2} = \frac{\cos\varphi}{\sin\psi}[\cos\varphi'\cos(\lambda'-\lambda) - \cos\varphi\sin^2\alpha'\cos\psi]$$

$$\frac{\partial^2 \psi}{\partial \varphi \partial \lambda} = \sin\alpha'(\sin\varphi - \cos\varphi\cos\alpha'\cot\psi)$$

where $\alpha$ is the azimuth as defined in Heiskanen and Moritz (1967). After some operations, the derivatives of the closed Stokes kernel with respect to $r$ and $\psi$ can be derived, see Table 1 ($P_{l,1}$ and $P_{l,2}$ are unnormalized Legendre functions for order 1 and 2, respectively). Some of them can already be found in Reed (1973) and Witte (1970). The formulas have been checked successfully against the spectral kernel functions by calculating the kernel function for various $r$ and $x$. Analogously, the kernel functions for the extended Hotine formula are listed in Table 2. They may be used for the upward continuation and transformation of airborne gravity disturbances.

Gravity gradients can be obtained using the following steps:

*i)* Remove long-wavelength components from the local data using a global geopotential model.

*ii)* Compute first (Eq. (9)) and second derivatives (Eq. (10)) of the disturbing potential for the area where data are available (near zone). Determine estimates of the far-zone contribution as described in Thalhammer (1994).

*iii)* Assemble the gravity tensor $\boldsymbol{T}^{\text{NEU}}$ in the NEU system using Eq. (8).

*iv)* Restore the long-wavelength components.

*v)* If necessary, rotate the tensor into a different coordinate system $\boldsymbol{T} = \boldsymbol{R}\boldsymbol{T}^{\text{NEU}}\boldsymbol{R}^T$, where $\boldsymbol{R}$ is a rotation matrix.

## 4 Numerical results

The following numerical tests provide insights into the numerical accuracy of the integral formulas. All tests are performed in a closed-loop based on gravity data from the global geopotential model GPM98A (Wenzel 1998). Gravity anomalies and gravity disturbances, complete up to degree and order 1800, have been computed at the Bjerhammer sphere and the full tensor at 250 km height. A regular grid of $0.25° \times 0.25°$ has been used. Note that this grid is coarse and some aliasing might occur. The test area is given as:

$$\sigma_0 : (\varphi, \lambda) \in [38°, 54°] \times [0°, 28°] \quad (11)$$

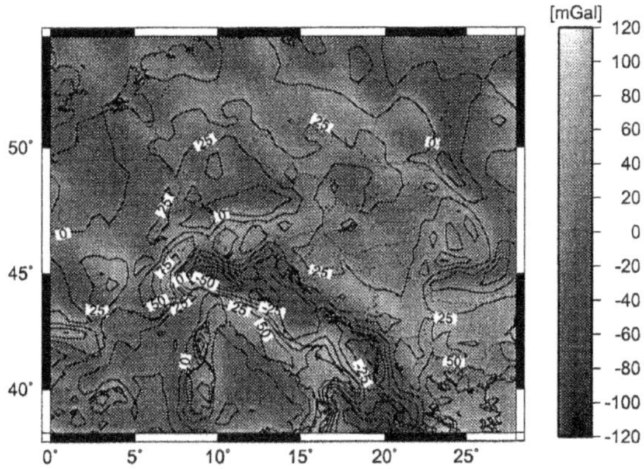

**Fig. 1** Gravity anomalies from GPM98A at the reference sphere [mGal]

It has relatively large gravity variations (RMS = 29.7 mGal), see Fig. 1. Since the long-wavelengths are not well represented by the local data, the long-wavelength components (using GPM98A) up to degree and order $l = 90$ have been removed. After the estimation of the gradients at satellite height, the long-wavelengths are restored again. A spherical cap of 6° has been used. Results of this test are summarized in Table 3. Differences of the estimated tensor components to the ones directly obtained from GPM98A are listed. The values in brackets are for the Hotine integration using gravity disturbances. Note that the differences are only computed for the output area $\sigma_1$:

$$\sigma_1 : (\varphi, \lambda) \in [44°, 48°] \times [6°, 22°] \quad (12)$$

All components can be computed with a numerical accuracy of about 1.3-2.5 mE (Std). Best results are obtained for the radial component. The larger errors on the other components could be due to some aliasing. However, this has to be studied in more detail. The largest errors occur at the edges of the computational area where missing data deteriorate the solutions.

The differences are shown graphically in Fig. 2. Due to the symmetry of the tensor, the components $yx$, $zx$ and $zy$ are equivalent to the components $xy$, $yz$ and $yz$, respectively. Therefore, they are shown in grey.

As mentioned before, the long-wavelength gravity field is not well represented by the local data

**Table 1** Extended Stokes kernel function and its derivatives. The summation of the series is taken from two up to infinity

| Kernel | Spectral form | Closed kernel expression |
|---|---|---|
| $S$ | $\sum_l \frac{2l+1}{l-1} t^{l+1} P_l(x)$ | $t\left[\frac{2}{D}+1-3D-tx\left(5+3\ln\frac{1-tx+D}{2}\right)\right]$ |
| $S_r$ | $-\sum_l \frac{2l+1}{l-1} t^{l+1}\left(\frac{l+1}{r}\right)P_l(x)$ | $-\frac{t^2}{R}\left[\frac{1-t^2}{D^3}+\frac{4}{D}+1-6D-tx\left(13+6\ln\frac{1-tx+D}{2}\right)\right]$ |
| $S_\psi$ | $-\sum_l \frac{2l+1}{l-1} t^{l+1} P_{l,1}(x)$ | $-t^2\sin\psi\left[\frac{2}{D^3}+\frac{6}{D}-8-3\frac{1-tx-D}{D\sin^2\psi}-3\ln\frac{1-tx+D}{2}\right]$ |
| $S_{rr}$ | $\sum_l \frac{2l+1}{l-1} t^{l+1}\frac{(l+1)(l+2)}{r^2}P_l(x)$ | $\frac{t^3}{R^2}\left[(1-tx)\left(\frac{3(1-t^2)}{D^5}-\frac{4}{D^3}\right)-\frac{(1+t^2)}{D^3}-\frac{10}{D}-18D+2\right.$ $\left.-3tx\left(15+6\ln\frac{1-tx+D}{2}\right)+\frac{24}{D}+\frac{4(1-t^2)}{D^3}\right]$ |
| $S_{r\psi}$ | $\sum_l \frac{2l+1}{l-1} t^{l+1}\left(\frac{l+1}{r}\right)P_{l,1}(x)$ | $\frac{t^3\sin\psi}{R}\left[\frac{3(1-t^2)}{D^5}+\frac{4}{D^3}+\frac{6}{D}-13-6\ln\frac{1-tx+D}{2}+\right.$ $\left.\frac{6tx(D+1)}{D(1-tx+D)}\right]$ |
| $S_{\psi\psi}$ | $\sum_l \frac{2l+1}{l-1} t^{l+1}\left(-\cot\psi P_{l,1}(x)+P_{l,2}(x)\right)$ | $-t^2x\left[\frac{2}{D^3}+\frac{6}{D}-8-3\frac{1-tx-D}{D\sin^2\psi}-3\ln\frac{1-tx+D}{2}\right]$ $+t^3\sin^2\psi\left[\frac{6}{D^5}+\frac{6}{D^3}+3\frac{(D-1)}{D^2\sin^2\psi}+3\frac{(D+1)}{D(1-tx+D)}-\right.$ $\left.3\frac{1-tx-D}{D\sin^2\psi}\left(\frac{2x}{t\sin^2\psi}+\frac{1}{D^2}\right)\right]$ |

**Table 2** Extended Hotine kernel function and its derivatives. The summation of the series is taken from zero up to infinity

| Kernel | Spectral form | Closed kernel expression |
|---|---|---|
| $H$ | $\sum_l \frac{2l+1}{l+1} t^{l+1} P_l(x)$ | $\frac{2t}{D}-\ln\left(\frac{v}{1-x}\right)$ |
| $H_r$ | $-\sum_l \frac{2l+1}{l+1} t^{l+1}\left(\frac{l+1}{r}\right)P_l(x)$ | $-\frac{2R}{r^2D}-\frac{2t^2(x-t)}{rD^3}-\frac{t(x-t)}{Drv}+\frac{R}{r^2v}$ |
| $H_\psi$ | $-\sum_l \frac{2l+1}{l+1} t^{l+1} P_{l,1}(x)$ | $-\frac{2t^2\sin\psi}{D^3}-\frac{t\sin\psi}{Dv}-\frac{\sin\psi}{v}+\frac{\sin\psi}{1-x}$ |
| $H_{rr}$ | $\sum_l \frac{2l+1}{l+1} t^{l+1}\frac{(l+1)(l+2)}{r^2}P_l(x)$ | $\frac{t}{r^2D}\left(4-\frac{2t^2}{D^2}-\frac{t}{v}\right)+\frac{t}{r^2v}\left(\frac{t}{v}-2\right)$ $+\frac{t(x-t)}{Dr^2}\left(\frac{8t}{D^2}+\frac{6t^2(x-t)}{D^4}+\frac{2}{v}+\frac{t(x-t)}{D^2v}+\frac{t(x-t)}{Dv^2}-\frac{2t}{v^2}\right)$ |
| $H_{r\psi}$ | $\sum_l \frac{2l+1}{l+1} t^{l+1}\left(\frac{l+1}{r}\right)P_{l,1}(x)$ | $\frac{4t^2\sin\psi}{rD^3}+\frac{6t^3\sin\psi(x-t)}{D^5r}+\frac{t\sin\psi}{rDv}+\frac{t^2\sin\psi(x-t)}{rD^3v}$ $+\frac{t^2\sin\psi(x-t)}{rD^2v^2}-\frac{t^2\sin\psi}{rDv^2}+\frac{t\sin\psi(x-t)}{rDv^2}-\frac{t\sin\psi}{rv^2}$ |
| $H_{\psi\psi}$ | $\sum_l \frac{2l+1}{l+1} t^{l+1}\left(-\cot\psi P_{l,1}(x)+P_{l,2}(x)\right)$ | $\frac{-2t^2x}{D^3}+\frac{6t^3\sin^2\psi}{D^5}-\frac{tx}{Dv}+\frac{t^2\sin^2\psi}{D^3v}+\frac{t^2\sin^2\psi}{D^2v^2}+\frac{t\sin^2\psi}{Dv^2}$ $-\frac{x}{v}+\frac{t\sin^2\psi}{Dv^2}+\frac{\sin^2\psi}{v^2}+\frac{x}{1-x}-\frac{\sin^2\psi}{(1-x)^2}$ |

and the contribution of a global gravity field is subtracted. After the estimation of the gravity tensor, the removed effects are restored. The previous tests involved long-wavelength information up to degree and order 90. For a comparison, a reference field of degree and order 60 produces a RMS error of about 2.6 mE and a 360 field an error of about 0.7 mE for the radial component. Thus in this comparison the use of a higher reference field model improves the solution. In any case, the results are optimistic since

**Table 3** Differences between $T_{ij}$ directly from GPM98A and estimates from terrestrial data [mE]. Values in brackets are from the extended Hotine formula

| $ij$ | Min | Max | Mean | Std |
|---|---|---|---|---|
| $xx$ | -6.1 (-5.9) | 8.7 (8.5) | 0.0 (0.0) | 2.5 (2.4) |
| $xy$ | -5.0 (-4.7) | 4.3 (4.1) | -0.2 (-0.2) | 1.9 (1.8) |
| $xz$ | -5.5 (-4.9) | 5.4 (4.5) | 0.0 (0.0) | 2.1 (1.8) |
| $yy$ | -8.2 (-7.9) | 5.5 (5.7) | 0.0 (0.0) | 2.3 (2.1) |
| $yz$ | -5.6 (-4.8) | 4.4 (3.9) | -0.1 (0.1) | 1.8 (1.6) |
| $zz$ | -3.4 (-3.6) | 3.8 (3.9) | 0.0 (-0.0) | 1.3 (1.4) |

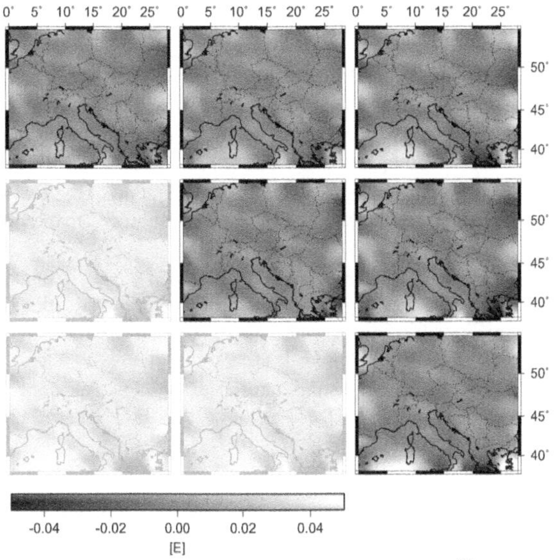


**Fig. 2** Gravity gradient differences ($T_{xx} - T_{xx}^{gpm98a}$, $T_{xy} - T_{xy}^{gpm98a}$, $T_{xz} - T_{xz}^{gpm98a}$, $T_{yx} - T_{yx}^{gpm98a}$, $T_{yy} - T_{yy}^{gpm98a}$, $T_{yz} - T_{yz}^{gpm98a}$, $T_{zx} - T_{zx}^{gpm98a}$, $T_{zy} - T_{zy}^{gpm98a}$, $T_{zz} - T_{zz}^{gpm98a}$ from left to right, top to bottom) in [E], grey areas are equivalent components

the long-wavelengths are assumed to be errorless. Improved results may be possible by introduction of kernel modifications and/or the use of weighting functions. This is, however, outside the scope of this investigation.

In the previous tests a spherical cap of 6° was used, which requires a large data area such as Europe, North America or Australia. The question arises if a smaller integration cap, in combination with a smaller data area, can produce similar results. Six test runs have been done using spherical caps of 1° – 6°. The differences to the values directly obtained from the GPM98A model are computed and summarized in Fig. 3. In all cases, the output area is $\sigma_1$. Clearly, smaller spherical caps, such as 1° or 2°, produce large errors (up to 38 mE) and cannot be used for the problem at hand. A spherical cap of 3° is the minimum in order to keep the error well below 5 mE. Obviously, this also limits the selection of data areas as they have to be sufficiently large.

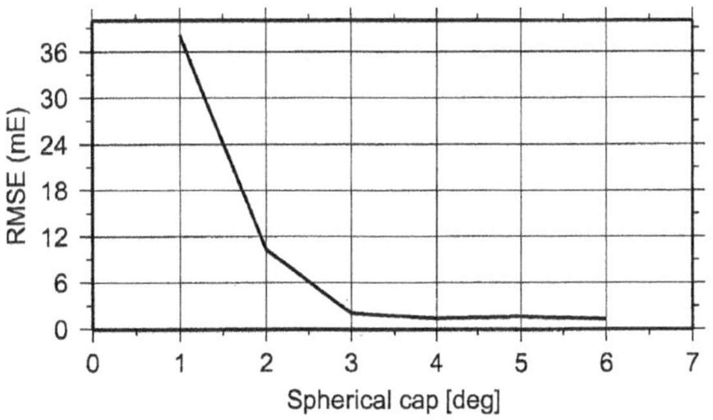


**Fig. 3** RMS differences using different spherical cap sizes for the radial component (extended Stokes function is used)

The last test involves errors in the input data. In particular, the effect of white noise and a regional bias is studied. Noise of 2 mGal standard deviation and a bias of 8 mGal on Austria is shown in Fig. 4. It is added on the input data and an integration with a spherical cap of 6° is performed. The results are shown in Fig. 5. Clearly, the bias directly affects the radial component $T_{rr}$. It is also visible, yet with different sign and half of the magnitude, in the other diagonal components. Compared with the other components, the horizontal component $T_{xy}$ is less affected by the bias. It is worth noting that the effect of the noise (up to 2 mGal Std) is small due to the fact that the kernel functions have low-pass filter characteristics (large upward continuation height).

## 5 Summary and conclusions

Formulas have been presented for the estimation of the full gravity tensor at satellite height using gravity anomalies and gravity disturbances. They allow for an estimation of the gravity tensor in the North-East-

**Fig. 4** Noise and bias [mGal]

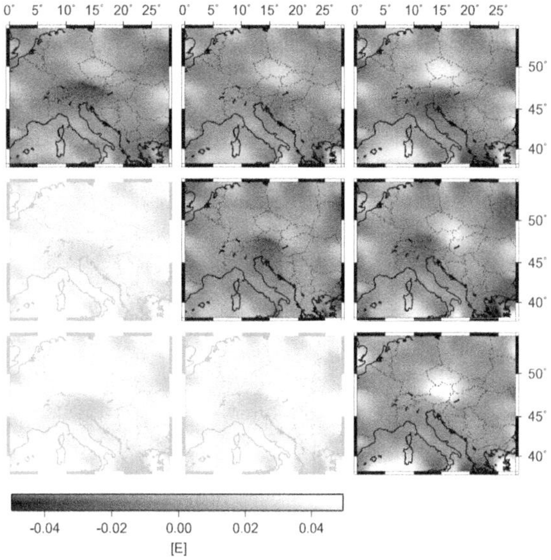

**Fig. 5** Gravity gradient differences [E], order as in Figure 2

Up coordinate system if a large data area is available. Closed-loop simulations have shown that best results are obtained with a spherical cap of 3° or larger. Furthermore, the use of a high-degree reference field (D/O 60 or higher) is advantageous. Then, the integration error is between 1.3 mE and 2.5 mE, but additional tests using a finer grid spacing are necessary. A bias in the input data affects all components. In a relative sense, however, the off-diagonal components are less affected. The effect of (white) noise in the input gravity data is small since the kernel functions have low-pass filter characteristics.

The closed-loop simulation has demonstrated that the presented integral formulas can be used for the computation of gravity gradients at satellite height. Future refinements of the method are the introduction of kernel modifications or weighting functions. Also, error propagation is necessary. The obtained reference gradients may then be used for calibration and validation purposes of the satellite mission GOCE. The actual calibration and validation procedure is outside the scope of this paper and will be addressed in an upcoming contribution.

## 6 Acknowledgement

Financial support for the first author came from an external ESA fellowship. This is gratefully acknowledged.

## References

Arabelos D, Tscherning CC (1998) Calibration of satellite gradiometer data aided by ground gravity data. Journal of Geodesy. 72: 617-625

Bouman J, Koop R, Tscherning CC, Visser P (2004) Calibration of GOCE SGG data using high-low SST, terrestrial gravity data and global gravity field models. Journal of Geodesy. 78, DOI 10.1007/s00190-004-0382-5

Cesare S (2002) Performance requirements and budgets for the gradiometric mission. Issue 2 GO-TN-AI-0027, Preliminary Design Review, Alenia, Turin

Denker H (2002) Computation of gravity gradients for Europe for calibration/validation of GOCE data. In: Proc of the 3rd IGGC, Thessaloniki, Greece, pp 287-292

Ecker E (1969) Sphärische Integralformeln in der Geodäsie. Deutsche Geodätische Kommission. Nr. 142. Munich.

Haagmans R, Prijatna K, Omang O (2002) An alternative concept for validation of GOCE gradiometry results based on regional gravity. In: Proc of the 3rd IGGC, Thessaloniki, Greece, pp. 281-286

Heiskanen WA, Moritz H (1967) Physical geodesy. WH Freeman and Company, San Francisco

Koop R, Stelpstra D (1989) On the computation of the gravitational potential and its first and second order derivatives. Manuscripta Geodaetica. 14: 373-382.

Müller J, Denker H, Jarecki F, Wolf KI (2004) Computation of calibration gradients and methods for in-orbit validation of gradiometric GOCE data. Paper published in the Proceedings of the 2nd Int. GOCE user workshop in Frascati.

Pail R (2002) In-orbit calibration and local gravity field continuation problem. In H. Sünkel (ed). From Eötvös to mGal+. Final Report. pp. 9-112.

Pick M, Picha J, Vyskočil V (1973) Theory of the Earth's gravity field. Elsevier, Amsterdam

Reed GB (1973) Application of kinematical geodesy for determining the short wave length components of the gravity field by satellite gradiometry. Technical Report No. 201. Department of Geodetic Science. The Ohio State University. Columbus, Ohio.

Thalhammer M (1994) The geographical truncation error in satellite gravity gradiometer measurements. Manuscripta Geodaetica. 19: 45-54.

Tscherning CC (1976) Computation of the second-order derivatives of the normal potential based on the representation by a Legendre series. Manuscripta Geodaetica. 1:72-92.

Tscherning CC (1993) Computation of covariances of derivatives of the anomalous gravity potential in a rotated reference frame. Manuscripta Geodaetica. 8(3): 115-123.

Visser P, Koop R, Klees R (2000) Scientific data production quality assessment. In: Sünkel H (ed). From Eötvös to mGal. WP 4. pp. 157-176

Wenzel G (1998) Ultra high degree geopotential models GPM98A, B and C to degree 1800. In: Proc Joint Meeting International Gravity Commission and International Geoid Commission, 7-12 September 1998. Trieste. Available at http://www.gik.uni-karlsruhe.de/~wenzel/gpm98abc/gpm98abc.htm

Witte B (1970) Die Bestimmung von Horizontalableitungen der Schwere im Aussenraum aus einer Weiterentwicklung der Stokesschen Funktion. Gerlands Beiträge zur Geophysik. Akademische Verlagsgesellschaft. Geest and Protig KG, Leipzig. East Germany. 79, No 2. 87-94.

# Evaluation of Airborne Gravimetry Integrating GNSS and Strapdown INS Observations

Ch. Kreye, G.W. Hein, B. Zimmermann
Institute of Geodesy and Navigation
University FAF Munich, Werner-Heisenberg-Weg 39, D-85579 Neubiberg, Germany

**Abstract.** Airborne gravimetry systems provide the most economical way to improve the spatial resolution of gravity data measured by satellite missions.

So the paper deals with the presentation of a modern airborne gravitymeter designed, developed and tested at the university FAF Munich. The specific forces are measured by a high precision strapdown INS and the kinematical accelerations are derived using numerous differential GNSS observations.

So the first part of the paper describes the system architecture, the test environment and the area of two finished flight test campaigns. The error models of GNSS and INS measurements are demonstrated and evaluated in regard to airborne gravimetry applications.

In this context the derivation of kinematical accelerations out of GNSS raw data is investigated. Thereby the additional performance potential of five GNSS receivers in the aircraft and twelve reference stations along the flight trajectory for acceleration determination is taken into account.

In the scope of integration filter design important aspects are emphasized concerning the low dynamic input data and the analogue processing of GNSS and INS data streams. Finally a first result of the observed gravity signal is presented.

**Keywords.** Airborne gravimetry, acceleration determination, strapdown inertial navigation system

## 1 Introduction

Information about the earth gravity field is used for many applications in geophysics and geodesy dealing with figure and structure of our planet. In the context of current or planned satellite missions methods for determination of the gravity field are in discussion today. Caused by thresholds in possible spatial resolution (50-100 km), however, applications using data of satellite based systems are restricted to global or regional investigations. The observation of gravity with wavelengths smaller than 1 km in an efficient way, especially important for economical applications, is only possible with airborne methods.

Today airborne gravitymeters basing on the platform design provide the amount of gravity in local to regional areas. Also strapdown systems are available today, but nevertheless the combination of modern INS technology basing on the strapdown principle and sophisticated processing methods should be able to improve these systems. Important advantages are the observation of the full gravity vector and the simpler system design that enables more efficient airborne gravimetry campaigns. But it must be approved if it is possible to reach the often postulated goal: an accuracy of 1 mGal with a spatial resolution of 1 km.

## 2 Fundamentals

The goal of airborne vector gravimetry is to provide the gravity disturbance vector $\boldsymbol{\delta g}$ as the difference between the measured gravity $\mathbf{g}$ and the normal gravity $\boldsymbol{\gamma}$ in the same observation point. Following Newton's second law of motion the gravity $\mathbf{g}^i$ in an inertial coordinate system can be calculated by the difference between the specific force $\mathbf{f}^i$ and the kinematical acceleration $\mathbf{a}^i$ (= second derivative of the position) .

**Fig. 1** Principle of airborne gravimetry

In a strapdown INS approach the specific forces **f** are provided by an accelerometer and gyro triad permanently fixed to the body of the aircraft. The inertial data can be transformed using the following formulas:

$$\mathbf{f}^i = \mathbf{C}_b^i \cdot \mathbf{f}^b \tag{1}$$

with

$$\mathbf{C}_b^i = \mathbf{C}_{b0}^i + \int \mathbf{C}_b^i \cdot \mathbf{\Omega}_{ib}^b dt \tag{2}$$

where $\mathbf{\Omega}_{ib}^b$ skew-symmetric matrix of gyro rates
$\mathbf{C}_b^i$ rotation matrix between body- and inertial frame

The error models for specific forces $\mathbf{f}^b$ and gyro rates $\boldsymbol{\omega}_{ib}^b$ are described and evaluated for airborne gravimetry, e.g. in Kreye et al. (2003). Critical elements are especially the time variations of systematic errors, unmodelled influences and sensor noise.

In order to calculate the kinematical acceleration of the aircraft GNSS observations in a DGPS configuration are used.

In the traditional approach first the phase solution for the aircraft position $\mathbf{x}^e$ is computed. The next step is the derivation of $\mathbf{x}^e$ to the earth-fixed velocity $\mathbf{v}^e$ and acceleration $\mathbf{a}^e$. Neglecting the variation of earth rotation rate (Euler term) the kinematical acceleration in inertial frame $\mathbf{a}^i$ can be generated by

$$\mathbf{a}^i = \mathbf{C}_e^i \left( \mathbf{a}^e + 2\mathbf{\Omega}_{ie}^e v^e + \mathbf{\Omega}_{ie}^e \mathbf{\Omega}_{ie}^e \mathbf{x}^e \right) \tag{3}$$

where $\mathbf{\Omega}_{ib}^b$ skew-symmetric matrix of earth rotation rate
$\mathbf{C}_e^i$ rotation matrix between earth- and inertial frame

Another processing method is presented by Jekeli 1992. In this case the phase measurement itself and its derivations as well as the code solution is required to compute the kinematical acceleration $\mathbf{a}^i$ directly.

In opposite to the traditional method here the least squares estimation is followed by the low-pass filtering and derivation processing. At the same time it is not necessary to solve the phase ambiguities. Therefore it seems to be possible to allow longer baselines between reference station and aircraft. Furthermore only the L1 phase observation with a lower noise level than the ionospheric free linear combination can be used for acceleration determination. So from this point of view the direct method should be preferred, but the particular performance level must be investigated in practical tests (see below).

Both algorithms, however, are based on GNSS phase observations $\Theta_A^i$ with the following general error model:

$$\theta_A^i = S_A^i + N_A^i \cdot \lambda + \delta S_{A[CLK]}^i + \delta S_{A[PCV]}^i + \delta S_{A[REL]}^i + \delta S_{A[ION]}^i + \delta S_{A[TROP]}^i + \varepsilon_A^i \tag{4}$$

where
$S_A^i$ range between satellite i and receiver A
$N_A^i$ ambiguity term
$\delta S_{A[CLK]}^i$ clock error including satellite and receiver clock, hardware biases and synchronisation errors
$\delta S_{A[PCV]}^i$ phase center variation
$\delta S_{A[REL]}^i$ relativistic errors
$\delta S_{A[ION]}^i$ ionosspheric error
$\delta S_{A[TROP]}^i$ tropossheric error
$\varepsilon_A^i$ phase noise

The influence of measurement errors presented in equation 4 is well-known for the position determination using GNSS phase observations. Airborne gravimetry, however, is nearly the only application where GNSS measurements should provide accurate, mGal-level acceleration data. In this case the error influences must be evaluated in a completely different way. It has to be taken into account, that the process of differentiating amplifies these errors as function of increasing frequency, causing them to be larger as the upper edge of the bandwidth is increased.

In general the spectral window of airborne gravimetry is between 0.02 and 0.002 Hz. So the performance of an airborne gravimetry system depends on the ability to estimate the systematic errors and to suppress or model the stochastic error components. Both terms are influenced on the one hand by sensors on the other hand by the used processing algorithms. In chapter 4 and 5 the performance of the system described below is evaluated using laboratory investigations and practical flight test.

## 3 System design

Beside of resolution and accuracy also the economic efficiency of a sensor system decides on its future applications. In case of an airborne gravimetry system this means that it must be as cheap, small and light as possible. Following these arguments in the context of a new airborne gravimetry project we try to investigate the performance of an integrated sensor system basing on GNSS receivers and a commercial high precision strapdown INS. As it is demonstrated in figure 2 the specific forces are measured by a SAGEM Sigma 30 INS fitted with triads of ring-laser gyros and pendulous accelerometers. Using special interfaces both the raw data and the navigation data are available for the gravity calculation.

**Fig. 2** Components of the airborne gravimetry system

In order to derive the kinematic acceleration on the one hand an ASHTECH L1/L2 receiver with a raw data rate of 10 Hz is used. Additionally the L1-observations of four other GNSS antennas with fixed baselines are generated by two NOVATEL Beeline multi-antenna systems. The integration of the L1/L2 observations with this data at first should provide better performance of the phase ambiguity determination using the information of known baseline lengths on the aircraft. Secondly the redundant determination of the airplane dynamics should increase the accuracy of acceleration computation and a better evaluation of GNSS results. It must be pointed out that the multi-antenna configuration is not used for attitude determination of the aircraft. Therefore variations in the corresponding baseline lengths up to 1 cm are tolerable. GNSS reference ground stations along the flight track guarantee differential observations. Finally a central PC provides the sensor controlling during the flight periods, the time synchronization of GNSS and INS data sets and the storage capabilities for all observations.

## 4 Static lab tests

Before practical flight tests are carried out lab tests can provide a first impression of the system performance. Beginning with the inertial sensor first of all specific forces and gyro rates with a data rate of 100 Hz are observed in order to evaluate their time response. A frequency analysis of specific force measurements in static case over 30 minutes is carried out. The amplitude of accelerometer errors like time variations of bias and scale factors or noise effects in the relevant spectral area smaller than 0.04 Hz is at 1 mGal level.

In a second test the static inertial data are analysed using equations 1 and 2 to investigate the overall INS performance including the gyro rates. After an alignment phase the gyro rates are used to transform the specific forces in the navigation frame. After low pass filtering to a time resolution of 25 s the derived INS measurements for the down component representing a time span of 1 hour behave like it is shown in figure 3. The linear trend caused by sensor errors compensated in a real observation situation must be separated from the data. The analysis of the deviations yields to a RMS error of 1.5 mGal. So the INS data seems to be in the required accuracy range. A final evaluation also fo the dynamic case is possible not until the flight test data sets are completely processed.

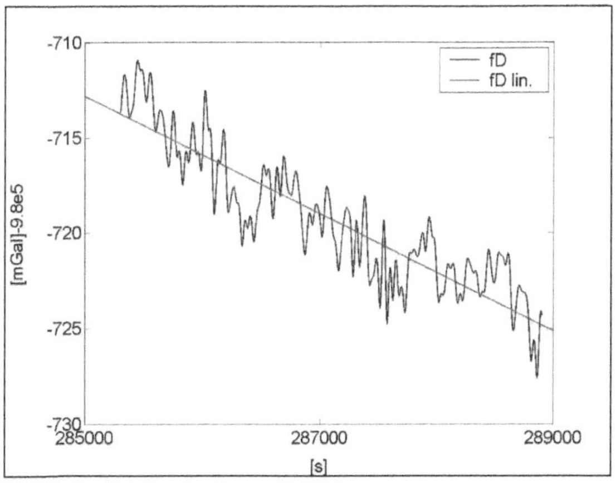


**Fig. 3** Static performance of inertial data

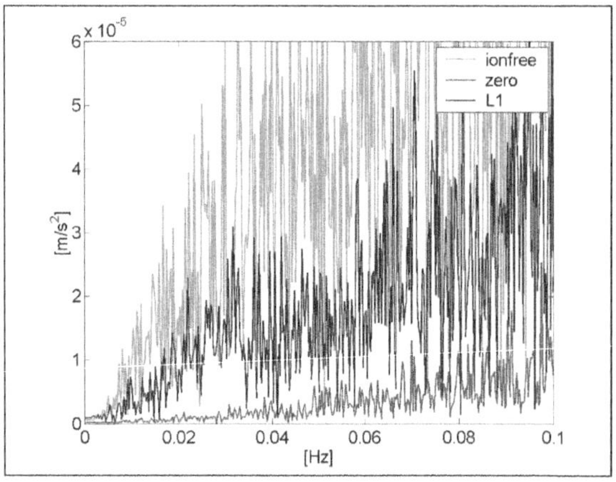


**Fig. 4** Frequency spectrum of kinematical accelerations

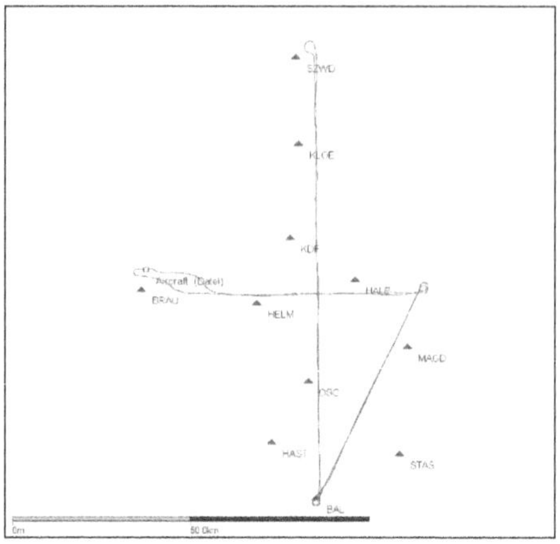

**Fig. 5** Trajectory of flight tests with reference stations

Also the kinematical accelerations based on GNSS observations are firstly investigated in static tests. The measurements of two geodetic GNSS receivers with data rate of 1 Hz are used to generate a phase solution first in a zero baseline configuration then with a baseline length of 50 m in a standard multipath environment. These results basing on L1 only and ionospheric free data are differentiated twice to get the required accelerations. The corresponding frequency spectra allow an evaluation of different error sources. The L1 phase noise as the sole error source in the zero baseline test cause only very small discrepancies from the static observation condition. Also the L1 solution with atmospheric and multipath errors can fulfil the proposed requirements of airborne gravimetry. The amplitudes are smaller than 2 mGal up to the frequency threshold of 0.04 Hz. The tenfold higher noise level of the ionospheric free linear combination, however, causes a significant decrease in accuracy (see figure 4). If the filtered acceleration values itself are analysed a RMS of 1.5 mGal can be computed for the L1 solution, a value of 5 mGal for the ionospheric-free result. So from this point of view the single frequency observations must be preferred, but the influences of ionospheric errors should be studied using practical test data.

## 5 Practical flight tests

In order to approve the performance of the described airborne gravimetry system practical flight test are carried out with a Do 128-6 aircraft in March 2004 in the middle of Germany near Magdeburg. The average speed was 70 m/s with height over ground of 300 m.

The flight path in an area of 120 to 80 km with a gravity disturbance of 60 mGal is presented in figure 5. During the observation periods specific forces up to 0.2 $m/s^2$ reduced to 1 km spatial resolution can be measured. The data of one testflight with 12 reference stations along the trajectory should be used to stress the results of some further investigations concerning the GNSS error influences on the derived kinematical accelerations. It must be pointed out that the following results refer to a spatial resolution of 1 km generated with a Tchebychev FIR-Filter.

According to equation 4 the first error source is the ambiguity term. An undetected cycle slip of only 2 phases leads to a difference in the aircraft acceleration of more than 100 mGal, but it is not critical to detect its occurrence.

Caused by its short term behaviour, the clock error of receiver and satellite affect the pseudorange accelerations in a dominant way.

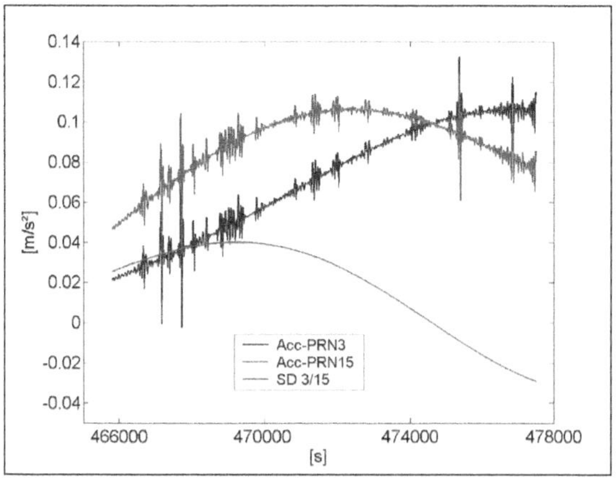


**Fig. 6** Influence of receiver clock error

An example is shown in figure 6 during a static observation case. The high correlation in the high frequency signal, eliminated by calculating the single difference, is demonstrated significantly. Therefore double difference observations must be the input data for kinematical acceleration processing.

The influence of the atmosphere for this application is discussed in many papers. Using real GNSS observations it is very difficult to separate ionospheric and tropospheric errors from other influences. So we use a GNSS signal simulator to recreate the satellite signals during the real flight test received on the aircraft and on defined reference stations in combination with the same GNSS receiver installed in the aircraft. The real and simulated aircraft dynamics differ only between 1 and 10 mGal. Using the signal simulator it is possible to define different scenarios with and without modelled atmospheric errors. The result is shown in figure 7.

A comparison of two identical simulations without any modelled errors leads to a standard deviation of 0.2 mGal, caused by receiver noise. The value for the difference between error free and atmospheric error solution is 0.8 mGal. Consequently is can be pointed out, that standard atmospherical conditions, cause only small acceleration errors. In this case it is not required to use two frequency observations or a very dense reference network. This result can be emphasized if the computed kinematical accelerations of the aircraft in regard to different reference station are compared to each other. There are no significant variations between two reference stations with 20 km and 90 km baseline (see figure 8).

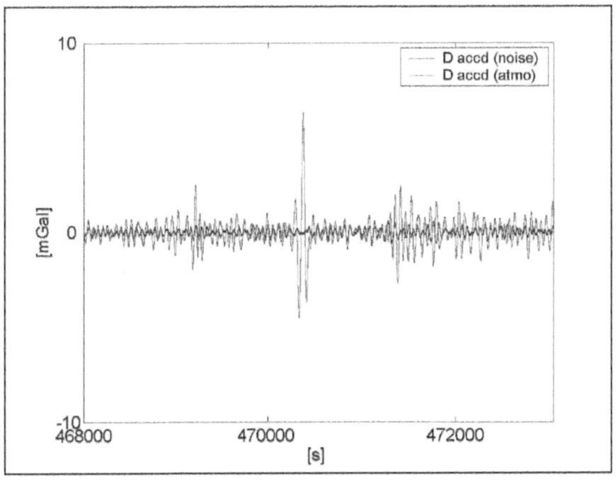


**Fig. 7** Influence of atmospherical error

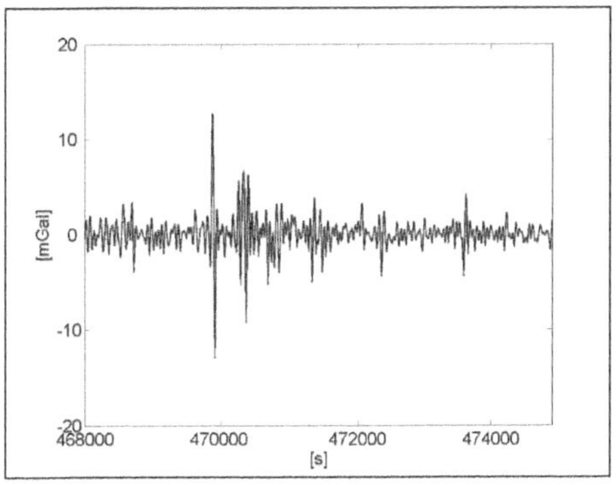


**Fig. 8** Accuracy level of kinematical accelerations

Errors depending on the baseline length, like e.g. ionospherical influences, affect the acceleration solution only in a secondary dimension. Definitely The more long term character of these errors is definitely a reason for this result.

The differences of the aircraft accelerations using two reference stations during one test flight are presented in figure 8. The RMS error, as a first indicator for the accuracy level of derived kinematical accelerations, is 2.1 mGal.

A multi antenna configuration on the aircraft allows simultaneous 1 Hz observations of L1 GNSS measurements with fixed baselines between 1 m and 13 m. Using the gyro rates of the INS the lever arm effects can be corrected. The results of these single antennas are correlated by these corrections and the same satellite constellation, but averaging calculations can reduce the important influence of receiver noise. The differences in aircraft acceleration derived by two GNSS receivers are presented in figure 9.

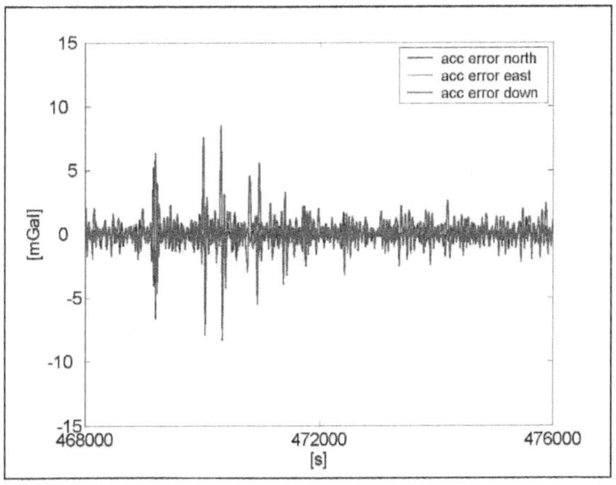


**Fig. 9** Difference of aircraft accelerations between different GNSS aircraft antennas

## 6 Integration Processing

Concerning the integration processing only some discussing points should be highlighted. In order to calculate the full gravity vector a combination on acceleration level is taken into account. In a first approach we use a modified Kalman filter algorithm presented in Kwon and Jekeli 2001 avoiding form filters to separate the gravity signal.

In this case the state vector consists of accelerometer, gyro and orientation errors. The observation vector is directly the difference between measured specific forces, kinematical accelerations and normal gravity in an inertial coordinate system. This term contains the gravity disturbance as well as systematic and stochastic sensor errors. An accurate estimation or suppression of these errors during the filtering process is required to provide the gravity information. As investigated in the previous processing important aspects are also the time synchronisation and the equal filter algorithms of GNSS and INS data, the reduced error estimation performance of the Kalman Filter caused by the low dynamics of the input signal and the correlated dynamic reduction of specific forces and gyro rates. The last point is taken into account by low-pass filtering of direction cosine elements.

Figure 10 shows the observed gravity signal for the down component in the observation vector before the estimation of systematic INS errors. Especially using the blue line, representing the required spatial resolution for airborne gravimetry (1 km), the correlation of the gravity signal between identical flight legs can be pointed out. Further processing should be able to evaluate the performance of the described airborne gravimetry system.

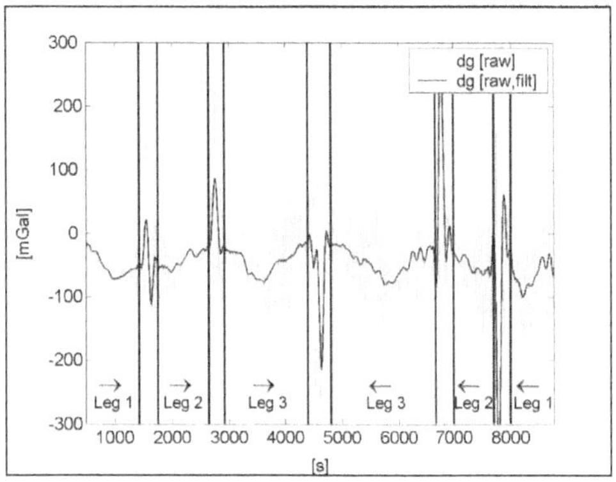


**Fig. 10** Gravity signal in raw sensor data

## 7 Acknowledgments

The investigations and developments during the described airborne gravimetry project are in the context of the "GEOTECHNOLOGIEN" program "observation of the system earth from space" founded by the German Federal Ministry of Education and Research (BMBF) and the German Research Council (DFG). In cooperation with two other research institutes and two industrial partners within the project the performance of different airborne gravimetry systems should be investigated.

## 8 References

Bruton A.M., Schwarz K.P., Ferguson S., Kern M., Wei M. (2002). Deriving acceleration from DGPS: Towards higher resolution applications for airborne gravimetry, GPS Solutions, Vol. 5 No. 3.

Czombo J. (1994). GPS accuracy test for airborne gravimetry, Proceedings of ION-GPS-1993 Technical Meeting, Salt Lake City, Utah

Eissfeller B., Spietz P.(1989). Basic filter concepts for the integration of GPS and an inertial ring laser gyro strapdown system, Manuscripta Geodetica 14: 166-182

Hehl K. (1992). Bestimmung von Beschleunigungen auf einem bewegten Träger durch GPS und digitale Filterung, Schriftenreihe des Studienganges Vermessungswesen, University FAF Munich, Heft 43, 1992

Jekeli C. (1994). On the computation of vehicle accelerations using GPS phase measurements, International Symposium on Kinematic Systems in Geodesy, Geomatics and Navigation, Banff, Alberta, Canada, 1994

Kleusberg A., Goodacre A., Beach R.J. (1989). On the use of GPS for airborne gravimetry, 5th Int. Geodetic Symposium on Satellite Positioning, Las Cruces, New Mexico, March 1989

Kreye Ch., Hein G.W. (2003) GNSS based kinematic acceleration determination for airborne vector gravimetry –methods and results-, Proceedings of ION GPS/GNSS 2003 Portland, Oregon, September 2003

Kwon J. H., Jekeli C. (2001). A new approach for airborne vector gravimetry using GPS/INS, Journal of Geodesy 74, 690-700

Wei M., Schwarz K.-P. (1994). An error analysis of airborne vector gravimetry, International Symposium on Kinematic Systems in Geodesy, Geomatics and Navigation, Banff, Alberta, Canada, 1994

# Network Approach versus State-space Approach for Strapdown Inertial Kinematic Gravimetry

Assumpció Térmens. Institut Cartogràfic de Catalunya.
Parc de Montjuic. 08038-Barcelona. Spain

Ismael Colomina. Institute of Geomatics.
Parc Mediterrani de la Tecnologia - Campus de Castelldefels. 08860-Castelldefels. Spain

**Abstract.** The extraction of gravity anomalies from airborne strapdown INS gravimetry has been mainly based on state-space approach (SSA), which has many advantages but displays a serious disadvantage, namely, its very limited capacity to handle space correlations (like the rigorous treatment of cross-over points). This paper examines an alternative through the well known geodetic approach, where the INS differential mechanization equations are interpreted as a least-squares network parameter estimation problem. The authors believe that the above approach has some potential advantages that are worth exploring. Mainly, that modelling of the Earth gravity field can be more rigorous than with SSA and that external observation equations can be better exploited.

**Keywords.** INS/GPS, airborne gravimetry, kinematic gravimetry, geoid determination, INS calibration, network approach (NA), state-space approach (SSA).

## 1 Motivation

A relatively recent technique in the field of airborne kinematic gravimetry is the combined use of strapdown inertial navigation systems (strapdown INS or SINS) —or inertial measurement units (IMU)— and the Global Positioning System (GPS) —Schwarz (1985). We will refer to it as INS/GPS-gravimetry. INS/GPS-gravimetry uses the differences between the linear accelerations measured by the accelerometers of an IMU and the accelerations derived from GPS. INS/GPS-gravimetry is mainly affected by two error sources: short term GPS-derived acceleration errors and long term INS inertial sensor errors —Schwarz and Li (1995).

For geoid determination applications, short term errors —i.e., the noise of GPS-derived accelerations— have been identified as one of the limiting factors of the technique. Fortunately, the situation is likely to improve significantly with the advent of the European global navigation satellite system Galileo because of its higher signal-to-noise ratio and with the subsequent use of hybrid Galileo/GPS receivers.

On the long wavelength side of the problem, the correct measurement of gravity —or, rather, of the anomalous gravity field— with INS/GPS-gravimetry depends on the correct separation of the INS/GPS errors from the actual variations of the gravity field itself (now, the long-wavelength bias stability is the limiting factor). This separation is, in principle, feasible because of the different characteristics of the two signals: errors of the inertial sensors can be reasonably modeled as time functions, whereas the variations of the gravity field are, strictly, spatial functions. (Understandably, so far, most of the research has focused on the INS/GPS short wavelength errors as the practical use of the technique and its competitiveness with traditional terrestrial gravimetry is bounded by, moderate to high, precision and resolution thresholds.) An improvement of the calibration of inertial sensors may be seen as an improvement of the long wavelength errors of INS/GPS-gravimetry. By doing so, we are not only achieving an overall improvement of INS/GPS-gravimetry but, in particular, we are extending its spectral window of applicability. This extension might be instrumental to the integrated use of GOCE gravimetry and INS/GPS-gravimetry as the sole means of gravimetry for geoid determination.

In this paper, we investigate algorithms to better calibrate the systematic errors of the inertial sensors. More specifically, we investigate an alternate procedure to the traditional Kalman filtering and smoothing. The advantage of the "new" procedure is that it can assimilate all the information available in a gravimetric aerial mission; from ground gravity control to the cross-over conditions, among other observational information types. The proposed procedure is nothing else than *geodesy as usual* in that we redefine the INS/GPS-gravimetry problem as a network adjustment problem —early studies can be seen at Forsberg (1986) for least-squares methods in land-based and helicopter-based inertial gravimetry.

Last, we note that better and more reliable algorithms for inertial sensor calibration may allow the use of low noise inertial sensors even if they suffer from large drifts. This, in turn, has a positive impact on the low frequency end of the INS/GPS-gravimetry spectral window.

## 2 INS/GPS-gravimetry: geodesy as usual

So far, extraction of gravity anomalies from INS/GPS-gravimetry has been mainly based on a state-space approach (SSA): the output of the stochastic dynamical system defined by the INS mechanization equations is Kalman-filtered and -smoothed with the GPS-derived positions and/or velocities – see Schwarz (1985), Wei and Schwarz (1990), Schwarz and Li (1995), Tomé (2002).

In INS/GPS-gravimetry, the separation of the INS/GPS errors from the variations of the gravity field is obtained by the use of appropriate models —e.g., stochastic differential equations— for the IMU sensor systematic errors and for the gravity field anomalies. Given the INS mechanization equations, the IMU calibration equations and the gravity field variation equations (sic), the SSA generates "optimal" estimates for the IMU trajectory (position, velocity and attitude), for the IMU errors and for the gravity field differences with respect to some reference gravity model.

In INS/GPS-gravimetry, the SSA is essentially given —Wei and Schwarz (1990)— by

$$\begin{aligned} \dot{r}^e &= v^e \\ \dot{v}^e &= R^e_b(f^b + w^b_f) - 2[\omega^e_{ie}\times]v^e + g^e \\ \dot{R}^e_b &= R^e_b\,[(\omega^b_{ib} + w^b_\omega)\times] - [\omega^e_{ie}\times]R^e_b \end{aligned} \qquad (1)$$

where $r^e$ and $v^e$ are the position and velocity vectors in the Conventional Terrestrial frame ($e$); $R^e_b$ is the transformation matrix from the body frame ($b$) to the $e$-frame; $\omega^e_{ie} = (0, 0, \omega_e)^T$ where $\omega_e$ is the rate of Earth rotation; $g^e$ is the gravity vector as a function of $r^e$; $w^b_f$ and $w^b_\omega$ are the generalized white-noise processes of the specific force ($f^b$) and angular velocities ($\omega^b_{ib}$), inertial observations respectively.

The numerical solution of this system can take many different forms which may be model-based or not, see Hammada and Schwarz (1997). It should be noted that hardly any of the active groups working on these problems uses Kalman filtering as a standard procedure today. Typically a two-step procedure is employed: in a first stage, FIR filtering or something similar to take care of time-dependent errors, and in a second stage, a cross-over adjustment to take care of the spatial structure of the gravity field.

The key to overcome SSA limitations is to look at the system equations (1) as a stochastic differential equations (SDE) that, through discretization, leads us to a time dependent geodetic network as discussed in —Térmens and Colomina (2003), Colomina and Blázquez (2004)— for geodetic, photogrammetric and remote sensing applications. A time dependent network is a network such that some of its parameters are time dependent or, in other words, stochastic processes. A time dependent network can be seen as a classical network that incorporates stochastic processes and dynamic models. A classical network can be seen as a particular case of a time dependent network.

To solve a time dependent network is to perform an optimal estimation of its parameters which may include some stochastic processes. As usual, the solution of the network will end in a large, single adjustment step where all parameters, time dependent and independent, will be simultaneously estimated.

In a time dependent network we may have static and dynamic observation models. A static observation model is a traditional observation equation. A dynamic observation model —or a stochastic dynamic model— is an equation of the type

$$f(t, \ell(t) + w(t), x(t), \dot{x}(t)) = 0 \qquad (2)$$

where $f$ is the mathematical functional model, $t$ is the time, $\ell(t)$ is the time dependent observation vector, $w(t)$ is a white-noise generalized process vector, $x(t)$ is the network parameter vector and $\dot{x}(t)$ the time derivative of $x(t)$. Note that $x(t)$ contains stochastic processes that, in particular, may be random constants thus including traditional time independent parameters. The discretization of the dynamic observation models together with the static observation models and further network least-squares adjustment will be referred to as the network approach (NA).

In general, NA has many potential advantages compared to SSA: parameters may be related by observations regardless of time; networks can be static and/or dynamic; covariance information can be computed selectively; and variance component estimation can be performed. In the context of INS/GPS-gravimetry the authors believe that some of the NA potential advantages are significant: modeling of the Earth gravity field can be more rigorous than with the SSA; external observational information can be better exploited; and more information for further geoid determination is produced. The main drawback of NA is that it cannot be applied to real-time INS/GPS navigation but this is certainly not an issue for a geodetic gravimetric task.

# 3 INS/GPS-gravimetry models for the NA

In this section we review the dynamic and static observation models that can be assimilated by the NA for INS/GPS-gravimetry. We note that the set of dynamic observation models corresponds to what is called *the system* in stochastic modeling and estimation. Analogously, the set of static observation models corresponds to what is called *the observations*. In the context of time dependent networks —Colomina and Blázquez (2004)— the names dynamic and static observation models are used to highlight the fact that we build our network from observations that contribute to the estimation of parameters either through dynamic or static equations.

## 3.1 Dynamic observation models

The dynamic observation models are, essentially, two. One model is the set of the INS mechanization equations and the other model expresses the "continuity" of gravity along the aircraft trajectory.

The mathematical model associated to SINS navigation is given by the well-known mechanization equations (1), that are usually extended with the angular rate sensors and accelerometers calibration states and models. The choice of these models has to guarantee that the estimated calibration parameters will not absorb other kind of effects, specially anomalous gravity. Investigations published in Nassar et al (2003) show that a linear calibration model is not sufficient, but in this paper, to fix the ideas and for the sake of simplicity we restrict intentionally the calibration states to time dependent biases:

$$\begin{aligned}\dot{r}^e &= v^e \\ \dot{v}^e &= R_b^e(f^b + w_f^b + a^b) - 2[\omega_{ie}^e \times]v^e + g^e \\ \dot{R}_b^e &= R_b^e[(\omega_{ib}^b + w_\omega^b + o^b)\times] - [\omega_{ie}^e\times]R_b^e \\ \dot{o}^b &= F_{gyr}(o^b) \\ \dot{a}^b &= F_{acc}(a^b)\end{aligned} \tag{3}$$

where $F_{gyr}$ and $F_{acc}$ are the calibration model functions of the angular rate sensors ($o^b$) and accelerometers biases ($a^b$). (Needless to say, the calibration functions and the calibration states depend on the type of sensors.)

The system (3) can be extended with a new mathematical model —GDT model— that shows the changes of the gravity disturbance along the trajectory of a moving vehicle with respect to time. The changes of the gravity vector $g^e$ along the trajectory with respect to time can be given —Jekeli (2001)and Schwarz and Wei (1995)— by

$$\dot{g}^e = (\bar{G}^e - [\omega_{ie}^e\times][\omega_{ie}^e\times])v^e = \Delta G^e v^e, \tag{4}$$

where $\bar{G}^e$ is the gravitational gradient tensor. For the gravity disturbance vector, similar differential equations — $\dot{\delta g}^e = \Delta G^e v^e$ — are obtained. If no gravity gradiometer measurements are available $\Delta G^e v^e$ can typically be modelled by simple stochastic models. Then, to fix the ideas and to simplify the modeling, the gravity disturbance can be represented by a random walk.

Now the dynamic observation models formed by SINS mechanization equations (3) and GDT model are:

$$\begin{aligned} VEL: &\quad \dot{r}^e = v^e + w_0 \\ FB: &\quad \dot{v}^e = R_b^e(f^b + w_f^b + a^b) - 2[\omega_{ie}^e\times]v^e + \\ &\quad +\delta g^e + \gamma(r^e) \\ WIB: &\quad \dot{R}_b^e = R_b^e[(\omega_{ib}^b + w_\omega^b + o^b)\times] - [\omega_{ie}^e\times]R_b^e \\ OB: &\quad \dot{o}^b = F_{gyr}(o^b + w_o^b) \\ AB: &\quad \dot{a}^b = F_{acc}(a^b + w_a^b) \\ GDT: &\quad \dot{\delta g}^e = w_g^e \end{aligned}$$

These models are time dependent equations of the type (2), where $\ell(t) = (f^b, \omega_{ib}^b)^T$ and $x(t) = (r^e, v^e, R_b^e, \delta g^e, a^b, o^b)^T$.

## 3.2 Static observation models

The static observation models considered are: the coordinate update point (CUPT), the velocity update point (VUPT), the zero velocity update point (ZUPT), the gravity update point (GUPT), the static gravity update point (SGUPT) and the cross-over points (XOVER).

**CUPT model.** A coordinate update is a point where the position of the platform is known from an independent procedure (usually GPS). The CUPT equation is $p_0 + w_p = r^e$.

**VUPT model.** If instead of the position the speed is also known, the associated equation is $v_0 + w_v = v^e$.

**XOVER model.** Usually, the trajectory of gravimetric flight follows a regular pattern. The intersection points of the trajectory are known as cross-overs. Since, in practice, actual intersections are hard to materialize, horizontal observations with small height differences are allowed for the cross-over points. The cross-over observation equation imposes that gravity is the same in coincident points.

**ZUPT model.** The zero velocity update is based on $v^e = 0$ and it is widely used in terrestrial inertial surveying. In a gravimetric flight, it can only be applied at the beginning and at the end of the survey. This model is equivalent to a VUPT with $v = 0$.

**SGUPT model.** For every ZUPT equation, gravity can be considered as a constant function. It can be seen as a XOVER observation.

**GUPT model.** If gravity is known in some point of the trajectory, the following equation is obtained: $g_0 + w_g = \delta g^e + \gamma(r^e)$.

## 3.3 Discretization of the dynamic observation equations

The dynamic observations equations are SDE. SDE arise naturally from real-life ODE (ordinary differential equations) whose coefficients are only approximately known because they are measured by instruments or deduced from other data subject to random errors. The initial or boundary conditions may be also known just randomly. In these situations, we would expect that the solution of the problem be a stochastic process.

Like in ODE theory, certain classes of SDE have solutions that can be found analytically using various formulas, and others —most of them— have no analytic solution. There are several numerical techniques to solve SDE — Kloeden and Platen (1999). All of them are based on their correct stochastic discretization which is not a trivial issue.

Now, in this paper, we will limit the discussion to a simplest approximation method: the explicit midpoint method or leap-frog method. Consider for a function $x[n]$, $\dot{x}[n] = \Lambda(x, n) = (x[n+1] - x[n-1])/(2\delta t)$. This method is not generally acceptable, because the existence of weak stabilities. However, in this paper, it suffices to illustrate the use of NA for INS/GPS-gravimetry.

Then, the previous equations —the dynamic and the static— can be transformed into a finite set of observation equations. They can be discretized and afterwards written as $\ell + w = F(x)$, where $\ell$ are the observations (in our case $f^b$, $\omega^b_{ib}$), $w$ are the residuals of $\ell$ and $x$ are the parameters to be determined ($r^e$, $v^e$, $\delta g^e$, $a^b$, $o^b$, $q$):

$$\begin{array}{ll}
VEL: & 0 + w_0 = v^e[n] - \Lambda(r^e, n) \\
FB: & f^b[n] + w^b_f = -a^b[n] - \\
 & -R^b_e(q, n)\,\{\delta g^e[n] + \gamma(r^e[n]) - \\
 & -2[\omega^e_{ie}\times]\,v^e[n] - \Lambda(v^e, n)\} \\
WIB: & \omega^b_{ib}[n] + w^b_w = -o^b[n] + R^b_e(q, n)\,\omega^e_{ie} + \\
 & +2M_q(q, n)^T\,\Lambda(q, n) \\
OB: & 0 + w_0 = \Lambda(o^b, n) - F_{gyr}(o^b[n]) \\
AB: & 0 + w_0 = \Lambda(a^b, n) - F_{acc}(a^b[n]) \\
GDT: & 0 + w_0 = \Lambda(\delta g^e, n)
\end{array}$$

## 3.4 Discretization of the static observation

Following the same procedure as in the previous section —discretize and arrange to $\ell + w = F(x)$— for each static observation equations, we obtain:

$$\begin{array}{ll}
CUPT: & p_0 + w_p = r^e[n] \\
VUPT: & v_0 + w_v = v^e[n] \\
GUPT: & g_0 + w_g = \delta g^e[n] + \gamma(r^e[n]) \\
XOVER: & 0 + w_0 = \|\delta g^e[n] + \gamma(r^e[n])\| - \\
 & -\|\delta g^e[k] + \gamma(r^e[k])\|
\end{array}$$

## 3.5 Final INS/GPS-gravimetric network

As a result of the preceeding discussion, our problem can be reduced to the solution of the system of equations formed by the dynamic models —VEL, FB, WIB, AB, OB, GDT— and the static models —CUPT, GUPT, VUPT, XOVER. The above mathematical models have been implemented in the GeoTeX/ACX software system —Colomina et al. (1992)— developed at the ICC

| Model | $l$ | $x$ |
|---|---|---|
| IMU data | | |
| VEL | 0 | $r^e[n-1]$, $r^e[n+1]$, $v^e[n]$ |
| FB | $f^b[n]$ | $a^b[n]$, $q[n]$, $\delta g^e[n]$, $r^e[n]$ $v^e[n-1]$, $v^e[n]$, $v^e[n+1]$ |
| WIB | $\omega^b_{ib}[n]$ | $o^b[n]$, q[n-1], q[n], q[n+1] |
| OB | 0 | $o^b[n-1]$, $o^b[n]$, $o^b[n+1]$ |
| AB | 0 | $a^b[n-1]$, $a^b[n]$, $a^b[n+1]$ |
| GDT | 0 | $\delta g^e[n-1]$, $\delta g^e[n+1]$ |
| GPS data | | |
| CUPT | $p_0$ | $r^e[n]$ |
| VUPT | $v_0$ | $v^e[n]$ |
| Additional data | | |
| GUPT | $g_0$ | $\delta g^e[n]$, $r^e[n]$ |
| XOVER | 0 | $r^e[n]$, $r^e[k]$, $\delta g^e[n]$, $\delta g^e[k]$ |

Table 1: INS/GPS-gravimetry mathematical models for NA.

since 1988. The associated observations and parameters for each model are shown in Table 1.

Solving the system is to perform an optimal estimation of its parameters in the sense of least-squares; i.e. the expectation of the parameters and their covariance is known. We emphasize the possibility to compute the covariance of a limited number of selected parameters —perform a selective inversion of the normal matrix— and the variance component estimation. In our case, these selected parameters will be the gravity disturbances and their covariances for further geoid determination.

# 4 Implementation issues

First consider that IMU data is collected at *Hz(IMU)* frequency ($50Hz \leq Hz(IMU) \leq 200Hz$ for operational environments), the GPS data at *Hz(GPS)* frequency ($1Hz \leq Hz(GPS) \leq 5Hz$) and the additional data can be collected a different frequencies. Second, consider that we want to compute all the parameters at $Hz(IMU)$ frequency. Then the size of the system increase considerably. This can be shown in Table 2 for some operational environments.

Although the large size of the associated normal matrices, they are essentially of the band-bordered type and we can apply sparse matrix techniques, fill-in reduction techniques and memory-to-disk paging to solve the system of linear equations. If this is done, the computational load is comparable to that of SSA.

| | Case 1 | Case 2 | Case 3 | Case4 |
|---|---|---|---|---|
| $H_z(GPS)$ | 1 | 1 | 5 | 2 |
| $H_z(IMU)$ | 200 | 50 | 200 | 50 |
| $t$ (s) | 10 800 | 14 400 | 14 400 | 11 301 |
| $N_{XOVER}$ | | 30 000 | | 332 150 |
| $N_{IMU}$ | 2 160 000 | 720 000 | 2 880 000 | 565 050 |
| $N_{GPS}$ | 10 800 | 14 400 | 72 000 | 22 602 |
| $N_{GUPT}$ | | | | 1 907 |
| $N^*_{eq}$ | 38 912 400 | 13 033 200 | 52 056 000 | 10 572 763 |
| $N_{par}$ | 38 880 000 | 12 960 000 | 51 840 000 | 10 170 900 |
| $r_b$ | 0.0008 | 0.0056 | 0.0042 | 0.0380 |

Table 2: Information data of some operative environments at the ICC and UofC. Case 1: Applanix flight (ICC). Case 2: CASI flight (ICC). Case 3: GeoMobil (ICC). Case 4: Rockies'95 (UofC). $N_{eq} = N^*_{eq} + N_{al}$. $N^*_{eq} = 18N_{IMU} + 3N_{GPS}$. $r_b = \frac{N_{eq}-N_{par}}{N_{eq}}$ (average redundancy).

If it is still necessary to reduce the size of the system, it is possible to take into account that some of the parameters —or unknowns— have a slow variation in time and subsets of them can be grouped in a single one. This can considerably reduce the number of unknowns. For instance, it should be possible to compute $a^b$, $o^b$, $\delta g^e$ at $Hz(GPS)$ frequency and $r^e$, $v^e$ and $q$ at $Hz(IMU)$ frequency. In this case, $Npar$ is reduced to half (19537200 in Case 1, 6609600 in Case 2, 26568000 in Case 3 and 5288868 in Case 4 respectively).

Table 2 also shows the small average redundancy of the systems to be solved ($0.3 \leq r_b \leq 0.9$ for geodetic and photogrammetric bundle networks). It is well-known that accurate least-squares adjustment needs high redundancy. Looking at Table 2, the number of IMU data and GPS data cannot be increased without increasing the number of parameters to be determined. Then to increase the redundacy of the system, we have to increase the number of additional equations. So, it is advisable to use all the existing observational information to increase the redundancy of the system.

It is under investigation the study of how to plan the surveys to increase them: cross-overs (repetition or/and intersection of flight lines), ZUPT periods at the beginning and at the end of the survey, upwarded gravity points, etc.

# 5 Conclusions, ongoing work and further research

It has been seen that the determination of the anomalous gravity by inertial techniques is critically employee of the capacity to separate the errors of the system of the effects of the gravitational field. This separation is based mainly on the different characteristics from both signals: the errors of the inertial sensors (INS) can be reasonably considered like time function, whereas the variations of the gravitational field are only function of the position.

Actually, SINS airborne gravimetry has been mainly based on Kalman filtering (SSA approach). In Kalman, the error separation is obtained, at first, by the use of different correlations of the bias and the variations of the gravitational field. The advantage of Kalman filter is the good physical description of the instruments errors, but it displays a serious disadvantage by the incapacity of handling space correlations, like the condition of cross-over points.

It has been presented that the development of an adjustment method in genuinely geodetic post-process with the explicit purpose to determine precise gravity anomalies taking advantage at maximum the space characteristics of the gravitational field. This method tries to jointly deal with the bias like function the time and the anomalous gravity like function of the position, by means of the resolution of the corresponding system of equations.

This system of equations can be, at first, very large and its redundancy small. It is under investigation some methods (numerical and geodetic) to handle with and to increase its redundancy.

# 6 Acknowledgements

The second author of this paper has been supported in his research by the OTEA-g project of the Spanish National Space Research Programme (reference: ESP2002-03687) of the Spanish Ministry of Education and Science.

# References

Colomina,I., Navarro,J., Térmens,A., 1992. GeoTeX: a general point determination system. International Archives of Photogrammetry and Remote Sensing, Vol. 29 - B3, International Society of Photogrammetry and Remote Sensing, pp. 656–664.

Colomina,I., Blázquez,M., 2004. A unified approach to static and dynamic modelling in photogrammetry and remote sensing. ISPRS International Archives at Photogrammetry, Remote Sensing and Spatial Information Sciences, Vol. 35 - B1, Comm. I, pp. 178–183.

Forsberg,R., 1986. Inertial Geodesy in Rough Gravity Field. UCSE Report N.30009, University of Calgary, 1986, 71 pages.

Hammada,Y., Schwarz,K.P., 1997. Airborne Gravimetry: Model-based versus Frequency-domain Filtering Approaches. Proc. of the Int. Symp. on Kinematic Systems in Geodesy, Geomatics and Navigation, Banff, Canada, June 3-6, 1997. pp. 581–595.

Jekeli,C., 2001. Inertial Navigation Systems with Geodetic Applications. de Gruyter.

Kloeden,P.E., Platen,E., 1999. Numerical solution of Stochastic Differential Equations, Springer Verlag, New York, US.

Nassar,S., Schwarz,K.P., Noureldin,A., El-Sheimy,N., 2003. Modeling Inertial Sensor Errors using Autoregressive (AR) Models. Proc. ION NTM-2003. Annaheim, USA, January 22-24, 2003. pp. 116–125.

Schwarz,K.P., 1985. A Unified Approach to Post-mission Processing of Inertial Data. Bulletin Géodésique Vol. 59, N. 1, pp. 33–54.

Schwarz,K.P., Li,Y.C., 1995. What can airborne gravimetry contribute to geoid determination? IAG Symposium 64 "Airborne Gravimetry," IUGG XXI General Assembly, Boulder, CO, US pp. 143–152.

Schwarz,K.P., Wei,M., 1995. Inertial Geodesy and INS/GPS Integration (partial lecture notes for ENGO 623). Department of Geomatics Engineering. University of Calgary.

Térmens,A., Colomina,I., 2003. Sobre la corrección de errores sistemáticos en gravimetría aerotransportada. Proceedings of the 5. Geomatic Week, Barcelona, ES.

Tomé,P., 2002. Integration of Inertial and Satellite Navigation Systems for Aircraft Attitude Determination. Ph.D. Thesis. Department Applied Mathematics. Faculty of Sciences. University of Oporto.

Wei,M., Schwarz,K.P., 1990. A strapdown inertial algorithm using an Earth-fixed cartesian frame. *Navigation*, Vol. 37, No. 2, pp. 153–167.

# The Airborne Gravimeter CHEKAN-A at the Institute of Flight Guidance (IFF)

T. H. Stelkens-Kobsch
Institute of Flight Guidance,
Technical University of Braunschweig, Hermann-Blenk-Strasse 27, 38108 Braunschweig, Germany

**Abstract.** The Institute of Flight Guidance (IFF) of the Technical University of Braunschweig (TU BS) is involved in the development of airborne gravimetry since 1985. Fundamental examinations of airborne gravimeters were carried out between 1991 and 1993. In 1998 a high-precision two-frame inertial platform and a gravimeter sensor were purchased and modified for airborne application in cooperation with the Russian manufacturer ELEKTROPRIBOR.

Successful flight tests have been executed in the recent years. So far the resolution achieved is 2 km with a standard deviation of 3 mGal.

The two-frame inertial platform was extended to a three-frame INS by mounting a ring laser gyro on top of the platform. The gyro provides an additional degree of freedom (yaw) around the vertical axis.

Since 2001 the IFF is involved in a project funded by the German Federal Ministry of Education and Research. The goal of the project is to develop an airborne gravimetry system with a resolution of 1 mGal.

**Keywords.** Airborne gravimetry, gravimeter, inertial platform

## 1 Introduction

Due to the relatively high speed and low costs at which measurements can be made, an airborne gravimetry system provides an attractive alternative to conventional terrestrial and space based methods to determine the gravity field of the earth. An aircraft provides access to difficult terrain and uniform sampling, but the success of land and sea gravimetry cannot be transferred to airborne gravimetry easily. Airborne gravimetry is a big challenge on flight measurement technology and filtering techniques. It is of particular importance for the future of airborne gravimetry to increase accuracy and resolution. The IFF is involved in the development of a gravity measurement system using an inertial platform, a gravity sensor, a barometric sensor and kinematic differential carrier phase based GPS positioning.

The gravity disturbance signal may be obtained from the time-synchronised difference between the measurements of a precise inertial accelerometer signal (gravity sensor) and an altimeter signal (GPS differential carrier phase, barometric sensor). The gravity sensor measures specific force, which is composed of a gravity component and an aircraft acceleration component. The Global Positioning system (GPS) measures the aircraft motion only.

The two basic approaches for specific force measurement use strapdown inertial systems [Wei, Schwarz] or platform inertial systems [Olesen, Forsberg, Gidskehaug]. Damped inertial platforms are typical for commercial airborne gravimetry systems. The direction stabilisation of the gravity sensor is one of the fundamental problems, which has to be solved in all approaches of airborne gravimetry. The accuracy of the vertical specific force measurement is dependent on the performance of the vertical accelerometer and the levelling accuracy of the inertial platform. Aircraft dynamics and flight manoeuvres affect the accuracy of the vertical specific force measurement via platform levelling errors.

The inertial platform used for an airborne gravimetry system can be a three-axes or a two-axes gyro stabilised platform. For the two-axes inertial platform, the azimuth axis can be calculated by a strapdown calculation.

To carry out the airborne gravimetry research at IFF, a Russian sea gravimeter (CHEKAN-A) was acquired, modified and successfully put into operation onboard the experimental twin-engine turboprop aircraft Dornier Do 128-6. The platform was originally designed to be installed in submarines and was developed at the Central Scientific & Research Institute "ELEKTROPRIBOR".

In the first work step the focal points were the adjustment of the technology in order to satisfy the demanded high accuracy and the verification of the specifications. Many difficulties were to be overcome up to the operational readiness of the airborne gravimeter and the first flight tests with meaningful results. A detailed description of the executed work can be found in the interim report SFB 420 [Schänzer, Abdelmoula].

In this study the evolution of the gravimetry system will be described.

## 2 Platform description

Beside the sensors for altitude measurement the inertial platform is the decisive item of the sensor package necessary for airborne gravimetry. The platform supplies both the acceleration signal necessary for the determination of gravity anomalies and data necessary for different corrections. The measuring system supplied from the Russian manufacturer "ELEKTROPRIBOR" consists of two main parts: a control unit and a gyro-stabilised platform. In Fig. 1 the two main components are shown mounted in the experimental aircraft of the Institute of Flight Guidance of the Technical University of Braunschweig.

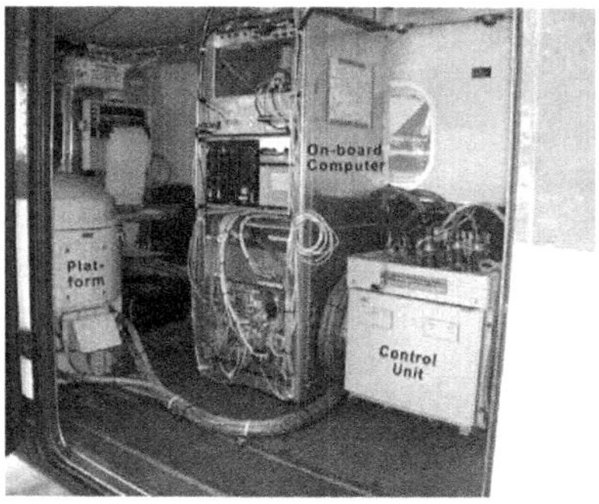

**Fig. 1** Gravity measurement system in the test aircraft

The inertial platform serves as the stabilisation of the gravity sensor's sensitivity axis in vertical direction. It is designed as a two-axes gyro stabiliser with a two-degree-of-freedom gyro and a gearless servo drive. Outer and inner rotating gimbals are supported by precise bearings and their axes are aligned parallel to the longitudinal and lateral vehicle axes correspondingly. A set of sensors is installed on the platform. A highly sensitive accelerometer (gravity sensor, Fig. 2) is located in the centre and is responsible for measuring the vertical acceleration. Two further accelerometers are mounted in the horizontal level and provide data for platform levelling. Each gimbal axis is provided with an angular position transducer (rotary transformer) and a brushless torque motor. The angular range of inner gimbal tilt is ±15° and of the outer gimbal ±30°.

The control unit contains the electronics for amplification, control and indication of functional efficiency.

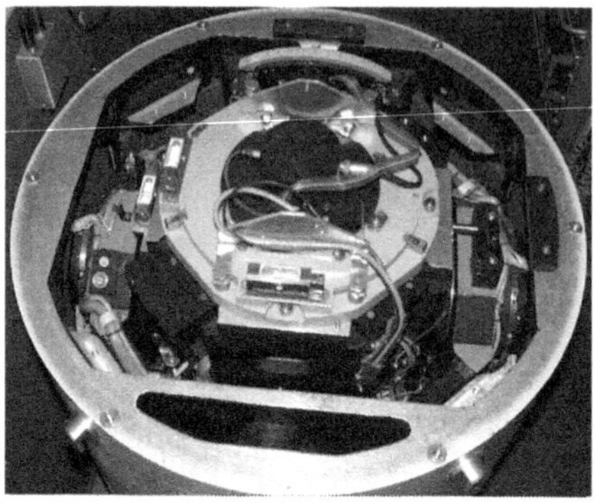
**Fig. 2** Inertial gravimeter platform

To compensate the influence of Earth rotation $\Omega_E$ and the relatively large drift ($D$=3°/$h$) of the stabilising gyro, the platform was originally controlled with an analogue internally damped stabilisation loop. However, this kind of levelling loop does not meet the requirements for an airborne gravity measurement system. For movements of submarines this kind of stabilisation is appropriate but after high dynamic course manoeuvres like turning flights, the analogue controller needs unacceptable long to realign the platform. Therefore this control was upgraded with a digital controller, which creates an additional degree of freedom (yaw) for the platform around the vertical axis. The azimuth information is acquired from a ring laser gyro, which is installed on top of the platform. In other words, by use of GPS-measurements and a heading gyro a three-axis-platform is established with a strapdown calculation.

## 3 Modelling of the gravimeter

The central element within the system is the gravity sensor (Fig. 3). The sensing element consists of two identical torsion frames with pre-stressed quartz filaments. On each filament a pendulum with a mass and a reflector are mounted. The pendulums are inversely arranged to minimize cross coupling effects. Under normal gravity conditions the pendulums are horizontally aligned.

The torsion frames (Double Quartz Elastic System) are mounted on a carrier which also holds a reference prism whose reflecting areas have the same orientation as the pendulum reflectors. A change of gravity causes a change of the pendulum angle according to $\xi = K \cdot \delta g$. $K$ describes the sensitivity of the quartz filament and is in the range of $0.36\ ^{arcsec}/_{mGal}$.

To enhance attenuation of the vibratory pendulum system the entire construction is installed in a case which is filled with a viscous fluid.

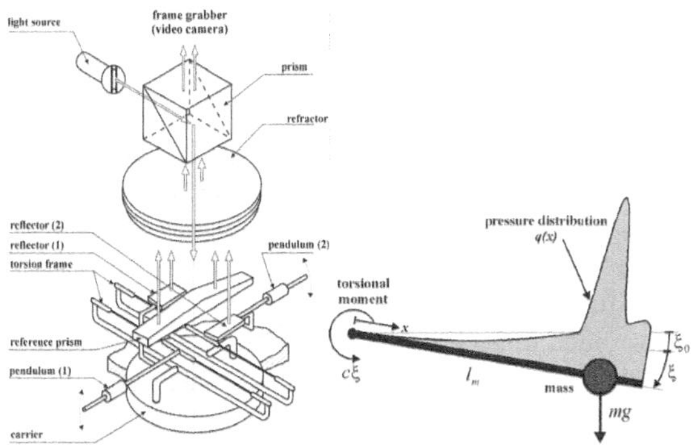


**Fig. 3** Construction and modelling of the gravity sensor

As the fluid used to attenuate the movement of the pendulums is viscous, the occurrence of friction between the pendulums and the fluid has to be considered. The basic approach is here the Newtonian law of friction [Kaufmann]. Besides the friction also buoyancy of pendulums in the fluid occurs. Steady state equilibrium results from the torsional moment of the quartz filaments, the weight of the mass and the buoyancy inside the fluid. The angle $\xi_0$ is the steady state deflection of the pendulum.

The mathematical description for a pendulum in viscous fluid shown in Fig. 3 is:

$$\begin{aligned} ml_m\ddot{\xi} &= -c\xi - \int_0^l q(x)x\,dx \\ &+ m(\delta g - a_z)l_m \cos(\xi + \xi_0) \\ &+ m a_x l_m \sin(\xi + \xi_0) \end{aligned} \tag{1}$$

The fluid attenuates motion. According to the specification of viscous fluids follows:

$$\int_0^l q(x)x\,dx = \left(d_1\mu + d_2\left|\dot{\xi}\right|\right)\dot{\xi}\,. \tag{2}$$

In equation (2) parameter $d_1$ describes friction and parameter $d_2$ describes drag. With equation (1) equation (2) can be written as:

$$\begin{aligned} &\ddot{\xi} + \frac{1}{ml_m^2}\left(d_1\mu + d_2\left|\dot{\xi}\right|\right)\dot{\xi} + \frac{c}{ml_m^2}\xi \\ &= \frac{\cos(\xi + \xi_0)}{l_m}(\delta g - a_z) + \frac{\sin(\xi + \xi_0)}{l_m}a_x \end{aligned} \tag{3}$$

In steady state the deflection comes to

$$\xi_\infty = \frac{ml_m \cos(\xi_\infty + \xi_0)}{c}\delta g_\infty\,. \tag{4}$$

The sensitivity of the sensor can be adjusted with $l_m$. The maximum deflection $\xi_{max}$ for the whole measurement range is approximately 1°.

Experiments demonstrated that pressure drag can be neglected in comparison to friction drag. This means $d_2\left|\dot{\xi}\right| \ll d_1\mu$ and therefore $d_2\left|\dot{\xi}\right|$ will be neglected in the following.

Equation (3) can now be written as:

$$\begin{aligned} &\ddot{\xi} + \frac{d_1\mu}{ml_m^2}\dot{\xi} + \frac{c}{ml_m^2}\xi \\ &= \frac{\cos(\xi + \xi_0)}{l_m}(\delta g - a_z) + \frac{\sin(\xi + \xi_0)}{l_m}a_x \end{aligned} \tag{5}$$

The system has two cutoff frequencies. As the system cannot sample very high frequencies the influence of $\ddot{\xi}$ can be neglected. This leads to a simplified equation which describes the sensor as a low pass:

$$\begin{aligned} &\frac{d_1\mu}{c}\dot{\xi} + \xi \\ &= \frac{ml_m \cos(\xi + \xi_0)}{c}(\delta g - a_z) + \frac{\sin(\xi + \xi_0)}{c}a_x \end{aligned} \tag{6}$$

In Fig. 4 the output signal of the gravity sensor is shown in comparison to the developed model.

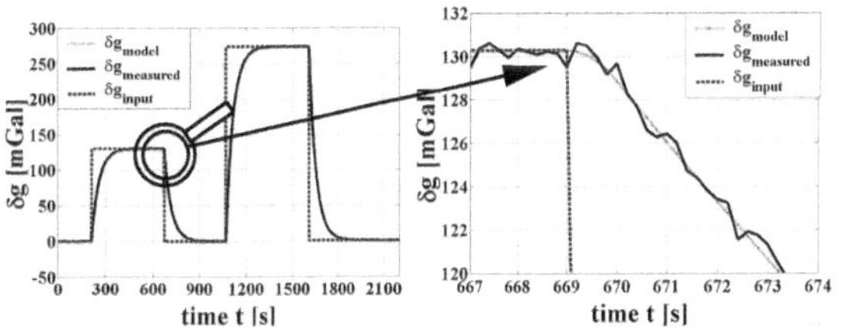


**Fig. 4** gravity sensor output signal

## 4 Platform stabilisation

To achieve a gravity measurement accuracy of 1 mGal onboard a light fixed-wing aircraft under the influence of horizontal accelerations (turbulence of air) a sophisticated platform levelling is required. This requirement can hardly be met by commercially available inertial platforms. The aircraft would furthermore have to maintain constant velocity with very high accuracy, which is quite difficult for light fixed-wing aircraft. Even the magnitude of horizontal acceleration is about 1 m/s$^{-2}$, dependent on air turbulence conditions. For a Schuler-tuned inertial platform accelerometer biases must be less than $5 \cdot 10^{-5}\,\mathrm{ms}^{-2}$ in order to achieve the demanded levelling accuracy. Gyro drifts must be less than $0.001°/h$ [Zhang]. This grade of inertial sensors is either not commercially available or too expensive to be used in civilian applications. To solve this problem, new techniques of using GPS measurements for damping the platform oscillation must be developed.

### 4.1 Velocity-aided platform

Inertial platform levelling errors can be physically damped by feeding measurements of a reference system into the platform levelling loop. The advantage of this approach over internally damped platforms is to provide the vertical accelerometer with a more stable environment and to physically reduce influences of horizontal accelerations on the measurement of vertical specific force.

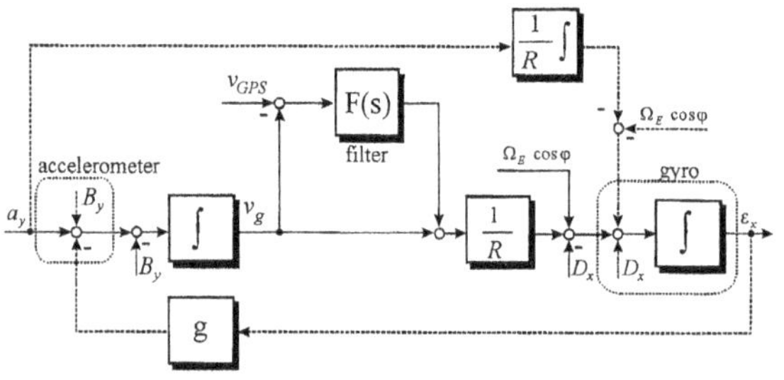


**Fig. 5** Levelling loop of the platform stabilised by GPS-velocity (East-channel)

In Fig. 5 the diagram of a levelling loop is shown, which is physically stabilised by GPS-velocity since there is no better reference system to date. Dashed lines in the diagram represent a physical feedback path of the gravity signal due to platform mislevelling and a forward path of local level frame rotation due to system movement and Earth rotation. The other part of the diagram is feeding velocity back into the platform levelling loop. The levelling error $\varepsilon_x$ is defined as the platform mislevelling with respect to reference ellipsoid. The Laplace transformation of the levelling error can be found from Fig. 5 as:

$$\hat{\varepsilon}_x = \frac{\hat{D}s - \frac{\hat{B}}{R}\left(1 + F(s)\right) + \frac{s}{R}F(s)\delta\hat{V}_{ref}}{s^2 + \frac{g}{R}\left(1 + F(s)\right)}, \qquad (7)$$

$$F(s) = k\frac{T_D s + 1}{T_I s + 1}$$

where $s$ is the operator of the Laplace transformation, $g$ is the magnitude of gravity and $R$ is the average Earth radius.

Equation (7) shows that the levelling error is not affected by system acceleration and the levelling loop can be stabilised by choosing values of $T_D$, $T_I$ and $k$. The characteristic polynomial $n(s)$ can be written in dependency of controller gains and Schuler frequency $\omega_S$:

$$n(s) = T_I s^3 + s^2 + \omega_s^2\left(T_I + kT_D\right)s + \left(k+1\right)\omega_s^2 \qquad (8)$$

This equation has three Eigenvalues. The aim is to determine the controller gains to get an accurate levelling loop. Many simulations and flight tests have been done in order to get the according values. These tests showed that the optimal behaviour of the platform is achieved if the three poles of the characteristic polynomial are located on the left part of a circle with the radius $\omega_0$.

The levelling loop is a low pass filter for accelerometer biases and a band-pass filter for gyro drift and GPS velocity. Errors within the frequency range close to the natural frequency have major effects on platform levelling errors.

### 4.2 Position-aided platform

In normal operation GPS velocity shows a time delay in relation to INS velocity due to complex signal processing in the GPS receiver. An inaccurate determination of this delay causes an excitation of the platform, which in turn causes an enlargement of platform misalignment. Despite these deficits the controller on velocity basis impresses by its simple practicability and its satisfying behaviour during a quasi-stable measuring process.

The time delay problem is basically treated, as the position information of GPS is used as reference for the platform. The position information received from GPS has a time delay of an order of magnitude smaller than the velocity information. Furthermore it is exactly assignable.

In order to carry out a control based on position information, the control loop must be changed as indicated in Fig. 6 (north channel). The problem of platform control on position basis is the stabilisation of the control loop. Now three integrations have to be performed in the loop and therefore it tends to be instable. The influence of the additional integrator must be compensated by a suitable selection of the filter $G(s)$. For the represented block diagram the following transfer function results for levelling errors of the platform:

$$\hat{\varepsilon}_y = \frac{\hat{D}s^2 - \frac{\hat{B}}{R}\left(s + \frac{G(s)}{R}\right) + s^2 \frac{G(s)}{R}\delta\hat{\varphi}_{ref}}{s^3 + \frac{g}{R}\left(s + \frac{G(s)}{R}\right)} . \quad (9)$$

The characteristic polynomial $n(s)$ becomes:

$$n(s) = (s + \omega_0)\left(s^2 + 2\zeta_1\omega_0 s + \omega_0^2\right)\left(s^2 + 2\zeta_2\omega_0 s + \omega_0^2\right) \quad (10)$$

The conditions regarding the critical band close to the natural frequency are almost the same as for velocity based control, but the delay of the system is now much smaller. Therefore the control on basis of position information outperforms the velocity based control.

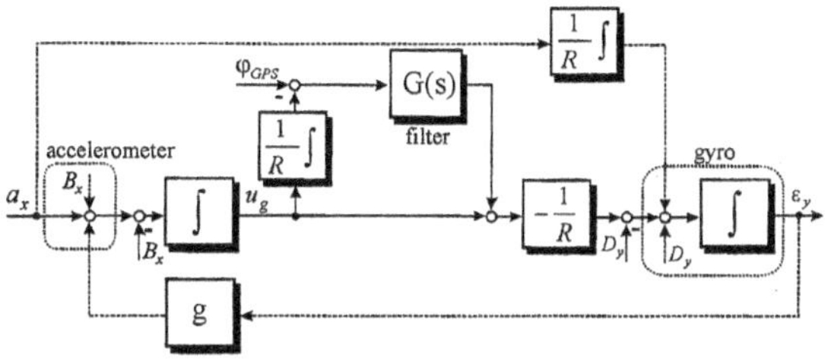


**Fig. 6** Levelling loop of the platform stabilised by GPS-position (North-channel)

### 4.3 Initial alignment

Initial alignment and navigation algorithms are fundamental to operate an inertial platform. The task of initial alignment is to level the platform and to find the azimuth angle. In addition, gyro drifts can be calibrated during alignment. After initial alignment, the platform is switched to navigation mode and is ready for gravity survey. Using the accelerometer outputs as input, the navigation algorithm computes system position and velocity as well as command rates for the mechanical gyro in real-time.

To obtain a highly accurate alignment result, a fine alignment and gyro calibration has to be performed. In a first step the controller gains are selected to obtain a narrow system bandwidth. The computed velocity can be determined using equation (11):

$$u = \frac{g(\Omega_E \cos\varphi(\sin\psi - \sin\psi_s) + D_y)}{s^2 + \frac{g}{R}(1 + F(s))}$$
$$v = \frac{g(\Omega_E \cos\varphi(\cos\psi - \cos\psi_s) - D_x)}{s^2 + \frac{g}{R}(1 + F(s))} \quad (11)$$

where $\varphi$ is the geodetic latitude, $\Omega_E$ is the Earth rotation, $u$ and $v$ are the horizontal velocity components, $D_x$ and $D_y$ are the components of gyro drift.

In steady state and under the assumption of constant gyro drift during fine alignment velocity signals are recorded as

$$u_\infty = \frac{R}{1+k}(\Omega_E \cos\varphi(\sin\psi - \sin\psi_s) + D_y)$$
$$v_\infty = \frac{R}{1+k}(\Omega_E \cos\varphi(\cos\psi - \cos\psi_s) - D_x) \quad (12)$$

where $\psi$ and $\psi_s$ are true azimuth and estimated azimuth in coarse alignment, respectively.

In a second step the platform is turned by a certain angle $\Delta\psi$ in azimuth and the levelling step is repeated. Using the velocity data obtained from the two orientations ($u_{1\infty}$, $v_{1\infty}$ and $u_{2\infty}$, $v_{2\infty}$), both the azimuth at the two orientations and the constant components of gyro drift can be estimated. The azimuth of the second orientation can be computed using equation (13)

$$\begin{bmatrix} \sin\psi \\ \cos\psi \end{bmatrix} = \begin{bmatrix} 1 & \frac{-\sin\Delta\psi}{1-\cos\Delta\psi} \\ \frac{\sin\Delta\psi}{1-\cos\Delta\psi} & 1 \end{bmatrix} \begin{bmatrix} b_1 \\ b_2 \end{bmatrix} \quad (13)$$

where

$$\begin{bmatrix} b_1 \\ b_2 \end{bmatrix} = \frac{1+k}{2R\Omega_E \cos\varphi} \begin{bmatrix} u_{2\infty} - u_{1\infty} \\ v_{2\infty} - v_{1\infty} \end{bmatrix} \quad (14)$$

If the constant components of gyro drift were determined beforehand and then compensated, the azimuth can be obtained after setting the first orientation:

$$\sin\psi = \frac{(1+k)u_\infty}{R\Omega_E \cos\varphi} + \sin\psi_s$$
$$\cos\psi = \frac{(1+k)v_\infty}{R\Omega_E \cos\varphi} + \cos\psi_s \quad (15)$$

The relations between azimuth and constant gyro components are shown in equation (11) Using the azimuth obtained from equation (15) the constant gyro components can be obtained with

$$D_x = \Omega_E \cos\varphi(\cos\psi - \cos\psi_s) - \frac{(1+k)v_{2\infty}}{R}$$
$$D_y = \Omega_E \cos\varphi(\sin\psi - \sin\psi_s) - \frac{(1+k)u_{2\infty}}{R} \qquad (16)$$

The algorithms introduced above have been used for initial alignment of the inertial platform and gyro calibration. The levelling accuracy is mainly limited by accelerometer biases. If gyros and accelerometers are well calibrated, the platform can be levelled within 2 arcseconds. At the end of the initial alignment the constant components of the gyro drift can be calibrated with an accuracy of 0.03 °/h and the azimuth angle can be calculated with an accuracy of 30 arcseconds. The limiting factor for the accuracy of gyro calibration and azimuth angle estimation is the change in mean values of gyro drifts during fine alignment. Therefore the fine alignment time should be designed as short as possible.

## 5 Further development of the system

During the actual project work the demand for an even better platform levelling came up in order to achieve the high goal of 1 mGal accuracy. On this account a Kalman filter for platform levelling is under development. As the platform has been extended by an additional degree of freedom around the vertical axis it is possible to implement a levelling loop similar to those used in integrated navigation systems. At the IFF a simulation of the entire system has been developed. This system contains a 13 state variables Kalman filter. The principle of the levelling loop of the simulation is shown in Fig. 7.

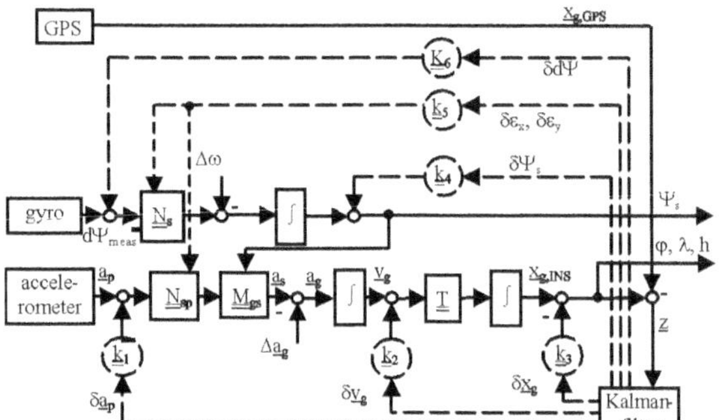


**Fig. 7** Scheme of the Kalman filter

The results of the simulation show, that a much better performance is to be expected with a levelling algorithm that makes use of a Kalman filter. The comparison between the control mentioned in chapter 4 and the Kalman filter control are shown in Fig. 8. The levelling errors of the Kalman filter are magnitudes smaller than them of the control loop.

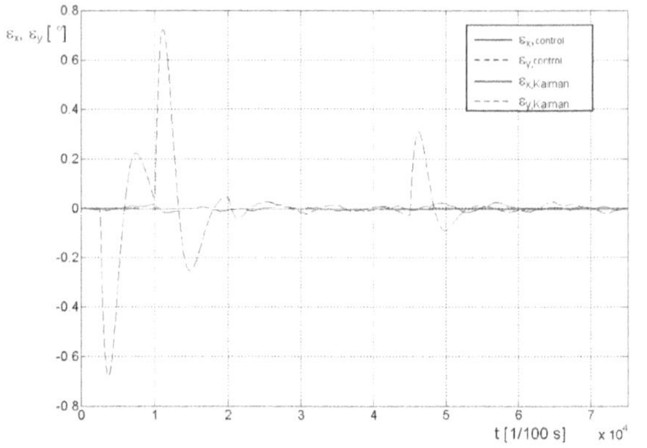


**Fig. 8** comparison of developed levelling methods

## 6 Summary and outlook

There are two major error sources which influence vertical specific force measurements. One is the gravity sensor error, the other is the platform levelling error. The first is mainly dependent on dynamic performance of the gravity sensor and aircraft vertical accelerations. The second affects vertical force measurements only when horizontal aircraft acceleration exists. To achieve an overall gravity measurement accuracy of 1 mGal the platform must be levelled with high accuracy under dynamic conditions typical for small aircraft. For this reason the existent two-frame inertial platform was upgraded by a ring laser gyro to provide an additional degree of freedom. Furthermore high performance levelling algorithms combined with a Kalman filter are under development.

## References

Abdelmoula., Ein Beitrag zur Bestimmung der Erdbeschleunigungsanomalien an Bord eines Flugzeuges; Verlag Shaker, 2000.

Kaufmann., Technische Hydro- und Aerodynamik; Springer-Verlag, Berlin Göttingen Heidelberg, 1963.

Olesen, Forsberg, Gidskehaug., Airborne Gravimetry Using the LaCoste and Romberg Gravimeter – An Error Analysis. In: Proc. of International Symposium on Kinematic Systems in Geodesy, Geomatics and Navigation (KIS97), Banff, Canada, June 3-6, pp. 613-618; 1997.

Schänzer, Abdelmoula., Fluggravimeter; Zwischenbericht Sonderforschungsbereich Flugmesstechnik SFB 420, 1999.

Wei, Schwarz., Flight Test Results from a Strapdown Airborne Gravity System. Journal of Geodesy, 72, pp. 323-332, 1998.

Zhang., Development of a GPS-aided Inertial Platform for an Airborne Scalar Gravity System; Calgary, 1995.

# Numerical investigation of downward continuation methods for airborne gravity data

I.N. Tziavos, V.D. Andritsanos
Aristotle University of Thessaloniki, Department of Geodesy and Surveying, University Box 440, 54124, Thessaloniki, Greece, E-mail: tziavos@olimpia.topo.auth.gr

R. Forsberg, A.V. Olesen
Geodynamics Department, National Survey and Cadastre (KMS), Denmark, Rentemestervej 8, DK-2400, Copenhagen, Denmark, E-mail: rf@kms.dk

## Abstract

Two airborne surveys were carried out in the Crete region of Greece. A first airborne gravity survey was done in February 2001 to cover the southern part of the Aegean sea and the island of Crete in the frame of the European Community CAATER (Coordinated Access to Aircraft for Transnational Environmental Research) project, primarily aiming at the establishment of the gravity information needed to connect the existing gravity data on the island of Crete to the altimetric gravity field in the open sea. A second airborne gravity survey was carried out in January 2003 over the island of Gavdos and the surrounding sea areas south of Crete, in the frame of the GAVDOS project, an ongoing European Community-funded project aiming at the development of a calibration site in the island of Gavdos for altimetric satellites. The main goal of this paper is the evaluation of the airborne measurements from the above mentioned campaigns through a comparison of the downward continuation methods in the space and frequency domain. Different downward continuation schemes are evaluated, with comparisons to satellite altimetry showing accuracies close to 4 mGal in terms of standard deviation of the differences between the downward continued gravity anomalies and the altimetry derived gravity anomalies. Finally, some remarks are presented with regards to the construction of a detailed gravity anomaly grid based on all the available satellite and surface gravity data sources.

**Keywords.** Airborne gravimetry, airborne gravity data, downward continuation methods

## 1 Introduction

Airborne gravimetry is an efficient method for fast and economic gravity measurements. The main advantages of the method are the uniform distribution of the data and the quick coverage of mountainous and sea areas.

Ground gravity data are usually collected on a very complicated surface. On the contrary, in airborne gravimetry the flight trajectories are known and smooth, making the use of fixed boundary value problem feasible. The boundary value problem of airborne gravimetry is discussed in detail in Li (2000) and Novak and Heck (2002). In real-world applications of airborne gravity, the typical situation is that both airborne and surface gravity data are available and should be utilized together for an optimal solution. Here least- squares collocation is an obvious method, allowing the combination of precise point observations at the surface, with airborne gravity data at altitude, which typically is along-track filtered at resolutions of 5-10 km and accuracy of 1-3 mGal for modern data (Wei, 1999). A fundamental disadvantage of the airborne measurement is the attenuation of the high wave numbers due to the flight altitude. For this reason the downward continuation procedure is introduced, with special treatments to stabilize the solution and consequently avoiding the amplification of short-wavelength noise in the airborne data. Nevertheless, the method is suitable for geoid determination due to the geoid low-pass filtering scheme, which offsets noise amplification.

There are two main techniques for the evaluation of the downward continuation problem. The first one is a purely deterministic approach using the integral Poisson formula. Various implementations of the method have been presented during the past years, including pure integral solutions, e.g., Novak et al. (2003), Kern (2003), as well as their spectral equivalents, e.g., Forsberg (1987). The second method is based on collocation theory as described in Moritz (1980). Only the latter is able to use surface and airborne data jointly.

In the present paper some comparisons between

different methodologies for downward continuation are presented, and some numerical tests and comparisons with altimetry-gravity data are carried out.

## 2 Theory

### 2.1 Collocation

Least-squares collocation can be efficiently used to downward continue airborne gravity data. The gravity anomaly signal "s" at a ground grid point is estimated from a vector "x" containing all available airborne data according to the formula:

$$\hat{s} = C_{sx}\left[C_{xx} + D\right]^{-1} x \qquad (1)$$

Covariance matrices, $C_{xx}$ and $C_{sx}$, are taken from a full self-consistent spatial covariance model such as the spherical earth model of Tscherning and Rapp (1974) or the flat earth model for airborne gravimetry developed by Forsberg (1987); $D$ is the noise matrix.

### 2.2 FFT methods

If the flight altitude is constant, downward continuation can be carried out by frequency domain methods. Let $\Delta g$ and $\Delta g^*$ be the gravity anomalies at altitude $h$ and at the geoid, respectively. Taking the 2D Fourier transform of $\Delta g$

$$F(\Delta g) = \Delta g(x, y) e^{-i(k_x x + k_y y)} dx dy, \qquad (2)$$

gravity anomalies at the geoid are obtained by

$$F(\Delta g^*) = e^{kh} F(\Delta g), \qquad k = \sqrt{k_x^2 + k_y^2}, \qquad (3)$$

where $k_x$, $k_y$ are the wave numbers in north and east directions. Assuming a Kaula rule decay of the power spectral density (PSD) of $\Delta g$ and white noise for the data, the optimal Wiener filter downward continuation operator becomes

$$\Delta g^* = \Delta g \frac{e^{kh}}{1 + ck^4}, \qquad (4)$$

where $c$ is a resolution parameter depending on the ratio of noise to gravity signal covariance (Forsberg 2002).

### 2.3 Downward continuation stabilization schemes

The downward continuation of airborne gravity data is a high-pass filtering operation, which mainly amplifies short-wavelength noise in the airborne data. A major advantage in geoid determination by using airborne gravity measurements is the low-pass geoid filtering operation, which will offset the amplification of short-wavelength noise.

As far as the noise reduction of the airborne gravity is concerned, this can be achieved by the use of proper data reductions and filtering of the final products (Olesen et al., 2001). Recent investigations have shown that observation noise has almost a constant power over the frequency spectrum currently considered in airborne gravimetry. Therefore, the observation noise of airborne gravity may be approximated by white noise models (Bruton, 2000), except of the shortest wavelengths (5-10 km). In practical applications stabilization is implicitly obtained by gridding airborne gravity data onto a regular grid by interpolation (by collocation, etc.). The interpolation operator generates a smooth function by applying a low-pass filter to the original data, with care matching the along-track filtering, inherent in all airborne gravity data. A remove – restore scheme can also be applied to the airborne geoid determination procedure:

$$\Delta g_{red} = \Delta g_{obs} - \Delta g_{GM} - \Delta g_h, \qquad (5)$$

where $\Delta g_{red}$ are the reduced (smoothed) airborne gravity anomalies at flight level, $\Delta g_{GM}$ is the contribution of a global high degree geopotential model and $\Delta g_h$ are the terrain effects at flight altitude. The use of terrain effects may generate additional noise if terrain and density models are not sufficiently accurate. A consistent application of the remove-restore principle will, however, limit these errors significantly. It should also be noted that since the airborne gravity data are along-track filtered, the terrain effects computed at altitude must be filtered by the same filter algorithm that is used in the airborne processing; otherwise, short-wavelength terrain-effects may actually produce noise.

## 3 Data analysis and results

The need for an accurate geoid solution for the GAVDOS project (GAVDOS, 2004), and the lack of gravity data near the coast of Crete led us to establish the airborne campaign. This campaign was divided into two phases, both using a LaCoste and Romberg S-meter as the primary gravity sensor. The first phase involved the aero-gravity survey performed in the frame of the EU CAATER project, using a large F-27 aircraft at a flight altitude of 3.3

km. Flight conditions were quite turbulent, and data thus quite noisy. Descriptions of the survey instrumentation and data preprocessing are provided in Olesen et al. (2002). The final product was free-air gravity anomalies at flight altitude. The measurement error was estimated at the level of 3 mGal for the track data.

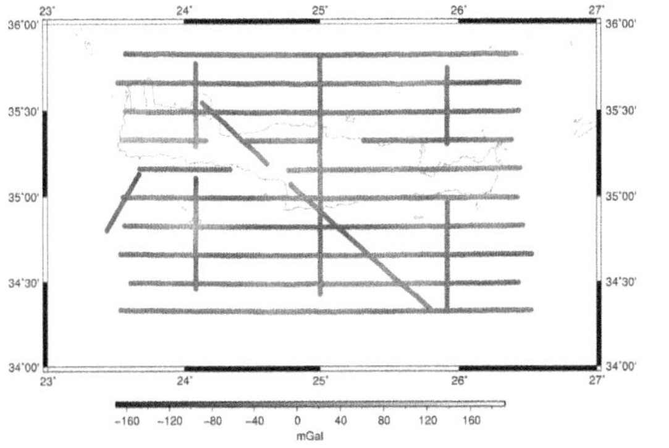


**Fig. 1** Reduced CAATER gravity anomalies at nominal 3.3. km flight level (mGal)

The final gravity anomalies at flight level, after the removal of EGM96 contribution and the terrain effects (RTM effects), are shown in Figure 1. In the second phase, airborne operations were performed with the HB-LID Twin-Otter airplane of the Swiss Federal Office of Topography in the period from January 9 - 15, 2003. These operations were performed in the frame of the GAVDOS Project and were divided into low altitude (300 m) and high altitude (3500 m) campaigns. The tracks were flown at two elevations due to the high topography of western Crete.

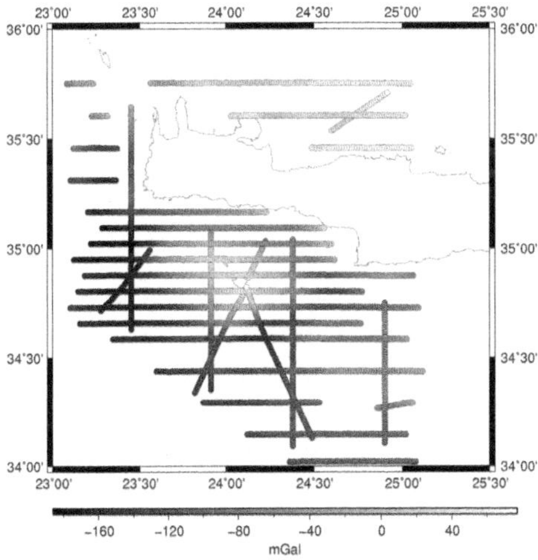


**Fig. 2** Free-air anomalies at GAVDOS low level (300 m) flights (mGal). Estimated accuracy of data 2.4 mGal.

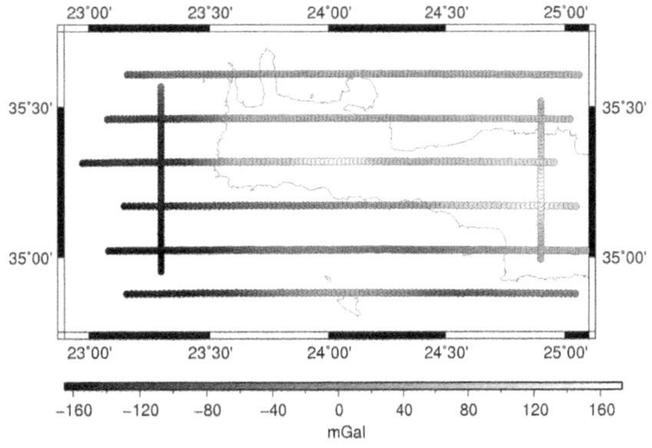


**Fig. 3** Free-air anomalies at GAVDOS high flight level (mGal). Estimated accuracy of data 2.6 mGal.

Flight trajectories of the low and high campaigns are shown in Figures 2 and 3, respectively. The estimated accuracy of the low and high level flights are 2.4 and 2.6 mGal, based on cross-over analysis.

### 3.1 Phase one: CAATER data

Terrain effects were computed by introducing a surface of 30 arcsec grid spacing from newly available SRTM (Satellite Radar Topography Mission) and estimating the residual terrain effects (Forsberg, 1987). A symmetric second order Butterworth filter with half power point at 200 seconds was used to filter efficiently the computed terrain effects. The same filter, corresponding to a resolution of 5.5 km (half-wavelength), was applied to the airborne observations (Olesen, 2003). The contribution of EGM96 was calculated at flight level. The statistics are presented in Table 1.

**Table 1.** The statistics of CAATER airborne gravity data (mGal).

| | mean | sd | min | max |
|---|---|---|---|---|
| $\Delta g$ | -3.002 | 71.340 | -169.400 | 192.300 |
| $\Delta g - \Delta g_{EGM96}$ | 0.945 | 32.293 | -71.099 | 154.040 |
| $\Delta g_{red}$ | 0.994 | 30.825 | -75.921 | 136.350 |

In order to achieve a smooth measurement field, point airborne data were used to predict free-air gravity anomalies at a mean flight level onto a regular grid of 5′×5′ resolution. During the gridding procedure an error of 3 mGal for each observation was adopted. The final grid statistics refer to the region $34.3 \leq \varphi \leq 35.75$ and $23.5 \leq \lambda \leq 26.5$, and are given in Table 2.

**Table 2.** The statistics of the grid airborne data at mean flight of 3 km ( mGal).

| | mean | sd | min | max |
|---|---|---|---|---|
| $\Delta g_{red}$ (grid) | 1.007 | 30.627 | -69.283 | 118.287 |

Using the gridded data, three downward continuation schemes were followed. At first, least squares collocation was used choosing a planar model of covariance function (PLSC) for fitting empirical data. In the planar model, gravity covariances between gravity anomalies at two altitudes ( $h_1$ , $h_2$ ) are of the form (Forsberg, 1987):

$$C\left(\Delta g^{h_1}, \Delta g^{h_2}\right) = -\sum_n a_n \log\left(D_n + \sqrt{s^2 + (D_n + h_1 + h_2)^2}\right), \quad (6)$$

where $a_n$ are weight factors combining terms relating to two depth value terms ( $D_n = D + nT$ ), with the "free parameters" D and T taking the role analogous to the Bjerhammar sphere depth of the spherical collocation and a "compensating depth" attenuation factor, correspondingly. Alternatively, the classical spherical Earth covariance model of Tscherning and Rapp (1974) was employed (SLSC). Finally, the fast Fourier harmonic continuation was used as described in section 2.2. The statistical analysis of the downward continued reduced gravity anomalies is presented in Table 3.

**Table 3.** The statistics of the downward continued CAATER gravity data (mGal).

| | mean | sd | min | Max |
|---|---|---|---|---|
| PLSC | 1.068 | 42.561 | -112.008 | 178.173 |
| SLSC | 0.974 | 43.249 | -116.024 | 204.388 |
| FFT | 1.043 | 40.576 | -103.890 | 185.590 |

The differences between the downward continuation methods are presented in Table 4. It is seen that the high level – necessitated by topography and weather – give quite large differences between methods.

**Table 4.** Differences of the downward continued methods (mGal).

| | mean | sd | min | max |
|---|---|---|---|---|
| PLSC – SLSC | 0.094 | 3.684 | -26.215 | 21.952 |
| PLSC – FFT | 0.025 | 5.873 | -34.816 | 47.915 |
| SLSC – FFT | -0.068 | 5.375 | -34.412 | 45.117 |

The major differences between planar and spherical collocation are located in cells near the coast-sea boundary and over land. Furthermore, the comparison between the collocation and the FFT method presents some discrepancies at the border of the area due to spectral edge effects. For an independent evaluation we compare the direct derived downward continued airborne gravity anomalies and the free-air gravity anomaly grid KMS02. The latter has been constructed using altimetry-derived gravity (Andersen and Knudsen, 1998). In order to have consistent quantities to compare, a restore procedure of EGM96 and terrain effects was performed. The resulted free-air gravity anomalies were interpolated at the grid points and the differences are tabulated in Table 5. The differences between downward continued free-air gravity anomalies derived from planar collocation and the KMS02 gravity database are depicted in Figure 4. A standard deviation of 9 mGal and 6 mGal has been found in the northern and southern parts of the area, respectively.

**Table 5.** Differences between airborne and altimetry-derived gravity data (mGal).

| | | pts | mean | sd |
|---|---|---|---|---|
| Whole area | PLSC | | 1.747 | 10.794 |
| $34.33 \le \varphi \le 35.75$ | SLSC | 3065 | 1.912 | 11.127 |
| $23.5 \le \lambda \le 26.5$ | FFT | | 1.340 | 10.965 |
| North of Crete | PLSC | | 0.011 | 9.121 |
| $35.5 \le \varphi \le 35.75$ | SLSC | 688 | 0.392 | 9.391 |
| $23.5 \le \lambda \le 26.5$ | FFT | | -0.118 | 9.471 |
| South of Crete | PLSC | | -1.066 | 5.675 |
| $34.33 \le \varphi \le 34.75$ | SLSC | 1092 | -1.025 | 5.722 |
| $23.5 \le \lambda \le 26.5$ | FFT | | -1.222 | 6.621 |

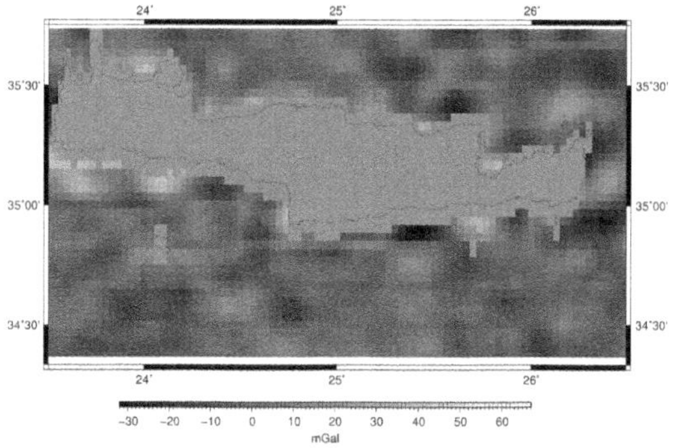


**Fig. 4** Differences between downward continued CAATER data derived from planar collocation and the KMS02 gravity database.

## 3.2 Phase two: GAVDOS airborne campaign

### Low altitude data

Following the same procedure as in phase one, low altitude campaign data were reduced and gridded onto a regular grid of 5′×5′ resolution. The area under study was bounded by $34 \le \varphi \le 35.75$ and $23 \le \lambda \le 25$. The statistics are presented in Table 6. Following the three downward continuation schemes of the previous section, reduced free-air gravity anomalies at sea level were computed and the statistical results are shown in Table 7.

As expected planar and spherical collocation gave approximately the same results, with differences having a 2 mGal standard deviation compared to the FFT solution.

**Table 6.** The statistics of GAVDOS low altitude airborne gravity reductions (mGal).

| | max | min | mean | sd |
|---|---|---|---|---|
| $\Delta g$ | 76.700 | -185.300 | -71.885 | 68.460 |
| $\Delta g - \Delta g_{EGM96}$ | 106.368 | -109.037 | -12.193 | 38.102 |
| $\Delta g_{red}$ | 105.798 | -108.777 | -11.872 | 37.922 |
| $\Delta g_{red}$ (grid) | 97.112 | -110.348 | -3.540 | 31.413 |

**Table 7.** Results and differences between various downward continuation methods - low flight level (mGal).

| | max | min | mean | sd |
|---|---|---|---|---|
| PLSC | 99.215 | -112.828 | -3.545 | 32.107 |
| SLSC | 99.494 | -112.919 | -3.536 | 32.115 |
| FFT | 99.370 | -113.880 | -3.891 | 32.074 |
| PLSC – SLSC | 1.154 | -0.985 | -0.009 | 0.231 |
| PLSC – FFT | 22.918 | -17.131 | 0.346 | 1.978 |
| SLSC – FFT | 22.919 | -17.082 | 0.355 | 1.930 |

With the low altitude (~ 300 m flight level) all methods gave very close results, which illustrates the flight level sensitivity to the downward continuation operation.

**Table 8.** Comparison of downward continued data derived from planar collocation and KMS02 - low flight level (mGal).

| | pts | mean | sd |
|---|---|---|---|
| Whole area ($34 \le \varphi \le 35.75$, $23 \le \lambda \le 25$) | 2798 | 3.450 | 13.030 |
| South part ($34 \le \varphi \le 34.75$, $24 \le \lambda \le 25$) | 713 | -0.016 | 4.565 |
| Open sea part ($34 \le \varphi \le 34.5$, $24 \le \lambda \le 25$) | 496 | -0.543 | 3.861 |

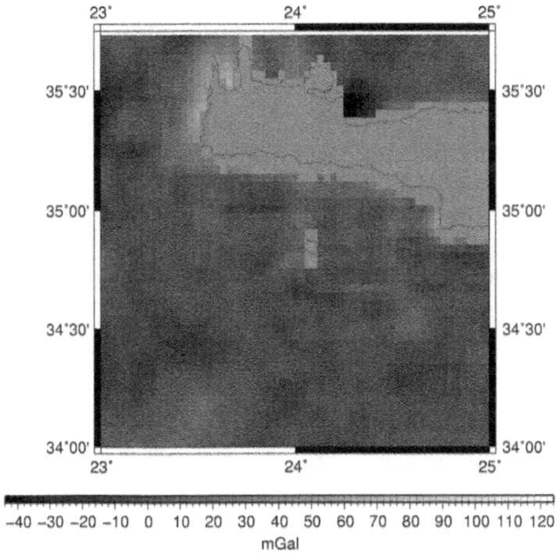


**Fig. 5** Differences between downward continued data derived from planar collocation and the KMS02 altimetric gravity database.

After restoring EGM96 and terrain contributions, a new comparison with KMS02 altimetry data was carried out. The results of this comparison are summarized in Table 8 and presented in Figure 5. As expected, major differences arise near the boundary between sea and land because of the problematic altimetry-to-gravity conversion there. In the open sea area, a standard deviation of 3.9 mGal is achieved, comparable with the 2.4 mGal measurement error of the airborne data.

## High altitude data

The same procedure as before was followed for the high altitude airborne gravity data. The reduction and gridding results are presented in Table 9. The statistics of the resulted downward continuation data and the differences between each reduction technique can be found in Table 10. It is obvious that the high flight level affected the stability of the methods. Thus, discrepancies between collocation methods reach a standard deviation of 2 mGal and between collocation and spectral method 9.5 mGal. A mean accuracy of 12 mGal in terms of standard deviation was found over the whole area between the downward continued airborne gravity and the KMS02 altimetric grid, as it is shown in Table 11.

**Table 9.** The statistics of GAVDOS high altitude airborne gravity reductions (mGal).

| | max | min | mean | sd |
|---|---|---|---|---|
| $\Delta g$ | 172.800 | -165.500 | -11.159 | 85.921 |
| $\Delta g - \Delta g_{EGM96}$ | 136.962 | -66.433 | 0.346 | 38.169 |
| $\Delta g_{red}$ | 104.684 | -66.323 | 0.193 | 35.579 |
| $\Delta g_{red}$ (grid) | 104.199 | -63.247 | 2.748 | 36.222 |

**Table 10.** Results and differences between various downward continuation methods - high flight level (mGal).

| | max | min | mean | sd |
|---|---|---|---|---|
| PLSC | 161.424 | -87.447 | 3.384 | 51.639 |
| SLSC | 161.771 | -88.905 | 3.328 | 52.067 |
| FFT | 148.300 | -88.790 | 3.120 | 47.724 |
| PLSC – SLSC | 6.998 | -7.195 | 0.056 | 2.105 |
| PLSC – FFT | 50.952 | -48.125 | 0.208 | 9.867 |
| SLSC – FFT | 49.092 | -46.667 | 0.264 | 9.611 |

**Table 11.** Comparison of downward continued data derived from planar collocation and KMS02 grid - High flight level (mGal)

| | Pts | Min | max | mean | sd |
|---|---|---|---|---|---|
| PLSC | 812 | -28.090 | 80.012 | 8.041 | 12.496 |
| SLSC | 812 | -31.099 | 80.502 | 8.240 | 12.582 |
| FFT | 812 | -32.966 | 77.979 | 7.533 | 13.387 |

From Table 11, one can see the agreement of the collocation methods and the relatively poorer performance of the FFT method due to some edge effects. It should be pointed out, however, that the high level flight data only covers the coast-near sea region, where the KMS02 is inaccurate.

## 4 Conclusions

In the frame of the GAVDOS Project, airborne gravity measurements were performed in the wide area of Crete Island. Three methods were used in order to downward continue the airborne data. The first two were based on collocation theory and the third one on the spectral implementation of the integral formulation. The variation in flight altitude between campaigns led to some remarks concerning the dependence of the downward continuation operator with height. As evidenced from the results, the planar covariance model in the collocation method gave the closest comparisons with respect to the altimetric grid. However, high altitude campaigns are obviously more noisy due to the unstable downward continuation operator. An accuracy of 0.2 – 5 mGal was estimated in the differences between each method result. Collocation with planar and spherical models gave approximately equivalent fields and both were close enough to the spectral solution. An external comparison with KMS02 altimetric data gave an accuracy of 3.8 – 12.5 mGal, depending on the area of comparison: over open sea the accuracy reach the level of 3.8 mGal at worst while over land-sea boundaries major discrepancies increase the agreement to 12.5 mGal. This can be explained by the well-known weakness of satellite altimetry data near coastlines. Low altitude measurements gave consistent results between downward continuation methods and minor differences with the altimetric database over open sea. A thorough comparison in terms of geoid height will lead to more reliable conclusions on the use of GAVDOS airborne campaigns to the establishment of an ultra-high resolution geoid in the area. Such geoid models should use all available data (high, low airborne and surface), and would thus implemented by collocation.

## Acknowledgement

Funding for this research was provided from the EU under contract GAVDOS: EVR1-CT-2001-40019 in the frame of the EESD-ESD-3 Fifth Framework Program. We thank Arne Gidskehaug, University of Bergen, Norway, for providing the airborne gravimeter.

## References

Andersen OB and P Knudsen (1998): Global marine gravity field from the ERS-1 and Geosat geodetic mission altimetry, J. Geophys. Res. Vol. 103 , No. C4 , p. 8129.

Bruton AM (2000): Improving the accuracy and resolution of SINS/DGPS airborne gravimetry. Ph.D. Thesis, Dept. of Geomatics Engineering, University of Calgary.

Forsberg R (1987): A new covariance model for inertial gravimetry and gradiometry. Journal of Geophysical Research, vol. 92, pp. 1305 – 1310.

Forsberg R (2002): Downward continuation of airborne gravity data – an Arctic case story. In Proceedings of the 3rd Meeting of the International Gravity and Geoid Commission. IN Tziavos (ed.). Thessaloniki, Aug. 26 – 30.

GAVDOS (2004). Internet url: http://www.gavdos.tuc.gr.

Kern M (2003): An Analysis of the Combination and Downward Continuation of Satellite Airborne and Terrestrial Gravity Data. Ph.D. Thesis, Dept. of Geomatics Engineering, University of Calgary.

Li YC (2000): Airborne Gravimetry for Geoid Determination. Ph.D. Thesis. Dept. of Geomatics Engineering, University of Calgary.

Moritz H (1980): Advanced Physical Geodesy. Wichmann, 2nd edition, Karlsruhe.

Novak P and B Heck (2002): Downward continuation and geoid determination based on band limited airborne gravity data. Journal of Geodesy, vol. 76, pp. 269 – 278.

Novak, P., M. Kern, K.-P. Schwarz, M.G. Sideris, B. Heck, S. Ferguson, Y. Hammada and M. Wei (2003): On geoid determination from airborne gravity. Journal of Geodesy, vol. 76, pp. 510 – 522.

Olesen AV, R Forsberg, K Keller, AHW Kearsley (2001): Error sources in airborne gravimetry employing spring-type gravimeter. In: K.P. Schwarz (ed.): Vistas for Geodesy in the new Millennium. Proc. IAG General Assembly, Budapest, Sept. 2001, Springer Verlag IAG Series 125, pp. 205 – 210.

Olesen, AV, IN Tziavos and R Forsberg (2002): New Airborne Gravity Data Around Crete – First Results from CAATER Campaign. In Proceedings of the 3rd Meeting of the International Gravity and Geoid Commission. IN Tziavos (ed.). Thessaloniki, Aug. 26 – 30.

Olesen, A.V. (2003): GAVDOS airborne gravity acquisition and processing report. GAVDOS project annual report, December 2003.

Tscherning CC and RH Rapp (1974): Closed Covariance Expressions for Gravity Anomalies, Geoid Undulations, and Deflections of the Vertical implied by Anomaly Degree Variance Models. Rep. No. 208, Dept. of Geodetic Science and Surveying, Ohio State University, Ohio.

Wei M (1999): From airborne gravimetry to airborne geoid mapping. Report of the IAG SSG 3.164. In Determination of the gravity field, R Forsberg (ed.), pp. 25 – 32.

# Status of the European Gravity and Geoid Project EGGP

H. Denker [(1)], J.-P. Barriot, R. Barzaghi, R. Forsberg, J. Ihde, A. Kenyeres, U. Marti, I.N. Tziavos

[(1)] Institut für Erdmessung, Universität Hannover, Schneiderberg 50, D-30167 Hannover, Germany
E-mail: denker@ife.uni-hannover.de; Fax: +49-511-7624006

**Abstract.** The European Gravity and Geoid Project (EGGP) is a project within IAG Commission 2, reporting to Sub-commission 2.4. The main goal of the project is to compute an improved European geoid and quasigeoid model. This appears to be possible now because significant new and improved data sets have become available since the last computation in 1997 (EGG97). These improvements include better global geopotential models from the CHAMP and GRACE missions, better digital elevation models (DEMs) in some regions (e.g., new national DEMs, SRTM3, GTOPO30), updated gravity data sets for selected regions, updated ship and altimetric gravity data including improved merging procedures, the use of GPS/levelling data, as well as improved modelling and computation techniques.

An overview is given on the project structure, the computation strategy, the available data sets, the expected accuracies, the time schedule, and the work done so far. The primary input data sets are high-resolution gravity and terrain data supplemented by a state-of-the-art global geopotential model. The general computation strategy is the remove-restore procedure. The initial computations are based on the spectral combination approach with integral formulas evaluated by 1D FFT. First results based on an updated terrestrial gravity data set and GRACE geopotential models show significant improvements (up to 60 %) as compared to GPS/-levelling. Moreover, also the tilts, existing in previous solutions, have been reduced to typically below 0.1 ppm.

**Keywords.** Geoid, quasigeoid, gravity field modelling, GPS/levelling, EGGP, CHAMP, GRACE

## 1 Introduction

The latest high-resolution European geoid and quasigeoid models (EGG97) were computed at the Institut für Erdmessung (IfE), University of Hannover, acting since 1990 as the computing center of the International Association of Geodesy (IAG) Sub-commission for the European Geoid (predecessor of the EGGP), for details cf. Denker and Torge (1998). EGG97 is based on high-resolution gravity and terrain data in connection with the global geopotential model EGM96. The evaluation of EGG97 by GPS/levelling data revealed the existence of long wavelength errors at the level of 0.1 to 1 ppm, while the agreement over distances up to about 100 km is at the level of 0.01 m in many areas with a good quality and coverage of the input data (Denker and Torge, 1998; Denker, 1998).

Since the development of EGG97, significant new or improved data sets have become available, including strongly improved global geopotential models from CHAMP and GRACE, new national and global terrain data sets, new or updated gravity data sets, improved altimetric results, as well as new GPS/levelling campaigns. Furthermore, also the combination of ship and altimetric data has been refined, and new gravity field modelling techniques, e.g., wavelet techniques, fast collocation, etc., have become operational.

Considering all these advancements, a complete re-computation of the European geoid/quasigeoid is appropriate and promises significantly improved accuracies, especially at long wavelengths. Therefore, after the IUGG General Assembly in Sapporo in 2003, it was decided to support this task in the form of an IAG Commission 2 Project, named CP2.1 and entitled "European Gravity and Geoid Project EGGP". The project is reporting to Sub-commission 2.4 and has strong connections to the International Gravity Field Service (IGFS), with its centers Bureau Gravimétrique International (BGI), International Geoid Service (IGeS), National Geospatial-Intelligence Agency (NGA), and GeoForschungsZentrum Potsdam (GFZ), as well as to several other IAG bodies, e.g., EUREF. The EGGP is running within the 4-year period from 2003 to 2007 until the next IUGG General Assembly. The project is organised by a steering committee (H. Denker (Chair), J.-P. Barriot, R. Barzaghi, R. Forsberg, J. Ihde, A. Kenyeres, U. Marti, I.N. Tziavos) and has about 50 national delegates (project members) from most of the countries in Europe. The EGGP terms of reference can be found in EGGP (2003).

This contribution gives an overview on the general computation strategy and on the progress in the collection of gravity data, terrain data, and global geopotential models from the new space missions CHAMP and GRACE. First updated geoid/quasigeoid solutions are presented based on the new global geopotential models from CHAMP and GRACE. Moreover, results from an improved terrestrial gravity data set, including reprocessed ship gravity and new altimetric anomalies, are presented.

## 2 Computation Strategy

The basic computation strategy is based on the remove-restore technique, considering high-resolution terrestrial gravity and terrain data in combination with a state-of-the-art global geopotential model (probably based on the GRACE mission). Terrain reductions will be applied to smooth the data and to avoid aliasing effects. At present, the residual terrain model (RTM) technique according to Forsberg and Tscherning (1981) is favoured. Bathymetry and density data may be considered in special test areas. Moreover, GPS/levelling data will be used for control purposes, and may also be used for a combined solution (e.g., Denker et al., 2000), depending on the quality and availability of data. All data sets will be referred to uniform horizontal, vertical, and gravity reference systems. The collection of the relevant data sets is pursued by the steering committee and the members of the project.

A significant problem with high-resolution gravity and terrain data is the confidentiality of data, which must be assured to most of the data owners. For this reason, it was decided to have only one data and computation center at the Institut für Erdmessung (IfE), University of Hannover. In addition, a second confidential gravity data center is setup at Bureau Gravimétrique International (BGI) to use the expertise of BGI in the validation and cleaning of large gravity data sets. The inclusion of data in the confidential BGI project database requires separate agreements between the data owners and BGI, and there will be no connection to the BGI public database.

The primary gravity field quantity to be computed will be the height anomaly or the quasigeoid undulation, with the advantage that only gravity field observations at the Earth's surface and in its exterior enter into the calculation, avoiding assumptions about the Earth's interior gravity field. A geoid model is then derived by introducing a density hypothesis, which should be identical to the one used for the computation of corresponding orthometric heights.

Initially, the gravity modelling at IfE will be based on the spectral combination technique with integral formulas (e.g., Wenzel, 1982). In this method, the combination of terrestrial gravity data and a global geopotential model is done by means of spectral weights, which depend on the accuracy of the input data sets. Due to the high accuracy of the global models at long wavelengths, the terrestrial data mainly contribute the shorter wavelength components. Lateron, IfE may also test other modelling techniques, e.g., least squares collocation or wavelets. Moreover, it is planned to use the fast collocation approach developed by the Milan group (e.g., Sansò and Tscherning, 2003). Regarding the time frame, it is planned to have the final geoid/quasigeoid models in 2007 and preliminary solutions in 2005 and 2006.

The final goal is to strive for an accuracy of 0.01 m for the computed geoid/quasigeoid models (for distances up to some 100 km). Obviously, this is only possible in areas with a good coverage and quality of the input gravity and terrain data. The data requirements can be derived from theoretical and numerical studies including spectral analysis. With respect to the gravity data, a spacing of 2 to 5 km and an accuracy at the level of 1 mgal (white noise) is sufficient (Denker, 1988; Forsberg, 1993; Grote, 1996), but on the other hand even small systematic gravity errors affecting large regions may integrate up to significant geoid errors. For the elevation models, a resolution of roughly 100 m to 1000 m is adequate for alpine to low relief, respectively, with an accuracy at the level of some ten meters.

## 3 Recent Progress in Data Collection

### 3.1 Gravity Data

Since the start of the project, significant improvements of the gravity database have been made, including new data sets for several countries, e.g., Belgium, Luxemburg, Germany, Slovenia, Switzerland, and Netherlands. Moreover, positive responses, indicating a data update in the near future, were received from Austria, the Baltic States, Croatia, France, Greece, Poland, Serbia, Russia, the Scandinavian countries, etc. In addition to this, also the public domain data set from the Arctic Gravity Project became available (Forsberg and Kenyon,

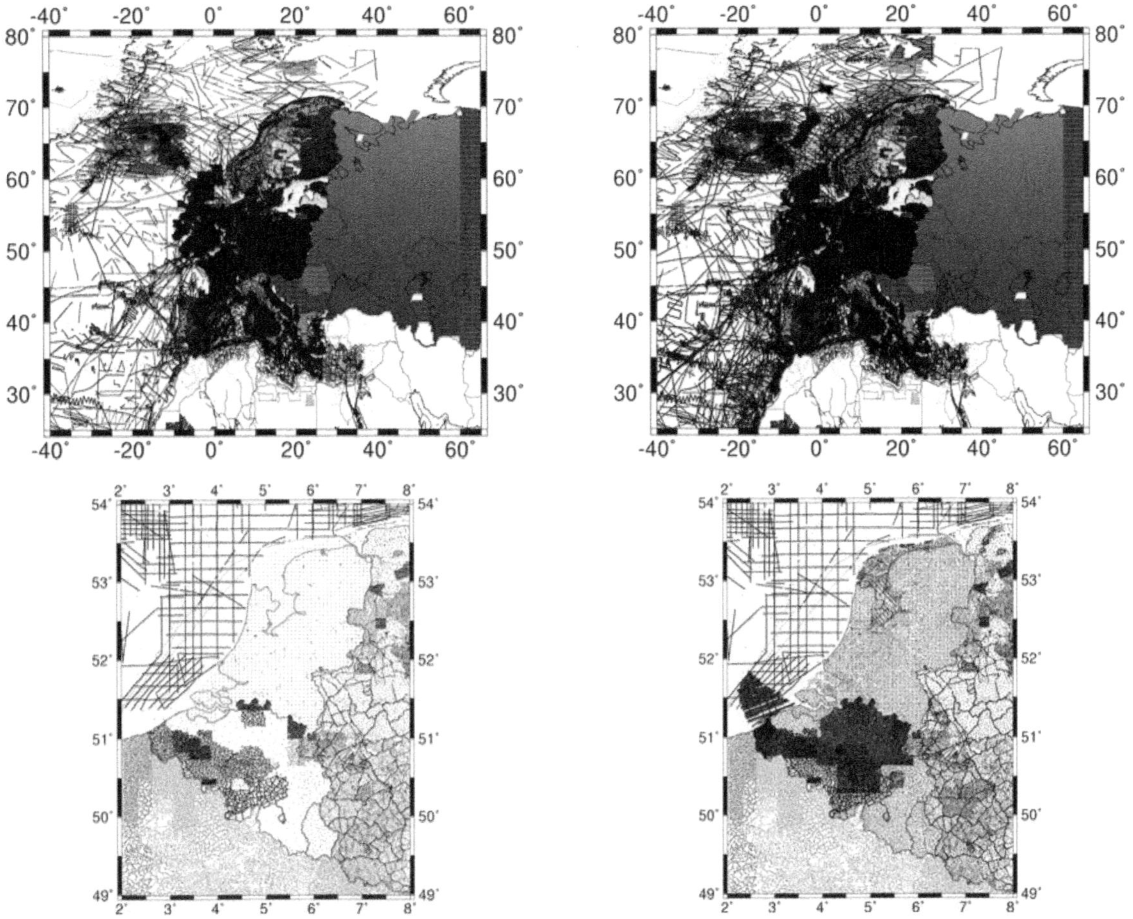


**Fig. 1.** Locations of terrestrial gravity data for entire Europe (top) and a sub-area (bottom). The left part shows the status for EGG97 and the right part shows the status of July 2004.

2004). As one example to document the progress in the collection of gravity data, Fig. 1 (bottom) depicts the old status (EGG97, Denker and Torge, 1998) and the new status in July 2004 for an area covering Belgium, Luxemburg, Netherlands, as well as parts of France and Germany.

In addition, the older gravity data sets were revised regarding the underlying reference systems, the target systems being ETRS (European Terrestrial Reference System), UELN (United European Levelling Network) and absolute gravity. Within the EGGP, only data which can be related without any doubts to the target reference systems will be included in the primary data base.

Significant progress was also made in the collection and reprocessing of ship gravity data (e.g., at IfE and other institutions). The ship gravity data, collected from several institutions for the European Seas, were crossover adjusted using a bias per track error model in order to reduce instrumental and navigational errors, incorrect ties to harbour stations, etc. (for details see Denker and Roland, 2003). An "original" and an "edited" data set were considered, where the edited data set excluded data affected by ship turns, data in the Red Sea, data from short tracks (< 3 points), and tracks with large crossover differences. Table 1 shows the crossover statistics for both data sets before and after the adjustment. The table clearly shows that the editing of some very bad observations resulted already in an improvement of the crossover differences by a factor of two, while the crossover adjustment again reduced the crossovers by a factor of two. Before the adjustment, the RMS crossover difference is 15.5 mgal for the original and 8.4 mgal for the edited data set; the corresponding values after the adjustment are 7.0 mgal and 4.7 mgal, respectively.

**Table 1.** Statistics of crossover differences from ship gravity observations. Units are mgal.

| data set<br>adjustment | original<br>before | original<br>after | edited<br>before | edited<br>after |
|---|---|---|---|---|
| # | 89,328 | 89,328 | 78,929 | 78,929 |
| Mean | 0.20 | -0.02 | 0.04 | -0.01 |
| RMS | 15.48 | 7.01 | 8.37 | 4.69 |
| Min | -258.43 | -204.98 | -109.91 | -48.56 |
| Max | +259.54 | +198.37 | +128.40 | +49.16 |

The improvement of the ship data by editing and crossover adjustment was also illustrated by comparisons with altimetric anomalies from the KMS02 model (Andersen et al., 2003), giving a RMS difference of 18.0 mgal for the original data set and 10.2 mgal for the edited data set, both before the adjustment. The crossover adjustment further reduced the RMS difference to 7.8 mgal for the edited data set, proving the effectiveness of the entire procedure. In sub-areas, e.g., around Iceland, the RMS difference between the ship and KMS02 data is only 4.2 mgal.

Fig. 1 (top) depicts the locations of gravity data for entire Europe; the left part shows the status for EGG97, and the right part shows the status for July 2004 including the reprocessed ship gravity data.

### 3.2 Digital Elevation Models (DEMs)

For the EGG97 model, digital elevation models (DEMs) were available with a resolution of about 200 m for most countries in Central and Western Europe, while coarser grids with a resolution of 0.5 km to 10 km had to be used for the remaining parts of Europe. For EGG97, only in Germany a DEM with a very high resolution of 1" × 1" (approx. 30 m) was available. Meanwhile, also Switzerland has released a 1" × 1" DEM, and Austria has indicated the release of a corresponding model. However, especially in Eastern Europe and some other areas, fill-ins from global public domain databases have to be used, either because high-resolution DEMs do not exist or are not released for confidentiality reasons. For this purpose, the SRTM3 model with a resolution of 3" × 3" (JPL, 2004) and the public domain global model GTOPO30 with a resolution of 30" × 30" (LP DAAC, 2004) can be used. The SRTM3 model has been released recently from the analysis of the Shuttle Radar Topography Mission as a preliminary and "research-grade" model, covering the latitudes between 60°N and 54°S. On the other hand, the GTOPO30 model has global coverage and was derived already in 1996 from several raster and vector sources of topographic information (LP DAAC, 2004).

The SRTM3 and GTOPO30 DEMs were evaluated at IfE by comparisons with national DEMs for Germany, based on 1" × 1" data (Denker, 2004a). The comparisons of the national and SRTM3 models revealed that one of the national models contained less accurate fill-ins in some areas outside of Germany. After excluding these areas, the differences between the best national model and the SRTM3 DEM show a standard deviation of 7.9 m with maximum differences up to about 300 m. The largest differences are located in opencast mining areas and result from the different epochs of the data. Histograms of the differences show a clear deviation from the normal distribution with a long tail towards too high SRTM3 elevations. Moreover, the presently available SRTM3 "research-grade" models contain numerous data voids (undefined elevations), which cause significant problems. The filling of these data gaps by interpolation must be handled with care, especially for larger gaps in mountainous areas (Denker, 2004a).

The evaluation of the GTOPO30 model by national and SRTM3 DEMs demonstrated that in Germany the longitudes of GTOPO30 should be increased by 30" (one block). The longitude shift reduced the standard deviation of the differences to the national and SRTM3 models by roughly 75 %, yielding final values of about 6.8 m and 11.5 m for the national and SRTM3 models, respectively.

Thus, the national DEMs, augmented by the SRTM3 and GTOPO30 data will allow the creation of DEMs for entire Europe with a resolution of at least 30" × 30", which is a significant improvement compared to the previous EGG97 computation.

### 3.3 Global Geopotential Models

The CHAMP and GRACE missions have led to significant improvements in the modelling of long wavelength gravity signals. This is documented, e.g., by the accumulated formal geoid error, which does not exceed 0.01 m for spherical harmonic degrees up to about 25 for the CHAMP models (e.g., Reigber et al., 2004a) and 75 for the GRACE models (e.g., Reigber et al., 2004b). On the other hand, the limit of 0.01 m is already exceeded at degree 8 for the EGM96 model. Correspondingly, the limit of 0.05 m is exceeded at about degree 20 for EGM96, 40 for the CHAMP models, and 90 for the GRACE models.

The new geopotential models from the CHAMP and GRACE missions, in combination with terrestrial gravity data of good quality (±1 mgal) and coverage, allow the computation of significantly improved continental-scale geoid and quasigeoid models. Error estimates based on the degree variance approach result in standard deviations of about 0.02 m to 0.03 m for solutions based on the GRACE models, with the largest contribution (about 0.02 m) coming from the degree range 90 to 360. The corresponding values for geoid solutions based on the CHAMP models and EGM96 are about 0.04 m and 0.06 m, respectively.

## 4 First Results

Updated European geoid/quasigeoid models were computed based on the new CHAMP and GRACE geopotential models. The computations were done using the EGG97 terrestrial gravity data set as well as an updated data set (section 3.1). The computations were done using the remove-restore technique in connection with the least squares spectral combination method. The spectral weights were derived from the error estimates of the global models and the terrestrial data. Terrain reductions were done using the RTM method. The computation area is 25°N – 77°N and 35°W – 67.4°E. The grid spacing is 1' × 1.5', yielding 3,120 × 4,096 grid points. The GRS80 constants, the zero degree undulation terms, and the zero-tide system were used throughout all computations (for details see Denker, 2004b).

All computed quasigeoid models were evaluated by GPS/levelling data from the European EUVN data set (Ihde et al., 2000) and by national campaigns. Fig. 2 shows the differences (after subtracting a common bias) between 166 stations of the EUVN GPS/levelling data set (only stations with UELN normal heights were used) and the EGG97 gravimetric quasigeoid based on EGM96 (left part), as well as a new solution (right part) based on the EIGEN-GRACE02S geopotential model (Reigber et al., 2004b); the terrestrial gravity data are identical in both solutions (status of EGG97). Fig. 2 shows clearly that the long wavelength discrepancy over Central Europe almost disappears for the GRACE solution; the largest discrepancies remain at coastal stations, especially around the Mediterranean Sea where the gravity data quality is weak. The RMS difference is 0.262 m for EGG97 and reduces to 0.230 m for the EIGEN-GRACE02S solution (12 % improvement). Correspondingly, a solution based on the CHAMP EIGEN-3 model (Reigber et al., 2004a) gives a RMS difference of 0.238 m (9 % improvement). When using the updated terrestrial gravity data set from 2004 in combination with the EIGEN-GRACE02S model, the RMS difference reduces to 0.203 m (23 % improvement compared to EGG97). Furthermore, when transforming the GPS results (according to Poutanen et al., 1996) from the non-tidal to the zero-tide system, which is used for the quasigeoid solutions, another slight reduction of the RMS difference to 0.197 m can be observed (25 % total improvement versus EGG97). Tilt parameters were also computed, but not considered any further as they were less than 0.1 ppm in all cases.

**Table 2.** RMS differences from comparisons of GPS/levelling with EGG97 and a new quasigeoid based on EIGEN-GRACE02S. A constant bias is subtracted. Units are m.

| Country | # pts. | EGG97/ EGM96 | EGG04/ EIGEN-GRACE02S | Improvement |
|---|---|---|---|---|
| Belgium | 31 | 0.061 | 0.046 | 25 % |
| France | 965 | 0.128 | 0.084 | 34 % |
| Germany | 678 | 0.107 | 0.041 | 62 % |
| Hungary | 299 | 0.089 | 0.057 | 36 % |
| Netherlands | 84 | 0.035 | 0.031 | 11 % |
| Switzerland | 147 | 0.084 | 0.063 | 24 % |
| EUVN | 166 | 0.262 | 0.230 | 12 % |

Additional comparisons of EGG97 and the new quasigeoid solution based on EIGEN-GRACE02S

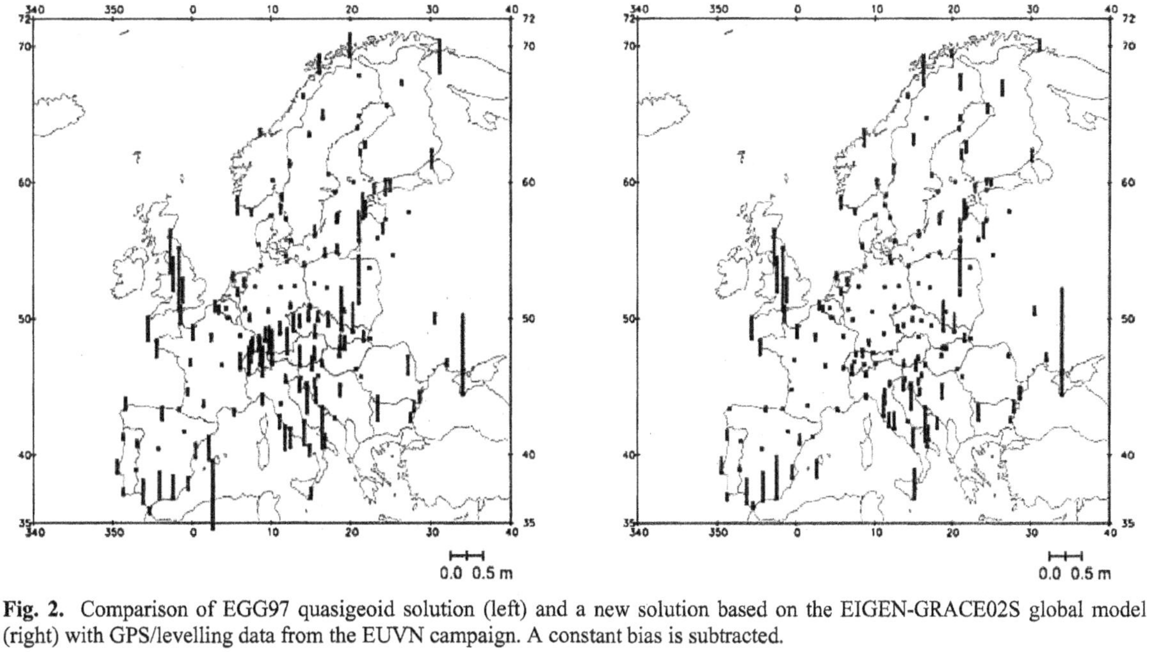


**Fig. 2.** Comparison of EGG97 quasigeoid solution (left) and a new solution based on the EIGEN-GRACE02S global model (right) with GPS/levelling data from the EUVN campaign. A constant bias is subtracted.

with several national GPS/levelling data sets are shown in Table 2. Again, both solutions use identical terrestrial gravity data (status EGG97). The table provides the RMS differences for both solutions after subtracting a common bias. The table clearly shows that in all cases the use of the new GRACE model improves the geoid/quasigeoid modelling significantly. The maximum improvement is more than 60 % for the German data set. A more detailed analysis shows that the tilts, existing in EGG97, are reduced to typically below 0.1 ppm, i.e. by one order of magnitude in some cases.

Furthermore, with the updated solutions based on the GRACE models, accurate determinations of $W_0$ (reference geopotential of the vertical datum) and vertical datum unifications become possible. When considering the 2004 terrestrial gravity data set, the EIGEN-GRACE02S model, the EUVN GPS/levelling data, and a transformation of the GPS heights to the zero-tide system, an estimate of $W_0$ (Europe) of 62,636,857.02 ±0.15 $m^2s^{-2}$ is obtained. The corresponding value from the German GPS/-levelling data is 62,636,856.91 ±0.02 $m^2s^{-2}$. Both values are in good agreement with the value 62,636,857.25 $m^2s^{-2}$ published for Europe by Burša et al. (2002). However, the European values deviate by about 1.0 $m^2s^{-2}$ from the global best estimates (e.g., Burša et al., 2002).

## 5 Conclusions

Significant progress was made within the framework of the European Gravity and Geoid Project EGGP regarding the collection and homogenization of high-resolution gravity and terrain data. Several new data sets became available, and especially the new geopotential models from the CHAMP and GRACE missions improved the geoid/quasigeoid modelling very much. In the GPS/levelling comparisons, the RMS differences reduced up to about 60 % when using the GRACE models and up to 30 % for the solutions based on CHAMP, as compared to the previous EGG97 model relying on EGM96. In addition, the tilts, existing in EGG97, were also reduced to typically below 0.1 ppm. Due to the support with data by numerous people and agencies, further improvements are to be expected in the future.

## References


Andersen, O.B., P. Knudsen, S. Kenyon, and R. Trimmer (2003). KMS2002 Global Marine Gravity Field, Bathymetry and Mean Sea Surface. Poster, IUGG2003, Sapporo, Japan, June30-July11, 2003.

Burša, M., et al. (2002). World height system specified by geopotential at tide gauge stations. IAG Symposia 124: 291-296, Springer Verlag.

Denker, H. (1988). Hochauflösende regionale Schwerefeldmodellierung mit gravimetrischen und topographischen Daten. Wiss. Arb. Fachr. Verm.wesen, Univ. Hannover, Nr. 156.

Denker, H. (1998). Evaluation and Improvement of the EGG97 Quasigeoid Model for Europe by GPS and Leveling Data. Reports of the Finnish Geodetic Institute, 98:4, 53-61, Masala, 1998.

Denker, H. (2004a). Evaluation of SRTM3 and GTOPO30 terrain data in Germany: Proceed. GGSM2004, Porto.

Denker, H. (2004b). Improved European geoid models based on CHAMP and GRACE results. Proceed. GGSM2004, Porto.

Denker, H., M. Roland (2003). Compilation and evaluation of a consistent marine gravity data set surrounding Europe. Proceed., IUGG2003, Sapporo, Japan, June30-July11, 2003.

Denker, H., W. Torge (1998). The European gravimetric quasigeoid EGG97 – An IAG supported continental enterprise. IAG Symposia 119:249-254, Springer Verlag.

Denker, H., W. Torge, G. Wenzel, J. Ihde, U. Schirmer (2000). Investigation of Different Methods for the Combination of Gravity and GPS/Levelling Data. IAG Symposia 121:137-142, Springer-Verlag.

EGGP (2003). Commission Project 2.1, European Gravity and Geoid, Terms of Reference and Objectives, http://www.ceegs.ohio-state.edu/iag-commission2/cp2.1.htm.

Forsberg, R. (1993). Modelling the fine-structure of the geoid: methods, data requirements and some results. Surveys in Geophys. 14: 403-418.

Forsberg, R., S. Kenyon (2004). Gravity and geoid in the Arctic Region – The northern GOCE polar gap filled. Proceed. 2nd Internat. GOCE Workshop, Esrin, March 8-10, 2004.

Forsberg, R., C.C. Tscherning (1981). The use of height data in gravity field approximation by collocation. Journal of Geophys. Research 86: 7843-7854.

Grote, T. (1996). Regionale Quasigeoidmodellierung aus heterogenen Daten. Wiss. Arb. Fachr. Verm.wesen, Univ. Hannover, Nr. 212.

Ihde, J. et al. (2000). The height solution of the European Vertical Reference Network (EUVN). Veröff. Bayer. Komm. für die Internat. Erdmessung, Astronom. Geod. Arb., Nr. 61: 132-145, München.

JPL (2004). SRTM – The Mission to Map the World. Jet Propulsion Laboratory, California Inst. of Techn., http://www2.jpl.nasa.gov/srtm/index.html.

LP DAAC (2004). Global 30 Arc-Second Elevation Data Set GTOPO30. Land Process Distributed Active Archive Center, http://edcdaac.usgs.gov/gtopo30/gtopo30.asp.

Poutanen, M., M. Vermeer, J. Mäkinen (1996). The permanent tide in GPS positioning. J. of Geodesy 70: 499-504.

Reigber, Ch., et al. (2004a). Earth gravity field and seasonal variability from CHAMP. Earth Observation with CHAMP: 25-30, Springer Verlag.

Reigber, Ch., et al. (2004b). An Earth gravity field model complete to degree and order 150 from GRACE: EIGEN-GRACE02S. J. of Geodynamics, accepted June 25, 2004.

Sansò, F., C.C. Tscherning (2003). Fast spherical collocation: theory and examples. Journal of Geodesy 77:101-112.

Wenzel, H.-G. (1982). Geoid computation by least squares spectral combination using integral kernels. Proceed. IAG General Meet., 438-453, Tokyo.

# Merging a Gravimetric Model of the Geoid with GPS/Levelling data : an Example in Belgium

H. Duquenne
Laboratoire de Recherche en Géodésie, IGN/ENSG, 6 et 8 avenue Blaise Pascal, 77455 Marne-la-Vallée (France), henri.duquenne@ensg.ign.fr

M. Everaerts
Royal Observatory of Belgium, Av. Circulaire 3, 1180 Brussels (Belgium), everaerts@oma.be

P. Lambot
National Geographic Institute, 13 Abbaye de la Cambre, 1050 Brussels (Belgium), pla@ngi.be

**Abstract.** A new model of the gravimetric quasi-geoid covering Belgium was computed in 2003. In order to adapt it to the needs of levelling by GPS, it was necessary to fit it to a set of levelled GPS points embodying the Belgian geodetic and levelling reference systems. A dense network was available, allowing significant methodological and statistical studies on how to model differences between the gravimetric geoid and the levelled GPS points. A method using kriging with the outlier detection in GPS-levelling data was applied. Several models of trend and covariance functions were tested. A reference surface dedicated to levelling by GPS in Belgium is presented, the accuracy of which is about 2~2.5 cm.

**Keywords.** Geoid, levelling by GPS, kriging, Belgium.

## 1 Introduction

Several methods have been proposed to combine a gravimetric geoid and a set of levelled GPS points in order to derive a height reference surface suitable for levelling by GPS. Kotsakis and Sideris (1999) gave a review of publications on this topic and presented two approaches to estimate the accuracy of a geoid model and to model systematic distortions affecting data sets. Marti et al. (2001) proposed a common adjustment of GPS, levelling and the geoid which needed the total variance-covariance matrix of the data and led to a change in the height reference system. Computation of a corrector to the geoid by collocation was used for instance by Smith and Milbert (1999) for the USA, Duquenne (1999) for France, Forsberg et al. (2002) for the British Isles. This method has the advantage of producing by itself a stochastic model of discrepancies between the geoid model and the levelled GPS points, while leaving the height reference system unchanged. The area under investigation is Belgium, where a dense set of levelled points is available. This allows for interesting statistical studies, especially on covariance modelling.

## 2 Data description

A height reference system (also called height datum) is defined by its fundamental benchmark and a height system (orthometric, normal). Thanks to the smooth and low topography and the small magnitude of the gravity anomalies over the territory of Belgium, there is no appreciable difference between height systems. So the height datum was materialized by a levelling network without orthometric or normal correction. This network is known as "Deuxième Nivellement Général" (DNG).

The new reference geodetic network of Belgium (BEREF) consists of 36 points measured by GPS. It was adjusted in one block connected to 7 stations of the European GPS Permanent Network (EPN). The standard deviation of the coordinates was evaluated at about 5 mm. Referenced to the BEREF, a dense geodetic network of 4024 points covering an area of 30500 km$^2$ (1 point per 8 km$^2$) was set up by GPS. All the baselines were adjusted in a single process, leading to a standard deviation of the coordinates of 8 to 17 mm. These points were linked to the height reference DNG, using precise levelling with a tolerance of 1 cm/km.

The first gravity measurement in Belgium was carried out at the Royal Observatory (ORB) in 1884. The first Belgian gravity network was observed in 1924. It included 381 stations, the precision of which was better than 0.7 mGal. Since 1948 the Belgian National Geographical Institute (IGN-B) and the ORB have been closely collaborating to make the gravity coverage denser. This was achieved in 2002, when more than 30000

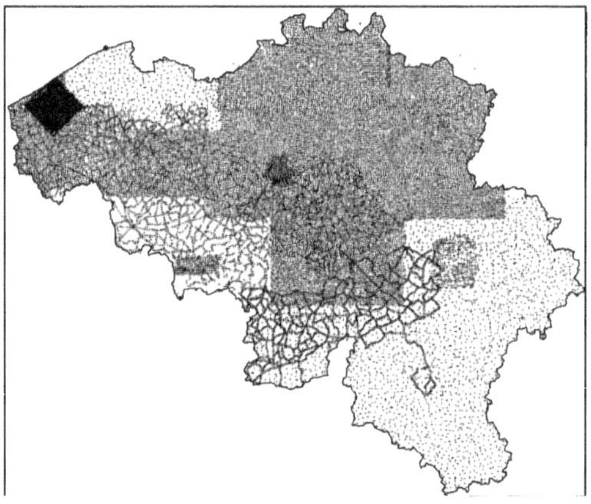

**Fig 1** Gravity coverage of Belgium.

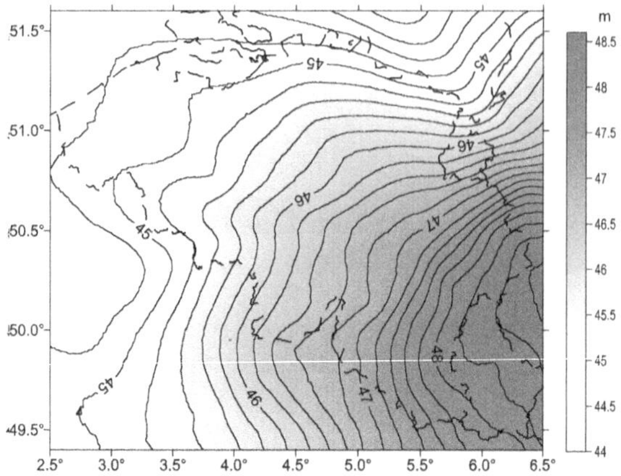



**Fig 2** The quasigeoid model BG03.

measurements were completed (figure 1). Their density is less in the south-west part of the country (1 station per 2.5 to 5 $km^2$) than in the rest (1 station per $km^2$). Their precision reaches 0.1 mGal. The data base of the ORB contains presently more than 250000 gravity points covering the whole of Belgium and surrounding countries. Gravity data were supplied by the Bureau de Recherche Géologiques et Minières for France, the British Geological Survey for the United Kingdom, the Rijkswaterstaat for the Netherlands, and H. G. Wenzel (personal communication) for Germany. All these data were used to compute the quasigeoid BG03.

A first estimation of the Belgian geoid model was carried out in 1996 (Pâquet et al. 1996). It was computed by Stokes's method and by collocation. The model was compared with the 36 levelled GPS points which were available at that time. Its precision was 3 to 4 cm in areas sufficiently covered with gravity data.

The quasigeoid BG03 (figure 2) was computed at the International Geoid Service at DIAAR in Milan (Barzaghi et al. 2004). Two models were computed, the first one by rapid collocation, the second one by Stokes's method and FFT. The results were similar. Compared to BG96, BG03 is improved due to denser gravity data in the south-east part of the country, the use of the high-resolution geopotential model GPM98CR (Wenzel 1998) and a digital terrain model including bathymetry.

## 3 Elaboration of a height reference surface

### 3.1 Goals and methodology

For ten years or so, precise models of the geoid or quasigeoid have been used for levelling by GPS. Nevertheless it is well known that a model of the geoid (or the quasigeoid) cannot be used directly as the reference frame, as it is done with the levelling network and today with levelled GPS points. This is due to various errors in the data used to compute the geoid, the orthometric and ellipsoidal heights of the GPS points (see for example Duquenne 1999). Furthermore when working out orthometric or normal heights from GPS measurements, users expect to find the same results as if they had used classical levelling. Therefore the quasigeoid model must be adapted through a process which ensures that:

- the height reference system of the classical levelling is virtually unchanged,
- outliers in the quasigeoid and/or levelled GPS points are detected and removed,
- precision is evaluated.

Merging the gravimetric quasigeoid with levelled GPS points brings a new realization of the height datum with the following advantages:

- a higher density, thanks to the gravimetric quasigeoid,
- a well known geodetic reference, due to the levelled GPS points,
- the numerical realization of the datum via a grid, which makes its interpolation easy. In that sense one can talk about a *height reference surface*.

Collocation or kriging were proposed as tools to meet these requirements. $\zeta_{Grv}$ refers to the quasigeoid undulation derived from gravimetry, whereas $\zeta_{GPS-Lev}$ is the quasigeoid undulation computed from GPS and levelling (geometric quasigeoid). Let $\{P_i, i=1,\ldots,n\}$ be a set of levelled GPS points where $\zeta_{Grv,i}$ and $\zeta_{GPS-Lev,i}$ are known, and $(\varphi_i, \lambda_i)$ are the latitude and longitude of $P_i$.

The differences $\delta_i = \zeta_{GPS-Lev,i} - \zeta_{Grv,i}$ are split in two parts:

$$\delta_i = s_i + n_i \tag{1}$$

From a statistical point of view, $s_i$ represents the correlated part of $\delta_i$, while $n_i$ is the uncorrelated part. From a geodetical point of view, $s_i$ is mainly due to errors in gravity data the integration of which causes correlation between $\zeta_{Grv,i}$'s. GPS or levelling errors are partly correlated. $n_i$ contains errors due to heights of GPS antennae, unexpected vertical motion of benchmarks, etc. In the field of signal processing, $s_i$ is called the signal and $n_i$ the noise. We continue with these usual terms, although they are not fully suitable here. The signal and noise can be calculated by collocation (Moritz 1989, eq. 14-23 and 14-43), also presented in (Chilès and Delfiner 1999) as a particular case of *co-kriging*:

$$\mathbf{s} = \mathbf{C}_{ss}\mathbf{C}_{\delta\delta}^{-1}\delta \tag{2}$$

$$\mathbf{n} = \delta - \mathbf{s} \tag{3}$$

where $\delta$, $\mathbf{s}$, $\mathbf{n}$ are the vectors of the difference, signal and noise, respectively, $\mathbf{C}_{ss}$ and $\mathbf{C}_{\delta\delta}$ are the covariance matrices of the signal and raw differences, respectively. Once computed, the signal and noise can be used to retrieve outliers. More specifically, a large noise level reveals an error in a particular levelled GPS point, whereas the large signal indicates a possible error in gravity survey. The signal can also be used to correct the geoid model since it can easily be interpolated. The simple kriging equation (2) holds as long as the difference $\delta = \zeta_{GPS-Lev} - \zeta_{Grv}$ is a stationary random function (Chilès and Delfiner 1999). If not, alternative solutions must be considered:

- detrending the differences, for instance, with a low degree polynomial of the coordinates, then kriging the residuals,
- universal kriging that involves simultaneous determination of the trend and signal,
- use of a variogram instead of the covariance function.

In this limited study and owing to lack of space we focus on the first method.

## 3.2 Covariance modelling

Generally, the covariance matrices in Eq. (2) are not known a priori. Especially in our case study, the covariance matrix of heights of the levelled GPS points is not known. Then the covariance must be computed by fitting a covariance model to the empirical covariance of the data. Afterwards several models commonly used in geostatistics are tested to model the covariance of the signal. Let $f$ be the covariance function of the signal and $\sigma_n^2$ the variance of the noise (a constant). The covariance of the noise is assumed to be zero. Since signal and noise are not correlated, the covariance of the signal and differences takes the form:

$$C_s(P_i, P_j) = f(P_i, P_j) \tag{4}$$

$$C_\delta(P_i, P_j) = \begin{cases} f(P_i, P_j) & \text{if } P_i \neq P_j \\ \sigma_n^2 + f(P_i, P_j) & \text{if } P_i = P_j \end{cases} \tag{5}$$

### 3.2.1 *Isotropic models*

With these models, the covariance depends only on the spherical distance $r$ between points. The parameters of the model are:

- $\sigma_n^2$, the noise variance
- $C_0$, the signal covariance for $r = 0$, and
- $d_{half}$, the correlation length, i.e. the distance for

**Table 1** Isotropic covariance models, mainly drawn from (Chilès and Delfiner 1999) slightly modified: the special parameter $\alpha$ is introduced here so that the correlation length $d_{half}$ has the same meaning for all the models, see text.

| Name | Function | Special parameter |
|---|---|---|
| Exponential | $C(r) = C_0 \exp(-r \ln 2 / d_{half})$ | none |
| Markov 2° order | $C(r) = C_0(1 + r/(\alpha\, d_{half}))\exp(-r/(\alpha\, d_{half}))$ | $\alpha$ : $\alpha = 1/\ln(2 + 2/\alpha)$, $\alpha \approx 0.59582$ |
| Triangular | $C(r) = C_0(1 - r/(\alpha\, d_{half}))$ | $\alpha = 2$ |
| Spherical | $C(r) = C_0(1 - 0.5\, r/(\alpha\, d_{half}) + 0.5\,(r/\alpha\, d_{half})^3)$ | $\alpha$ : $\alpha^3 - 3\alpha^2 + 1 = 0$, $\alpha \approx 2.87939$ |
| Cardinal sine | $C(r) = C_0(\alpha\, d_{half}/r)\sin(r/(\alpha\, d_{half}))$ | $\alpha$ : $\alpha \sin(1/\alpha) = 1/2$, $\alpha \approx 0.52756$ |

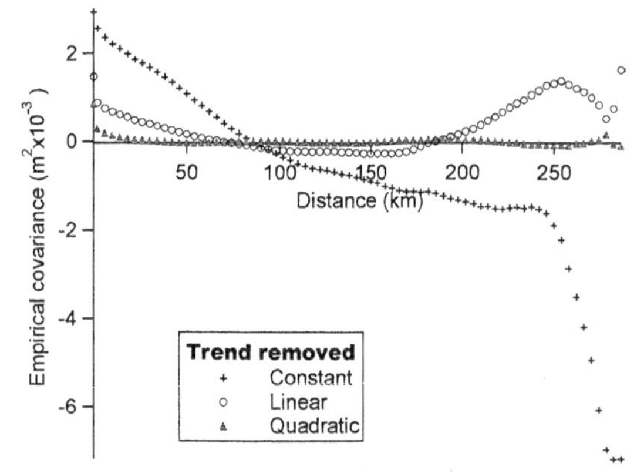


**Fig 3** Empirical covariance of the differences $\delta$ for 4024 levelled GPS points. Classes of distance are 4 km wide.

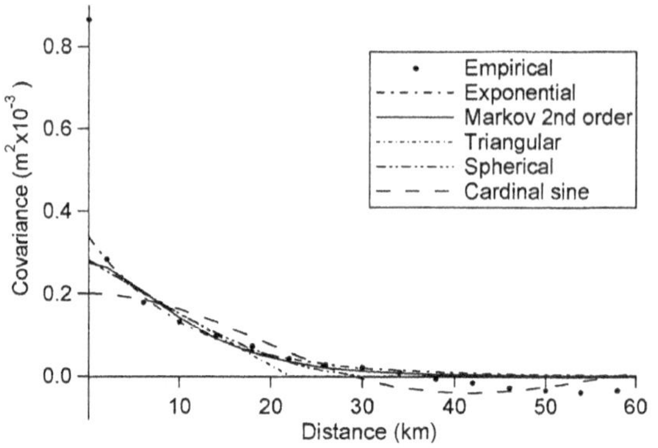


**Fig 5** Adjustment of isotropic covariance models to the empirical covariance of the differences $\delta$ up to 38 km (quadratic trend removed).

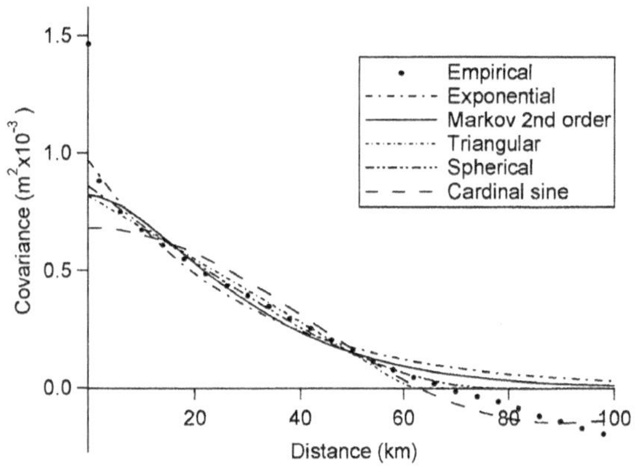


**Fig 4** Adjustment of isotropic covariance models to the empirical covariance of the difference $\delta$ up to 78 km (linear trend removed).

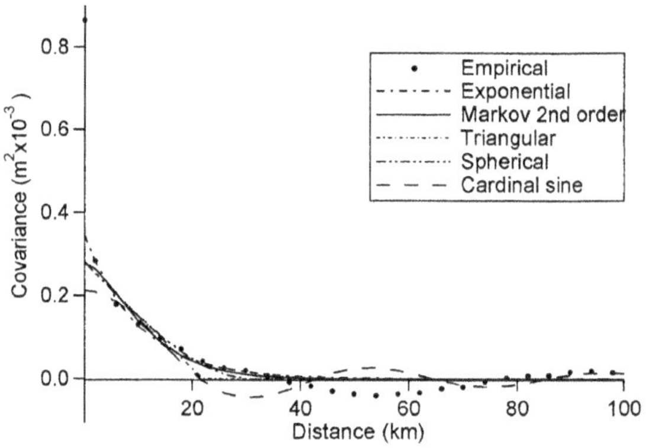


**Fig 6** Adjustment of isotropic covariance models to the empirical covariance of the differences $\delta$ up to 78 km (quadratic trend removed).

**Table 2** Adjusted parameters and statistics of isotropic covariance models. Note that the trend variance is not included in $C_0$. The standard deviations of the residuals are computed up to the distance of adjustment.

| Trend removed | Model | Distance of adjustment | $C_0$ ($10^{-6}$ $m^2$) | $d_{half}$ (km) | $\sigma_n$ ($10^{-3}$ m) | Residual of covariance ($10^{-6}$ $m^2$) Min | Max | Std. dev. | Figure # |
|---|---|---|---|---|---|---|---|---|---|
| linear | exponential | 78 km | 967.9 | 20.06 | 22.24 | -49.0 | 101.4 | 48.6 | 4 |
| linear | Markov 2nd order | 78 km | 819.2 | 27.02 | 25.36 | -65.5 | 71.8 | 38.2 | 4 |
| linear | triangular | 78 km | 815.1 | 30.51 | 25.44 | -90.4 | 35.5 | 33.1 | 4 |
| linear | spherical | 78 km | 857.4 | 27.18 | 24.60 | -54.3 | 29.0 | 20.8 | 4 |
| linear | cardinal sine | 78 km | 679.6 | 38.03 | 27.98 | -200.3 | 72.0 | 67.7 | 4 |
| quadratic | exponential | 38 km | 337.0 | 7.28 | 22.96 | -10.7 | 17.3 | 9.0 | 5 |
| quadratic | Markov 2nd order | 38 km | 275.4 | 10.25 | 24.27 | -19.6 | 25.9 | 14.5 | 5 |
| quadratic | triangular | 38 km | 278.4 | 11.00 | 24.21 | -41.7 | 24.0 | 25.9 | 5 |
| quadratic | spherical | 38 km | 280.8 | 10.82 | 24.16 | -29.2 | 22.2 | 18.3 | 5 |
| quadratic | cardinal sine | 38 km | 199.9 | 17.45 | 25.78 | -84.8 | 35.9 | 41.0 | 5 |
| quadratic | exponential | 78 km | 346.3 | 6.87 | 22.76 | -15.2 | 42.4 | 22.6 | 6 |
| quadratic | Markov 2nd order | 78 km | 277.7 | 10.06 | 24.22 | -17.8 | 41.2 | 23.0 | 6 |
| quadratic | triangular | 78 km | 278.4 | 11.00 | 24.21 | -41.7 | 40.9 | 26.8 | 6 |
| quadratic | spherical | 78 km | 280.8 | 10.83 | 24.16 | -29.2 | 40.9 | 23.9 | 6 |
| quadratic | cardinal sine | 78 km | 212.2 | 12.99 | 25.54 | -73.88 | 67.9 | 44.7 | 6 |

which the signal equal to $C_0/2$.

Details on tested models are given in Table 1. Figure 3 shows the empirical covariance of the residuals of the differences $\delta_i = \zeta_{GPS-Lev,i} - \zeta_{Grv,i}$ after removing a trend. Clearly with a constant trend, empirical covariances are too large at long distances and the stationarity hypothesis has to be rejected. Figures 4 to 6 and Table 2 depict results of the covariance model fitting after subtracting a linear or quadratic trend from the $\delta$ 's. From this test we can conclude:

- removing the quadratic trend allows reducing the adjustment range and improves the fit by a factor of 1.1 to 5 in terms of the standard deviation of the residuals;
- the Markov and cardinal sine models do not fit well especially for short distances, due to their zero derivative at the origin;
- triangular model lacks the curvature at the middle distances;
- best results are globally obtained by the exponential model;
- standard deviation of the noise $\sigma_n$ computed from the variance of the residuals and $C_0$, is stable enough (about 22~25 mm).

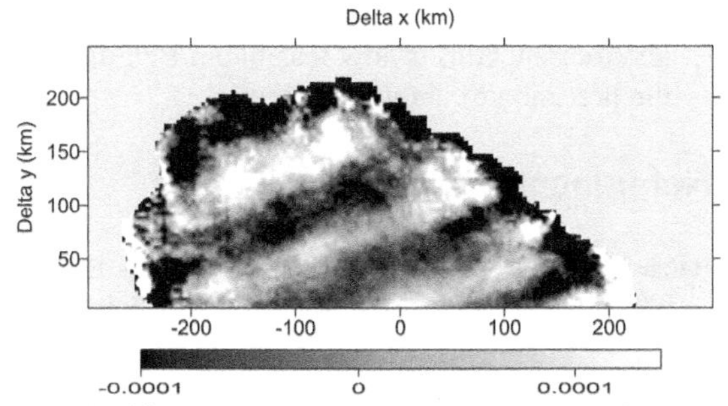


**Fig 7** 2D histogram of the empirical covariance of the differences $\delta$ for 4024 levelled GPS points. The quadratic trend is subtracted. Classes of coordinate differences are 4 km wide. Unit: $m^2$.

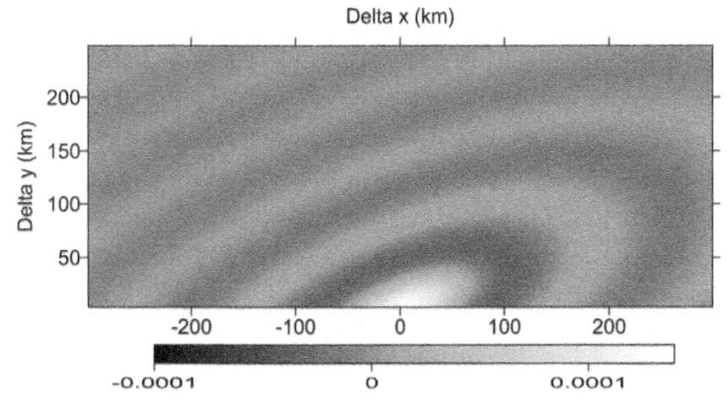


**Fig 8** 2D histogram of the cardinal sine covariance model. Distance of adjustment 78 km. Unit: $m^2$.

### 3.2.2 *Anisotropic models*

Anisotropic covariance models depend on the distance and direction. Starting from an isotropic model, the most simple way to bring in anisotropy is to define a new coordinate system in the tangent plane to the sphere. Let $\Delta x$ and $\Delta y$ be the components of $\overrightarrow{P_i P_j}$ in that plane towards east and north, respectively. The distance $r = \left(\Delta x^2 + \Delta y^2\right)^{1/2}$ in the equation of the isotropic model is changed into:

$$r' = \left[\left(k\left(\Delta x \cos\theta + \Delta y \sin\theta\right)\right)^2 + \left(\left(\Delta x \sin\theta - \Delta y \cos\theta\right)/k\right)^2\right]^{1/2}$$

where $\theta$ is the main angle of anisotropy and $k$ the anisotropy ratio. Figure 7 shows the 2D empirical covariance after a quadratic detrending, whereas figure 8 represents the cardinal sine model. Table 3 summarizes the results of the adjustments. Significantly better results are obtained with the quadratic detrending. The tested anisotropic models work worse than the isotropic ones. This might be due to the inability of all the models but the cardinal sine one to render the "hole effect" (i.e. the negative values of the covariance), and the misfit of the cardinal sine model for the short distances. More anisotropic models should be experienced.

**Table 3** Adjusted parameters and statistics of anisotropic covariance models.

| Trend removed | Model | Distance of adjustment | $C_0$ ($10^{-6}$ $m^2$) | $d_{half}$ (km) | $k$ (aniso. ratio) | $\theta$ (aniso. angle) | $\sigma_n$ ($10^{-3}$ m) | Residual of covariance ($10^{-6}$ $m^2$) Min | Max | Std. dev. |
|---|---|---|---|---|---|---|---|---|---|---|
| linear | exponential | 78 km | 1069.2 | 18.64 | 0.71 | 101.15° | 19.83 | -251.5 | 464.5 | 124.3 |
| linear | Markov 2nd order | 78 km | 795.3 | 28.70 | 0.73 | 101.35° | 25.82 | -264.3 | 457.1 | 118.6 |
| linear | triangular | 78 km | 692.4 | 36.44 | 0.76 | 98.80° | 27.75 | -268.2 | 445.4 | 78.7 |
| linear | spherical | 78 km | 786.4 | 30.79 | 0.75 | 99.92° | 26.00 | -272.1 | 445.4 | 79.9 |
| quadratic | exponential | 78 km | 287.2 | 10.66 | 2.49 | 110.68° | 24.02 | -72.5 | 112.2 | 43.0 |
| quadratic | Markov 2nd order | 78 km | 203.9 | 17.24 | 2.42 | 110.74° | 25.70 | -78.0 | 111.9 | 42.7 |
| quadratic | triangular | 78 km | 171.0 | 22.72 | 2.28 | 110.61° | 26.33 | -107.7 | 111.7 | 41.1 |
| quadratic | spherical | 78 km | 193.3 | 19.32 | 2.30 | 110.64° | 25.91 | -94.1 | 111.7 | 42.0 |
| quadratic | cardinal sine | 78 km | 139.7 | 29.71 | 1.63 | 111.93° | 26.92 | -122.9 | 95.6 | 30.0 |

### 3.3 Data validation

The validation is obtained through the analysis of noise in an iterative process. At each step, the trend and covariance parameters are updated, signal and noise are computed for each validated points, as well as the standard deviation of noise and a tolerance (2.57 times the standard deviation). Then the noise is compared with the tolerance. Figure 9 shows how the process is sensitive to a change of the covariance model: the exponential model, which fits better the empirical covariance, appears to be more selective than the triangular one. The former leads to a noise standard deviation of 1.6 cm after 6 iterations, against 2.2 cm after 9 iterations for the latter.

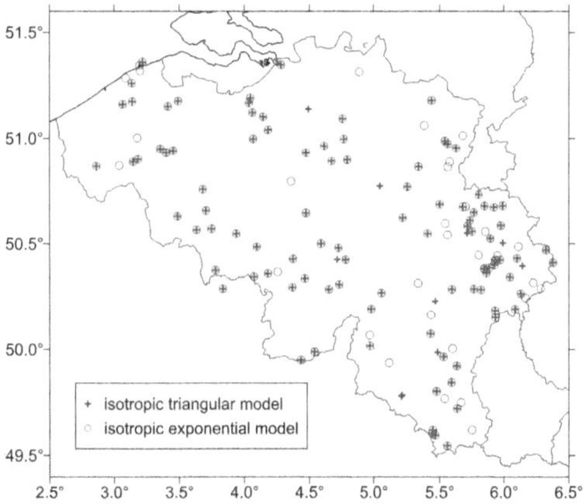


**Fig 9** Point rejection using two different covariance models.

### 3.4 The Belgian height reference surface

As stated above, the height reference surface is obtained by combining a grid of the quasigeoid model with the trend and interpolated values of the signal computed from all the accepted GPS points. Obviously interpolation is carried out using kriging. In order to evaluate the precision, a degraded test surface was produced from a subset of 594 points using the isotropic exponential covariance model. The heights of the remaining points were computed from the GPS measurements, then compared to the heights from levelling. The standard deviation of the differences reaches 2.9 cm. Besides the errors of the degraded height reference surface itself, this figure includes the errors of levelling and GPS measurements. The standard deviation of the noise (1.6 cm in the best case) is certainly a too optimistic estimator of the precision of the final reference surface (the one computed using 4024 points), whereas 2.9 cm is certainly pessimistic.

## 4 Conclusions

In order to adapt the new Belgian quasigeoid model BG03 to the requirements of levelling by GPS, the kriging method was employed. Particularly the problems of finding a suitable covariance model and its fit it to the empirical covariance were addressed. From this study, we can conclude:

- it is desirable to consider covariance models like the exponential one, which in our case worked better than the commonly used Gaussian or Markovian models;
- although the anisotropic models were tested, their relevance was not obvious;
- outlier rejection is very sensitive to the choice of the covariance model;
- levelling by GPS is now feasible in Belgium with the precision of about 2~2.5 cm.

## References

Barzaghi R., A. Borghi, B. Ducarme and M. Everaerts (2004). Quasi-geoid BG03 computation in Belgium. In: *Newton's Bulletin Nr 1, BGI & IGeS.*

Chilès J.P. and P. Delfiner (1999). Geostatistics. Wiley & Sons, New York, USA.

Duquenne H. (1999). Comparison and Combination of a Gravimetric Quasigeoid with a Levelled GPS Data Set by Statistical Analysis. Phys. Chem. Earth (A) 24:1, 79-83.

Forsberg R., G. Strykowski, J.C. Iliffe, M. Ziebart, P.A. Cross, C.C. Tscherning, P. Cruddace, K. Stewart, C. Bray, O. Finch (2002). OSGM02: a new geoid model of the British Isles. In: *Gravity and Geoid 2002, 3rd Meeting of the IGGC, Ziti Edition, Thessaloniki, Greece.*

Kotsakis J. and M.G. Sideris (1999). On the adjustment of combined GPS/levelling/geoid networks. J Geod 73: 412-421.

Marti U., A. Schlatter, E. Brockmann and A. Wiget (2001). The Way to a Consistent National Height System for Swizerland. In: *IAG Symposia 125: Vistas for Geodesy in the New Millennium, Springer 2002.*

Moritz H. (1989). Advanced Physical Geodesy. Wichmann, Karlsruhe, Germany.

Pâquet P., Z. Jiang and M. Everaerts (1996). A new Belgian geoid determination BG96. In: *IAG Symposia 117: Gravity, Geoid and Marine Geodesy, Springer 1997.*

Smith D.A. and D.G. Milbert (1999). The GEOID96 high-resolution geoid height model for the United States. J Geod 73: 219-236.

Wenzel, G. (1998). Ultra high degree geopotential model GPM3E97A to degree 1800 tailored to Europe. In: *M. Vermeer and J. Adam (Editors): Proceedings Second Continental Workshop on the Geoid in Europe. Reports of the Finnish Geodetic Institute 98:4, pp. 71-80, Marsala, Finland.*

# The Antarctic Geoid Project: Status Report and Next Activities

Mirko Scheinert
TU Dresden, Institut für Planetare Geodäsie, D-01062 Dresden, Germany

**Abstract.** The Antarctic Geoid Project (AntGP) forms the Commission Project 2.4 within the IAG Subcommission 2.4 *Regional Geoid Determination*. The main scientific goal of AntGP is to compile gravity data for the entire Antarctic, thus to fill in the southern polar gap and to improve the terrestrial gravity data coverage. A completed Antarctic gravity dataset will substantially contribute to the determination of the global gravity field in combination with the new satellite missions and will serve as an excellent basis for regional and continental geoid improvement.

Therefore, efforts have to be made to compile already existing data and to promote new gravity observation campaigns. Under Antarctic conditions, airborne gravimetry has proved the most powerful technique. Also, the connection of the different gravity surveys to a common gravity datum has to be made. Reference stations observed with absolute gravity meters should play an important role for this task.

The paper will review the status of AntGP and the work done so far. Special attention will be given to pointing out the liaison to initiatives within the Scientific Committee on Antarctic Research (SCAR). The activities to be undertaken within the near future will be outlined as well.

**Keywords.** Gravity surveys, geoid determination, Antarctica.

## 1 Introduction

The *Antarctic Geoid Project* (AntGP) was adopted a project within the new IAG structure after the IUGG General Assembly in Sapporo in July 2003. AntGP can be found as Commission Project 2.4 within Commission 2 *Gravity Field*, Sub-Commission 2.4 *Regional Geoid Determination*. This paper reviews the status of the project – among other considerations the current knowledge of gravity in Antarctica, possibilities and limits of gravity surveys and the linkage to other programmes and initiatives.

## 2 Goals of the Antarctic Geoid Project

In contrast to other regions of the world, in Antarctica there is no obligation to fulfill a state's responsibility for national surveying, for instance to determine a geoid in order to provide a reference for the national height system. On the contrary, due to the Antarctic treaty, Antarctica is open to scientific research in accordance with environmental protection issues. Regarding the rationale of a geoid determination in Antarctica, three arguments should be given:

*First argument: It is necessary to densify and to improve global gravity field models in Antarctica.*

Looking at the new satellite gravity missions CHAMP (launched July 15, 2000), GRACE (launched March 17, 2002) and GOCE (to be launched 2006) there still remains the polar gap problem. This problem can be seen in an area (described by a spherical cap of about 1 to 3 degrees radius) which will not be covered by satellite observations, which influences the determination especially of the zonal coefficients of the spherical harmonic expansion of the Earth's gravitational field. Such global models describe the longer wavelengths of the field, i.e., up to degree and order 200 to 360 (spatial resolution about 180 to 100 km). The current knowledge of the Earth's gravity field in Antarctica without incorporating the new satellite data can be illustrated by the model EGM96 (Lemoine et al., 1996). New, up-to-date global models, determined by a combination of CHAMP, GRACE and terrestrial data are GGM01C resp. GGM02C (Center of Space Research, University of Texas, (CSR, 2004), complete up to degree and order 200) and EIGEN-CG01C (GFZ Potsdam, (Schwintzer et al., 2004), complete up to degree and order 360). The differences between EGM96 and the new combination models like GGM01C reach magnitudes of about 4 meters. These differences illustrate mainly the improvement of the models by the new satellite data (esp. GRACE). Nevertheless, for the terrestrial data incorporated in these models there are no or only minor improvements in Antarctica. Therefore, it is necessary to improve the global models by providing a better coverage of gravity data as well as to carry out an improvement on regional scales (e.g., regional geoid determination).

*Second argument: Satellite-derived gravity field models should be validated in Antarctica.*

Although the new satellite missions improve the accuracy of the global models, it is necessary to validate these results. This could be carried out utilizing independent data, e.g. terrestrial, airborne and shipborne gravity data. Validation in this context means *product validation*, i.e. the gravity field model yielded by a satellite solution will be compared with a model resulted from independent data, mostly for a specific region. This is nicely demonstrated for instance by D. McAdoo (McAdoo et al., 2004), who showed a high correlation between the surface gravity field of the Arctic Gravity Project (Kenyon and Forsberg, 2002) and that of the GRACE model GGM01S (both filtered with a 2.5° Gaussian). To follow a similar procedure in Antarctica, one has to fill in a sufficient number of gravity tracks and points, respectively.

*Third argument: A high-resolution, high-accuracy gravity/geoid model in Antarctica should be determined for further studies in geodesy, geophysics, glaciology and oceanography.*

Using terrestrial (shipborne, airborne) gravity data, regional and continental geoid models in Antarctica should be determined. This would serve as a densification of global models (down to scales of the order of 10 km). Such models could be used for further studies in geodesy and neighbouring disciplines, e.g. for the determination of so-called free-board heights of sea ice or for serving as reference in ice mass balance studies. In geophysics a dense grid of free-air or Bouguer anomalies will be used for inversion, i.e., the investigation of the inner structure of the Earth.

Following these three main arguments, the terms of reference emphasize as objectives of AntGP the compilation of a completed gravity dataset for Antarctica (AntGP, 2004). Activities should be focussed on collecting existing gravity data as well as performing new (especially airborne) gravity surveys. Precise gravity ties between reference stations and old and new traverses and airborne surveys should be carried out. For the observation of reference gravity stations absolute gravity meters should be used. It should also be emphasized, that AntGP should serve as a liaison to similar initiatives in geodesy, geophysics and glaciology.

## 3 Gravity determination in Antarctica

Discussing possible techniques to measure gravity in Antarctica, one has to deal with specific conditions. Antarctica presents a harsh and remote environment. Sea ice, ice shelves, glaciers and the continental ice sheet cause limits to otherwise unproblematic techniques. Bad weather conditions (especially the wind when exceeding a certain velocity) could prevent measurements altogether. Most activities are restricted to the Antarctic summer season. Altough there exists a number of wintering stations, the working range during the winter is very much limited. Regarding observation campaigns, one has to state that there is no (gravity) survey especially dedicated to geodesy. It is absolutely necessary to work interdisciplinarily, especially with geophysicists and glaciologists. Finally, one has to acknowledge the huge logistic effort to carry out any survey in Antarctica.

One possibility to differentiate between observation techniques are *trackwise* and *pointwise* measurements, resp. Shipborne and airborne measurements represent the first group, whereas absolute and relative gravity observations represent the latter one. Especially airborne surveys combine several different geophysical observation methods, with gravity being only one of them. Furtheron, standard aerogeophysical techniques comprise magnetics, radio echo sounding (RES), laser profiler or laser scanner, GPS and INS.

Reviewing the current status of gravity data coverage in Antarctica, one has to be aware that a) data were surveyed by different nations and programmes, b) data could complement or overlap each other, and c) data are measured at different observation epochs and with different precision and resolution.

There are already some projects to compile gravity data of different sources. One of them is *Antarctic Digital Gravity Synthesis* (ADGRAV), which originates from an initiative of the SCAR Working Group on Solid Earth Geophysics at the SCAR Conference in Concepcion, in July 1998. The project was carried out at the Lamont-

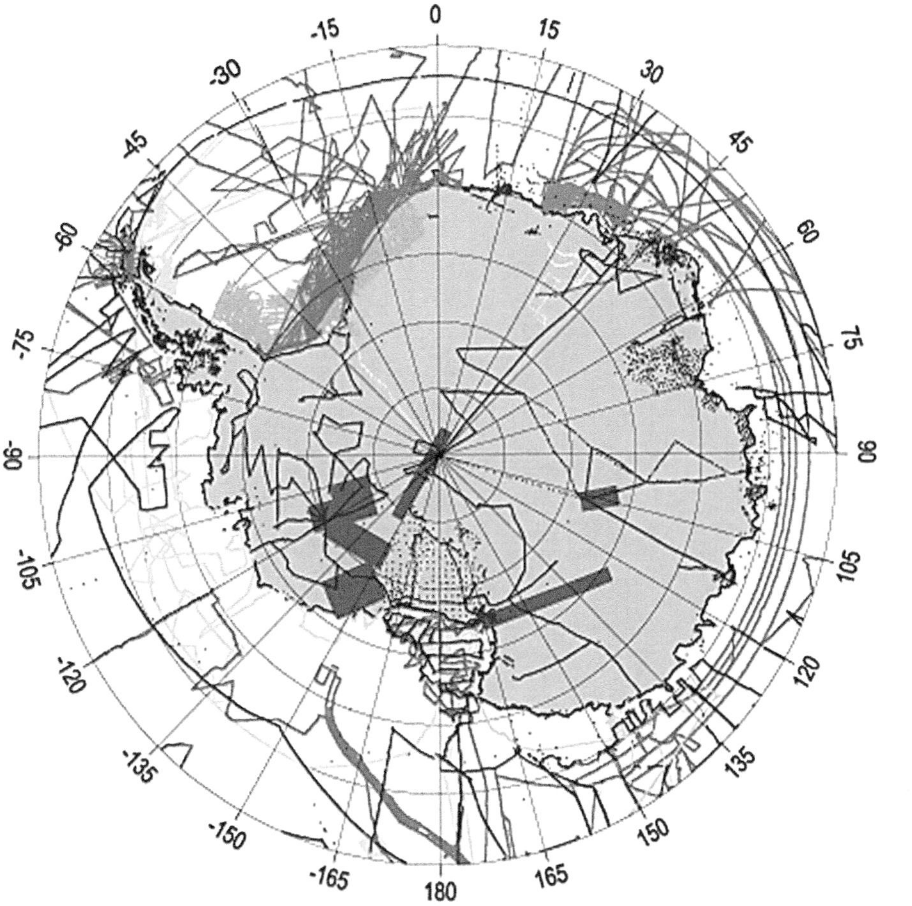


Figure 1: Gravity data coverage at ADGRAV (status of 02/2002) (ADGRAV, 2002)

Doherty Earth Observatory of Columbia University (Palisades, USA) (ADGRAV, 2002). The compilation (see Fig. 1) comprises a number of ship tracks, point measurements as well as some airborne surveys mostly in West Antarctica (Bell et al., 1999; Studinger et al., 2003; Studinger et al., 2004).

A further compilation is hosted by VNII-Okeangeologia (St. Petersburg, Russia). This dataset has been compiled since 1970 and comprises about 50,000 km marine profiles, ground observations (Filchner-Ronne ice shelf, Ross ice shelf, coastal Enderby Land, Amery ice shelf) and airborne tracks (southern Weddell Sea, western Dronning Maud Land, East Antarctica between 30° and 60° E) (Aleshkova et al., 2004; Leitchenkov, 2004), see Fig. 2. Additionally, data available through ADGRAV could be included.

Extensive airborne surveys were carried out also by German institutions. The Alfred Wegener Institute for Polar and Marine Research (AWI Bremerhaven, Germany) has conducted aerogeophysical campaigns in Dronning Maud Land. They cover marine and continental regions (EMAGE and VISA projects, (Nixdorf et al., 2004)). The Federal Agency for Geosciences and Resources (BGR Hannover, Germany) surveyed the region of the Lambert Glacier during a joint programme with Australian partners in 2002/03 (PCMEGA, (Damaske and McLean, 2004)).

Regarding pointwise observations, there are a lot of relative gravimetry campaigns. Relative gravimetry measurements were made either on traverses or on grids, using snow mobiles or helicopters as means of transport. Examples can be found throughout Antarctica, in Central Dronning Maud Land (Germany) or in Victoria Land (New Zealand, Germany and Italy), to mention just two regions. Absolute gravity measurements are mainly restricted to Antarctic stations at coastal regions, Fig. 3, see e.g.

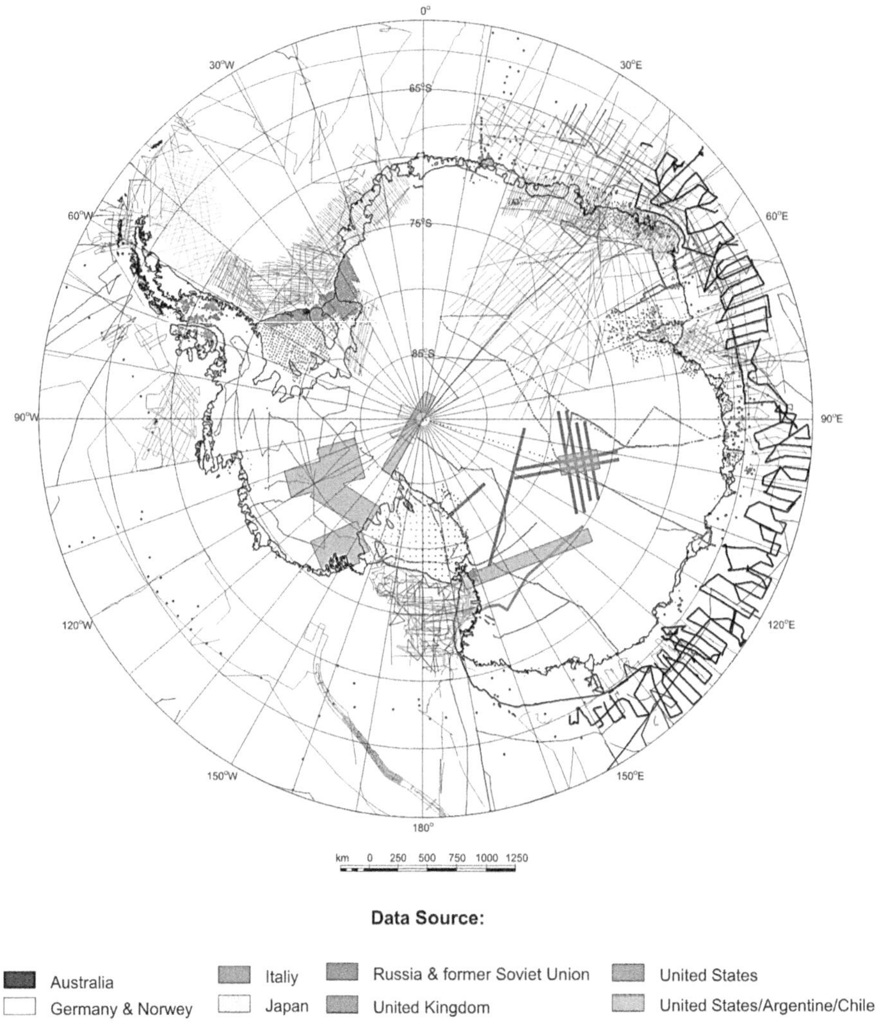


Figure 2: Russian Antarctic Gravity Dataset (Aleshkova et al., 2004; Leitchenkov, 2004)

(Almavict, 2004). There, the absolute gravity meter FG-5 turned out to be an observation standard in the last years.

When carrying out and analysing gravity surveys in Antarctica there are still a lot of problems which have to be solved or where (substantial) improvements have to be made. Precision and resolution have to be checked, e.g. by strategies like a cross-over adjustment. Older data points or tracks could be repeated to evaluate their precision. Most of the surveys are by their nature relative gravity observations, therefore a proper reference to a gravity datum has to be made. For instance, gravity ties have to be measured between absolute gravity stations (which provide a datum) and aero-gravimetry. Especially for the airborne surveys, the kinematic positioning solutions have to be improved. This is a crucial point because the surveys could rarely allow an optimum set-up of ground GPS reference stations, so that very often long baselines have to be solved. One has to look for a proper combination of GPS and INS. For a validation of geoid models, it will be necessary to get a linkage to tide gauge stations.

Furthermore, topographic data are needed to perform the data analysis and to determine improved geoid models. In Antarctica one has to face a unique situation: this is the densitiy contrast between ice and bedrock. Hence, measurements and/or models of the ice surface height and the ice thickness are needed in order to also infer the bedrock topography. A favoured observation method is airborne RES (see above). A continental model of bedrock topography and ice thickness in Antarctica is

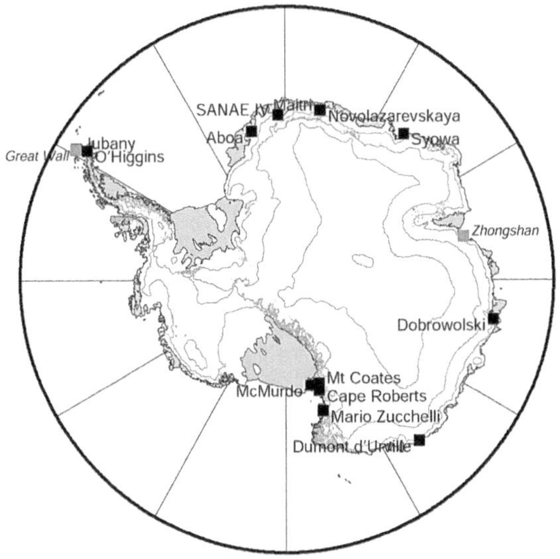


Figure 3: Absolute gravity stations in Antarctica

provided by BEDMAP, Figs. 4, 5. BEDMAP was a project sponsored by SCAR and by EISMINT (European Ice Sheet Modelling Initiative), maintained by the British Antarctic Survey (BAS) (BEDMAP, 2001).

## 4 Linkage to other programmes

AntGP should maintain close links to similar initiatives in geodesy, geophysics and glaciology. The Scientific Committee on Antarctic Research (SCAR) has to be mentioned first. The SCAR XXVIII conference took place in Bremen, in July 2004, for the first time including an Open Science Conference. The work plan 2004 – 2006 of the SCAR Geoscience Standing Scientific Group's (GSSG) sub-programme *Geodetic Infrastructure in Antarctica* (GIANT) includes project 3 *Physical Geodesy* which deals also with Antarctic gravity. Furtheron, several data initiatives exist: ADGRAV (already mentioned), ADMAP (magnetic field in Antarctica), BEDMAP (see above), and also ADD (Antarctic Digital Database (ADD, 2004)). It was agreed, that these inititatives should be continued and that especially new data should be incorporated.

The *International Polar Year 2007/08* (IPY) (IPY, 2004) will provide a platform for a challenging "international programme of coordinated, interdisciplinary, scientific research and observations in the Earth's polar regions" (Bell, 2004). It was adopted by the *International Council for Science* (ICSU) and the *World Meteorological Organization* (WMO). In autumn 2004, the work of the IPY planning group will be finished and a new structure will be installed (IPY, 2004). Under the project headline "*Remote Antarctic Exploration*" programmes for joint geophysical, geodetic and glaciological investigations should be focussed. Thus, the IPY offers excellent chances for geodesy – in connection with aerogeophysics – to close the polar gap in gravity.

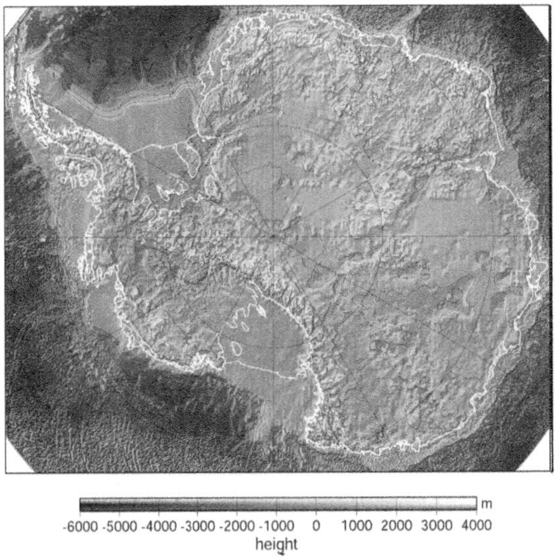


Figure 4: Antarctic bedrock topography (BEDMAP, 2001)

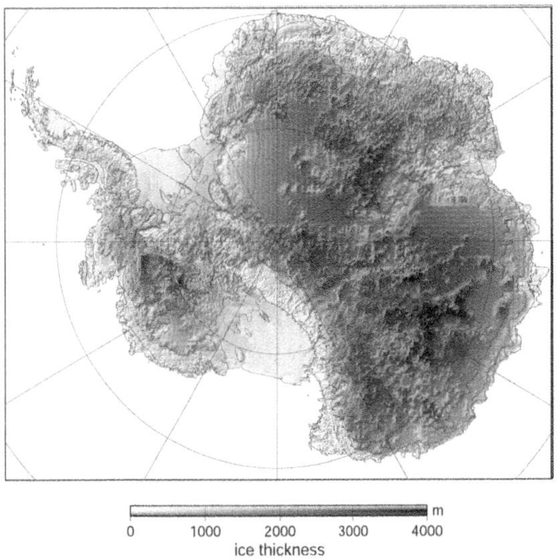


Figure 5: Antarctic ice thickness (BEDMAP, 2001)

## 5 Summary

For the north polar region, the *Arctic Gravity Project* (Kenyon and Forsberg, 2002; ArcGP, 2002) succesfully closed the polar gap and thus provided an incomparable gravity dataset for the Arctic. AntGP is pursueing a corresponding strategy for the Antarctic, thus contributing to fulfill the goals of the IAG. AntGP was put into action after its confirmation by the IAG in autumn 2003. Communication between people and communities has been established and will be maintained through circular letters, emails and while attending conferences. A website hosted by TU Dresden (AntGP, 2004) has been set up and will be updated regularly. There, information can be retrieved on the terms of reference, list of members and current status, as well as on further links. A close communication with other programmes within SCAR and IPY, but also with data or processing centres for global gravity field models has to be maintained. One major goal of AntGP, even at the current state, is to encourage scientists to use AntGP as one reference when applying for projects in geodesy and aerogeophysics in Antarctica.

## Acknowledgement

The constructive collaboration with all members of AntGP is gratefully acknowledged. Special thanks for discussions and/or providing information and material to: René Forsberg, German Leitchenkov, Uwe Nixdorf, Detlef Damaske, Jaakko Mäkinen, Michael Studinger, Gernot Reitmayr, Alessandro Capra, Reinhard Dietrich, Steve Kenyon and Dave McAdoo.

## References

ADD (2004). `www.add.scar.org`.

ADGRAV (2002). `www.marine-geo.org/antarctic/gravity/`.

Aleshkova, N., Masolov, V., Leitchenkov, G., Mandrikov, V., Alyavdin, S., Golynski, A., and R, K. (2004). Russian Antarctic Gravity Dataset. Poster presentation at SCAR XXVIII Open Science Conference, Bremen, July 26-28, 2004.

Almavict, M. (2004). Vertical displacement and variation of gravity predicted and observed in Antarctica. Poster presentation at SCAR XXVIII Open Science Conference, Bremen, July 26-28, 2004.

AntGP (2004). `www.tu-dresden.de/ipg/antgp/antgp.html`.

ArcGP (2002). `earth-info.nga.mil/GandG/agp/`.

BEDMAP (2001). `www.nerc-bas.ac.uk/public/aedc/bedmap`.

Bell, R. (2004). International Polar Year 2007-2008. Presentation at SCAR XXVIII Open Science Conference, IPY Discussion Forum, Bremen, July 28, 2004.

Bell, R. et al. (1999). Airborne gravity and precise positioning for geologic applications. *Journ. Geophys. Res.*, 104(B7):15281–15292.

CSR (2004). www.csr.utexas.edu/grace.

Damaske, D. and McLean, M. (2004). Airborne Geophysical Data Acquisition South of the Prince Charles Mountains, East Antarctica. Poster presentation at SCAR XXVIII Open Science Conference, Bremen, July 26-28, 2004.

IPY (2004). `www.ipy.org`.

Kenyon, S. and Forsberg, R. (2002). Arctic Gravity Project - A Status. In Sideris, M., editor, *Gravity, Geoid and Geodynamics 2000, IAG Symposium 123, Springer*, pages 391–395.

Leitchenkov, G. (2004). (pers. comm.).

Lemoine, F. et al. (1996). The Development of the NASA GSFC and NIMA Joint Geopotential Model. In *Proc. GRAGEOMAR, Tokyo, Sep. 30 - Oct. 5. 1996.*

McAdoo, D., Farrell, S., Laxon, S., Wagner, C., and Childers, V. (2004). GRACE-based Arctic Ocean Geoids for Oceanographic and Sea-Ice Investigations. Joint CHAMP/GRACE Science Meeting, July 6-8, 2004, Potsdam.

Nixdorf, U. et al. (2004). Airborne Gravity Measurements in Dronning Maud Land for the Validation, Densification and Interpretation of New Satellite Data. Presentation at SCAR XXVIII Open Science Conference, Bremen, July 26-28, 2004.

Schwintzer, P. et al. (2004). A High Resolution Global Gravity Field Model Combining CHAMP/GRACE Satellite Missions and Altimetry/Gravimetry Surface Gravity Data. Presentation at 1st EGU General Assembly, Nice, April 25-30, 2004.

Studinger, M., Bell, R., Buck, W., Karner, G., and Blankenship, D. (2004). Sub-ice geology inland of the Transantarctic Mountains in light of new aerogeophysical data. *Earth Plan. Sci. Letters*, 220:391–408.

Studinger, M., Karner, G., Bell, R., Levin, V., Raymond, C., and Tikku, A. (2003). Geophysical models for the tectonic framework of the Lake Vostok region, East Antarctica. *Earth Plan. Sci. Letters*, 216:663–677.

# First Results from new High-precision Measurements of Deflections of the Vertical in Switzerland

Anna Müller, Beat Bürki, Hans-Gert Kahle
Geodesy and Geodynamics Laboratory, ETH Zurich, CH-8093 Zurich, Switzerland
email : anna.mueller@geod.baug.ethz.ch

Christian Hirt
Institut für Erdmessung, Universität Hannover, Schneiderberg 50, 30167 Hannover, Germany

Urs Marti
Federal Office of Topography (swisstopo), Seftigenstrasse 264, CH-3084 Wabern, Switzerland

**Abstract.** In October 2003 two modernized digital zenith camera systems have been deployed in Switzerland during the campaign CHGeo2003. The mission was carried out under the auspices of the Swiss Geodetic Commission of the Swiss Academy of Sciences (SAS) and coordinated by the Swiss Federal Office of Topography (swisstopo). The goal of the campaign was to provide additional highly accurate deflections of the vertical in order to contribute to an improvement of the presently used Swiss geoid CHGeo98. The observations were carried out at 68 selected stations covering regions with old or inadequate data. Further reasons of the campaign were the proof of the field capability and the comparison of both systems concerning their accuracy by observing at several stations simultaneously. The paper describes the realization of the campaign CHGeo2003 and the results obtained with the digital systems.

**Keywords.** Digital zenith camera, deflection of the vertical, campaign CHGeo2003, geoid model

## 1 Introduction

During the last 3 years considerable work has been invested at the Geodesy and Geodynamics Laboratory (GGL) of ETH Zurich and the Institut für Erdmessung (IfE) of University Hanover to develop two digital zenith cameras up to field capability. The DIADEM system (**D**igital **A**stronomical **De**flection **M**easuring system) of ETH Zurich and the TZK2-D (**T**ransportable **Z**enith **C**amera **2** – **D**igitalsystem) of University Hanover are both based on the same construction which was initially invented at the IfE, University Hanover. Both systems have been upgraded to the state-of-the-art using modern CCD technology and GPS equipment instead of analogue photographic media and long wave time signal receivers. CCD sensors provide substantially higher sensitivities than analogue media. In addition, they offer the advantage of an instantaneous image processing by means of the software package AURIGA (**Au**tomatic **R**eal-Time **I**mage Processing System for **G**eodetic **A**stronomy), developed at IfE. Since the complete data acquisition process is controlled by a dedicated computer the digital systems can be used for a fast and direct determination of deflections of the vertical.

## 2 The digital zenith cameras: DIADEM and TZK2-D

### 2.1 Principle

While satellite-based methods like GPS provide geometrically defined ellipsoidal coordinates ($\varphi,\lambda$), zenith cameras determine the local direction of the gravity vector (local plumb line) expressed by astronomical parameters ($\Phi,\Lambda$). The difference between both directions is called deflection of the vertical ($\xi,\eta$). It reflects anomalies of the Earth's gravity field, caused by inhomogeneous mass distributions and is used for local geoid determination. The main idea in determining astronomical parameters ($\Phi,\Lambda$) is to interpolate the direction of the rotational axis of the zenith camera into the zenithal field of stars imaged on a CCD sensor.

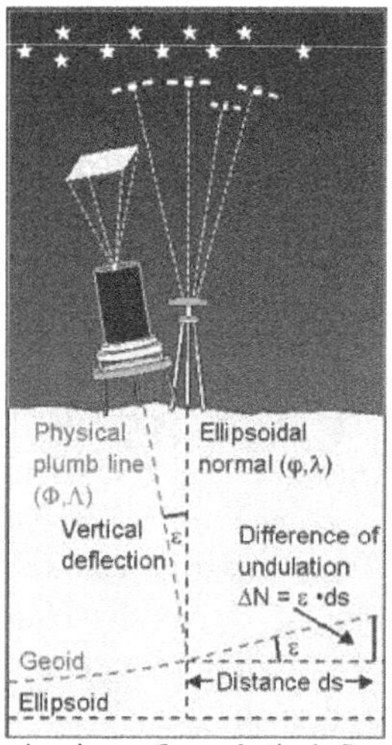


**Fig. 1** Determination of vertical deflections using a digital zenith camera and GPS equipment.

The positions of stars on the celestial sphere are defined by equatorial coordinates ($\delta,\alpha$) which are obtainable from high-precision star catalogues Tycho-2 (Høg et al., 2000) and UCAC (Zacharias et al., 2004). The equatorial coordinates can be linked with the astronomical parameters by GAST (Greenwich Apparent Siderial Time) (Eq.1).

$$\Phi = \delta \; ; \; \Lambda = \alpha - GAST \tag{1}$$

A GPS receiver, part of the digital zenith camera system, provides the exact exposure epochs and ellipsoidal coordinates ($\varphi,\lambda$), respectively. Thus deflections of the vertical can be calculated by Eq. 2.

$$\xi = \Phi - \varphi \; ; \; \eta = (\Lambda - \lambda)\cos\varphi \tag{2}$$

Information on how to apply deflections of the vertical for geoid determination can be found in (Torge, 2003).

## 2.2 Instrumentation

In general, the zenith camera is divided into two parts: a turnable superstructure and a fixed substructure mounted on a tripod. Both parts are separated by a special ball bearing. The superstructure can be rotated 180 degrees azimuthally for two opposite camera positions in order to eliminate the eccentricity of the CCD sensor and the zero offsets of the electronic tiltmeters.

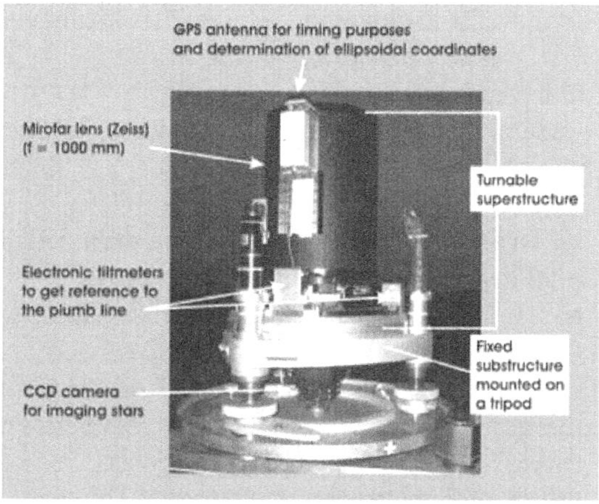


**Fig. 2** The main components of a digital zenith camera system: Mirotar lens, CCD camera, orthogonally mounted electronic tiltmeters and GPS equipment.

The turnable part consists of a highly light-sensitive Mirotar lens (Zeiss), a CCD camera for imaging stars and 2 pairs of orthogonally mounted high-resolution electronic tiltmeters to determine the inclination of the camera system. Furthermore, a GPS equipment is integrated in the zenith camera system to provide highly precise time information and ellipsoidal coordinates, respectively (Fig. 2). In addition, the DIADEM system uses 5 servo motors for levelling, focussing and azimuthal turning in order to enable a fully-automated measuring process. All these processes are controlled by an appropriate industrial computer (Müller, 2002; Hirt and Bürki, 2002). More technical details can be found in (Bürki et al., 2004).

# 3 Astro-Campaign CHGeo2003

## 3.1 Motivation

In the year 2003 the project CHGeo2003 has been initiated by the Swiss Federal Office of Topography (swisstopo) aiming at the improvement of the presently used geoid model CHGeo98 of Switzerland (Marti, 1997). Since this model still suffers from long wavelength errors it is mainly responsible for discrepancies between the orthometric heights from the new national height system of Switzerland (LHN95) and the orthometric heights derived by GPS and the geoid (Fig. 3). These discrepancies are in the order of several centimeters up to decimeters in some regions (Marti, 2002; Brockmann et al., 2003).

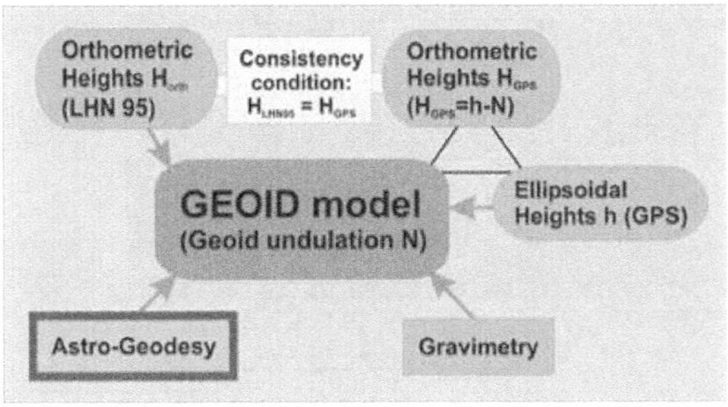


**Fig. 3** Orthometric heights from LHN95 and orthometric heights derived by GPS and the geoid should fulfill the consistency condition H-h+N=0. The geoid model is a combination of different data like GPS/-levelling, deflections of the vertical (Astro Geodesy) and gravity values (Gravimetry).

For a recalculation of the geoid of Switzerland the project CHGeo2003 included new and improved measurements (GPS/levelling, deflections of the vertical, gravity values) as well as the use of advanced mass models and calculation methods. The project also contributes to the European Combined Geodetic Network (ECGN) and the European Unified Vertical Network – Densification Action (EUVN-DA) of EUREF (Brockmann et al., 2003; Brockmann et al., 2004).

## 3.2 Realization

According to the aims of the CHGeo2003 project the two digital zenith camera systems have been deployed in October 2003 in Switzerland to provide highly precise deflections of the vertical. Other dedicated goals were to prove the field capability of the enhanced instruments and the comparison of both systems regarding their accuracy.

The zenith camera observations have been carried out in regions where the presently used geoid model showed uncertainties, that means larger discrepancies between GPS/levelling and geoid undulations derived from the model. Furthermore measurements have been performed at old astro-stations which had an inadequate accuracy (> 0.5 arcsec) and in regions with only few astro-stations to densify the existing network. Unlike former zenith camera deployments it is no longer required to perform the observations at geodetic reference points since a GPS receiver enables the determination of geodetic coordinates. To provide the geodetic coordinates with an adequate accuracy the GPS measurements have been realized in differential mode based on carrier phase corrections using the GSM capability from the swiss positioning service (SWIPOS). Thus, the coordinates could be provided with an accuracy of less than 10 cm in real-time. During the campaign, a total number of 68 stations have been measured in 16 observation nights. That means 1-5 stations per night and team depending on weather, location (mountains, valley) and distance between the stations. Four stations have been observed simultaneously, among them the geodetic reference station in Zimmerwald. In addition, this station has been measured repeatedly during different nights.

Due to instrumental effects the observations have been performed in 2 to 4 different azimuthal camera orientations. Thus the average time needed per station is about 45 minutes. Finally, 80 to 120 single solutions per station could be used for the determination of the direction of the local plumb line. The data analysis have been carried out with the software package AURIGA, a real-time data processing software for zenith camera images, developed at IfE, University Hanover (Hirt, 2001; Hirt and Bürki, 2002; Hirt, 2004).

### 3.3 Results

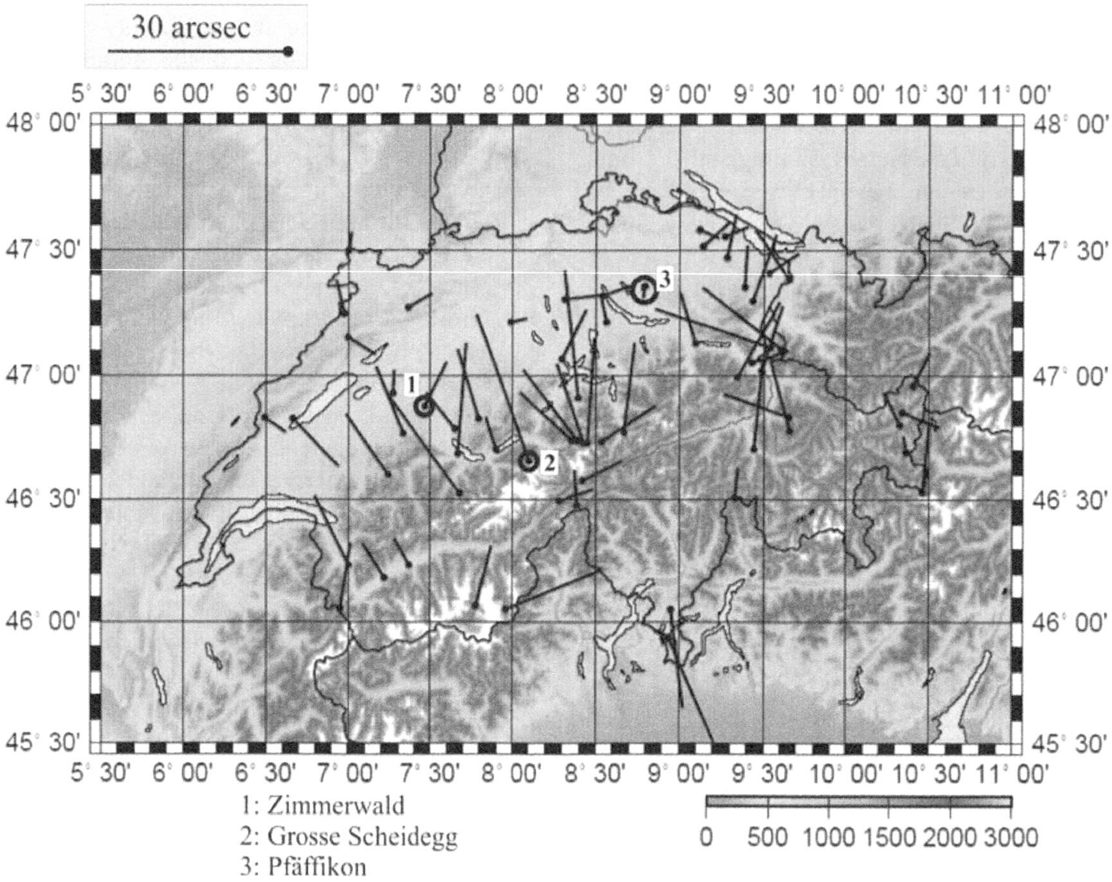


**Fig. 4** Measured deflections of the vertical in Switzerland during the CHGeo2003 campaign. 1: Zimmerwald, 2: Grosse Scheidegg, 3: Pfäffikon.

The CHGeo2003 campaign offered the possibility to deploy for the first time two modernized digital zenith cameras. It was concluded that both systems worked reliably and efficiently, also under very harsh conditions like in high mountains with temperatures at –15 °C. Since the whole acquisition process and the analysis of the data is steered and controlled by a PC, the work and time effort is significantly decreased. This holds true mainly for the fact that the evaluation work of the old analogue systems was very time consuming.

The measured deflections of the vertical at 68 stations in Switzerland are shown in Fig. 4. The maximum value was observed at the station "Grosse Scheidegg", south of Grindelwald, Bernese Oberland, with 27". The smallest vertical deflection was measured in Pfäffikon near Zurich with 2".

The points of the vectors show in direction to the attracting masses. Thus, the influence of the Alpine chain is well recognizable since the deflections of the vertical measured to the north of the alps point in direction to the mountains. Also the effect of the Jura mountains (north-west) is well identifiable in 4 stations measured in this region. The two observed stations in the canton Tessin (south) express the influence of the Ivrea body, a density anomaly in the Earth's crust.

Regarding the internal accuracy obtainable with a digital zenith camera, the standard deviation of about 100 single deflections of the vertical per station is between 0.1" to 0.3". The station mean has a precision of less than 0.1". The standard deviation of repeated measurements during different nights at the geodetic reference station Zimmerwald is 0.1" for each system (Fig. 5).

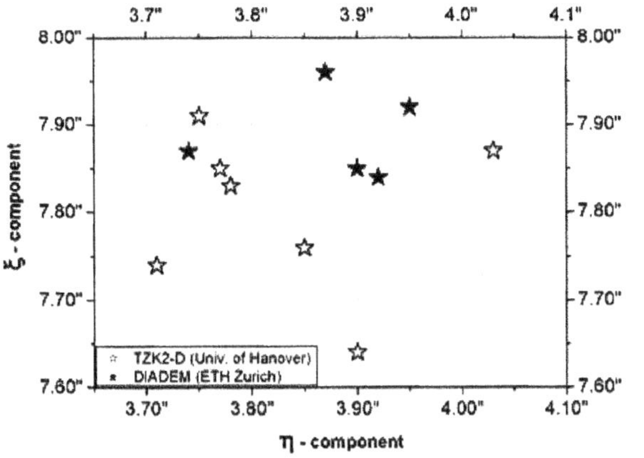


**Fig. 5** Results of repeated measurements with both systems at the geodetic reference station in Zimmerwald.

The results of the two digital systems show a good agreement within 0.3" for both deflection components. The difference between the calculated station means per system gives a first idea about the external accuracy which is on the order of 0.1"-0.15" for both ξ and η. Compared with the formerly used analogue zenith camera systems, the digital instruments achieve an accuracy which is 3-5 times higher. Further information concerning the accuracy of the TZK2-D system can be found in (Hirt, 2004; Hirt et al., 2004).
The 68 measured deflections of the vertical were used together with other 690 available deflections for a first computation of a purely astrogeodetic geoid. By using older data sets systematic offsets in some regions could be detected. Therefore only measurements since 1980 were used. The final solution showed that it was possible to eliminate the long wavelength errors of the formerly used geoid model (Marti, 2004). A first combined solution of deflections of the vertical and GPS/levelling showed significantly improved GPS/levelling residuals. They are now in the order of some millimeters in flatter areas and up to some centimeters in Alpine regions (Brockmann et al., 2004).

## 4 Conclusions

The results obtained with the digital zenith camera systems are promising and show their high potential for local gravity field and precise geoid determination. Compared with the formerly used analogue zenith camera systems, the digital instruments achieve an accuracy which is 3-5 times higher. The improved accuracy is due to the increased number of single solutions per station, a better modelling of systematic errors, the use of highly precise electronic tiltmeters and the application of modern high-precision star catalogues (Hirt, 2004). Concerning the efficiency and handling of the system the time needed for preparation, data acquisition and analysis could be significantly reduced. This is mainly attributed to the use of modern CCD technology which enables a fast acquisition and processing of the images. For data analysis the software package AURIGA has been applied. Thus, an immediate control and validation of the measurements on site is possible. A presently remaining point of uncertainty is the dependence of azimuthal corrections, which require observations in different azimuthal directions. If this problem can be solved, then it will be possible to decrease the time needed from 45 minutes in average to 20 minutes for preparation, data acquisition and analysis.
The main goal of the CHGeo2003 campaign was to improve the used geoid model. The procedure for a recalculation of the geoid has been applied by U. Marti, swisstopo. First solutions showed that it was possible to eliminate the long wavelength errors of the geoid model and to decrease the GPS/levelling residuals significantly. Detailed information concerning the new geoid model CHGeo2003 can be found in (Marti, 2004).

## References

Brockmann, E., Marti, U., Schlatter, A. and Schneider, D. (2003). *CHCGN activities in Switzerland. Activities for a Swiss Combined Geodetic Network.* EUREF'03 Symposium, Toledo. Appeared also as report 03-41 from the Swiss Federal Office of Topography (swisstopo), Wabern.

Brockmann, E., Becker, M., Bürki, B., Gurtner, W., Haefele, P., Hirt, C., Marti, U., Müller, A., Richard, Ph., Schlatter, A., Schneider, D. and Wiget, A. (2004). *Realization of a Swiss Combined Geodetic Network (CH-CGN).* EUREF'04 Symposium of the IAG Commission 1 – Reference Frames, Subcommission 1-3a Europe (EUREF), Bratislava, Slovakia.

Bürki, B., Müller, A. and Kahle, H.-G. (2004). *DIADEM: The New Digital Astronomical Deflection Measuring System for High-precision Measurements of Deflections of the Vertical at ETH Zürich.* Proceedings IAG GGSM symposium 2004 (CD), Porto. Portugal.

Hirt, C. (2001). *Automatic Determination of Vertical Deflections in Real-Time by Combining GPS and Digital Zenith Camera for Solving the GPS-Height-Problem.* Proceedings of the 14th International Technical Meeting of the Satellite Division of the Institute of Navigation: 2450-2551, Alexandria, VA.

Hirt, C. and Bürki, B. (2002). *The Digital Zenith Camera – A New High-Precision and Economic Astrogeodetic Observation System for Real-Time Measurement of Deflections of the Vertical.* Proceedings of the 3rd Meeting of the International Gravity and Geoid Commission of the International Association of Geodesy, Thessaloniki (ed. I. Tziavos): 161-166.

Hirt, C. (2004). *Entwicklung und Erprobung eines digitalen Zenitkamerasystems für die hochpräzise Lotabweichungsbestimmung.* Wissenschaftliche Arbeiten der Fachrichtung Vermessungswesen der Universität Hannover, Nr. 253, Dissertation.

Hirt, C., Reese, B. and Enslin, H. (2004). *On the Accuracy of Vertical Deflection Measurements Using the High-Precision Digital Zenith Camera System TZK2-D.* Proceedings IAG GGSM symposium 2004, Porto. Portugal.

Høg, E., Fabricius, C., Makarov, V.V., Urban, S., Corbin, T., Wycoff, G., Bastian, U., Schwekendiek, P. and Wicenec, A. (2000). *The Tycho-2 Catalogue of the 2.5 Million Brightest Stars.* Astronomy and Astrophysics 355: L27-L30.

Marti, U. (1997). *Geoid der Schweiz 1997.* Geodätisch-geophysikalische Arbeiten in der Schweiz Nr. 56, Schweizerische Geodätische Kommission.

Marti, U. (2002). *Modelling of Differences of Height Systems in Switzerland.* Proceedings of the 3rd Meeting of the International Gravity and Geoid Commission of the International Association of Geodesy, Thessaloniki (ed. I. Tziavos): 378-383.

Marti, U. (2004). *High Precision combined geoid determination in Switzerland.* Proceedings IAG GGSM symposium 2004 (CD), Porto. Portugal.

Müller, A. (2002). *Umbau des bestehenden Zenitkamera-Messsystems auf digtiale Bildtechnik.* Unveröffentlichte Diplomarbeit am Institut für Photogrammetrie und Fernerkundung, Technische Universität Dresden.

Torge, W. (2003). *Geodäsie.* 2. Auflage. W. de Gruyter, Berlin, New York.

Zacharias, N., Urban, S.E., Zacharias, M.I., Wycoff, G.L., Hall, D.M., Monet, D.G. and Rafferty, T.J. (2004). *The Second US Naval Observatory CCD Astrograph Catalog (UCAC2). The* Astronomical Journal 127: 3043-3059.

# Error Propagation with Geographic Specificity for Very High Degree Geopotential Models

N.K. Pavlis and J. Saleh
Raytheon ITSS Corporation, 1616 McCormick Drive, Upper Marlboro, Maryland 20774, USA
Nikolaos_Pavlis@raytheon.com Fax: +301-883-4140

**Abstract.** Users of high-resolution global gravitational models require geographically specific estimates of the error associated with various gravitational functionals (e.g., $\Delta g, N, \xi, \eta$) computed from the model parameters. These estimates are composed of the commission and the omission error implied by the specific model. Rigorous computation of the commission error implied by any model requires the complete error covariance matrix of its estimated parameters. Given this matrix, one can compute the commission error of various model-derived functionals, using covariance propagation. The error covariance matrix of a spherical harmonic model complete to degree and order 2160 has dimension ~4.7 million. Because the computation of such a matrix is beyond the existing computing technology, an alternative method is presented here which is capable of producing geographically specific estimates of a model's commission error, *without* the need to form, invert, and propagate such large matrices. The method presented here uses integral formulas and requires as input the error variances of the gravity anomaly data that are used in the development of the gravitational model.

**Keywords**. Geopotential, high-degree spherical harmonic models, error propagation, convolution.

---

## 1 The Main Idea

Geopotential models like EGM96 are composite solutions. A low degree *comprehensive* solution (e.g., Nmax=70 for EGM96) employing complete normal matrices and least-squares adjustment techniques combines the satellite-only information with surface gravity and satellite altimetry data. The higher degree and order part of the model (e.g., from n=71 to 360 for EGM96) is determined *solely* from a complete, global grid of $\Delta g$.

Beyond the maximum degree and order of the available satellite-only solution, there is little need to form complete normal matrices, since no "adjustment" takes place within this degree range. The merged (terrestrial plus altimetry-derived) $\Delta g$ are the only data whose signal and error content determine the model's signal and error properties in this degree range. This fact enables high-degree error propagation, *with* geographic specificity, through the use of integral formulas with band-limited kernels, *without* the need to form, invert, and propagate extremely large matrices.

## 2 An Example Illustrating the Principle

The gravity anomaly computed from a composite model is ($L$ and $H$ stand for *Low-* and *High-degree*):

$$\hat{\Delta g} = \hat{\Delta g}_L + \hat{\Delta g}_H = \sum_{n=2}^{L} \hat{\Delta g}_n + \sum_{n=L+1}^{H} \hat{\Delta g}_n \quad . \tag{1}$$

The corresponding geoid undulation is:

$$\hat{N} = \hat{N}_L + \hat{N}_H = \sum_{n=2}^{L} \hat{N}_n + \sum_{n=L+1}^{H} \hat{N}_n \quad , \tag{2}$$

and can be written as (Heiskanen and Moritz, 1967):

$$\hat{N} = \frac{R}{4\pi\gamma} \iint_{\sigma} \hat{\Delta g} S(\psi) d\sigma \quad , \quad S(\psi) = \sum_{n=2}^{H} \frac{2n+1}{n-1} P_n(t) \tag{3}$$

where $t = \cos(\psi)$. We define:

$$S_L(\psi) = \sum_{n=2}^{L} \frac{2n+1}{n-1} P_n(t) \quad , \quad S_H(\psi) = \sum_{n=L+1}^{H} \frac{2n+1}{n-1} P_n(t) \tag{4}$$

Wong and Gore (1969) used a similar separation of harmonic components to modify Stokes' kernel in the context of truncation theory. Equations (2), (3), and (4) and the orthogonality of spherical harmonics imply that:

$$\hat{N} = \frac{R}{4\pi\gamma} \iint_{\sigma} (\hat{\Delta g}_L + \hat{\Delta g}_H)[S_L(\psi) + S_H(\psi)] d\sigma \quad \Rightarrow$$

$$\hat{N} = \frac{R}{4\pi\gamma} \iint_{\sigma} \hat{\Delta g}_L S_L(\psi) d\sigma + \frac{R}{4\pi\gamma} \iint_{\sigma} \hat{\Delta g}_H S_H(\psi) d\sigma \, . \tag{5}$$

Therefore, a strict, degree-wise separation of spectral components can be achieved by restricting the spectral content of the kernel function accordingly, *as long as* the integration is performed globally. The band-limited version of Stokes' equation:

$$\hat{N}_H = \frac{R}{4\pi\gamma} \iint_{\sigma} \hat{\Delta g}_H \, S_H(\psi) d\sigma \tag{6}$$

implies, for uncorrelated errors of $\hat{\Delta g}_H$, the error propagation formulas:

$$\sigma^2(\hat{N}_H) = \left(\frac{R}{4\pi\gamma}\right)^2 \iint_\sigma \sigma^2(\hat{\Delta g}_H) S_H^2(\psi) d\sigma \qquad (7a)$$

$$\sigma_{12}(\hat{N}_H) = \left(\frac{R}{4\pi\gamma}\right)^2 \iint_\sigma \sigma^2(\hat{\Delta g}_H) S_H(\psi_1) S_H(\psi_2) d\sigma \quad (7b)$$

Discretized versions of equations (7a, b) allow the computation of $\sigma^2(\hat{N}_H)$ and $\sigma_{12}(\hat{N}_H)$ from $\sigma^2(\hat{\Delta g}_H)$ through global convolutions. We implement (7a) using 1D FFT (Haagmans et al., 1993), with $H$ covering the degree range where the merged (terrestrial plus altimetry-derived) $\Delta g$ define solely the solution. The geoid error covariances from equation (7b) are also computed using global convolution, although with much less efficiency compared to the computation of error variances for points on regular grids. This approach is applicable to any functional related to $\Delta g$ by an integral formula. Equations (7a, b) employ the spherical approximation, which we consider adequate for error propagation work. Apart from this, (7a, b) are rigorous, and their numerical implementation is only subject to discretization errors. Finally, the band limiting of integration kernels removes the singularity at the origin of kernels like Stokes' and Vening Meinesz's, therefore the innermost zone effects require no special treatment.

If we assume that the error correlation between $\hat{\Delta g}_L$ and $\hat{\Delta g}_H$ is negligible due to orthogonality, then the total error variance of a field functional, $f$, at the geographic location $(R,\varphi,\lambda)$, as computed from a specific geopotential model, can be written as:

$$\begin{aligned}\sigma_f^2(R,\varphi,\lambda) \approx\ & \sigma_f^2(R,\varphi,\lambda)\_commission\_L \\ & +\sigma_f^2(R,\varphi,\lambda)\_commission\_H \\ & +\sigma_f^2(R,\varphi,\lambda)\_omission\end{aligned} \qquad (8)$$

where $\sigma_f^2(R,\varphi,\lambda)\_commission\_L$ is computed by propagation of the complete error covariance matrix of the comprehensive solution employing 2D FFT (Haagmans and Van Gelderen, 1991) (see Figure 1 for EGM96), $\sigma_f^2(R,\varphi,\lambda)\_commission\_H$ is computed by global convolution based on an integral formula, and $\sigma_f^2(R,\varphi,\lambda)\_omission$ may be estimated using local covariance models. This approach circumvents the need to form, invert, and propagate extremely large matrices.

## 3 Functional Relations

We present the functional relations between gravity anomalies and other gravity field functionals of interest. From these relations, the corresponding error propagation formulas can be derived, in exactly the same fashion as with equations (6) and (7).

***Gravity Anomaly*** ($\Delta g$)***:*** To propagate gravity anomaly errors we use Poisson's integral (Heiskanen and Moritz, 1967):

$$\Delta g_p = \frac{1}{4\pi}\iint_\sigma \Delta g\, D(\psi,r)\, d\sigma$$

$$D(\psi,r) = \frac{R^2}{r}\left(\frac{r^2-R^2}{l^3} - \frac{1}{r} - \frac{3R}{r^2}t\right)$$

$$D_{N_1}^{N_2}(\psi,r) = \sum_{n=N_1}^{N_2}(2n+1)\left(\frac{R}{r}\right)^{n+1} P_n(t) \ . \qquad (9)$$

When $r = R$, this kernel can be computed by:

$$D_{N_1}^{N_2}(\psi) = \sum_{n=N_1}^{N_2}(2n+1)P_n(t) = \begin{cases} \dfrac{(N_2+1)(P_{N_2}-P_{N_2+1}) - N_1(P_{N_1-1}-P_{N_1})}{(1-t)} & t \neq 0 \\ (N_2+1)^2 - N_1^2 & t = 0 \end{cases} \qquad (10)$$

The kernel (10) has interesting filtering properties (see also Jekeli, 1981). As $N_2 \to \infty$ this kernel becomes the spherical equivalent of the Dirac delta function. As long as the convolution with this kernel is global (as is the case in our application), the result is an ideal filtering (apart from discretization errors) of the function convolved with this kernel, since the eigenvalues of this operator are:

$$\lambda_n = \begin{cases} 1 & N_1 \le n \le N_2 \\ 0 & otherwise \end{cases} \qquad (11)$$

This property allows us to extract from a given set of $\sigma^2(\Delta g)$ the portion of the error variance that corresponds to a given degree range $[N_1, N_2]$, which enables the computation of $\sigma^2(\hat{\Delta g}_H)$.

***Geoid Undulation*** ($N$)***:*** The band-limited version of Stokes' kernel has the series form:

$$S_{N_1}^{N_2}(\psi,r) = \sum_{n=N_1}^{N_2}\frac{(2n+1)}{(n-1)}\left(\frac{R}{r}\right)^{n+1} P_n(t) \ . \qquad (12)$$

***Gravity Disturbance*** ($\delta g$)***:*** We use the integral:

$$\delta g_p = -\frac{R}{4\pi}\iint_\sigma \Delta g \frac{\partial S(\psi,r)}{\partial r} d\sigma \ ,$$

whose kernel is given by:

$$\frac{\partial S(\psi,r)}{\partial r} = -\frac{R(r^2-R^2)}{rl^3} - \frac{4R}{rl} - \frac{R}{r^2} + \frac{6Rl}{r^3} + \frac{R^2}{r^3}t(13 + 6\ln\frac{r-Rt+l}{2r}) \ , \quad l = \left(r^2+R^2-2rRt\right)^{1/2}$$

which can be written in series form as:

$$R\frac{\partial S_{N_1}^{N_2}}{\partial r}(\psi,r) = \sum_{n=N_1}^{N_2}\frac{(2n+1)(n+1)}{(n-1)}\left(\frac{R}{r}\right)^{n+2} P_n(t) \ . \quad (13)$$

***N-S and E-W Deflections of the Vertical*** ($\xi,\eta$)***:*** We use the Vening Meinesz integral formula:

$$\begin{Bmatrix} \xi \\ \eta \end{Bmatrix}_p = \frac{1}{4\pi\gamma}\iint_\sigma \Delta g \frac{\partial S(\psi,r)}{\partial \psi}\begin{Bmatrix} \cos\alpha \\ \sin\alpha \end{Bmatrix} d\sigma$$

whose kernel is:

$$\frac{\partial S(\psi,r)}{\partial \psi} = \sin\psi \times$$

$$\left[-\frac{2R^2 r}{l^3} - \frac{6R^2}{rl} + \frac{8R^2}{r^2} + \frac{3R^2}{r^2}\left(\frac{r-Rt-l}{l\sin^2\psi} + \ln\frac{r-Rt+l}{2r}\right)\right]$$

and takes the series form:

$$\frac{\partial S_{N_1}^{N_2}}{\partial \psi}(\psi,r) = \sum_{n=N_1}^{N_2} \frac{(2n+1)}{(n-1)}\left(\frac{R}{r}\right)^{n+1} \frac{\partial P_n(t)}{\partial \psi} \quad . \qquad (14)$$

## 4 Verification

First we verified that the computation of functionals using global convolution (equations like (6)) gives results that agree with those computed using harmonic synthesis, over the same degree range. Geoid undulations computed from EGM96 via harmonic synthesis, for the degree range (n=71 to 360), are shown in Figure 2. Their RMS value is ±1.01 m. The difference between these values and those computed via convolution (using 30′ cells) has an RMS value of ±4 mm (Figure 3). This indicates a level of agreement between the two methods that we consider more than adequate for error propagation work. The discrepancies between the two estimates are affected primarily by the cell size used in the convolution approach (discretization error), becoming exceedingly small for 5′ cells.

We also verified that error propagation based on global convolutions gives results that agree well with those computed using rigorous error covariance matrix propagation (linear algebra). Obviously, this can only be done for relatively low degree expansions, where the sizes of the matrices involved allow the rigorous approach to be implemented. To this end we used the 1°x1° terrestrial (i.e., no altimetry) $\Delta g$ standard deviations used in the development of EGM96, and computed propagated errors for $N$, in two ways. First through rigorous covariance matrix propagation (Figure 4), and second based on a global convolution (Figure 5). In both cases the degree range was n=2 to 90. The first method takes 90 minutes of execution time, while the second takes 15 seconds. The computation was done on a SunFire v480 server with four 1.2GHz Ultra SPARC III processors. Percentage wise, the maximum difference between the two estimates is ~14%, while the RMS difference is ~3.6%.

Finally, Figures 6 and 7 present 5′x5′ commission error maps for the deflections of the vertical, computed from PGM2004A to degree 2159 (Pavlis et al., this issue). These figures demonstrate clearly that our technique produces propagated errors that preserve the geographic variations of the $\sigma^2(\Delta g)$ values used in the development of the model with a high degree of fidelity. Step discontinuities and certain areas with minimal geographic variation of the propagated errors (e.g., Antarctica) reflect certain shortcomings of the $\sigma^2(\Delta g)$ values assigned to the gravity data. A corresponding map for geoid undulations is shown in (Pavlis et al., *ibid.*), together with comparisons of the propagated errors versus the *observed* performance of PGM2004A, as obtained from independent data tests. It takes about 36 minutes of execution time to compute a global 5′x5′ error map, with the method presented here, on the same Sun server.

## 5 Summary

We have developed and verified a method for error propagation *with* geographic specificity, from very high degree spherical harmonic gravitational models. This approach is very efficient and yields results that are accurate enough to be useful. The approach uses global convolutions with band-limited kernels to isolate and compute the error contribution of the harmonics beyond the maximum degree of the comprehensive solution. These developments open up new possibilities for the application of optimal $\Delta g$ weighting by degree (or degree range), $\Delta g$ error calibration using locally available independent data, and permit the examination of the implications of $\Delta g$ weights by region.

## References


Haagmans, R., E. de Min, M. Van Gelderen (1993). Fast evaluation of convolution integrals on the sphere using 1D FFT, and a comparison with existing methods for Stokes' integral. *manusc. geod.*, ***18***, 227-241.

Haagmans, R.H.N., M. Van Gelderen (1991). Error variances-covariances of GEM-T1: Their characteristics and implications in geoid computation. *J. Geophys. Res.*, ***96*** *(B12)*, 20011-20022.

Heiskanen, W.A. and H. Moritz (1967). *Physical Geodesy*. W.H. Freeman, San Francisco.

Jekeli, C. (1981). Alternative Methods to Smooth the Earth's Gravity Field. *Rep. 327*, Dept. of Geod. Sci. and Surv., The Ohio State University, Columbus, Ohio.

Wong, L. and R. Gore (1969). Accuracy of Geoid Heights from Modified Stokes Kernels. *Geophys. J. R. Astr. Soc.*, ***18***, 81-91.

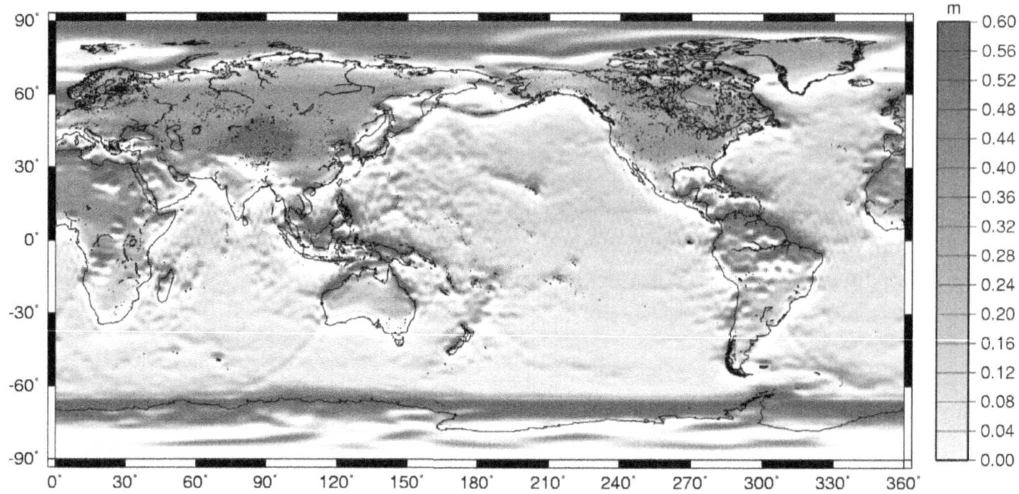


**Fig. 1** Geoid commission error for EGM96 (n=2 to 70) from its full error covariance matrix. RMS = ±0.18 m.

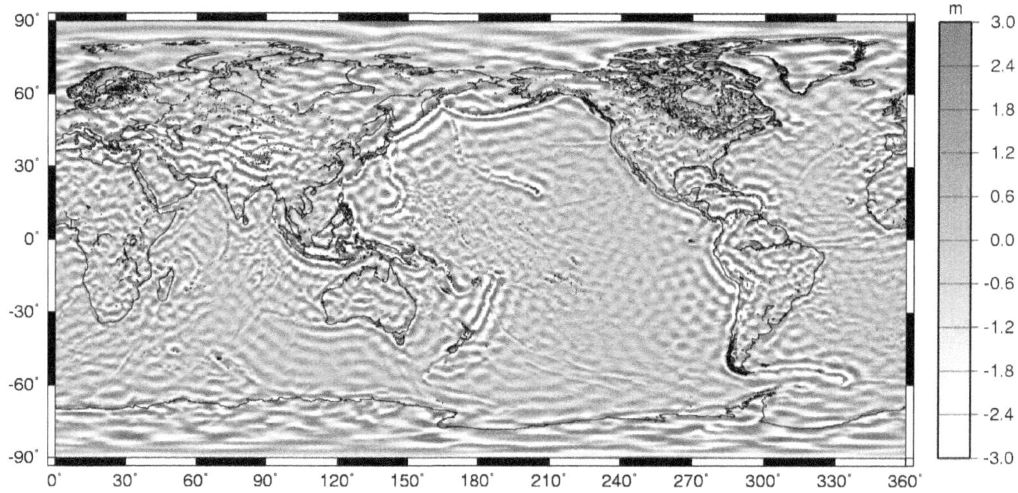


**Fig. 2** 30´x30´ synthetic geoid undulations from EGM96 (n=71 to 360). RMS = ±1.01 m.

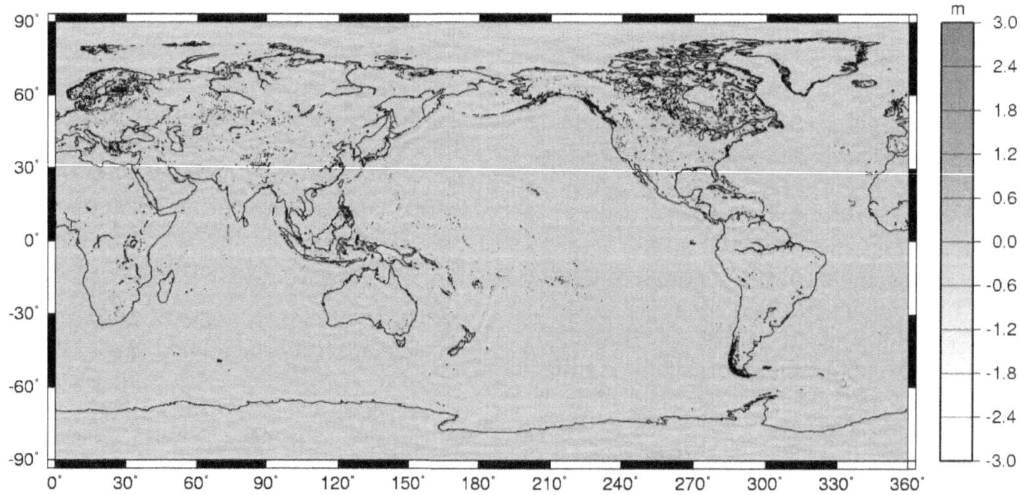


**Fig. 3** 30´x30´ geoid undulation differences: [synthetic Δg (n=2 to 360)]*[Stokes' kernel (n=71 to 360)] minus the synthetic undulations of Figure 2. RMS = ±0.004 m.

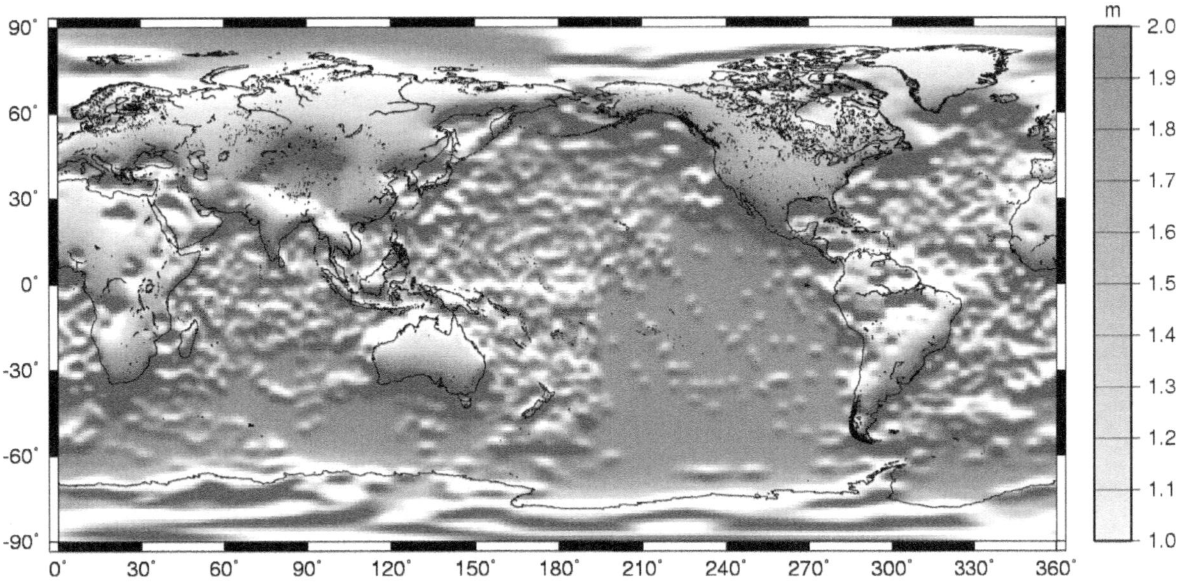


**Fig. 4** 1°x1° geoid undulation commission error computed by propagating the full error covariance matrix. RMS = ±1.64 m.

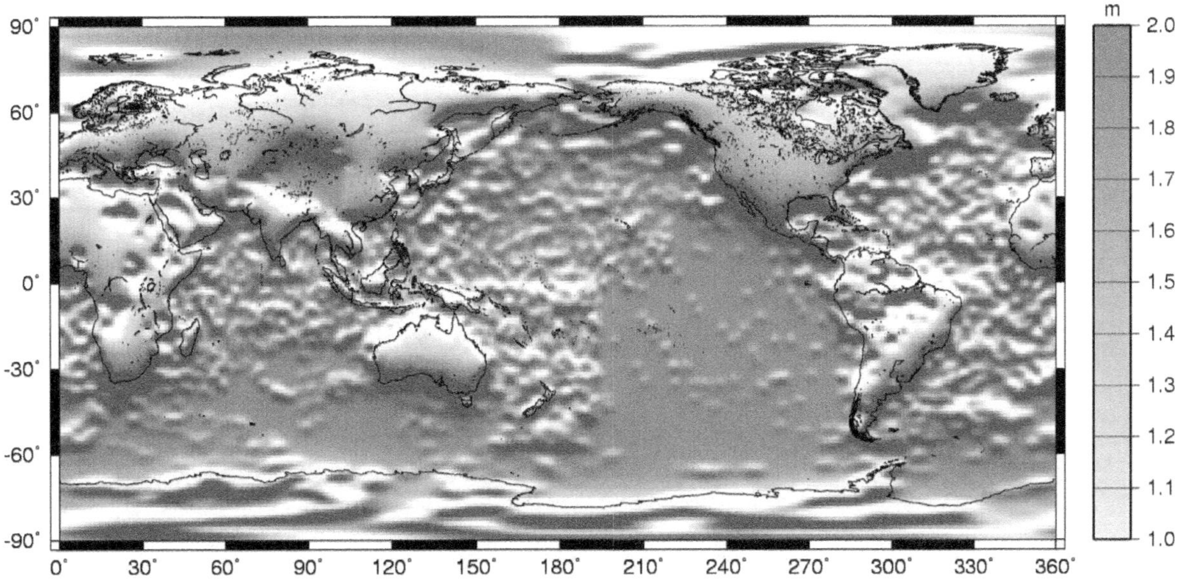


**Fig. 5** 1°x1° geoid undulation commission error computed using Stokes' integral formula. RMS = ±1.69 m.

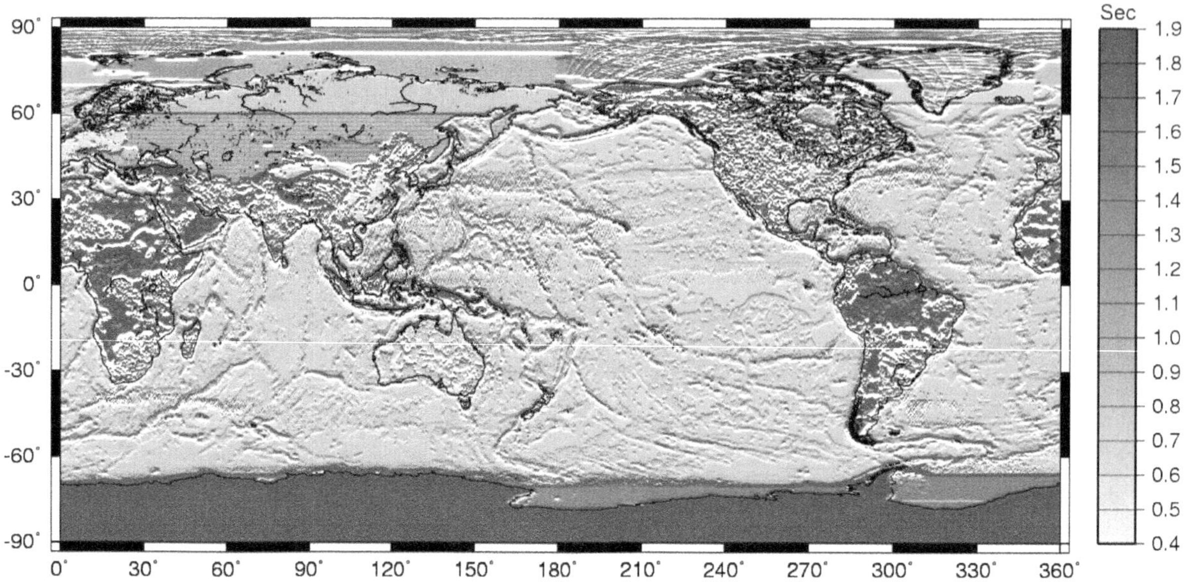


**Fig. 6** 5′x5′ $\xi$ commission error (arc seconds) computed from PGM2004A (n=2 to 2159) using convolution (Vening Meinesz's formula). RMS = ±1.047″.

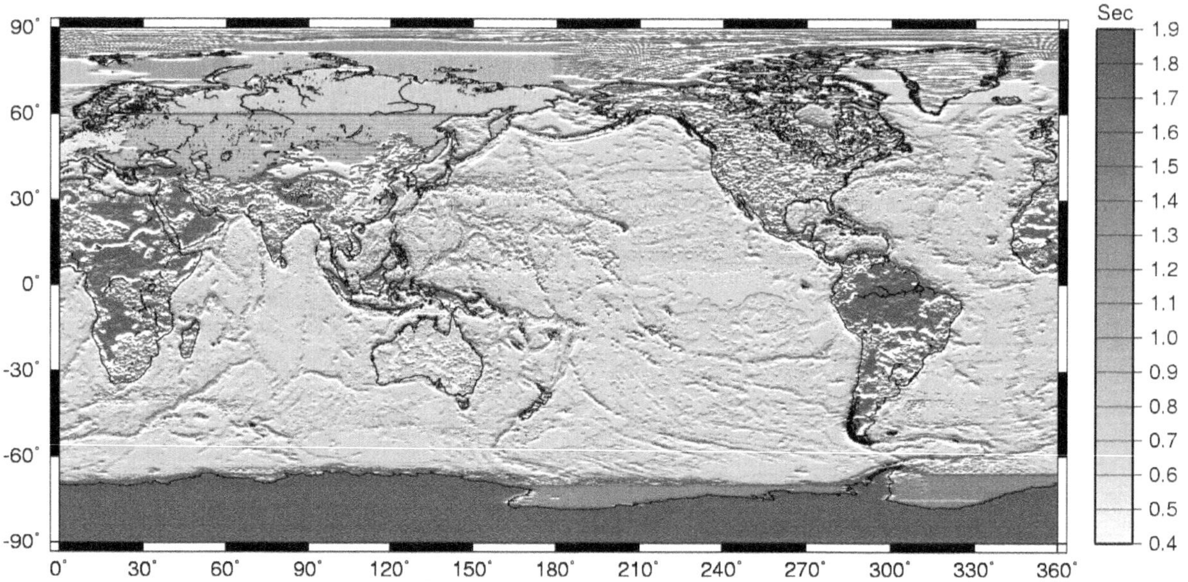


**Fig. 7** 5′x5′ $\eta$ commission error (arc seconds) computed from PGM2004A (n=2 to 2159) using convolution (Vening Meinesz's formula). RMS = ±1.057″.

# Gravity Data Base Generation and Geoid Model Estimation Using Heterogeneous Data

G.S. Vergos✉, I.N. Tziavos, V.D. Andritsanos
Department of Geodesy and Surveying, Aristotle University of Thessaloniki, University Box 440, 541 24, Thessaloniki, Greece, Fax: +30 31 0995948, E-mail: vergos@topo.auth.gr.

**Abstract.** The computation of high-resolution and high-precision geoid models in the Eastern part of the Mediterranean Sea usually suffers from the few gravity observations available. In the frame of the EU-sponsored GAVDOS project, a systematic attempt has been made to collect all available gravity data for an area located in the Southern part of Greece and determine new and high-resolution geoid models. Thus, all available gravity data have been collected for both land and marine regions and an editing/blunder-removal processing scheme has been followed to generate an optimal gravity dataset for use in geoid determination. The basic analysis and validation of the gravity databank was based on a gross-error detection visualization and collocation scheme. The Least Squares Collocation (LSC) method was employed to predict gravity at known stations and then validate the observations and detect blunders. The finally generated gravity database presents a resolution of 1 arcmin in both latitude and longitude while its external and internal accuracies were estimated to about ±5 mGal and ±0.2 – ±0.4 mGal, respectively. Based on the derived gravity database a gravimetric geoid model was developed using the well-known remove-compute-restore method with an application of a 1D Fast Fourier Transform (FFT) to evaluate Stokes' integral. Altimetric geoid solutions have been also determined from the GEOSAT and ERS1 geodetic mission altimetry data. Finally, combined geoid models have been computed using the FFT-based Input Output System Theory (IOST) and the LSC methods. The consistency of the geoid models estimated was assessed by comparing the geoid height value at the Gavdos Tide Gauge (TG) station on the isle of Gavdos. Their accuracy was determined through comparisons with stacked T/P sea surface heights. From the comparisons performed it was found that the accuracy of the gravimetric, altimetric and combined models was at the ±14.5 cm, ±8.6 cm and ±12.5 cm level, and their consistency at about ±2 cm.

**Keywords.** Gravity database, least squares collocation, geoid.

## 1 Introduction

The determination and availability of a high-resolution and high-accuracy geoid model is nowadays a necessity in a large number of geo-sciences, since it serves as the reference surface where other measurements and phenomena of the Earth system are related. This paper summarizes the results obtained during the last two years towards the generation of a consistent and accurate gravity database for Southern Greece and the estimation of a high-accuracy and high-resolution geoid model in support of the EU funded GAVDOS project. GAVDOS project focuses on the establishment of a sea level monitoring and altimeter calibration site on the isle of Gavdos, Greece. To achieve the objectives of the project, the estimation of a geoid model was necessary to serve as a reference surface for the oceanographic, sea level monitoring, tectonic and other studies. Our group was responsible for the collection of all available gravity data for the area under study and the estimation of gravimetric and combined, with altimetric Sea Surface Heights (SSHs), geoid models. The structure of the procedure followed can be summarized as: a) collection and unification of all available gravity data, b) blunder removal and editing, c) gravity database generation and gravimetric geoid model estimation, d) combination of heterogeneous data to estimate a combined geoid model.

## 2 Gravity database generation

### 2.1 Gravity data collection

For the generation of the gravity database a total number of 103289 point and mean gravity data were collected from different data sources. Various databases including absolute gravity measurements, relative marine, land and airborne observations were collected for the area of Gavdos bounded between $33° \leq \varphi \leq 38°$ and $20° \leq \lambda \leq 28°$. The distribution of the gravity data is presented in Figure 1.

The entire database was divided into separate files and the observations were encoded according to their origin. In addition, they were reformulated in the classical format of "station ID, latitude, longitude, height and observation". The original databases used were: a) 33560 point marine and land free-air gravity anomalies collected during international campaigns and international projects (Andritsanos and Tziavos 2002; Casten and Makris 2001; Lagios et al. 1996), b) 9322 marine free-air gravity anomalies derived from satellite altimetry (Andersen and Knudsen 1998), which were basically used to fill several gaps in the Eastern Mediterranean Sea, c) 21372 marine free-air gravity anomalies from GEODAS (NGS 2001), d) 1680 airborne free-air gravity anomalies from the CAATER, and e) 39647 marine gravity data from the digitisation of Morelli's maps (Behrend et al. 1996). All observation were referred to IGSN71, the gravity anomalies were computed using the International Gravity Formula of 1980 for the normal gravity and the geographical coordinates were transformed to the GRS80 ellipsoid.

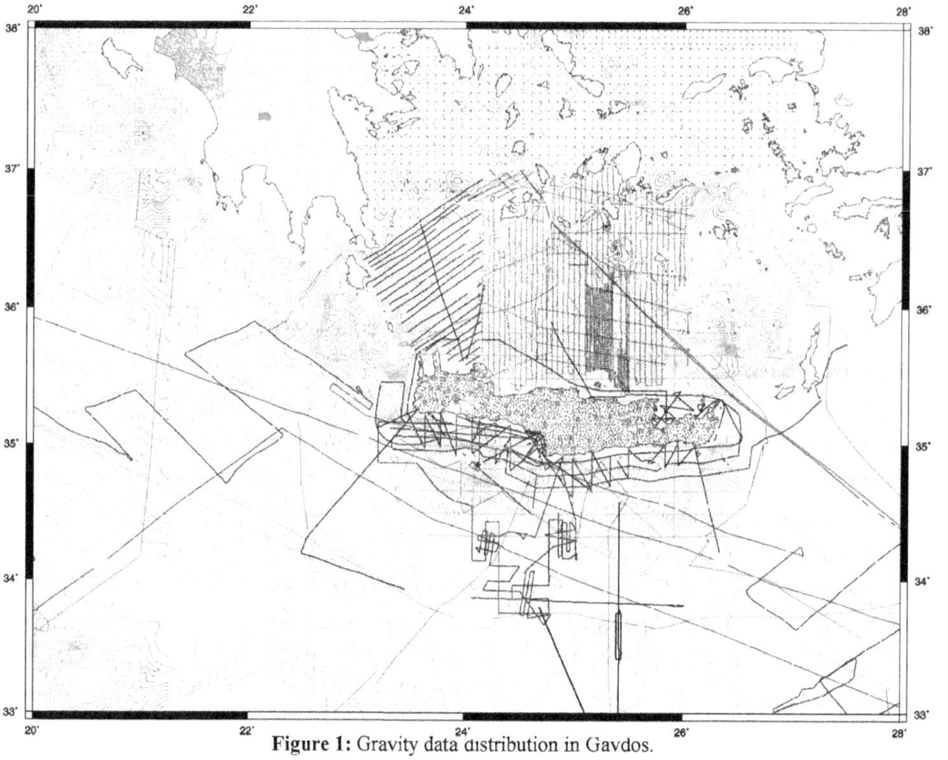


**Figure 1:** Gravity data distribution in Gavdos.

## 2.2 Methodology for blunder detection

For the identification and removal of blunders a two-step procedure was followed, i.e., a) visual inspection and b) least-squares collocation.

Taking into account that data related to the gravity field are spatially correlated, gravity quantities of the same type and not far apart will be very similar. Especially, after the removal of a highly expanded geopotential model and of the effect of the neighboring masses, the distribution of the data should be close to normal. According to BGI (1992) an effective check can be done by 2D contouring the data. Thus, following this method, a map of residual, i.e., EGM (Earth Gravity Model) and topographically reduced gravity anomalies, was generated and deep holes and steep spikes were considered to indicate suspicious observations. Since the smoothness of the field is the highest one after the removal of high and low frequency information, large discrepancies can be identified as blunders.

Least squares collocation (LSC) was also used to remove any existing outliers that were not removed during the preceding visual check. A gravity anomaly $y$ was predicted from a set of values $x$, in neighboring points, spaced as evenly as possible in all directions according to the well-known collocation formula (Tscherning 1991)

$$\tilde{y} = \mathbf{C}_y \bar{\mathbf{C}}^{-1}\mathbf{x} \quad (1)$$

where $C_y$ is the vector of covariances between $y$ and the $x_i$ values and $\bar{\mathbf{C}} = \mathbf{C} + \mathbf{D}$ is the sum of the covariance matrix of the $x_i$ quantities and the variance-covariance matrix of the noise (error) associated with the quantities. An error estimate was also computed for the difference $|y - \tilde{y}|$ as

$$\sigma^2(y - \tilde{y}) = C_o - \mathbf{C}_y^T \bar{\mathbf{C}}^{-1}\mathbf{C}_y \quad (2)$$

where $C_o$ is the variance of the gravity values. A gross-error was then detected when

$$|y_{obs} - \tilde{y}| > k\sqrt{\sigma^2(y - \tilde{y}) + \sigma_y^2} \quad (3)$$

where $k$ is a constant generally having the value 3 to 5 depending on the check strictness and $\sigma_y^2$ is the error variance of the observation $y_{obs}$. From the above equations it is obvious that gross-errors are most easily found if $C_o$ is as small as possible. Thus it is obvious that the removal of the long and short wavelengths of the gravity field is necessary for the outlier detection to lead to rigorous results.

## 2.3 Results of validation

### 2.3.1 *Visual inspection*

The total number of free-air gravity anomalies were collected and reformulated in a single file giving as station number a characteristic code for each dataset, so they could be distinguished at a later step. This set formulated the initial set of point free-air gravity anomalies which were then used for the construction of the GAVDOS project gravity database. For the visual inspection, the contribution of the GPM98b EGM (Wenzel 1999) was removed from the raw data while the topographic effects were taken into account through a simple Bouguer reduction. The statistics of the reduced gravity anomalies are presented in Table 1.

**Table 1**. Statistics of point $\Delta g_f$ before and after the reduction to GPM98b; Bouguer gravity anomalies ($\Delta g_B$) before and after the visual inspection test. Unit: [mGal].

| | max | min | mean | σ |
|---|---|---|---|---|
| $\Delta g_f$ | 270.60 | -247.76 | -34.46 | ±82.27 |
| $\Delta g_{f\,red}$ | 149.03 | -111.87 | -2.83 | ±14.34 |
| $\Delta g_B$ (before) | 159.74 | -247.76 | -36.52 | ±80.41 |
| $\Delta g_B$ (after) | 122.40 | -136.93 | -4.89 | ±80.38 |

Employing the so-derived reduced anomaly field, a contour map of the area was generated. Some outliers were identified considering that spikes and holes in the gravity field do not describe local irregularities since the main topographic signal from a Bouguer plate was removed. After this visual inspection test, 94 gravity anomaly observations have been identified as blunders and were subsequently removed from the global database.

### 2.3.2 Collocation scheme

The procedure described in the previous section was followed to eliminate any existing gross-errors that passed the visual inspection test. Due to the large number of observations, the area under study was divided in 20 sectors of 1°×2° in latitude and longitude each. That was necessary in order to be able to handle the large amount of data and preserve the homogeneity of the field. The total number of observations in each compartment was then divided in two files with equal and homogeneously distributed data points. The data were then reduced to the EGM96 geopotential model and an empirical covariance function was computed and fitted to the Tscherning and Rapp (1974) analytical model from the observations of the first data file. Using the parameters of this model, predictions at the locations of the points of the second file were then estimated. Due to the unavailability of proper measurement error and the ambiguous quality of the data, an error of ±5 mGal was assigned to each observation. A rejection criterion with a parameter k=2 was followed, which was stricter compared to that used in earlier studies (Vergos et al. 2003), so as to remove more suspicious observations and generate a more accurate gravity database. This parameter in conjunction with the overestimated $\sigma$ of the observations ensured the removal of the largest blunders. A total number of 5729 points were rejected as suspicious gross-errors. The points removed represent a 5.6425% of the total database, while those remaining were 95804 gravity observations. The statistics before and after the blunder removal are tabulated in Table 2.

**Table 2**. Statistics of reduced to EGM96 Bouguer gravity anomalies before and after the gross-error removal test with LSC. Unit: [mGal].

| | max | min | mean | σ |
|---|---|---|---|---|
| $\Delta g_{B\,red}$ (before) | 118.45 | -115.23 | -5.34 | ±26.26 |
| $\Delta g_{B\,red}$ (after) | 118.45 | -113.40 | -5.60 | ±26.01 |

## 2.4 Gravity database estimation

The remaining 95804 point Bouguer gravity anomalies were transformed to free-air anomalies by restoring the effect of the Bouguer plate. Finally, and in order to fill gaps in the database over Turkey, gravity anomalies from GPM98b were estimated, whilst the CAATER airborne gravity data were also implemented in the database (Olesen et al. 2003). Thus, a total number of 97466 reduced to EGM96 point free-air gravity anomalies were available.

Most databases are usually given as regular grids of mean values rather than irregularly distributed point values. The former representation is preferred due to a) the smaller size of the data files, b) the easier manipulation of the data with spectral methods, c) restrictions in the availability of the point data, etc. Therefore, the final step for the construction of the database refers to the estimation of gravity anomalies on a regular grid. For the prediction results to be more rigorous, regardless of the gridding algorithm used, the field to be interpolated has to be as smooth as possible, thus the effect of the topographic masses was removed from the data through an RTM reduction (Forsberg 1984). The Digital Terrain and Depth Model (DTDM), whose statistics are given in Table 3, has a resolution of 1 km in both latitude and longitude and was constructed by the authors at an earlier phase of the project (details available at www.gavdos.tuc.gr).

**Table 3**. Statistics of the 1-km DTM used for the RTM effects. Unit: [m].

| | max | min | mean | σ |
|---|---|---|---|---|
| **DTM** | 2394.790 | -5065.950 | -1562.476 | ±1272.917 |

The removal of the RTM-effects from the EGM96 reduced gravity anomalies resulted in a residual gravity anomaly field. From Table 4, which summarizes the statistics of the free-air gravity anomalies before and after the RTM reduction, it is evident that the residual field is indeed smoother since the range is reduced by 45% (139.4 mGal), the mean by 83.20% (2.82 mGal) and the σ by 40% (10.87 mGal).

To construct the final gravity grid, different gridding algorithms such as spline interpolation and weighted means were tested. But, for the gridding procedure to be rigorous we chose to grid the data using collocation. This method is obviously more time consuming compared to the other two, but provides statistically optimal results. To grid the data using LSC the correlation length and the variance of the residual field had to be computed, thus the empirical covariance function of the data has been computed and fitted to the Tscherning and Rapp model (see Figure 2). Figure 2 presents the empirical covariance functions of the gravity data before (cross) and after the RTM reduction (asterisk) as well as the one fitted to the analytical model (dot). Figure 2 strengthens the previous conclusion that the data after the RTM reduction are indeed smoother, since the variance of the data reduced and the correlation length of the field increased.

**Table 4**. Statistics of reduced to EGM96 free-air gravity anomalies ($\Delta g_f$) before and after the RTM reduction. Unit: [mGal].

| | max | min | mean | σ |
|---|---|---|---|---|
| $\Delta g_{f\,red}$ (before) | 206.08 | -108.17 | -3.39 | ±27.13 |
| $\Delta g_{f\,red}$ (after) | 78.99 | -95.87 | -0.57 | ±16.26 |

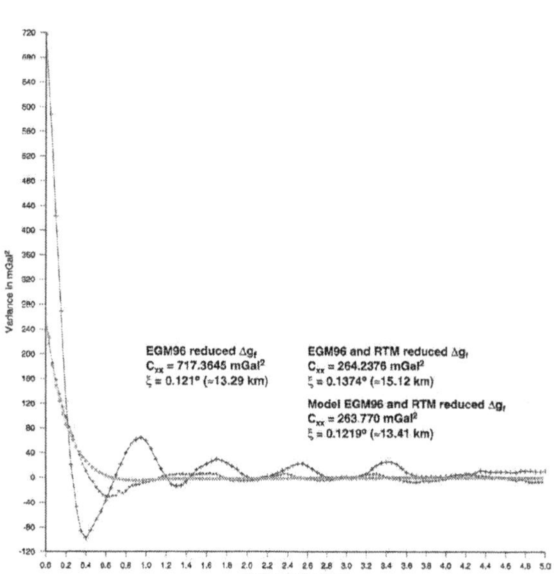


**Figure 2:** Empirical covariance functions of the reduced (cross) and residual (asterisk) gravity anomalies and the fitted analytical model (dot).

In this way the final residual (RTM and EGM96 reduced) free-air gravity anomaly grid has been estimated. Then, adding back the effect of the topography/bathymetry and that of the geopotential model resulted in the final 1′×1′ (corresponding to about 1.7 km spatial resolution) gravity database. Figure 3 depicts the final gravity database in the area under study, while the statistics of the gravity field over Gavdos are presented in Table 5.

**Table 5**. Statistics of reduced to EGM96 and final free-air gravity anomaly grid (the GAVDOS project database). Unit: [mGal].

| | max | min | mean | rms | σ |
|---|---|---|---|---|---|
| $\Delta g_{f\,red}$ | 186.17 | -104.09 | -0.85 | 25.14 | ±25.12 |
| $\Delta g_f$ | 230.24 | -237.81 | -22.16 | 86.07 | ±83.17 |

## 3 Geoid determination and validation

Using the estimated gravity database, as well as satellite altimetry data from the geodetic missions of ERS1 and GEOSAT, gravimetric, altimetric and combined geoid models have been determined for the area. In all cases the well-known remove-compute-restore method was employed, while the estimation of the gravimetric geoid was carried out using the 1D-FFT spherical Stokes convolution (Haagmans et al. 1993). The methodology followed to process the altimetric SSHs and estimate the combined ERS1 and GEOSAT altimetric geoid model is described in detail in Vergos and Sideris (2003) and Vergos et al. (2003) and will not be discussed here.

The determination of a combined gravimetric and altimetric geoid solution was performed with two methods, i.e., conventional LSC in the space domain (Moritz 1980) and the FFT-based Multiple Input Multiple Output System Theory (MIMOST) in the frequency domain (Andritsanos and Tziavos 2002). Due to a) the high-resolution of the geoid models, b) the extension of the area, and c) the fact that our main interest was over the isle of Gavdos, the combined solutions have been estimated in the inner parts of the wider area under study, to speed up the computations. Table 6 presents the statistics of the four estimated geoid models, where the gravimetric and altimetric solutions refer to the entire area under study and the combined ones to a smaller part of it. Since the LSC method is the most time-consuming one, it was limited to the area bounded between $34° \le \varphi \le 35°$ and $23° \le \lambda \le 24°$. Figures 4 and 5 depict the LSC, gravimetric and MIMOST geoid models.

**Table 6**. Statistics of the final geoid models for the area of Gavdos. Unit: [m].

| MODEL | max | min | mean | σ |
|---|---|---|---|---|
| $N^{gravimetric}$ | 39.813 | 0.780 | 21.185 | ±10.352 |
| $N^{altimetric}$ | 40.206 | 1.057 | 21.376 | ±10.484 |
| $N^{MIMOST}$ | 37.733 | 6.168 | 2.899 | ±9.127 |
| $N^{LSC}$ | 25.638 | 9.857 | 16.867 | ±3.951 |

The validation of the estimated geoid models was performed through comparisons with stacked T/P SSHs

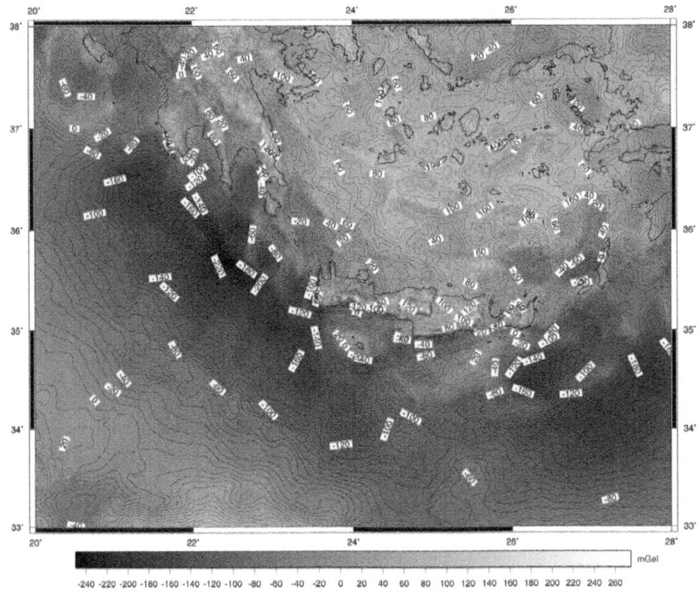

**Figure 3:** The final free-air gravity database in the area under study.

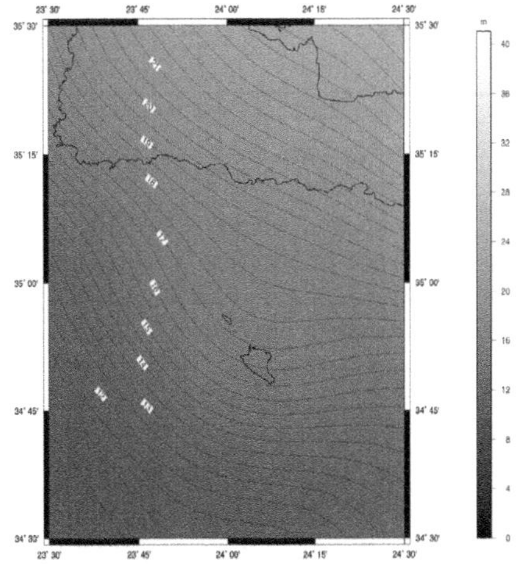

**Figure 4:** LSC combined geoid model for the area of Gavdos. (the asterisk shows the Karave TG station)

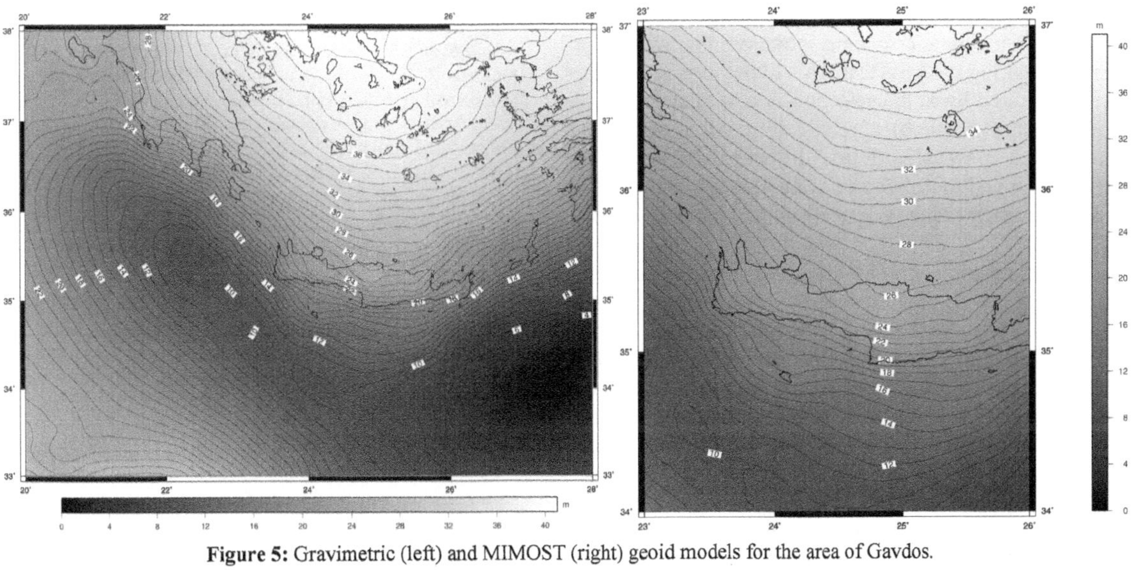

**Figure 5:** Gravimetric (left) and MIMOST (right) geoid models for the area of Gavdos.

spanning over nine years of the satellite mission (1993 – 2001). Furthermore, geoid heights from each model have been predicted for the Gavdos (KARAVE) tide gauge (TG) station. Table 7 summarizes the statistics of the comparisons with T/P (standard deviations only) after the fit of the differences with a 4-parameter transformation model as well as the estimated geoid heights at the TG station. No results for the comparison between the LSC geoid and T/P are presented, since the satellite points within the LSC solution are very few and would lead to over-optimistic outcomes. In Table 7, $N^i$ refers to the geoid heights from the different models estimated, $N^{TG}$ to the predicted height at the Karave TG station and $\sigma_{N^{TG}}$ to the accuracy of the latter. From the statistics, it can be concluded that all estimated geoid models are consistent to each other since the geoid height at the TG varies within 4 cm in the worst case. As far as the models developed are concerned, it should be noted that the altimetric geoid is superior over purely marine areas (homogeneous coverage and high accuracy), the gravimetric should be used over land areas and close to the coastline, while the combined solutions offer the advantages of both "worlds" since they represent accurately the geoid over both marine and land areas. The MIMOST combined model, provides an agreement with the T/P SSHs at the ±12.5 cm level (1σ), which is about 6 cm better compared to the previous combined solution for the area (Andritsanos et al. 2001).

**Table 7**. Predicted geoid height at the Gavdos TG station from the different models and comparisons with T/P SSHs.

| model | $\sigma_{N^{TP}-N^{i}}$ (cm) | $N^{TG}$ (m) | $\sigma_{N^{TG}}$ (cm) |
|---|---|---|---|
| $N^{grav}$ | ±14.5 | 16.70 | ±1.41 |
| $N^{altim}$ | ±8.60 | 16.70 | ±0.91 |
| $N^{MIMOST}$ | ±12.5 | 16.68 | ±1.19 |
| $N^{LSC}$ | ---------- | 16.72 | ±0.40 |

## 4 Conclusions

A gravity anomaly database has been created from marine, land and airborne data, towards the determination of a high-accuracy and high-resolution geoid model in Gavdos, Greece. The methodology employed was a two-step procedure, i.e., a visual inspection test followed by a least squares collocation blunder detection and removal scheme. Both tests are highly objective, since in the former, holes and spikes not so deep or steep respectively can be either removed as blunders or remain in the database. In the latter, the removal of an observation as erroneous depends solely on the selection of the data error (±5 mGal in our case) and the rejection constant *k* (2 in our case). As far as the visual inspection test is concerned, only observations that could be clearly distinguished as blunders were removed, considering that any remaining erroneous observations would be removed during the LSC test. This hypothesis was more or less ensured by a) the selection of a very strict rejection constant *k* and b) a relatively small observation error. Thus practically, if the difference between the observation and the LSC prediction was larger than ~10 mGal, the observation was removed as blunder (see Eq. 3).

For the determination of the geoid models the well-known remove-compute-restore method was employed to estimate gravimetric, altimetric and combined solutions. From the results obtained and the comparisons with stacked T/P SSHs, we can conclude that the altimetric geoid gives the most precise results and outperforms the gravimetric solution by about 6 cm. One should take into account that the estimated altimetric geoid is highly correlated with the T/P data, therefore their comparison is expected to give good results. A much better validation dataset would be GPS/Leveling geoid heights at the TG station, but such information was and still is unavailable. This difference of 6 cm is probably the quasi-stationary sea surface topography in the area, which was not removed from the altimetric observations due to the unavailability of a local QSST model and the inappropriateness of the global models in closed sea areas. Of course, the altimetric geoid is prone to errors close to the coastline and unavailable on land, therefore combined gravimetric and altimetric models were determined for the area. The latter agree at about ±12.5 cm with the T/P SSHs which is a major improvement compared to the previous geoid models for the area. Finally, the consistency between the geoid models is at the ±2 – ±4 cm level, which is considered as satisfactory.


## Acknowledgement

Funding for this research was provided from the EU under contract GAVDOS: EVR1-CT-2001-40019 in the frame of the EESD-ESD-3 Fifth Framework Program (www.gavdos.tuc.gr).


## References


Andritsanos VD, Tziavos IN (2002) Estimation of gravity field parameters by a multiple input/output system. Phys and Chem of the Earth, Part A 25(1): 39-46.

Andritsanos VD, Vergos GS, Tziavos IN, Pavlis EC and Mertikas SP. (2001) A High Resolution Geoid for the Establishment of the Gavdos Multi-Satellite Calibration Site. In: Sideris MG (ed) Proc of International Association of Geodesy Symposia "Gravity Geoid and Geodynamics 2000", Vol. 123. Springer - Verlag Berlin Heidelberg, pp 347-354.

Andersen OB, Knudsen P (1998) Global gravity field from ERS1 and Geosat geodetic mission altimetry. J Geophys Res 103(C4): 8129-8137.

Behrent D, Denker H, Schmidt K (1996) Digital gravity data sets for the Mediterranean Sea derived from available maps. BGI Bulletin d' information 78: 31-39.

BGI (1992) BGI Bulletin d' Information Vol 70, 71.

Casten U, Makris J (2001) Erkundung der Krustenstruktur von Kreta durch detaillierete Schwere- und Magnetfeldmessunggen. Project Report DFG: Ca 83/8-1 bis 3 Ma 719/54-1 bis 3.

Forsberg R (1984) A study of terrain corrections, density anomalies and geophysical inversion methods in gravity field modeling. Rep of the Dept of Geodetic Sci and Surv No 355 The Ohio State Univ, Columbus, Ohio.

Haagmans R, de Min E, van Gelderen M (1993) Fast evaluation of convolution integrals on the sphere using 1D FFT, and a comparison with existing methods for Stokes' integral. Manuscr Geod 18: 227-241.

Lagios E, Chailas S, Hipkin RG (1996) Newly compiled gravity and topographic data banks of Greece. Geophys J Int 126: 287-290.

Moritz H (1980) Advanced Physical Geodesy. 2nd Ed Wichmann, Karlsruhe

National Geophysical Data Center (2001) GEODAS Marine trackline geophysics – Gravity, bathymetry, seismic, geophysical data.

Olesen AV, Tziavos IN, Forsberg R (2003) New Airborne Gravity Dta Around Crete – First results from the CAATER Campaign. In: Tziavos (ed) Proc of the 3rd Meeting of the Gravity and Geoid Commission "Gravity and Geoid 2002", pp 40-44.

Tscherning CC (1991) The use of optimal estimation for gross-error detection in databases of spatially correlated data. BGI, Bulletin d' Information 68: 79-89.

Tscherning CC, Rapp (1974) Closed Covariance Expressions for Gravity Anomalies, Geoid Undulations, and Deflections of the Vertical Implied by Anomaly Degree-Variance Models. Rep of the Dept of Geodetic Sci and Surv No 208 The Ohio State Univ, Columbus, Ohio.

Vergos GS, Sideris MG (2003) Estimation of High-Precision Marine Geoid Models Off-shore Newfoundland, Eastern Canada. In: Tziavos (ed) Proc of the 3rd Meeting of the Gravity and Geoid Commission "Gravity and Geoid 2002", pp 126-131.

Vergos GS, Tziavos IN, Andritsanos VD (2003) On the Determination of Marine Geoid Models by Least Squares Collocation and Spectral Methods using Heterogeneous Data. Presented at Session G03 of the 2004 IUGG General Assembly, Sapporo, Japan, July 2-8, 2003. (accepted for publication to the conference proceedings)

Wenzel HG (1999) Global models of the gravity field of high and ultra-high resolution. In: Lecture Notes of IAG's Geoid School, Milano, Italy.

# A new strategy for processing airborne gravity data

B.A. Alberts, R. Klees and P. Ditmar
Delft Institute of Earth Observation and Space Systems (DEOS),
Delft University of Technology, Kluyverweg 1, PO box 5058, 2600 GB Delft, The Netherlands.

**Abstract.** We present a new approach for regional gravity field analysis from airborne gravity data, which combines separate processing steps into one inversion scheme. The approach uses a spectral representation of the Earth's gravity field in terms of a series of harmonic base functions, expressed in a local Cartesian reference frame. The parameters of this representation are estimated using least-squares techniques. Special emphasis is put on the proper modeling of data noise. Frequency-dependent data weighting is applied to handle colored noise, which replaces the traditional method of low-pass filtering and cross-over adjustment. To suppress model errors at the highest spatial frequencies, first-order Tikhonov regularization is applied. The performance of the developed technique is assessed for the inversion of gravity disturbances into the disturbing potential, using simulated data.

**Keywords:** Airborne gravimetry, airborne gravity data processing, regional gravity field analysis, frequency-dependent weighting.

## 1 Introduction

The processing of airborne gravity data traditionally consists of several independent steps such as filtering, adjustment of cross-over misfits and gridding. The resulting estimates of gravity disturbances at flight level can be downward continued and inverted into related gravity functionals at ground level (e.g. gravity anomalies). For an overview of methods for the inversion of airborne gravity data see Alberts and Klees (2004). Each of the various processing steps may, however, introduce errors, which accumulate in the course of processing and limit the accuracy and resolution of the final solution. The objective of this paper is to present an alternative approach for the inversion of airborne gravity data, which combines the different processing steps into one inversion scheme.

For the analytical representation of the disturbing potential, we choose a spectral representation; the disturbing potential is expressed as a series of harmonic base functions. The expressions for the base functions can be selected depending of the coordinate system under consideration. Here we use Cartesian coordinates, and the base functions are fundamental solutions of Laplace' equation in Cartesian coordinates. The basic methodology of the developed technique, which is sketched in section 2, is closely related to the approach that has been developed for the computation of the Earth's gravity field from satellite data acquired by the new satellite gravity missions (Ditmar and Klees, 2002; Klees et al., 2003). An important aspect of this approach is that we use frequency-dependent data weighting to handle colored observation noise, instead of the traditional low-pass filtering. This is discussed in section 3, whereas section 4 shows that the same concept can be used for the handling of a bias and tilt in the data, which replaces the traditional cross-over adjustment. Because the downward continuation amplifies model errors, a stabilization technique must be applied. Here we use first-order Tikhonov regularization, of which the derivation is given in section 5.

We investigated the performance of the new approach for the inversion of gravity disturbances at flight level into the disturbing potential at ground level. As the area size is limited, the long wavelength part has been removed from the gravity signal. To assess the accuracy of the method, several tests were done with simulated data corrupted by white and colored noise. The results of the numerical tests are presented in section 6.

## 2 Inversion methodology

The inversion technique is based on the spectral representation of the Earth's gravity field. The gravity potential is expressed in a local Cartesian reference frame as a series of harmonic base functions. Therefore, the coordinates of the observation points are first transformed to a local Cartesian coordinate system $\mathbf{x}_{lc} = (x, y, z)$. For geocentric coordinates $\mathbf{x}_{gc} = (X, Y, Z)$ this transformation is given as

$$\mathbf{x}_{lc} = \mathbf{P}\mathbf{R}_Z(\lambda_c)\mathbf{R}_Y(\varphi_c)\,\mathbf{x}_{gc} + \mathbf{t}, \tag{1}$$

where $\mathbf{P}$ is a permutation matrix to set the $y$-axis pointing north and the $x$-axis pointing east, $\mathbf{R}_Y$ and $\mathbf{R}_Z$ denote the rotation matrices, $(\varphi_c, \lambda_c)$ is the coordinate centre of the area considered and $\mathbf{t} = (t_x, t_y, t_z)$ is a translation in the corresponding directions.

The gravitational potential is expressed in a series of fundamental solutions of Laplace's equation in Cartesian coordinates, which is given as

$$V(x,y,z) = \sum_{l=0}^{\infty}\sum_{m=0}^{\infty} (A_{lm}\cos\alpha_l x \cos\beta_m y + B_{lm}\cos\alpha_l x \sin\beta_m y + C_{lm}\sin\alpha_l x \cos\beta_m y + D_{lm}\sin\alpha_l x \sin\beta_m y)\, e^{-\gamma_{lm} z}, \quad (2)$$

with

$$\alpha_l = \frac{2\pi l}{D_x}, \quad \beta_m = \frac{2\pi m}{D_y},$$

$$\gamma_{lm} := \sqrt{\alpha_l^2 + \beta_m^2} = 2\pi\sqrt{\left(\frac{l}{D_x}\right)^2 + \left(\frac{m}{D_y}\right)^2},$$

where $D_x$ and $D_y$ determine the size of the area in $x$ and $y$ direction, respectively. It is important to note that this approach does not use a planar Earth approximation, as the observations remain at the positions where they are taken (e.g. on a sphere when a constant flight level is assumed).

With the chosen representation it is implicitly assumed that the signal is periodic. For regional computations this assumption clearly does not hold; there will be a discontinuity at the edge of the area, causing oscillations in a neighborhood of the boundaries of the area. These edge effects can, however, be reduced by extending the area with zero observations (zero-padding) or by using some data tapering.

The relation between observed gravity disturbances $V_z$ and the unknown coefficients is found by applying the derivative operator in $z$-direction to Eq. (2)

$$V_z(x,y,z) := \frac{\partial V}{\partial z} = \sum_{l=0}^{\infty}\sum_{m=0}^{\infty} -\gamma_{lm} e^{-\gamma_{lm} z} \times (A_{lm}\cos\alpha_l x \cos\beta_m y + B_{lm}\cos\alpha_l x \sin\beta_m y + C_{lm}\sin\alpha_l x \cos\beta_m y + D_{lm}\sin\alpha_l x \sin\beta_m y)\,. \quad (3)$$

Note that if $l = m = 0$, $\gamma_{lm}$ is equal to zero. As a consequence the coefficient $A_{00}$ (the constant part of the potential) cannot be determined from airborne gravimetry measurements.

Eq. (3) can be used as the functional part of a standard Gauss-Markov model

$$\mathbf{y} = \mathbf{A}\mathbf{x} + \mathbf{e}, \quad E\{\mathbf{e}\} = 0, \quad E\{\mathbf{e}\mathbf{e}^T\} = D\{\mathbf{y}\} = \mathbf{Q_y}.$$

Here, $\mathbf{y}$ is the $n \times 1$ observation vector containing the gravity disturbances $V_z$, $\mathbf{e}$ is the vector of measurement errors, $\mathbf{x}$ is the $r \times 1$ vector of unknown coefficients and $\mathbf{Q_y}$ is the covariance matrix of the observations. The $n \times r$ design matrix $\mathbf{A}$ describes the functional relations between $\mathbf{y}$ and $\mathbf{x}$. The optimal solution is given by the best linear unbiased estimator (BLUE)

$$\hat{\mathbf{x}} = (\mathbf{A}^T \mathbf{Q_y}^{-1}\mathbf{A})^{-1}\mathbf{A}^T\mathbf{Q_y}^{-1}\mathbf{y}. \quad (4)$$

The solution vector $\hat{\mathbf{x}}$ contains the estimated coefficients $\hat{A}_{lm}$, $\hat{B}_{lm}$, $\hat{C}_{lm}$, $\hat{D}_{lm}$. Then, Eq. (2) is used to compute the disturbing potential at given positions.

## 3 Frequency-dependent weighting

Airborne gravity measurements are acquired in a very dynamic environment, resulting in strong noise in the observations, which is often colored, i.e. non-uniformly distributed over the frequencies. To reduce the effects of the noise, an aggressive low-pass filtering is usually applied. For a discussion on low-pass filtering in airborne gravimetry see Childers et al. (1999). Low-pass filtering, however, deteriorates the gravity signal. Therefore, the concept of filtering is replaced by the concept of frequency-dependent data weighting. Frequencies at which the noise is large get a lower weight than frequencies at which the noise is small, but the signal is preserved.

When we are dealing with colored noise, the covariance matrix of the observations will be non-diagonal. The inverse $\mathbf{Q_y}^{-1}$ cannot be easily computed when the number of observations is large. However, when we consider the product $\mathbf{Q_y}^{-1}\mathbf{A}$ as the application of $\mathbf{Q_y}^{-1}$ to the columns of $\mathbf{A}$, the problem changes to solving a system of linear equations with equation matrix $\mathbf{Q_y}$, which can efficiently be solved using the pre-conditioned conjugate gradients (PCCG) method. If it is assumed that the noise is stationary and if there are no data gaps in the time series of measurements, the covariance matrix is Toeplitz. An approach to handle stationary colored noise in the presence of data gaps is presented in (Klees and Ditmar, 2002). An efficient method to solve Toeplitz systems has been proposed by Klees et al. (2003). The numerical complexity of this method is $O(n)$. It uses an ARMA (autoregressive moving

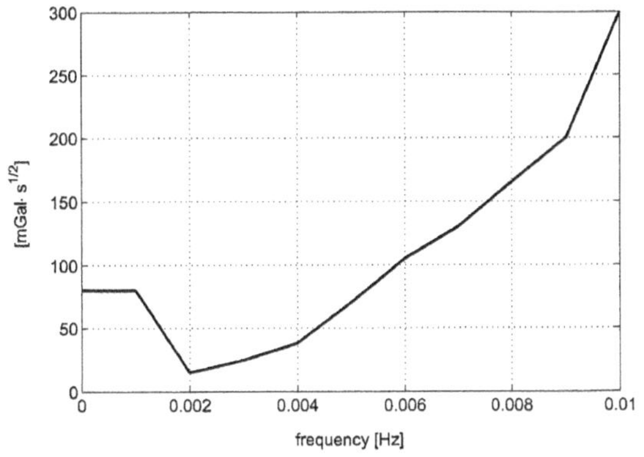


**Fig. 1.** Amplitude spectrum of vertical accelerations (taken from Childers et al., 1999).

average) model of the colored noise:

$$\xi_n = \varepsilon_n - \sum_{k=1}^{p} a_{p,k}\xi_{n-k} + \sum_{i=1}^{q} b_{q,i}\varepsilon_{n-i}, \quad n \in \mathcal{Z}, \tag{5}$$

where $\xi_n$ is the colored noise process, $\varepsilon_n$ is a white noise process with zero mean and variance $\sigma_\varepsilon^2$, and $\mathcal{Z}$ denotes the set of integer values (for more details see e.g. Brockwell and Davis (1991)). The coefficients $a_{p,k} : k = 1, \ldots, p$ and $b_{q,i} : i = 1, \ldots, q$ are the model parameters and the pair $(p, q)$ defines the order of the ARMA model. The ARMA model can be computed from a power spectral density function (PSD) or a noise realization, if this information is available. Figure 1 shows an example of a PSD of vertical aircraft accelerations, averaged for 33 flights (Childers et al., 1999). Because most noise of the gravimeter scales with the amplitude of the aircraft motion, the PSD in figure 1 may be used to obtain an ARMA model.

In practice, however, the noise characteristics are different for each flight due to the sensor used and the flight conditions. Then, a different approach must be used to obtain a noise model. For instance, one may use the residuals of a preliminary least-squares solution to estimate a noise model. This model could then be used as input to the frequency-dependent weighting scheme.

## 4 Handling of bias and tilt

The traditional processing of gravity data includes a cross-over adjustment to deal with remaining low-frequency errors. This is done by minimizing gravity differences at cross-over locations between crossing flight lines. See Glennie and Schwarz (1997) for an example of cross-over adjustment applied to SINS gravity measurements. This method of adjustment of cross-over misfits requires a close coincidence of measurement points at crossing lines, which means that observations must be acquired at about the same altitude.

We propose to handle bias and tilt together with the estimation of the gravity field parameters. The bias and tilt parameters can be included in the functional model, Eq. (3), and estimated jointly with the potential coefficients. Alternatively, they can be dealt with using frequency-dependent data weighting, by setting up a covariance matrix that filters out the low-frequency noise. It can easily be shown that both methods, estimation and filtering, lead to exactly the same results.

When the bias and tilt parameters are estimated explicitly, the functional model changes to

$$\mathbf{y} = \mathbf{Ax} + \mathbf{Bq} + \mathbf{e}, \tag{6}$$

where the vector $\mathbf{q} = (a \ b)^T$ contains the tilt and bias parameters $a$ and $b$, respectively, and

$$\mathbf{B} = \begin{pmatrix} t_1 \ t_2 \ \ldots \ t_n \\ 1 \ \ 1 \ \ldots \ 1 \end{pmatrix}^T .$$

In this case only one bias and tilt parameter are estimated, but the model can easily be extended to estimate one bias and tilt parameter per flight line.

Alternatively we can construct a filter matrix that removes a bias and tilt from the data. Consider a linear filter matrix $\mathbf{F}$ with the property $\mathbf{FB} = 0$. Then, applying this matrix to Eq. (6) gives

$$\mathbf{Fy} = \mathbf{FAx} + \mathbf{FBq} + \mathbf{Fe} = \mathbf{FAx} + \mathbf{Fe}.$$

Let $\mathbf{F}_1 = \mathbf{B}(\mathbf{B}^T\mathbf{Q_y}^{-1}\mathbf{B})^{-1}\mathbf{B}^T\mathbf{Q_y}^{-1}$ be the projection matrix onto the column space of $\mathbf{B}$. Then $\mathbf{F} := \mathbf{I} - \mathbf{F}_1$ satisfies $\mathbf{FB} = 0$. It can be shown that, using the generalized inverse of $\mathbf{FQ_yF}^T$ and observing that $\mathbf{F}^2 = \mathbf{F}$, the least squares solution $\hat{\mathbf{x}}$ is obtained as

$$\hat{\mathbf{x}} = (\mathbf{A}^T\tilde{\mathbf{Q}}_\mathbf{y}^{-1}\mathbf{A})^{-1}\mathbf{A}^T\tilde{\mathbf{Q}}_\mathbf{y}^{-1}\mathbf{y}, \tag{7}$$

with

$$\tilde{\mathbf{Q}}_\mathbf{y}^{-1} = \mathbf{F}^T\mathbf{Q_y}^{-1}\mathbf{F} = \mathbf{Q_y}^{-1}\mathbf{F}.$$

Again it is not necessary to compute the new covariance matrix explicitly as it can be considered as a series of matrix multiplications which are solved in the same manner as discussed in section 3.

For the most simple case, when the data contain only a bias and are contaminated by white noise, we have $\mathbf{B} = (1\ 1\ \ldots\ 1)^T$ and the filter matrix simplifies to

$$\mathbf{F} = \begin{pmatrix} 1-1/n & -1/n & \ldots & -1/n \\ \vdots & \vdots & & \vdots \\ -1/n & -1/n & \ldots & 1-1/n \end{pmatrix}.$$

For subtracting a bias (and tilt) from each flight line, the derivation of the filter matrix is the same as above, resulting in a block-diagonal matrix $\mathbf{F}$, with the number of blocks equal to the number of profiles.

## 5 Regularization

The inversion of airborne gravity data into gravity functionals at ground level, is an ill-posed problem. The LS solution strongly oscillates, because data errors and unmodeled signal are strongly amplified due to the ill-conditioned normal matrix. This problem can be solved using a regularization method. One of the best-investigated methods is Tikhonov regularization, which is often used in geodetic inverse problems. The regularized solution can be obtained as

$$\hat{\mathbf{x}} = (\mathbf{A}^T\tilde{\mathbf{Q}}_{\mathbf{y}}^{-1}\mathbf{A} + \alpha\mathbf{R})^{-1}\mathbf{A}^T\tilde{\mathbf{Q}}_{\mathbf{y}}^{-1}\mathbf{y}, \qquad (8)$$

with the regularization parameter $\alpha$ and a symmetric positive definite regularization matrix $\mathbf{R}$. In many inverse problems the regularization matrix is set equal to unity. In our case this is equivalent to minimizing the $L_2$-norm of the disturbing potential at $z = 0$ and is called zero-order Tikhonov (ZOT) regularization. Alternatively, we can derive the expression for first-order Tikhonov (FOT) regularization, analogously to the derivation in (Ditmar et al., 2003). Instead of the gravity potential itself, we minimize the horizontal gradient of the gravity potential, i.e. the functional to be minimized is

$$\Phi^{\mathrm{FOT}} = \iint_D (\nabla_H V(x,y,z))^2 \mathrm{d}x\,\mathrm{d}y, \qquad (9)$$

where $\nabla_H$ is the surface gradient operator

$$\nabla_H = \begin{pmatrix} \frac{\partial}{\partial x} \\ \frac{\partial}{\partial y} \\ 0 \end{pmatrix}.$$

Inserting the expression for the gravity potential, Eq. (2), into the regularization condition of Eq. (9), and using the orthogonality properties of the sine and cosine functions, we obtain

$$\Phi^{\mathrm{FOT}} = \frac{D_x D_y}{4}\mathbf{x}^T\mathbf{R}^{\mathrm{FOT}}\mathbf{x}, \qquad (10)$$

where $D_x$ and $D_y$ denote the area size (see Eq.(2)). The elements of the regularization matrix are given as

$$\mathbf{R}_{ij}^{\mathrm{FOT}} = \begin{cases} 2\,\delta_{ij}(\alpha_{l_i}^2 + \beta_{m_i}^2)\,e^{-2\gamma_{l_i m_i} z}, & \text{if } l_i = 0 \\ & \text{or } m_i = 0 \\ \delta_{ij}(\alpha_{l_i}^2 + \beta_{m_i}^2)\,e^{-2\gamma_{l_i m_i} z}, & \text{otherwise.} \end{cases} \qquad (11)$$

Another important aspect of regularization is the selection of the regularization parameter $\alpha$. For the numerical experiments discussed in section 6, the regularization parameter is determined empirically. In the future we plan to implement a parameter choice rule, such as the L-curve method (Hansen and O'Leary, 1993), variance component estimation (Koch, 1990) or generalized cross-validation (Golub et al., 1979).

## 6 Computational experiment

The performance of the developed technique is assessed using simulated data from the global geopotential model GPM98b (Wenzel, 1998), complete up to degree and order 1800. Gravity disturbances $V_z$ have been computed at a flight level of $h = 4$ km, with the long-wavelength signal (up to degree 360) removed from the data. The area for which the gravity disturbances have been computed is situated in the Rocky Mountains with the coordinate centre $(\varphi_c, \lambda_c)$

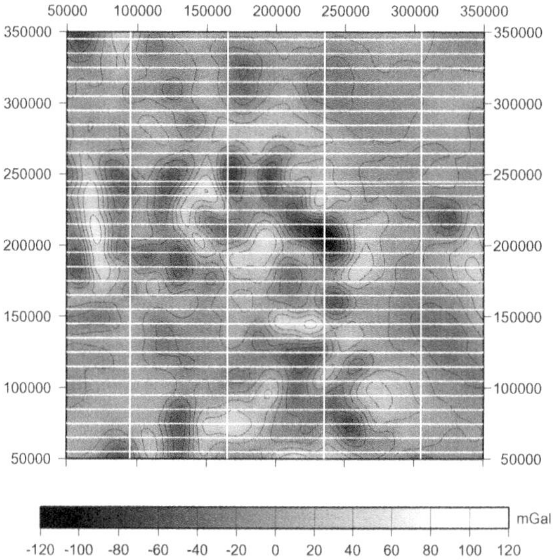


**Fig. 2.** Gravity disturbances at h=4 km and the simulated flight pattern (white lines); units at the axes in m.

at a Northern latitude of 45° and an Eastern longitude of 250°. The size of the area is $D_x = D_y = 300$ km. The data have been computed at 30 East-West oriented flight lines, with a distance of 10 km between the lines and at 4 North-South oriented flight lines at distances of 70 km apart. The spacing of the points along track was 500 m, resulting in 20434 observations. Figure 2 shows the simulated gravity signal at flight level together with the simulated flight pattern. At the stage of data processing, the area was extended by 50 km at all sides and filled with zeroes to reduce edge effects. Potential coefficients have been estimated complete to degree $l_{max} = 40$ and order $m_{max} = 40$, corresponding to a half-wavelength resolution of 5 km. The total number of potential coefficients was 6560.

Several tests were done to assess the accuracy of the method. For the first test a random bias of 4.0 mGal was added to the flight lines. Figure 3 shows for each flight line the bias added and the bias estimated, where the lines 31–34 denote the N-S profiles. The figure shows that the biases are estimated very well, except for the flight lines that are situated near the Northern (lines 1–3) and Southern (lines 28–30) boundaries of the area, which is caused by edge effects. The maximum difference is 0.5 mGal, whereas the RMS difference is only 0.2 mGal.

The influence of noise on the solution was investigated for both white and colored noise. The observations were first corrupted by random white noise of $\sigma = 2$ mGal. The solution was computed using first-order regularization, with the regularization parameter derived empirically, resulting in a RMS geoid height error of 1.4 cm for the inner area of 200 × 200 km (the maximum error is 4.4 cm).

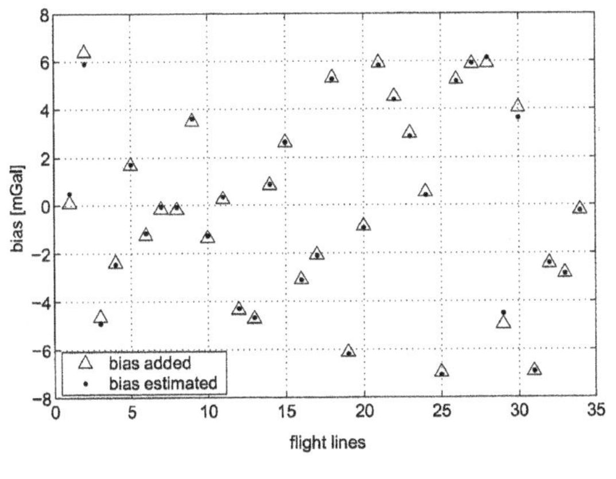


**Fig. 3.** Results of bias estimation where a random bias of $\sigma = 4$ mGal was added to the flight lines. The lines 31–34 denote the N-S flight lines.

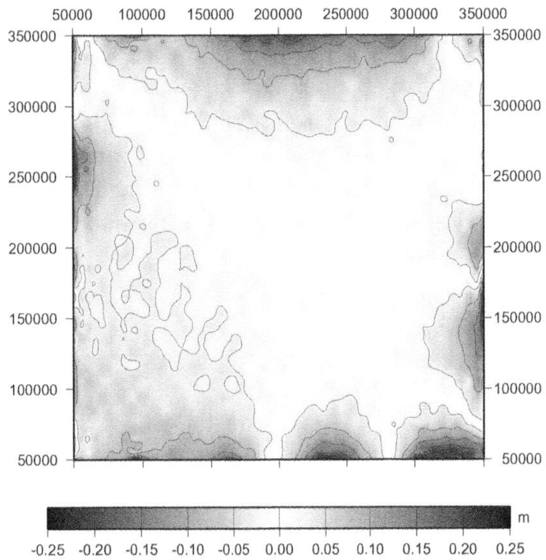


**Fig. 4.** Geoid height errors for the inversion of gravity disturbances corrupted by random white noise of $\sigma = 2$ mGal.

The errors are shown in figure 4. Although edge effects were reduced, there are still large errors up to 25 cm close to the boundary of the area. For the test of frequency-dependent data weighting the PSD of figure 1 was used to compute a noise realization ($\sigma = 17.2$ mGal), which was added to the observations. The same PSD was used to fit an autoregressive model of the order 100 using the method proposed in (Klees and Broersen, 2002). Again, FOT regularization was applied and the regularization parameter was derived empirically.

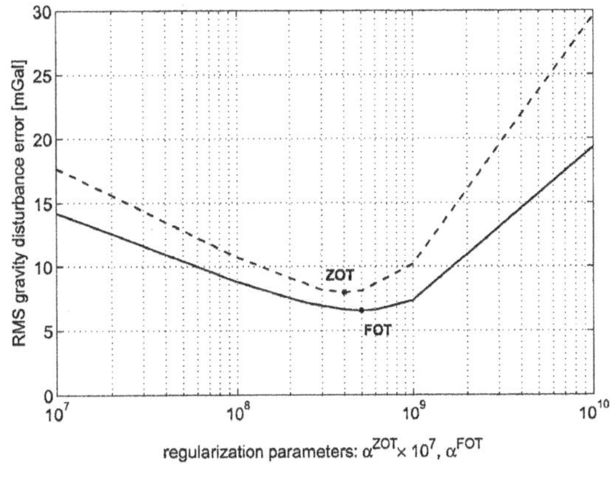


**Fig. 5.** The RMS error of gravity disturbances at ground level as function of the regularization parameter for ZOT (dashed line) and FOT (solid line). Gravity disturbances at $h = 4$ km were corrupted by colored noise.

According to figure 5, FOT regularization provides smaller RMS gravity disturbance errors than ZOT regularization. Figure 6 shows the resulting geoid height errors when FOT regularization is applied. The RMS error in the inner 200 × 200 km is 3.2 cm and the maximum error is 9.6 cm. The errors are larger than in the presence of white noise due to a larger noise level.

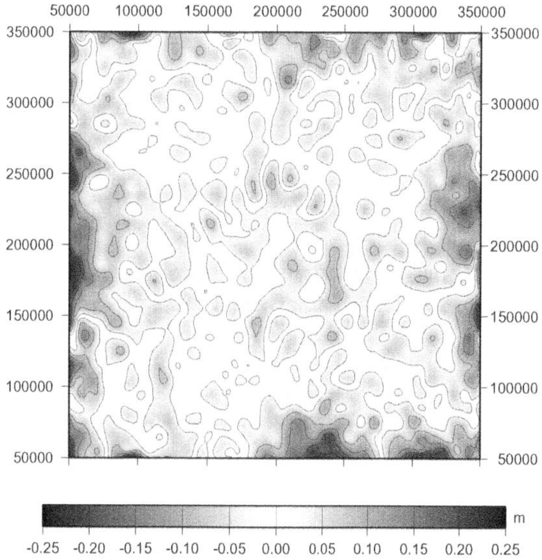


**Fig. 6.** Geoid height errors for the inversion of gravity disturbances, corrupted by colored noise. Frequency-dependent data weighting is applied.

## 7 Summary

We presented a new strategy for processing airborne gravity data, which combines separate processing steps, such as filtering, cross-over adjustment and downward continuation into one algorithm. The performance of the method was investigated for the downward continuation of gravity disturbances, computed from a global geopotential model. We have shown that the approach can efficiently invert airborne gravity measurements into gravity functionals on the Earth's surface. Although the developed technique provides accurate results, several aspects need further research.

First of all, in this paper we determined the regularization parameter empirically, which is not possible when processing real data. Therefore, we plan to implement a parameter choice rule to determine the regularization parameter from the data only. Furthermore, a noise model to be used in the frequency-dependent data weighting will usually not be available, and a method to estimate the noise in the data must be developed. Another issue that must be addressed is the handling of edge effects. Although they can be reduced by extending the area with zero observations or using a data tapering, the errors are very large near the boundaries of the area.

## References

Alberts B, Klees R (2004) *A comparison of methods for the inversion of airborne gravity data.* J Geodesy 78: 55–65

Brockwell PJ, Davis RA (1991) *Time series: theory and methods.* Second edition. Springer Series in Statistics, New York

Childers VA, Bell RE, Brozena JM (1999) *Airborne gravimetry: An investigation of filtering.* Geophysics 64(1): 61–69

Ditmar P, Klees R (2002) *A method to compute the Earth's gravity field from SGG/SST data to be acquired by the GOCE satellite.* Delft University Press (DUP) Science, Delft

Ditmar P, Kusche J, Klees R (2003) *Computation of spherical harmonic coefficients from gravity gradiometry to be acquired by the GOCE satellite: regularization issues.* J Geodesy 77: 465–477

Glennie C, Schwarz KP (1997) *Airborne Gravity by Strapdown INS/DGPS in a 100 km by 100 km Area of the Rocky Mountains.* Proc Int Symp Kinematic Systems in Geodesy, Geomatics and Navigation (KIS97), June 3–6, Banff, Canada, pp 619–624

Golub GH, Heath M, Wahba G (1979) *Generalized cross-validation as a method for choosing a good ridge parameter.* Technometrics 21: 215–223

Hansen P, O'Leary DP (1993) *The use of the L-curve in the regularization of discrete ill-posed problems.* SIAM J Sci Comput 14: 1487–1503

Klees R, Broersen P (2002) *How to handle colored noise in large least-squares problems - Building the optimal filter.* Delft University Press (DUP) Science, Delft

Klees R, Ditmar P (2004) *How to handle colored noise in large least-squares problems in the presence of data gaps.* Sansó F (ed.), Proc Vth Hotine-Marussi Symposium on Mathematical Geodesy, June 17–21 2002, Matera, Italy, pp 39–48

Klees R, Ditmar P, Broersen P (2003) *How to handle colored observation noise in large least-squares problems.* J Geodesy 76: 629–640

Koch KR (1990) *Bayesian inference with geodetic applications.* Springer Berlin, Heidelberg, New York

Wenzel G (1998) *Ultra high degree geopotential models GPM98A, B and C to degree 1800.* Proc Joint Meeting International Gravity Commission and International Geoid Commission, 7–12 September 1998, Trieste, Italy

# Multiresolution representation of a regional geoid from satellite and terrestrial gravity data

M. Schmidt(1), J. Kusche(2), J.P. van Loon(2), C.K. Shum(3), S.-C. Han(3), O. Fabert(4)

(1) Deutsches Geodätisches Forschungsinstitut (DGFI), Marstallplatz 8, 80539 Munich, Germany
(2) Delft Institute of Earth Observation and Space Systems (DEOS), TU Delft
(3) Laboratory for Space Geodesy and Remote Sensing, OSU Columbus
(4) Differential Geometry and Topology Group, LMU Munich

**Abstract.** In this paper we present results from modeling the Earth's gravitational field over the northern part of South-America using spherical wavelets. We have applied our analysis to potential data that we derived from CHAMP using the energy balance method, and to terrestrial gravity anomalies. Our approach provides a regional correction to the EGM96 reference gravity field, expressed in various detail levels, which are partly determined by the satellite data and partly by the terrestrial data.

**Keywords.** Multiresolution representation, gravity data, CHAMP, terrestrial gravity.

## 1 Introduction

Current gravity models of the Earth, such as EGM96 or the new CHAMP- and GRACE-derived models, are based on a series expansion of the harmonic potential in terms of spherical harmonics. This approach is well-established, but it meets its limitations if one is interested in very high resolution, in the combination of data sets of spatially heterogeneous distribution and quality, or in the treatment of geographically restricted features (aliasing along continent boundaries). It is therefore, that we are interested in alternative representations of the gravity field. Spherical wavelets have been considered for long as a candidate representation due to their flexible filtering and localization characteristics and the existence of fast pyramidal algorithms for wavelet decomposition and reconstruction (Freeden et al. 1998, Schmidt et al. 2002), but until now they have rarely been applied to real gravity data (Fengler et al. 2004).

Here we present results from modeling the Earth's gravitational field over the northern part of South-America using spherical wavelets of Blackman type. We have applied our analysis to potential data that we derived from CHAMP kinematic orbits and accelerometry data using the energy balance method covering an area slightly larger than our recovery region, and to gravity anomalies derived from terrestrial, airborne, and altimetric data. Our approach provides a regional correction to the EGM96 reference gravity field, expressed in various detail levels, which are partly determined by the satellite data and partly by the terrestrial data.

## 2 Multiresolution Representation

The potential $V(\mathbf{r})$ of the Earth's gravitational field ($\mathbf{r}$ being the geocentric position vector) fulfills the *Laplacian differential equation* $\Delta V = 0$ outside the generating masses. On a reference sphere, $V$ is typically expressed by the *spherical harmonic expansion*

$$V(\mathbf{r}) = \sum_{n=0}^{\infty} \sum_{m=-n}^{n} V_{nm} Y_{nm}(\mathbf{r}) \qquad (1)$$

with $Y_{nm}$ being the spherical harmonics of degree $n$ and order $m$. The $Y_{nm}$ share the useful property that subsequent detail spaces generated by spectral bands $n = n_1, \ldots, n_2$ are mutually orthogonal. An alternative concept, which goes back on Freeden et al. (1998), is to exploit a scale discrete *wavelet multiresolution*, i.e.

$$V(\mathbf{r}) = (\phi_{i'} \star V)(\mathbf{r}) + \sum_{i=i'}^{I} (\psi_i^{(2)} \star V)(\mathbf{r}) \,. \qquad (2)$$

In this theory, $\phi_i(\mathbf{r}, \cdot)$ and $\psi_i(\mathbf{r}, \cdot)$ are rotationally symmetric harmonic *scaling* and *wavelet function* of resolution level (scale) $i \in \{i', \ldots, I\}$, the symbol '$\star$' denotes convolution on the sphere, and $\psi^{(2)} \star V$ means $\psi \star (\widetilde{\psi} \star V)$ where $\widetilde{\psi}$ is the dual wavelet function. Whereas the first term, acting as a *low-pass filter*, represents the coarse spatial scales, the sum in (2) accumulates subsequent detail spaces each related to a certain frequency band, i.e. the spherical convolution $\psi_i^{(2)} \star V$ provides the band-pass filtered *detail signal*

$$v_i(\mathbf{r}) := (\psi_i^{(2)} \star V)(\mathbf{r}) \tag{3}$$

of level $i$. The two representations (1) and (2) can be combined (which is not what we do in this study) by truncating (1) at a degree $N$, and adding the sum in (2) over the detail spaces only, see Freeden et al. (1998) or Schmidt et al. (2004). In contrast, because we will compute a correction to a reference gravity model based on new regional data sets, we use (2) exclusively. The frequency behaviour of this representation is controlled by the choice of the scaling kernel. The wavelet kernels are then derived from the idea that they should represent differences of subsequent levels of low-pass filters. A rotationally symmetric scaling function $\phi_i$ of resolution level $i$ can be defined as a *Legendre series*, i.e.

$$\phi_i = \sum_{n=0}^{\infty} \frac{2n+1}{4\pi} \, \Phi_i(n) \, P_n \, , \tag{4}$$

wherein $\Phi_i(n)$ are the Legendre coefficients and $P_n$ the Legendre polynomials of degree $n$. The Legendre coefficients $\Psi_i(n)$ of the wavelet function

$$\psi_i = \sum_{n=0}^{\infty} \frac{2n+1}{4\pi} \, \Psi_i(n) \, P_n \, , \tag{5}$$

of level $i$ follow from the *two-scale relation*, either defined as

$$\Psi_i(n) \, \widetilde{\Psi}_i(n) = \Phi_{i+1}(n) - \Phi_i(n) \tag{6}$$

or

$$\Psi_i(n) = \sqrt{\Phi_{i+1}(n) - \Phi_i(n)} \tag{7}$$

if $\widetilde{\psi}_i = \psi_i$ holds. In this study we use the *Blackman scaling function*, defined by the Legendre coefficients

$$\Phi_i(n) = \begin{cases} 1 & n = 0, \ldots, 2^{i-1}-1 \\ (A_i(n))^2 & n = 2^{i-1}, \ldots, 2^i - 1 \\ 0 & n = 2^i, \ldots, \infty \end{cases} \tag{8}$$

with

$$A_i(n) = 0.42 - 0.50 \cos\left(\frac{2\pi n}{2^i}\right) + 0.08 \cos\left(\frac{4\pi n}{2^i}\right) ,$$

and the *Blackman wavelet function* computed by means of Eq. (7) and shown in Fig. 1 for different levels. As can be seen from Fig. 1c this wavelet

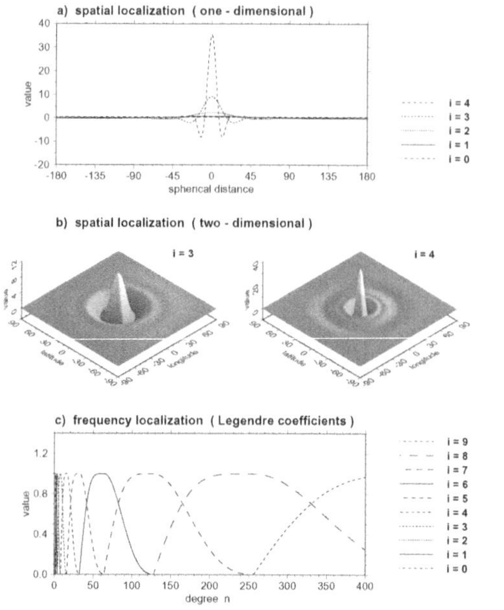


**Fig. 1**: Blackman wavelet of different resolution levels $i$ in the spatial and in the frequency domain.

function is *strictly band-limited*, i.e. only the Legendre coefficients $\Psi_i(n)$ within the frequency band

$$B_i := \{n \,|\, 2^{i-1} \leq n < 2^{i+1}\} \tag{9}$$

are not equal to zero. Hence, the spatial resolution for the evaluation of the spherical convolution in Eq. (3) depends on the corresponding frequency band $B_i$. In additon the Blackman wavelet shows less oscillations in the spatial domain than e.g. the Shannon wavelet. This is an important asset in regional applications, because the convolutions are then restricted to the domain where we have data.

## 3 Analysis of CHAMP Data and Terrestrial Data for South-America

For the preparation of a regional satellite geopotential data set, we have analyzed kinematic CHAMP orbits that were kindly provided by D. Svehla and M. Rothacher, TU Munich. These orbits cover two years (days 70/2002–70/2004), and were processed following the zero-differencing strategy, see Svehla and Rothacher (2003). Kinematic positions with an a-priori sigma exceeding a threshold were removed. We have then converted these orbits via the *energy balance approach* into a time series of residual geopotential values, using GFZ's ISDC accelerometry and attitude data products in our analysis, which essentially follows Jekeli (1999) or Han et al. (2002). EGM96 complete to degree $n$ = 120 was used as a *reference*. For details see van Loon and Kusche (2005). We believe the accuracy of the resulting data is at the 0.8 $m^2/s^2$ level. For the region under consideration, orbital arcs were then selected, and a degree-1 polynome was fitted through the data per arc to estimate and apply a correction to the a-priori energy constant and a linear drift caused by unmodeled accelerometer bias. The residual geopotential was then approximately continued down- or upward to a mean orbital sphere using EGM96. The maximum distance between true orbit and mean sphere did not exceed 30 km. These values provide the input for the subsequent multiresolution analysis and downward continuation.

The satellite data were complemented by a high-resolution ($2' \times 2'$) data set containing Faye anomalies, derived from terrestrial, airborne and altimetric observations, and kindly provided to us by L. Sánchez from the Instituto Geográfico Augustin Codazzi (IGAC), Bogotá. Again, EGM96 complete to degree 120 was used as a reference field.

In summary, the following two figures show the data we used as input for the multiresolution analysis of the regional geoid, namely the *residual geopotential* from CHAMP (Fig. 2) and the *residual gravity anomalies* computed from terrestrial data (Fig. 3).

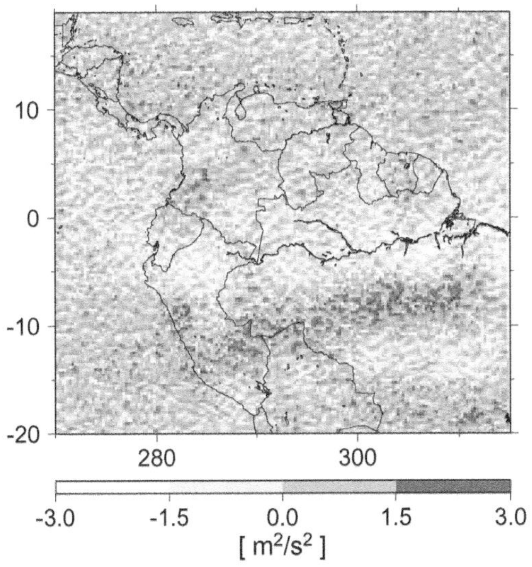


**Fig. 2**: Satellite input data: Residual geopotential at mean orbital altitude over northern part of South-America, computed from CHAMP; EGM96 up to degree $n = 120$ is subtracted; data in [$m^2/s^2$]

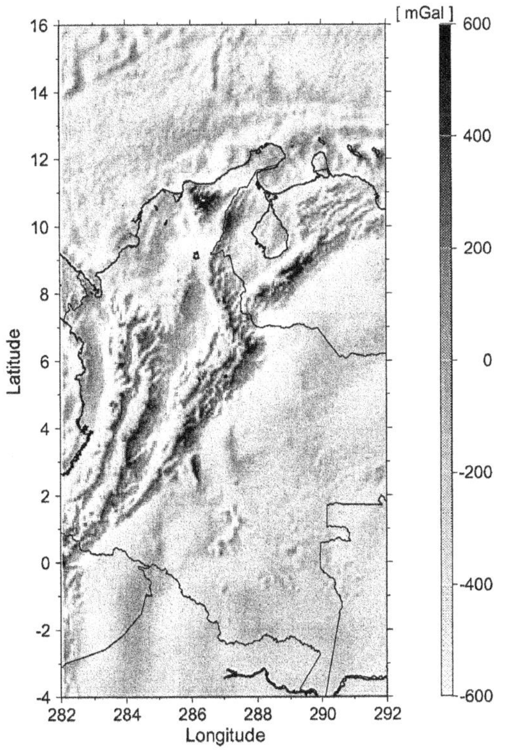


**Fig. 3**: Terrestrial input data: Residual gravity (Faye) anomalies of Colombia; EGM96 up to degree $n = 120$ is subtracted; data in [mGal]

## 4 Results

Our strategy aims at constructing a multiresolution representation of the regional geoid by exploiting the properties of a reference spherical harmonic model, satellite data, and terrestrial data in an optimal way. To be more specific, our objective is to approximate *high-resolution geoid undulations N* by the decomposition

$$
\begin{aligned}
N_I(\mathbf{r}) &= N^{EGM}_{2,\ldots,120}(\mathbf{r}) + \underbrace{\sum_{i=i'}^{i''} n_i(\mathbf{r})} + \underbrace{\sum_{i=i''+1}^{I} n_i(\mathbf{r})} \\
&= N^{EGM}_{2,\ldots,120}(\mathbf{r}) + \delta N^{wav}_{i',\ldots,i''}(\mathbf{r}) + \delta N^{wav}_{i''+1,\ldots,I}(\mathbf{r}) \,, \qquad (10)
\end{aligned}
$$

wherein $N^{EGM}_{2,\ldots,120}$ means the geoid undulations from the reference model EGM 96 up to degree $n = 120$. The detail signals $n_i$ for levels $i = i', \ldots, i''$ are computed from satellite data, the remaining detail signals for levels $i = i''+1, \ldots, I$ are calculated from the terrestrial data. Note, that in this study we do not deal with an overlapping part, i.e.the computation of certain detail signals from both satellite and terrestrial data is not considered. With respect to the spectral domain the overlap between satellite and terrestrial data is fixed by the Legendre coefficients of the Blackman wavelet; see Fig. 1c. Recall, that the model parts $\delta N^{wav}_{i',\ldots,i''}$ and $\delta N^{wav}_{i''+1,\ldots,I}$ in Eq. (10) are *corrections* to the reference model $N^{EGM}_{2,\dots,120}$, which has in fact no physical meaning anymore.

It is instructive to look first at the regional geoid construction from satellite data. In our present wavelet analysis, the detail spaces from level $i = i'$ up to level $i = i''$ are derived from the CHAMP residual geopotential data shown in Fig. 2. The following two figures display the wavelet representations $\delta N^{wav}_{2,\dots,5}$ and $\delta N^{wav}_{2,\dots,6}$, i.e. we chose $i' = 2$ and $i'' = 5$ in Fig. 4 as well as $i' = 2$ and $i'' = 6$ in Fig. 5, according to Eq. (10).

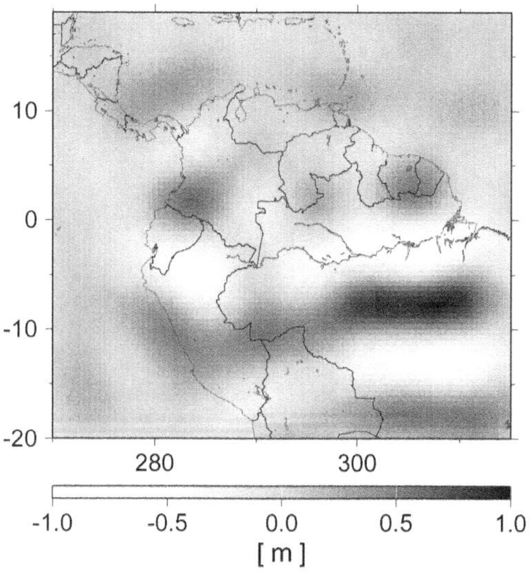


**Fig. 4**: Residual geoid wavelet model $\delta N^{wav}_{2,\dots,5}$ according to Eq. (10), computed from CHAMP; data in [m]

Some spurious patches appear in the level-6 representation $\delta N^{wav}_{2,\dots,6}$ which we attribute to the fact that there is no regularization present in our detail solutions. This is a problem we will consider in future work. The satellite-only construction, after restoring the reference model, i.e.

$$N^{sat}_{i''}(\mathbf{r}) := N^{EGM}_{2,\dots,120}(\mathbf{r}) + \delta N^{wav}_{2,\dots,i''}(\mathbf{r}) \,, \tag{11}$$

can be compared to recent GRACE-derived global gravity models. Fig. 6 shows the corresponding residual geoid undulations

$$\delta N^{GRACE}_{2,\dots,120}(\mathbf{r}) := N^{GRACE}_{2,\dots,120}(\mathbf{r}) - N^{EGM}_{2,\dots,120}(\mathbf{r}) \tag{12}$$

derived from GFZ's EIGEN-GRACE02S model. To be more specific, for an oceanic area ($\lambda = 270° \dots 280°, \phi = -20° \dots 10°$) we find as rms differences of $\delta N^{wav}_{2,\dots,i''}$ with respect to $\delta N^{GRACE}_{2,\dots,120}$ the values 0.17 m for level $i'' = 5$ and 0.68 m for level $i'' = 6$, respectively. For a continental area ($\lambda = 280° \dots 300°$,

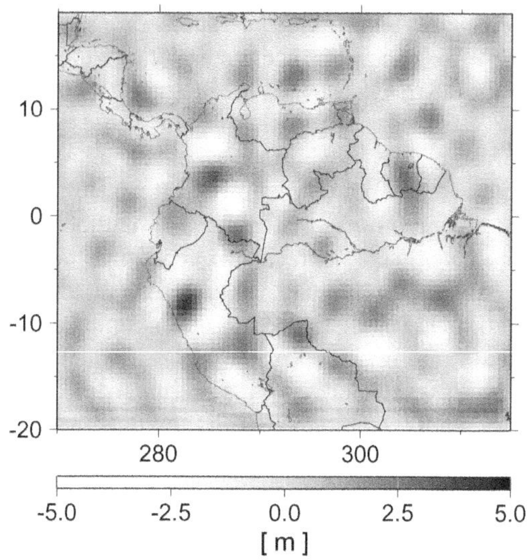


**Fig. 5**: Residual geoid wavelet model $\delta N^{wav}_{2,\dots,6}$ according to Eq. (10), computed from CHAMP; data in [m]

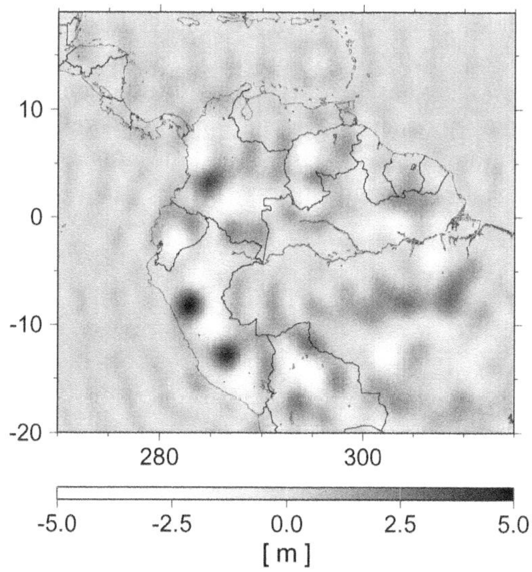


**Fig. 6**: Residual geoid GRACE model $\delta N^{GRACE}_{2,\dots,120}$ according to Eq. (12), up to degree $n = 120$; data in [m]

$\phi = -15° \dots 10°$) the rms becomes 0.99 m (level 5), 0.92 m (level 6). If we consider the detail signal of level 6 only for the continental area, which is perfectly reasonable in a wavelet multiresolution analysis, the rms drops to 0.65 m for the overall region shown in the figures. Obviously there is a good agreement between this *multi-level wavelet model*

$$\delta N^{wav}_{2,\dots,5/6}(\mathbf{r}) := \sum_{i=2}^{5} n_i(\mathbf{r}) + \underbrace{n_6(\mathbf{r})}_{\text{over land}} \,, \tag{13}$$

shown in Fig. 7, and the corresponding residual

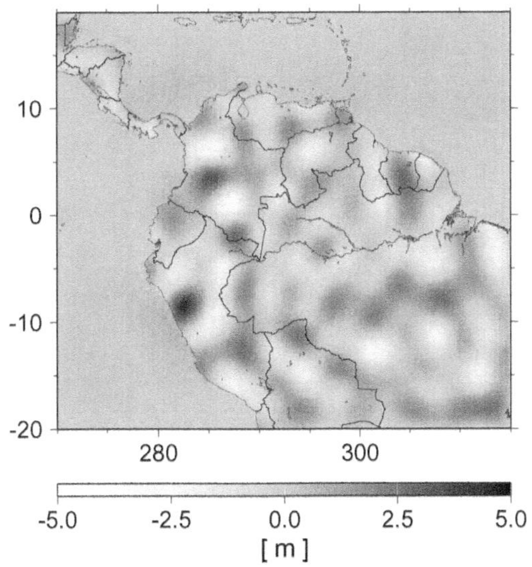


**Fig. 7**: Residual geoid wavelet model $\delta N^{wav}_{2,\ldots,5/6}$ according to Eq. (13), up to level $i'' = 5$ over the oceans and up to level $i'' = 6$ over land, computed from CHAMP; data in [m]

geoid GRACE model $\delta N^{GRACE}_{2,\ldots,120}$, defined in Eq. (12) and displayed in Fig. 6 (correlation $\approx 0.65$).

For a comparison, GFZ's EIGEN-CHAMP03S and EIGEN-GRACE02S differ by about 0.31 m (ocean region), 0.33 m (continental region) at degree 75 and 0.87 m (ocean region), 0.94 m (continental region) at degree 120. Hence, we can conclude that within the chosen region of South-America our regional multi-level wavelet model

$$N^{sat}_{5/6}(\mathbf{r}) := N^{EGM}_{2,\ldots,120}(\mathbf{r}) + \delta N^{wav}_{2,\ldots,5/6}(\mathbf{r}) \qquad (14)$$

of the geoid, computed from CHAMP data, fits better to the global EIGEN-GRACE02S model than the global EIGEN-CHAMP03S model does for the same region (all models up to degree $n = 120$). This is partly due to the damping of the Blackman Legendre coefficients defined in Eq. (8) and shown in Fig. 1c, in connection with the fact that CHAMP data does not contain sufficient information to determine spherical harmonics at this degree without strong regularization.

Any higher-level reconstruction from the satellite data exhibits almost only noise. We have therefore decided to extend the geoid multiresolution model from level 7 onwards on the basis of the residual gravity (Faye) anomalies shown in Fig. 3. The following figures show the representation $N^{sat}_{5/6}$ (Fig. 8) and according to Eq. (10) the combined satellite-terrestrial representations

$$N_I(\mathbf{r}) = N^{sat}_{5/6}(\mathbf{r}) + \sum_{i=7}^{I} n_i(\mathbf{r}) \qquad (15)$$

for $I = 8$ (Fig. 9) and $I = 11$ (Fig. 10). Since Faye anomalies are related to the quasigeoid, an additional correction has to applied to the corresponding detail signals $n_i$ (Torge 2001). Note, that the Blackman scaling coefficients drop to 0.5 at about harmonic degree 330 (level 8) and degree 2650 (level 11). The desired high-resolution gravity model of Colombia shown in Fig. 10 contains, however, frequencies up to degree 4095.

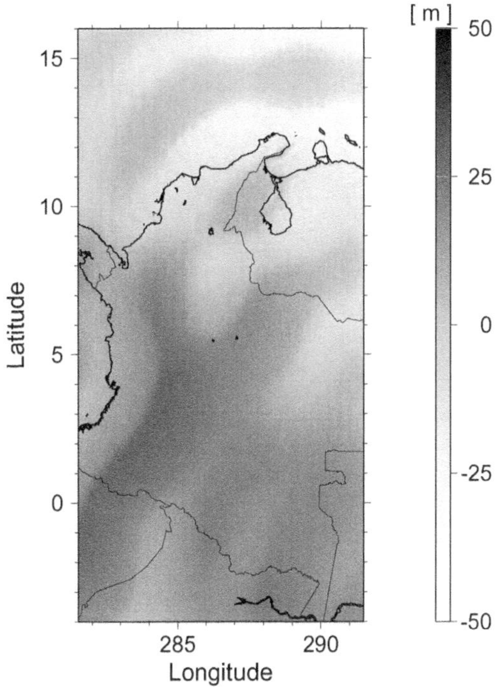


**Fig. 8**: Colombian geoid in detail space construction from CHAMP data up to level $i'' = 5$ over the oceans and up to level $i'' = 6$ over the continents ($= N^{sat}_{5/6}$, see Eq. (14)); data in [m]

## 5 Conclusions

We have demonstrated that multiresolution techniques can be successfully applied to real satellite and terrestrial gravity data sets. A regional geoid model of the northern part of South America has been derived by combination of CHAMP-derived data covering two years, and a high-resolution gravity anomaly data set. It is based on a spherical wavelet multiresolution, using the Blackman scaling function. Corrections to the reference spherical harmonic model (EGM96 complete to degree 120) are at the at the 1 m niveau for level 5 and 3 m to 5 m for level 6. Whereas these levels are from CHAMP only, all higher detail levels have been derived from the gravity anomaly data set only.

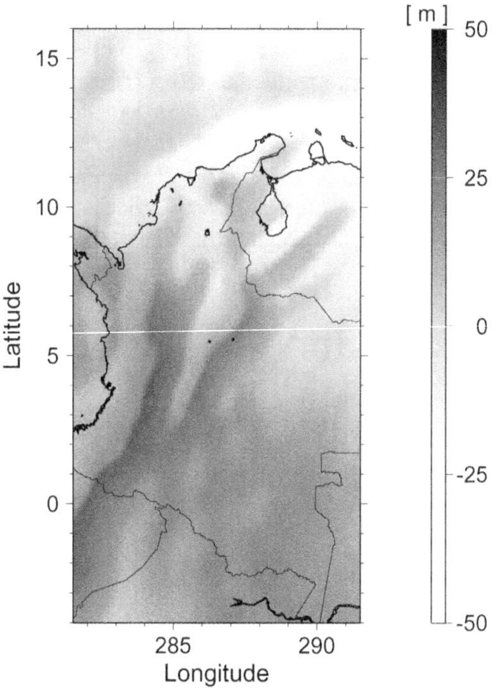


**Fig. 9**: Colombian geoid in detail space construction from CHAMP up to level $i'' = 5$ over the oceans and up to level $i'' = 6$ over the continents and from terrestrial data for levels 7 and 8 ($= N_8$, see Eq. (15) with $I = 8$); data in [m]

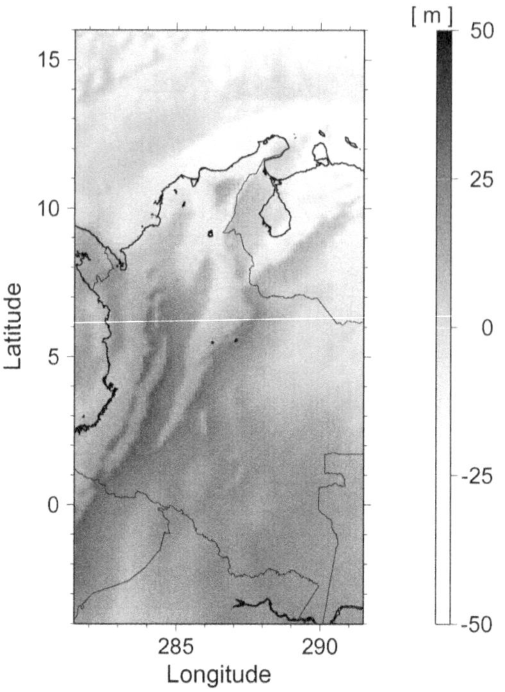


**Fig. 10**: Colombian high-resolution geoid in detail space constructions from CHAMP up to level $i'' = 5$ over the oceans and up to level $i'' = 6$ over the continents and from terrestrial data for levels 7 to 11 ($= N_{11}$, see Eq. (15) with $I = 11$); data in [m]

Future research will in particular consider regularization in the process of satellite gravity multiresolution and downward continuation, statistical approaches that estimate wavelet coefficients by exploiting error statistics of the input data, and the optimal combination of terrestrial and satellite data sets (including GRACE). Furthermore, we will consider validation using GPS leveling data sets.

## Acknowledgements


We are grateful to GFZ Potsdam for providing CHAMP data through their ISDC. Thanks go also to Drazen Svehla (TUM) for providing CHAMP kinematic orbits, and to Laura Sánchez (IGAC) for contribution gravity anomaly data. JPvL acknowledges funding by the dutch GO-2 program (SRON EO-03/057).


## References


Fengler M, Freeden W, Kusche J (2004) *Multiscale geopotential solutions from CHAMP orbits and accelerometry*. Reigber et al. (eds.) Earth Observation with CHAMP – Results from three years in Orbit, Springer

Freeden W, Gervens T, Schreiner M (1998) *Constructive Approximation on the Sphere (with Applications to Geomathematics)*. Clarendon Press, Oxford

Han S–C, Jekeli C, Shum CK (2002) *Efficient gravity field recovery using in situ disturbing potential observables from CHAMP*. GRL 29(16), DOI 10.1029/2002GL015180

Jekeli C (1999) *The determination of gravitational potential differences from satellite–to–satellite tracking*. Cel Mech Dyn Astr 75: 85–100

van Loon J, Kusche J (2005) *Stochastic model validation of satellite gravity data: a test with CHAMP pseudo–observations*. Gravity, Geoid and Space Missions, Porto, August 29 - September 4, 2004, accepted, this volume

Schmidt M, Fabert O, Shum CK (2002) *Multi–resolution representation of the gravity field using spherical wavelets*. Weikko A. Heiskanen Symposium, Ohio State University, Columbus

Schmidt M, Fabert O, Shum CK, Han S–C (2004) *Gravity field determination using multi–resolution techniqes*. 2nd International GOCE User Workshop, Frascati, March 8–10, 2004

Svehla D, Rothacher M (2003) *Kinematic and reduced–dynamic precise orbit determination of low earth orbiters*. Adv Geosciences 1: 47–56

Torge W (2001) *Geodesy*. de Gruyter, Berlin

# A Study on Two-boundary Problems in Airborne Gravimetry and Satellite Gradiometry

P. Holota [1] and M. Kern [2]
[1] Research Institute of Geodesy, Topography and Cartography, 250 66 Zdiby 98, Praha-vychod, Czech Republic
e-mail: holota@pecny.asu.cas.cz, Tel.: +420 323649235, Fax: +420 284890056
[2] Institute of Navigation and Satellite Geodesy, Graz University of Technology, Steyrergasse 30/III,
A-8010 Graz, Austria e-mail: kern@geomatics.tu-graz.ac.at, Tel.: +430 3168736349, Fax:+430 3168736845

**Abstract.** Current gravity data, which are used for refined studies of the Earth's gravity field, result from cutting-edge technologies and advanced measuring techniques. They often refer to some exceptional surfaces or manifolds and can lead to interesting mathematical problems. In this paper, three cases are considered where airborne gravimetry is combined with terrestrial gravity data, airborne measurements are collected at two flight levels, and satellite gradiometry is combined with terrestrial gravity data. The three cases are studied mathematically and the disturbing potential is expressed explicitly. The spectral representation based on spherical harmonics is used. Subsequently, an optimum procedure is discussed in order to treat the overdetermined problems and to keep the typical regularity of harmonic functions at infinity. Some practical aspects are also mentioned.

**Keywords.** Earth's gravity field, geodetic boundary-value problems, overdetermined problems, optimization

---

## 1 Introduction

One of the challenges in gravity field modeling in the near future will be the combination of heterogeneous gravity data. As more data become available, certain kinds of overdetermined problems have to be solved. In this paper, theoretical studies are performed for the combination of terrestrial gravity data and airborne gravity data. Also, the possible use of airborne data measured at two height layers is considered. Finally, in view of the satellite mission GOCE and the so-called space-wise approach, investigations are done into the combination of terrestrial gravity data and (satellite) gradiometry.

The problems may be systematically approached by confining ourselves to a layer $\Omega$ bounded by two surfaces. With some degree of simplification we can even suppose that $\Omega$ is bounded by two spheres of radius $R_i$ and $R_e$, respectively, assuming that $R_i < R_e$.

We will refer our considerations to Euclidean three-dimensional Euclidean space $\boldsymbol{R}^3$ with rectangular Cartesian coordinates $x_i, i = 1,2,3$ and the origin at the center of gravity of the Earth. Then $\boldsymbol{x} \equiv (x_1, x_2, x_3)$ is a general point in $\boldsymbol{R}^3$. We will assume that the Earth is a rigid body and that the system of coordinates rotates together with the Earth with a known constant angular velocity $\omega$ around the $x_3$-axis.

In addition, we will also use the spherical coordinates $r$ (radius vector), $\varphi$ (geocentric latitude) and $\lambda$ (geocentric longitude). They are related to $x_1, x_2, x_3$ by the equations

$$x_1 = r\cos\varphi\cos\lambda\,, \quad x_2 = r\cos\varphi\cos\lambda \tag{1}$$

$$x_3 = r\sin\varphi \tag{2}$$

Finally, we will identify with $W$ and $U$ the gravity and a normal potential of the Earth, so that $T(\boldsymbol{x}) = W(\boldsymbol{x}) - U(\boldsymbol{x})$ is the disturbing potential. $T$ is a harmonic function in the exterior of the Earth and the quantity to be determined from the available data. In the first part of the paper, three two-level boundary value problems using terrestrial, airborne and satellite data are discussed. In the second part, an optimum procedure for the combination of the overdetermined problems is considered.

## 2 Terrestrial and airborne gravimetry

Consider the case that

$$\Delta T = 0 \quad \text{in} \quad \Omega \tag{3}$$

$$\frac{\partial T}{\partial r} + \frac{2}{R_i} T = -\Delta g \quad \text{for} \quad r = R_i \tag{4}$$

$$\frac{\partial T}{\partial r} = -\delta g^{(e)} \quad \text{for} \quad r = R_e \tag{5}$$

where $\Delta$ is Laplace's operator, $\Delta g$ stands for the usual gravity anomaly and $\delta g^{(e)}$ is the gravity disturbance that is given at $r = R_e$. Recall that $\delta g(\boldsymbol{x}) = g(\boldsymbol{x}) - \gamma(\boldsymbol{x})$, where $g = |\,\boldsymbol{grad}\,W\,|$ is the measured gravity and $\gamma = |\mathbf{grad}\, U|$ is the normal gravity. We will assume that our input data are corrected for gravitational interaction with the Moon, Sun and the planets, for the precession and nutation of the Earth and so on. (Note that $\Delta g$ does not mean Laplace's operator applied on $g$, but represents a standard symbol in physical geodesy.)

Recall that the representation of $T = T(r,\varphi,\lambda)$ in $\Omega$ means to write

$$T = \sum_{n=0}^{\infty}\left(\frac{R_i}{r}\right)^{n+1} T_n^{(i)}(\varphi,\lambda) + \sum_{n=0}^{\infty}\left(\frac{r}{R_e}\right)^{n} T_n^{(e)}(\varphi,\lambda) \quad (6)$$

where $T_n^{(i)}$ and $T_n^{(e)}$ are the respective Laplace surface spherical harmonics, cf. also Grafarend and Sansò (1984). In order to find the solution of our problem we can express $\Delta g$ and $\delta g^{(e)}$ in spherical harmonics

$$\Delta g(\varphi,\lambda) = \sum_{n=0}^{\infty} \Delta g_n(\varphi,\lambda) \quad (7)$$

and

$$\delta g^{(e)}(\varphi,\lambda) = \sum_{n=0}^{\infty} \delta g_n^{(e)}(\varphi,\lambda) \quad (8)$$

Inserting Eqs. (6) - (8) into Eqs. (4) - (5) and setting

$$q = \frac{R_i}{R_e} \quad (9)$$

we (in view of the orthogonality of spherical harmonics) obtain the following system

$$(n-1)T_n^{(i)} - (n+2)q^n T_n^{(e)} = R_i\,\Delta g_n \quad (10)$$

$$(n+1)q^{n+1}T_n^{(i)} \;-\; nT_n^{(e)} = R_e\,\delta g_n^{(e)} \quad (11)$$

which is valid for any individual $n$. Its determinant

$$D_n = (n+1)(n+2)q^{2n+1} - n(n-1) \quad (12)$$

is a function of $n$ and $q$. Obviously, the question whether there exist couples of $n$ and $q$ such that $D_n = 0$ is of particular interest. This equation means that

$$q^{2n+1} = \frac{n(n-1)}{(n+1)(n+2)} \quad (13)$$

Rigorously, Eq. (13) cannot be satisfied for all possible $n$ and $q$ since the right hand side of Eq. (13) is a rational number whereas the left hand side is more general and can be represented by rational as well as irrational numbers. Therefore, we can expect that in the majority of cases $D_n$ will differ from zero, though it cannot be not avoided that $D_n$ will have very small values or will even be equal to zero for some special couples of $n$ and $q$. Obviously, for $D_n \neq 0$ one obtains

$$T_n^{(i)} = -[R_i n\Delta g_n - R_e(n+2)q^n \delta g^{(e)}]/D_n \quad (14)$$

and

$$T_n^{(e)} = -[R_i(n+1)q^{n+1}\Delta g_n - R_e(n-1)\delta g_n^{(e)}]/D_n \quad (15)$$

Note also that $q \to 0$ as $R_e \to \infty$ and that at the same time the boundary-value problem discussed goes over into Stokes's problem, $D_n \to n(n-1)$ and obviously $D_1 \to 0$. Fig. 1 illustrates further details.

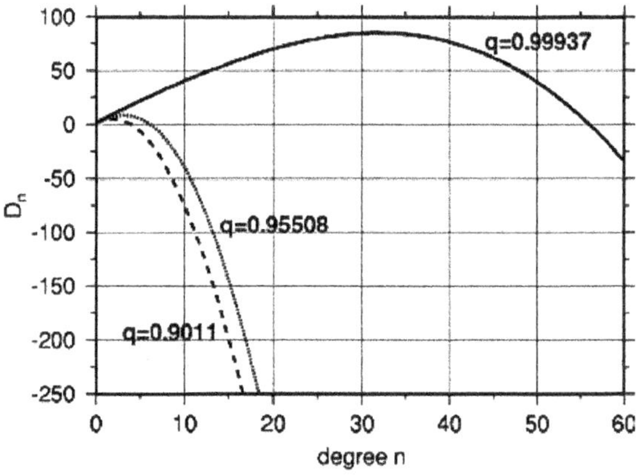


**Fig. 1.** $D_n$ for $R_i = 6378\,km$ and some selected values of $R_e$: $q = R_i/6382\,km = 0.99937$, $q = R_i/6678\,km = 0.95508$ and $q = R_i/7078\,km = 0.90110$

## 3 Two level airborne gravimetry

A similar situation arises when considering the gravity disturbance measured at two height levels. In this case we have the following problem

$$\Delta T = 0 \quad \text{in} \quad \Omega \quad (16)$$

$$\frac{\partial T}{\partial r} = -\,\delta g^{(i)} \quad \text{for} \quad r = R_i \quad (17)$$

$$\frac{\partial T}{\partial r} = -\,\delta g^{(e)} \quad \text{for} \quad r = R_e \quad (18)$$

where $\delta g^{(i)}$ and $\delta g^{(e)}$ are the gravity disturbances at $r = R_i$ and $r = R_e$, respectively.

The problem can be treated directly. However, for reasons that will be clear in Sect. 6, we prefer a slightly different approach. It can be useful in case at the focus is on the determination of the radial derivative of $T$. Therefore, we put

$$w = r\,\frac{\partial T}{\partial r} \quad (19)$$

Thus our problem is

$$\Delta w = 0 \quad \text{in} \quad \Omega \tag{20}$$

$$w = -R_i\, \delta g^{(i)} \quad \text{for} \quad r = R_i \tag{21}$$

$$w = -R_e\, \delta g^{(e)} \quad \text{for} \quad r = R_e \tag{22}$$

Similarly to Eq. (6), we will suppose that

$$w = \sum_{n=0}^{\infty} \left( \frac{R_i}{r} \right)^{n+1} w_n^{(i)}(\varphi, \lambda) + \sum_{n=0}^{\infty} \left( \frac{r}{R_e} \right)^{n} w_n^{(e)}(\varphi, \lambda) \tag{23}$$

where $w_n^{(i)}$ and $w_n^{(e)}$ are the respective Laplace surface spherical harmonics and that together with Eq. (8), valid for $\delta g^{(e)}$, also

$$\delta g^{(i)}(\varphi, \lambda) = \sum_{n=0}^{\infty} \delta g_n^{(i)}(\varphi, \lambda) \tag{24}$$

The apparatus of spherical harmonics then yields for any individual $n$ the following system

$$w_n^{(i)} + q^n w_n^{(e)} = -R_i\, \delta g_n^{(i)} \tag{25}$$

$$q^{n+1} w_n^{(i)} + w_n^{(e)} = -R_e\, \delta g_n^{(e)} \tag{26}$$

Its determinant

$$D_n' = 1 - q^{2n+1} \tag{27}$$

differs from zero, provided that $R_i \neq R_e$. Hence

$$w_n^{(i)} = -(R_i\, \delta g_n^{(i)} - R_e q^n \delta g_n^{(e)}) / D_n' \tag{28}$$

and

$$w_n^{(e)} = (R_i\, q^{n+1} \delta g_n^{(i)} - R_e\, \delta g_n^{(e)}) / D_n' \tag{29}$$

which solves our problem.

**Remark 1.** (*On the solution of Neumann's problem*) As mentioned above, the result we obtained, yields also the solution of the original problem given by Eqs. (16) - (18), but after some additional computation. Indeed, from Eq. (19) we have

$$T(r, \varphi, \lambda) = C(\varphi, \lambda) + \int \frac{w(r, \varphi, \lambda)}{r}\, dr \tag{30}$$

where $C$ is an arbitrary function, which does not depend of the radial distance $r$. Now inserting from Eq. (23) and integrating, we obtain

$$T = C(\varphi, \lambda) - \sum_{n=0}^{\infty} \frac{1}{n+1} \left( \frac{R_i}{r} \right)^{n+1} w_n^{(i)} + w_0^{(e)} \ln r + \sum_{n=1}^{\infty} \frac{1}{n} \left( \frac{r}{R_e} \right)^{n} w_n^{(e)} \tag{31}$$

Moreover, recalling that $T$ has to be a harmonic function, we see that necessarily

$$w_0^{(e)} = 0 \quad \text{and} \quad \Delta C(\varphi, \lambda) = 0 \tag{32}$$

In view of Eq. (29), the first equation means that the following condition has to be met

$$R_e^2\, \delta g_0^{(e)} = R_i^2\, \delta g_0^{(i)} \tag{33}$$

Similarly, the second equation results in

$$\frac{\partial^2 C}{\partial \varphi^2} + \tan \varphi \frac{\partial C}{\partial \varphi} + \frac{1}{\cos^2 \varphi} \frac{\partial^2 C}{\partial \lambda^2} = 0 \tag{34}$$

a special case of Legendre's differential equation (for $n = 0$). Hence $C = c\, P_{00}(\sin \varphi)$, where $P_{00}$ is Legendre's function and $c$ is an arbitrary constant. In consequence $C = \text{const.}$ Summing up, we have

$$T = C - \sum_{n=0}^{\infty} \frac{1}{n+1} \left( \frac{R_i}{r} \right)^{n+1} w_n^{(i)} + \sum_{n=1}^{\infty} \frac{1}{n} \left( \frac{r}{R_e} \right)^{n} w_n^{(e)} \tag{35}$$

which is in full analogy with the result obtained in another way in Holota (1995). Eq. (35) has the structure of the solution of Neumann's problem in a bounded domain, provided the condition (33) is met.

## 4 Gravimetry and gradiometry

The combination of satellite gravity gradiometry with terrestrial gravimetry data may be approached as follows

$$\Delta T = 0 \quad \text{in} \quad \Omega \tag{36}$$

$$\frac{\partial T}{\partial r} + \frac{2}{R_i} T = -\Delta g \quad \text{for} \quad r = R_i \tag{37}$$

$$\frac{\partial T^2}{\partial r^2} = G \quad \text{for} \quad r = R_e \tag{38}$$

Here $G = G(\varphi, \lambda)$ symbolizes the input coming from satellite gradiometry. We will suppose that

$$G(\varphi, \lambda) = \sum_{n=0}^{\infty} G_n(\varphi, \lambda) \tag{39}$$

where $G_n$ are the respective Laplace surface spherical harmonics. Hence, in view of the orthogonality of the spherical harmonics it is again not difficult to deduce for any individual $n$ the following system

$$(n-1) T_n^{(i)} - (n+2) q^n T_n^{(e)} = R_i\, \Delta g_n \tag{40}$$

$$(n+1)(n+2) q^{n+1} T_n^{(i)} + n(n-1) T_n^{(e)} = R_e^2\, G_n \tag{41}$$

The determinant of this system is

$$D_n'' = n(n-1)^2 + (n+1)(n+2)^2 q^{2n+1} \tag{42}$$

and always differs from zero. Thus

$$T_n^{(i)} = [R_i n(n+1) \Delta g_n + R_e^2 (n+2) q^n G_n] / D_n'' \tag{43}$$

and

$$T_n^{(e)} = -[R_i (n+1)(n+2) q^{n+1} \Delta g_n - R_e^2 (n-1) G_n] / D_n'' \tag{44}$$

## 5 Optimization

In all three cases we found a solution which is a harmonic function in the bounded domain $\Omega$. In the following we will symbolize this solution by the letter $u$. It can be written as

$$u = \sum_{n=0}^{\infty} \left( \frac{R_i}{r} \right)^{n+1} u_n^{(i)}(\varphi, \lambda) + \sum_{n=0}^{\infty} \left( \frac{r}{R_e} \right)^{n} u_n^{(e)}(\varphi, \lambda) \quad (45)$$

where $u_n^{(i)}$ and $u_n^{(e)}$ are the respective Laplace surface spherical harmonics. The problem, however, is that in general the continuation of $u$ for $r > R_e$ need not be a regular function at infinity, in other words it is not guaranteed that for $r \to \infty$ the continuation decreases as $c/r$, where $c$ is a constant. This can be considered a consequence of the fact that on the boundary of $\Omega$ (two parts) the input data are contaminated by some measurement errors and thus do not exactly represent the same function.

In fact the data given on the sphere of radius $R_i$ are enough to uniquely determine (possibly apart from first degree harmonics) a harmonic function in the whole domain $\Omega_{ext} \equiv \{ \boldsymbol{x} \in \boldsymbol{R}^3 ; r > R_i \}$ and thus also in $\Omega$, which is a part of $\Omega_{ext}$. For this reason the data on $R_e$ have the nature of excess data and in general (when some measurement noise is present) give rise to the ("internal") terms $(r/R_e)^n u_n^{(e)}$ that are not regular at infinity. This calls for a regularization that at the same time can reduce errors. The problem has been treated in literature as an overdetermined problem, see e.g. Sacerdote and Sansò (1985) and Rummel et al. (1989), but here, i.e. in the sequel, we will approach it through analytical regularization.

In solving the overdetermined problem as above, it is tempting to look for a harmonic function $f$ which is regular at infinity and minimizes the following functional

$$\Phi(f) = \int_{\Omega} (f-u)^2 \, dx \quad (46)$$

where $dx = dx_1 dx_2 dx_3$ is the volume element in Cartesian coordinates. We will suppose that $f \in H_2(\Omega_{ext})$, where $H_2(\Omega_{ext})$ is the space of harmonic functions endowed with inner product

$$(f,g) \equiv \int_{\Omega_{ext}} fg \frac{1}{r^2} \, dx \quad (47)$$

This inner product induces also the norm $\| f \| = (f,f)^{1/2}$. Roughly speaking, $H_2(\Omega_{ext})$ is a space of harmonic functions, which are defined on $\Omega_{ext}$ and square integrable under the weight $r^{-2}$.

However, an application of the optimization described above to the individual cases discussed in Sects. 2, 3 and 4 requires some care. We will return to it in the following.

Here we will pay some attention to the properties of the functional $\Phi$. It is not extremely difficult to show that for $f \in H_2(\Omega_{ext})$

$$\lim \Phi(f) = \infty \quad \text{as} \quad \| f \| \to \infty \quad (48)$$

which (using the standard terminology) means that $\Phi$ is a coercive functional on $H_2(\Omega_{ext})$. Take now an arbitrary $v \in H_2(\Omega_{ext})$ and form the function $\Phi(f+tv)$ of $t \in (-\infty, \infty)$. We easily compute that

$$\frac{d}{dt} \Phi(f+tv) \Big|_{t=0} = 2 \int_{\Omega} (f-u) v \, dx \quad (49)$$

and it is clear that

$$D\Phi(f,v) \equiv \frac{d}{dt} \Phi(f+tv) \Big|_{t=0} \quad (50)$$

is a bounded linear functional of the variable $v$ (Gâteaux' differential of $\Phi$ at the point $f$). Moreover, we can show that for all $f, v \in H_2(\Omega_{ext})$

$$D\Phi(f+v,v) - D\Phi(f,v) \geq 0 \quad (51)$$

The last inequality and the coerciveness of $\Phi$ makes it possible to conclude that $\Phi$ attains its minimum in $H_2(\Omega_{ext})$, for proof based on the theory of non-linear functionals see e.g. Necas and Hlavacek (1981).

On the contrary suppose that at a point $f \in H_2(\Omega_{ext})$ the functional $\Phi$ has its local minimum. Thus,

$$D\Phi(f,v) = 0 \quad (52)$$

for all $v \in H_2(\Omega_{ext})$, which immediately yields

$$\int_{\Omega} fv \, dx = \int_{\Omega} uv \, dx \quad (53)$$

for all $v \in H_2(\Omega_{ext})$.

**Remark 2.** In calculus of variations the identity (53) represents Euler's necessary condition for the functional $\Phi$ to have a minimum at the point $f$.

The integral identity represented by Eq. (53) is a natural starting point for a numerical solution. First, however, we put for the indices $n = 0,1,2,\ldots,\infty$ and $m = -n, -n+1, \ldots, -1, 0, 1, \ldots, n-1, n$

$$Y_{nm}(\varphi, \lambda) = P_{n|m|}(\sin \varphi) \begin{cases} \cos m\lambda & \text{for } m \geq 0 \\ \sin |m| \lambda & \text{for } m < 0 \end{cases} \quad (54)$$

where $P_{n|m|}$ is the usual (associated) Legendre function. Then

$$v_{nm} = \left(\frac{R_i}{r}\right)^{n+1} Y_{nm}(\varphi,\lambda) \tag{55}$$

are the solid spherical harmonics and it is known that in general

$$f = \sum_{n=0}^{\infty} \sum_{m=-n}^{m=n} f_{nm} v_{nm} \tag{56}$$

where $f_{nm}$ are scalar coefficients. In consequence, taking into account the orthogonality of spherical harmonics, Eq. (53) transforms into the following system of equations for the coefficients $f_{nm}$:

$$f_{nm} \int_{\Omega} v_{nm}^2 \, dx = \int_{\Omega} u v_{nm} \, dx \tag{57}$$

Here

$$\int_{\Omega} v_{nm}^2 \, dx = \int_{R_i}^{R_e} \int_{\sigma} \left(\frac{R_i}{r}\right)^{2n+2} Y_{nm}^2(\varphi,\lambda) r^2 d\sigma dr = \frac{R_i^3}{2n-1}(1-q^{2n-1}) \int_{\sigma} Y_{nm}^2(\varphi,\lambda) d\sigma \tag{58}$$

and $d\sigma$ is the surface element of the unit sphere $\sigma$.

With respect to the integral on the right hand side of Eq. (57), we first recall that in general in Eq. (45)

$$u_n^{(i)} = u_n^{(i)}(\varphi,\lambda) = \sum_{m=-m}^{m=n} a_{nm}^{(i)} Y_{nm}(\varphi,\lambda) \tag{59}$$

and

$$u_n^{(e)} = u_n^{(e)}(\varphi,\lambda) = \sum_{m=-m}^{m=n} a_{nm}^{(e)} Y_{nm}(\varphi,\lambda) \tag{60}$$

where $a_{nm}^{(i)}$ and $a_{nm}^{(e)}$ are the respective coefficients. Thus we obtain

$$\int_{\Omega} u v_{nm} \, dx = \int_{R_i}^{R_e} \int_{\sigma} \left(\frac{R_i}{r}\right)^{2n+2} u_n^{(i)} Y_{nm} r^2 d\sigma dr + \int_{R_i}^{R_e} \int_{\sigma} \left(\frac{r}{R_e}\right)^{n} \left(\frac{R_i}{r}\right)^{n+1} u_n^{(e)} Y_{nm} r^2 d\sigma dr = \tag{61}$$

$$= a_{nm}^{(i)} \int_{\Omega} v_{nm}^2 \, dx + a_{nm}^{(e)} \frac{R_i R_e^2}{2} q^n (1-q^2) \int_{\sigma} Y_{nm}^2 d\sigma$$

and subsequently

$$f_{nm} = a_{nm}^{(i)} + \alpha_n a_{nm}^{(e)} \tag{62}$$

where

$$\alpha_n = \frac{(2n-1)(1-q^2)}{2(1-q^{2n-1})} q^{n-2} \tag{63}$$

Hence, inserting into Eq. (56), we get

$$f = \sum_{n=0}^{\infty} \left(\frac{R_i}{r}\right)^{n+1} [u_n^{(i)} + \alpha_n u_n^{(e)}] \tag{64}$$

and one can show that $\alpha_0 = (1+q)/2q$ and that

$$\lim \alpha_n = 0 \quad \text{as} \quad n \to \infty \tag{65}$$

see also Fig. 2.

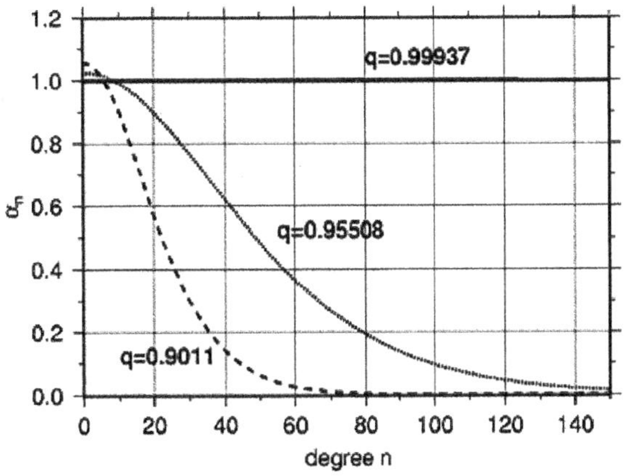


Fig. 2. Some values of $\alpha_n$ ($q$ as in Fig. 1)

(Note. Further values of $\alpha_n$ for $q = 0.99937$ are: $\alpha_{360} = 0.9919$, $\alpha_{720} = 0.9672$, $\alpha_{1440} = 0.8763$.)

## 6 Final remarks and Outlook

In order to apply our optimization to the problems discussed in Sects. 2, 3 and 4 we still have to return to Laplace's surface spherical harmonics $u_n^{(i)}$ and $u_n^{(e)}$, as they were derived in the individual cases.

The structure of the functional $\Phi$ offers a reasonable interpretation of our minimization concept especially in the case of the problem discussed in Sect. 3. Indeed, here $u = w$ and due to Dirichlet's principle among all functions $\tilde{u}$ that are defined on $\Omega$ and that for $r = R_i$ and $r = R_e$ attain the same values as $u$, the function $u$ (as a harmonic function) minimizes the Dirichlet's integral:

$$\int_{\Omega} \sum_{i=1}^{3} (\partial \tilde{u} / \partial x_i)^2 \, dx \tag{66}$$

see e.g. Rektorys (1977) or Kellogg (1953). In this sense $u$ is also the smoothest function among the functions $\tilde{u}$. [Dirichlet's principle belongs to variational methods, in geodesy see e.g. Holota (2004).]

Thus, inserting from Eqs. (28) and (29) into Eq. (64), we obtain

$$f = \sum_{n=0}^{\infty} \left(\frac{R_i}{r}\right)^{n+1} [A_n^{(i)} R_i \delta g_n^{(i)} + A_n^{(e)} R_e \delta_n^{(e)}] \tag{67}$$

where

$$A_n^{(i)} = -(1-q^{n+1}\alpha_n)(1-q^{2n+1})^{-1} \quad (68)$$

and

$$A_n^{(e)} = (q^n - \alpha_n)(1-q^{2n+1})^{-1} \quad (69)$$

It is interesting that up to a relatively high degree $n$ the coefficients $A_n^{(i)}$ and $A_n^{(e)}$ are constant (about $-0.5$), especially for $q = 0.99937$, see Fig. 3.

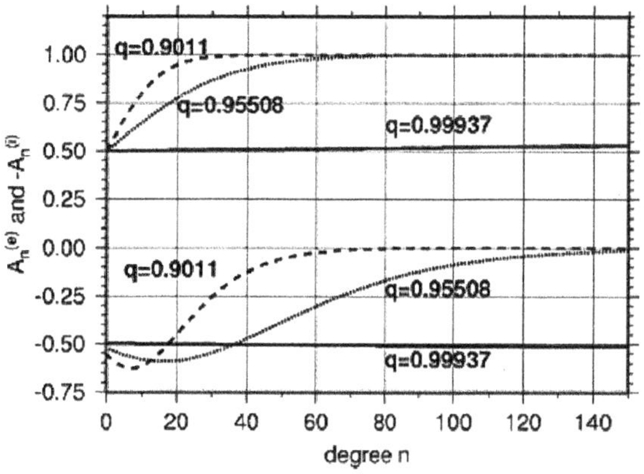


**Fig. 3.** Values of the coefficients $A_n^{(i)}$ and $A_n^{(e)}$

For $q = 0.99937$, we have for the higher degrees:

$$A_{360}^{(i)} = -0.5748, \quad A_{360}^{(e)} = -0.5333$$
$$A_{1440}^{(i)} = -0.7128, \quad A_{1440}^{(e)} = -0.5657$$
$$A_{2880}^{(i)} = -0.9248, \quad A_{2880}^{(e)} = -0.4582$$

Hence, we can conclude that our optimization actually offers a natural concept for weighting the input data (neglecting any data stochastics).

The problem in Sect. 2 has a slightly different nature and its solution deserves more attention, in particularly due to the fact that for some values of $n$ the system given by Eqs. (10) and (11) is nearly singular or even singular. For these reasons and also the page limit we avoid a straightforward application of our optimization concept, leaving the discussion of the problem for a new paper in preparation.

Finally, in case of the problem described in Sect. 4 the optimization leads to

$$f = \sum_{n=0}^{\infty} \left(\frac{R_i}{r}\right)^{n+1} \frac{1}{n+1} [A_n^{(i)} R_i \Delta g_n + A_n^{(e)} R_e^2 G_n] \quad (70)$$

with

$$A_n^{(i)} = (n+1)^2 [n - \alpha_n (n+2) q^{n+1}] / D_n'' \quad (71)$$

and

$$A_n^{(e)} = (n+1) [(n+2) q^2 + \alpha_n (n-1)] / D_n'' \quad (72)$$

The coefficients $A_n^{(i)}$ and $A_n^{(e)}$ for $q = 0.95508$ and $q = 0.90110$ are shown in Fig 4.

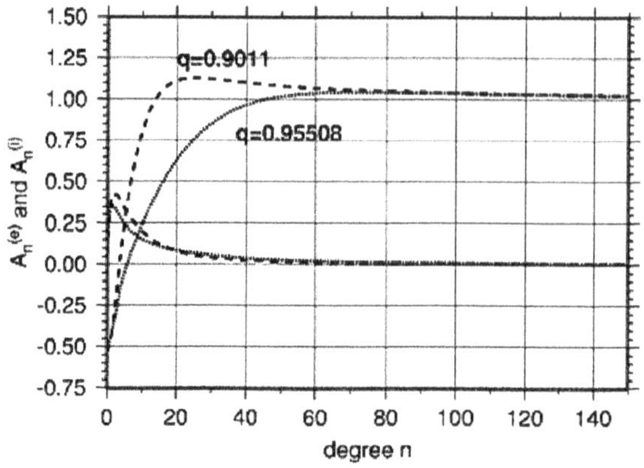


**Fig. 4.** Values of the coefficients $A_n^{(i)}$ and $A_n^{(e)}$

The figure shows that a good balance between the terrestrial and gradiometry data is a challenging problem which needs further investigations. Furthermore, the concepts are extended by taking the data stochastics into account. Finally, issues related to the distance zones and the topography will be considered in an upcoming contribution.

**Acknowledgements.** The work on this paper started during the stay of the first author at the Department of Geomatics Engineering of the University of Calgary and was stimulated by discussions with Prof. Klaus-Peter Schwarz. At the home institute, the first author was supported by the Grant Agency of the Czech Republic through Grant No. 205/04/1423 and partly by the Ministry of Education, Youth and Sports of the Czech Republic through Projects No. LN 00A005. The second author is financially supported by an external ESA fellowship. All this support is gratefully acknowledged.

## References


Grafarend E and Sansò F (1984) The multibody space-time geodetic boundary value problem and the Honkasalo term. Geophys. J.R. astr. Soc. 255-275

Holota P (1995) Boundary and Initial Value Problems in Airborne Gravimetry. In: Proc. IAG Symp. on Airborne Gravity Field Determination, IUGG XXI Gen. Assembly, Boulder, CO, USA, Jul 2-14, 1995, (Conv.: Schwarz K-P, Brozena JM, Hein GW). Special rep. No. 60010, Dept. of Geomatics Eng., The Univ. of Calgary, Calgary, 67-71

Holota P (2004) Some topics related to the solution of boundary-value problems in geodesy. Intl. Assoc. of Geodesy Symposia, Vol. 127, Springer, Berlin etc., 189-200

Kellogg OD (1953) Foundations of potential theory. Dover Publications, Inc., New York

Necas J and Hlavacek I (1981) Mathematical theory of elastic and elasto-plastic bodies: An introduction. Elsevier Sci. Publ. Company, Amsterdam-Oxford-New York

Rektorys K (1977) Variational methods. Reidel Co., Dordrecht-Boston

Rummel R, Teunissen P and Van Gelderen M (1989) Uniquely and over-determined geodetic boundary value problem by least squares. Bull. Géodésique, Vol. 63, 1-33

Sacerdote F and Sansò F (1985) Overdetermined boundary value problems in physical geodesy. Manuscripta Geodaetica, Vol. 10, No. 3, 195-207

# Local Geoid Computation by the Spectral Combination Method

O. Gitlein, H. Denker, J. Müller
Institut für Erdmessung, University of Hannover, Schneiderberg 50, D-30167 Hannover, Germany

**Abstract.** The spectral combination method is investigated for the determination of a local geoid in Lower Saxony (Germany) using a global geopotential model, radial gravity gradients from the upcoming GOCE satellite mission, and gravity anomalies. The main goal of this study is to test the method with regard to the validation of GOCE data products. In order to prove the effectiveness and numerical accuracy of the method, the computations are done in a closed-loop simulation based on the EGM96 geopotential model. The gravity field signal is decomposed into long, medium and short wavelength components using corresponding spectral weight functions. The long wavelength information up to about degree $n = 30$ is computed from a global geopotential model, the medium wavelength part ($n = 30 - 130$) is taken from radial gravity gradients at GOCE altitude (250 km), and the high frequency part ($n = 130 - 360$) is derived from terrestrial gravity anomalies. The modified spherical Butterworth filter and a cosine filter are tested as spectral weighting functions. The results from the closed-loop simulation are discussed, and an error analysis is done considering the commission and omission errors of the input data sets.

**Keywords.** Spectral combination method, radial gravity gradients, downward continuation

## 1 Introduction

From the analysis of the upcoming satellite mission GOCE, scheduled for launch in 2006, a precise static global Earth gravity field model with a spatial resolution of about 100 km and a geoid accuracy of 0.01 m is expected (ESA, 1999). GOCE will operate a gradiometer to measure gravity gradients at an altitude of about 250 km. For the verification of the high geoid precision, the GOCE products have to be validated. The main goal of this study is to contribute to the validation of the GOCE products.

A promising method for the validation of GOCE data is the use of ground gravity data from some well-surveyed areas (e.g., Arabelos and Tscherning, 1998; Haagmans, 2001). In this contribution, the spectral combination method is investigated for the determination of a local geoid in Lower Saxony using a global geopotential model, radial gravity gradients from the GOCE satellite mission, and gravity anomalies. The method allows the determination of the complete geoid spectrum as well as band-limited estimates, which can then be used for the validation and calibration of the GOCE data according to Haagmans et al. (2002). In order to prove the effectiveness and numerical accuracy of the method, the computations are done in a closed-loop simulation with synthedite data generated from the EGM96 geopotential model.

The methodology is described in section 2. The results based on different empirical spherical filters (modified spherical Butterworth and cosine) are discussed in section 3. The study also considers gravity anomalies located on a sphere and on an ellipsoid. Moreover, an error analysis is presented considering the commission and omission errors of the input data set.

## 2 The spectral combination method

The spectral combination method allows the computation of regional geoid models by closed integral formulas. In this approach, a spectral decomposition of the disturbing potential is done and the components, derived from different input data sets, are combined using spectral weights (Wenzel, 1981, 1982). In this study, the long wavelength components are computed from a global geopotential model (GGM), and the medium and short wavelength components are derived from radial gravity gradients from the GOCE mission and gravity anomalies.

### 2.1 Mathematical model

The geoid height $N$ and the disturbing potential $T$ at the computation point $P$ are related by Bruns' formula $N = T/\gamma$, where $\gamma$ is the normal gravity at the computation point $P(\theta, \lambda, r)$. For this reason, mainly the disturbing potential $T$ will be discussed below as the basic gravity field quantity. $T$ can be decomposed into its spectral components and expressed as an infinite series of spherical harmonics:

$$T(\theta, \lambda, r) = \sum_{n=2}^{\infty} \left(\frac{a}{r}\right)^{n+1} T_n, \qquad (1)$$

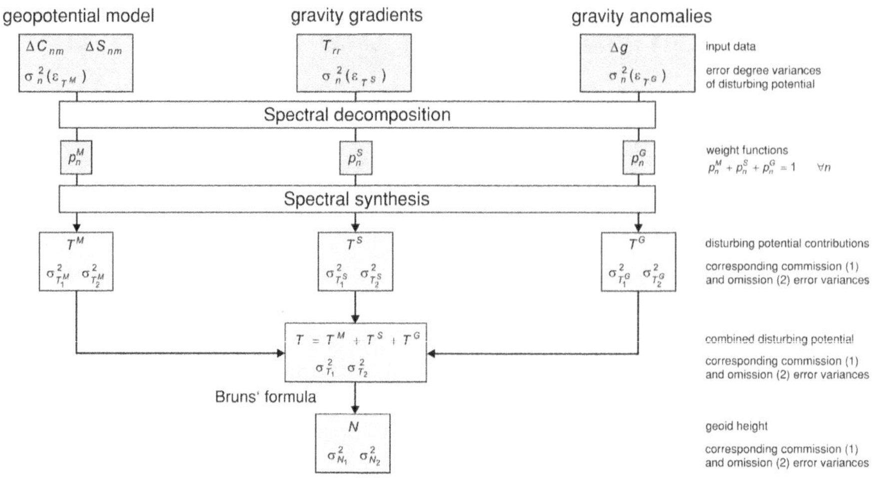


**Fig. 1.** Processing scheme for geoid determination from a geopotential model, radial gravity gradients and gravity anomalies.

where $T_n$ are the Laplace surface harmonics of degree $n$, and $a$ is the semi-major axis of the reference ellipsoid GRS 80. This expansion can be divided into the three parts $T^M$, $T^S$ and $T^G$ (see also Fig. 1):

- **long wavelength** contribution from the geopotential model up to degree and order $n^M_{max}$, $T^M$,
- **medium wavelength** contribution from the radial gravity gradients $T_{rr}$ (2nd radial derivatives of $T$) at satellite altitude, $T^S$, and
- **short wavelength** contribution from the terrestrial free-air gravity anomalies $\Delta g$, $T^G$.

The contributions $T^M$, $T^S$ and $T^G$ to the final combined disturbing potential $T$ are controlled by spectral weight functions $p^M_n$, $p^S_n$ and $p^G_n$ for the long, medium and short wavelength part, respectively. The spectral weights (Fig. 3) depend on the spherical harmonic degree $n$, and for each degree the sum of the spectral weights is one (Wenzel, 1981, 1982; Haagmans et al., 2002). The combined disturbing potential surface harmonic of degree $n$ can be computed by

$$T_n = p^M_n T^M_n + p^S_n T^S_n + p^G_n T^G_n . \tag{2}$$

The weights can be derived on the basis of statistical principles using the error information of the individual data sets (e.g. Wenzel, 1981; Denker, 2002) or by empirical procedures. In this study, the $p^M_n$, $p^S_n$ and $p^G_n$ are computed from two low pass filters $w_1$ and $w_2$:

$$p^M_n = w_1, \; p^S_n = w_2 - w_1, \; p^G_n = 1 - w_2. \tag{3}$$

This procedure is suggested in Haagmans et al. (2002), considering the spectral properties of the observations (e.g., measurement bandwidth of GOCE). The generation of $w_1$ and $w_2$ is done by empirical functions like the modified spherical Butterworth filters (MSB; Haagmans et al., 2002) or cosine filters (Denker, 2002).

The individual components of the disturbing potential can be computed by:

$$T^i = \sum_{n=2}^{n^i_{max}} \left(\frac{a}{r}\right)^{n+1} p^i_n T^i_n, \tag{4}$$

where $i = M, S, G$. The corresponding surface harmonics of degree $n$ are given by

$$T^M_n = \frac{GM}{a}\sum_{m=0}^{n}(\Delta\overline{C}_{nm}\cos m\lambda + \Delta\overline{S}_{nm}\sin m\lambda)\overline{P}_{nm}(\cos\theta),$$

$$T^S_n = \frac{R_S^2}{4\pi}\frac{2n+1}{(n+1)(n+2)}\left(\frac{R_S}{a}\right)^{n+1}\iint_{\sigma_0} T_{rr}P_n(\cos\psi)d\sigma,$$

$$T^G_n = \frac{R}{4\pi}\frac{2n+1}{n-1}\left(\frac{R}{a}\right)^{n+1}\iint_{\sigma_1}\Delta g P_n(\cos\psi)d\sigma. \tag{5}$$

In the above Eqs. (4) and (5), the following symbols are used (see also Fig. 2): $n, m$ = degree and order of the expansion, $GM$ = geocentric gravitational constant, $R, R_S$ = mean Earth radius and GOCE radius, $\Delta\overline{C}_{nm}$,$\Delta\overline{S}_{nm}$ =normalised potential coefficients, $\overline{P}_{nm}(\cos\theta)$=normalised Legendre functions, $P_n(\cos\psi)$=Legendre polynomials, $n^M_{max}$=maximum degree of the geopotential model, $n^S_{max}, n^G_{max}$ = maximum degree of the expansions for $T_{rr}$ and $\Delta g$ according to the data resolution, $\sigma_0, \sigma_1$ =spherical caps used for $T_{rr}$ and $\Delta g$, $\psi$ =spherical distance between the computation point $P\,(\theta, \lambda, r)$ and the source point $Q\,(\theta', \lambda', r')$, $d\sigma = sin\theta d\theta d\lambda$ = surface element on the unit sphere.

## 2.2 Reformulation of the model

For the practical calculation of the disturbing potential $T$, the formulas in section 2.1 are rewritten in this section, considering that for each degree the sum of the spectral weights is

$$p_n^M + p_n^S + p_n^G = 1. \tag{6}$$

Introducing Eq. (6) into Eq. (2) gives a numerically more advantageous formulation:

$$T_n = T_n^M + p_n^S \left(T_n^S - T_n^M\right) + p_n^G \left(T_n^G - T_n^M\right),$$
$$T_n = T_n^M + p_n^S T_n'^S + p_n^G T_n'^G. \tag{7}$$

From the comparison of Eqs. (2) and (7) it is obvious that the reformulation leads to the remove-restore procedure. The disturbing potential components from the $T_{rr}$ and $\Delta g$ observations (i.e. $T_n^S$ and $T_n^G$) are reduced for the effect of the geopotential model ($T_n^M$), respectively, yielding the residual quantities $T_n'^S$ and $T_n'^G$ (remove step). The advantage is that the residual quantities are smaller in magnitude than the original data, which leads to reduced truncation errors from a limited integration. Furthermore, if the geopotential model parts are computed by exact formulas or ellipsoidal approximations, the effect of spherical approximations, used in the handling of the residual gravity field quantities, is also diminished. Finally, the complete effect of the global geopotential model up to degree and order $n_{max}^M$ is added back (restore step).

The combined disturbing potential $T$ follows from Eqs. (1) and (7):

$$T = \sum_{n=2}^{\infty} \left(\frac{a}{r}\right)^{n+1} \left(T_n^M + p_n^S T_n'^S + p_n^G T_n'^G\right), \tag{8}$$

$$T = \widehat{T}^M + \widehat{T}^S + \widehat{T}^G. \tag{9}$$

The components in the above equation can be computed as follows:

**a) $\widehat{\mathrm{T}}^{\mathrm{M}}$ from GGM**

$$\widehat{T}^M = \sum_{n=2}^{n_{max}^M} \left(\frac{a}{r}\right)^{n+1} T_n^M. \tag{10}$$

**b) $\widehat{\mathrm{T}}^{\mathrm{S}}$ from the radial gravity gradients**

$$\widehat{T}^S = \frac{R_S^2}{4\pi} \iint_{\sigma_0} F(\psi, R_S, r) \left(T_{rr} - T_{rr}^M\right) d\sigma, \tag{11}$$

with the modified integration kernel for the radial gravity gradients

$$F = \sum_{n=2}^{n_{max}^S} \frac{2n+1}{(n+1)(n+2)} \left(\frac{R_S}{r}\right)^{n+1} p_n^S P_n(\cos\psi), \tag{12}$$

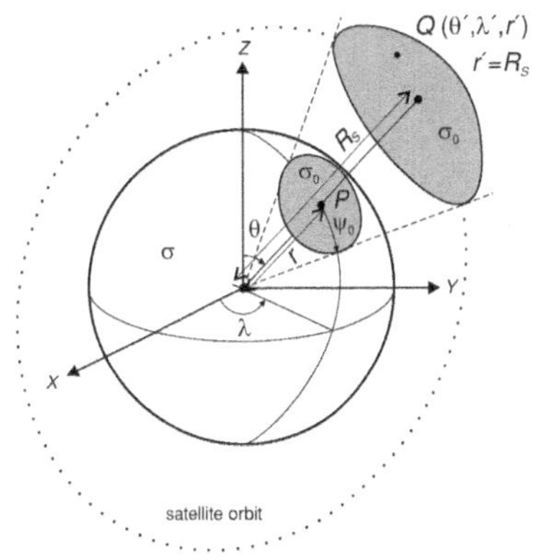


**Fig. 2.** Coordinate system.

and the radial gravity gradients $T_{rr}^M$ from the geopotential model up to degree and order $n_{max}^M$ at satellite altitude:

$$T_{rr}^M = \sum_{n=2}^{n_{max}^M} \left(\frac{a}{r'}\right)^{n+3} \frac{(n+1)(n+2)}{a^2} T_n^M. \tag{13}$$

**c) $\widehat{\mathrm{T}}^{\mathrm{G}}$ from the gravity anomalies**

$$\widehat{T}^G = \frac{R}{4\pi} \iint_{\sigma_1} S(\psi, R, r) \left(\Delta g - \Delta g^M\right) d\sigma, \tag{14}$$

with the modified integration kernel for gravity anomalies

$$S = \sum_{n=2}^{n_{max}^G} \frac{2n+1}{n-1} \left(\frac{R}{r}\right)^{n+1} p_n^S P_n(\cos\psi), \tag{15}$$

and the gravity anomalies $\Delta g^M$ from the geopotential model up to degree and order $n_{max}^M$:

$$\Delta g^M = \sum_{n=2}^{n_{max}^M} \left(\frac{a}{r'}\right)^{n+2} \frac{n-1}{a} T_n^M. \tag{16}$$

The integration in Eqs. (11) and (14) can be limited to a spherical cap $\sigma_0$ with a spherical radius $\psi_0$ (Fig. 2), because the observations are used in a remove-restore procedure in combination with a GGM. However, the limited integration of the observations causes a truncation or omission error ($\sigma_{N_2}$ for geoid undulations), which can be estimated using the anomaly degree variance model from Tscherning and Rapp (1974). On the other hand, the commission error ($\sigma_{N_1}$) is depending on the accuracy of the input data sets and can be derived by error propagation. The derivation of the complete formulas for the commission and omission errors is given in Haagmans et al. (2002), Denker (2002) and Heck (1979).

**Table 1.** Geoid contributions from global model EGM 96 with $n^M_{max} = 90$, radial gravity gradients at satellite altitude and gravity anomalies. The computations are based on the MSB filter using different cap sizes $\psi_0$ and $\psi_1$.

| | GGM | Gradients with $\psi_0$ | | | Anomalies with $\psi_1$ | | | Comb. Geoid |
|---|---|---|---|---|---|---|---|---|
| | | $5°$ | $10°$ | $15°$ | $5°$ | $10°$ | $15°$ | $13°$ and $10°$ |
| MIN [m] | 38.103 | -0.583 | -0.571 | -0.571 | -0.678 | -0.672 | -0.672 | 37.707 |
| MAX [m] | 48.896 | 0.706 | 0.701 | 0.701 | 0.796 | 0.825 | 0.827 | 49.111 |
| MEAN [m] | 43.381 | 0.040 | 0.039 | 0.039 | 0.003 | 0.001 | 0.001 | 43.422 |
| RMS ±[m] | 43.507 | 0.291 | 0.286 | 0.286 | 0.255 | 0.256 | 0.256 | 43.550 |
| $\sigma_{N_1^i}$ ±[cm] | 25.6 | 1.1 | 1.1 | 1.1 | 2.7 | 2.7 | 2.7 | 25.8 |
| $\sigma_{N_2^i}$ ±[cm] | 0 | 1.6 | 1.2 | 0.5 | 2.2 | 0.3 | 0.1 | 0.8 |

## 3 Numerical experiments/results

The results presented below are based on synthetic observations $T_{rr}$ and $\Delta g$, which are computed from EGM 96 up to degree and order 360 in a regular $30' \times 30'$ grid. The comparison of the combined geoid with the geoid determined directly from EGM 96 up to degree and order 360 serves as a consistency check. All radial gravity gradients are located on a sphere with radius $R_S = R + 250$ km (GOCE satellite altitude). Two data sets of gravity anomalies $\Delta g$ are tested. The first one contains gravity anomalies, which are located on a sphere with a mean Earth radius $R$, and the second data set is computed on a mean Earth ellipsoid to simulate approximately the real case for terrestrial data. For the long wavelength part, EGM96 with a maximum degree of expansion $n^M_{max} = 90$ was chosen in the remove-restore procedure, because actual GGM's are accurately known up to this degree.

The spectral weights are defined empirically using the modified spherical Butterworth filter (MSB) and a cosine filter (see Fig. 3). Due to the used weighting procedure, the geoid computation performed in this study is a band-limited estimation to a maximum degree, which is controlled by the maximum degree considered for the terrestrial gravity anomalies. In the tests presented here, this maximum degree is restricted to 360. Other weighting schemes were also tested, where information up to higher degrees was extracted from the GOCE gradients. However, these cases gave slightly larger differences in the closed-loop computations (see below), which is due to the low signal to noise ratio of the gravity gradients at high degrees, affecting the downward continuation of the GOCE data.

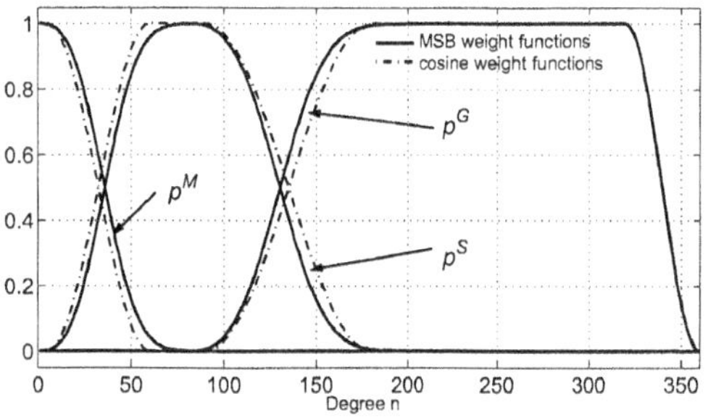


**Fig. 3.** Weight functions $p^M_n$, $p^S_n$ and $p^G_n$ computed from MSB and cosine filters.

### 3.1 Geoid based on MSB filter

This part of the investigation is based on a modified spherical Butterworth filter (MSB). Tests with the spherical Butterworth (SB) filter (Haagmans, 2001) led to unsatisfactory results due to big oscillations of the integration kernels (see also Haagmans et al., 2002). Hence the SB filter was modified, and the following two low pass filters are used in this study:

$$w_1 = \left[1 + \left(\frac{n}{n_b}\right)^{2k}\right]^{-\frac{1}{2}n} \quad \text{with}\; n_b = 180,\; k = 1, \qquad (17)$$

$$w_2 = \begin{cases} 1 & \text{for}\; 2 \le n \le 80, \\ \left[1 + \left(\frac{n-80}{n_b}\right)^{2k}\right]^{-\frac{1}{2}(n-80)} & \text{with}\; n_b = 300,\; k = 1. \end{cases}$$

From $w_1$ and $w_2$, the spectral weights $p^M_n$, $p^S_n$ and $p^G_n$ are derived according to Eq. (3). Furthermore, the $p^G_n$ values are multiplied with a cosine window at the end of the spectral range of interest (see Fig. 3) in order to reduce oscillations of the kernel $S$.

The resulting integration kernels $F$ and $S$ according to Eq. (12) and (15) are shown in the upper part of Fig. 4. The figure clearly shows that the integration can be limited to spherical cap sizes of about 15°. An additional advantage of the modified integration kernels is, that they are finite in the computation point due to the finite summation limits $n^S_{max}$ and $n^G_{max}$. A special consideration of the inner zone is therefore not necessary.

Table 1 shows the contributions from the individual data sources. The contributions from the radial gravity gradients and the gravity anomalies are provided for different spherical cap sizes. The commission errors $\sigma_{N_1^i}$ are computed with 1 mE uncorrelated gradient errors, 1 mGal uncorrelated gravity anomaly errors, and the EGM96 coefficient errors. The final geoid for Lower Saxony (see Fig. 5) was computed with cap sizes of 13° and 10° for the gravity gradients and gravity anomalies, respectively. The cap sizes were chosen in order to keep the omission errors below 1 cm for each of the input data sets.

As the results are based on simulated observations with a priori known ground truth values, the combined geoid from the spectral combination approach can be compared directly with the geoid determined from EGM96 up to degree and order 360. In this closed-loop computation the two geoid models show differences in the range of -0.5 to +0.4 cm (±0.2 cm RMS), which is in good agreement with the truncation error $\sigma_{N_2} = \pm 0.8$ cm from Table 1.

### 3.2 Geoid based on cosine filter

In this section, the cosine filter is applied in the computations. Although the spectral weights $p_n^M$, $p_n^S$ and $p_n^G$ based on the cosine low pass filters are very similar to the ones from the MSB filters (see Fig. 3), the corresponding integration kernels show significant differences (see Fig. 4). The figure clearly shows bigger oscillations of the integration kernels based on the cosine filters, especially for the gravity gradients. Consequently, larger cap sizes are required for the integration. The omission error of the geoid contribution from the radial gravity gradients is 36.2, 5.3, 4.0, and 1.8 cm for integration cap sizes of 5°, 10°, 15° and 20°, respectively.

On the other hand, the integration kernels for gravity anomalies based on the MSB and cosine filters show a similar behaviour (see Fig. 4). This is also confirmed by the estimated omission errors of 1.6, 0.4 and 0.2 cm for cap sizes of 5°, 10° and 15°, respectively. These figures are similar to the values in Table 1. Here it should be noted that both weight functions $p_n^G$ (MSB and cosine filter) use the cosine filter at the end of the spectrum (Fig. 3, degree range $n = 320 - 360$).

### 3.3 Band-limited geoid contributions

This section discusses the computation of geoid contributions from radial gravity gradients and gravity anomalies for the same bandwidth, followed by a comparison of the two estimates. The aim of this investigation is to contribute to the validation and calibration of the GOCE products, where the band-limited GOCE results are to be directly compared to the corresponding components from the terrestrial gravity data (see also Haagmans et al., 2002). The computation of the band-limited geoid signals $\widehat{T}^S$ and $\widehat{T}^G$ is realized by setting $p_n^G$ equal to $p_n^S$. The calculations are done with the MSB and cosine filters. The resulting integration kernels for the radial gravity gradients are identical to the ones shown in Fig. 4. For the gravity anomalies, the integration kernels are shown in Fig. 6. Here, the integration kernel based on the MSB filters again converges slightly faster to zero than the kernel based on the cosine filters.

For the calculations based on the MSB filters, an integration cap size of 13° and 10° is used as before for the radial gravity gradients and the gravity anomalies, respectively, leading to a geoid omission error of 0.7 cm for both components. For the integra-

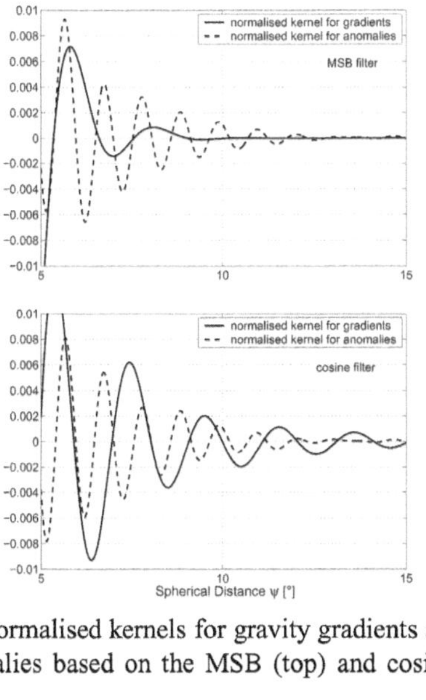


Fig. 4. Normalised kernels for gravity gradients and gravity anomalies based on the MSB (top) and cosine filters (bottom).

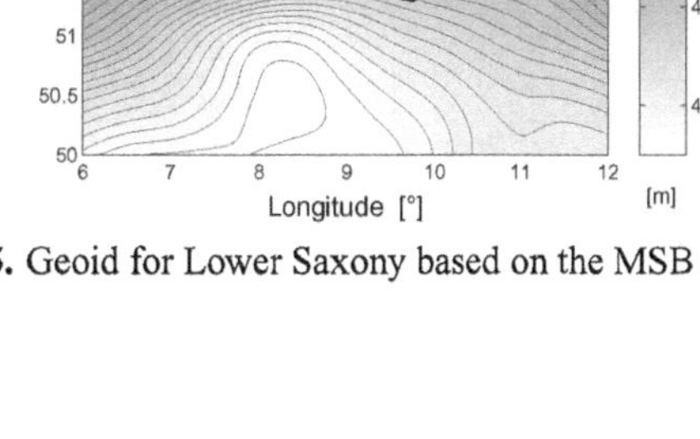


Fig. 5. Geoid for Lower Saxony based on the MSB filter.

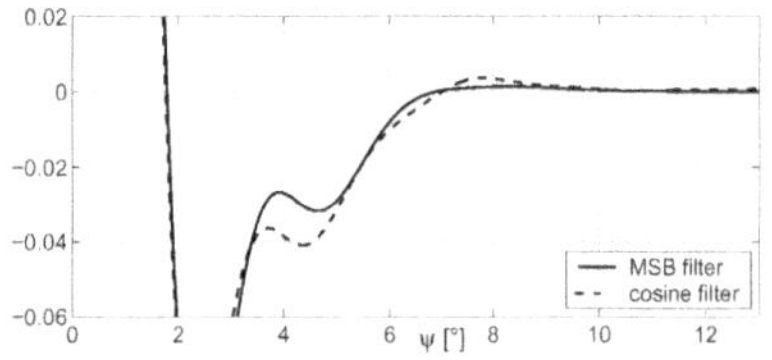


**Fig. 6.** Normalised kernels $\bar{S}$ for gravity anomalies based on MSB and cosine filters for the case $p_n^G = p_n^S$.

tion kernel based on the cosine filter, the cap size is limited to 15°, resulting in omission errors of 4 cm for both cases.

The choice of other spectral weight functions with a MSB filter, extracting signals from the GOCE gravity gradients up to degree $n = 300$, leads to band-limited geoid differences (see above) of $\pm 2$ cm RMS between the components from the gravity gradients and the gravity anomalies, respectively. This is caused by the low signal to noise ratio of the gravity gradients at higher degrees, affecting the downward continuation implicitly used in the computations. On the other hand, the differences between the components derived from the global geopotential model and the gravity anomalies show only a RMS difference of $\pm 0.1$ cm.

### 3.4 Gravity anomalies on the ellipsoid

Because a sphere is only a rough approximation of the Earth, some tests with gravity anomalies located on an ellipsoid were performed. Fig. 7 (left part) shows the differences between the two combined geoids calculated from gravity anomalies on a sphere and on an ellipsoid, both in connection with GOCE gradients and the GGM up to $n = 90$. Fig. 7 (right part) depicts the differences between the band-limited geoid contributions from $T_{rr}$ and $\Delta g$ on the ellipsoid (see also previous section). In both cases, the values are in the range of -0.8 and +0.6 cm with a RMS of $\pm 0.2$ to 0.3 cm. Hence, the procedure is also suited for the application to real terrestrial data.

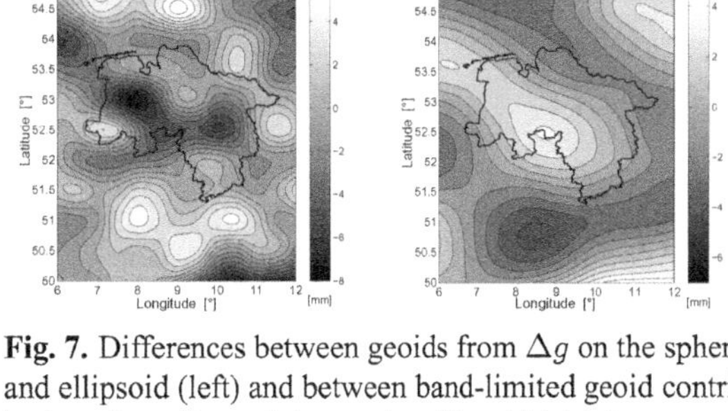


**Fig. 7.** Differences between geoids from $\Delta g$ on the sphere and ellipsoid (left) and between band-limited geoid contributions from $T_{rr}$ and $\Delta g$ on the ellipsoid (right).

## 4 Conclusions

A local geoid for Lower Saxony was computed based on the spectral combination method, combining a global geopotential model, radial gravity gradients from GOCE and terrestrial gravity anomalies. The effectiveness and numeriacal accuracy of the procedure is tested in a closed-loop simulation based on EGM96. For the spectral weighting, modified spherical Butterworth (MSB) and cosine filters were used. The MSB filters gave superior results associated with smaller omission errors. The differences between the computed and ground truth values show a RMS of about $\pm 0.2$ to 0.3 cm for all cases, including the cases with gravity anomalies on the sphere and the ellipsoid. Hence, the method is also suited for the application to real data, but further studies are needed to investigate this case.

## References

Arabelos, D. and Tscherning, C. C. (1998). *Calibration of satellite gradiometer data aided by ground gravity data.* Journal of Geodesy, 72:617-625.

Denker, H. (2002). *Computation of Gravity Gradients Over Europe for Calibration/Validation of GOCE Data.* Proceed. of GG2002 Meeting in Thessaloniki, 287-292.

ESA (1999). *Gravity Field and Steady-State Ocean Circulation - The Four Candidate Earth Explorer Core Missions, SP-1233(1).*

Haagmans, R. (2001). *GOCE The First Core Mission of the Earth Observation Envelope Programme and a Concept for Applications at Regional Scale.* ESA, recitation held in ECGS Luxembourg, JGL 89th, Nov. 2001.

Haagmans, R., Prijatna, K., and Dahl-Omang, O. (2002). *An Alternative Concept for Validation of GOCE Gradiometry Results Based on Regional Gravity.* Proceed. of GG2002 Meeting in Thessaloniki, 281-286.

Heck, B. (1979). *Zur lokalen Geoidbestimmung aus terrestrischen Messungen vertikaler Schweregradienten.* DGK, Reihe C, Heft 259, München.

Tscherning, C. C. and Rapp, R. H. (1974). *Closed covariance expressions for gravity anomalies, geoid undulations, and deflections of the vertical implied by anomaly degree variance models.* OSU Rep. 208, Columbus.

Wenzel, H. G. (1981). *Zur Geoidbestimmung durch Kombination von Schwereanomalien und einem Kugelfunktionsmodell mit Hilfe von Integralformeln.* Zeitschr. f. Verm.wesen, 106:102-111.

Wenzel, H. G. (1982). *Geoid Computation by Least Squares Spectral Combination using Integral Kernels.* Proceed. of the General Meeting of the IAG, Tokyo, 438-453.

# On the incorporation of sea surface topography in establishing vertical control

G. Fotopoulos[1], I.N. Tziavos[2], and M.G. Sideris[1]
[1]University of Calgary, Department of Geomatics Engineering, 2500 University Drive N.W., Calgary, AB, Canada, T2N 1N4

[2]Aristotle University of Thessaloniki, Department of Geodesy and Surveying, University Box 440, 54124, Thessaloniki, Greece.

## Abstract

One of the major sources of distortion in vertical control networks is caused by neglecting sea surface topography (SST) at tide gauge stations. Often, the orthometric height is fixed to zero at these stations without applying proper corrections for the deviation of the mean sea surface from the equipotential surface represented by the geoid. In view of the significant improvements in SST determination made in the past decade (particularly the low-to-medium frequencies) and the expected improvement in global gravity field models in the near future, it is appropriate to consider practical methods for the incorporation of SST into establishing vertical control. The purpose of this paper is to develop a consistent procedure for incorporating the mean SST values into the combined height network adjustment of terrestrial GPS-on-benchmark data and GPS-on-tide gauge data typically located in coastal areas, harbours, estuaries and/or river mouths. Two main issues that arise for the proper incorporation of SST information into the optimal heterogeneous height network adjustment include (i) the modelling of systematic errors and datum discrepancies among the height data types (ellipsoidal, orthometric, geoid and SST) using a corrector surface and (ii) the separation of random errors for estimating variance components for each height type. The limiting factor in all of these studies is data availability or rather lack of quality data and obtaining reliable initial covariance matrices for the height data in a particular region. However, in lieu of the increased need for cm-level accurate vertical control it is expected that this situation will be significantly improved in the near future.

## 1 Introduction: Traditional approach to establishing vertical control

Figure 1 depicts a typical scenario for the establishment of a reference benchmark to define a regional vertical datum used for vertical control. The tide gauge records the instantaneous sea level height $H_{ISL}$ and these values are averaged over a long term in order to obtain the mean value of the local sea level $H_{MSL}$. The height of the tide gauge is also measured with respect to a reference benchmark that is situated on land preferably (although not always) a short distance from the tide gauge station. Then the height of the reference benchmark above mean sea level $H_{BM}$ is computed by:

$$H_{BM} = H_{MSL} + \Delta H_{BM-TG} \tag{1}$$

Levelling begins from this benchmark and reference heights are accumulated by measuring height differences along levelling lines. The accuracy of the reference benchmark height derived in this manner is dependent on the precision of the height difference $\Delta H_{BM-TG}$ and the value for mean sea level $H_{MSL}$. If one assumes that the value for mean sea level is computed over a sufficiently long period of time which averages out all tidal period components and any higher frequency effects such as currents, then the accuracy depends on $\Delta H_{BM-TG}$.

For highly accurate heights as those needed for a cm-level vertical datum, the tide gauges cannot be assumed to be vertically stable. It is well known that land motion at tide gauges is a source of systematic error, which causes distortion in the height network if it is not corrected for. Land motion at tide gauges and reference benchmarks may be caused abruptly by earthquakes or by erosion or more subtle changes such as post-glacial rebound and land subsidence. The solution to this problem is to include independent space-based geodetic techniques such as GPS (or DORIS, GLONASS and in the future GALILEO) and satellite altimetry in order to estimate the land motion at these tide gauges (Kuo et al., 2004). An additional advantage of introducing satellite-derived heights is that it relates the regional vertical datum to a global vertical reference surface.

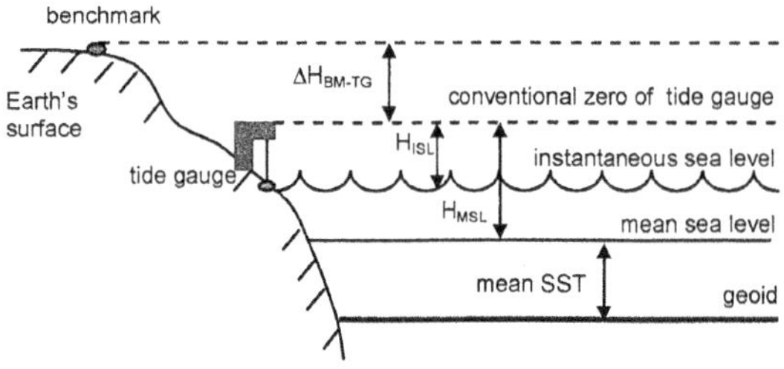


**Fig. 1** Establishment of vertical reference benchmark

As new methodologies and techniques evolve to the point where cm-level (and even sub-cm-level) accurate coordinates are needed, the distortions in traditionally-defined regional vertical networks are no longer acceptable. Modern alternatives to this problem have been proposed over the years (see e.g., Heck and Rummel, 1990). The following section outlines a method for combining the heterogeneous height data types in order to "modernize" vertical control.

## 2 Introducing GNSS: Optimal combination method

Descriptions of optimal combination schemes for incorporating global navigation satellite system (GNSS) derived ellipsoidal heights into vertical control are exhaustive in current geodetic literature. The procedure followed in this paper will be based on the optimal combination scheme for ellipsoidal ($h$), orthometric ($H$), and geoid ($N$) heights that has been thoroughly discussed in Fotopoulos (2003). A summary of the optimal combination approach is provided in Figure 2. The method addresses two main issues that arise for the proper adjustment of the heterogeneous height network, namely (i) the modelling of systematic errors and datum discrepancies among the height data types (ellipsoidal, orthometric, geoid) using a corrector surface and (ii) the separation of random errors for estimating variance components for each height type.

As input the required information includes the triplet of height values ($h, H, N$) at *each* GPS-levelling benchmark and the associated initial covariance (CV) matrices ($\mathbf{C}_h, \mathbf{C}_H, \mathbf{C}_N$). Given this network of points a combined least-squared adjustment can be performed in order to estimate the parameters of a corrector surface model, denoted by the bilinear term $\mathbf{a}^T\mathbf{x}$. This term is incorporated in order to deal with numerous datum inconsistencies and systematic effects inherent among the heterogeneous height data. There are several different types of models that can be used and numerous investigations have been presented in literature and will not be dwelled on herein. It is, however, important to stress that the selection procedure used to determine the best corrector surface model should be carefully evaluated/executed to avoid mis-interpretation of actual model performance.

The next step in the combined adjustment involves the estimation of variance components, denoted by $\boldsymbol{\theta} = [\sigma_h^2 \;\; \sigma_H^2 \;\; \sigma_N^2]^T$, for each of the height data types in order to re-evaluate/re-scale the input CV matrices. The variance component estimation (VCE) schemes used throughout are iterative versions of the minimum norm quadratic unbiased estimation (MINQUE; Rao, 1971) and the almost unbiased estimation (AUE; Horn and Horn, 1975) schemes. Details can also be found in Fotopoulos (2003).

This paper aims to address one of the omissions/approximations often made in establishing vertical control, namely that of sea surface topography. Although there are many other approximations/assumptions that should also be considered (i.e. instability of reference station due to land motion, neglecting river discharge corrections on tide gauge measurements, inexact normal/orthometric height corrections, using approximate/normal gravity values instead of actual surface gravity values for computing orthometric heights, etc.) it is appropriate to address this particular issue in view of the potential benefits from recent (and future) dedicated gravity field satellite missions to both geodesists and oceanographers. In the following section, a consistent procedure for incorporating the mean SST values into the combined height network adjustment problem described herein is developed. Emphasis is also placed on the evaluation of the accuracy of SST models, which is often overshadowed by, although not strictly independent of, geoid accuracy evaluation. As in the case with terrestrial height data, variance component estimation can play a vital role in improving/calibrating SST covariance information (from altimetric, oceanographic or assimilated schemes) at tide gauge stations and surrounding areas.

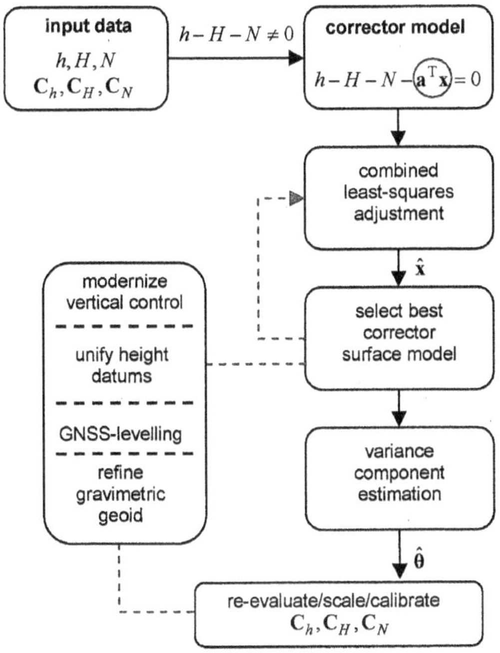


**Fig. 2** Simplified flowchart of main components for optimal combination of ellipsoidal, orthometric and geoid height values

## 3 Incorporating sea surface topography

A major source of distortion in vertical control networks is caused by neglecting sea surface topography at tide gauge stations. Often, the orthometric height is fixed to zero at these stations, without applying proper corrections for the deviation of the mean sea surface from the equipotential surface represented by the geoid (known as sea surface topography which is approximately ± 2 metres). Over the past decade or so, numerous advances have been made in modelling the low-to-medium frequency components of the sea surface, particularly through satellite altimetry measurements, which provide accurate and global uniform coverage (Andersen and Knudsen, 2000). Over the open seas, altimetric-derived stationary SST (the equivalent of GNSS-levelling at sea) is theoretically given by the following relationship:

$$SST = h_{alt} - N_m \qquad (2)$$

where $h_{alt}$ is the mean sea surface height with respect to geocentric reference ellipsoid derived from satellite altimetry measurements along the satellite tracks and the geoid heights, $N_m$, are interpolated at track points from a marine geoid (or global geopotential model). The principle satellite altimetry measurements and the relationship between the various height reference surfaces of interest is approximately illustrated in Figure 3.

In the past, the utility of this approach has been hindered by uncertainties in the global gravity field models and errors in the altimetric measurements (radial orbit error, atmospheric effects, tides, electromagnetic bias and measurement noise). Most of the errors affecting the altimetric ranges can be modelled and corrected leaving only residual effects at the cm-level. Thus, the dominant error source affecting the mean SST computed via Eq. (2) is due to the global gravity field model, which is expected to greatly improve. Although the performance of satellite altimetry is very good over open seas (i.e., 2 to 4 cm for TOPEX/POSEIDON), the measurement accuracy significantly deteriorates along the coast, over the shelf (which varies from tens to hundreds of kilometers off the coasts), in shallow depths, and fresh water inflows (Hipkin, 2000). Improvements in global SST models (including in coastal areas) are expected with the increase in accuracy of the satellite-only global geopotential models from the new and upcoming dedicated low-earth orbiting satellite gravity missions. Already, with preliminary results from the CHAMP and more recently the GRACE data, dramatic improvements in the long wavelength features of the SST models have been observed (Gruber and Steigenberger, 2000). The meso-scale features are also expected to be enhanced with the availability of the GOCE mission data. Furthermore, the accuracy of oceanographic methods for SST will also improve from the assimilation of the improved gravity field solutions into the estimation scheme. Therefore, it is likely that SST models will rely heavily on the altimetric method in the future. Along the coasts, it is most likely that this information will be combined with oceanographic methods, such as steric levelling and global ocean circulation models, for determining SST (Hipkin, 2000).

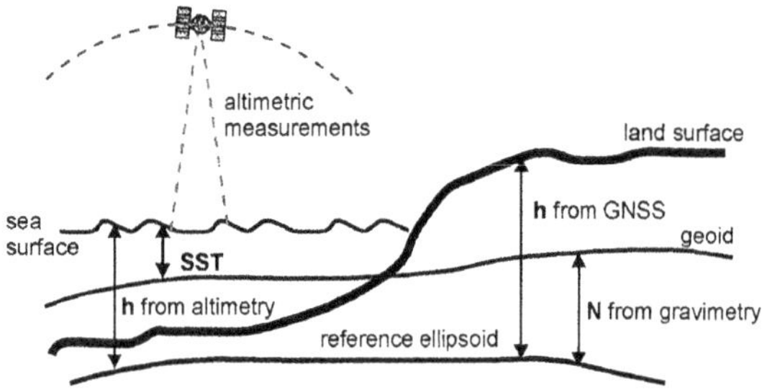


**Fig. 3** Satellite altimetry measurements and the relationship between various height reference surfaces

In view of the expected improvement in global gravity field models in the near future, it is appropriate to consider the incorporation of SST into establishing vertical control. The proposed scheme involves the combined adjustment of terrestrial GPS-on-benchmark data, as described in the previous section, in addition to GPS-on-tide gauge data typically located in coastal areas, harbours, estuaries, and/or river mouths. Obtaining reliable and accurate (mean) SST values at these tide gauge locations remains a challenge.

### 3.1 General methodology

To facilitate the discussion, consider a typical vertical control network consisting of $m$ terrestrial in-land stations (GPS-on-benchmarks) and $n$ stations on the coasts at tide gauges (GPS-on-tide gauges). It should be noted that in the algorithms provided below, the ellipsoidal height information at the tide gauges is typically determined through GPS measurements and the SST values may be interpolated from oceanographic, altimetric or a combined model at the tide gauge stations.

The new observation equation model for this *mixed* (in terms of input data types) adjustment is described as follows:

$$\ell_i = h_i - \widetilde{H}_i - N_i = \mathbf{a}_i^{\mathrm{T}} \mathbf{x} + v_i\,, \quad i = 1,2,\ldots,p \tag{3}$$

where $p = m + n$ is the total number of stations, $\widetilde{H}_i$ for terrestrial points refers to the orthometric height $\widetilde{H}_i = H_i$ (see Heiskanen and Moritz, 1967 for details). At the tide gauge stations the observation equation model is modified to accommodate SST values such that $\widetilde{H}_i = SST_i$. The new vector of observations consists of two types of observations as expressed by

$$\ell = \begin{bmatrix} \ell_I \\ \ell_{(I)} \end{bmatrix} = \begin{bmatrix} h_j - H_j - N_j \\ h_k - SST_k - N_k \end{bmatrix}, \quad \begin{matrix} j = 1,2,\ldots,m \\ k = 1,2,\ldots,n \end{matrix} \tag{4}$$

where the terrestrial points are contained in the $(m \times 1)$ subvector $\ell_I$ and the observations at the tide gauge stations comprise the $(n \times 1)$ subvector $\ell_{(I)}$. In virtually all practical network situations, there will be fewer tide gauge stations than benchmarks on land, $n < m$.

The solution for the unknown parametric model coefficients is a straight-forward application of a combined least-squares adjustment and is given by

$$\hat{\mathbf{x}} = \left[\mathbf{A}^{\mathrm{T}} \mathbf{C}^{-1} \mathbf{A}\right]^{-1} \mathbf{A}^{\mathrm{T}} \mathbf{C}^{-1} \ell \tag{5}$$

It should be noted from the above equation that the corrector surface parameters are computed and applied to all stations (terrestrial and tide gauge) and therefore remains common throughout. The CV matrix for the disjunctive observations is formulated as follows:

$$\mathbf{C} = \mathbf{C}_h + \mathbf{C}_{\widetilde{H}} + \mathbf{C}_N \tag{6}$$

which is shown in block-diagonal form

$$\mathbf{C} = \begin{bmatrix} \mathbf{C}_{h_I} + \mathbf{C}_{H_I} + \mathbf{C}_{N_I} & \mathbf{0} \\ \mathbf{0} & \mathbf{C}_{h_{(I)}} + \mathbf{C}_{SST_{(I)}} + \mathbf{C}_{N_{(I)}} \end{bmatrix}$$

The individual CV matrices are defined as follows:

$$\begin{aligned} &\mathbf{C}_{h_I} = \sigma_h^2 \mathbf{Q}_{h_I}, \mathbf{C}_{h_{(I)}} = \sigma_h^2 \mathbf{Q}_{h_{(I)}} \\ &\mathbf{C}_{H_I} = \sigma_H^2 \mathbf{Q}_{H_I} \\ &\mathbf{C}_{N_I} = \sigma_N^2 \mathbf{Q}_{N_I}, \; \mathbf{C}_{N_{(I)}} = \sigma_N^2 \mathbf{Q}_{N_{(I)}} \\ &\mathbf{C}_{SST_{(I)}} = \sigma_{SST}^2 \mathbf{Q}_{SST_{(I)}} \end{aligned} \tag{7}$$

where $\mathbf{Q}_{h_I}, \mathbf{Q}_{H_I}, \mathbf{Q}_{N_I}$ are the positive-definite cofactor matrices for the ellipsoidal, orthometric and geoid heights, respectively, on land. $\mathbf{Q}_{h_{(I)}}$ and $\mathbf{Q}_{N_{(I)}}$ are the positive-definite cofactor matrices for the ellipsoidal and geoid height data, respectively, at the tide gauge stations. It is worth noting that Eq. (6) is a simplified form of the stochastic model, which assumes no correlation among the different height types. For instance, $H$ and $N$ are correlated through the mapping of the height of the gravity stations into the gravity anomaly and therefore the geoid height. However, the treatment of these cross-correlations among the heterogeneous height types is beyond the scope of this paper.

The vector of random errors is modified to include the new observation type as follows:

$$\mathbf{v} = \left[\mathbf{v}_h^{\mathrm{T}} \quad \mathbf{v}_{\widetilde{H}}^{\mathrm{T}} \quad \mathbf{v}_N^{\mathrm{T}}\right]^{\mathrm{T}} \tag{8}$$

where $\mathbf{v}_{(\cdot)}$ is a vector of random errors for each of the $h, H\ (SST), N$ data types. The corresponding covariance matrix is described by the general formulation $E\{\mathbf{v}\mathbf{v}^T\} = \mathbf{C}_\mathbf{v}$, where $E\{\cdot\}$ is the mathematical expectation operator.

Finally, the separate adjusted residuals, according to height data type can be resolved as shown in Kotsakis and Sideris (1999). In this particular case, the explicit formulations for the ellipsoidal and geoid height residuals are somewhat modified to accommodate the $\mathbf{C}_{\widetilde{H}}$ term and is provided by

$$\hat{\mathbf{v}}_h = \mathbf{C}_h(\mathbf{C}_h + \mathbf{C}_{\widetilde{H}} + \mathbf{C}_N)^{-1}\mathbf{M}\ell \tag{9a}$$

$$\hat{\mathbf{v}}_N = \mathbf{C}_N(\mathbf{C}_h + \mathbf{C}_{\widetilde{H}} + \mathbf{C}_N)^{-1}\mathbf{M}\ell \tag{9b}$$

where, $\mathbf{M} = \mathbf{I} - \mathbf{A}(\mathbf{A}^{\mathrm{T}}\mathbf{C}^{-1}\mathbf{A}^{\mathrm{T}}\mathbf{C}^{-1})$. A slight alteration is applied to the separate adjusted residuals corresponding to $\widetilde{H}$ and is computed as follows:

$$\hat{\mathbf{v}}_{\widetilde{H}} = \begin{bmatrix} \hat{\mathbf{v}}_H \\ \hat{\mathbf{v}}_{SST} \end{bmatrix} = -\begin{bmatrix} \mathbf{C}_H & 0 \\ 0 & \mathbf{C}_{SST} \end{bmatrix} \cdot \left\{\begin{bmatrix} \mathbf{C}_H & 0 \\ 0 & \mathbf{C}_{SST} \end{bmatrix} + \mathbf{C}_h + \mathbf{C}_N\right\}^{-1} \mathbf{M} \begin{bmatrix} \ell_I \\ \ell_{(I)} \end{bmatrix} \tag{9c}$$

where $\mathbf{C}_{SST} = E\{\mathbf{v}_{SST}\mathbf{v}_{SST}^{\mathrm{T}}\}$ is the covariance matrix for the SST data at the tide gauge stations in the network. Past studies on constructing CV matrices for the SST values involve applying error propagation to the spherical harmonic series expansion of SST and incorporating degree variances (see Levitus, 1982, Hwang, 1996 and Lemoine et al., 1998).

### 3.2 Estimating variance components

The convenience and flexibility of the MINQUE variance component estimation procedure is demonstrated through the effective incorporation of a fourth variance component, $\sigma_{SST}^2$, for the sea surface topography values at the tide gauge stations in the network. We start with the general form of the stochastic model for the observations given by

$$\mathbf{C} = \sum_{\alpha=1}^{u} \theta_\alpha \mathbf{T}_\alpha \tag{10}$$

where $\boldsymbol{\theta} = \left[\sigma_h^2 \quad \sigma_H^2 \quad \sigma_N^2 \quad \sigma_{SST}^2\right]^{\mathrm{T}}$ contains the four unknown variance components and the $\mathbf{T}_i$ matrix for each height data type is explicitly stated by

$$\mathbf{T}_h = \begin{bmatrix} \mathbf{Q}_{h_I} & \mathbf{Q}_{h_{I(I)}} \\ \mathbf{Q}_{h_{(I)I}} & \mathbf{Q}_{h_{(I)}} \end{bmatrix}, \ \mathbf{T}_H = \begin{bmatrix} \mathbf{Q}_{H_I} & 0 \\ 0 & 0 \end{bmatrix},$$

$$\mathbf{T}_h = \begin{bmatrix} \mathbf{Q}_{N_I} & \mathbf{Q}_{N_{I(I)}} \\ \mathbf{Q}_{N_{(I)I}} & \mathbf{Q}_{N_{(I)}} \end{bmatrix}, \ \mathbf{T}_{SST} = \begin{bmatrix} 0 & 0 \\ 0 & \mathbf{Q}_{SST_{(I)}} \end{bmatrix} \tag{11}$$

Since a common adjustment is performed for **all** of the ellipsoidal and the geoid heights (regardless if they are situated at terrestrial benchmarks or at tide gauge stations), the cross-covariance information between the two sets of points for the ellipsoidal and geoid heights is included and denoted by $\mathbf{Q}_{h_{I(I)}} = \mathbf{Q}_{h_{(I)I}}^{\mathrm{T}}$ and $\mathbf{Q}_{N_{I(I)}} = \mathbf{Q}_{N_{(I)I}}^{\mathrm{T}}$, respectively. This also results in a common variance factor to be estimated for each of the ellipsoidal and geoid heights. The positive-definite cofactor matrix for the sea surface topography values is given by $\mathbf{Q}_{SST_{(I)}}$. The decomposition of the stochastic model through the use of the $\mathbf{T}_{(\cdot)}$ matrices in Eq. (10) allows for unbalanced data to be used for the heterogeneous height data types and for more than one variance component to be estimated for the group of heights collectively denoted by $\widetilde{H}$, which may prove to be a valuable characteristic when dealing with real data of various sources.

Substituting the above equations into the general rigorous MINQUE formulation, the four unknown variance components are estimated by solving the following system:

$$\mathbf{S}\hat{\boldsymbol{\theta}} = \mathbf{q} \tag{12}$$

where each element, $s_{\alpha\beta}$, in the $\mathbf{S}$ matrix is computed from

$$s_{\alpha\beta} = tr(\mathbf{RQ}_{\alpha}\mathbf{RQ}_{\beta}), \quad \alpha, \beta = h, H, N, SST \tag{13}$$

and the vector $\mathbf{q}$ is composed of the quadratic forms as follows:

$$\mathbf{q} = \begin{bmatrix} \ell^{\mathrm{T}}\mathbf{R}^{\mathrm{T}}\mathbf{T}_{h}\mathbf{R}\ell \\ \ell^{\mathrm{T}}\mathbf{R}^{\mathrm{T}}\mathbf{T}_{H}\mathbf{R}\ell \\ \ell^{\mathrm{T}}\mathbf{R}^{\mathrm{T}}\mathbf{T}_{N}\mathbf{R}\ell \\ \ell^{\mathrm{T}}\mathbf{R}^{\mathrm{T}}\mathbf{T}_{SST}\mathbf{R}\ell \end{bmatrix} \tag{14}$$

The matrix $\mathbf{R}$ is computed from $\mathbf{R} = \mathbf{C}^{-1}\mathbf{M}$ by substituting the appropriate form of the covariance matrix for the observations as given in Eq. (6). The above formulation is applied iteratively until all unknown variance components converge satisfactorily.

## 4 Remarks and future work

The use of SST values at tide gauge stations used to establish vertical control is still plagued with practical limitations. Tide gauges are situated near coastal areas and even with the use of satellite altimetry, which has revolutionized sea surface observations and greatly improved SST models in open oceans, the performance in coastal areas is still quite poor. Global ocean circulation models derived from satellite altimetry data and hydrostatic models may reach accuracies of 2-3 cm in the open oceans, but the models fall apart in shallow coastal areas giving uncertainties on the order of tens of centimetres. As discussed in Kearsley (2004), the accuracy of the $N$ values at the tide gauge stations still suffers from deficiencies in off-shore gravity field information, which can be improved through airborne gravimetry or by connecting to a first order benchmark located sufficiently inland where local gravity is adequate for a precise geoid solution. Therefore, with significant problems still looming in the coastal regions, distortions will be evident in heights referred to a vertical datum that is defined with low accuracy SST models. However, as realistic estimates of the accuracy of these models becomes available and a better understanding of the process of tide gauge observations (e.g., reference benchmark stability), estimates of the accuracy of the observations can be made. Future work will involve implementing the procedure using real data and testing SST error models derived from altimetric, oceanographic or hybrid models.

## References


Andersen O.B. and Knudsen P. (2000) The role of satellite altimetry in gravity field modelling in coastal areas. *Physics and Chemistry of the Earth*, vol. 25, no. 1, pp. 17-24.

Fotopoulos G. (2003) An analysis on the optimal combination of geoid, orthometric and ellipsoidal height data. *Ph.D. Thesis*, University of Calgary, Department of Geomatics Engineering Report Number 20185.

Gruber T. and Steigenberger P. (2002) Impact of new gravity field missions for sea surface topography determination. *Proceedings of the 3rd Meeting of the International Gravity and Geoid Commission,* Tziavos (Ed.), Thessaloniki, Greece, Aug. 26-30, pp. 320-325.

Heck B. and Rummel R. (1990) Strategies for solving the vertical datum problem using terrestrial and satellite geodetic data. *IAG Symposia*, vol. 104, Sünkel H and Baker T (Eds.), Springer-Verlag, pp. 116-128.

Heiskanen W.A. and Moritz H. (1967) Physical Geodesy. W.H. Freeman and Company San Francisco.

Hipkin R. (2000) Modelling the geoid and sea surface topography in coastal areas. *Physics and Chemistry of the Earth*, vol. 25, no. 1, pp. 9-16.

Horn S.D. and Horn R.A. (1975) Comparison of estimators of heteroscedastic variances in linear models. *Journal of the American Statistical Association*, vol. 70, Issue 352, pp. 872-879.

Hwang C. (1996) A study of the Kuroshio's seasonal variabilities using an altimetric-gravimetric geoid and TOPEX/POSEIDON altimeter data. *Journal of Geophysical Research*, Vol. 101, pp. 6313-6335.

Kearsley A.H.W. (2004) Unification of vertical datums. Report on IAG Commission X Working Group 3 on the Worldwide Unification of Vertical Datums 1999 to 2003, UNSW, Sydney, Australia.

Kotsakis C. and Sideris M.G. (1999) On the adjustment of combined GPS/levelling/geoid networks. *Journal of Geodesy*, vol. 73, no. 8, pp. 412-421.

Kuo C.Y., Shum C.K., Braun A. and Mitrovica J.X. (2004) Vertical crustal motion determined by satellite altimetry and tide gauge data in Fennoscandia, *Geophys. Res. Lett.*, 31, L01608, doi:10.1029/2003GL019106.

Lemoine F.G., et al. (1998) The development of the joint NASA GSFC and the National Imagery and Mapping Agency (NIMA) geopotential model EGM96. NASA Technical Publication -1998 - 206861, July 1998.

Levitus S. (1982) Climatological Atlas of the World Ocean. NOAA/ERL GFDL Professional Paper 13, Princeton, N.J., 173 pp.

Rao C.R. (1971) Estimation of Variance Components - MINQUE Theory. *Journal of Multivariate Statistics*, vol.1, pp.257-275.

# A New Gravimetric Geoidal Height Model over Norway Computed by the Least-Squares Modification Parameters

H. Nahavandchi, A. Soltanpour, E. Nyrnes

Norwegian University of Science and Technology, Division of Geomatics, 7491 Trondheim, Norway

**Abstract.** Over the last decade, there has been an increased interest in the determination of the geoid. This is mainly due to the demands for height transformation from users of GPS. Costly conventional levelling operations can be replaced with quicker and cheaper GPS-levelling surveys, as long as the geoidal height is computed to a high accuracy. Therefore, there is a common goal among geodesists to determine "the 1-cm geoid model". This study uses a least-squares procedure to compute the modification parameters for the geoidal height determination over Norway. For the computation of the long-wavelength contribution, the new Global Geopotential Models (GGM) GGM01s from GRACE twin-satellites is used.

**Keywords.** Stokes' formula, geoid, global gravity model, modification, topography

## 1 Introduction

The geoid is the equipotential surface of the Earth's gravity field (more or less) coinciding with mean sea level. Civil engineers use it as the reference surface for elevations while oceanographers use it for studies of ocean circulation, currents and tides. It is also valuable to geophysicists for geodynamics studies, geophysical interpretation of the Earth's crust, and prospecting. These types of applications require knowledge of the geoid with a precision of ±1-10 cm. The new technologies of satellite altimetry and satellite positioning have also placed a high demand on precise geoid determination research. To obtain such high accuracy, terrestrial (or airborne) gravity observations in a combination with a GGM must be employed and the modification procedures have to be realized.

An accurate solution of the boundary-value problem in physical geodesy has usually been found using Stokes' well-known formula (1849) for the anomalous gravity potential, with the geoidal height calculated through Bruns' formula. During the past decade, gravimetric geoid computations have been an active research topic in most countries: in USA (Milbert 1995); in Canada (Sideris and She 1995) (Vanicek et al. 1996); in Europe (Denker et al. 1997); in Scandinavia (Forsberg et al. 1996) are among them.

## 2 Geoidal Height Determination

Removing the effect of external masses to the geoid (topographic and atmospheric masses) or reducing them inside the geoid (the direct effect) is a requirement of Stokes' formula. This correction applies to both long- and short-wavelength parts of the final geoid. The effects of masses are then restored after applying Stokes' integral (primary indirect effect). There is also another indirect effect resulting from a free-air correction of gravity from geoid to co-geoid, i.e. secondary indirect effect.

Gravimetric geoidal height can then be computed by the following formula (see e.g. Heiskanen and Moritz 1967, Nahavandchi 2004):

$$N(R,\varphi,\lambda) = N_1(R,\varphi,\lambda) + N_2(R,\varphi,\lambda) + \delta N_{\mathrm{Corr}} \quad (1)$$

where

$$\delta N_{\mathrm{Corr}} = \delta N^{\mathrm{atm}}_{\mathrm{total}}(R,\varphi,\lambda) + \delta N^{\mathrm{top}}_{\mathrm{sec\ ind}}(R,\varphi,\lambda) + \delta N^{\mathrm{top}}_{\mathrm{ind}}(R,\varphi,\lambda) + \delta N^{\mathrm{top}}_{\mathrm{dir}}(R,\varphi,\lambda) \quad (2)$$

N is the final gravimetric geoid computed at a point on the geoid with latitude φ, longitude λ and mean radius R, and

$\delta N^{\mathrm{atm}}_{\mathrm{total}}(R,\varphi,\lambda)$ is the total effect (direct plus indirect effects ) of atmospheric masses,

$\delta N^{\mathrm{top}}_{\mathrm{sec\ ind}}(R,\varphi,\lambda)$ is the secondary indirect effect due to the topography,

$\delta N^{\mathrm{top}}_{\mathrm{ind}}(R,\varphi,\lambda)$ is the primary indirect effect on geoid due to topography,

$\delta N^{\mathrm{top}}_{\mathrm{dir}}(R,\varphi,\lambda)$ is the direct effect of topography on the long-wavelength part of geoidal heights

$N_1(R,\varphi,\lambda)$ is the short-wavelength contribution to the final geoid computed from Stokes' integral

$N_2(R,\varphi,\lambda)$ is the long-wavelength contribution to the final geoid determined from a GGM.

The modification methods differ from each other by the choice of the modification parameters and modification procedure. Next section discusses a least-squares method.

## 3 Least-Squares Modification

Sjöberg (1984) proposes the truncation error due to the incomplete integration area, erroneous terrestrial gravity data and potential harmonic errors be reduced in a least-squares sense. Such a modification can be written (Sjöberg, 1984):

$$N_1 + N_2 = N^{\mathrm{Sj}} = \frac{R}{4\pi\gamma}\iint_{\sigma_0} S_M(\psi)\Delta g(R,\varphi,\lambda)d\sigma + c\sum_{n=2}^{M}(Q_{Nn} + s_n)\Delta g_n \quad (3)$$

where

$$Q_{Nn} = Q_n - \sum_{k=2}^{N}\frac{2k+1}{2}s_k e_{nk} \quad (4)$$

and

$$S_M(\psi) = S(\psi) - \sum_{k=2}^{M}\frac{2k+1}{2}s_k p_k(\cos\psi) \quad (5)$$

where $S(\psi)$ is the original Stokes' formula, ψ is the spherical distance between computation and running points, Δg is the gravity anomaly on the geoid, $\gamma$ is normal gravity on the ellipsoid, $\sigma$ is the unit sphere and $\sigma_0$ is the cap of integration.

The modification parameters $s_n$ in the least-squares model are computed from the system of linear equation (Sjöberg, 1984):

$$\sum_{r=2}^{M} a_{kr}s_r = h_k \qquad k = 2,3,\cdots,M \quad (6)$$

where

$$a_{kr} = (\sigma_r^2 + dc_r)\delta_{kr} - \frac{2k+1}{2}\sigma_r^2 e_{kr} - \frac{2r+1}{2}\sigma_k^2 e_{rk} + \frac{2k+1}{2}\frac{2r+1}{2}\sum_{n=2}^{M} e_{nk}e_{nr}(\sigma_n^2 + c_n) \quad (7)$$

and

$$h_k = \frac{2\sigma_k^2}{k-1} - Q_k\sigma_k^2 + \frac{2k+1}{2}\sum_{n=2}^{M}\left[Q_n e_{nk}(\sigma_n^2 + c_n) - \frac{2}{n-1}e_{nk}\sigma_n^2\right] \quad (8)$$

Here

$$e_{kn}(\psi_0) = \int_{\psi_0}^{\pi} P_k(\cos\psi) P_n(\cos\psi) \sin\psi \, d\psi \quad (9)$$

and

$$Q_k(\psi_0) = \int_{\psi_0}^{\pi} S(\psi) P_k(\cos\psi) \sin\psi \, d\psi \quad (10)$$

are the Molodensky truncation coefficients. $\delta_{kr}$ is Kronecker's delta symbol. The gravity anomaly degree variance $c_n$ can be computed from

$$c_n = \frac{(GM)^2}{a^4} (n-1)^2 \sum_{m=0}^{n} (C_{nm}^2 + S_{nm}^2) \quad (11)$$

using potential coefficients $C_{nm}$ and $S_{nm}$ from a GGM. GM is the product of the universal gravitational constant G and the mass of the Earth M and a is the equatorial radius of the reference ellipsoid. The gravity anomaly error degree variance, due to erroneous potential coefficients is computed from

$$dc_n = \frac{(GM)^2}{a^4} (n-1)^2 \sum_{m=0}^{n} (\delta_{C_{nm}}^2 + \delta_{S_{nm}}^2) \quad (12)$$

where $\delta_{Cnm}$ and $\delta_{Snm}$ are the standard deviations of potential coefficients taken from a GGM. The error degree variances for the terrestrial gravity anomalies ($\sigma_n^2$) can be computed from (Sjöberg, 1986)

$$C(\psi) = c_1 \left[ \frac{1-\Omega}{\sqrt{1-2\Omega\cos\psi+\Omega^2}} - (1-\Omega) - (1-\Omega)\cos\psi \right] \quad (13)$$

where $\sigma_n^2$ are expressed from

$$\sigma_n^2 = c_1 (1-\Omega)\Omega^n \quad (14)$$

The parameters $c_1$ and $\Omega$ are determined from the values of C(0)=10 $\mathrm{mGal}^2$ and a correlation length of $0.1^\circ$ in this study.

## 4 Corrections

### 4.1 Direct Topographic Correction

The classical integral formula for direct effect on gravity anomaly is (Moritz 1980)

$$\delta\Delta g_{\mathrm{dir}}^{\mathrm{top}}(R,\varphi,\lambda) = \frac{\mu R^2}{2} \iint_{\sigma_0} \frac{(H-H_P)^2}{\ell_0^3} d\sigma \quad (15)$$

where H and $H_p$ are the heights of the running and computation points, respectively, $\mu = G\rho_0$ where $\rho_0$ is the density of topography, and

$$\ell_0 = R\sqrt{2(1-\cos\psi)} = 2R\sin\frac{\psi}{2} \quad (16)$$

In determining the long-wavelength part of the geoidal heights from a GGM, one must expect a bias of the external harmonic series when applied at the geoid within the topographic masses. This bias can be corrected by removing the effects of topographical masses using following formula (Nahavandchi and Sjöberg 1998):

$$\delta N_{\mathrm{direct}}^{\mathrm{topography}} = -\frac{2\pi\mu}{\gamma} \sum_{n,m} \frac{n+2}{2n+1} (H^2)_{nm} Y_{nm}(P) - \frac{2\pi\mu}{R\gamma} \sum_{n,m} \frac{(n+2)(n+1)}{3(2n+1)} (H^3)_{nm} Y_{nm}(P) \quad (17)$$

where

$$(H^\upsilon)nm = \frac{1}{4\pi} \iint_{\sigma} H_P^\upsilon Y_{nm} d\sigma, \qquad \upsilon = 2, 3 \quad (18)$$

and $Y_{nm}$ are fully normalized spherical harmonics.

### 4.3 Primary Indirect Topographic Correction

The classical formula for determining the indirect topographic effect on the geoid is (Wichiencharoen 1982)

$$\delta N_{\mathrm{ind}}^{\mathrm{top}}(R,\varphi,\lambda) = -\frac{\pi\mu H^2}{\gamma} - \frac{\mu R^2}{6\gamma} \iint_{\sigma_0} \frac{H^3 - H_P^3}{\ell_0^3} d\sigma \quad (19)$$

## 5 Numeical Investigations

### 5.1 Data Sources

The area of study is entire Norway. The intention is to determine a new gravimetric geoid in this area.

The gravity data were provided by Norwegian Mapping Authority (NMA). The height data with 1 km resolution were available from National Imagery and Mapping Agency (NIMA). The mean free-air anomalies are derived in 3 x 6 minutes geographic cells, ranging from -141.170 mGal to 193.908 mGal with an average of 1.927 mGal, and a standard deviation of ± 24.708 mGal. The height coefficients $(H^{\upsilon})_{nm}$ (to degree and order 360) were determined using Eq (18). For this, a 30' × 30' Digital Terrain Model (DTM) was generated using the GETECH 5' ×5' DTM (GETECH 1995) and averaged using area weighting. The parametric definitions are: G=6.673 $\times 10^{-11}$ $m^3kg^{-1}s^{-2}$, $\rho_0$ = 2670 $kgm^{-3}$, $R$=6371 km, and $\gamma$ = 981 Gal. In all the integral formulas, the 1 × 1 km DTM (NIMA) is used. The numerical values of the coefficients $Q_n$ and $e_{kj}$ are computed using the recursive algorithms given by Paul (1973) and Hagiwara (1976).

### 5.2 Corrections

The direct topographic effect on gravity in Stokes' formula, computed from Eq. (15), is plotted in Figure 1. This effect (in the study area) is significant.

The direct topographic effect on geopotential [Eq. (17)], to degree 360, is directly computed on the geoid and illustrated in Figure 2. This effect is smaller than the topographic effect in Stokes' integral. It varies from -79 mm to +43 mm.

The primary indirect topographic effect, computed from Eq. (19), is illustrated in Figure 3. This effect ranges to a value of -22 cm.

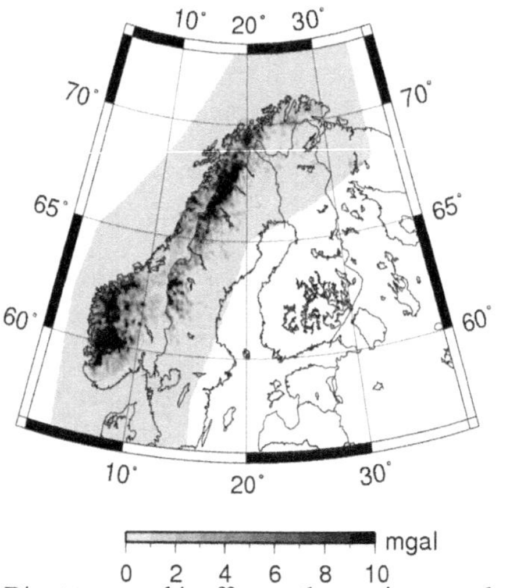


**Fig. 1** Direct topographic effect on the gravity anomaly

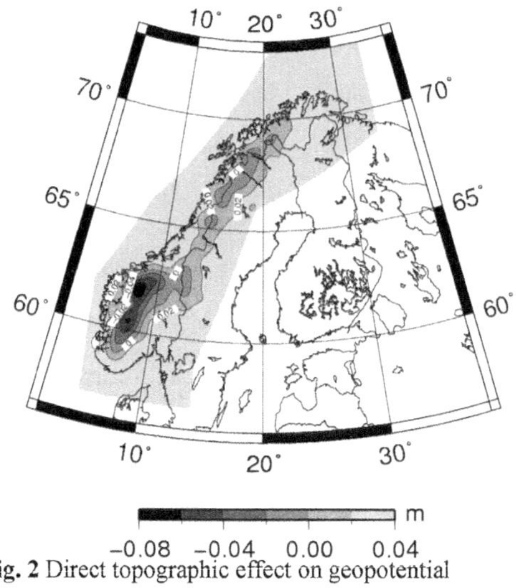


**Fig. 2** Direct topographic effect on geopotential

Three less significant corrections are also applied in this study: Secondary indirect topographic effect (Nahavandchi and Sjöberg 1998), the total atmospheric effect (Sjöberg and Nahavandchi 2000) and ellipsoidal effect (Moritz 1980). The first correction varies from -20 to 30 mm, the total atmospheric correction varies from -30 to 32 mm and the third correction has a maximum value of 83 μGal to the gravity anomaly.

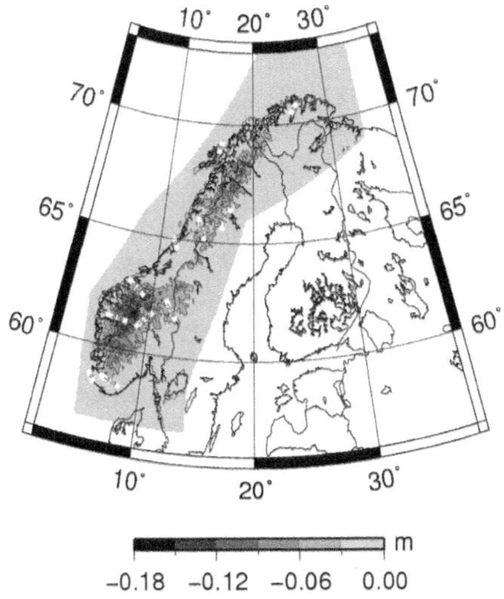


**Fig. 3** Primary indirect topographic effect

### 5.3.2 Comparisons and Final Computations

Comparison of the gravimetric geoid with a sufficient GPS-levelling data provides a means of estimating, and removing, possible systematic error sources in the gravity and levelling data as well as GPS measurements. The geoid computed by GPS-levelling data is here called geometric geoid. The differences between the geometric and gravimetric geoids in this study have an average of 0.23 m, and a standard deviation of ±0.36 m. 344 GPS-levelling stations are used for the comparison. Figure 4 shows the distribution of the GPS-levelling stations. The final gravimetric geoid is shown in Figure 5. It is referenced to the geocentric GRS80 ellipsoid.

In addition, the offsets between the gravimetric and geometric geoidal heights are minimized by introducing the four-parameter regression model. To realize this regression formula, the final gravimetric geoid model is fitted to the same 344 GPS-levelling points. The evaluation of the suitability of a gravimetric geoid model can also be based on the post-fitting residuals. The RMS of post fitting residuals is 32.6 cm. The post-fitting residuals are demonstrated in Figure 6, ranging from -78.9 cm to 87.9 cm. For the most of Norway (80 percent), the absolute value of residuals is within 40 cm.

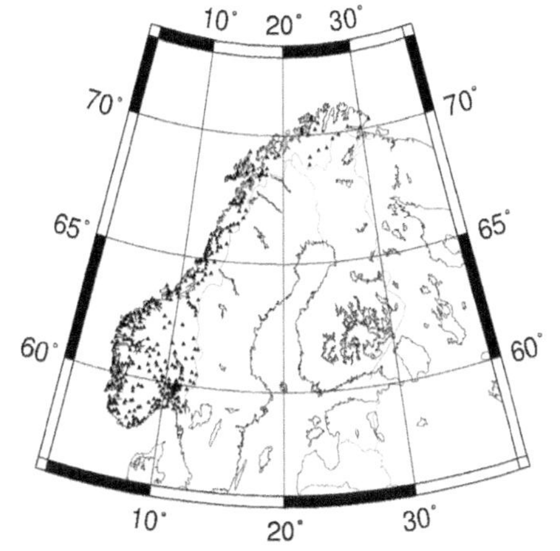


**Fig. 4** Distribution of the GPS-levelling stations

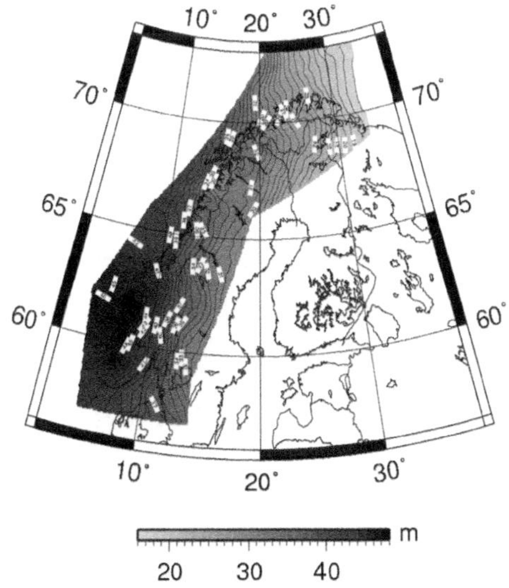


**Fig. 5** Final geoid of Norway with the least-squares modification parameters

Figure 6 also shows that the largest differences between the gravimetric and geometric geoid heights are located in the area above the parallel 65. This area has the least density of gravity data in this study. We believe that using a denser data set (DTM and gravity data) will decrease the pre- and post- fitting residuals, significantly, to 1-dm level. It should also be mentioned that part of these discrepancies could be attributed to the hypothesis for the density of the topographic masses.

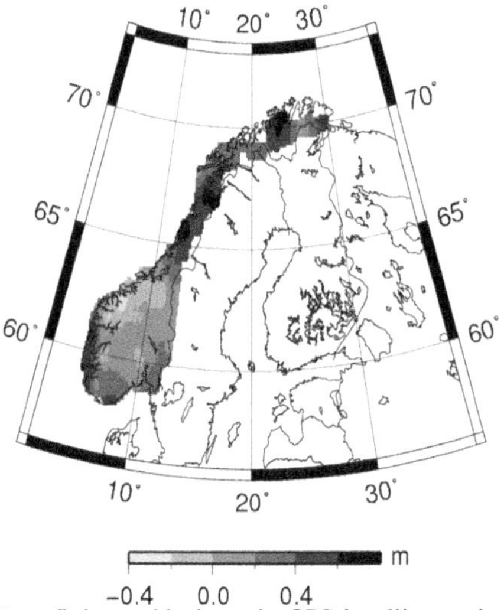

**Fig. 6** Post-fitting residuals on the GPS-levelling stations

## 6 Discussions

A new high-resolution geoid model for Norway has been computed. The least-squares modification parameters are used in the geoid computations. This study also employs the new GRACE model GGM01s. The comparison of the gravimetric geoid with GPS-levelling-derived geoid provides an agreement in a few decimeter levels. However, these somehow large discrepancies are mostly resulted from the incomplete data sets. It is believed this procedure, improved with the use of recent techniques for topographic corrections, would result to a 1-dm gravimetric geoid model if all available data were provided.

## References


Denker H, Behrend D, Torge W (1997) The European gravimetric quasi-geoid EGG96. In: Gravity, Geoid and Marine Geodesy. Springer, Berlin, pp. 532-540.

Forsberg R, Kaminskis J, Solheim D (1996) Geoid for the Nordic and Baltic Region from Gravimetry and Satellite Altimetry. Proceeding of International Symposium on Gravity, Geoid and Marine geodesy (GRAGEOMAR), Tokyo, IAG SympVol. 117, pp 540-547. Berlin:Springer-verlag.

GETECH (1995) Global DTM5. Geophysical Exploration Technology (GETECH), University of Leeds, Leeds.

Hagiwara Y (1976) A new formula for evaluating the truncation error coefficients. Bull Geod, 50:131-135.

Heiskanen WA, Moritz H (1967) Physical Geodesy. W H Freeman and Company, San Francisco.

Milbert DG (1995) Improvement of a high resolution geoid model in the United States by GPS height on NAVD88 benchmarks. IGeS Bulletin 4:13-36.

Moritz H (1980) Advanced physical geodesy. Herbert Wichman Verlag, Karlsruhe

Nahavandchi H (2004) The quest for a precise geoidal height model. Kart og Plan 1: 46-56.

Nahavandchi H, Sjöberg LE (1998) Terrain correction to power $H^3$ in gravimetric geoid determination. Journal of Geodesy 72: 124-135.

Paul NK (1973) A method of evaluating the truncation error coefficients for geoidal height. Bulletin Geodesique 110: 413-425.

Sideris MG, She BB (1995) A new high resolution geoid for Canada and part of the US by the 1D-FFT method. Bulletin Geodesique 69:107-118.

Sjöberg LE (1984) Least squares modification of Stokes's and Vening Meinesz' formulas by accounting for errors of truncation, potential coefficients and gravity data, Department of Geodesy, University of Uppsala, No. 27, Uppsals, Sweden.

Sjöberg LE (1986) Comparison of some methods of modifying Stokes's formula. Bollettino di Geodesia e Scienze Affini 3: 229-248.

Sjöberg LE, Nahavandchi H (2000) The atmospheric geoid effects in Stokes formula. Geophysical Journal International 140: 95-100.

Stokes GG (1849) On the variation of gravity on the surface of the Earth. Trans Camb Phil Soc 8: 672-695.

Vanicek P, Kleusberg A, Martinec Z, Sun W, Ong P, Najafi M, Vajda P Harrie L, Tomasec P, Ter Horst B (1996a) Compilation of a precise regional geoid. Department of Geodesy and Geomatics Engineering, Technical Report 184, University of New Brunswick, Fredericton.

Wichiencharoen C (1982) The indirect effects on the computation of geoid undulations. Rep 336, Department of Geodetic Science, The Ohio State University, Columbus.

# On the Accuracy of Vertical Deflection Measurements Using the High-Precision Digital Zenith Camera System TZK2-D

Christian Hirt
Institut für Erdmessung, Universität Hannover, Schneiderberg 50, 30167 Hannover, Germany
E-mail: hirt@mbox.ife.uni-hannover.de Fax: +49 511 762 4006

Birger Reese
Institut für Photogrammetrie und GeoInformation, Universität Hannover,
Nienburger Str. 1, 30167 Hannover, Germany
E-mail: reese@ipi.uni-hannover.de Fax: +49 511 762 2483

Heinz Enslin
Alstedder Str. 180, 44534 Lünen, Germany, Fax: +49 2306 50021

**Abstract.** In this paper the accuracy of vertical deflections $(\xi, \eta)$ provided by the Digital Zenith Camera System TZK2-D is comprehensively analysed. During 2003 and 2004 various vertical deflections have been measured repeatedly at particular sites with different types of comparative data available. Emphasis is placed on the presentation of comparative measurements with highly accurate reference data at the Hamburg PZT station. As main result, the external accuracy level of the deflection data has been found to be about $0''.10$ - $0''.15$ as such exceeding considerably the accuracy of the formerly used photographic zenith cameras. The high accuracy level makes the Digital Zenith Camera System TZK2-D a very powerful system for highly accurate geoid determination in local areas.

**Keywords.** Digital Zenith Camera System, Photographic Zenith Tube (PZT), vertical deflection, accuracy

## 1 Introduction

At the University of Hannover, the Digital Zenith Camera System TZK2-D (Transportable Zenitkamera 2 - Digitalsystem, Figure 1 and 5) has been constructed for the automated determination of vertical deflections $(\xi, \eta)$ and is described in Hirt (2004, 2003); Hirt and Bürki (2002); Hirt and Seeber (2002); Hirt (2001). In the course of instrumental implementation attention has been laid on the development of appropriate calibration methods. After completing the development and calibration phase, various vertical deflection measurements have been carried out at about 100 stations in Northern Germany and Switzerland (cf. Brockmann et al. 2004; Hirt 2004; Hirt and Reese 2004; Müller et al. 2004; Reese 2004). An important issue is the estimation of the accuracy level of the observed deflections. Therefore, comparative and repeated measurements have been extensively performed at selected stations in Hamburg, Hannover and Benthe. In this contribution some exemplary results are presented, demonstrating the accuracy level reached for the deflection data $(\xi, \eta)$. The deflection data listed in this paper is on the reference system ETRS89. For precise ellipsoidal position data of the measurement stations presented in this contribution the reader is referred to Hirt (2004).

## 2 Comparative Measurements at the Hamburg PZT station

The site of the Photographic Zenith Tube (PZT) in Hamburg is considered as one of the most precise astronomical reference stations in Germany. The Hamburg PZT (see Figure 2) has been operated within the framework of the international services for monitoring Earth rotation, the Bureau International de l'Heure and the International Polar Motion Service, over a period of almost three decades. (e.g. Enslin 1964, 1972; BIH 1984; IPMS 1984). In 1986, the operational use of the PZT signed off. Due to the high-precision of astronomical PZT-observations being in the order of $0''.1$, the site of the former PZT is well-suited for comparisons with modern astrogeodetic instrumentations like the Digital Zenith Camera TZK2-D. On the one hand comparative measurements are used to estimate the external accuracy of the deflection data. On the other hand larger remaining systematic errors may be revealed.

At the Hamburg PZT station, the system TZK2-D has been extensively used for comparative measurements over five nights during winter 2003 and spring 2004. The internal accuracy (precision) is obtained

**Figure 1.** The Digital Zenith Camera System TZK2-D (during observation at the former PZT site in Hamburg). The depiction shows the complete system including electronic components for data acquisition and processing. All parts of the system are compactly mounted on a vehicle simplifying the astrogeodetic field work.

by repeated measurements during the same night. Table 1 exemplarily shows results obtained on 9th December 2003. Considering the spreading of single solutions, the internal accuracy is found to be about 0.″08. Vertical deflection measurements carried out during other nights show comparable internal accuracy estimates for the mean values (Hirt 2004).

**Table 1.** TZK2-D measurements at Hamburg PZT station on 20031209

| Data set | $\xi$ [″] | $\eta$ [″] | $r_\xi$ [″] | $r_\eta$ [″] |
|---|---|---|---|---|
| 20031209_2000 | 1.88 | -0.94 | -0.08 | 0.01 |
| 20031209_3000 | 1.72 | -0.88 | 0.08 | -0.05 |
| 20031209_4000 | 1.87 | -1.07 | -0.07 | 0.14 |
| 20031209_5000 | 1.78 | -0.87 | 0.02 | -0.06 |
| 20031209_6000 | 1.76 | -0.91 | 0.04 | -0.02 |
| Mean | 1.80 | -0.93 | | |
| Std.dev. | | | **0.07** | **0.08** |

**Figure 2.** The formerly used Hamburg Photographic Zenith Tube (photo courtesy of Bundesamt für Seeschifffahrt und Hydrographie, H. Besser)

The mean $(\xi, \eta)$ values obtained in five different nights are listed in Table 2, the nominal (reference) values provided by PZT observations are given by Table 3. Assuming that the PZT numbers correspond to the true values, the external accuracy of both components is 0.″14. The remaining, non-significant differences of 0.″11 in $\xi$ and 0.″06 in $\eta$ underline the high degree of correspondence between both types of deflection data.

**Table 2.** TZK2-D measurements at Hamburg PZT station in Winter 2003 and Spring 2004. The last columns contain the differences $(\varepsilon_\xi, \varepsilon_\eta)$ between the TZK2-D observations and the PZT reference values.

| Date | $\xi$ [″] | $\eta$ [″] | $\varepsilon_\xi$ [″] | $\varepsilon_\eta$ [″] |
|---|---|---|---|---|
| 20031209 | 1.80 | -0.93 | -0.01 | 0.17 |
| 20040325 | 2.01 | -0.99 | 0.20 | 0.11 |
| 20040413 | 2.04 | -1.10 | 0.23 | 0.00 |
| 20040415 | 1.89 | -1.26 | 0.08 | -0.16 |
| 20040424 | 1.87 | -0.93 | 0.06 | 0.17 |
| Mean | 1.92 | -1.04 | | |
| Std.dev. | | | **0.14** | **0.14** |

Table 3. Comparison at PZT station Hamburg. The last columns contain the differences $\Delta\xi$, $\Delta\eta$ with respect to the TZK2-D mean values.

| Data set | $\xi$ [″] | $\eta$ [″] | $\Delta\xi$ [″] | $\Delta\eta$ [″] |
|---|---|---|---|---|
| TZK2-D Mean | 1.92 | -1.04 | | |
| PZT Values | 1.81 | -1.10 | **0.11** | **0.06** |
| EGG97 | 2.15 | -1.12 | **-0.23** | **0.08** |

At the same site computed deflections based on the European Gravimetric Geoid EGG97 (Torge and Denker 1999) could also be used for a comparison (Table 3). The mean accuracy of the computed deflections is specified to be in the order of 0″.2 (Denker 2004, 1988). The comparison shows non-significant differences of -0″.23 ($\xi$) and 0″.08 ($\eta$) between observed and computed vertical deflections.

A remarkable, larger difference of 0″.34 in the $\xi$-component between the PZT and EGG97 values reveals a discrepancy of the comparison data sets. Due to their accuracy specification, the difference is believed to come for the most part from the computed deflections based on EGG97. It should be positively noticed that the vertical deflection values provided by the TZK2-D are located within the range formed by the comparison data (component $\xi$). Although the reference data is not of higher-level accuracy in comparison to those of the TZK2-D, the results clearly indicate an external accuracy level of at least 0″.15.

## 3 Comparative Measurements in Hannover

In order to determine the repeatability and to obtain very precise reference coordinates, repeated measurements over 14 nights over a period of 13 months have been carried out at a reference station at the University of Hannover between February 2003 and March 2004. As reference station, a stable pillar located on the roof of the geodetic institute has been selected.

The deflection data acquired covers a wide spectrum of environmental conditions. Besides daytime and season-dependent residual effects, most different weather situations (temperature, air pressure) and resulting refraction anomalies are implicitly contained in the deflection data as well as the impact of various star fields taken from star catalogues Tycho-2 (Høg et al. 2000) and UCAC (Zacharias et al. 2004) used for astrometric data reduction.

Table 4. Comparison at station Hannover

| Data Set | $\xi$ [″] | $\eta$ [″] | $\Delta\xi$ [″] | $\Delta\eta$ [″] |
|---|---|---|---|---|
| TZK2-D Mean | 6.47 | 1.15 | | |
| EGG97 | 6.60 | 1.15 | **-0.13** | **0.00** |

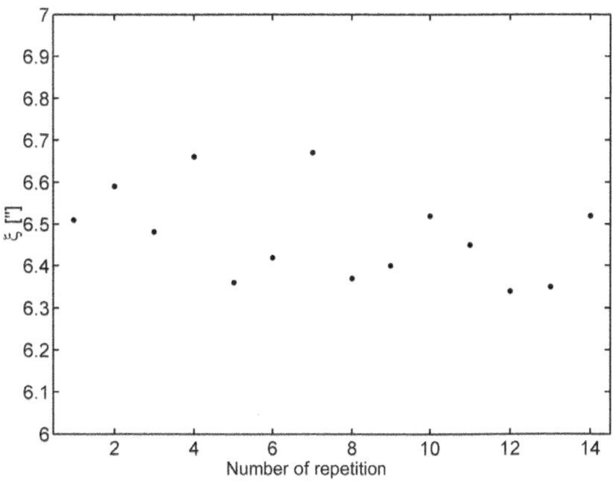


Figure 3. TZK2-D observations of $\xi$ at station Hannover

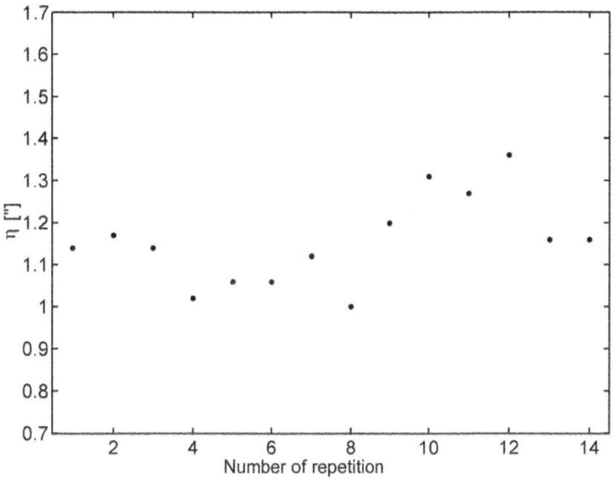


Figure 4. TZK2-D observations of $\eta$ at station Hannover

Figures 3 and 4 show the observation results for the Hannover station. The standard deviation of the $(\xi, \eta)$ mean values obtained at different nights is 0″.11 for both components; the scattering interval obtained as max-min values of the observation series is less than 0″.35. In conclusion, the observation of vertical deflections using the system TZK2-D reveals a high repeatability under changing environmental conditions. A comparison between measured and computed deflections (from EGG97) shows an excellent consistency with non-significant differences of 0″.13 in $\xi$ and 0″.00 in $\eta$ (Table 4).

**Figure 5.** The Digital Zenith Camera System TZK2-D at station Hannover

## 4 Comparative Measurements in Benthe

At Benthe near Hannover, 10 different stations have been occupied twice in various nights. Regarding the residuals of the double measurements, standard deviations below 0.″10 are obtained (cf. Hirt and Reese 2004; Hirt 2004). At one station, measured and computed deflections were compared (Table 5). In both components, the discrepancies are less than 0.″10 as such being non-significant.

**Table 5.** Comparison at station Benthe

| Data Set | $\xi$ [″] | $\eta$ [″] | $\Delta\xi$ [″] | $\Delta\eta$ [″] |
|---|---|---|---|---|
| TZK2-D Mean | 7.07 | 2.71 | | |
| EGG97 | 7.01 | 2.80 | **0.06** | **-0.09** |

## 5 Discussion

The accuracy of vertical deflection measurements using the Digital Zenith Camera TZK2-D has been comprehensively investigated at different sites in Northern Germany. Based on the results obtained both by means of numerous repeated and comparative measurements, a reasonable external accuracy estimate is 0.″10 to 0.″15.

Compared to the accuracy level of the Digital Zenith Camera System TZK2-D, the accuracy of formerly used analogue (photographic) zenith cameras has been found to be in the order of 0.″5 (cf. Seeber and Torge 1985; Wissel 1982; Bürki 1989). Hence, the accuracy level of the Digital Zenith Camera TZK2-D exceeds those of the analogue zenith cameras considerably by a factor of 3 to 5.

Some application fields significantly benefit by the accuracy improvement achieved. Highly precise deflection data provided by the system TZK2-D opens a new level of accuracy for local astrogeodetic geoid determinations. Even the millimeter-range can be attained economically in local areas by means of astronomical levelling. Typical applications are, for example, the validation of geoid models, the analysis of the fine structure of the gravity field and the localization of density anomalies below the surface (cf. Hirt and Reese 2004). In mountainous regions such as the European Alps highly precise determined vertical deflections serve as valuable observation of the Earth's gravity field for geoid computations (cf. Brockmann et al. 2004; Müller et al. 2004; Marti 2004).

A couple of reasons in combination are responsible for this significant accuracy improvement compared to analogue zenith cameras. Above all, the high-precision star catalogues Tycho-2 and UCAC enable the highly accurate determination of vertical deflections. They provide access to the celestial reference system with external accuracies for star positions of a few 0.″01 (cf. Zacharias et al. 2000). Furthermore, a high number of repeated observations (usually at least 60 single solutions per station) reduces residual error sources significantly. But also the application of state-of-the-art digital technology for data acquisition and a refined modelling of instrumental systematics contribute to the increased accuracy level.

The difference between the internal accuracy, found to be about 0.″08, and external accuracy level (0.″10 to 0.″15) most likely comes from zenithal refraction anomalies resulting in tilt of atmospheric layers (cf. Dimopoulos 1982, Ramsayer 1967). Such anomalies can falsify the measured direction of the plumb line and thus the vertical deflection data. This effect is considered to limit the accuracy of the vertical deflection measurements. In order to overcome these restrictions future research should deal with the analysis of atmospheric data sets (e.g. digital weather models) to correct for zenith refraction.

## Acknowledgement

The development of the Digital Zenith Camera System has been financed by the DFG (Deutsche Forschungsgemeinschaft, German National Research Foundation). The field work has been supported by Dipl.-Ing. René Käker and cand.-geod. Tobias Krömer. The authors would like to thank the Landesbetrieb Geoinformation und Vermessung Hamburg (LGV) for recovering the site of the formerly used PZT.

## References

BIH (1984). *Annual Report for 1983.* Bureau International de l'Heure, Paris.

Bürki, B. (1989). *Integrale Schwerefeldbestimmung in der Ivrea-Zone und deren geophysikalische Interpretation.* Geodätisch-geophysikalische Arbeiten in der Schweiz, Nr. 40, Schweizerische Geodätische Kommission.

Brockmann, E., Becker, M., Bürki, B., Gurtner, W., Haefele, P., Hirt, C., Marti, U., Müller, A., Richard, P., Schlatter, A., Schneider, D. and Wiget, A. (2004). *Realization of a Swiss Combined Geodetic Network (CH-CGN).* EUREF'04 Symposium of the IAG Commission 1 - Reference Frames, Subcommission 1-3a Europe (EUREF), Bratislava, Slovakia.

Denker, H. (1988). *Hochauflösende regionale Schwerefeldbestimmung mit gravimetrischen und topographischen Daten.* Wissenschaftliche Arbeiten der Fachrichtung Vermessungswesen der Universität Hannover Nr. 156.

Denker, H. (2004). Personal communication.

Dimopoulos, T. (1982). *Untersuchungen über die Genauigkeit der Ermittlung der astronomischen Refraktion.* Dissertation, Fakultät Bauingenieur- und Vermessungswesen der Universität Stuttgart.

Enslin, H. (1964). *Der Breitendienst des Deutschen Hydrographischen Institutes.* Zeitschrift für Vermessungswesen 89: 266-279.

Enslin, H. (1972). *The Variation of Mean Latitude of the Hamburg PZT.* Colloquium Nr. 1 of the IAU, La Plata, Argentina: 71-76.

Hirt, C. (2001). *Automatic Determination of Vertical Deflections in Real-Time by Combining GPS and Digital Zenith Camera for Solving the GPS-Height-Problem.* Proc. 14th International Technical Meeting of The Satellite Division of the Institute of Navigation: 2540-2551, Alexandria, Virginia.

Hirt, C. (2003). *The Digital Zenith Camera TZK2-D - A Modern High Precision Geodetic Instrument for Automatic Geographic Positioning in Real-Time.* Astronomical Data Analysis Software and Systems XII, Astronomical Society of the Pacific Conference Series Vol. 295, San Francisco: 155-159.

Hirt, C. (2004). *Entwicklung und Erprobung eines digitalen Zenitkamerasystems für die hochpräzise Lotabweichungsbestimmung.* Wissenschaftliche Arbeiten der Fachrichtung Geodäsie und Geoinformatik der Universität Hannover Nr. 253.

Hirt, C. and Bürki, B. (2002). *The Digital Zenith Camera - A New High-Precision and Economic Astrogeodetic Observation System for Real-Time Measurement of Deflections of the Vertical.* Proc. of the 3rd Meeting of the International Gravity and Geoid Commission of the International Association of Geodesy, Thessaloniki, Greece (ed. I. Tziavos): 161-166.

Hirt, C. and Reese, B. (2004). *High-Precision Astrogeodetic Determination of a Local Geoid Profile Using the Digital Zenith Camera System TZK2-D.* Electronic Proc. IAG GGSM2004 Symposium, Porto, Portugal.

Hirt, C. and Seeber, G. (2002). *Astrogeodätische Lotabweichungsbestimmung mit dem digitalen Zenitkamerasystem TZK2-D.* Zeitschrift für Vermessungswesen 127: 388-396.

Høg, E., Fabricius, C., Makarov, V. V., Urban, S., Corbin, T., Wycoff, G., Bastian, U., Schwekendiek, P. and Wicenec, A. (2000). *The Tycho-2 Catalogue of the 2.5 Million Brightest Stars.* Astronomy and Astrophysics 355: L27-L30.

IPMS (1984). *Annual Report of the International Polar Motion Service for the year 1983.* Central Bureau of the IPMS, Mizusawa.

Marti, U. (2004). *High Precision combined geoid determination in Switzerland.* Proc. IAG GGSM2004 Symposium, Porto, Portugal.

Müller, A., Bürki, B., Hirt, C., Marti, U. and Kahle, H.-G. (2004). *First Results from New High-precision Measurements of Deflections of the Vertical in Switzerland.* Proc. IAG GGSM2004 Symposium, Porto, Portugal.

Ramsayer, K. (1967). *Investigation on Errors in the Determination of Astronomical Refraction.* Proc. Intern. Symp. Figure of the Earth and Refraction, Wien. Published in: Österreichische Zeitschrift für Vermessungswesen, Sonderheft 25: 260-269.

Reese, B. (2004). *Untersuchungen zur hochpräzisen und wirtschaftlichen Lotabweichungsbestimmung mit dem digitalen Zenitkamerasystem TZK2-D.* Diploma thesis at the Institut für Erdmessung of the Universität Hannover, unpublished.

Seeber, G. and Torge, W. (1985). *Zum Einsatz transportabler Zenitkameras für die Lotabweichungsbestimmung.* Zeitschrift für Vermessungswesen 110: 439-450.

Torge, W. and Denker, H. (1999). *Zur Verwendung des Europäischen Gravimetrischen Quasigeoids EGG97 in Deutschland.* Zeitschrift für Vermessungswesen 124: 154-166.

Wissel, H. (1982). *Zur Leistungsfähigkeit von transportablen Zenitkameras bei der Lotabweichungsbestimmung.* Wissenschaftliche Arbeiten der Fachrichtung Vermessungswesen der Universität Hannover Nr. 107.

Zacharias, N., Urban, S. E., Zacharias, M. I., Wycoff, G. L., Hall, D. M., Monet, D. G. and Rafferty, T. J. (2004). *The Second US Naval Observatory CCD Astrograph Catalog (UCAC2).* The Astronomical Journal 127: 3043-3059.

Zacharias, N., Zacharias, M. I., Urban, S. E. and Høg, E. (2000). *Comparing Tycho-2 astrometry with UCAC1.* The Astronomical Journal 120: 1148-1152.

# TERRA: A feasibility study on local geoid determination in Bolivia with strapdown inertial airborne gravimetry.

M. Giménez, I. Colomina, J. J. Rosales, M. Wis
Institute of Geomatics, Generalitat de Catalunya & Universitat Politècnica de Catalunya
Av. del Canal Olímpic, ES-08860 Castelldefels, Spain

C. C. Tscherning
Department of Geophysics
University of Copenhagen, Juliane Maries Vej 30, DK-2100 Copenhagen, Denmark

E. Vásquez
Instituto Geográfico Militar
Estado Mayor del Ejército Av. Saavedra 2303, La Paz, Bolivia

**Abstract.** Strapdown inertial airborne gravimetry as a practical and feasible tool for local and regional geoid determination is under research at the Institute of Geomatics (IG). Within the framework of the Spanish-Bolivian cooperation project TERRA, between the Instituto Geográfico Militar of Bolivia (IGM) and the IG, a feasibility analysis on geoid determination in Bolivia by means of INS/GNSS integration is being carried out. TERRA includes, among other goals, the specification of the geoid model that best fits the needs of Bolivia from a global point of view. Simulations in the spatial and spectral domain for GNSS and INS data will allow for an assessment of the performance of the technology. In addition, an empirical covariance model of the gravity anomalies has been derived from CHAMP-data, so that realistic simulations will be carried out. The contribution of other new satellite gravity missions, such as —GOCE—, will be evaluated as well. Actual gravity data will be obtained from a test flight over various topographic conditions to analyze the system response to a highly variable gravity signal. The data will be also processed to empirically verify the simulation results. The aim of this paper is to introduce the project and present some preliminary results.

**Keywords.** Strapdown inertial airborne gravimetry, INS/GNSS integration, geoid specification, simulation, CHAMP, GOCE

## 1 Introduction

Around the middle of the last decade, the Instituto Geográfico Militar of Bolivia implemented a qualitative improvement of the national geodetic infrastructures. This general aim included the establishment of a GNSS-based geodetic network, with passive and active stations and coordinates in the SIRGAS (Sistema de Referencia Geocéntrico para América del Sur) reference system, the determination of a low-medium resolution digital elevation model (DEM) for the whole country, the revision of the leveling network and its extension to new lines and to the permanent stations, the expansion of the gravimetric network and, finally, the determination of a high resolution local geoid for Bolivia.

Within the framework of this main project, the Institute of Geomatics is conducting a feasibility study on the use of strapdown inertial airborne gravimetry for geoid determination, under both technical and economical points of view.

## 2 State-of-the-art

Since the early stages of gravity measurement for geoid determination, airborne gravimetry has been considered as the optimum technology for covering a huge area with a reasonable dense set of values. Stable platform-based modified gravity meters have been tested in some experimental flights (Brozena et al., 1997) and even for production-level missions (Klingelé et al., 1994).

Strapdown inertial airborne gravimetry has been a matter of research since the last decade, see (Schwarz and Wei, 1998) and (Forsberg et al.,

1996). Both approaches show apparently similar performances (Glennie et al., 2000). However, to the best knowledge of the authors, very few geoid determination campaigns have up to now been conducted with this technology (Li and Schwarz, 2000), although it has slight operational advantages and a lower cost of purchase —not operating— costs (Bruton et al. 2001). Those test flights have demonstrated an accuracy of to 2-3 cm in the geoid undulations, limited to the high frequencies (Schwarz and Wei, 1998); the low and medium frequency content are usually derived from global geopotential models, which adds several decimeters to the error budget.

Gravity-devoted satellite missions are intended to partially fill the gap between the accurate long wavelength provided by global models and airborne-derived data. The CHAMP mission will provide data with a maximum resolution of 250 km (Reigber et al., 2003); GRACE-derived data will have reasonable accuracy for resolutions below 150 km (García, 2002) and the future European mission GOCE (ESA, 1999) will recover wavelengths even below 100 km.

# 3 The TERRA project

The aim of TERRA is to give answers to the following questions: What kind of geoid (i.e. specification of the geoid) fits the general needs of Bolivia? What budget of geodetic infrastructures are required for the determination of the geoid by means of the technology under research? Which should be the features —specification— of the inertial measurement unit (IMU) or combination of them to be used in order to meet the required resolution and accuracy? To what extent will the combination of airborne and satellite gravimetry avoid or minimize the need for terrestrial campaigns? And, last but not least: What kind of earth observation mission should be planned in order to make the economical effort become affordable for the country?

The key factor in the TERRA project is the comparison between the spectral characteristics of the INS/GNSS assembly-derived geoid and its combination with satellite data and the geoid specification for Bolivia.

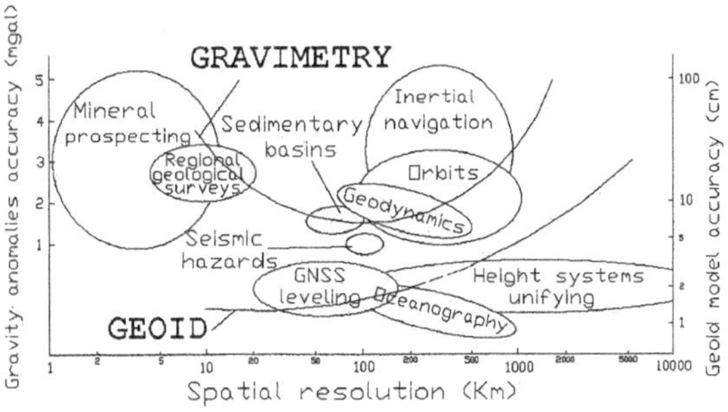


Figure 1: Preliminary geoid specification.

## 3.1 Geodetic and social context analysis

One of the main objectives of the study is to establish the features of the geoid model for Bolivia. This model should meet the needs of the country from a general point of view, that is, not only technical but also socioeconomic reasons.

The geoid specification will take into account the opinion of local potential users and will be supervised by Prof. Carl Christian Tscherning. The requirements for the various depicted applications are taken from (ESA, 1999).

Documentary research and personal talks with members of the Marine and Photogrammetric Service of the Bolivian Army lead to a preliminary draft of the geoid specification shown in Figure 1. Natural resources have a strategic interest for the country; therefore, special attention is paid to the suitability of gravity data for mineral prospecting from local to basin scales. The geoid specification reflects the need of Bolivia for a unified physical height system —a must in hydrographic engineering, i.e. for channel planning—. The proposal takes also into account the advantages of turning GNSS into a leveling tool for engineering issues.

## 3.2 Test flight

In order to empirically assess the results that come from simulations, two experimental complementary flights will be conducted between August and October this year. The airplane, a Beechcraft King C-90 is owned by the Aviation Company of the Bolivian Army.

The first one is intended to cover a reasonably extensive area and will follow a photogrammetric path near La Paz, with main direction N - S; it will allow for cross-over checking and for external comparison with upward-continued in-

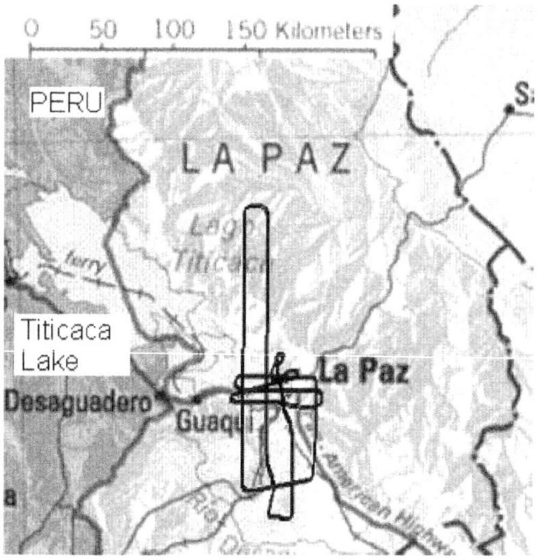


Figure 2: Trajectory of flight 1.

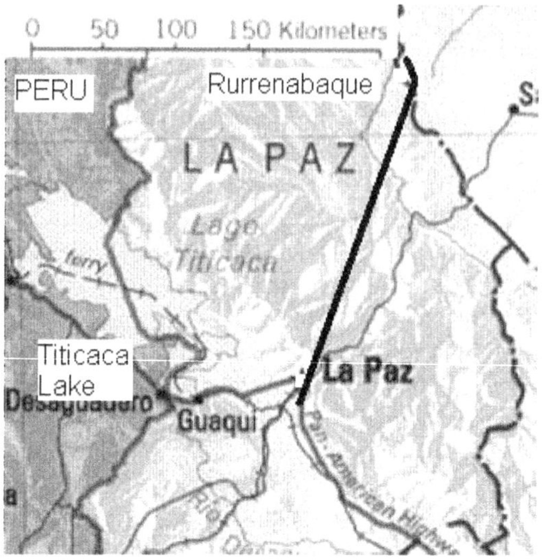


Figure 3: Trajectory of flight 2.

dependent data. Although the straight lines are no longer than 200 km, the topographic variation of the overflown terrain will be significant. Its trajectory can be seen in Figure 2.

The second flight has a different purpose. It will consist of a long straight line between La Paz and Rurrenabaque, in the Amazonic rainforest. Its length, about 250 km, will allow an empirical estimation of the maximum wavelength that can be recovered with the Institute of Geomatics' INS/GNSS system, within a certain range of accuracy. The final purpose is to evaluate how would this value combine with the current satellite gravimetry resolution capacities. An added value is that it will overfly the whole range of topographic signatures, from 4000 m height in La Paz airport (El Alto) to more than 5000 in the Andes, north from La Paz, and then going down to 300 in the rainforest. Some interesting conclusions are therefore expected from this flight, whose trajectory is shown in Figure 3.

### 3.2.1 The INS/GPS system

The device to be flown is called TAG (Trajectory, Attitude and Gravimetry), that mainly consists of an IMU/GNSS assembly with fully operational capabilities. Figure 4 shows the system, with the IMU inside the small metallic case in close-up. The IMU will be installed in an Aluminium bench (Figure 5 shows an assembly of the present IMUs owned by the IG, actually placed in the bench). The inertial sensor is a Northrop Grumman —former Litton— LN-200. For more information about the system the reader is referred to (Wis et al., 2004).

### 3.2.2 Data processing

A preliminary processing of the trajectory will be carried out by means of an extended Kalman Filter with smoothing. The extension accounts for a state for the gravity disturbance.

Nevertheless, a new geodetic approach is under development at the IG. It is intended to avoid the problems of Kalman Filter when dealing with gravity, whose correlation is basically space- and not time- dependent. The new approach takes advantage of cross-over data, ZUPT (Zero-velocity Update Point), CUPT (Coordinate Update Point) and any other information that may be supplied to the system. More information about this topic can be found in (Colomina and Blázquez, 2004) and (Térmens and Colomina, 2004).

## 3.3 Simulations

An IMU-simulator is under development at the IG. It will allow for measuring the impact of the sensor features in the final error estimates and, for this reason, recommendations will be made about the kind of sensor assembly (accelerometers, gyros, number of IMUs or maybe the use of redundant units) to be used if a certain level of accuracy has to be reached.

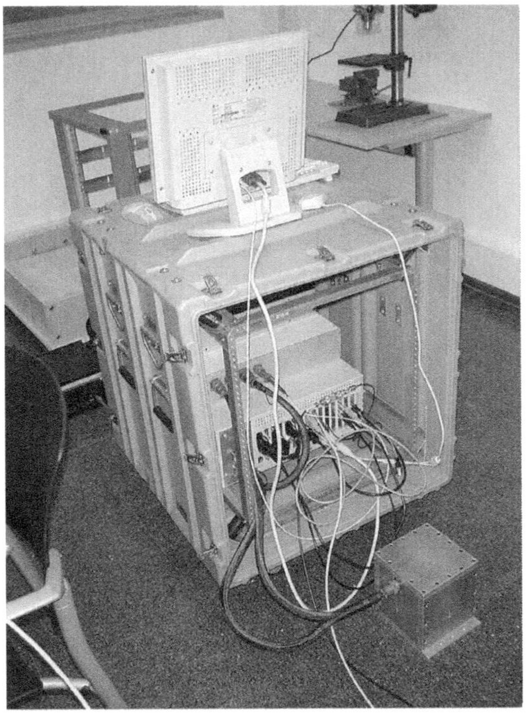

Figure 4: The TAG System

Figure 5: IMU Bench.

Meanwhile, by using least-squares collocation (LSC), see (Moritz, 1980) it is possible to obtain error estimates for the quality of height anomalies (quasi-geoid heights) from various combinations of data and from various types of gravity field variations. It is only necessary to know the position, type and error estimate of the data and its statistical characteristics (covariance function). The GRAVSOFT package (Tscherning et al., 1992) has been used to execute the computations described below.

Files concerning the simulations can be found in http://cct.gfy.ku.dk/bolivia/bolivia.htm.

In the preliminary stages of the project, we have computed error-estimates for a part of SW of Bolivia. The topography has been obtained from DTM2002 (Saleh and Pavlis, 2002).

CHAMP-data (Howe and Tscherning, 2004), were used to generate gravity anomalies at an altitude of 300 m above the terrain in order to simulate 32400 airborne gravity data. The creation of this data-set was in principle not necessary since the simulations —as mentioned above— may be carried out without using real data. Only positions and error-estimates of the data are needed.

As a reference field, EGM96 to degree 24 was used. The effect of the residual terrain was also subtracted from the CHAMP data.

This data was used to generate empirical covariance functions in two sub-areas,(-24 deg. - -13 deg. in latitude and -70.0 deg. -63 deg. in longitude) one with high mountains and one in the lowland, (-24 deg. to -13 deg. in latitude and -63 deg. to -55 deg. in longitude).

The main result is that a gravity field variance of 413.5 $mgal^2$ and 1226.6 $mgal^2$ were found in the low and in the high area, respectively, for data from which EGM96 to deg. 24 and residual topographic effects have been subtracted. The value for the high area is probably underestimated. Real data is necessary in order to obtain a better value.

These values were then used in the computation of simulated error estimates of height anomalies in two sub-areas within the two areas, bounded by (-24.0 -13.0) in latitude and (-63.0 -55.0) in longitude, and (-24.0 -13.0) in latitude and (-70.0 -63.0) in longitude. About 3000 CHAMP height anomalies and 2000 gravity anomalies were used in each case.

The subtraction of EGM96 to degree 24 from the data, results in a height anomaly standard deviation of about 3 m. Using the CHAMP data without topography then permits the prediction of height anomalies with overall errors around 1 m. The error will be larger in the high area and lower in the low area.

Using gravity data with a 5' spacing in the above mentioned sub- areas, the error decreased to between 0.14 m and 0.43 m for the low area and between 0.18 m and 0.56 for the high area, see Figure 6 and Figure 7.

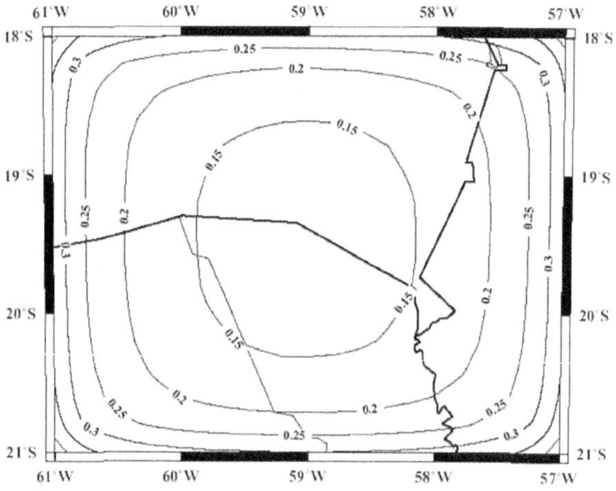


Figure 6: Error estimate for a low-height area.

### 3.4 The contribution of the new satellite mission GOCE

Some studies have already shown that the expected resolution of the ESA GOCE mission for a centimetric geoid accuracy is at the 70 km level (Suenkel, 2002). If the current limitation of airborne gravimetry can be extended up to this level, the combination of data from both sources might be a definite solution for production- level gravimetric geoid determination with no need for terrestrial campaigns.

Within the TERRA project, the performance of the ESA GOCE mission in combination with airborne gravimetry will be estimated by means of both an empirical and a numerical-stochastic analysis based on simulations.

## 4 Final remarks and further developments

A local geoid model is nowadays a necessary infrastructure. Bolivia aspires to have such a model with a certain accuracy in the following years, as a tool for development.

Given the special conditions of Bolivia in terms of size, orographic complexity, logistic problems and difficulty of access to extensive parts of the country, the Instituto Geográfico Militar of Bolivia has seen in strapdown inertial airborne gravimetry a suitable solution, which is being evaluated by the Institute of Geomatics.

The first task will be the specification of the geoid model in Bolivia.

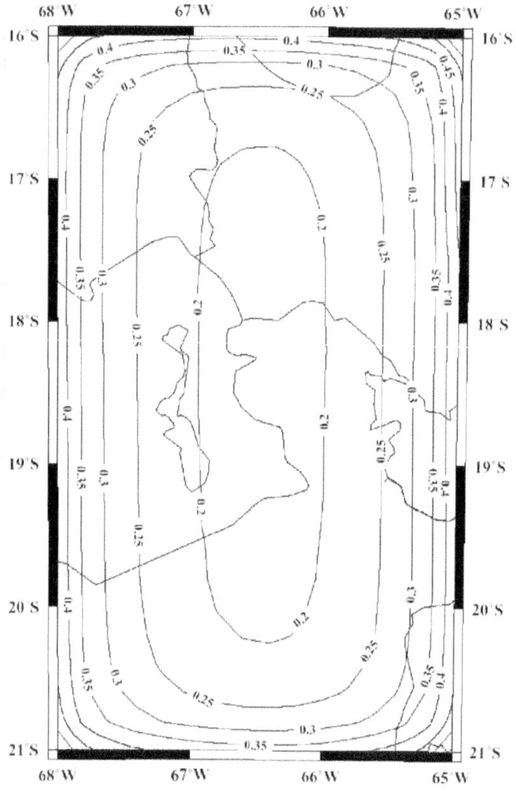


Figure 7: Error estimate for mountainous area.

Simulations of gravity anomaly errors from an INS/GNSS assembly in Bolivia are being conducted at the IG. The final aim is to establish the features of the system that should be used in a production flight.

Until full development of an own IMU-simulator, some performance analysis have been carried out from CHAMP-data. It can be up to now concluded that using CHAMP data combined with airborne gravity data (with 2.0 mgal mean error) spaced 5' apart it is possible to obtain height anomalies with mean errors of 0.14 m in low areas and 0.18 m in high areas. Improvements may be obtained using a DTM of higher resolution than the one used here and more dense gravity data. The use of a complete higher order reference field like EGM96 to degree 360 is not expected to give much improvement because the regional data used to construct the model is not of high quality.

Two planned experimental flights will provide realistic error estimates. They will include some cross-overs, where intrinsic accuracy of the data can be assessed; moreover, an area of Bolivia with some terrestrial gravity covering will be overflown, that will allow for external checking of the data.

Finally, the combination of GOCE and airborne gravimetry data will be evaluated. The aim is to establish the spectral validity window where both technologies can avoid the cumbersome terrestrial gravity campaigns.

## 5 Acknowledgements

The research reported in this paper has been funded by the Spanish Ministry of Education and Science, through the OTEA-g project of the Spanish National Space Research Programme (reference: ESP2002-03687), and the Spanish Ministry of Economy, through the TERRA project of the FAD (Fund for Development Aid) cooperation programme.

## References

Brozena, J.M., Peters, M.F., Childers, V.A., 1997. The NRL airborne gravimetry program. In: Cannon, M.E. and Lachapelle, G. (eds.), Proceedings of the International Symposium on Kinematic Systems in Geodesy, Geomatics and Navigation, pp. Banff, Canada, pp. 553-556.

Bruton, A.M., Hammada, Y., Ferguson, S., Schwarz, K.P., Wei, M., Halpenny, J., 2001. A Comparison of Inertial Platform, Damped 2-axis Platform and Strapdown Airborne Gravimetry. Proceedings of the International Symposium on Kinematic Systems in Geodesy, Geomatics and Navigation, Banff, Canada, pp. 542-550.

Colomina, I. and Blázquez, M., 2004. A unified approach to static and dynamic modeling in photogrammetry and remote sensing. In: Altan, O. (ed.), Proceedings of the XXth ISPRS Congress, Istanbul, Turkey, pp. 178-183.

ESA, 1999. Gravity Field and Steady-State Ocean Circulation Mission. ESA SP-1233 (1) — The Four Candidate Earth Explorer Core Missions. ESA Publications Division, Noordwijk, The Netherlands.

Forsberg, R., Hehl, K., Meyer, U., Gidskehaug, A., Bastos L., 1996. Development of an airborne geoid mapping system for coastal oceanography —AGMASCO. In: Geoid and Marine Geodesy, Vol. 117, IAG Symp. Series. 163-170.

García, R., 2002. Local Geoid Determination from GRACE Mission. Dept. of Civil and Environmental Engineering and Geodetic Science. The Ohio State University. Report 460.

Glennie, C. L., Schwarz, K. P., Bruton, A. M., Forsberg, R., Olesen, A. V., Keller, K., 2000. A comparison of stable platform and strapdown airborne gravity, Journal of Geodesy, Vol. 74, pp. 383-389.

Howe, E., Tscherning, C. C., 2004. Gravity field model UCPH2004 from one year of CHAMP data using energy conservation. Presented at the IAG GGSM2004 conference, August 30 - Sept. 3, 2004, Porto, Portugal.

Klingelé, E. E., Bagnaschi, L., Halliday, M., Cocard, M., Kahle, H.G., 1994. Airborne Gravimetric Survey of Switzerland. Final Results. Institut fuer Geodaesie und Photogrammetrie, E.T.H. Zuerich. Report 239.

Li, Y.C. and Schwarz, K.P., 2000. Accuracy and resolution of the local geoid determination from airborne gravity data. In: Gravity, geoid and geodynamics 2000, IAG Vol. 123, pp. 241-246.

Moritz, H., 1980. Advanced Physical Geodesy. H.Wichmann Verlag, Karlsruhe.

Reigber, C., Balmino, G., Schwintzer, P., Biancale, R., Bode, A., Lemoine, J. M., Knig, R., Loyer, S., Neumayer, H., Marty, J. C., Barthelmes, F., Perosanz, F., Zhu, S. Y., 2003. New Global Gravity Field Models from Selected CHAMP Data Sets. In: First CHAMP Mission Results for Gravity, Magnetic and Atmospheric Studies. Ed. Springer, pp. 120-127.

Schwarz, K.P. and Wei, M., 1998. Flight test results from a strapdown airborne gravity system. Journal of Geodesy, Vol. 72, pp. 323-332.

Saleh, J. and Pavlis, N.K., 2002. The Development and Evaluation of the Global Digital Terrain Model DTM2002. In: Tziavos, I.N. (Ed. ZITI), Proceedings of the 3rd Meeting of the International Gravity and Geoid Comission. Thessaloniki, Greece, pp. 207-212.

Suenkel, H., 2002. From Eötvös to Milligal. Final Report. ESA Contract No. 14287/00/NL/DC. Graz, Austria.

Térmens, A., Colomina, I., 2004. Network Approach versus State-Space Approach for Strapdown Inertial Kinematics Gravimetry. Presented at the IAG GGSM2004 conference, August 30 - Sept. 3, 2004, Porto, Portugal.

Tscherning, C. C. , Forsberg, R., Knudsen, P., 1992. The GRAVSOFT package for geoid determination. Proc. 1. Continental Workshop on the Geoid in Europe, Prague, May 1992, pp. 327-334. Research Institute of Geodesy, Topography and Cartography, Prague.

Wis, M., Samsó, L., Aigner, E., Colomina, I., 2004. Present Achievements of The experimental navigation system TAG. In: Altan, O. (ed.), Proceedings of the XXth ISPRS Congress in Istanbul, Turkey, pp. 136-141

# Artificial Neural Network: A Powerful Tool for Predicting Gravity Anomaly from Sparse Data

A.R. Tierra
Geodetic Laboratory, Department of Geography and Environmental Engineering, Army Polytechnic School, Av. El Progreso s/n, Sangolquí, Ecuador, atierra@espe.edu.ec

S.R.C. de Freitas
Graduate Course of Geodesy, Federal University of Paraná, Centro Politécnico, P.O. Box 19001, ZC. 81531-990, Curitiba, Brazil. sfreitas@ufpr.br

**Abstract.** Generally, the gravity surveys are developed along roads or other communication ways. This leads to an irregular space distribution and lacking data in large areas, like that containing high mountains, wetlands, lakes and forests. The usual methods for geoid computation from gravity data need a regular grid of gravity anomalies. Numerous methods have been developed for gravity anomalies interpolation at regular distribution. This paper reports on implementation of an interpolation method by using techniques for learning and training of Artificial Neural Networks (ANN) in predicting both free-air and Bouguer gravity anomalies from irregular and sparse data. The method was applied for a region in the Ecuador (5°S - 1°N and 75°W – 81°W) that has strong variations in crustal density and morphology. The free-air gravity anomalies prediction results were compared with the method of Kriging interpolation. The ANN method presented better results in predicting gravity anomalies in the considered region.

**Keywords.** Free air and Bouguer anomalies, Predicting, Artificial Neural Network

## 1 Introduction

The gravity anomalies obtained from gravity terrestrial data are fundamental for high frequency geoid computation based on the solution of the Geodetic Boundary Value Problem. The basic condition for using Stokes' Integral formula in this solution is the knowledge of gravity anomalies of regular distribution. However, in practice, terrestrial gravity observations are carried mainly along roads or in areas easily accessible. Then it is usual to have large areas without terrestrial observations, like that of dense forests, wet terrains or high mountains. These kinds of landscape are usual in the Amazonian region and Andean region. Because of this reason, high frequency gravity anomalies data obtained directly from terrestrial gravity observation with good spatial distribution are not available for several regions like that comprised among 5°S - 1°N and 75°W – 81°W.

Several techniques exist to carry out the interpolation gravity anomalies, for example: Kriging, based on geostatistics concepts, the Nearest Neighbors based on spatial distribution of data, Minimal Curvature based on tension factors, Least-Squares collocation (LSC) based on determination of one empirical covariance function, among others. A discussion about these methods can be found respectively in Cressie (1993), Watson (1992), Smith and Wessel (1990), Moritz (1980). The choice of such a method depends on the distribution and frequency of data. The main problem is to determine the optimal interpolator when different kinds of data and several structural and morphological aspects are present in a region under consideration. Tierra and De Freitas (2002) proposed the use of ANN for predicting gravity anomalies in regions with different kinds of gravity data distribution because this offers the possibility to integrate different sources of gravity data together gravity, like Earth crust structural and morphologic data.

In this paper are presented ANN trained for predicting free-air and Bouguer anomalies from sparse gravity data. The Kriging method was also applied for such prediction. The results are compared for control points.

## 2 Gravity Anomalies

The gravity anomaly $\Delta g$ represents the disagreement between the gravity at the geoid and the normal gravity in a related point at the reference ellipsoid. It is fundamental to establish the geometrical relationship between these surfaces. The gravity anomaly is defined as the difference between the gravity in P at the Earth's surface reduced to the geoid (i.e. $g_0$) and the normal gravity ($\gamma$) in the corresponding point on the surface of the reference ellipsoid obtained by normal projection of P. The value of the gravity anomaly is given by (Heiskanen and Moritz, 1967, p.83):

$$\Delta g = g_0 - \gamma \quad (1)$$

Because the measuring point at the Earth's surface is not on the geoid, the value of $g_0$ is obtained by reduction. The gravity reductions must observe (Heiskanen and Moritz, 1967, p. 290): geophysical meaning; easy computation; interpolation possibilities.

A basic kind of gravity reduction is the so-called free-air reduction $C_F$ that considers only the variation of gravity with the height and given by (Torge, 1989, p. 89):

$$C_F = -\frac{\partial g}{\partial H} H \quad (2)$$

where $\partial g/\partial H$ is the vertical gradient of gravity, and $H$ is the orthometric height.

Because the vertical gradient of gravity and $H$ are usually unknown, they are replaced by the normal gradient ($\partial\gamma/\partial h$) and by the leveled height ($H_L$) in the computations. Then $C_F$ can be given by:

$$C_F \cong -[\ (\frac{\partial \gamma}{\partial h}) H_L + \frac{1}{2}(\frac{\partial^2 \gamma}{\partial h^2}) H_L^2\ ] \quad (3)$$

The free-air anomaly $\Delta g_F$ is obtained after the application of the free-air correction $C_F$ on the gravity value $g$ measured in P. Then (1) becomes:

$$\Delta g_F = g + C_F - \gamma \quad (4)$$

The Bouguer anomaly $\Delta g_B$ is defined with the aim to eliminate most of the effect from the crust mass between the point P and the geoid. The $\Delta g_B$ is obtained in the following analytic form:

$$\Delta g_B = \Delta g_F - 2\pi G \bar{\rho} H_L - B + C_T \quad (5)$$

where $2\pi G \bar{\rho} H_L$ is the so-called Bouguer correction that corresponds to the vertical component of gravitational force produced by an infinite plate of thickness $H_L$ and density $\bar{\rho}$ between P and the geoid; $B$ is the Earth surface curvature correction; and $C_T$ is the terrain correction.

In general, for computing Bouguer anomaly the density of the slab is considered constant and equal to 2.67 g/cm$^3$. However, in some regions lateral variations of density may appear and they can introduce errors in the gravity anomaly computed values.

## 3 Artificial Neural Networks -ANN

The ANN's were conceived from the observation of the human brain behavior in comparison with digital computers. Even nowadays the digital computers can execute several millions of operations by second, the human brain can do parallel tasks more efficiently than computers. The reason is the ability to activate millions of neuron-cells simultaneously.

It can be established that the human brain is a non-linear complex parallel structure that can process and operate information stored in the connections among neurons (Haykin, 2001, p. 28), (Muñoz, 1996, p. 105).

Loesch and Sari (1996, p. 5) define the ANN as a computer system able to emulate some abilities of a biologic nervous system by using a certain number of artificial interconnected neurons.

### 3.1 Characteristics of Artificial Neurons

An Artificial Neuron (AN) is a logic mathematic structure that tries to emulate the functions and behavior of a biologic one. An input layer whose links with the artificial cell body are made by connection weights develops the actions of dendrites. These weights correspond to the synapses in biologic cells. Also, each of the neurons is associated with a bias weight shown in figure 1.

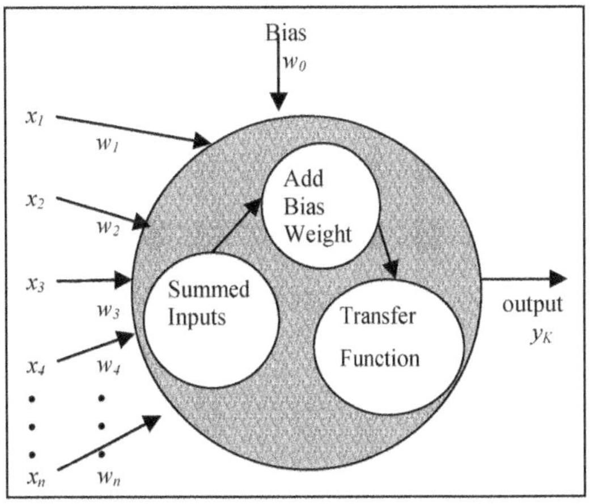


**Fig. 1.** Typical processing steps inside an artificial neuron

Inside each neuron, a weighted sum of the inputs is calculated, a bias weight is added, and this value is transformed by a transfer function that permits to limit the output signal in a finite interval (Tierra, 2003). The transformed result is sent to neurons in the next layer.

Then, an AN structure can be modeled in the following way (Li and Bridgwater, 1999, p. 67), (Loesch and Sari, 1996, p. 28), (Haykin, 2001, p. 38):

$$S = \sum_{i=1}^{n} x_i w_i + w_0 \qquad (6)$$

where:

- $x_i$ - are the input signals;
- $w_i$ - are the connection weights from the previous layer of neurons;
- $w_0$ - is the bias weight that corresponds to an additional independent input in an AN.

The result from equation 6 is applied in a transfer function *f(S)* that gives an output $y_k$ according the model:

$$y_k = f(S) \qquad (7)$$

Although many different functions could be successful transfer function, usually a differentiable and bounded function is used. The used transfer functions in this work were the linear:

$$f(S) = S \qquad (8)$$

and the hyperbolic tangent:

$$f(S) = \frac{e^S - e^{-S}}{e^S + e^{-S}} \qquad (9)$$

### 3.2 Some Structures for ANN

An ANN can learn or improve its behavior from external information. The learning is obtained by iterative process by adjustment of bias and connection weights. The learning or training process can be with supervision. In this case known values are related to the patterns of input and output. In this process each input produces an output that is compared with known values, and can indicates how adequate the obtained output is. If the answer is not in accordance with the ideal pattern, an error signal is used in an iterative form to adjust the connection weight of the ANN until an adequate value of output is obtained.

An ANN can be structured in different forms depending on the array of neurons with flux of signals among AN in lateral, backward, and forward. The usual Artificial Neural Network multi-layers form (ANNM) permits only backward and forward flux, with different forms of connection. The ANNM with supervision and training by an algorithm called back-propagation was developed in the 1980s, and has been used to solve successfully several problems. The neurons in an ANNM are connected with all the neurons of neighbor layers but without connections in the same layer. A typical three-layer network structure is displayed in figure 2. This kind of structure with progressive signal flux without direct or indirect lateral connections is called feed forward.

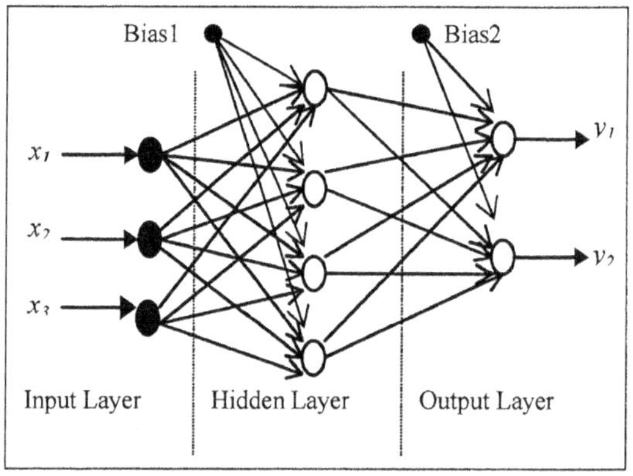


**Fig. 2.** Artificial Neural Multi-layer Network

An ANNM is formed by three kinds of layers: **Input layer** formed by *n* input units with the function of distributing the external signals to the next layer; **Hidden layer** formed by the processing elements (neurons) without external contacts. The number of hidden layers is variable for each kind of ANNM; **Output layer** is singular and formed by *m* processing elements (neurons) from that emerge output vectors to form the output of the ANNM.

### 3.3 ANNM Training

The ANNM is trained by using an algorithm called back-propagation that is based on learning by error correction (Haykin, 2001, p. 183). The back-propagation is used as base for supervised learning. In this sense, it is necessary to have input data for training and output reference data as basis for controlling the connection weights. In synthesis, the learning by back-propagation consists in finding adequate weights to solve the problem. The training is developed in two steps through the network:

1) The input vector crosses the network in the forward sense and emerges from the output layer as a true answer of the ANNM. In this process all connection weights are constant. Each neuron in the hidden layer receives the input signal with a pattern $p$ and operates them according equation (6). The input function generates an output signal for $p$ when applied in the transfer function. These output signals in this hidden layer are applied as input in the next layer following in the same way until the output layer.

2) In the process of back-propagation the connection weights receive adjustment by a rule obtained from the observed errors in the output. The propagation across the ANNM in this process is in the backward sense. The error $\varepsilon_j$ in the output of a neuron $j$ is defined as the difference between the ideal value $d_j$ and the obtained value $y_j$ in the first step (Haykin, 1999, p. 161). It is given by:

$$\varepsilon_j = d_j - y_j \tag{10}$$

The error $\varepsilon_j$ active a control structure aiming to apply a sequence of correction to the connection weights in the neuron $j$. The adjustments are planned to be done by steps aiming to approximate $y_j$ with the ideal answer $d_j$. This goal is achieved by minimization of the cost function based on the total error of the network given by:

$$SQE = \frac{1}{2}\sum_{j=1}^{nn} \varepsilon_j^2 \tag{11}$$

where:

$nn$ – is the number of neurons in the output layer.

In synthesis, the aim of back-propagation is to obtain adequate values for connection and bias weights $w_{ij}$ achieving the minimization of the function $SQE$. Then, the learning process demands the correct mapping of input vectors and respectively outputs and can be expressed like a problem for minimizing the cost function $SQE$ defined in the space W or observing the condition:

$$\frac{\partial SQE}{\partial w_{ij}} = 0 \tag{12}$$

where:

$w_{ij}$ - is the connection weight between nodes $j$ and $i$.

The weights are refreshed and corrected by one factor $\Delta w_{ij}$ since the output layer to the input layer. The adaptation of the weights follows model for a the iteration $n$:

$$w_{ij}(n+1) = w_{ij}(n) + \Delta w_{ij}(n) \tag{13}$$

The training of ANNM by forward and backward propagation is done until a specified convergence or number of iterations is achieved. The $SQE$ function allows to evaluate and to control the process when an epoch or error level is defined in the beginning of the training.

As mentioned above the supervised learning (training) was employed in our research. Therefore, the values of input units and the corresponding output units are known. The set of known input and output values is termed as input-output pair.

All input-output pairs are usually divided into two sets. The first is termed as learning or training set which is used to determine the connection weights. The second, named validation set, is used to choose the optimal parameters of neural network. When the training is complete, the network can be used for prediction.

## 4 Used ANNM and Data Base for Predicting Free- Air Anomalies

After several tests with different ANN and composition of each layer including number of neurons, the ANNM general structure used for predicting free-air anomalies in this work was composed by the input layer, one hidden layer, and one output layer with different numbers of neurons for different regions. The transfer functions in the hidden layer was a sigmoid hyperbolic tangent, and linear in the output layer. The referred selected area for testing in Ecuador was divided in three regions: Coastal; Andean; and Amazonian. The total number of gravity data points available was 15187. The figure 3 shows the training points (points) sufficient for learning and 31 validation points (cross) where the values of free-air and Bouguer anomalies were known, these points are not used in the training of the ANNM. The validation points were selected to have a distribution with different heights, variations in crustal density and located in areas without data.

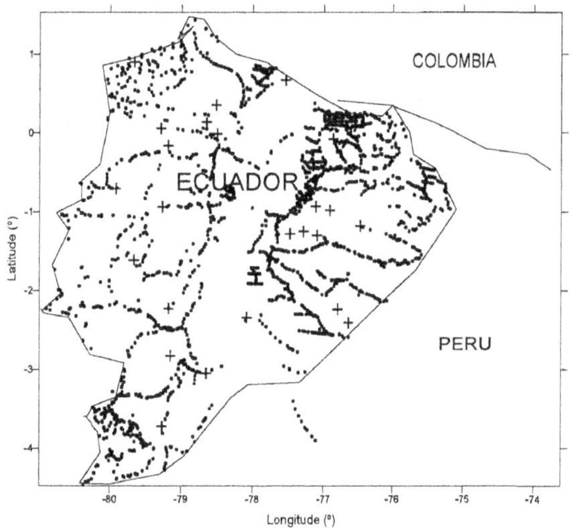


**Fig. 3.** Points used for training and evaluation of the ANNM.

The best final structure for the ANMM in the input, hidden and output layers were respectively: Coastal region [4, 18, 1]; Andean region [4, 5, 1]; Amazonian region [4, 5, 1]. These structures correspond to 4 variables in the input layer $(\phi, \lambda, H_L, \gamma)$ respectively latitude, longitude, leveled height, and normal gravity. 18, 5, and 5 neurons respectively for each region composed the hidden layer. It was necessary only one neuron in the output layer for all regions. An adequate training was achieved for 5153 or more data points. The figure 4 shows the ANNM for predicting free-air anomalies in the Andean region.

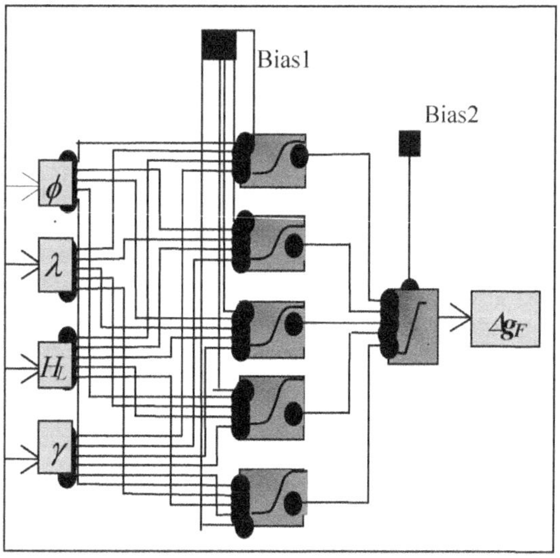


**Fig. 4.** Structure of the ANNM [4, 5, 1] applied for predicting free-air anomalies in the Andean region.

## 5 Used ANNM and Data Base for Predicting Bouguer Anomalies

In the same form, different tests were done to found a good structure for ANNM to be applied for predicting Bouguer anomalies at the 3 named regions in the section 4. The final structures were [5, 15, 1] for the coastal region, [5, 5, 1] for the Andean region, and [5, 6, 1] for the Amazonian region. For predicting Bouguer anomalies five input variables were used for each point $(\phi, \lambda, H_L, \bar{\rho}, \gamma)$ where $\bar{\rho}$ is the crust density. The only variation applied in the structure was the number of neurons in the hidden layer, respectively 15, 5, and 6 for coastal, Andean and Amazonian regions. The considered minimal number of points for training was also 5153 for this case. The figure 5 shows the used structure for predicting Bouguer anomalies in the Andean region.

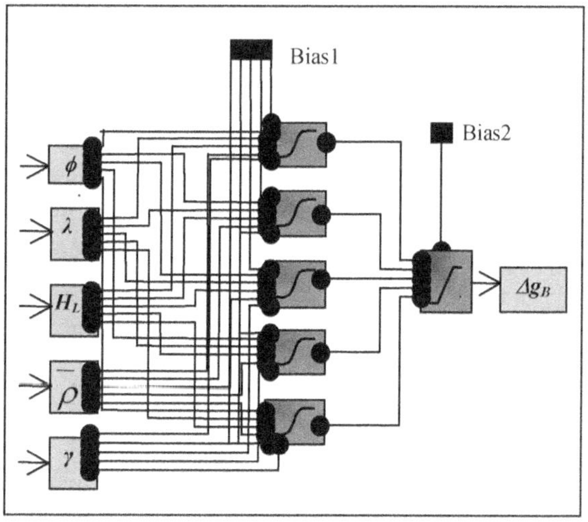


**Fig. 5.** Structure of the ANNM [5 5 1] applied for predicting Bouguer anomalies in the Andean region.

## 6 Results

The trained ANNM according sections 4 and 5, were used for predicting gravity anomalies in the 31 checkpoints. The free-air anomalies were obtained directly and free-air anomalies were obtained indirectly from the predicted Bouguer anomalies at the same checkpoints. In another step, grids of free-air anomalies for points respectively at 5 km, 10 km, 15 km, and 20 km were establishes, applying the Kriging method. From these grids were interpolated free-air anomalies for the checkpoints.

The obtained results for the three sets of free-air anomalies at the validation points were compared with the known values. Their differences are summarized in the table 1 respectively by mean, standard deviation, and maximum and minimum differences. The row FREE-ANNM is related to the free-air anomalies obtained directly, the row FREE-B-ANNM is related to the free-air anomalies obtained from the predicted Bouguer anomalies. The process KRIGING is related with the free-air anomalies interpolated from the grids.

**Table 1.** Observed differences among known values and predicted values for free-air anomalies in Ecuador.

| PROCESS | | Results in mGal | | | |
|---|---|---|---|---|---|
| | | **Mean** | **Std. Deviat.** | **Min. Value** | **Max. Value** |
| FREE-ANNM | | 2,54 | 6,21 | -15,65 | 12,67 |
| FREE-B-ANNM | | -0,55 | 6,91 | -15,75 | 13,72 |
| KRIGING | 5km | 2,17 | 17,85 | -64,88 | 29,00 |
| | 10km | 1,72 | 18,32 | -66,59 | 28,56 |
| | 15km | 0,19 | 19,60 | -70,09 | 28,06 |
| | 20km | 0,81 | 20,95 | -82,46 | 28,30 |

## 7 Conclusions and Remarks

Considering the obtained results for the study area in Ecuador can be concluded that:

- From the table 1 it is possible to infer that there are no differences from the ANNM predicted free-air anomalies directly or that obtained from Bouguer anomalies;
- The prediction of gravity anomalies with ANN presents good results even for sparse data or non-regular distributions. This fact is probably related to the ANN ability to integrate heterogeneous data in the learning process;
- In the study area, even with sparse data it was possible to train the ANNM with a third-part of the available data point.

**Acknowledgments.** The authors would like to thank CNPq (Brazilian National Council of Research) for the financial support of this work.

## 8 References

Cressie, N (1993). Statistics for Spatial Data. ISU, New York.

Li, Y., and Bridgwater, J (2000). Prediction of Extrusion Pressure Using Artificial Neural Network. Powder Technology 108: 65-73.

Haykin, S (1999). Neural Networks: A Comprehensive Foundation. 2. ed. New Jersey: Prentice Hall.

Haykin, S (2001). Redes Neurais: Princípios e Prática. 2. ed. Porto Alegre: Bookman.

Heiskanen, W., and Moritz, H (1967). Physical Geodesy. H. Freeman and Company.

Forsberg, R (1994). Terrain Effects in Geoid Computations. In: Lectures Notes of the International School for the Determination and Use of the Geoid. International Geoid Service, Milan.

Moritz, H (1980). Advanxed Physical Geodesy. H. Wichmann Verlag, Karlsruhe.

Muñoz, A (1996). Aplicación de Técnicas de Redes Neuronales Artificiales al Diagnóstico de Procesos Industriales. Ph.D. Thesis, Dept. de Electrotécnia y Sistemas, Universidad Pontificia Comillas de Madrid.

Loesch, C., and Sari, S (1996). Redes Neurais Artificias. Fundamentos e Modelos. Blumenau, Brazil.

Smith, W., and Wessel, P (1990). Gridding with Continuous Curvature Splines in Tension. Geophysics 55(3):293-305.

Tierra, A., and De Freitas, S (2002). Predicting Free-Air Gravity Anomaly Using Artificial Neural Network. International Association of Geodesy Symposia: Vertical Reference Systems. Springe 124: 215-218.

Tierra, A (2003). Metodologia para a Geração da Malha de Anomalias Gravimétricas para Obtenção do Geoide Gravimétrico Local a partir de Dados Esparsos. Ph.D. Thesis Department of Geomatics, Federal University of Paraná.

Torge, W (1989). Gravimetry. New York: de Gruyter, New York.

Watson, D (1992). Contouring: A Guide to the Analysis and Display of Spatial Data. Pergamon.

# Photon Counting Airborne Laser Swath Mapping (PC-ALSM)

W. E. Carter, R. L. Shrestha, and K.C. Slatton
Department of Civil and Coastal Engineering, University of Florida, Gainesville, FL 32611, USA
Fax: 352-392-5032, bcarter@ce.ufl.edu

**Abstract.** Commercially marketed airborne laser swath mapping (ALSM) instruments currently use lasers with sufficient energy per pulse, in combination with optics of sufficient aperture, to work with return signals of thousands of photons per shot. The resulting high signal to noise level virtually eliminates spurious range values caused by noise, such as background solar radiation and sensor thermal noise. However, the high signal level approach requires laser repetition rates of hundreds of thousands of pulses per second to obtain contiguous coverage of the terrain at sub-meter spatial resolution, and with currently available technology, affords little scalability for significantly downsizing the hardware, or reducing the costs.

University of Florida (UF) researchers are developing an ALSM unit based on a different paradigm, referred to as photon counting ALSM, or PC-ALSM. In the PC-ALSM approach, relatively low energy laser pulses are transmitted, and are used to illuminate a surface 'patch' of terrain a few meters in extent. The returning photons are detected by a multi-channel photomultiplier tube, which separately senses the returns from an array of groundals comprising each patch, providing high (few decimeter) resolution contiguous coverage of the terrain. A multi-channel multi-stop timing unit records both noise and signal events within a range gated window, which enables noise to be filtered out of the data during post flight processing.

Researchers at NASA GSFC have already tested a first generation system based on this new paradigm. The NASA system operated from a high altitude aircraft, to obtain proof of concept data, prior to the development of a satellite based instrument. Details of the preliminary UF design for a second generation system that will operate from a light aircraft flying less than 1000 meters above local ground level and providing contiguous coverage of the terrain with 30 cm spatial resolution are reviewed.

**Keywords**. Airborne, laser, terrain mapping.

## 1 Introduction

Commercially marketed airborne laser swath mapping (ALSM) instruments currently use lasers with sufficient energy per pulse (typically about 100 micro-joules) in combination with optics of sufficient aperture, to obtain return signals of thousands of photons per shot. Strong returns make it possible to achieve high signal-to-noise ratio (SNR) using standard noise suppression techniques such as temporal (range gate), spatial (field stop) and spectral (narrow band-pass optical) filters. However, to keep the peak power levels low enough to avoid damaging optical components and comply with eye-safety regulations, the laser pulse widths are typically about 10 nanoseconds, equivalent to about 3 meters. To achieve a range resolution of a few centimeters, the round trip travel time of the laser pulse must be measured to about one percent of the pulse length - a challenge when the return pulses vary in amplitude by 2 to 3 orders of magnitude and the shapes of many are distorted by reflection from uneven surfaces.

An alternative approach to ALSM is to design the receiver to work with a much lower signal level – perhaps as low as a single photon. Potential advantages of the photon counting approach to ALSM (PC-ALSM) include reductions in the size, weight, power consumption and cost of the instrumentation. In addition, because the illumination per unit surface area can be significantly lower, the laser footprint can be increased and a multi-channel sensor can be used to achieve contiguous surface coverage with improved spatial resolution. The primary disadvantage attendant to working with a low SNR is that some percentage of the events recorded will be noise, which must be identified and removed during the reduction and processing of the data.

## 2 Low SNR Laser Ranging

Beginning with the Apollo 11 lunar mission, in 1969, a number of retro-reflector arrays were placed on the moon by American astronauts and USSR un-manned missions. The lasers available in that era made it very difficult to design a system, even using a very large aperture telescope, that resulted in return signals above a single to a few photons. By using tight spatial, spectral and temporal filters researchers were successful in obtaining accurate earth-moon ranges despite the very low SNR.

More recently, researchers at NASA Goddard Space Flight Center (GSFC) began to investigate using the low SNR approach to develop a satellite based laser topographic mapping system [Degnan and McGarry, 1998; Degnan et. al., 1998 and 2001]. Based on promising results derived from theoretical and engineering calculations, the NASA team constructed a "brass-board" system to test the approach from a high altitude aircraft. Early results proved that the low SNR unit could produce high-resolution topographic coverage, even in haze and light fog, and also could penetrate coastal waters to depths of a few meters [Degnan, 2002].

Based on the NASA results, UF researchers concluded that low SNR laser ranging could be used to develop a small, lightweight, low power consumption ALSM instrument that could operate from an unmanned aerial vehicle (UAV). A number of potential military applications for such a system were immediately obvious, including the mapping of coastal areas prior to amphibious attack.

After a review of readily available commercial components, a preliminary design was made for the Coastal Area Tactical-mapping System (CATS). A proposal was submitted to the Office of Naval Research (ONR), and was approved with a starting date of March 15, 2004.

## 3 Overview of CATS Sensor Design

Our immediate goal is to build a sensor head that can be used to demonstrate, study, and evaluate the strengths and weaknesses of PC-ALSM. We decided to minimize the time and costs of developing a first generation CATS sensor head, and devote the largest possible portion of the resources to the collection, study and evaluation of PC-ALSM observations in varied settings and conditions. To do this, we are building the CATS sensor so it can be mounted rigidly to and operate simultaneously with our Optech Inc. model 1233 ALTM unit. This approach means that the inertial measurement unit (IMU) and GPS data collected routinely by our current system can be used to reduce and analyze data from both units.

Our decision to integrate the new CATS sensor with our existing system immediately set certain design parameters, among them the nominal altitude (600 to 1000 meters) and ground speed (50 to 70meters per second) at which it must operate. With these parameters set simple scaling of the GSFC unit yields first estimates of the laser energy (3 to 6 micro-joules) per pulse and the aperture (5 to 8 centimeter diameter) of the optics required.

Even with a range gate, narrow band-pass optical filter, and tight spatial filter, it is still likely that there typically will be sufficient noise events from atmospheric backscattering of the laser and sun light, to warrant the use of a multiple stop event timer. And, the repetition rates (5 to 10 KHz) of currently readily available micro-lasers make it necessary to illuminate a "patch" of terrain a few meters on a side to achieve contiguous coverage of the terrain on a single pass. Taking these factors into consideration, two components comprise the heart of the CATS sensor: a multi-channel photomultiplier tube (PMT), to achieve a sub-meter lateral spatial resolution, and a multi-channel multi-stop event timer, to minimize the effects of noise and surface clutter.

We selected a Hamamatsu model R4110U-74-M100D PMT, (see Figure 1.) The PMT has a metal envelope, which is approximately 5.3 x 5.3 cm on a side, and 3.1 cm in height. The 10 x 10 array of photo-cathodes (100 channels) is 1.8 cm. by 1.8 cm. The photocathode material is GaAsP, which has a typical quantum efficiency of 40% at 0.5320 micrometers. The dark noise for each channel is only a few hundred counts per second, at 25 degrees centigrade. The PMT has 2 stages of filmless micro-channel plate internal

Figure 1. This image shows a Hamamatsu Electronics Inc. model R4110U-74-M100D photomultiplier tube, next to a size AA battery.

amplification. The total effective quantum efficiency of the detector is expected to be better than 25%. The typical rise time for a single photo-electron event is 200 picoseconds. The recovery time before another event (in the same channel) can be detected has not yet been determined, but is expected to be less than one nanosecond.

The CATS detector package, comprised of the PMT and power supply, multi-channel multi-stop event timer, and controller, is being built by Fibertek Inc. The event timer will initially have 64 channels, with provisions for upgrading to 96 channels. The timing for all channels of the event timer will be closely aligned and driven from a single oscillator, operating at an effective frequency of at least 1 GHz, with a design goal of 2 GHz. Each clock cycle will represent a "timing bin" and bins containing events will be recorded for all channels after each laser shot. The range resolution is one-half the laser roundtrip timing resolution, corresponding to 15 cm at 1 GHz, and 7.5 cm at 2 GHz. The event timer is designed to record events in contiguous bins, if the sensor signal is narrow enough and is capable of responding to such closely spaced events, for laser pulse rates up to 10,000 pps.

Other major components of the CATS sensor head include the micro-laser and power supply, as well as the optical and mechanical components that make up the transmit and receive paths. Sigma Space Inc. has been contracted to design and build the sensor head, including interfacing for the detector. Figure 2 is a sketch showing the preliminary conceptual design of CATS sensor head.

There is no intent to push the state-of-the-art of micro-lasers for the CATS project. A commercially marketed laser will be purchased, which we expect will provide 5 to 10 thousand pulses per second, with 3 to 6 microjoules per pulse, at a wave length of 0.5320 micrometers. The output of the laser will likely have a circular cross section, with Gaussian distribution of energy across the aperture (heavily peaked in the center). The laser light will pass through a beam expander and then a holographic element to change the shape of the cross section to a square, and to distribute the energy as uniformly across the aperture as possible - referred to as a "top hat" distribution.

The CATS sensor head will utilize a scanner developed under the NASA GSFC program, which consists of two clear aperture "Risley" prisms that can be driven independently, in both

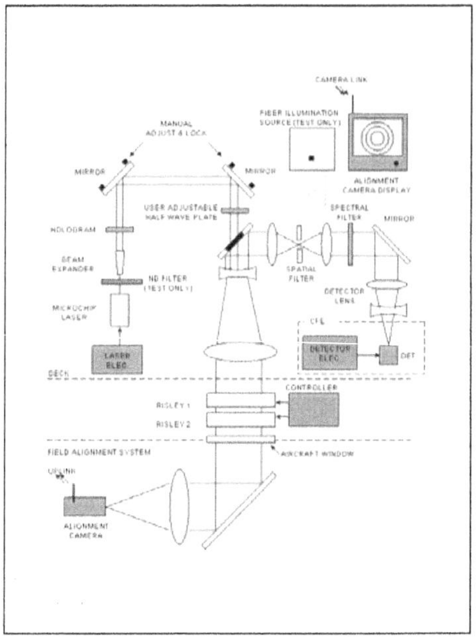


Figure 2. Schematic showing the conceptual design of the CATS sensor head.

directions and over a range of angular velocities. The scanner design allows a number of scan patterns to be selected. For example, if both prisms are aligned and rotated in the same direction and at the same angular velocity, the scan pattern will be the same as that of a classical conical scanner. However, if the prisms are offset in phase and rotated in opposite directions, an inclined raster pattern can be generated [Degnan, 2002].

## 4 Importance of Laser Pulse Length

The nominal pulse length of the micro-laser selected for CATS will be a few hundred picoseconds – equivalent to a fraction of a meter. The pulse length is actually a more important parameter of the laser than the number of pulses per second or the energy per pulse, because it strongly affects the ability of the system to collect high resolution full 3 dimensional data, a major advantage of PC-ALSM over classical high SNR systems

High SNR ALSM units typically operate with laser pulse energies of tens to hundreds of micro-joules. To obtain these energy levels the laser pulse widths are typically 10 nanoseconds or more, at full-width-half-maximum (FWHM), corresponding to some 3 meters or longer in units of length. Identifying and separating returns from objects spaced less than a pulse length apart is difficult.

Consider, for example, a hard object such as a military vehicle parked under a net, with a clearance of one meter. The net may reflect only a small fraction of the laser energy, but that may be sufficient to trigger the event timer, and the subsequent receiver "dead time" will prevent detection of the larger return from the vehicle. Even if the timing electronics are capable of handling multiple events, the laser pulse length sets a lower limit on the spacing of objects that can be resolved.

At best, even if "long-pulse" systems are capable of recording multiple stops, the point clouds produced are dominated by first-surface returns. For example, in a densely forested area, there will be many points at the tops of the trees, and relatively few at ground level. In between the returns will tend to be quantized, at intervals roughly equal to the pulse length. Data of this type is sometimes referred to as having 2.5 dimensions, because the point clouds do not generally map intermediate structure (between the first and last surfaces) well.

One approach to improving the ability of "long pulse" systems to detect objects spaced by less than one pulse length is to record the shape of the return signal - using a waveform digitizer. Unfortunately, that approach generally results in an order of magnitude or more increase in the observational data, and still there is no way to uniquely deconvolve the various components contained in the return signal. Nonetheless, work on this approach continues, and enjoys some success.

The micro laser purchased for CATS will have a pulse length approximately a factor of 30 shorter than conventional ALSM units. When combined with the fast rise time of the PMT, and the high resolution multiple event interval timer, CATS will capture nearly full 3 dimension datasets.

## 5 Concluding Remarks

The CATS design uses leading edge commercial components such as the Hamamatsu 100 channel PMT and the Fibertek multi-channel multi-stop event timer, which are built to order. There are substantial waiting times for these components, which will result in a build time of at least 10 months, after the notices for the contractors to proceed. Our current schedule is aimed at collecting the first observations sometime during the summer of 2005. However, UF graduate students have already begun research on various aspects of the data collection, reduction and analysis, focusing on the efficient extraction of information from PC-ALSM observations. Consideration is also being given to the next generation system, with the goal of substantially reducing the size, weight, and power consumption to enable the system to operate from a moderate sized UAV.

## References


Carter, W. E., R.L. Shrestha and S. P. Leatherman; "Airborne Laser Swath Mapping: Applications to Shoreline Mapping," Proceedings of International Symposium on Marine Positioning (INSMAP ?98), Melbourne, FL, Nov. 30 - Dec. 4, pp. 323-333, 1998.

J. J. Degnan and J. F. McGarry, "Feasibility study of multikilohertz spaceborne microlaser altimeters", Proc. European Geophysical Society (EGS) Annual Symposium, Nice, France, April 20-24, 1998.

J. J. Degnan, McGarry, T. Zagwodzki, P. Dabney, J. Geiger, R. Chabot, C. Steggerda, J. Marzouk and A. Chu, "Design and Performance of an Airborne Multikilohertz Photon-Counting, Microlaser Altimeter," Proceedings of Land Surface Mapping and Characterization Using Laser Altimetry, Vol. XXXIV 3-W4 International Archives of Photogrammetry and Remote Sensing, Annapolis, MD, Oct22-24, pp. 9-16, 2001.

J. .J. Degnan, "Photon-Counting Multikilohertz Microlaser Altimeters for Airborne and Spaceborne Topographic Measurements", Journal of Geodynamics (Special Issue on Laser Altimetry), 2002.

J. J. Degnan, "A Conceptual Design for a Spaceborne 3D Imaging LIDAR," Elektrotechnik und Informationtionstechnik, heft. 4, pp. 99 – 106, April, 2002.

# Evaluation of SRTM3 and GTOPO30 Terrain Data in Germany

H. Denker
Institut für Erdmessung, Universität Hannover, Schneiderberg 50, D-30167 Hannover, Germany
E-mail: denker@ife.uni-hannover.de; Fax: +49-511-7624006

**Abstract.** High-resolution terrain data are crucial for gravity field modelling in mountainous regions. In areas without national digital elevation models (DEMs) available, fill-ins from global models have to be used. For this purpose, the global models GTOPO30 (30" resolution) and SRTM3 (3" resolution) are considered. The SRTM3 model has been released recently from the analysis of the Shuttle Radar Topography Mission and covers the latitudes between 60°N and 54°S, while the GTOPO30 model is a global public domain data set completed already in 1996.

In this contribution, 1" x 1" national DEMs for Germany are used to evaluate the global models. The differences between the best national models and the SRTM3 data show a standard deviation of 7.9 m with maximum differences up to about 300 m. The largest differences are located in opencast mining areas and result from the different epochs of the DEMs. Histograms of the differences reveal a clear deviation from the normal distribution with a long tail towards too high SRTM3 elevations. The evaluation of GTOPO30 shows that the longitudes should be increased by 30" (one block) in Germany. For the shifted GTOPO30 DEM, the standard deviation of the differences with respect to the best national model is 6.8 m, roughly 75 % smaller than for the original model.

**Keywords.** Digital elevation model, DEM, terrain data, DEM evaluation, SRTM3, GTOPO30

## 1 Introduction

High resolution digital elevation models play an important role in gravity field modelling, as the short wavelength gravity field variations are highly correlated with the topography. The modelling is usually based on the remove-restore procedure, where the terrain data are used to smooth the gravity field observations in order to avoid aliasing effects and to facilitate gridding and field transformations, for details see, e.g., Forsberg and Tscherning (1981), Denker (1988), Forsberg and Sideris (1989), and Sideris and Forsberg (1991).

In view of continental geoid computations, e.g., for Europe (Denker and Torge 1998), it has to be considered that digital elevation models (DEMs) are not available for some countries, either because they have not been created or because of confidentiality reasons. Therefore, in these areas fill-ins from global models have to be used. For this purpose, the SRTM3 model with a resolution of 3" × 3" (JPL, 2004) and the public domain global model GTOPO30 with a resolution of 30" × 30" (LP DAAC, 2004) can be used.

In this contribution, 1" x 1" national DEMs for Germany are used to evaluate the global models SRTM3 and GTOPO30. The differences between the national DEMs and the SRTM3 and GTOPO30 models are analyzed and the statistics are provided for different relief types. Moreover, the terrain models are compared to elevations from gravity stations.

## 2 Digital Elevation Models (DEMs)

A short description of all digital elevation models (DEMs) used in this study is given in Table 1. The area of investigation is between 47° - 56° north latitude and 5° - 16° east longitude.

The Shuttle Radar Topography Mission (SRTM) occurred Feb. 11-22, 2000 and successfully fulfilled all mission objectives. The SRTM data covers most of the land surfaces between 60° north latitude and 54° south latitude (targeted land coverage was 80% of the Earth's landmass). Following the calibration and validation phase, the raw data were processed continent by continent into digital elevation models. Details on the SRTM mission and concepts are explained, e.g., in Bamler (1999). So far, an unedited data set with a resolution of 3 arc second (SRTM3) was released to the public domain. This product is preliminary and is distributed for evaluation by the research and applications user community. It can be downloaded from ftp://edcsgs9.cr.usgs.gov/pub/data/srtm/ (USGS, 2004). The National Geospatial-Intelligence Agency

**Table 1.** Digital elevation models (DEMs) for the area 47°N – 56°N and 5°E – 16°E.

| DEM | Resolution | Accuracy | # elev. | # undefined elev. |
|---|---|---|---|---|
| SRTM3-1 | 3" x 3" | 16 m | 142,584,001 | 329,304 |
| SRTM3-2 | 3" x 3" | 16 m | 142,584,001 | 0 |
| GTOPO30 | 30" x 30" | 30 m | 1,425,600 | 256,077 |
| FRG-1A | 1" x 1" | 20 m | 1,283,040,000 | 613,792,787 |
| FRG-1B | 1" x 1" | 20 m | 1,283,040,000 | 619,754,163 |
| FRG-2A | 1" x 1" | 20 m | 1,283,040,000 | 612,159,380 |
| FRG-2B | 1" x 1" | 20 m | 1,283,040,000 | 618,606,658 |

(NGA) is currently editing and verifying the SRTM data to bring them into conformance with map accuracy standards, and these "finished" data will then be released to the public by the end of 2005 (JPL 2004). According to the accuracy specifications, the absolute vertical accuracy (90% linear error) is 16 meters and the absolute horizontal accuracy (90% circular error) is 20 meters for SRTM3 (Bamler, 1999; JPL, 2004). The elevations are given relative to the EGM96 geoid, and the horizontal datum is WGS84. Furthermore, it should also be noted that the SRTM is a "first return system" which provides elevations based on whatever the radar has bounced off from. While in many instances the elevations may be referring to actual ground level, this is not the case in dense forests, built-up areas, etc. (Showstack, 2003).

The presently available unedited "research-grade" SRTM3 data in particular may contain numerous voids (regions with no data) and other spurious points, and in addition to this, water bodies are not well-defined because they produce very low radar backscatter (JPL, 2004). Table 1 gives the statistics for the original (downloaded) SRTM3 data set, which is denoted as SRTM3-1. The number of undefined elevations in the study area is 329,304, i.e. 0.23% of all values. The SRTM3-1 model is also depicted in Fig. 1, where the undefined elevations are shown as black dots. In flat areas, most of the undefined values are associated with water bodies, e.g., along the Rhine and Danube river and in lake districts. Moreover, a significant number of undefined elevations are located in the Alps area with very high mountains and narrow gorges. As a complete DEM is required in gravity field modelling applications, a second version, denoted as SRTM3-2, was created, where the undefined elevations were replaced with interpolated values from neighbouring data (weighted mean).

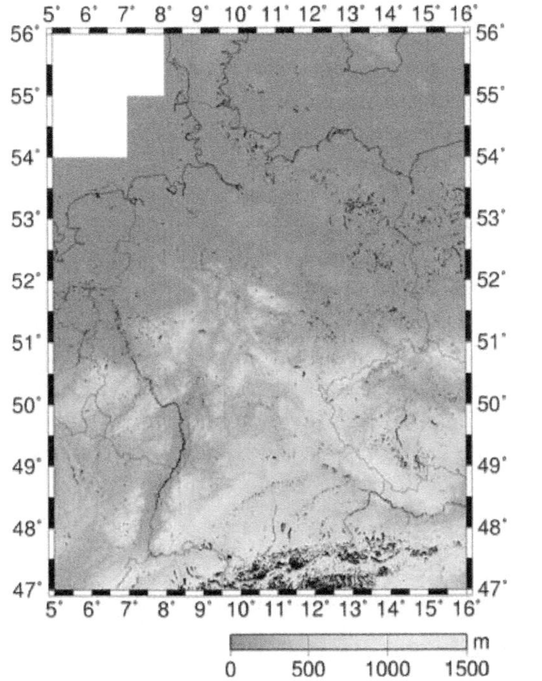


**Fig. 1.** SRTM3-1 digital elevation model. Data voids are marked as black dots.

Moreover, also the GTOPO30 data with a resolution of 30" × 30" (LP DAAC, 2004) were considered in this investigation, mainly because the SRTM data does not cover Northern Europe, which is part of our target area for geoid computations. The GTOPO30 data were downloaded by ftp from LP DAAC (2004). The DEM has global coverage and was derived from several raster and vector sources of topographic information. The horizontal coordinates refer to WGS84 and the elevations are referenced to mean sea level (MSL). The accuracy varies by location according to the source data. In Germany, the major data source is the digital terrain elevation data (DTED) from NGA (e.g., NGA, 1996) and the vertical accuracy is specified as 30 meters (90% linear error). The GTOPO30 DEM is also listed in Table 1; the undefined values are all located in ocean areas and were replaced by zero values for the subsequent comparisons.

The national DEMs originate from the German Military (AMilGeo, 1992) and have a resolution of 1" × 1". The models cover the territory of Germany and were derived by digitization of 1:50,000 maps in the years 1985 to 1990. The horizontal datum of the models used in this study is WGS84. The elevations are referenced to MSL. The absolute vertical accuracy (90% linear error) is specified as 20 meters and the absolute horizontal accuracy (90% circular error) is 26 meters (AMilGeo, 1992). Several versions of the original DEM were derived (see Table 1). The original DEM is FRG-1A, while the model FRG-2A contains some updates south of 49.5°N latitude. In addition, the versions FRG-1B and FRG-2B were derived from the corresponding "A" versions by excluding data in two sub-areas outside of Germany, located in the Alps Mountains (Austria) and Ore Mountains (Czech Republic); this was done because in these areas obviously less accurate fill-ins were used in the national DEMs (see also below). Table 1 summarizes the main features of the national DEMs.

For the evaluation of the SRTM3 and GTOPO30 DEMs, the 1" × 1" elevations from the national DEMs were averaged to 3" × 3" and 30" × 30" grids, respectively. During this step, also a re-interpolation was necessary due to the underlying different grid coordinate systems.

## 3 Evaluation of SRTM3

The SRTM3 DEMs were evaluated by comparisons with the national models. The statistics of the differences are provided in Table 2. Moreover, the differences between the original SRTM3-1 model and the national models FRG-1A and FRG-2A are depicted in Figs. 2 and 3 for the complete study area (top) as well as for a sub area in southern Germany (middle). In addition, Figs. 2 and 3 (bottom) also contain histograms of the corresponding differences for the complete study area.

From Figs. 2 and 3 it is clear that the largest differences between the SRTM3-1 and national DEMs are located in southern Germany in the mountainous Alps region, in the Czech Republic, as well as in the opencast mining districts around Leipzig and Halle, north of Dresden (Lausitz) and west of Cologne. The discrepancies in the mining districts are clearly related to the different epochs of the DEMs, i.e. the SRTM data are up-to-date, while the national DEMs were created in the 1980s. From Figs. 2 and 3 it is also clear that the agreement between the SRTM3-1 DEM and the national FRG-2A model is superior as compared to the FRG-1A model. Moreover, Fig. 3 documents that FRG-2A contains low quality fill-ins in some areas in southern Germany and to some extent also in the Czech Republic. This was the main reason for removing the two sub-areas shown in Figs. 2 and 3 (marked by dotted patterns), leading to the corresponding FRG-1B and FRG-2B DEMs, respectively.

The statistics of the differences, shown in Table 2, also prove that the largest differences are located in the two sub-areas mentioned above. In the comparisons of the original SRTM3-1 DEM with the FRG-1B and FRG-2B models, the standard deviations reduce by about 3% as compared to the corresponding FRG-1A and FRG-2A models, and the maximum differences reduce from more than 800 m for the "A" models to about 350 m for the "B" models. The comparison with the national DEM FRG-2B yields a standard deviation of the differences of 7.9 m with maximum discrepancies up to 324 m (0.08 % of the differences are larger than 50 m and 0.01 % are larger than 100 m). For FRG-1B the corresponding figures are slightly larger, i.e. the standard deviation is 8.3 m and the maximum discrepancy is 365 m (0.14 % of the differences are larger than 50 m and 0.02 % are larger than 100 m).

In all comparisons of the SRTM3-2 models, where the data voids have been filled by a simple weighted mean prediction, the discrepancies with the national DEMs deteriorate as compared to the original SRTM3-1 data set. The relevant standard deviations increase by about 10 % and the maximum differences go up to about 900 m for both the "A" and "B" versions of the national DEMs (see Table 2). A more detailed inspection of the results

**Table 2.** Differences between 3" × 3" DEMs. Units are m.

| Difference | # | Mean | Stddev | Min | Max |
|---|---|---|---|---|---|
| SRTM3-1 - FRG-1A | 74,234,512 | +2.69 | 8.56 | -447.0 | +848.0 |
| SRTM3-1 - FRG-1B | 73,589,462 | +2.68 | 8.30 | -365.0 | +339.0 |
| SRTM3-1 - FRG-2A | 74,405,567 | +2.74 | 8.16 | -421.0 | +848.0 |
| SRTM3-1 - FRG-2B | 73,710,998 | +2.72 | 7.90 | -324.0 | +258.0 |
| SRTM3-2 - FRG-1A | 74,373,176 | +2.67 | 9.42 | -818.0 | +848.0 |
| SRTM3-2 - FRG-1B | 73,708,498 | +2.66 | 9.02 | -818.0 | +848.0 |
| SRTM3-2 - FRG-2A | 74,554,083 | +2.69 | 9.48 | -941.0 | +918.0 |
| SRTM3-2 - FRG-2B | 73,835,063 | +2.69 | 8.95 | -941.0 | +918.0 |

**Table 3.** Differences SRTM3-1 minus FRG-2B (3" x 3" DEMs) for different relief types. Units are m.

| Relief | # | Mean | Stddev | Min | Max |
|---|---|---|---|---|---|
| low | 38,417,922 | +0.53 | 6.00 | -320.0 | +162.0 |
| medium | 34,342,758 | +5.02 | 8.84 | -213.0 | +230.0 |
| alpine | 950,318 | +8.10 | 12.27 | -324.0 | +258.0 |

**Table 4.** Differences between 3" × 3" DEMs and gravity station heights (gelev). Units are m.

| Difference | # | Mean | Stddev | Min | Max |
|---|---|---|---|---|---|
| FRG-1A - gelev | 247,017 | +0.14 | 6.37 | -270.1 | 378.4 |
| FRG-1B - gelev | 247,017 | +0.14 | 6.37 | -270.1 | 378.4 |
| FRG-2A - gelev | 247,017 | +0.17 | 6.24 | -149.1 | 378.4 |
| FRG-2B - gelev | 247,017 | +0.17 | 6.24 | -149.1 | 378.4 |
| SRTM3-1 - gelev | 246,900 | +1.87 | 6.94 | -210.2 | 389.5 |
| SRTM3-2 - gelev | 247,017 | +1.85 | 7.95 | -795.2 | 594.4 |

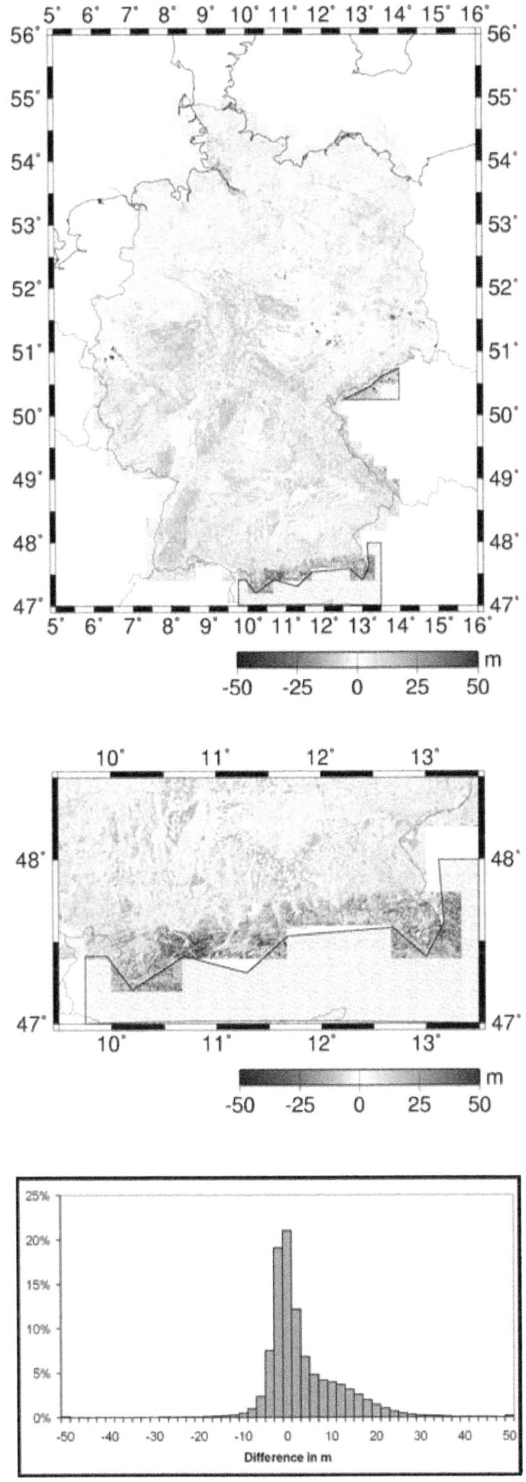


**Fig. 2.** Differences SRTM3-1 minus FRG-1A for complete study area (top) and a sub area in southern Germany (middle), supplemented by a histogram of all differences (bottom). The two polygon sub-areas excluded in the corresponding FRG-1B model are marked by dotted patterns.

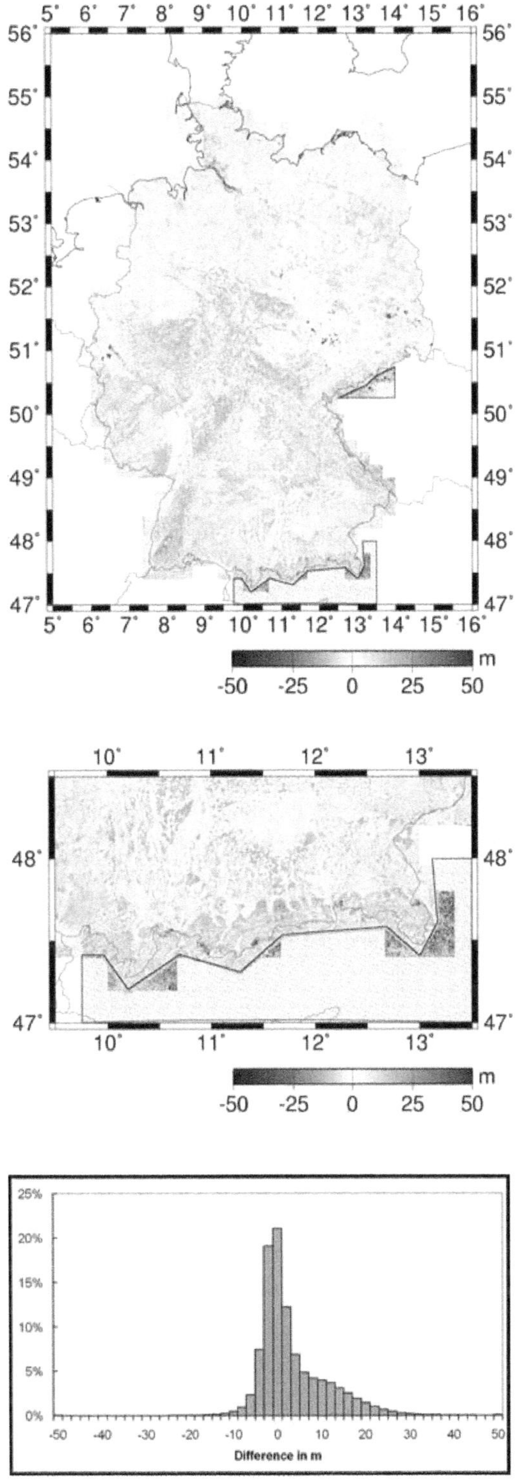


**Fig. 3.** Differences SRTM3-1 minus FRG-2A for complete study area (top) and a sub area in southern Germany (middle), supplemented by a histogram of all differences (bottom). The two polygon sub-areas excluded in the corresponding FRG-2B model are marked by dotted patterns.

shows that small data voids can be filled by interpolation, but larger data voids in mountainous areas should not be filled by interpolation, as this may lead to large errors. Thus the data voids in the SRTM data pose a significant problem for a number of applications.

Table 3 provides the statistics of the differences between the original SRTM3-1 DEM and the national FRG-2B model for different relief types. While for all three relief types the maximum discrepancies go up to about 300 m, the standard deviations are varying (6.0 m for low, 8.8 m for medium, and 12.3 m for alpine relief).

Furthermore, the histograms of the differences between SRTM3-1 and FRG-1A and FRG-2A (Figs. 2 and 3) show a quite obvious deviation from the normal distribution. There is a long tail towards too high elevations of the SRTM3 model, which is expected due to the fact that SRTM is a "first return system", providing elevations of whatever the radar has bounced off from, and in many instances this is above the actual ground level, e.g., in dense forests, built-up areas, etc. (Showstack, 2003).

Another evaluation of the 3" × 3" DEMs was done by comparisons with the elevations from gravity stations in Germany. The statistics of the differences are provided in Table 4. The standard deviations of the differences are 6.4 m and 6.2 m for the national models FRG-1A/B and FRG-2A/B, respectively and the maximum discrepancies go up to 378 m. The comparisons with the "A" and "B" DEM versions yield identical results, because there are no gravity stations located in the two excluded areas (see Figs. 2 and 3). For the SRTM3-1 and SRTM3-2 DEMs, the standard deviations increase to 6.9 m and 8.0 m, respectively. For SRTM3-2, with the data voids filled by interpolation, some very large differences up to about 800 m show up. This confirms the above conclusion, that the interpolation of larger data voids must be handled with care.

## 4 Evaluation of GTOPO30

The GTOPO30 DEM was evaluated by comparisons with the national and SRTM3 models. For this purpose, the latter models were averaged using all defined elevations within a 30" × 30" cell. Table 5 shows the statistics for selected comparisons. In addition to the original GTOPO30 model, also a shifted version (GTOPO30-S) was considered, where the longitudes were increased by 30" (one block). This was suggested by a correlation analysis between the original GTOPO30 and national DEMs.

Table 5 clearly proves that the shifted version GTOPO30-S yields a significantly better agreement with the national and SRTM3 data than the original model. The standard deviations of the differences between the original GTOPO30 model and the national and SRTM3 models are about 27 m and

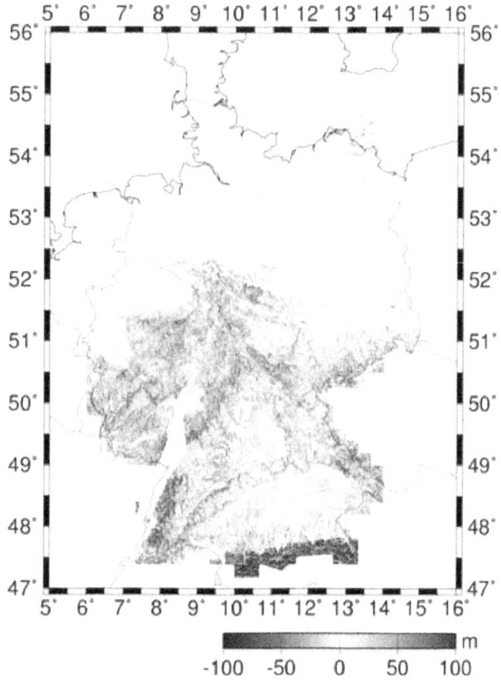


Fig. 4. Differences GTOPO30 minus FRG-2A.

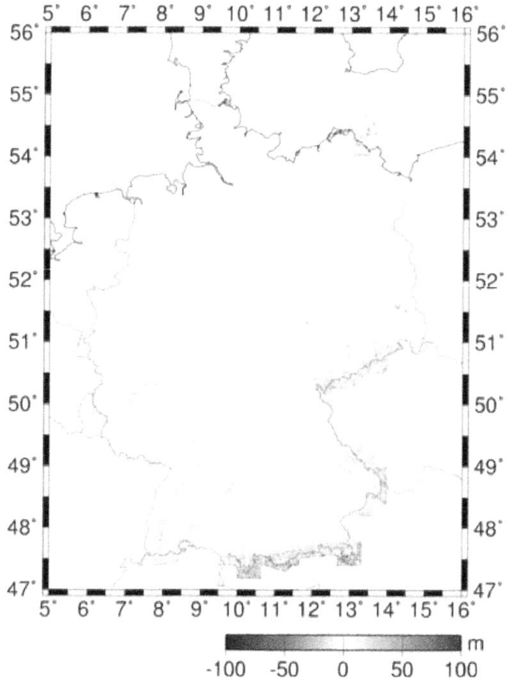


Fig. 5. Differences GTOPO30-S minus FRG-2A.

**Table 5.** Differences between 30" × 30" DEMs. Units are m .

| Difference | # | Mean | Stddev | Min | Max |
|---|---|---|---|---|---|
| GTOPO30 - FRG-1B | 705,414 | -0.37 | 26.93 | -743.0 | +567.0 |
| GTOPO30 - FRG-2B | 706,627 | -0.33 | 27.68 | -811.0 | +619.0 |
| GTOPO30 - SRTM3-1 | 1,169,356 | -3.65 | 42.26 | -912.0 | +852.0 |
| GTOPO30 - SRTM3-2 | 1,169,523 | -3.69 | 42.00 | -769.0 | +702.0 |
| GTOPO30-S - FRG-1B | 705,337 | -0.59 | 6.86 | -688.0 | +434.0 |
| GTOPO30-S - FRG-2B | 706,550 | -0.56 | 6.77 | -674.0 | +460.0 |
| GTOPO30-S - SRTM3-1 | 1,169,197 | -3.75 | 11.57 | -610.0 | +730.0 |
| GTOPO30-S - SRTM3-2 | 1,169,364 | -3.79 | 11.30 | -797.0 | +550.0 |

42 m, respectively. The corresponding values for GTOPO30-S are about 6.8 m and 11.5 m, respectively. Thus the longitude shift reduces the differences by roughly 75 %, which is a very significant improvement. The improvement of GTOPO30-S versus the original version is also quite obvious from Figs. 4 and 5, showing the differences between the two GTOPO30 versions and the national model FRG-2A. The maximum differences up to almost 800 m occur in the mountainous parts of the study area (in the Alps). Moreover, some patterns are visible around the Ore Mountains, related probably to the compilation of the GTOPO30 data.

The GTOPO30 DEMs were also compared with the elevations from gravity stations in Germany. The standard deviation of the differences is 30.3 m for the original version and 19.2 m for GTOPO30-S (max. 480 m), and accordingly for the 30" × 30" SRTM and national DEM versions standard deviations from 19.9 m to 20.9 m are obtained.

## 5 Conclusions

The SRTM3 digital elevation models (DEMs) were evaluated by comparisons with national models for Germany. Two SRTM3 versions were considered. The first model (SRTM3-1) consists of the original data, while in the second model (SRTM3-2) the undefined elevations (data voids) were replaced by interpolated values from neighbouring data (weighted mean). The comparisons revealed that one of the national models contained less accurate fill-ins in some areas outside of Germany. After excluding these areas, the differences between the best national model (FRG-2B) and the SRTM3-1 DEM show a standard deviation of 7.9 m with maximum differences up to about 300 m. The largest differences are located in opencast mining areas and result from the different epochs of the data. The differences were also analyzed for different relief types, yielding standard deviations of 6.0 m for low, 8.8 m for medium, and 12.3 m for alpine relief. Furthermore, histograms of the differences show a clear deviation from the normal distribution with a long tail towards too high SRTM3 elevations. For the SRTM3-2 model, the comparison results deteriorate, i.e. the standard deviation with respect to the best national model is 9.0 m and the maximum differences go up to about 940 m, showing that the filling of data voids by interpolation must be handled with care. Additional comparisons with elevations of gravity stations in Germany gave a standard deviation of the differences of 6.9 m for SRTM3-1 and 8.0 m for SRTM3-2. To sum up, the SRTM3 DEM fully complies with the accuracy specifications.

The evaluation of the GTOPO30 model by national and SRTM3 DEMs demonstrated that in Germany the longitudes of GTOPO30 should be increased by 30" (one block). The longitude shift reduced the standard deviation of the differences to the national and SRTM3 models by roughly 75 %, yielding final values of about 6.8 m and 11.5 m for the national and SRTM3 models, respectively.

## References


AMilGeo (1992). Elevation Model DHM/M745. Amt für Militärisches Geowesen, Euskirchen, pers. comm..

Bamler, R. (1999). The SRTM Mission: A World-Wide 30 m Resolution DEM from SAR Interferometry in 11 Days. In: D. Fritsch and R. Spiller (eds.): Photogrammetric Week 99, Wichmann Verlag Heidelberg: 145-154.

Denker, H. (1988). Hochauflösende regionale Schwerefeldmodellierung mit gravimetrischen und topographischen Daten. Wiss. Arb. Fachr. Verm.wesen, Univ. Hannover, Nr. 156.

Denker, H., W. Torge (1998). The European gravimetric quasigeoid EGG97 – An IAG supported continental enterprise. IAG Symposia, Vol. 119:249-254, Springer Verlag.

Forsberg, R., M.G. Sideris (1989). On topographic effects in gravity field approximation. In: E Keylso, K. Poder, C.C. Tscherning (eds.): Festschrift to Torben Krarup, Geodaetisk Institut, Meddelelse No. 58: 129-148, Copenhagen.

Forsberg, R., C.C. Tscherning (1981). The use of height data in gravity field approximation by collocation. Journal of Geophys. Research 86: 7843-7854.

JPL (2004). SRTM – The Mission to Map the World. Jet Propulsion Laboratory, California Inst. of Techn., http://www2.jpl.nasa.gov/srtm/index.html.

LP DAAC (2004). Global 30 Arc-Second Elevation Data Set GTOPO30. Land Process Distributed Active Archive Center, http://edcdaac.usgs.gov/gtopo30/gtopo30.asp.

NGA (1996). Performance specification Digital Terrain Elevation Data (DTED). National Geospatial-Intelligence Agency, Document MIL-PRF-89020A.

Sideris, M.G., R. Forsberg (1991). Review of geoid prediction methods in mountainous regions. IAG Symposia, Vol. 106: 51-62, Springer Verlag.

Showstack, R. (2003). Digital Elevation Maps Produce Sharper Image of Earth's topography. EOS, Transactions, American Geophysical Union, Vol. 84, No. 37: 363.

USGS (2004). US Geological Survey, ftp://edcsgs9.cr.usgs.gov/pub/data/srtm/.

# Multiscale Estimation of Terrain Complexity Using ALSM Point Data on Variable Resolution Grids

K.C. Slatton[1, 2], K. Nagarajan[1], V. Aggarwal[1], H. Lee[1], W. Carter[2], R. Shrestha[2]
[1] Department of Electrical and Computer Engineering
[2] Department of Civil and Coastal Engineering
University of Florida; PO Box 116130; Gainesville, FL 32611
Tel: 352.392.0634, Fax: 352.392.0044, E-mail: slatton@ece.ufl.edu

**Abstract.** Multiscale Kalman smoothers (MKS) have been previously employed for data fusion applications and estimation of topography. However, the standard MKS algorithm embedded with a single stochastic model gives suboptimal performance when estimating non-stationary topographic variations, particularly when there are sudden changes in the terrain. In this work, multiple MKS models are regulated by a mixture-of-experts (MOE) network to adaptively fuse the estimates. Though MOE has been widely applied to one-dimensional time series data, its extension to multiscale estimation is new.

## 1 INTRODUCTION

Airborne Laser Swath Mapping (ALSM) and Interferometric Synthetic Aperture Radar (InSAR) have emerged as important technologies for remotely sensing topography. Traditionally, InSAR has been employed for mapping extended areas with moderate resolution, while ALSM has been used to map local elevations at high resolution over smaller regions. A multiscale estimation framework can then be used to fuse ALSM and InSAR data having different resolutions to produce improved estimates over large coverage areas while maintaining high resolution locally.

Multiscale Kalman smoothers (MKS) [1] that are modeled on fractional Brownian motion (fBm) stochastic models have been used previously to estimate elevations [2]. The MKS-based estimation provides a dense set of estimates and estimate error variances at multiple resolutions. However, the standard MKS algorithm employs a single stochastic model and therefore produces suboptimal estimates when presented with spatial variations in the elevation statistics. For example, rough undulating terrain yields an elevation surface with a shorter correlation length than flat smooth terrain. The inclusion of multiple models in the estimation framework has been employed in this work and is found to give better results over non-stationary terrain. A mixture-of-experts (MOE) network [3] is then employed to weigh the filter outputs based on the Kalman filter residuals (innovations).

This work also makes use of fractal theory concepts for estimating the parameters of the prior models by finding the fractal dimension of the image over different regions. A brief description of the test area is presented in Section 2. Fractal based segmentation and the multiscale estimation algorithm are described in Sections 3 and 4, respectively. The mixture of experts adaptation is given in Sections 5 and 6. We then summarize the results obtained on a set of ALSM data acquired over an urban forest in Section 7.

## 2 STUDY AREA

The study area analyzed for this work is a mixed coniferous and deciduous forest in North-central Florida, USA. The forest is part of the Hogtown Greenway within the city of Gainesville, so urban development (roads and houses) appear along the forest edges, as shown in Fig. 1.

Fig. 1 Near-infrared digital image of study site acquired coincidently with ALSM data.

The study site was imaged with an Airborne Laser Swath Mapping (ALSM) sensor owned by the University of Florida (UF) from an above ground altitude of 600m. The UF ALSM is an Optech 1233 system capable of recording first stop and last stop returns for each laser pulse. The first stop corresponds to the elevation of the first, therefore highest, object that intercepts the laser beam. The last stop corresponds to the last, therefore lowest, object that intercepts the laser beam. For ground surfaces, first and last stop elevations are generally the same, but they can be separated by many meters over forest canopies.

For this analysis, the last stop returns were used due to a recording error in the first stop returns (see Fig. 2). The laser beam is approximately 15 cm in diameter when it intercepts the terrain and land cover. The site was imaged with two flight lines, resulting in an average of 2-3 laser returns per square meter. Thus the last stop returns sense the tops of opaque targets, such as roads and rooftops, but represent some penetration on targets that have sparse discrete structure at the scale of tens of centimeters, such as the forest canopy.

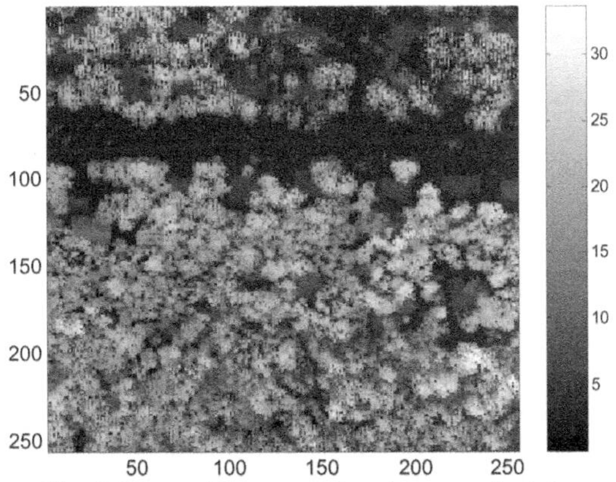

Fig. 2 Mean of last stop elevations acquired from two overflights of the study area. Axis and colorbar values are in meters.

## 3 FRACTAL-BASED DATA SEGMENTATION

In order to determine a natural partitioning for the imaged scene, the ALSM point data were gridded into an image array with a spatial resolution of 1m × 1m. The fractal dimension was computed, as shown if Fig. 3, the histogram of which was found to exhibit a bimodal structure. The fractal dimension was calculated using a 9m × 9m window with the Triangular Prism Surface Area (TPSA) method [4, 5].

For two-dimensional data, the fractal dimension $D$ is related to the Hurst exponent $\mathbf{H}$ as [6, 7]. $\mathbf{H}$ can in turn be related to a slope parameter $\mu$ of the power spectrum density through [8, 9].

$$\mathbf{H} = 3 - D. \tag{1}$$

$$\mu = 2\mathbf{H} + 1. \tag{2}$$

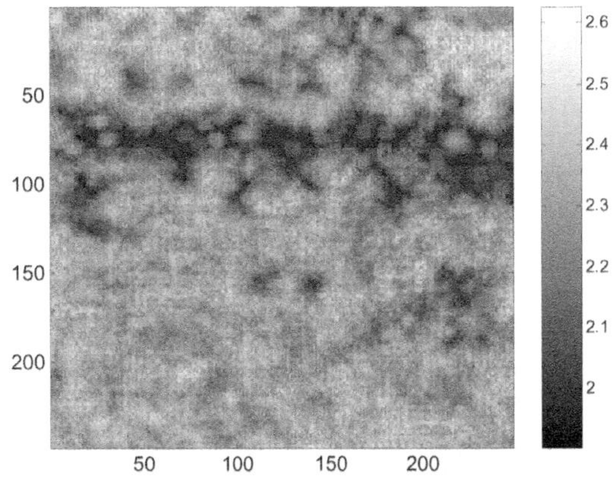

Fig. 3 Fractal dimension $D$.

The TPSA method was used to generate two (due to the bimodal distribution of $D$) fractal models. The total variance of the models at the 256 × 256 support level was set equal to the sample variance computed for the "smooth" road and the "rough" forest in the ALSM elevation data. By virtue of using the multiscale fractional Brownian motion model, the incremental variances between scales can be expressed as

$$Var(W) = \sigma^2 \left[ 1 + \left(\frac{1}{2}\right)^{\mathbf{H}} + \left(\frac{1}{2}\right)^{2\mathbf{H}} + \cdots + \left(\frac{1}{2}\right)^{7\mathbf{H}} \right], \tag{3}$$

where $W$, represents the windowed data used in the sample variance calculation. Using (3), we solve for the variance scale factor $\sigma$.

## 4 MULTISCALE ESTIMATION FRAMEWORK

MKS estimation of two-dimensional data was proposed in [1] and provides a scale recursive framework for estimating topographies at multiple resolutions. These models are defined on index sets, organized as quadtrees as shown in Fig. 4.

The multiscale estimation is initiated with a fine-to-coarse sweep up the quadtree that is analogous to Kalman filtering with an added merge step. The upward sweep is followed by a coarse-to-fine sweep down the quadtree that corresponds to Kalman smoothing.

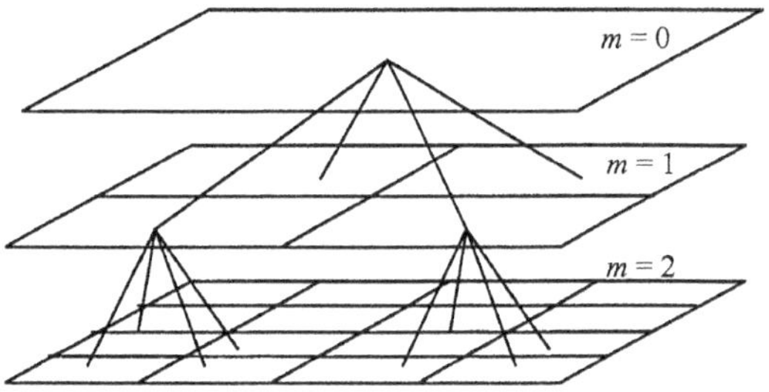

Fig. 4 Quadtree data structure where $m$ denotes the scale

The linear coarse-to-fine model in [1] is given as,

$$\begin{aligned} x(s) &= A(s)x(\gamma s) + B(s)w(s) \\ y(s) &= H(s)x(s) + v(s) \end{aligned} \tag{4}$$

where $s$ is the node index on the tree, and $\gamma$ is a backshift operator in scale such that $\gamma s$ is one scale coarser than $s$. Here, $x(s)$ is the state variable (elevations), and $y(s)$ represents the ALSM observations. $w(s)$ is assumed to be a Gaussian white noise process with identity variance, and the measurement error $v(s)$ is a Gaussian white noise process with variance $R(s)$. $A(s)$ is the coarse-to-fine state transition operator, $B(s)$ is the coarse-to-fine process noise standard deviation, and $H(s)$ is the measurement-state relation. The scale is represented by the level in the quadtree, and is denoted by $m$. The support of the image at level $m$ is $2^m \times 2^m$ where $m = \{0, \cdots, M\}$.

The standard Kalman formulation provides optimal estimates (in the mean squared sense) when there is perfect *a priori* knowledge of the state and measurement model parameters $A(s)$, $B(s)$, $H(s)$, and $R(s)$. The state process is assumed to follow a $1/f^{\mu}$ model known as fractional Brownian motion (fBm). Fractional Brownian motion models can be used to represent a wide range of natural process such as topography and atmospheric turbulence. Using this model, the power spectrum of the state variable $x(s)$ can be represented by the multiscale model in (4) by using $A(s) = 1$ and $B(s) = B_0 2^{\frac{(1-\mu)m}{2}}$ [1]. The values of $B_0$ and $\mu$ are determined using the technique outlined in Sections 3 and 6.

The MKS approach provides reasonable estimates for the evolution of the state process in scale $\hat{x}(s)$. However, the resulting coarse-to-fine process noise variance $B^2(s)$ is constant at each scale; hence, it cannot accommodate non-stationarities in the imagery at a particular scale. The fine-to-coarse process noise variance $Q(s)$ used in the Kalman filter is a function of $B^2(s)$ and is therefore also spatially uniform. Elevations from more than one type of terrain or landcover comprise a non-stationary $2D$ process. Therefore, using a data model that is variable in scale but uniform in space will lead to suboptimal estimates in general.

## 5 MIXTURE OF EXPERTS

The MOE used in this work is a single-layer network that consists of multiple experts and a gating network that arbitrates among the expert estimators. The MOE assigns weights to the experts in an unsupervised fashion as they compete for the desired response. In this work, each expert system is a multiscale Kalman filter with parameter vector $\alpha_i$, as shown in Fig. 5. The most likely model among these filters for a particular input is assigned a higher weight by the gating network. The weights $g_i$ satisfy

$$\begin{aligned} &0 < g_i < 1 \quad \forall g_i \\ &\sum_{i=0}^{N} g_i = 1 \end{aligned} \quad . \tag{5}$$

The input to the estimator is the observation data set, which in this case consists of ALSM point data aggregated (gridded) to different resolutions. The desired outputs are the elevation estimates of the terrain and landcover.

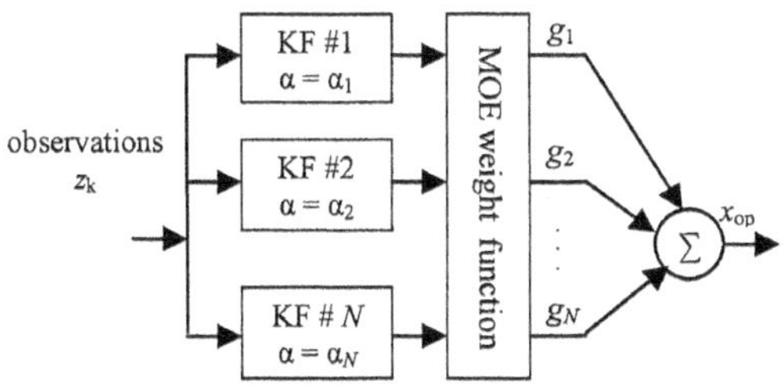


Fig. 5: Mixture-of-experts network blending of Kalman filter bank estimates.

The weights can be interpreted as *a priori* probabilities for the corresponding experts given the current input. A softmax transformation forces the condition on the weights given in (5)

$$g_i = \frac{e^{u_i}}{\sum_{j=0}^{N} e^{u_j}} \quad . \tag{6}$$

Here, $u_i = z^T a_i$ and $a_i$ is the weight matrix of the $i^{th}$ filter in the modular network. The residuals (innovations) $r_k$ are used to calculate the Gaussian conditional probability $f$ of the measured observations at time $k$, given the filter realization and the measurements up to index $k-1$ [10, 11].

$$f\left(z_k \middle| \alpha_i, z_{k-1}\right) = \frac{1}{\sqrt{2\pi |W_k|}} e^{-\frac{1}{2} r_k^T W_k^{-1} r_k}, \tag{7}$$

where

$$W_k = H_k P_k^- H_k^T + R_k$$
$$r_k = z_k - H_k \hat{x}_k^-$$

The probability distribution of the filter bank is given as the weighted sum of the individual conditional distributions of the filters

$$f(z_k) = \sum_{i=1}^{N} f(z_k | \alpha_i) g_i \cdot \tag{8}$$

Equation (8) can be considered as a likelihood function, which upon maximization gives the optimal estimates. The *a posteriori* probabilities $h_i$ are defined as,

$$h_i = \frac{f(z_k | \alpha_i) g_i}{\sum_{j=1}^{N} f(z_k | \alpha_j) g_j} \cdot \tag{9}$$

The weight matrix $a_i$ is updated by maximizing (9) using a gradient ascent procedure, yielding an update $a_i \leftarrow a_i + \eta(h_i - g_i) z_k$, where $\eta$ is a learning rate parameter [3].

Each filter is embedded with its own $1/f^{\mu}$ model for a particular class of topography. In the case of MKS, the filter recursion steps represent different scales and therefore cannot exceed the base-2 logarithm of the finest-scale image support.

## 6 MULTIPLE MODEL SELECTION

Data input into the MKS algorithm are generally used only at one scale, which corresponds to their maximum resolution [2]. However, having data at only one or two scales does not always allow sufficient recursion steps for the multiple model weights to converge. Because the original ALSM data in this study are composed of discrete points in three-dimensional space, the points are gridded to several different resolutions to represent terrain and landcover phenomenology at different scales. In this work, the ALSM data were thus gridded to populate the quadtree at scales $m = \{8, 7, 6, 5, 4, 3, 2\}$.

The Kalman model parameters must be specified before the multiple model MKS estimator can be implemented. The fractional Brownian stochastic model dictates that $A(s) = 1$. The value for the mapping matrix $H(s)$ is unity in grid cells containing ALSM data and zero in empty grid cells. For quadtree levels containing observations, the measurement noise variance $R$ was specified empirically for each scale, yielding rms values of {10cm, 32cm, 71cm, 100cm, 120cm, 140cm, 220cm}. The *a priori* values for the state and error variance were set to zero and the sample variance of the image, respectively.

Finally, the stochastic detail $B(s)$ terms are determined. Corresponding to the bimodal nature of the fractal dimension values, a two-model filter bank was chosen. One model represents low variance (smooth) terrain, while the other represents high variance terrain and landcover. Parts of the forest encompassing sudden changes from tall trees to bare ground (forest gaps) exemplified the highly varying region (variance = 16.96). Road surfaces represented the other extreme (variance = 0.3666). These variances are then used to find the value of $\sigma$ in (3) using the corresponding Hurst exponents calculated from the local fractal dimension. The value of $\sigma$ can then be used to specify the $B$ terms as

$$B_m = \sigma^2 \left(\frac{1}{2}\right)^{(m-1)\mathbf{H}}. \quad (10)$$

where $m$ ranges from 1 to 8 for a quadtree with 256 × 256 leaf nodes. Since the variance decays linearly in log space for the fractal models, the value of $B_0$ is found through extrapolation. Hence with two different sets of $\sigma$ and $\mathbf{H}$, we obtain two different sets of stochastic detail functions corresponding to the two different filter models.

## 7 RESULTS

The proposed two-model estimator was applied to ALSM data acquired by the UF ALSM system over the study area in Fig. 1. A region 256m × 256m was extracted from the ALSM point cloud data. The estimation algorithm starts with the upward sweep wherein the estimates calculated by the individual filters are employed to calculate the MOE weights. The filter and smoother then continue to run independently of the weighing network. The weights are then used to fuse the Kalman filter estimates in the Kalman smoothing downward sweep. The presence of observations at multiple scales provides several recursion steps for the model weights to adapt. *A priori* (scale 8) probabilities were set to 0.5 for each model.

The relative model weighting was found to be sensitive to the $\sigma$ values used in the stochastic detail functions. When $\sigma$ for Model 1 ($\sigma_1$) and Model 2 ($\sigma_2$) are both large relative to the standard deviation of the image, the model with the smaller $\sigma$ value receives the higher weight most often, regardless of terrain. When both $\sigma_1$ and $\sigma_2$ are small, the model with the larger $\sigma$ is selected most often. If both $\sigma$ values are very close to the standard deviation of the image, both models receive weights close to 0.5. However, if $\sigma_1$ and $\sigma_2$ are sufficiently different from each other, while remaining reasonably close to the image standard deviation, the models are well-matched to the data and their weights react strongly to terrain and landcover differences. This was the case when the $\sigma$ values were calculated from the sample variance of the forest and road regions, as described in Section 6.

The stochastic detail values used for each model are shown in Table 1A. Average weights assigned to each model for different terrain and landcover regimes are listed in Tables 1B, 1C, and 1D. Notice that Model 1 receives high weights over the smooth road, low weights over the forest with high fractal dimension, and intermediate weights over the more uniform forest. The converse is true for Model 2. The relative weighting remains consistent across scales.

Table 1: (A) Incremental variances for the two prior models . (B): average model weights for road. (C): average model weights for high fractal dimension forest. (D): average model weights for intermediate fractal dimension forest.

| A | $B(0)$ / $B(5)$ | $B(1)$ / $B(6)$ | $B(2)$ / $B(7)$ | $B(3)$ / $B(8)$ | $B(4)$ |
|---|---|---|---|---|---|
| Model 1 | 0.723 | 0.366 | 0.185 | 0.094 | 0.047 |
| | 0.024 | 0.012 | 0.006 | 0.003 | |
| Model 2 | 26.79 | 16.96 | 10.73 | 6.793 | 4.299 |
| | 2.721 | 1.722 | 1.089 | 0.689 | |

| B | Scale 7 | Scale 6 | Scale 5 | Scale 4 |
|---|---|---|---|---|
| Model 1 | 0.8400 | 0.8582 | 0.8817 | 0.8948 |
| Model 2 | 0.1600 | 0.1418 | 0.1183 | 0.1052 |

| C | Scale 7 | Scale 6 | Scale 5 | Scale 4 |
|---|---|---|---|---|
| Model 1 | 0.4001 | 0.3164 | 0.3156 | 0.3140 |
| Model 2 | 0.5999 | 0.6836 | 0.6844 | 0.6860 |

| D | Scale 7 | Scale 6 | Scale 5 | Scale 4 |
|---|---|---|---|---|
| Model 1 | 0.5807 | 0.4890 | 0.4932 | 0.5001 |
| Model 2 | 0.4193 | 0.5110 | 0.5068 | 0.4999 |

In Fig. 6 and 7, weights for Model 1 and Model 2 are shown. The weights over the road and the high variance part of the forest (lower right) remain well correlated across all scales. The weights in the lower variance forest (lower left) initially have many high and low values, but tend to a more uniform 0.5 value at coarser scales. This seems to indicate that as measurement resolution decreases, the discrete three-dimensional structure of the forest medium begins to appear more uniform. This scale-wise phenomenology has important implications for inverse modeling and parameter estimation that use other types of remote sensing data over forested terrain, such as radar or multi-spectral data.

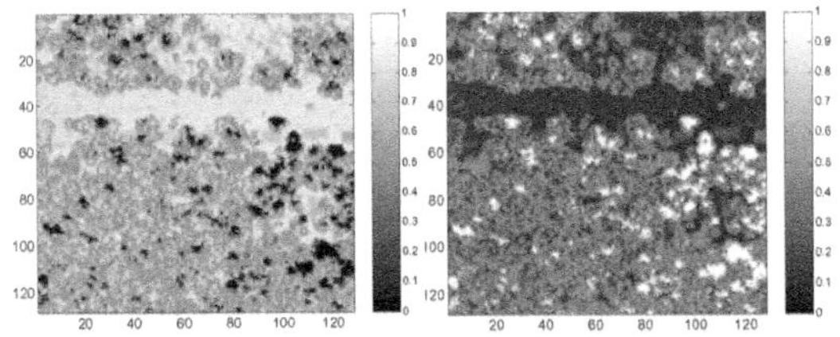

Fig. 6: Weights assigned to Model 1 (left) and Model 2 (right) by the mixture of experts at scale 7.

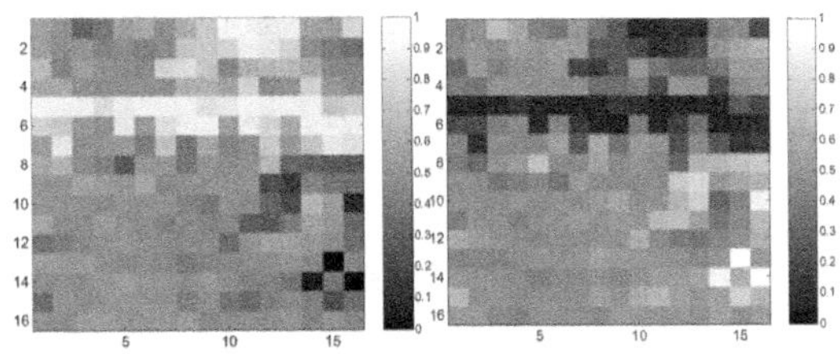

Fig. 7: Weights assigned to Model 1 (left) and Model 2 (right) by the mixture of experts at scale 4.

## 8 CONCLUSIONS

The inclusion of multiple models in the multiscale estimation framework provides an improvement over single model approaches. The results obtained show promise and warrant further analysis. Fractal based data segmentation provides insight into appropriate partitioning of the data and *a priori* stochastic models. Fast adaptation of weights makes the mixture of experts well suited for MKS, since the number of recursions in scale is generally limited to 10 or less.

Aggregating the ASLM point data into different cell sizes allows the observations to be "realized" at many resolutions, thus revealing the scale-wise phenomenology of different terrain and landcover. Providing data at multiple scales also allows the weights generated by the mixture of experts network to converge towards stable values. Ongoing work deals with the analysis of the weighting algorithm to generate more consistent weights.

## 9 REFERENCES


[1] P. W. Fieguth, W. C. Karl, A. S. Willsky, and C. Wunsch, "Multiresolution optimal interpolation and statistical analysis of TOPEX/POSEIDON satellite altimetry," *IEEE Trans. Geosci. Remote Sensing*, vol. 33, pp. 280–292, Mar. 1995.

[2] K. C. Slatton, M. M. Crawford, and B. L. Evans, "Fusing interferometric radar and laser altimeter data to estimate surface topography and vegetation heights," *IEEE Trans. Geosci. Remote Sensing*, vol. 39, pp. 2470–2482, November 2001.

[3] W. S. Chaer, R. H. Bishop, J. Ghosh, " A mixture of Experts framework for adaptive Kalman filtering," *IEEE Trans. Systems, Man and Cybernetics*, vol. 27, no. 3, June 1997.

[4] K. Jeyarani , and K. Jayaram Hebbar , " Fractal concept to the classification of crop and forest type in IRS data," *IEEE Geoscience and Remote Sensing Symposium*, vol. 1, pp.784 - 786, May 1996.

[5] K. C. Clarke, "Computation of the fractal dimension of topographic surfaces using the triangular prism surface area method," *Computers and Geoscience*, vol. 12 No 5, pp 713 – 722, May 1986.

[6] Manfred Schroeder, *Fractals, Chaos, Power Laws*, Freeman, 1992.

[7] J. L. Starck, F. Murtagh, and A. Bijaoui, *Image Processing and Data Analysis: the Multiscale Approach*, Cambridge, 1998.

[8] Donald L. Turcotte, *Fractals and Chaos in Geology and Geophysics*, 2nd ed., Cambridge, 1997.

[9] Heinz-Otto Peitgen, and Dietmar Saupe, *The Science of Fractal Images*, Springer-Verlag, 1988.

[10] P. S. Maybeck, P. D. Hanlon, "Performance enhancement of a multiple model adaptive estimator," *IEEE Trans. Aerospace and Electronic Systems*, vol. 31, no. 4, October 1995.

[11] D. T. Magill, "Optimal adaptive estimation of sampled stochastic processes," *IEEE Trans. Automatic Control*, vol. ac-10, no.4, October 1995.

# A comparison of different isostatic models applied to satellite gravity gradiometry

F. Wild, B. Heck
Geodetic Institute, University of Karlsruhe, Englerstr 7, D-76128 Karlsruhe, Germany
e-mail: wild@gik.uni-karlsruhe.de; Fax: +49-721-608-6808

**Abstract.** In satellite gradiometry, the gravitational signals originating from the Earth's topography and its isostatic compensation can be recognized in the gravity gradients observed along the satellite orbit. One general task should be the reduction of these effects to produce a smooth gravity field suitable for downward continuation. Based on different isostatic models such as the Airy-Heiskanen model, the Pratt-Hayford model, the combination of the Airy-Heiskanen model (land area) and the Pratt-Hayford model (ocean area), and the generalized Helmert model, the topographic-isostatic effects are calculated for a GOCE-like satellite orbit. For the second vertical (radial) derivative of the gravitational potential the order of magnitude of both topographic and isostatic components amounts to about 10 E.U. while the combined topographic-isostatic effect reduces to about 1 E.U.. In this paper, the focus lies on the comparison between the classical isostatic models and the generalized Helmert model, consistently using a rigorous spherical formulation for all models. By variation of the depth of the condensation layer, it is possible to demonstrate that the classical isostatic models become equivalent to the Helmert model related to a specific condensation depth d. E.g., the standard Airy-Heiskanen model related to a normal crustal thickness T = 25 km is best approximated using the compensation depth d = 24 km in the generalized Helmert model. Instead of the conventional remove-restore techniques which lead to high numerical efforts, the use of the generalized Helmert model is recommended.

**Keywords.** Satellite gravity gradiometry, topographic reduction, isostatic models (Airy-Heiskanen, Pratt-Hayford , Helmert)

## 1 Introduction

The topographic masses and their isostatic balance masses have effects on the outer gravity field of the Earth. This will be visible in satellite gravity gradiometry (SGG) observations, e.g. in the planned GOCE mission. The effects of the topographic-isostatic masses should be reduced by some kind of remove-restore technique in order to produce a smooth field which can be downward continued more easily.

In this paper, the topographic reduction of the SGG data is explained in chapter 2. The isostatic reductions, especially based on the classical Airy-Heiskanen model and the Pratt-Hayford model as well as the generalized Helmert model, are described in chapter 3. The comparison of the classical isostatic models with the generalized Helmert model is made in chapter 4 with the aid of a simulation. In the derivation of the formulae for the topographic and isostatic effects on the potential and its second derivatives a consistent spherical formulation is applied throughout the paper; this is in contrast to the conventional procedure (see e.g. Tsoulis, 1999) based on modelling of the masses by rectangular prisms.

## 2 Topographic Reduction of SGG Data

The approximation of the geoid as a sphere $\Sigma_g$ (radius R) enables the modelling of the gravitational potential of the topographic masses between sea level and the Earth's topographic surface. S is the irregular topographic surface of the Earth.

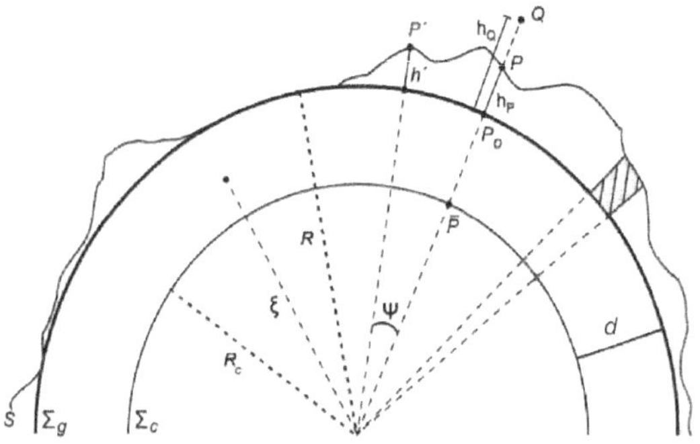


**Fig. 1** Geometry of the topography and the condensation model in spherical approximation.

The geocentric radius of the computation point Q is denoted by r, $r' = R + h'$ is the radius of the variable integration point P′ on the Earth's surface. $r_P$ is the radius of the point P which is placed below Q and upon the Earth's surface. $\psi$ is the angle between the radius vectors of Q and P′, G is the gravitational constant (see Fig. 1). The decomposition of Newton's integral into a two-dimensional spherical integral and a one-dimensional integral over the radial parameter $\xi$ results in a topographic potential, given by

$$V_t(Q) = G \cdot \rho \cdot \iint_\sigma \left[ \int_R^{r'} \frac{\xi^2}{\ell} d\xi \right] \cdot d\sigma, \qquad (1)$$

where

$$\ell = \sqrt{r^2 + \xi^2 - 2r\xi \cdot \cos\psi} \qquad (2)$$

and the density $\rho$ is assumed to be constant.

The effects of the topographic masses on the gravity gradient tensor are obtained from the second derivatives of the Newton integral (Eq. 1). The formula of the second radial derivative results in (see Heck and Wild, 2004)

$$\frac{\partial^2 V_t(Q)}{\partial r^2} = \frac{8\pi G\rho}{3r^3} \cdot \left(r_P^3 - R^3\right)$$
$$+ \frac{G\rho}{r^2} \iint_\sigma \left[ \begin{array}{l} -\frac{r'^3}{2\bar{\ell}'} + \frac{r_P^3}{2\bar{\ell}} + r'\bar{\ell}' - r_P\bar{\ell} \\ + \frac{r'^3 \cdot \left(r^2 - r'^2\right)}{2\bar{\ell}'^3} - \frac{r_P^3 \cdot \left(r^2 - r_P^2\right)}{2\bar{\ell}^3} \\ + 3r \cdot \cos\psi \cdot \left(\bar{\ell}' - \bar{\ell}\right) \\ + r^2 \cdot \left(3\cos^2\psi - 1\right) \\ \cdot \ell n \left| \frac{\bar{\ell}' + r' - r \cdot \cos\psi}{\bar{\ell} + r_P - r \cdot \cos\psi} \right| \end{array} \right] \cdot d\sigma, \qquad (3)$$

where

$$\bar{\ell}' = \sqrt{r^2 + r'^2 - 2rr' \cdot \cos\psi},$$
$$\bar{\ell} = \sqrt{r^2 + r_P^2 - 2rr_P \cdot \cos\psi}. \qquad (4)$$

Here, the first part of the second radial derivative of the topographic potential (out-of-integral term) represents the influence of a spherical shell of constant density $\rho$ and thickness ($r_P$ - R) on the vertical gravity gradient at Q. The second part, respectively the integral term, illustrates the terrain effect. This part would be zero, if there was no topography on the Earth.

# 3 Isostatic Reductions of SGG Data

## 3.1 Isostatic Model of Airy-Heiskanen

In order to model the effects of the isostatic masses, the compensation according to G.B. Airy and W.A. Heiskanen occurs in local, vertical columns by varying their thickness and leaving the density constant. The standard column has the constant density $\rho_0$ and the thickness T. Continental columns generate "roots" ($t' > 0$), whereas oceanic columns create "anti-roots" ($t' < 0$) (see Fig. 2).

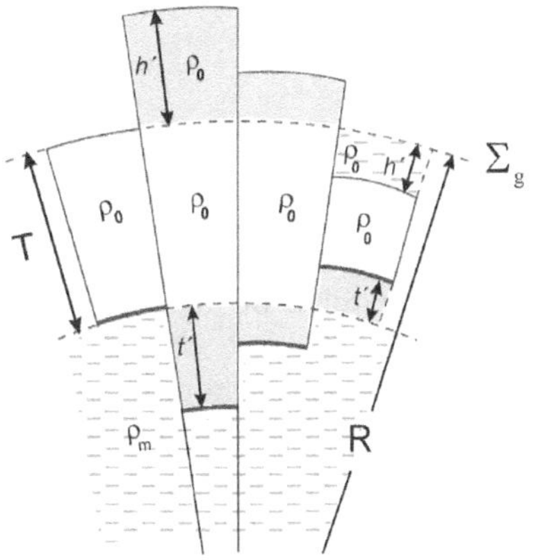


**Fig. 2** Isostatic model of Airy-Heiskanen in spherical approximation

The formula of the isostatic potential at the computation point Q is

$$V_{A-H}(Q) = G \cdot \Delta\rho \cdot \iint_\sigma \left[ \int_{R-T-t'}^{R-T} \frac{\xi^2}{\ell} d\xi \right] \cdot d\sigma, \qquad (5)$$

where $\Delta\rho = \rho_m - \rho_0$ is the difference between the density of the Earth's mantle and crust. In our case we use the JGP95E "rock-equivalent" digital terrain model for simulations which is based on a compression of water masses into a density $\rho_0 = 2.67\ g{\cdot}cm^{-3}$. In Kuhn (2003) the relation between the terrain height $h'$ and the root depth $t'$ has been derived, resulting in the formula

$$t' = (R - T) - \sqrt[3]{(R - T)^3 - \left(r'^3 - R^3\right) \cdot \mu}, \qquad (6)$$

where $\mu = \rho_0 / (\rho_m - \rho_0)$. The formula of the second radial derivative is (see Wild and Heck, 2004)

$$
\frac{\partial^2 V_{A-H}(Q)}{\partial r^2}=\frac{8\pi G\Delta\rho}{3r^3}\cdot\left((R-T)^3-(R-T-t_p)^3\right)
$$

$$
+\frac{G\Delta\rho}{r^2}\iint_\sigma\left[\begin{array}{l}
-\dfrac{(R-T-t_p)^3}{2\bar{\ell}_c}+\dfrac{(R-T-t')^3}{2\bar{\ell}'_c}\\
+(R-T-t_p)\cdot\bar{\ell}_c-(R-T-t')^3\cdot\bar{\ell}'_c\\
+\dfrac{(R-T-t_p)^3\cdot\left(r^2-(R-T-t_p)^2\right)}{2\bar{\ell}_c^3}\\
-\dfrac{(R-T-t')^3\cdot\left(r^2-r_P^2\right)}{2\bar{\ell}'^3_c}\\
+3r\cdot\cos\psi\cdot\left(\bar{\ell}'_c-\bar{\ell}_c\right)\\
+r^2\cdot\left(3\cos^2\psi-1\right)\\
\cdot\ell n\left|\dfrac{\bar{\ell}_c+(R-T-t_p)-r\cdot\cos\psi}{\bar{\ell}'_c+(R-T-t')^3-r\cdot\cos\psi}\right|
\end{array}\right]\cdot d\sigma, \tag{7}
$$

where

$$
\begin{aligned}
\bar{\ell}_c&=\sqrt{r^2+(R-T-t_p)^2-2r(R-T-t_p)\cdot\cos\psi},\\
\bar{\ell}'_c&=\sqrt{r^2+(R-T-t')^2-2r(R-T-t')\cdot\cos\psi},
\end{aligned} \tag{8}
$$

with the decomposition of the integration interval into two parts $[R-T-t', R-T-t_P]$, where $t_P$ is related to the "root" or "anti-root" of the point P. Combining the topographic and the isostatic potential, the out-of-integral term cancels out:

$$
\frac{\partial^2 V_t(Q)}{\partial r^2}-\frac{\partial^2 V_{A-H}(Q)}{\partial r^2}=
$$

$$
\frac{G\rho}{r^2}\iint_\sigma\left[\begin{array}{l}
\rho\cdot\left(-\dfrac{r'^3}{2\bar{\ell}'}+\dfrac{r_P^3}{2\bar{\ell}}+r'\bar{\ell}'-r_P\bar{\ell}\right.\\
+\dfrac{r'^3\cdot\left(r^2-r'^2\right)}{2\bar{\ell}'^3}-\dfrac{r_P^3\cdot\left(r^2-r_P^2\right)}{2\bar{\ell}^3}\\
+3r\cdot\cos\psi\cdot\left(\bar{\ell}'-\bar{\ell}\right)\\
+r^2\cdot\left(3\cos^2\psi-1\right)\\
\left.\cdot\ell n\left|\dfrac{\bar{\ell}'+r'-r\cdot\cos\psi}{\bar{\ell}+r_P-r\cdot\cos\psi}\right|\right)\\
-\Delta\rho\cdot\left(-\dfrac{(R-T-t_p)^3}{2\bar{\ell}_c}+\dfrac{(R-T-t')^3}{2\bar{\ell}'_c}\right.\\
+(R-T-t_p)\cdot\bar{\ell}_c-(R-T-t')^3\cdot\bar{\ell}'_c\\
+\dfrac{(R-T-t_p)^3\cdot\left(r^2-(R-T-t_p)^2\right)}{2\bar{\ell}_c^3}\\
-\dfrac{(R-T-t')^3\cdot\left(r^2-r_P^2\right)}{2\bar{\ell}'^3_c}\\
+3r\cdot\cos\psi\cdot\left(\bar{\ell}'_c-\bar{\ell}_c\right)\\
+r^2\cdot\left(3\cos^2\psi-1\right)\\
\left.\cdot\ell n\left|\dfrac{\bar{\ell}_c+(R-T-t_p)-r\cdot\cos\psi}{\bar{\ell}'_c+(R-T-t')^3-r\cdot\cos\psi}\right|\right)
\end{array}\right]\cdot d\sigma. \tag{9}
$$

## 3.2 Isostatic Model of Pratt-Hayford

To calculate the isostatic effect, J.H. Pratt divided the Earth's topography into local, vertical columns like G.B. Airy. He assumed a standard column of constant density $\rho_0$ and thickness D. The isostatic balance is achieved by varying the density of each column (see Fig. 3) postulating that the total mass in each column over the same surface element is constant. Continental columns in mountainous regions have a lower density, oceanic columns a higher density in comparison to $\rho_0$.

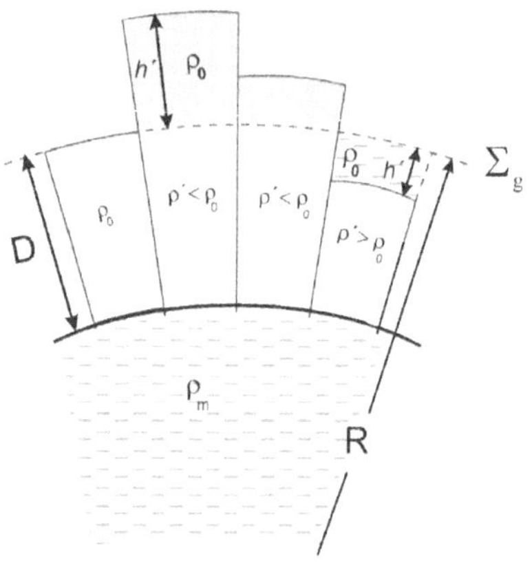


**Fig. 3** Isostatic model of Pratt-Hayford in spherical approximation

The formula of the isostatic potential at the computation point Q can be described analogously to the topographic potential by the Newton integral in spherical coordinates

$$
V_{P-H}(Q)=G\cdot\iint_\sigma\left[\int_{R-D}^{R}\Delta\rho'\cdot\frac{\xi^2}{\ell}d\xi\right]\cdot d\sigma. \tag{10}
$$

As a modification of the original Pratt-Hayford model we assume here that the topographic masses have the constant density $\rho_0$, consistent with the usual Bouguer reduction involving a constant crustal density, see e.g. Heiskanen and Moritz (1967) p.138. $\Delta\rho' = \rho_0 - \rho'$ is the density difference between the normal value $\rho_0$ and the actual density $\rho'$ of the isostatic masses below sea level, where

$$
\rho'=\rho_0\cdot\left[1-\frac{h'}{D}\cdot\frac{\left(1+\frac{h'}{R}\right)+\frac{1}{3}\left(\frac{h'}{R}\right)^2}{\left(1-\frac{D}{R}\right)+\frac{1}{3}\left(\frac{D}{R}\right)^2}\right]. \tag{11}
$$

By adding and subtracting the density $\rho'_{P0}$ of the point $P_0$ below Q the second radial derivative can be expressed by

$$\frac{\partial^2 V_{P-H}(Q)}{\partial r^2} = \frac{8\pi G \Delta\rho'_{P_0}}{3r^3}\cdot\left(R^3-(R-D)^3\right)$$
$$+\iint_\sigma \left(\rho'_{P_0}-\rho'\right)\cdot\left[\begin{array}{l} \frac{-R^3}{2\tilde{\ell}}+\frac{(R-D)^3}{2\tilde{\ell}_c}+R\cdot\tilde{\ell} \\ -(R-D)\cdot\tilde{\ell}_c+\frac{R^3\cdot\left(r^2-R^2\right)}{2\tilde{\ell}^3} \\ -\frac{(R-D)^3\cdot\left(r^2-(R-D)^2\right)}{2\tilde{\ell}_c^3} \\ +3r\cdot\cos\psi\cdot\left(\tilde{\ell}-\tilde{\ell}_c\right) \\ +r^2\cdot\left(3\cos^2\psi-1\right) \\ \cdot\ln\left|\frac{\tilde{\ell}+R-r\cdot\cos\psi}{\tilde{\ell}_c+(R-D)-r\cdot\cos\psi}\right| \end{array}\right]\cdot d\sigma,$$

where (12)

$$\tilde{\ell}=\sqrt{r^2+R^2-2rR\cdot\cos\psi},$$
$$\tilde{\ell}_c=\sqrt{r^2+(R-D)^2-2r(R-D)\cdot\cos\psi}.$$

The combination of the effects of the topographic and isostatic masses yields

$$\frac{\partial^2 V_t(Q)}{\partial r^2}-\frac{\partial^2 V_{P-H}(Q)}{\partial r^2}=$$
$$\iint_\sigma\left[\begin{array}{l} \left(-\frac{r'^3}{2\tilde{\ell}'}+\frac{r_P^3}{2\tilde{\ell}}+r'\tilde{\ell}'-r_P\tilde{\ell}\right. \\ +\frac{r'^3\cdot\left(r^2-r'^2\right)}{2\tilde{\ell}'^3}-\frac{r_P^3\cdot\left(r^2-r_P^2\right)}{2\tilde{\ell}^3} \\ +3r\cdot\cos\psi\cdot\left(\tilde{\ell}'-\tilde{\ell}\right) \\ +r^2\cdot\left(3\cos^2\psi-1\right) \\ \left.\cdot\ln\left|\frac{\tilde{\ell}'+r'-r\cdot\cos\psi}{\tilde{\ell}+r_P-r\cdot\cos\psi}\right|\right) \\ -\left(\rho'_{P_0}-\rho'\right)\cdot\left(\frac{-R^3}{2\tilde{\ell}}+\frac{(R-D)^3}{2\tilde{\ell}_c}\right. \\ +R\cdot\tilde{\ell}-(R-D)\cdot\tilde{\ell}_c \\ +\frac{R^3\cdot\left(r^2-R^2\right)}{2\tilde{\ell}^3}+3r\cdot\cos\psi\cdot\left(\tilde{\ell}-\tilde{\ell}_c\right) \\ -\frac{(R-D)^3\cdot\left(r^2-(R-D)^2\right)}{2\tilde{\ell}_c^3} \\ +r^2\cdot\left(3\cos^2\psi-1\right) \\ \left.\cdot\ln\left|\frac{\tilde{\ell}+R-r\cdot\cos\psi}{\tilde{\ell}_c+(R-D)-r\cdot\cos\psi}\right|\right) \end{array}\right]\cdot d\sigma. \quad (13)$$

The out-of-integral term cancels out similarly to $\partial^2 V_t(Q)/\partial^2 r^2$ - $\partial^2 V_{A-H}(Q)/\partial^2 r^2$ (see Eq. 9).

### 3.3 Combination of the Airy-Heiskanen model and the Pratt-Hayford model

To achieve an optimal assimilation of the classical isostatic models to the generalized Helmert model, the Airy-Heiskanen model is applied in the land area, the Pratt-Hayford model in contrast is used in the ocean area. By this mixture, one of the drawbacks of the original Airy-Heiskanen model – the fact that the anti-roots may rise above the ocean bottom in deep sea trough areas – can be avoided.

### 3.4 Generalized condensation model of Helmert

An alternative to the classical isostatic models – described in sections 3.1 and 3.2 – is the generalized condensation model of Helmert (1884). The potential of the masses condensed on the spherical surface $\Sigma_c$ (radius $R_c$) at the depth d below the geoidal surface $\Sigma_g$ can be expressed by the single layer potential

$$V_c = G\cdot\iint_\sigma\frac{\kappa'}{\ell_c}\cdot R_c^2\cdot d\sigma, \quad (14)$$

where

$$\ell_c=\sqrt{r^2+R_c^2-2rR_c\cdot\cos\psi}.$$

The surface density $\kappa'$ is calculated in the following way

$$\kappa'=\rho_0\cdot\frac{r'^3-R^3}{3R_c^2} \quad (15)$$

to fulfill the mass balance postulate. The standard depth $d = R - R_c$ of the condensation layer is 21 km in case of Helmert's first model and 0 km in case of Helmert's second model. The formula of the second radial derivative has been derived in Heck and Wild (2004). As a result of the evaluation of only a two-dimensional integral instead of a three-dimensional integral in the case of the classical isostatic models, there exist some numerical advantages.

## 4 Results

To compare the effects of various isostatic models on the gravity gradients at satellite height, the digital elevation model JGP95E – resolution 1° x 1° grid of "rock-equivalent" topography – is used. The mean radius R of the Earth is 6371 km, the satellite height is 260 km, i.e. the radius of the computation point Q is r = R + 260 km = 6631 km. The standard column of the Airy-Heiskanen model has the depth 25 km, $\Delta\rho$ amounts to 0.6 $g\cdot cm^{-3}$. For the Pratt-Hayford model D = 100 km has been used.

Replacing the integration by a summation over any spherical grid cell of size 1° x 1° yields

$$\iint_{\Delta\sigma_i} f(r,r',\psi)\cdot d\sigma\approx f(r,r'_m,\psi_m)\cdot\Delta\sigma_i, \quad (16)$$

where $r'_m$ and $\psi_m$ refer to the centre of each cell.

Fig. 4 displays the second radial derivative $\partial^2 V_{P\text{-}H}(Q)/\partial r^2$. The order of magnitude is about 10 E.U.. The maxima are for example in Greenland, the Andes and Antarctica. These results are almost identical to those of the topographic potential, the isostatic potential of Airy-Heiskanen and the isostatic potential of the mixed Airy-Heiskanen/Pratt-Hayford model.

The order of magnitude of the combined effect of the topographic and isostatic potential of Pratt-Hayford type is about 0.8 E.U. A similar order of magnitude is achieved with the difference of the topographic and isostatic potential of mixed Airy-Heiskanen/Pratt-Hayford type. This is equivalent to former results with regard to the Airy-Heiskanen model (see Wild and Heck, 2004) and the first Helmert model (see Heck and Wild, 2004). In all cases the out-of-integral terms cancel out.

By variation of the depth of the condensation layer it is possible to adapt the generalized Helmert condensation model to the classical isostatic models. For this reason, the depth of the condensation layer has been varied between d = 0 km (Helmert's second model) and d = 60 km in order to achieve an optimal approximation. As described in Wild and Heck (2004) the Airy-Heiskanen model is practically equivalent to the Helmert model for a depth of about 24 km. The Pratt-Hayford model in contrast is best approximated by the Helmert model fixing the condensation depth d at 48 km (see Tab. 1); the approximation error in terms of the rms difference amounts to 0.0044 E.U. only. In the case of the mixed Airy-Heiskanen/Pratt-Hayford model the optimal Helmert condensation depth is 41 km (see Fig. 5), related to a rms difference of 0.04 E.U.. In all comparisons of the classical isostatic models with the Helmert model, the largest discrepancy exists for the second Helmert model. This phenomenon is due to the fact that the balancing isostatic compensation masses in geophysical reality are much more distant from the Earth's surface or the satellite orbit than it is assumed in Helmert's second condensation model; for the same reason, the residual gravity field after subtracting the topographic-isostatic effects is much rougher for the second Helmert model than for the other – more realistic – isostatic models.

## 5 Conclusions

The numerical investigations demonstrate that the generalized Helmert model can be made equivalent to the Airy-Heiskanen model, the Pratt-Hayford model or the mixed Airy-Heiskanen/Pratt-Hayford model, when the depth of the condensation layer of Helmert is adapted in a suitable way. Among these alternatives, the Airy-Heiskanen model (T = 25 km), resulting in an optimal depth of d = 24 km, reflects the best approximation to geophysical reality – considering the fact that the isostatic masses compensating the Earth's topography are situated at a depth of about 20 - 30 km below sea level (see Moritz, 1990; Heck, 2002). Therefore the generalized Helmert model – calculated with a simpler integral kernel – is an actual alternative to the classical isostatic models, in particular the Airy-Heiskanen model. The application of these topographic-isostatic reductions leads to a smoothed residual field and therefore the downward continuation will be strongly simplified.

**Acknowledgements:** The authors gratefully acknowledge the valuable comments by J. Kirby and D. Tsoulis.

## References

Heck B (2002) On the use and abuse of Helmert's second method of condensation. In: Àdam J, Schwarz KP (eds) Vistas for Geodesy in the new millenium. Springer IAG Symposia, Vol. 125

Heck B, Wild F (2004) Topographic-isostatic reductions in satellite gravity gradiometry based on a generalized condensation model, In: Sansò F (ed) Proceedings of IAG General Assembly, Sapporo 2003, Springer IAG Symposia, (in print)

Heiskanen WA, Moritz H (1967) Physical Geodesy. Freeman and Company, San Francisco, 1967

Helmert FR (1884) Die mathematischen und physikalischen Theorieen der Höheren Geodäsie. II. Teil: Die physikalischen Theorien, B.G. Teubner, Leipzig (reprinted 1962)

Kuhn M (2003) Geoid determination with density hypothesis from isostatic models and geological information. Journal of Geodesy, Vol. 77, 50-65

Moritz H (1990) The figure of the earth. Herbert Wichmann Verlag, Karlsruhe, 1990

Tsoulis D (1999) Analytical and numerical methods in gravity field modelling of ideal and real masses. Deutsche Geodätische Kommission, Series C, No. 510

Wild F, Heck B (2004) Effects of topographic and isostatic masses in satellite gravity gradiometry, Proc. Second International GOCE User Workshop "GOCE, The Geoid and Oceanography", ESA-ESRIN, Frascati, Italy, 8-10 March 2004 (ESA SP-569, June 2004)

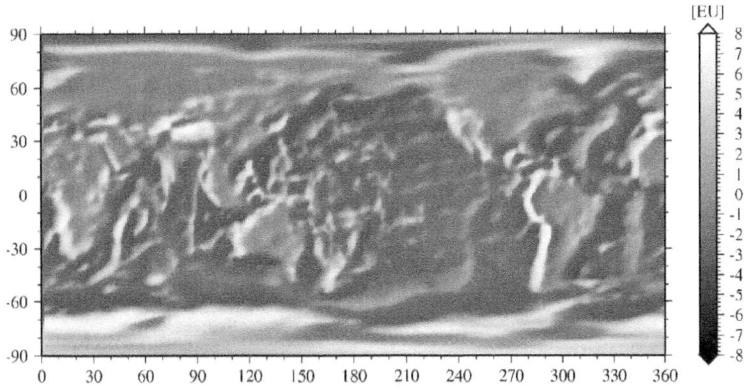


**Fig. 4** Isostatic reduction [ $\partial^2 V_{P\text{-}H}(Q)/\partial^2 r^2$]

**Tab. 1** Statistics of the difference [$\partial^2 V_c (Q)/\partial^2 r^2$ - $\partial^2 V_{P\text{-}H}(Q)/\partial^2 r^2$] (E.U.)

| Depth d of the Helmert model [km] | max | min | mean | rms |
|---|---|---|---|---|
| 0 | 1.358821 | -0.953901 | 0.006236 | 0.210743 |
| 21 | 0.707450 | -0.497449 | 0.003591 | 0.113168 |
| 25 | 0.592487 | -0.416584 | 0.003088 | 0.095618 |
| 30 | 0.452614 | -0.318124 | 0.002460 | 0.074117 |
| 35 | 0.316866 | -0.222517 | 0.001833 | 0.053096 |
| 40 | 0.185107 | -0.131558 | 0.001207 | 0.032581 |
| 45 | 0.062024 | -0.048697 | 0.000581 | 0.012849 |
| 46 | 0.041524 | -0.034358 | 0.000456 | 0.009196 |
| 47 | 0.028458 | -0.020472 | 0.000331 | 0.006003 |
| **48** | **0.026341** | **-0.027008** | **0.000206** | **0.004370** |
| 49 | 0.041008 | -0.043294 | 0.000081 | 0.005777 |
| 50 | 0.055552 | -0.066965 | -0.000044 | 0.008859 |
| 55 | 0.132676 | -0.187529 | -0.000669 | 0.027453 |
| 60 | 0.214990 | -0.304605 | -0.001292 | 0.046320 |

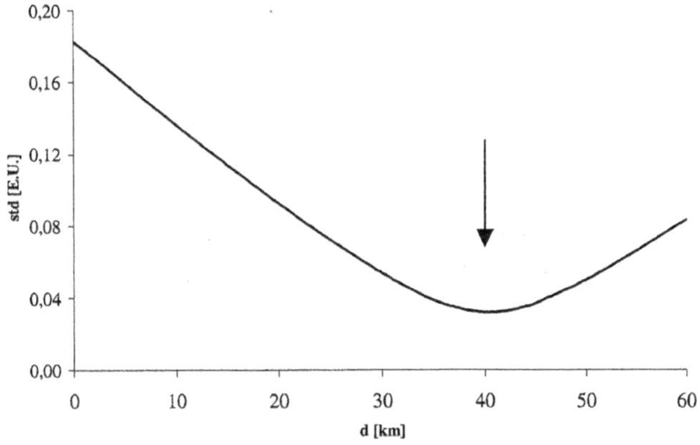


**Fig. 5** Difference of isostatic reductions [$\partial^2 V_c (Q)/\partial^2 r^2$ - $\partial^2 V_{A\text{-}H/P\text{-}H}(Q)/\partial^2 r^2$]

# Spectral Analysis of Mean Dynamic Ocean Topography From the GRACE GGM01 Geoid

Zizhan Zhang
Institute of Geodesy and Geophysics,
Chinese Academy of Science, 174 Xudong Road, Wuhan, China, 430077
The Graduate of School of the Chinese Academy of Sciences, 19A Yuquan Road, Beijing, China, 100039

Yang Lu
Institute of Geodesy and Geophysics,
Chinese Academy of Science, 174 Xudong Road, Wuhan, China, 430077
Unite Center for Astro-geodynamics Research, 80 Nandan Road, Shanghai, China, 200030

**Abstract.** With the development of the satellite gravity missions and the satellite altimetry missions, there have been recent improvements in geoid models and sea surface height (SSH) models, so now one can compute more accurate dynamic ocean topography (DOT). In this study, we computed a new mean DOT using a new geoid model (GGM01C) from GRACE and an improved mean SSH model (KMSS04) from a decade of multi-satellite altimetry. And DOT is computed using the GGM01C and EGM96 model to access the improvements and differences between these two geoid models. We find that the DOT is composed of long waves mainly, medium waves partially and short waves scarcely, and that the zonal spectrum of the DOT is different from its meridional spectrum. The differences in the DOTs from GGM01C and EGM96 in the test area indicate that great differences exist in these two gravity models, at least in equatorial area.

**Key words.** Mean dynamic ocean topography, GGM01, fast Fourier transform, wavelet analysis.

## 1 Introduction

The mean dynamic ocean topography (DOT) is usually defined as the difference between the mean sea surface height (SSH) and the geoid. Though its variation is only ±2~3 m, it is of high importance not only for oceanographic studies to directly monitor the ocean currents, do ocean forecasting etc, but also for geodetic studies to improve the geoid model, etc. However, it is difficult to estimate the DOT accurately. Many approaches were developed to estimate the DOT, such as the ocean climatology method, like the Levitus climatology [*Levitus 1982; Levitus et al., 1994*], mean circulating method, from ocean modeling (as the POCM series), synthetic climatology method [*Glenn et al., 1991; Ichikawa et al., 1995; Uchida et al., 1998*] and combining satellite altimetry with satellite gravimetry method, etc. However, these approaches are not efficient because of limitation of themselves or lacking data in some areas or bad precision of the data.

During the last two decades satellite altimetry has offered an abundance of measurement of the sea surface resulting in the improvement of marine gravity field and geoid determination, thus several high accuracy and resolution mean SSH models are developed (e.g. CLR, CLS01, KMSS1, KMSS04). Moreover, long-wavelength gravity models have been improved greatly in recent years. Such as the EIGEN-1S, EIGEN-2, EIGEN-3p from CHAMP, the GGM01S from GRACE etc., and especially the GGM01 gravity model, which represents a dramatic improvement over older geoid models [*Tapley et al., 2003*]. These breakthroughs make it possible to use the geodetic altimetry-derived products and satellite altimetry products to estimate a DOT accurately.

## 2 The Mean Dynamic Ocean Topography

### 2.1 Global Mean Sea Surface Height Model

At present, there are lots of mean SSH models. One of the latest mean SSH models is KMSS04, which is derived from a combination of altimetry from a total of 5 different satellites and a total of 8 different satellite missions like the T/P, T/P Tandem Mission, ERS1 ERM+GM, ERS2 ERM, GEOSAT GM, GFO-ERM data and new data from the JASON and ENVISAT data. The resolution of the mean sea surface is 2 minutes equivalent to 4 km at the Equator [*Andersen et al., 2004*].

### 2.2 High-accuracy GRACE Gravity Model GGM01

The GGM01 gravity model was determined from

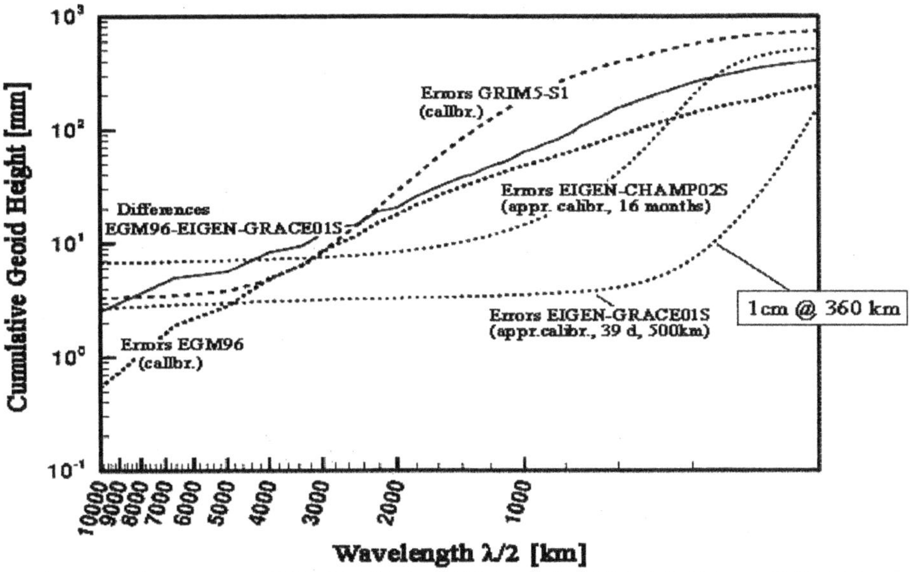


**Fig. 1** Cumulated geoid error for EIGEN-GRACE01S, EIGEN-CHAMP02S, EGM96 and GRIM5-S1 solutions and cumulated geoid differences between EIGEN-GRACE01S and EGM96. (For more details please see the web site: http://op.gfz-potsdam.de/grace/index_GRACE.html)

111 days of GRACE data spanning the months of April through November 2002, during the commissioning phase of the mission. It contains two models. One is the GGM01S obtained by GRACE satellite only, expanding to degree/order 120. The other is the GGM01C which is estimated by combining GGM01S model with historical multi-satellite tracking data, surface gravity data, altimetric sea surface heights, and so on, and it completes up to degree and order 200 [*http://www.csr.utexas.edu/grace/gravity/ggm01/GGM01_Notes.pdf*].

From Fig. 1, which shows the comparison of the precision of several gravity models as a function of wavelength, it can be seen that errors of EIGEN-GRACE01S are very small at wavelengths longer than 1000 km and the error is only one centimeter at wavelength 360 km, while errors of EGM96 model are up to 20 centimeters or even larger at the same length scale. Although GGM01C model is only the initial model derived from the GRACE mission, the precisions of its long wavelengths improved greatly even 10 to 50 times better than the corresponding wavelengths of all pre-existing models. The expected accuracy of final GRACE geoids is expected to be a few *mm* out to degree/order 70[*Tapley et al., 2003*]. So, it may be considered that the precision of long wavelength contents of the geoid estimated from the model is higher enough to compute the DOT accurately.

Circulation maps are very useful to evaluate the improvement of the geoid model, since small changes in the geoid can lead to significant changes in the circulation, especially in the tropics [*Tapley et al., 2003*]. The Center for Space Research of the University of Texas at Austin studied the ocean currents in Gulf Stream region of the North Atlantic. Their results show a circulation map obtained from the GRACE geoid combined with satellite altimetry and ship measurements of ocean temperature and salinity. This map matches well with the map measured by ship-deployed floats. But, in another circulation map derived from the same way as above, except that the best gravity model prior to GRACE was used, the implied currents are actually flowing in the wrong direction when compared to the above two maps [*http://www.csr.utexas.edu/grace/gravity/oceanographic_sciences.html*]. As a test of the GGM01 gravity model, large-scale zonal and meridional surface geostrophic currents are computed from the dynamic ocean topography estimated by it and are compared with those derived from a mean hydrographic surface. Reduction in residual RMS between the two by 30~60% (and increased correlation) indicates that the GGM01 geoid represents a dramatic improvement over older geoid models, which were developed from multiple satellite tracking data, altimetry, and surface gravity measurements [*Tapley et al., 2003*]. The results show that the GGM01 model is superior to all historical gravity models.

No 'Kaula' constraint, no other satellite information, no surface gravity information and no other a priori

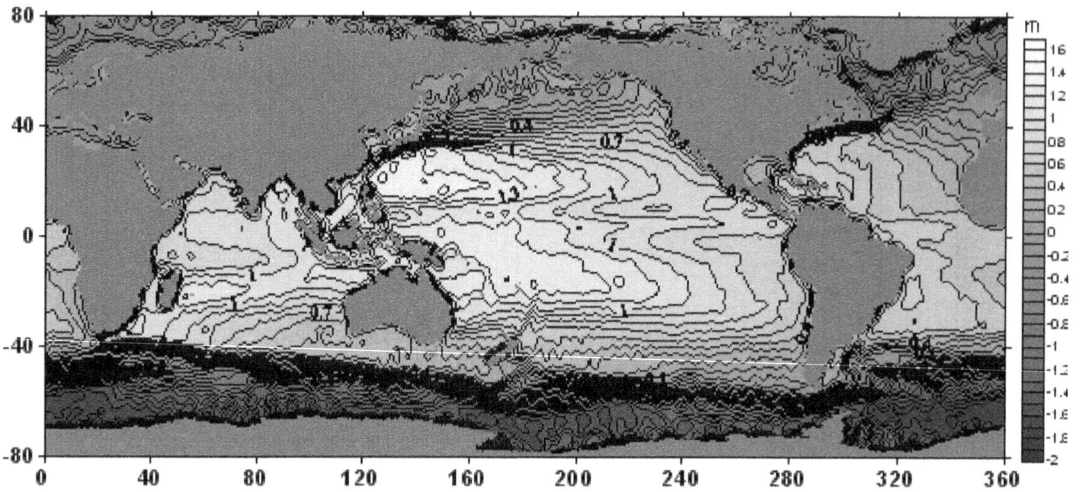


**Fig 2.** Global DOT estimated by the KMSS04 model minus the GGM01C geoid (units m)

conditioning were applied in generating the GGM01S solution. Tapley etc. suggested that on a global level, this solution has enough precision to show monthly variable gravity at spatial scales of ~1000km[*Tapley etc., 2004*]. From another point of view, the precision of long wavelength (larger than 1000km) of GGM01C model is higher than all old models, although wavelength shorter than 1000km may be contaminated by surface gravimetry data or other data used in generating it. [*http://www.csr.utexas.edu/grace/gravity/ggm01/GGM01_Notes.pdf*].

### 2.3 Global High Resolution DOT

Table 1 describes ratio of energy of EGM96 geoid spectrum. From this table, it could be known that the power spectrum of the geoid concentrates on the low frequency parts principally and those above 180 degree/order (equivalent wavelength is 200km) are less than 0.02% (table 1) [*Liu C. et al., 2004*], that's to say, a DOT based on the geoid expanded to degree/order 200 contains little the geoid signal. In this work, a global mean DOT (Fig. 2) was computed from the KMSS04 (spatial resolution equals to 0.5°×0.5°) and the GGM01C geoid to degree/order 200, and was filtered by Gauss filter with 1000km. The mean currents of the world, such as the Kuroshio Current, North Pacific Ocean flows, North Atlantic Ocean flow, California flow, South Equator flow, North Equator flow, Peru flow, etc., are displayed clearly in this plate.

| Frequency range | 2 ~ 36 | 37 ~ 180 | 181 ~ 360 | > 360 |
|---|---|---|---|---|
| Ratio (%) | 99.669 | 0.314 | 0.011 | 0.006 |

**Table 1** Ratio of energy of EGM96 geoid model spectrum

## 3 Spectral Analysis of the DOT

The mean DOT is successional in the global sea area and its wavelength contents are stationary relatively. The spectral characters of the DOT in an area should describe the global DOT structure approximately. A test area located in the Pacific with the bounds of (S30°~ N40°) (E160°~E230°) is selected for spectral analysis. The advantages of selecting this area come from the facts that there are only few lands and islands, but deep sea in this area, resulting in good quality of the altimetry data by reducing the influence of the shallow sea and land.

The use of spectral methods in physical geodesy has been developed during recent decades. The basic advantage of the analysis in the spectral domain is the algebraic simplicity of the convolution integrals. As it is well known, the convolution integrals are transformed to multiplications in the spectral domain, and the evaluation of some complicated formulas in gravity field modeling is speeded up [*Andritsanos et al., 2001*]. In this study, the two-dimensional Discrete Fast Fourier Transform (2DFFT) method is used [*Li J., 1993; Cadzow et al., 1981*].

Considering the initial data set, $f(x_k, y_l)$ and its Fourier transformed set, $F_F(u_m, v_n)$, then the relation between them would be

$$F_F(u_m, v_n) = \sum_{k=0}^{M-1}\sum_{l=0}^{N-1} f(x_k, y_l) e^{-i2\pi(u_m x_k + v_n y_l)} \Delta x \Delta y,$$

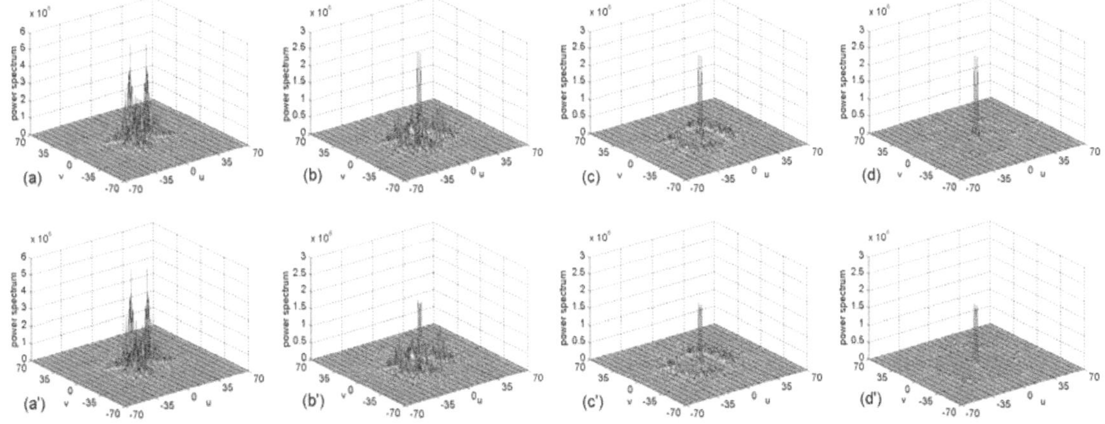

**Fig. 3** Power spectrum of the DOTs based on GGM01C (a-d) and EGM96 (a'-d') geoid model to 36,90,120,200

where $M, N$ denote the numbers of grids in $x, y$ direction respectively. $\Delta x, \Delta y$ denote the sampling distances (distance between two successive grids) in $x, y$ direction. Moreover, in order to reduce the influence of frequency alias and to process data easily, the zero mean value operation is applied before the 2DFFT processing. After this processing, the mean value of the initial data set would be zero.

### 3.1 Ensemble Spectral Feature of the DOT

The ensemble spectral characters of a signal show the overall reflection of its features in space domain after 2DFFT processing. Fig. 3 shows the general power spectral features of the DOTs estimated by two geoid models (e.g. EGM96 and GGM01C). In figure 3, u, v denote wave numbers, z-axle denotes power spectrum.

From Fig.3, the ensemble spectral features of the DOT are shown as follows: the energy spectrum of the DOT concentrates on the low frequency contents mainly, while those of the middle and high frequencies are small. The medium wavelengths of the geoid remain in the DOT partially when the geoid is expanded to low degree/order (see Fig.3 (a), (b), (a'), (b')). With the geoid model expanded to higher degree/order, the errors of the geoid are reduced, and the components between the medium and high frequencies are close to zero. As a result, the DOT is composed of low frequency mainly, medium frequency partially and high frequency scarcely.

### 3.2 Zonal and Meridional **Spectral Features of the DOT**

Fig. 4 shows the spectral features of the DOTs obtained from the GGM01C and EGM96 up to degree/order 200 in the zonal direction (Fig. 4(a)) and in the meridional direction (Fig. 4(b)). The magnified curves above the DOT-line correspond with the cures under the DOT-line.

From Fig.4 (a), it could be identified that some

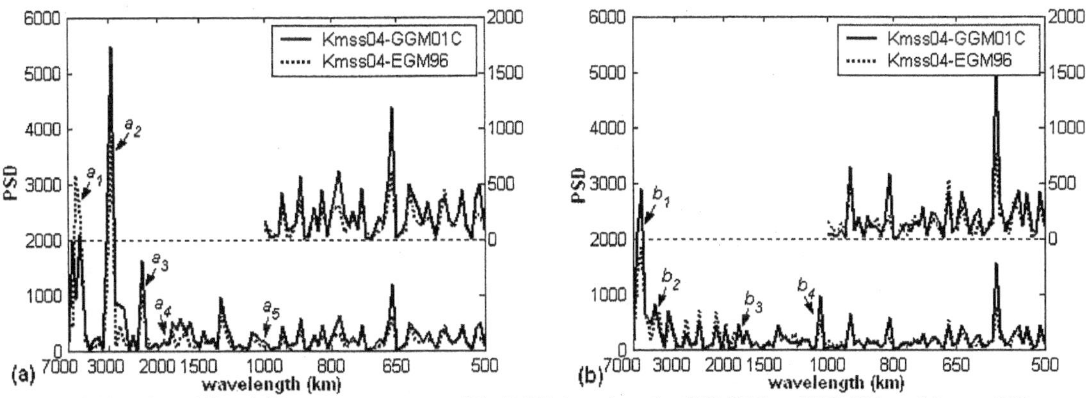


**Fig.4** Zonal (a) and meridional (b) power spectrums of the DOTs based on the GGM01C and EGM96 model up to 200

obvious differences exist between the power spectral maps of the DOTs from the EGM96 and GGM01C model. The main differences of the spectral features between the two DOTs exist in the long wavelength parts (see $a_1$ point and the range $a_2$~$a_3$ in Fig. 4(a)). The amplitude of the power spectrum of the DOT from the EGM96 model is larger than those from the GGM01C at wavelength longer than 5000km ($a_1$), while in the range of 2000km~ 3000km ($a_2$~$a_3$), the former is lower than the latter obviously. It could be also found that some slight differences exist at the medium wavelengths from 1000km to 2000km ($a_4$~$a_5$). From Fig. 4(b), it could be seen that the differences between the power spectral structure of the two DOTs computed from the EGM96 and GGM01C geoid are: (i) the amplitude of the former is larger than the latter at $b_1$ point, (ii) some slight differences exist at the range $b_2$~$b_3$, and (iii) almost no differences appear at the range $b_3$~$b_4$. The differences between wavelengths of 500 km and 1000 km in the zonal spectrum are lager than those of meridional spectrum (the curves shown in Fig. 4 above the DOT-line). These differences are real due to the way GRACE samples the gravity field in the N-S direction.

Comparing Fig4 (a) with (b), It could be seen that the amplitudes of wavelengths longer than 3000 km are much higher than those of shorter than 3000km in meridional spectrum, while the main amplitude is at 3000km in zonal spectrum, and wavelengths shorter than 3000km gives little contribution to the DOTs in both directions. The power spectral graph of the DOT from GGM01C geoid model shows more distinctive signals than that from the EGM96 geoid model.

## 4 Application of Wavelet Multi-scale Decomposition in the DOT

The wavelet transform introduces the idea of multi-scale analysis, which enjoys desirable properties of localization in both the space and the frequency domains. The wavelet transform can decompose signal into several components with different frequencies or scales and focus it into many specified details via dilation and shifting of the wavelet functions.

Following the principle of multi-scale analysis [*Osman et al., 2000; Hou Z., 2001*], we have

$$\Delta h = A_J H + D_1 H + D_2 H + \cdots + D_J H$$

Where $D_1 H$ is denoted by the first order wavelet detail of the DOT, $D_2 H$ by the second order, $D_J H$ by the J-th order, and $A_J H$ is denoted by the J-th order approximation of the DOT.

The first order wavelet detail of the DOTs reflects the noise signal in it. The second and third order wavelet details of the DOTs from the GGM01C and the EGM96 geoid mode look similar to each other, these details reflect seamounts, large-scale eddies and medium-scale ocean currents so on. Due to space limitations, these maps of the first to third order wavelet details are not shown in this paper. The plates (Fig. 5, (a) (b)) are the third-order approximation of the DOTs with wavelength longer than 800km, which reflect the large-scale deep ocean currents, sea bottom topography, etc. From this plate, it is easily found that there are obvious differences (especially in the area near the equator) between the DOTs from the GGM01C and the EGM96 geoid model. The equatorial flows in the former are displayed more clearly than those in the latter, which indicates that the GGM01C geoid

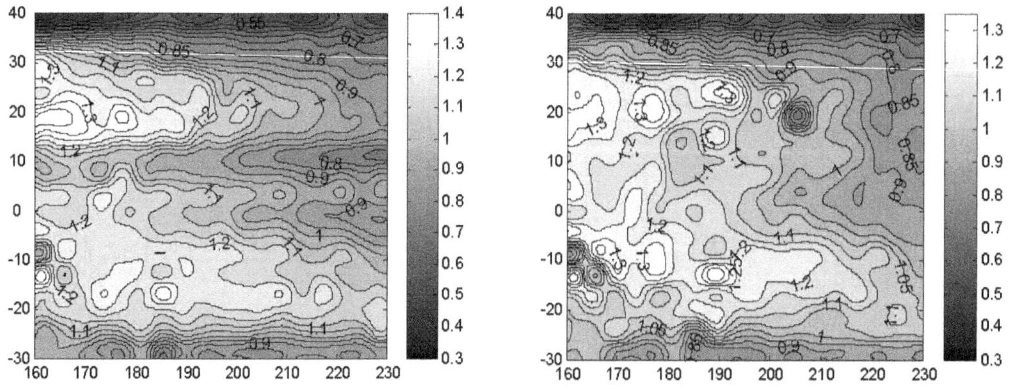


**Fig.5** The third-order approximations of the DOTs (a, b) based on the GGM01C and EGM96 geoid model to 200 (S30°~ N40°), (E160°~E230°)

model is superior to the EGM96 geoid model at this scale.

## 5 Summary and Conclusion

From this study, the following conclusions are drawn. The spectral structure of the DOT is very stationary, the energy concentrates on the low frequency contents mainly (the long wavelengths of the DOT), and medium and high frequency contents are lower; the long and medium waves of the geoid signal also remain in the DOT partially when the geoid is expanded to low degree/order.
The spectral plots of the DOTs are different between the zonal and meridional directions: the wavelengths longer than 3000km are the main components of the DOT in the meridional direction, while the main component of the DOT is the wavelength about 3000km in the zonal direction. Short waves contribute little to the DOT in both directions. The differences between the DOTs from GGM01C and EGM96 in the test area indicate that great differences exist in these two gravity models, and that the GGM01C model has been improved greatly, at least in equatorial area.

**Acknowledgements** The authors are grateful to professor David T. Sandwell and the reviewers of this paper for their helpful recommendations. The data used in this work were obtained from GeoForschungsZentrum Potsdam and the Center for Space Research of Texas University and Ole B. Andersen etc. in Danmark. This work was supported by the National Natural Science Foundation of China (Grant No. 40374007, 40234039), Foundations of Chinese Academy of Science (Grant Nos. KZCX2-SW-T1, KZCX3-SW-132), Foundation of marine 863 (Grant No. 2002AA639280) and the Informatization Construction of Knowledge Innovation Projects of the Chinese Academy of Sciences "Supercomputing Environment Construction and Application" (Grant No. INF105-SCE).

## References

Andritsanos, V. D., M. G. Sideris and I. N. Tziavos (2001). Quasi-stationary Sea Surface Topography Estimation by the Multiple Input/output Method. *Journal of Geodesy*, Vol.75: 216-226

Cadzow, J. A., and K. Ogino (1981), Two dimensional spectral estimation, *IEEE Trans., ASSP,* 29:396-401

CHRISTOPHER KOTSAKIS. Can We Filter Non-stationary Noise From Geodetic Data With Fast Spectral (FFT) Techniques? Bulletin of the International Geoid Service, 2001, No. 11, pp. 93—114

Glenn, S. M., D. L. Porter and A. R. Robinson (1991). A Synthetic Geoid Validation of Geosat Mesoscale Dynamic Topography in the Gulf Stream Region. *Journal of Geophysical Research,* 96, 7145-7166.

Hou Zunze (2001). Wavelet Transform and Its Application to Decomposition of Gravity Anomalies. WAA 201,LNCS 2251, pp. 404-410, 2001.

Knudsen, P., and O. B. Andersen (1997), Global Marine Gravity and Mean Sea Surface From Multimission Satellite Altimetry, in "Geodesy on the Move, Gravity, geoid, geodynamics and Antarctica", *Proceedings IAG scientific assembly*, Rio de Janeiro, sep.3-9, Eds, Forsberg, Feissel and Dietrich., *IAG symposia*, 119, 132-138, Springer, Berlin, 1998.

Ichikawa, K., S. Imawaki, and H. Ishii (1995). Comparison of surface velocities determined from altimeter and drifting buoy data. *Journal of Oceanography,* 51, 729-740.

Levitus, S. (1982). Climatological Atlas of the World Ocean NOAA, Geophysical Fluid Dynamics Laboratory, Professional Paper 13, Rockville, MD.

Levitus, S., R. Burgett, and T. P. Boyer (1994). World ocean atlas 1994 vol. 3: Salinity NOAA Atlas NESDIS 3, 99 pp. National Ocean and Atmosphere Administration USA.

Li Jiancheng (1993). The Spectral Methods in Physical Geodesy (in Chinese). Wuhan: Wuhan technical university of surveying and mapping. 25-27

Liu Changjian, Han Chunhao (2004). Instant Sea Level Model and Its Application in Positioning with Earth-synchronous Satellite (in Chinese). 9-16

Osman N. U., Serhat Seker, A. Muhittin Albora, and Atilla Ozmen (2000). Separation of Magnetic Fields in Geophysical Studies Using a 2-D Multi-resolution Wavelet Analysis Approach. Journal of the Balkan Geophysical Society, Vol. 3, No 3, p. 53-58

Tapley, B. D., D. P. Chambers, S. Bettadpur, and J. C. Ries (2003). Large-scale Ocean Circulation From the GRACE GGM01 Geoid. *Geophysical Research Letters,* Vol.30, No.22, 2163

Uchida, H., S. Imawaki, and J. H. Hu (1998). Comparison of Kuroshio surface velocities derived from satellite altimetry and drifting buoy data. *Journal of Oceanography*, 54, 115-122.

Ole B. Andersen, Anne L. Vest and P. Knudsen (2004). Mean sea surface determination and inter-annual ocean variability. *Observing and Understanding Sea Level Variations*, p.38

# Interannual to decadal sea level change in south-western Europe from satellite altimetry and in-situ measurements

L. Fenoglio-Marc
Institute of Physical Geodesy, Darmstadt University of Technology, Darmstadt, Germany

E. Tel, M.J. Garcia
Instituto Español de Oceanografía, Madrid, Spain

N. Kjaer
Geodetic Department, Kort og Matrikelstyrelsen, Copenhagen, Denmark

**Abstract.** Interannual to decadal sea level changes are investigated in the Iberian Peninsula in the interval 1993-2001 using satellite altimetry and tide gauge measurements. Eleven locations, six in the Mediterranean Sea and five in the Atlantic Ocean, are selected. Monthly de-seasoned sea level values are low-pass filtered to focus on signals with periods longer than one year. The correlation of altimetry and tide gauge monthly values is higher than 0.7 and significant at the 95% level at many of the locations, is regionally dependent and increases when using low-pass filtered data. The long-term components of the sea level height differences are mostly smaller than +/- 3 mm/yr. The sea level is inversely correlated to the North Atlantic Oscillation climatic index and the correlation increases at interannual time scales. A relative maximum in sea level in the years 1996-1997 coincides with a relative minimum of the North Atlantic Oscillation index. A maximum in the same time interval is observed in steric heights computed from hydrographic databases, that correspond to the change in sea level due to volume change. The correlation between steric heights and tide gauge yearly values in 1993-1998 is regionally dependent and significant at the 90% level at a few locations in the Mediterranean Sea. Highest values are reached with steric and thermo-steric heights of Medar/Medatlas in Malaga (0.7) and with thermo-steric heights of World Ocean Atlas 1998 in Ceuta and Algeciras (0.8).

**Keywords.** Sea level, altimetry, tide gauge, steric change

## 1 Introduction

A comparative analysis of sea level data from tide gauges and the Topex/Poseidon (T/P) altimeter is part of the European Sea Level Service – Research Infrastracture Project (ESEAS-RI). Data from more than 500 tide gauge stations of the PSMSL dataset (Permanent Service for Mean Sea Level - http://www.pol.ac.uk/psmsl/) are quality controlled and analysed (Tsimplis et al. 2004) to provide a basis for comparing and merging tide gauge data with T/P data. Analysis of tide gauge data per region is based on coherent sea level signal analysis (Shaw et al 2004). Spatial and temporal correlation with altimetry in 1993-2001 is analysed in Kjaer et al. (2004). The present study concentrates on the Iberian region with longitude 10°W-3.5°E and latitude 35-44°N. In addition to the monthly tide gauge records prepared within the ESEAS-RI project, hourly data from Spanish tide gauge stations are used to extend the series over the last decade until end of year 2001 as years 1999-2001 are often missed in the PSMSL database. Sea level variability at low and medium frequencies has been investigated in the late 20$^{th}$ century in this region by using tide gauge data and T/P satellite altimetry data. A general positive rate of sea level rise has been observed with a major increase in the last decade (Cazenave et al. 2001, Fenoglio-Marc 2001). A variation in the exchange flow through the strait of Gibraltar in the last decade was observed (Ross et al. 2000). The sea level in the North-Atlantic and in the Mediterranean Sea is strongly influenced by the North Atlantic Oscillation (NAO) (Woolf et al. 2003), which influences sea level anomalies in both freshwater flux (evaporation, precipitation and river runoff) and surface pressure (Tsimplis and Josey 2001). Changes in sea level heights arise from mass and volume changes. The mass change contributes to the gravity field variability and mainly arises from fresh water flux and net flow. The steric change corresponds to the volume change and is the portion of sea level change due to density variation, which is introduced by temperature and density variation and dominated by thermal effect. Its thermo-steric and halo-steric components are due respectively to changes in temperature and in salinity. Positive average expansion rates of 0.8 mm/yr for the 1000-2000 db layer and of 1.6 mm/yr for the full water column have been estimated for the interval 1950-1980 from temperature and density profiles near to the Iberian Peninsula (Arbic and Owens 2001). The values are comparable to the average value of 1.8 mm/yr estimated for the global sea level rise over the last century using tide gauge data, that include steric and other contributions (Douglas 2001). Regional temperature and salinity trends are detected from the hydrographic databases and from profiles collected in dedicated campaigns (Bethoux et al. 1998, Painter and Tsimplis 2003, Tsimplis and Rixen 2002). Warming trends in sea temperature have been observed in the last century with increase in the last decade: a warming trend of 0.02 °C/y in the temperature of the Alboran Sea in 1992-2001 in the 200 m deep layers of the Malaga Bay continental shelf and positive trends of 0.005 °C/y during the past century from the Medatlas dataset (Vargas et al. 2002). The Mediterranean Sea and the Atlantic Ocean present different characteristics of sea level variability mainly due to ocean tides, ocean circulation patterns and atmospheric conditions (Garcia-Lafuente et al. 2004). Studies based on monthly data (Zerbini et al. 1996, Fenoglio-Marc 2002, Mangiarotti 2003, Fenoglio-Marc 2004) show an agreement between seasonal and linear-term components of the sea level variability from tide gauges and altimetry in the Mediterranean Sea.

Interannual to decadal sea level changes measured by altimetry and tide gauge data are here investigated in the Iberian Peninsula in 1993-2001. In Section 2 a dataset is constructed that contains the hourly data collected from the local organisations after inspecting the differences with respect to the monthly ESEAS-RI dataset. In Section 3 altimetry and tide gauge sea level heights in 1993-2001 are compared. In Section 4 the relationship between sea level change and North Atlantic Oscillation is analysed using de-seasoned monthly averages and low-pass filtered data. In Section 5 interannual variability and steric heights from climatologic datasets and from hydrografic profiles are compared in the Mediterranean Sea.

## 2 Data preparation

Sea level monthly averages from altimetry and tide gauges are used. The Topex/Poseidon altimetry data between January 1993 and December 2001 are provided by the NASA Pathfinder Project (http://iliad.gsfc.nasa.gov, Version 9.2). The altimeter sea level anomalies are averaged in monthly grids with spacing in latitude and longitude of 1x1 degrees using a Gaussian weighted average method with half-weight parameter equal to 1 and search radius equal to 1.5 degrees (Fenoglio-Marc 2001). Alternatively time series at data points are treated.

The inverse barometric correction is applied to neither altimetry nor tide gauge data, the ocean tide correction given by the GOT002 model is applied only to the altimetry data before computing the monthly averages. This choice is based on comparisons between altimetry and tide gauge data (Fenoglio et al. 2004). Computing the monthly averages of hourly tide gauge data the short-period (diurnal, semi-diurnal) ocean tides average to zero at a given location, while the same does not apply for the ocean tide when the three T/P altimeter values for each month are averaged, that is why the ocean tide correction is applied to the altimeter data but not to the tide gauge data. Ocean tides of lower frequencies are neglected. The inverse barometer correction does not average to zero at a given location over one month using high frequency observations, that is why the inverted barometer correction is applied neither to altimeter nor to tide gauge data. The correlation of altimetry and tide gauge is higher when the inverse-barometric correction is not applied to the data as part of the correlation between the altimetric and tide gauge time-series is due to the inverse barometer response of the ocean, correcting the data reduces the correlation by less than 5% (Fenoglio et al., 2004). Monthly tide gauge data pre-selected by ESEAS-RI from the PSMSL (Permanent Service for Mean Sea Level - http://www.pol.ac.uk/psmsl/) and hourly data from local organisations are used. The PSMSL dataset contains 30 stations in the Iberian region in the last decade: 14 in the Mediterranean Sea and 16 in the Atlantic Ocean, but only 8 of those stations have complete coverage over 1993-2001, 4 are in the Mediterranean Sea (Ceuta, Algeciras, Malaga I and L'Estartit) and 4 in the Atlantic Ocean (Santander, La Coruña, Vigo, and Cadiz). The longer records at Cascais (1880-1993) and at Lagos (1900-1990) do not cover the altimetry period. The longest records including also the last decade are La Coruña, Santander, Vigo and Tarifa (1944-2001), Cadiz and Malaga (1960-2001). We have constructed an IBERIAN_LOCAL dataset containing tide gauge hourly data in the interval 1993-2001 provided by three Spanish organisations: the Instituto Español de Oceanografía (IEO), Puertos del Estado (PDE) and the Universitat Politecnica de Catalunya (Table 1). Figure 1 shows the geographical locations.

**Table 1.**Stations included in the Iberia_local dataset and 13 stations selected for the analysis in 1993-2001

| N | Ntg | PSMSL Code | Location | Source | Sea | Interval |
|---|---|---|---|---|---|---|
| 1 | 1 | 200006 | Bilbao | PDE | Atl | 1993-2001 |
| 2 | | 200011 | SantanderI | IEO | Atl | 1993-2001 |
| 3 | 2 | 200013 | SantanderII | PDE | Atl | 1993-2001 |
| 3 | | 200022 | GijonII | PDE | Atl | 1996-2001 |
| 4 | | 200030 | La CorunaI | IEO | Atl | 1993-2001 |
| 4 | 3 | 200031 | La CorunaII | PDE | Atl | 1993-2001 |
| 5 | | 200036 | Villagarcia | PDE | Atl | 1997-2001 |
| 6 | | 200041 | Vigo I | IEO | Atl | 1993-2001 |
| 6 | 4 | 200042 | Vigo II | PDE | Atl | 1993-2001 |
| 7 | | 220005 | Huelva | PDE | Atl | 1996-2001 |
| 8 | 5 | 220008 | Bonanza | PDE | Atl | 1993-2001 |
| 9 | 6 | 220003 | Cadiz III | IEO | Atl | 1993-2001 |
| 10 | 7 | 340008 | Ceuta | IEO | Atl | 1993-2001 |
| 11 | 8 | 220011 | Algeciras | IEO | Atl | 1993-2001 |
| 12 | 9 | 220021 | Tarifa | IEO | Atl | 1993-2001 |
| 13 | | 220031 | Malaga I | IEO | Atl | 1993-2001 |
| 13 | 10 | 220032 | Malaga II | PDE | Med | 1993-2001 |
| 14 | 11 | 220056 | Valencia | PDE | Med | 1993-2001 |
| 15 | | 225011 | Palma de Mallorca | IEO | Med | 1997-2001 |
| 16 | 12 | 220061 | Barcelona | PDE | Med | 1993-2001 |
| 17 | 13 | 220081 | L'Estartit | UPC | Med | 1993-2001 |

Stations used in the further analysis are in bold. Data are available at 21 tide gauge stations corresponding to 17 locations, as four locations (Santander, La Coruña, Vigo and Malaga) have two tide gauge stations of different institutions.

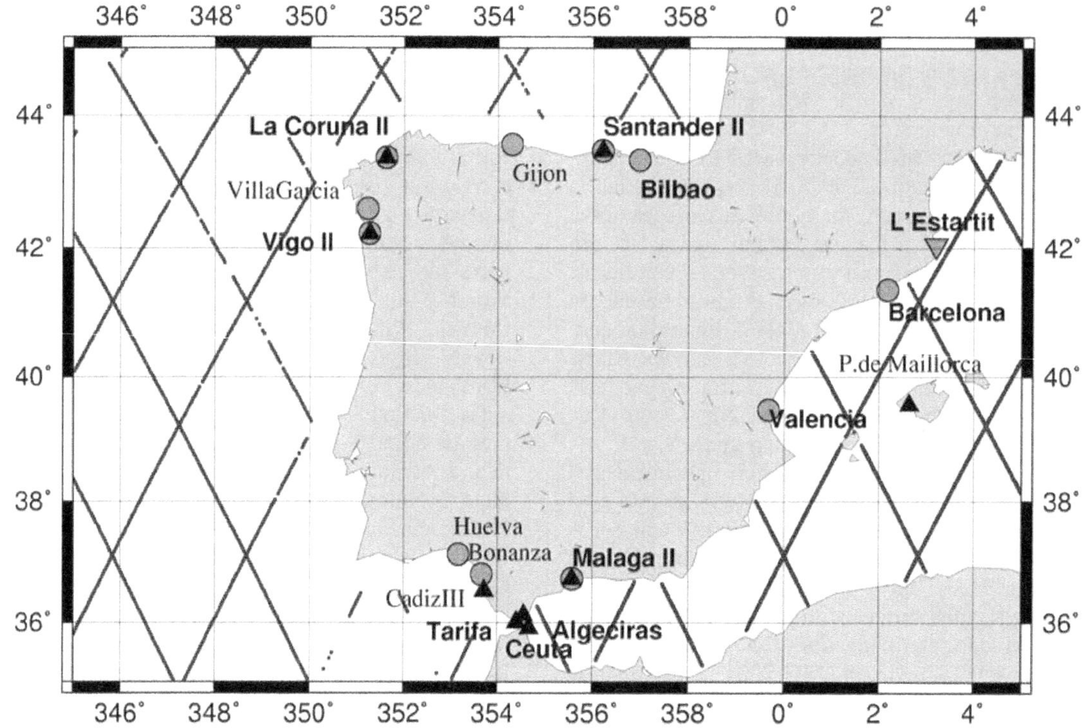


**Fig. 1** T/P ground-tracks and tide gauges with hourly data in 1993-2001: IEO (grey triangles), PDE (grey circles) and UPC (grey reversed triangle)

The IEO provides 9 stations, 5 in the Atlantic Ocean (Santander I, Coruña I, Vigo I, Tarifa and Cadiz) and 4 in the Mediterranean Sea (Ceuta, Algeciras, Malaga I and Palma de Maiorca), the PDE provides 11 stations of the REDMAR network, 8 in the Atlantic Ocean (Bonanza, Huelva, Santander II, Gijon, Bilbao, La Coruña II, Villagarcia, Vigo II) and 3 in the Mediterranean Sea (Barcelona, Valencia, Malaga II), the UPC provides 1 station (L'Estartit). At 13 locations (17 tide gauge stations) data cover the complete interval 1993-2001 with short or no gaps. Monthly means are computed from the hourly tide gauge data by averaging data over one month and are in good agreement with the PSMSL monthly data for all the 17 stations. The monthly averaged time-series of both altimetry and tide gauge data are detrended, de-seasoned and low-pass filtered using a simple 12-month window at 6 month intervals to focus on interannual signals. For the comparison with the yearly steric heights of the hydrographic datasets, yearly values are computed averaging the detrended, de-seasoned tide gauge and altimetric monthly time-series of a same year. The World Ocean Atlas 1998 (WOA98) contains annual grids of temperature at 14 standard levels in the interval 1945-1998. The Medar/Medatlas database (Medar 2002) contains annual grids of temperature and salinity at 46 standard levels in 1950-2002 in the Mediterranean Sea, grids are available corresponding to one, three and five year low-pass filtering, the filtering at one year is used. The spacing of the grids is 1 x 1 degree in latitude and longitude for WOA98 and 0.2 x 0.2 degree for Medar/Medatlas. Gridded time-series of the steric sea level anomalies are computed by evaluating first the density, the specific volume anomaly and the pressure at each level from temperature, salinity and depth data and then vertically integrating the specific volume anomalies over the water column (Gill 1982). Steric sea level anomalies are computed from temperature and salinity profiles collected by IEO four times a year during 1992-2001 at three stations near Malaga. The profiles are interpolated to every 5 db and time series constructed and the steric height time-series is obtained as an average of the three stations.

## 3 Sea level change from tide gauge and altimetry

Altimetry and tide gauge data are compared at the selected thirteen locations with data in the complete interval 1993-2001. The difference in the results obtained using gridded time-series or averaging over the nearest altimeter points is negligible. A set of parameters are relevant for the comparison between altimetry and tide gauge data (Fenoglio-Marc 2004). Table 2 shows the values of those parameters for the monthly records. Standard deviation and correlation of the records ($3^{rd}$ and $4^{th}$ columns), standard deviation and linear-term of the differences between altimetry and tide gauge station records ($2^{nd}$ and $7^{th}$ columns) are the relevant parameters. The $5^{th}$ and $6^{th}$ columns give the linear term of the altimetry and tide gauge station records. The standard deviation of the difference between the tide gauge and the nearest T/P point is higher (5-7 cm) for the Atlantic stations than for the Mediterranean stations (3-5 cm). The linear trend of sea level in the interval 1993-2001 evaluated from the tide gauge data along the Iberian coasts is positive almost everywhere, in agreement with altimetry (Fenoglio-Marc, 2004), it is significantly negative only in Cadiz. The correlation of monthly records is higher than 0.70 for most of the stations and is significant at the 95% level. Errors and problems in the tide gauge records are identified from the values of the relevant parameters. The linear-term of the monthly differences between altimetry and tide gauge stations is smaller than 3 mm/yr for most of the stations, but higher in Santander I, Cadiz, Malaga I, Bonanza while values in Santander II and Malaga II are low. Errors are identified in the records of the first three tide gauge stations, in Bonanza departures from the altimeter are expected as it is located near to a river estuary.

**Table 2.** Relevant parameters at 17 tide gauge stations

| TG | $\Delta_{at}$ mm | $\Delta_a / \Delta_t$ (mm) | $r_{at}$ | $b_a \pm \sigma_a$ (mm/yr) | $b_t \pm \sigma_t$ (mm/yr) | $b_{at} \pm \sigma_{at}$ (mm/yr) |
|---|---|---|---|---|---|---|
| Bilb | 46 | 68/80 | 0.82 | 5.9 +/- 2.5 | 6.5 +/- 2.9 | -0.6 +/- 1.7 |
| SanI | 57 | 64/88 | 0.76 | 5.6 +/- 2.5 | 11. +/- 3.2 | -5.8 +/- 2.1 |
| SanII | 46 | 64/82 | 0.73 | 5.6 +/- 2.4 | 5.4 +/- 3.0 | 0.2 +/- 1.7 |
| LCI | 67 | 69/95 | 0.70 | 3.8 +/- 2.7 | 2.4 +/- 3.6 | 1.4 +/- 2.6 |
| LCII | 61 | 81/87 | 0.73 | 3.0 +/- 2.9 | 1.2 +/- 3.4 | 1.8 +/- 2.4 |
| VigI | 63 | 66/95 | 0.75 | 3.9 +/- 2.5 | 0.3 +/- 3.6 | 3.6 +/- 2.4 |
| VigII | 51 | 66/90 | 0.83 | 3.6 +/- 2.5 | 2.0 +/- 3.4 | 1.6 +/- 1.9 |
| Bona | 58 | 60/77 | 0.66 | 3.6 +/- 2.2 | 11. +/- 2.7 | -7.7 +/- 2.1 |
| Cadi | 72 | 64/83 | 0.55 | 4.4 +/- 2.5 | -6.0+/- 3.1 | 10.4 +/- 2.6 |
| Ceut | 53 | 80/56 | 0.76 | 2.8 +/ 3.0 | -0.1+/- 2.1 | 2.9 +/- 2.0 |
| Alge | 60 | 80/61 | 0.67 | 3.0 +/- 3.0 | 5.8 +/- 2.2 | -2.7 +/- 2.2 |
| Tarif | 38 | 48/48 | 0.74 | 4.2 +/ 2.4 | 6.2 +/- 2.1 | -2.0 +/- 1.7 |
| MalI | 56 | 77/72 | 0.73 | 4.3 +/- 2.9 | 13. +/- 2.3 | -9.4 +/- 1.9 |
| MalI | 51 | 77/68 | 0.77 | 6.0 +/- 4.6 | 9.8 +/- 3.7 | -3.0 +/- 3.0 |
| Vale | 45 | 84/90 | 0.86 | 4.4 +/- 3.2 | 8.2 +/- 3.3 | -3.8 +/- 1.7 |
| Barc | 41 | 86/ 3 | 0.88 | 4.8 +/- 3.2 | 5.6 +/- 3.0 | -0.8 +/- 1.5 |
| L'Est | 28 | 77/ 8 | 0.94 | 4.8 +/- 2.8 | 4.4 +/- 2.9 | 0.4 +/- 1.0 |

Comparing results from tide-gauge stations at the same location (Santander, La Coruña, Malaga, Vigo), the PDE records agree better with the altimeter data in terms of correlation and trend of the sea level height differences (4th and 7th columns in Table 2).

**Table 3.** Correlations of tide gauge and T/P sea level

| | $r_{at}$ monthly | $r_{at}$ monthly deseasoned | $r_{at}$ interannual |
|---|---|---|---|
| Bilbao | 0.82 | 0.63 | 0.88 |
| Santander II | 0.73 | 0.74 | 0.87 |
| La Coruña II | 0.73 | 0.72 | 0.63 |
| Vigo II | 0.83 | 0.75 | 0.77 |
| Tarifa | 0.74 | 0.68 | 0.79 |
| Ceuta | 0.76 | 0.67 | 0.77 |
| Algeciras | 0.67 | 0.56 | 0.64 |
| Malaga II | 0.77 | 0.68 | 0.84 |
| Valencia | 0.86 | 0.67 | 0.60 |
| Barcelona | 0.88 | 0.80 | 0.56 |
| L'Estartit | 0.94 | 0.92 | 0.96 |

Further only the PDE stations are considered when both PDE and IEO are available at the same location, Bonanza and Cadiz are eliminated. Finally 11 stations are retained. Correlations of altimetry and tide gauge data computed from monthly, monthly de-seasoned and interannual (low-pass filtered) records are compared in Table 3. The higher correlation using interannual records shows that changes at long time-scales are associated to large spatial scales.

## 4 Sea level and NAO

In 1993-2001 the correlation between NAO and the tide gauge records is higher with interannual fields than with de-seasoned monthly fields. The correlation is higher if only Winter months are used, as NAO is mainly effective during Winter (Table 4). The de-seasoned monthly and interannual altimetry fields are decomposed in the Iberian region using the Principal Component Analysis method in dominant spatial and temporal components. The temporal patterns and the NAO climatic index are anti-correlated (Figure 2).

**Table 4.** Correlation between NAO and tide gauge stations in 1992-2001

| | Monthly deseasoned | Interannual | Winter-Monthly deseasoned |
|---|---|---|---|
| Santander II | -0.48 | -0.46 | -0.56 |
| Coruña II | -0.54 | -0.81 | -0.60 |
| Vigo | -0.57 | -0.67 | -0.65 |
| Cadiz | -0.56 | -0.65 | -0.68 |
| Algeciras | -0.59 | -0.70 | -0.71 |
| Tarifa | -0.59 | -0.56 | -0.73 |
| Malaga II | -0.61 | -0.77 | -0.73 |
| L'Estartit | -0.66 | -0.73 | -0.72 |

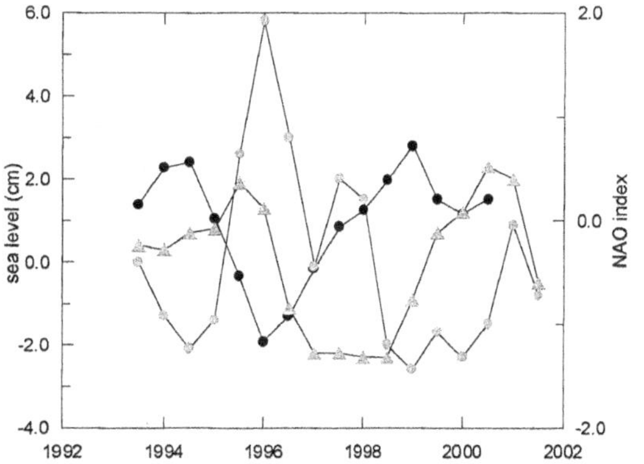

**Fig. 2** Temporal patterns of first two EOFs of low-pass filtered sea level altimetry in Iberian region in 1993-2001 (grey circles and triangles) and NAO index (black circle)

The spatial pattern of the first EOF corresponds to an oscillation in phase in the Mediterranean Sea and in the Atlantic Ocean, the second EOF to an oscillation out of phase. The first EOF of the interannual data explains 51% of the variability of the field and its temporal pattern has a correlation of –0.85 with the NAO index. A similar high correlation is observed between the temporal pattern of the first EOF of low-pass filtered altimetric sea level data and the NAO index in the Mediterranean Sea (-0.72) and in N-E Atlantic region (-0.86), defined as the region with longitude 24°W-5°E and latitude 25-55°N.

## 5 Steric heights and sea level change

The sea level change due to thermal expansion is evaluated in the Mediterranean Sea at the grid-point nearest to the last six stations of Table 3, located in the Mediterranean Sea. Deseasoned yearly means from tide gauges are compared to the steric heights evaluated from the WOA98 and Medar/Medatlas hydrographic databases. Table 5 shows the correlation between tide gauge and thermo-steric and steric heights at few stations for 1993-1998 and 1993-2001. WOA98 being only available until 1998, comparisons of results using thermo-steric Medar/Medatlas and WOA98 components are made in 1993-1998. The correlation is significant at the 80% level at a few locations. Significant values are obtained with both databases in Malaga and in Algeciras, where the highest correlations are found with Medar (0.7) and with WOA98 (0.84) respectively. In L'Estartit only the Medar database (0.52) and in Ceuta only the WOA98 (0.89) database gives high correlations. Values obtained with Medar/Medatlas over 1993-2001 are similar to the values obtained from the same dataset over 1993-1998. In Malaga and Algeciras the correlation with the steric height decreases due to a strong variation in the halo-steric component that occurs after 2000.

**Table 5.** Correlation and p-value in 1993-1998 (above) and in 1993-2001 (below) with tide gauge sea level

| | Thermo-steric WOA98 | Thermo-steric Medar/Medatlas | Steric Medar/Medatlas |
|---|---|---|---|
| Malaga II | 0.63 (0.18) | 0.70 (0.12) | 0.72 (0.12) |
| | | 0.73 (0.02) | 0.41 (0.27) |
| L'Estartit | -0.8 (0.02) | 0.52 (0.2) | 0.46 (0.3) |
| | | 0.59 (0.09) | 0,61 (0.08) |
| Ceuta | 0.89 (0.01) | 0.33 (0.5) | 0.03 (0.9) |
| | | 0.51 (0.16) | 0.21 (0.6) |
| Algeciras | 0.84 (0.03) | 0.61 (0.2) | 0.85 (0.03) |
| | | 0.75 (0.02) | 0.64 (0.06) |

In Malaga local hydrographic measurements are available from IEO. In 1993-1998 the highest correlation between the interannual tide gauge and thermo-steric heights is obtained with Medar/Medatlas (0.7), significant at the 90% level, while WOA98 gives a slightly lower correlation (0.63) (Figure 3). The amplitude of the interannual sea level signal at the tide gauge is 60 centimetres, which is comparable to the amplitude of the steric heights obtained from the Medar/Medatlas. The maximum is reached in 1996 using Medar/Medatlas and one year later using WOA98. Steric heights including the salinity contribution are computed over a longer interval (1993-2001) from Medar/Medatlas. The correlation between tide gauge sea level and thermo-steric component increases, while the correlation with the steric component decreases due to a strong salinity contribution after 2000. The interannual steric heights computed from the local hydrographic data have smaller amplitude and a different phase; the maximum occurs in 1998, i.e. two years later than the maximum obtained using Medar/Medatlas. Steric heights from Medar/Medatlas and interannual sea level from tide gauge and altimetry are compared in Figure 4. Tide gauge and altimetry data time-series are similar, with a phase difference of a few months with respect to the Medar/Medatlas steric heights. The low agreement between the steric height from hydrographic profiles off the coast and the observed sea level at the tide gauge in Malaga is partly due to the low-frequency (one per season) of the profiling.

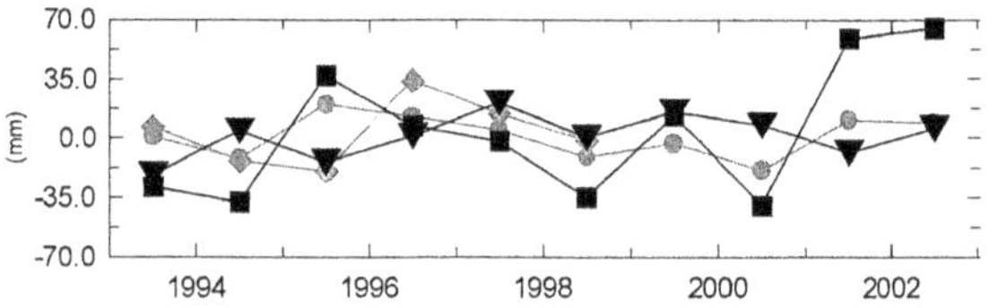


**Fig. 3** Steric sea level near Malaga from Medar/Medatlas (square) and campaign (reversed triangle), thermo-steric from WOA98 (diamond) and Medar/Medatlas (circle)

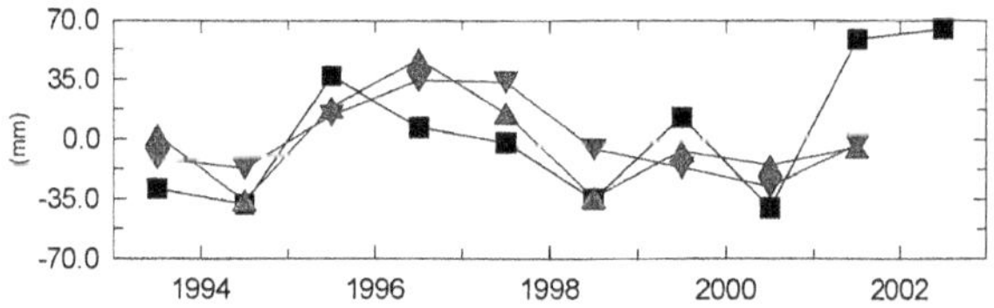


**Fig. 4** Steric sea level near Malaga from Medar/Medatlas (square) and interannual sea level from tide gauge (reversed triangle) and altimetry (triangle)

The expected annual cycle, i.e. a minimum in spring and a maximum in autumn, which is found in the tide gauge and altimetry data, is not present in the computed steric heights. Altimetry and tide gauge measurements have seasonal amplitudes of about 10 cm, while the seasonal signal detected in the profiles is out of phase and has an amplitude of 3 centimetres only. Currents and river discharges, that affect both salinity and temperature, and atmospheric forcing (pressure and associated wind) explain part of the difference between tide gauge and steric level computed from hydrographic data. Higher correlation is expected neglecting the surface layer in the computation of the steric anomaly, as temperature and salinity change very quickly in this layer due to atmospheric changes (pressure and temperature) and river runoff. Finally atmospheric condition on the day of sampling and sparse temporal sampling of hydrographical measurements, compared to the high frequency of tide gauge might have some influence.

## 6. Conclusions

The analysis is based on interannual time-series of sea level variability constructed by low-pass filtering the de-trended and de-seasonalised altimetric and tide gauge monthly averages. A good agreement is found over the last decade analysing four parameters: correlation and standard deviation of the time-series, standard deviation and trend of difference of the time-series. The correlation of altimetry and tide gauge monthly values is higher than 0.7 at many of the locations, is regionally dependent and increases when using low-pass filtered data. Sea level height differences of tide gauge and altimetry are regionally dependent too, mainly due to differences in tide and circulation patterns at the altimetry and tide gauge locations as well as due to wind and bathymetry effects. The long-term components of the sea level height differences are mostly smaller than +/- 3 mm/yr. In the Mediterranean Sea the correlations are generally higher, standard deviation and linear-term of the differences are smaller than along the Atlantic coasts. Using carefully checked tide gauge stations, the agreement between altimetry and tide gauge data is higher compared to the PSMSL dataset. Care in the station selection as well as updates of the PSMSL dataset and correction of IEO data is needed. The correlation between the NAO climatic index and the sea level height is negative and higher than 0.65 at most of the stations at interannual time-scales, the correlation is higher when only the Winter months are used. A relative maximum of the sea level variability in 1996-1997 coincides with a minimum in the NAO index. The identified coherent regional signals characterise regionally the interannual variability. The correlation between the thermo-steric heights computed in the Mediterranean Sea from the two databases Medar/Medatlas and WOA98 and the interannual sea level from tide gauge and altimetry in 1993-1998 show a regional dependence. In Malaga the correlation is significant for both datasets and higher with Medar/Medatlas (0.7). The opposite occurs in Algeciras. In a few locations only one dataset gives a significant correlations, while in others the correlation is low with both datasets. Steric heights including also the salinity contribution and computed over longer intervals (1993-2001) from Medar/Medatlas are in agreement with results using thermo-steric heights over the shorter period except in Malaga where the salinity contribution has to be further investigated after year 2000. The steric heights in Malaga computed by dedicated hydrographic campains have a lower agreement with the interannual sea level due to the low sampling of the hydrographic data. The maximum observed in 1996 by altimetry occurs a few months before in the steric heights evaluated from Medar/Medatlas in Malaga. The filtering used and the sampling are partly responsible for the different times at which the maximum occurs. Further data and analysis are needed to assess the percentage of variability due to mass and volume change both at seasonal and interannual time scales.

## References


Arbic, B.K. and W.Owens (2001). Climatic Warming of Atlantic Intermediate Waters. J. Climate 14: 4019- 4108.

Bethoux J.P., B. Gentili and D Tailliez (1998). Warming and freswater budget change in the Mediterranean since the 1940s, their possible relation to the greenhouse effect. Geoph. Res. Lett 25(7): 1023-1026.

Cazenave A., C. Cabanes, Dominh and S. Mangiarotti (2001). Recent sea level change in the Mediterranean Sea revealed by Topex/Poseidon satellite altimetry, Geoph. Res. Lett 28 (8): 1607-1610

Douglas B.C., (2001). Sea level change in the era of recording tide gauge, in Sea Level Rise history and consequences, 37-64, Academic Press

Fenoglio-Marc L. (2001). Analysis and representation of regional sea level variability from altimetry and atmospheric data, Geophysical Journal International, 145 (1): 1-18

Fenoglio-Marc L. (2002). Long-term sea level change in the Mediterranean Sea from multi-satellite altimetry and tide gauge data, , Physics and Chemistry of the Earth, 27: 1419-1431

Fenoglio-Marc L., C. Dietz and E. Groten (2004). Vertical land motion in the Mediterranean Sea from altimetry and tide gauge stations, Marine Geodesy 27 (3-4), in press

Garcia-Lafuente J., J. Del Rio, E. Alvarez Fanjiul, D. Gomis and J. Delgado (2004). Some aspects of the seasonal sea level variation around Spain. J. Geophys. Res. 109, C09008, doi:10.1029/2003JC002070

Gill A.E. (1982). Atmosphere Ocean-Dynamics, International Geophysics Series, Vol. 30, Academic Press, London

Kjaer N., O. Andersen, P. Knudsen and L. Fenoglio-Marc (2004). Report on sea level variations for thr European sea and the North Atlantic Ocean using altimetry and tide gauge data, First Annual Reports ESEAS-RI Project

Mangiarotti S., 2003. Les variations basse fréquence du niveau de la mer Méditerranée au cours de la deuxième moitié du XX siècle par altimétrie spatiale et maregraphie, Université Paul Sabatier Toulouse, PhD Thesis, 178 pages

MEDAR: Medar Group (2002). MEDATLAS 2002. Mediterranean and Black Sea database of temperatura, salinity and biochemical parameters climatological atlas, 4 CD-ROM. IFREMER. Ed (4 CD-ROM).

Painter and M. Tsimplis (2003). Temperature and salinity trends in the upper waters of the Mediterranean Sea as determined from the MEDATLAS dataset, Continental Science Res. 23: 1507-1522

Ross, Garrett and P.Y. Le Traon (2000). Western Mediterranean sea-level rise: changing exchange through the Strait of Gibraltar, Geoph. Res. Letters, 27 (18): 2949-2952

Shaw A. and Tsimplis M. and WP3.1 members (2004). ESEAS-RI Internal report, ESEAS-RI Project

Tsimplis M. and S.A. Josey (2001). Forcing of the Mediterranean Sea by atmospheric oscillations over the North Atlantic, Geoph. Res. Lett., 28 (5): 803-806

Tsimplis M. and M. Rixen (2002). Sea level in the Mediterranean Sea: the contribution of temperature and salinity changes, Geoph. Res. Lett. Vol. 29, 23, doi :10,1029/2002GL015870

Tsimplis M. and ESEAS-RI WP3.1 members (2004). Report on sea level variations for the European sea and the North Atlantic Ocean using tide gauge data, First Annual Reports ESEAS-RI

Vargas M., T. Raminez, D. Cortes, M. Sebastian and F. Plaza (2002). "Warming trends in the continental shelf of Malaga Bay (Alborán Sea)", Geoph. Res. Lett., 29 (22), 2082, doi:10.1029/2002 GL015306.

Woolf D.K., G.P. Shaw and M.N. Tsimplis (2003). The influence of the North Atlantic Oscillation on sea-level variability in the North Atlantic Region, The Global Atmosphere and Ocean system, 9 (4): 145-167

Zerbini S., Plag H.-P., Baker T. Becker M. et al., 1996: Sea level in the Mediterranean Sea: a first step towards separating crustal movements and absolute sea-level variations, Global and Planetary Change, 14:1-18.

# Space-time analysis of sea level in the North Atlantic from TOPEX/Poseidon satellite altimetry

S. M. Barbosa, M. J. Fernandes, M. E. Silva

Department of Applied Mathematics, Faculdade de Ciências, Universidade do Porto

Rua do Campo Alegre, 687, 4169-007 Porto, Portugal

**Abstract.** Spatial and temporal sea level variability in the North Atlantic is investigated from Topex/Poseidon (T/P) altimetry data. Time series of sea level anomalies on a regular 5° grid are analysed. Non-linear denoising through thresholding in the wavelet transform domain is carried out for each series in order to remove noise while preserving non-smooth features. Principal Component Analysis (PCA) is used to obtain a spatio-temporal description of the sea level field. To avoid modal mixing and improve interpretation of the principal modes, PCA is implemented separately for seasonal and trend components of the sea level field obtained from a wavelet-based multiresolution analysis. The leading pattern of the seasonal field reflects the dominance of a stable annual cycle over the study area and the change in the seasonal regime approaching the equator with contribution of the semi-annual cycle and phase-shift in the annual cycle in the tropical Atlantic. The leading pattern of the trend field is a broad spatial pattern associated with North Atlantic Oscillation (NAO), reflecting the influence of atmospheric conditions on interannual sea level variability.

**Keywords.** satellite altimetry, sea level change, discrete wavelet transform, PCA

---

## 1 Introduction

Recent interest in sea level change has been motivated by concerns related to the consequences of global climate change. Sea level is an important parameter to understand the connection between climatic and oceanographic phenomena.

Satellite missions with radar altimeters provide absolute sea level observations on a nearly global and spatially uniform scale. Topex/Poseidon mission achieved an unprecedented accuracy, yielding a high-quality space-time dataset of precise sea level measurements.

Principal Component Analysis (PCA) (Jollife (2002), von Storch and Zwiers (1999)) is an important tool to summarise and extract information from time-varying spatial fields. The field is decomposed into a sum of modes of variability, each of which is the product of a spatial pattern and a time-varying amplitude, so that the first few modes retain most of the variation in the original dataset. However statistical modes may not correspond to physically and dynamically independent modes, difficulting the interpretation of the resulting space-time patterns.

In this study the spatial and temporal variability of sea level in the North Atlantic is investigated from T/P altimetry data. Seasonal and interannual sea level variability in this region has been previously examined by Ferry et al. (2000), Esselborn and Eden (2001), Efthymiadis et al. (2002) and Volkov and van Aken (2003).

Estimation and interpretation of sea level principal modes is difficulted by the length of the altimetric records (~12 years) (Hendricks et al. (1996), Nerem et al (1997)). Here the discrete wavelet transform is used to enhance the sea level signal through denoising, and to avoid modal mixing and improve interpretation of principal modes.

## 2 Data

Topex/Poseidon analysed data are the GDR-Ms products provided by Aviso (AVISO (1996)) covering the North Atlantic from October 1992 to June 2003 (cycles 21-396).

Sea surface heights (SSH) have been derived by correcting data for all standard instrumental and geophysical effects, including the IB correction and filtering of the Topex dual frequency ionospheric correction (Fernandes et al. (2003)). Sea level anomalies (SLA) have been derived from the difference between the corrected SSH and the GSFC2000 mean sea surface (Wang, 2001).

For each cycle, a regular 5° grid of SLA values is obtained from the along-track data using an adjustable tension continuous curvature algorithm with a tension factor of 0.25 after block-median pre-processing to avoid spatial aliasing and eliminate redundant data (Smith and Wessel (1990)).

The resulting dataset comprises 153 time series of SLA values at each node of the grid and at approximately 10-days intervals (Fig. 1). The time series analysis performed on this grid is illustrated for point located at 65°W, 40°N.

## 3 Analysis

### 3.1 Signal enhancement

Time series of sea level anomalies are short, dominated by an irregular seasonal pattern and large high frequency variability. To retrieve the weak sea level change signal against this noisy background the noise contaminating the signal must be first removed.

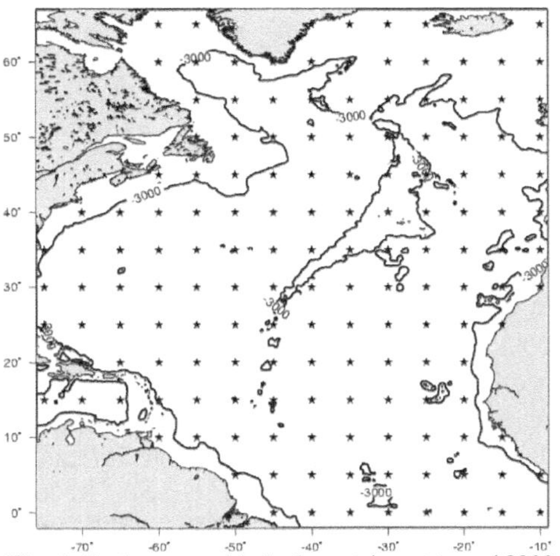


**Fig. 1** Study area with bathymetric contour (-3000 m) and time series locations (⋆)

Sea level records are usually smoothed through a linear filter to remove the high frequencies associated with mesoscale circulation. However, non-linear denoising may be a more efficient approach to deal with the temporal variability of sea level series. Unlike smoothing, denoising preserves non-smooth features and is therefore efficient for non-stationary signals and fractal noise.

Denoising of the SLA series has been carried out through non-linear thresholding in the wavelet transform domain. A level $J_0$ discrete wavelet transform of a time series X, $W=\mathcal{W}X$ uses the pyramid algorithm (Mallat (1989)) to obtain the transform coefficients (W) partitioned into $j=1,\dots,J_0$ individual vectors each associated with a specific time scale. Signal denoising via wavelet thresholding consists in the non-linear modification of the magnitude of the wavelet coefficients (through the application of a thresholding rule) and computation of the inverse wavelet transform of the thresholded coefficients (Donoho and Johnstone (1995)).

Wavelet coefficients are obtained from a level 5 Maximal Overlap Discrete Wavelet Transform, MODWT, (Percival and Walden (2000)) using the pyramid algorithm. The LA(8) filter (Daubechies Least Asymmetric filter of length 8), and periodic boundary conditions are considered. Universal level-dependent threshold is applied to wavelet coefficients up to the second level of the MODWT (Fig. 2). A denoised version of the time series is obtained through the inverse-MODWT of the wavelet coefficients surviving thresholding (Fig. 3).

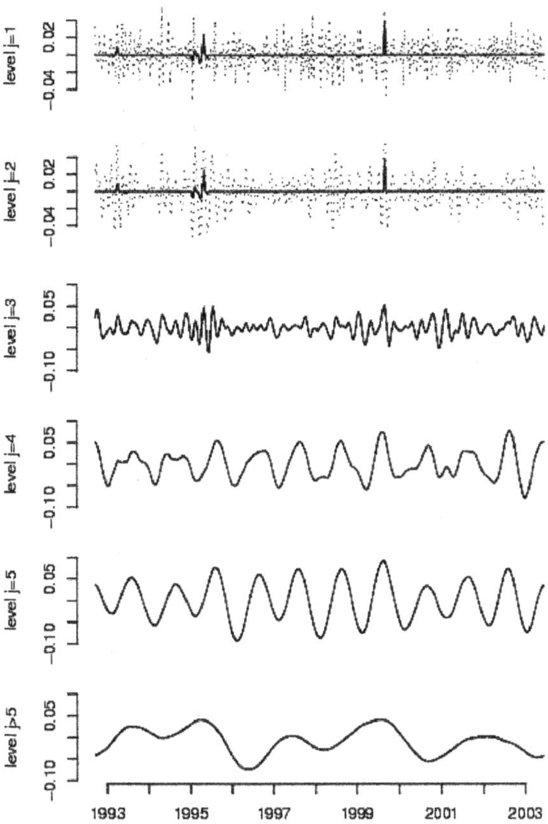


**Fig. 2** MODWT of a SLA time series (gridpoint 65°W, 40° N): wavelet coefficients (dotted) and coefficients surviving thresholding (solid)

### 3.2 PCA of SLA field

PCA is carried out to obtain a spatio-temporal description of the SLA denoised field.

The SLA dataset is previously standardised by subtracting the mean and dividing each time series by its standard deviation. Since the North Atlantic sea level field exhibits increased variability in the the Gulf Stream and northern regions, the normalisation of the SLA series allows to obtain spatial patterns without possible domination by such locations with larger variances. Although avoiding spatial variability to be driven by the most energetic gridpoints, normalisation yields coefficients corresponding to standardised sea level, and therefore less easy to interpret directly.

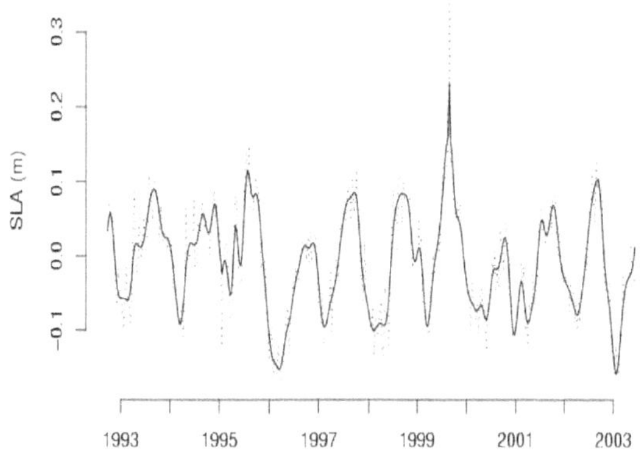


**Fig. 3** Original SLA time series (dotted) and denoised through thresholding (solid) - gridpoint 65°W, 40°N

The first mode corresponds to an annual oscillation, contaminated by a trend signal, and the second mode to a trend pattern with oscillatory contributions from both annual and semi-annual frequencies (Fig. 4.). The first mode accounts for 59% and the second mode for 8.5% of the overall SLA field variance.

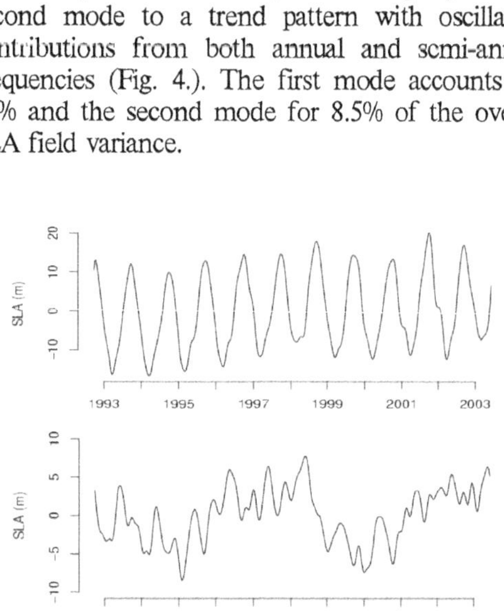


**Fig. 4** Temporal evolution of the first and second PCA modes for the SLA denoised field

The process of interpretation of principal modes may be difficult, and for non-stationary and cycle-stationary datasets such as the sea level anomalies field, traditional PCA may perform poorly and lead to modal mixing (Kim and Wu (1999)). To overcome these difficulties the sea level field may be decomposed into seasonal and inter-annual components prior to PCA implementation, but again linear and stationary decomposition methods may not be adequate for a non-stationary field. Due to its localisation properties in scale and time, the discrete wavelet transform is a suitable tool for decomposition of the SLA field.

### 3.3 Signal decomposition

From a level $J_0$ MODWT of a time series $X_t$, a scale-by-scale additive decomposition from fine to coarse scale is obtained from the inverse-MODWT for each individual level j, $j=1,\ldots,J_0$. Thus a multiresolution analysis (MRA) is defined, expressing each series $X_t$ as

$$X_t = \sum_{j=1}^{J_0} D_{j,t} + S_{J_0,t} \tag{1}$$

where each vector detail $D_j$ is associated with changes on physical scales of $2^{j-1} \times 10$ days and the smooth component $S_{J0}$ represents averages over physical scales larger than $2^{J0} \times 10$ days.

A multiresolution analysis based on a partial MODWT of level $J_0=5$ is carried out for each series of sea level anomalies (Fig. 5). Inter-annual variability is captured by the smooth series $S_5$, corresponding to the time series trend pattern. Detail $D_5$ is a non-stationary pure annual oscillation and detail $D_4$, a semi-annual oscillation with a small contribution at the annual frequency. Seasonality is thus captured by the sum of these two detail vectors.

### 3.4 PCA of SLA field components

PCA is implemented for the $D_5$ and $S_5$ fields corresponding to the seasonal and trend components obtained from the MRA for each gridpoint (Fig. 6 and Fig. 7). Again, the datasets are standardised to avoid domination of the analysis by the most energetic gridpoints. Again, the datasets are standardised to avoid domination of the analysis by the most energetic gridpoints.

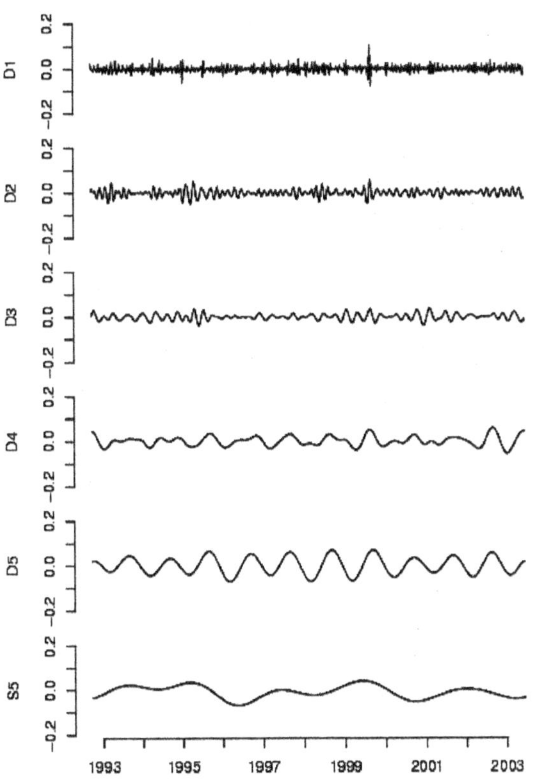


**Fig. 5** MODWT-based multiresolution analysis (MRA) – gridpoint 65°W, 40°N

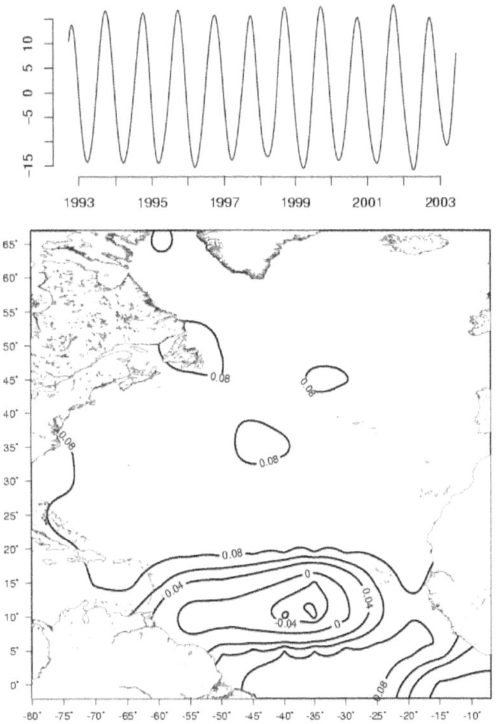


**Fig. 6** Temporal evolution and spatial pattern for the leading mode of the seasonal field ($D_5$)

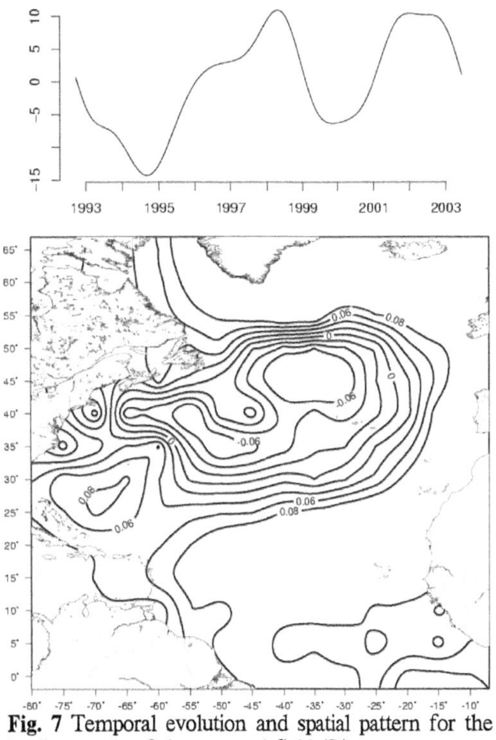


**Fig. 7** Temporal evolution and spatial pattern for the leading mode of the seasonal field ($S_5$)

The leading mode of the seasonal field is an annual oscillation explaining more than 75% of the overall seasonal variability. The temporal evolution is similar to the leading pattern of the total field although more stable in amplitude. Spatially the first seasonal mode exhibits a constant phase, as expected for a seasonal cycle, except in the tropical Atlantic region, reflecting a different seasonal regime with a strong semi-annual cycle and a phase-shift in the annual cycle.

The leading mode of the trend field corresponds to a broad spatial pattern explaining 37% of the overall field variance. The spatial pattern describes a coherent sea level variation in the northern and tropical Atlantic but opposing behaviour in the central North Atlantic, and seems to be associated with bathymetry (Fig. 1), with largest loadings east of the mid-Atlantic ridge. The temporal pattern exhibits strong fluctuations superimposed on a moderate increasing trend.

## 4. Discussion

Climate low frequency variability tends to occur in large spatial patterns associated with changes in circulation patterns. In the Northern Hemisphere the North Atlantic Oscillation (NAO) is one of the most important modes of low frequency variability with impacts reaching from the upper atmosphere to deep ocean (Hurrell and van Loon (1997), Marshall et al.. (2001)). NAO is characterised by a dipole of pressure anomalies most intense (both in intensity and area coverage) during winter. To investigate the relation between sea level inter-annual variability and NAO, the leading mode of the trend field is aggregated into average winter values for comparison with winter NAO index (Hurrell (1995)) (Fig. 8).

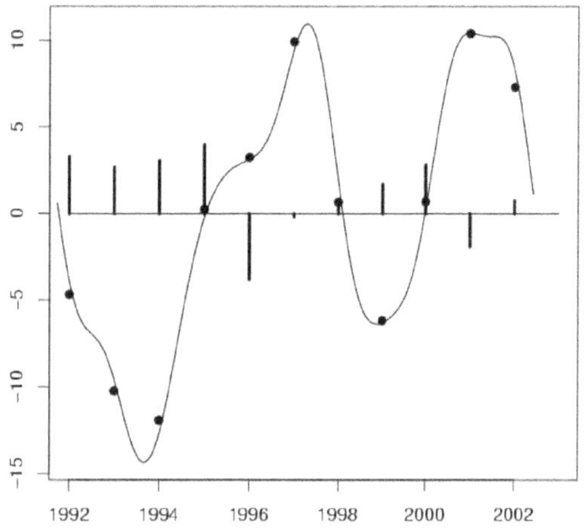


**Fig. 8** Leading mode of the trend field: 10-days (solid), winter (points) and NAO winter index (bars)

From the cross-correlation function between the two series (Fig. 9), the inter-annual leading mode and NAO index are anti-correlated, with maximum correlation for lag 1, corresponding to a Spearman rank correlation coefficient of -0.7.

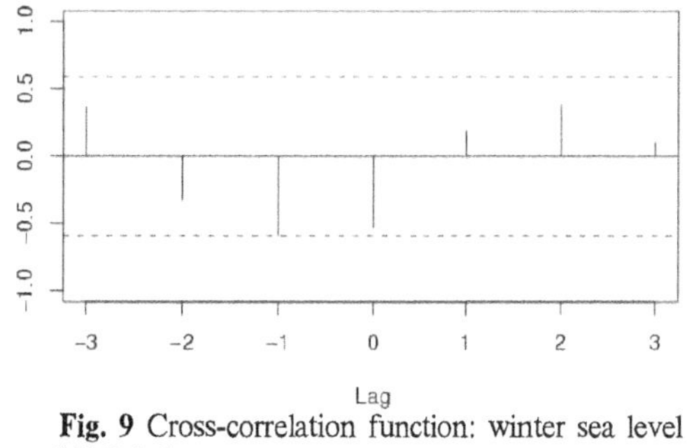


**Fig. 9** Cross-correlation function: winter sea level & NAO index

## 5. Conclusions

The properties of time and frequency localisation of the discrete wavelet transform have been used to pre-process the short and irregular time series of T/P sea level anomalies for the North Atlantic.

Denoising through thresholding in the wavelet domain has been carried out for flexible noise removal while preserving non-smooth features. A multiresolution analysis based on the maximal overlap version of the discrete wavelet transform has been used to decompose the sea level series into non-stationary seasonal and trend components. An interpretable spatio-temporal description of the SLA field has been obtained by separate principal component analysis of the seasonal and trend fields. The adopted methodology proved to be effective in avoiding modal mixing of principal modes. This allowed a more flexible and interpretable description of seasonal and inter-annual sea level variability in the study area.

The leading pattern of the seasonal field reflects the dominance of a stable annual cycle over the study area and the change in the seasonal regime approaching the equator with contribution of the semi-annual cycle and phase-shift in the annual cycle in the tropical Atlantic.

The leading pattern of the trend field reflects a broad pattern associated with the NAO. The interannual sea level change is anti-correlated with the NAO index with a maximum for lag 1, indicating that sea level responds to changes in the NAO index within 1 year (as suggested by Volkov and van Aken (2003) for the northern North Atlantic). Although the IB correction has been applied to the altimetry data, removing meteorological pressure effects, atmospheric conditions associated with long-term variability in the pressure field (such as the NAO pattern) influence sea level through heat and mass fluxes within the ocean-atmosphere system. Large scale ocean circulation responds to changes in atmospheric circulation, in particular changes in the wind stress curl, influencing oceanic heat transport (Esselborn and Eden (2001)).

A preliminary attempt to interpret the obtained spatial and temporal sea level patterns has been made. Further analysis is being carried out to improve the physical interpretation of sea level patterns and their correlation with atmosphere and ocean parameters.

**Acknowledgements**

Wavelet analysis was carried out with R package waveslim from B. Whitcher. (http://www.R-project.org)
Maps were created with GMT - Generic Mapping Tools (http://gmt.soest.hawaii.edu/)
Part of this work has been supported by program POCTI through the Centro de Investigação em Ciências Geo-espaciais (CICGE)

## References

AVISO (1996). User handbook merged TOPEX/ POSEIDON Products, AVI-NT-02-101-CN ed 3.0, 1996

Donoho, D. L. and I. M. Johnstone (1995). Adapting to unknown smoothness via wavelet shrinkage. *Journal of the American Statistical Association*, 90, pp. 1200-1224

Efthymiadis, D., F. Hernandez and P. Le Traon (2002). Large-scale sea level variations and associated atmospheric forcing in the subtropical North-East Atlantic Ocean. *Deep Sea Research II*, 42, pp. 3957-3981

Esselborn, S., and C. Eden (2001). Sea Surface Height changes in the North Atlantic Ocean related to the North Atlantic Oscillation. *Geophys. Res. Lett.*, 28, 3473-3476

Fernandes M. J., L. Bastos and M. Antunes (2003) Coastal Satellite Altimetry - Methods for Data Recovery and Validation, *Proceedings of the 3rd Meeting of the International Gravity & Geoid Commission (GG2002)*, pp. 02-307, Tziavos, I. N. (Ed.), Editions ZITI.

Ferry, N., G. Reverdin and A. Oschlies (2000). Seasonal sea surface height variability in the North Atlantic Ocean. *Journal of Geophysical Research - Oceans*, 105 (C3), pp. 6307-6326

Hendricks, J., R. Leben and G. Born (1996). Empirical Orthogonal Function analysis of global T/P altimeter data and implications for detection of global sea level rise. *Journal of Geophysical Research*, 101 (C6), pp. 14131-14145

Hurrell, J. (1995). Decadal trend in the NAO: regional temperatures and precipitation. *Science*, 269, pp. 676-679

Hurrel, J. and van Loon (1997) Decadal variations in climate associated with the NAO. *Climate Change*, 36, 301-326

Jollife, I. T. (2002). *Principal Component Analysis*, Springer

Kim, K. and Q. Wu (1999). A comparison study of EOF techniques: analysis of non-stationary data with periodic statistics. *Journal of Climate*, 12, pp. 185-189

Mallat, S. G. (1989). A theory for multiresolution signal decomposition: the wavelet representation. *IEEE Transactions on Pattern Analysis and Machine Intelligence*, 11, pp. 674-693

Marshall, J., Y. Kushnir, D. Battisti, P. Chang, A. Czaja, R. Dickson, J. Hurrell, M. McCartney, R. Saravanan, and M. Visbeck (2001). Review: North Atlantic climate variability: Phenomena, Impacts and Mechanisms. *Int. Journal of Climatology*, 21, pp. 1862-1898

Nerem ,R. S., K. E. Rachlin and B. D. Beckley (1997). Characterization of global mean sea level variations observed by TOPEX/POSEIDON using empirical orthogonal functions. *Surveys in Geophysics*, 18, pp. 293-302.

Percival, D. and A. Walden (2000). *Wavelet methods for time series analysis*, Cambridge University Press

Smith, W. H. F. and P. Wessel (1990). Gridding with continuous curvature splines in tension. *Geophysics*, 55, pp. 293-305

von Storch, H. and F. W. Zwiers (1999). *Statistical Analysis in Climate Research*, Cambridge University Press

Volkov, D. L. and H. M. van Aken. (2003). Annual and interannual variability of sea level in the northern North Atlantic Ocean, *Journal of Geophysical Research*, 108, C6, pp. 35-1 35-13, doi: 10.1029/2002JC001459

Wang, Y. M. (2001). GSFC00 Mean Sea Surface, gravity anomaly, and vertical gravity gradient from satellite altimeter data, *Journal of Geophysical Research*, 106, C12, pp.31167 -31174.

# Mean Sea Level and Sea Level Variation Studies in the Black Sea and the Aegean

I.N. Tziavos✉, G.S. Vergos
Department of Geodesy and Surveying, Aristotle University of Thessaloniki, University Box 440, 541 24, Thessaloniki, Greece, Fax: +30 31 0995948, E-mail: tziavos@olimpia.topo.auth.gr

V. Kotzev, L. Pashova
Central Laboratory of Geodesy, Bulgarian Academy of Sciences

**Abstract.** An analysis on the use of ERS1, ERS2 and TOPEX/Poseidon (T/P) data for the determination of a Mean Sea Surface (MSS) model in the Black Sea and the Aegean is presented. From all three satellites, the Exact Repeat Mission (ERM) data have been used spanning over a period of 12 years. The sea surface heights were edited for blunder detection and removal while an inter-comparison between the T/P data has been preformed as well. During the latter, a T/P validation data set was formed by stacking all 9 year of available T/P SSHs. Then, the differences between each year and the validation data set were estimated to identify patterns of sea level change during the period covered by the T/P data. This analysis was followed by the combination of the ERS1/2 and T/P data to form an MSS model. Three different gridding algorithms, i.e., ordinary weighted least squares, least squares collocation and interpolation using splines in tension were tested to determine the most appropriate. The gridded outputs were compared against the KMS2001 MSS model, which was used as a ground truth data set. Furthermore, T/P monthly, seasonal and annual SSHs have been compared with the best of the MSS model determined to identify patterns in sea level variation. Additionally, a comparison of the MSS model with gravimetric quasi-geoid heights at tide gauge (TG) stations across the Black Sea was conducted to assess their agreement and determine corrector surfaces for the transformation between altimetric and gravimetric geoid models.

**Keywords.** Mean sea surface, sea level variations, least squares collocation, corrector surfaces.

## 1 Introduction

During the last three decades altimeters on-board satellites have offered a tremendous amount of measurements of the ocean surface. This mapping of the oceans refers to the determination of the height separation between the sea surface and a reference ellipsoid. The measurements of sea surface, called sea surface heights (SSHs), are of main interest to both geodesy and oceanography. In geodesy, SSHs are commonly used to determine an altimetric marine geoid model and/or combined gravimetric and altimetric solutions. This is feasible, since the combination of multi-satellite SSHs leads to the determination of a Mean Sea Surface (MSS) model, which approximates the geoid very closely. It is then just a matter of reduction of the MSS heights, actually the removal of the quasi-stationary part of the sea surface topography, to determine a marine geoid model.

There have been many studies on the determination of MSS models (Andersen and Knudsen 1998; Cazenave et al. 1996; Yi 1995), global (Lemoine et al. 1998; Wenzel 1998) and regional geoid models (Arabelos and Tziavos 1996; Vergos et al. 2003) as well as on the recovery of gravity anomalies from altimetric measurements (Andersen and Knudsen 1998; Hwang et al. 1998; Tziavos et al. 1998). The main advantage of altimetric SSHs over shipborne gravity data can be viewed in terms of their high precision, resolution, homogeneity and global coverage. Finally, altimetric SSHs from the Exact Repeat Missions (ERMs) of various satellites are used in combination with MSS models to monitor sea level change and relate to historical and recent tide gauge (TG) records. A very good review on the applications of satellite altimetry to geodesy and sea level changes is given by Tapley and Kim (2001) and Nerem and Mitchum (2001) respectively.

The main purpose of the present study is to a) determine a MSS model for the Black Sea and the Aegean, b) investigate sea level changes in the area, c) validate the estimated MSS models with respect to quasi-geoid heights at TG stations, and d) estimate corrector surfaces between gravimetric and altimetric geoid models.

## 2 Mean Sea Surface Determination

To achieve the goals outlined above, altimetric data from the ERM missions of ERS1/2 and TOPEX/Poseidon (T/P) have been used (AVISO 1998). The ERS1 data (95576 point values) come from the 35-day ERM mission of the satellite from April 14, 1992 to December 13, 1993 and March 21, 1995 to May 16, 1995 phases *c* and *g* respectively. From ERS2, six years worth of data have been used (368617 point values) covering the period from April 21, 1995 to June 16, 2001. Finally, nine years of the T/P SSHs were employed (488634 point values) covering the period from October 2, 1992 to October 8, 2001. Fig. 1 depicts the area under study and the distribution of the altimetric data.

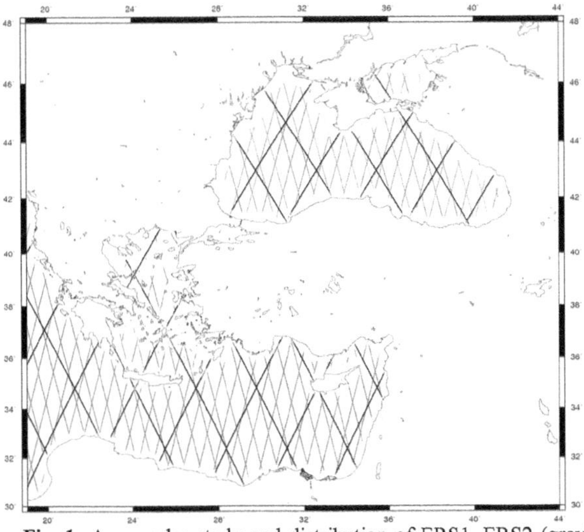

**Fig. 1:** Area under study and distribution of ERS1, ERS2 (gray) and T/P (black) tracks.

For the estimation of the MSS model, we have estimated single-satellite solutions, i.e., using T/P, ERS1 and ERS2 data only, while a combined model was determined as well employing all available SSHs. For each of these cases, three algorithms for the prediction on a grid have been tested, i.e., conventional weighted means, with the inverse of the square of the distance of each point serving as its weight, splines in tension (Smith and Wessel 1990) and least squares collocation (LSC) (Moritz 1980; Knudsen 1993). Therefore twelve solutions have been estimated and validated in total to decide on a) whether single or combined models are preferable and b) on the most appropriate algorithm. The prediction of the models using splines in tension was based on the algorithm incorporated in the Generic Mapping Tools software (Wessel and Smith 1998), while for the estimation of the LSC models a local covariance function for the area, determined by Vergos et al. (2003) from shipborne and airborne gravity data, has been used.

From the MSS models estimated it was concluded that the T/P only solutions (regardless of the algorithm) showed significant discontinuities within small areas, which are mainly attributed to the large cross-track spacing (~3°) of the T/P tracks. The ERS1 and ERS2 models had better behavior compared to T/P (reduced discontinuities) but once again there were many artifacts in the estimated field. Finally, all single-satellite MSS models did not manage to depict the short-wavelength features of the gravity field in the area due to the presence of, e.g., the Hellenic Trench. This was expected due to the large cross-track spacing of the data. On the other hand, the estimated combined models, regardless of the algorithm used, managed to give a more or less realistic picture of the gravity field of the area. From the comparisons between the algorithms, it was concluded that the weighted means solutions, at least with the particular weight used, does not manage to predict the short-wavelength details of the field, while spline interpolation produced some artifacts. Nevertheless, the combined model using splines in tension, managed to provide a very good picture of the MSS of the area, which was very close to the geoid signature of the underwater structures in the region. LSC is a method traditionally used in geodesy to predict gravity field related quantities or as a gridding algorithm. It is preferred due to its statistical rigorousness and the fact that it can take into account the local statistical characteristics of the area under study through the use of a local empirical covariance function. Therefore, it was expected that the LSC combined MSS model would give the best results. The estimated LSC combined model did not show any "trackiness" while no discontinuities were detected as well. Furthermore, it managed to depict very well all the structures of the gravity field of the area giving good detail.

But, in order to be confident on the selection of the collocation solution as the MSS model for the area, a comparison of all available combined models was performed w.r.t. the KMS01 MSS (Andersen and Knudsen 1998). The combined solutions were overwhelmingly better than single-satellite ones, thus the comparisons with KMS01 discussed below refer only to the combined models to assess the performance of the gridding algorithms. As an indication of the improvement that combined models offer, it should be mentioned that the standard deviation ($\sigma$) of differences between the LSC single-satellite ERS1, ERS2 and T/P MSS with KMS01 reached the ±0.303, ±0.214 and ±0.332 m level respectively. Table 1 presents the statistics of the differences between KMS01 and the estimated MSS models, before and after the minimization of the differences with a 3$^{rd}$ order polynomial model. In Table 1 SP denotes the spline, WM the weighted means and LSC the collocation MSS models respectively. It can be easily concluded that the LSC MSS model gives the best agreement w.r.t. KMS01 since the range is smaller by about 23% and the $\sigma$ by 16% compared to the SP model. The worst results are obtained when the WM model is compared with KMS01, showing that this gridding algorithm is inadequate. The last row in Table 1 gives the statistics of the LSC combined MSS model for the area under study, which is also depicted in Fig. 2. Thus, it was concluded that combined models are preferable to single-satellite ones, while least squares collocation was selected as the most rigorous algorithm to provide the final mean sea surface model for the area under study.

**Table 1.** Statistics of differences between KMS01 and the estimated MSS models. Unit: [m].

| | | max | min | mean | σ |
|---|---|---|---|---|---|
| SP | before | 1.505 | -1.990 | -0.086 | ±0.208 |
| | after | 1.679 | -1.854 | 0.000 | ±0.200 |
| WM | before | 1.622 | -3.121 | -0.094 | ±0.274 |
| | after | 1.790 | -3.079 | 0.000 | ±0.266 |
| LSC | before | 1.452 | -1.219 | -0.083 | ±0.177 |
| | after | 1.627 | -1.104 | 0.000 | ±0.168 |
| LSC MSS | | 42.159 | 0.840 | 21.607 | ±9.309 |

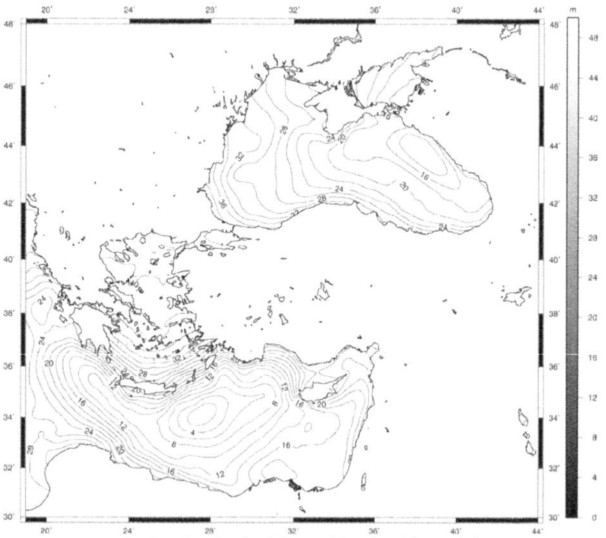

**Fig. 2:** The final LSC combined MSS model.

## 3 Sea Level Variations

The next goal of this study, as outline in Sec. 1, was the investigation for trends in the sea level change of the Black and Aegean Seas for the period covered by the data (1992-2001). To assess that, a two-fold procedure was followed. Firstly, a stacked dataset from all nine years of T/P data was constructed and differences between that and annually stacked T/P SSHs were formed. Stacking is a procedure of constructing means of repeated altimetric profiles, which successfully manages to remove effects influencing the data with period longer than the repeat period of the satellite (Knudsen 1992). Thus, for the 10-day repeat period of T/P, if an annual set of ERM data is stacked, then we manage to remove the ocean effects with period longer than 10 days and shorter than one year, influencing the ERM data. On the other hand, if a nine-year dataset is stacked, then we manage to remove effects with periods between 10 days and nine years. Thus, it can be assumed that the differences between an annual stacked dataset and a nine-year one, will be free of a) short-time (but longer than 10 days) phenomena and b) phenomena with a period between one and nine years. It should be noted that the effect of the dynamic ocean topography is not periodic thus it remains in the data. Therefore, what remains is the sea level variation between each year and a nine-year mean. Plotting these differences for all nine years and fitting a linear trendline will yield an annual rate of sea level change. The second approach was to generate monthly and annual datasets from the T/P SSHs for the period 1993-2001 and then compute their differences w.r.t. the final MSS model estimated for the area under study. In the latter approach, the differences presented are after the removal of a trend surface, so as to remove or at least reduce any mean trend between the datasets compared.

Fig. 3 presents a plot of the mean of the differences between the stacked annual T/P SSHs and the nine-year stacked dataset. From that figure a very small trend (-0.4 mm/y) in sea level change can be seen, which cannot be regarded as significant. Therefore we can conclude that there is no sea level change in the Black sea and the Aegean for the period 1993-2001.

Fig. 4 depicts the standard deviation of the differences between the monthly T/P datasets and the LSC combined MSS model for the period between January 1993 and January 2001. In all cases the differences are formed as $h^{monthly}$-*MSS*. The black circles represent the highest peak of each year, while the gray ones the lowest. It can be clearly seen that there is a very strong seasonal pattern in the sea level variation, with the highs and lows occurring in January and June to July of each year, respectively. Additionally, the highest peak in sea level in these nine years occurs in February 1998 and coincides with the El Niño event (~1 cm highest than the other yearly peaks), while La Niña in December 1998 does not seem to influence the sea level

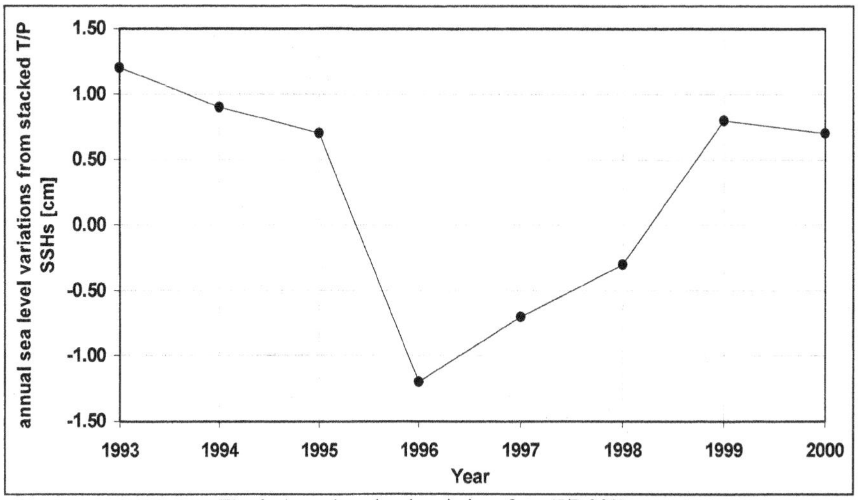


**Fig. 3:** Annual sea level variations from T/P SSHs.

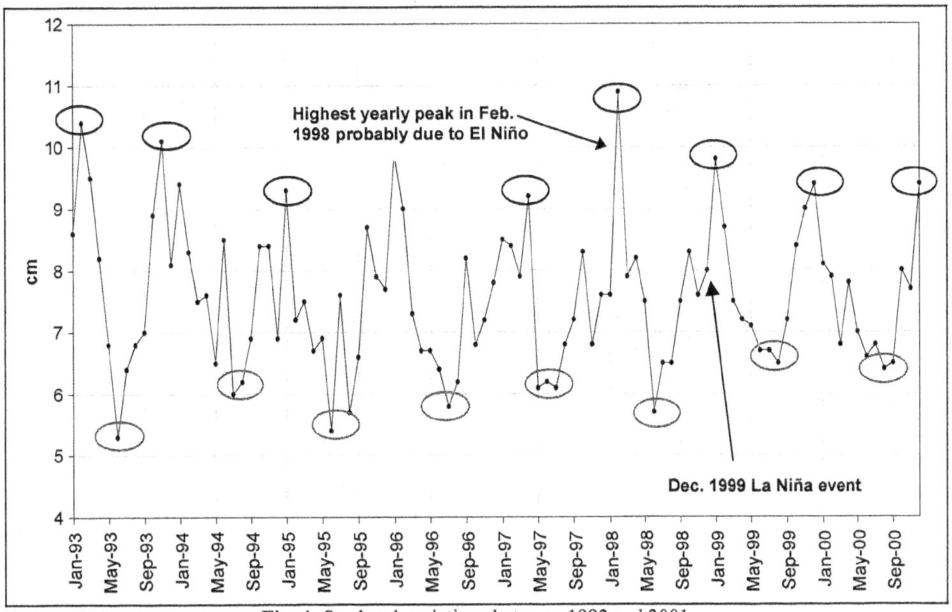


**Fig. 4:** Sea level variations between 1993 and 2001.

in the area significantly. This is probably expected since the Aegean and Black Seas are more or less closed, so they are less influenced by ENSO events, especially if the influence of the latter is not so strong (like in the case of La Niña). Finally, it is interesting to notice that even though the highs in each year occur in January, when El Niño occurs then the high appears in February. From the analysis of the differences between the annual T/P data sets and the estimated MSS, before the removal of the trend surface, it was found that for the period 1993-1995 there was of rate of sea level change of -0.33 mm/y while between 1995 and 2000 it reversed its sign and was rising by about +1.5 mm/y. But, when determining the rate of sea level change for the entire period (1993-2001), it was found to be at the sub-mm level, which agrees with what was found from the differences between the stacked T/P datasets. This is good evidence that the two approaches followed, i.e., a) comparisons between stacked T/P datasets and b) comparisons between a MSS model and monthly/annual T/P data can lead to reliable results. From the results obtained, it is evident that there is a very strong seasonal pattern in the sea level variation in the area under study while no sea level change can be identified.

## 4 Comparisons at TG stations

The final objective of the present study is the validation of the estimated MSS model w.r.t. TG records available for the region. There is only a small amount of TG stations (seven) available all of them located in the western part of the Black Sea (see Fig. 5). For these stations gravimetric quasi-geoid heights were available (Becker et al. 2002; Ihde et al. 2000; Kotzev et al. 1999) so that a comparison with the estimated MSS model and KMS01 was feasible. We assume that the QSST in the Black Sea is small in magnitude, so that the MSS and the geoid coincide, therefore such a comparison is feasible. Furthermore, since the TG stations are close to the coastline, it can be assumed that the geoid and the quasi-geoid coincide. The differences are minimized using a 2nd order polynomial models as

$$N^{TG} - N^{i} = x_1 + x_2\left(\phi_i - \bar{\phi}\right) + x_3\left(\lambda_i - \bar{\lambda}\right) + \\ + x_4\left(\phi_i - \bar{\phi}\right)^2 + x_5\left(\lambda_i - \bar{\lambda}\right)^2 + x_6\left(\phi_i - \bar{\phi}\right)\left(\lambda_i - \bar{\lambda}\right) + v \qquad (1)$$

where $N^{TG}$ is the TG quasi-geoid height, $N^{i}$ denotes the estimated MSS model or KMS01, and $x_1$, $x_2$, $x_3$, $x_4$, $x_5$, $x_6$ are parameters to be estimated using a least squares method. The results of these comparisons are summarized in Table 2 where the differences, both before (regular numbers) and after the fit (numbers in italics) of the polynomial model, are presented.

From Table 2 it can be concluded that the estimated MSS model gives smaller differences, w.r.t. the TG gravimetric geoid heights, compared to KMS01. These differences range from only -11 cm (for Bourgas) to almost -2.6 m for the TG station in Shabla. For the latter, KMS01 gives a very large difference as well (~-2 m), so it may be possible that the estimated gravimetric geoid height at that station is wrong. Apart from the differences at the Shabla station, the rest appear very satisfactory, since they reach a minimum of -68 cm. These large differences can be due to a) the fact that the altimetric data were not reduced for the sea surface topography and b) datum inconsistencies between the altimetric models and the gravimetric geoid estimated for the TG stations. This is especially evident for the

Varna TG station (difference of 14.2 cm after the fit), which was used for the definition of the Bulgarian height system and its connection to the Baltic system. This difference is very close to the one between the Baltic and Black Sea height systems. After the fit of the transformation model, the differences range between 9 mm and 35 cm for the estimated MSS model and 1.3 cm and 48 cm for KMS01. This is very encouraging for the methodology followed to determine the MSS model.

**Table 2**. Differences between the TG geoid height, the estimated MSS model and KMS01. Unit: [m].

| | $N^{TG}$-$N^{MSS}$ | | $N^{TG}$-$N^{KMS01}$ | |
|---|---|---|---|---|
| Constanza | -0.675 | *-0.029* | -0.445 | *-0.040* |
| Shabla | -2.590 | *-0.185* | -1.970 | *-0.251* |
| Varna | -0.588 | *0.142* | 0.112 | *0.192* |
| Irakli | -0.539 | *0.354* | 0.317 | *0.480* |
| Bourgas | -0.108 | *-0.278* | 0.549 | *-0.378* |
| Ahtopol | -0.465 | *-0.012* | 0.046 | *-0.017* |
| Mykolaiv | -0.324 | *0.009* | -0.245 | *0.013* |

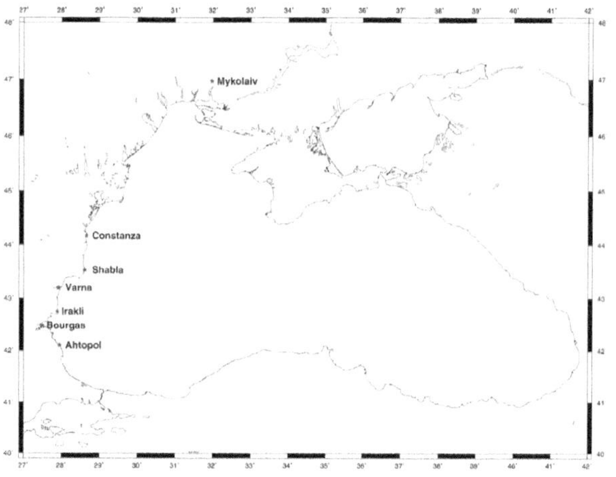


**Fig. 5:** Distribution of TG stations in the Black Sea.

From the minimization of the differences between the TG geoid heights and the MSS model a corrector surface has been determined (see Fig. 6) which corresponds to the 6 parameters estimated by Eq. 1. The translation parameter between the TG geoid heights and the MSS model estimated is 70 cm. This trend is very large to be attributed to either the height difference between the Black Sea and the Baltic Height systems, the QSST in the Black Sea and the datum inconsistencies between the MSS and quasi-geoid models. A first guess would be that the Shabla station influences the entire adjustment, therefore that TG was removed and the differences between the quasi-geoid heights and the estimated MSS were once again minimized. Of course, employing Eq. 1 with six observations, leads to a system with a unique solution, so no actual adjustment can take place. Therefore a 1st order polynomial was used, which is described by the first three $x_i$ parameters of Eq. 1. The differences after the fit at each station were now found to be -0.156 m, -0.085m, -0.088 m, 0.368 m, -0.107 m and 0.068 m for the Constanza, Varna, Irakli, Bourgas, Ahtopol and Mykolaiv TG station, which clearly depicts that the Shabla station influenced strongly the rest of the estimated residuals. The new corrector surface is depicted in Fig. 6 as well, while the estimated translation parameter ($x_1$) between the TG quasi-geoid heights and the MSS heights was now estimated at the 42.5 cm ±5.8 cm level, which corresponds very well to the combined effect of the height difference between the Black Sea and the Baltic Height systems (27 cm), the QSST in the Black Sea and the datum inconsistencies between the MSS and quasi-geoid models. This is good evidence, that at least for the western part of the Black Sea, a corrector surface has been determined for the transformation of altimetric geoid heights to gravimetric quasi-geoid heights and vice versa. This is especially important for Bulgaria and the connection between the Black Sea and Baltic Sea height systems. Finally, it should be pointed out that the estimated corrector surface is representative for the western part of the Black Sea only, where TG stations were available.

## 5 Conclusions

A method of estimating a mean sea surface model has been presented based on an analysis of single- and multi-satellite datasets employing various gridding algorithms. The combination of altimeter data from as many satellites as possible is preferable, since it manages to overcome the problems caused by the large cross-track spacing of ERM data, i.e., the discontinuities introduced in the estimated MSS model from the few observations available. Additionally, least squares collocation (employing a local covariance function) is preferred against weighted means and splines in tension to estimate the final grid, since it gives the more rigorous results without introducing artifacts in the final field. Furthermore, LSC manages to depict the short-wavelength features of the gravity field of the area while WM especially smoothes the estimated MSS model significantly.

From the analysis of the variations in the sea level, it was concluded that there is no detectable significant changes in the time period studied (1993-2001), since variations at the sub-mm level cannot be regarded as important. But, the sea level in the area under study has a seasonal pattern which is evident throughout the nine years covered by the present study. Additionally, it was possible to detect the response of the sea in the region to El Niño but not to La Niña, probably due to the small magnitude of the latter.

Finally, from the comparisons between the MSS model estimated and TG quasi-geoid heights it became evident that the model determined is comparable to KMS01 and in same cases better than that, since the differences it provides are smaller. Finally, after the fit

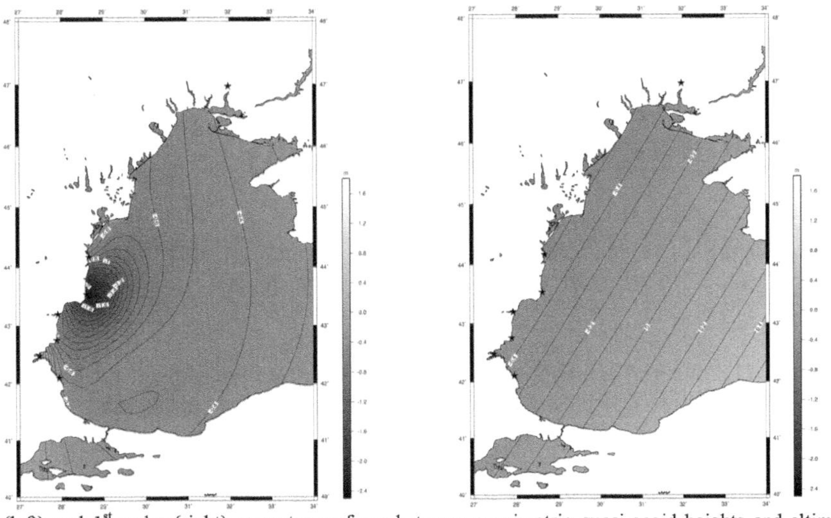

**Fig. 6:** $2^{nd}$ order (left) and $1^{st}$ order (right) corrector surfaces between gravimetric quasi-geoid heights and altimetric geoid heights.

of $2^{nd}$ and $1^{st}$ degree polynomials, corrector surfaces for the transformation between altimetric and quasi-geoid heights were determined.

## Acknowledgement

This research was funded from the Greek Secretariat for Research and Technology in the frame of the 3rd Community Support Program (Op. Sup. Pror. 2000 - 2006), Measure 4.3, Action 4.3.6 (International Scientific and Technological Co-operation), Bilateral co-operation between Greece and Bulgaria. Prof. Catalao, Joana Fernandes and an anonymous reviewer are gratefully acknowledged for their comments on earlier versions of the paper.

## References

Andersen OB, Knudsen P (1998) Global marine gravity field from the ERS1 and Geosat geodetic mission. J Geophys Res 103: 8129-8137.

Arabelos D, Tziavos IN (1996) Combination of ERS1 and TOPEX altimetry for precise geoid and gravity recovery in the Mediterranean Sea. Geophys J Int 125: 285-302.

AVISO User Handbook – Corrected Sea Surface Heights (CORSSHs) (1998). AVI-NT-011-311-CN, Edition 3.1.

Becker M, Zerbini S, Baker T, Burki B, Galanis J, Garate J, Georgiev I, Kozev V, Lobasov V, Marson I, Negusini M, Richter B, Veis G, Yuzefovich P (2002) Assessment of Height Variations by GPS at the Mediterranean and Black sea Coast Tide Gauges from the SELF Project. Global and Planetary Change 34: 1-58.

Cazenave A, Schaeffer P, Berge M, Brosier C, Dominh K, Genero MC (1996) High-resolution mean sea surface computed with altimeter data of ERS1 (geodetic mission) and TOPEX/POSEION. Geophys J Int 125: 696-704.

Hwang C, Kao EC, Parsons B (1998) Global derivation of marine gravity anomalies from Seasat, Geosat, ERS1 and TOPEX/POSEIDON altimeter data. Geophys J Int 134: 449-459.

Ihde J, Adam J, Gurtner W, Harsson B, Sacher M, Schlüter W, Wöppelmann G (2000) The Height Solution of the European Vertical Reference Network (EUVN). In: Torres J and Hornik H (eds) Reps of the EUREF Tech Working Group. Veröffentlichungen der Bayrischen Kommission für die Internationale Erdmessung, Heft N 61, München, 132-145.

Knudsen P (1992) Altimetry for Geodesy and Oceanography. In: Kakkuri J (ed) Geodesy and Geophysics. Finnish Geodetic Institute, pp 87-129.

Knudsen P (1993) Integration of Gravity and Altimeter Data by Optimal Estimation Techniques. In: Rummel R and Sanso F (eds) Satellite Altimetry for Geodesy and Oceanography, Lecture Notes in Earth Sciences. Springer-Verlag Berlin Heidelberg 50: 453-466.

Kotzev V et al. (1999) National vertical datum. Final report of the project SE-608/96 at the National Science fund, MES, Sofia.

Lemoine FG, Kenyon SC, Factor JK, Trimmer RG, Pavlis NK, Chinn DS, Cox C, Klosko SM, Luthcke SB, Torrence MH, Wang YM, Williamson RG, Pavlis EC, Rapp RH, Olson TR (1998) The development of the join NASA GSFC and NIMA geopotential model EGM96, NASA Technical Paper, 1998 – 206861.

Moritz H (1980) Advanced Physical Geodesy. $2^{nd}$ Ed Wichmann, Karlsruhe.

Nerem RS, Mitchum GT (2001) Sea Level Change. In: Fu L.-L. and Cazenave A (eds) Satellite Altimetry and Earth Sciences. Int Geophys Series Vol 69, Academic Press, San Diego California, 329-350.

Smith WHF, Wessel P (1990) Gridding with continuous curvature splines in tension. Geophysics 55: 293-305.

Tapley BD, Kim M.-G. (2001) Applications to Geodesy. In: Fu L.-L. and Cazenave A (eds) Satellite Altimetry and Earth Sciences. Int Geophys Series Vol 69, Academic Press, San Diego California, 371-406.

Tziavos IN, Sideris MG, Forsberg R (1998) Combined satellite altimetry and shipborne gravimetry data processing. Mar Geod 21: 299-317.

Vergos GS, Tziavos IN, Andritsanos VD (2003) On the Determination of Marine Geoid Models by Least Squares Collocation and Spectral Methods using Heterogeneous Data. Presented at Session G03 of the 2004 IUGG General Assembly, Sapporo, Japan, July 2-8, 2003. (accepted for publication to the conference proceedings)

Wenzel HG (1998) Ultra high degree geopotential model GPM3E97 to degree and order 1800 tailored to Europe. Presented at the $2^{nd}$ Continental Workshop on the geoid in Europe, Budapest, Hungary.

Wessel P, Smith WHF (1998) New improved version of Generic Mapping Tools released. EOS Trans 79(47): 579.

Yi Y (1995) Determination of gridded mean sea surface from altimeter data of TOPEX, ERS-1 and GEOSAT. Rep of the Dept of Geodetic Sci and Surv No 434 The Ohio State Univ, Columbus, Ohio.

# Modelling Future Sea-level Change under Greenhouse Warming Scenarios with an Earth System Model of Intermediate Complexity

O. Makarynskyy, M. Kuhn, W.E. Featherstone
Western Australian Centre for Geodesy, Curtin University of Technology, GPO Box U1987, Perth, WA 6845, Australia; Fax: +61 8 9266 2703; E-mail: O.Makarynskyy@curtin.edu.au

**Abstract.** Recently, a lot of effort has been put into estimating possible near-future changes (say, 10-100 years) in the Earth's abiotic system, especially changes induced by human activities. One of the most studied issues is the effect of greenhouse gases on global warming and the corresponding change in sea-level around the world due to the associated deglaciation. This study focuses at projections of global sea-level changes on geological time scales. The University of Victoria's (Canada) coupled Earth System Climate Model of intermediate complexity was implemented. Two different greenhouse-warming scenarios were studied on time-scales from hundreds to thousands years. The model was used to predict sea level variations under the combined influence of changes in sea ice coverage, global precipitation and evaporation, seawater salinity and temperature. Long-term projections show unequal water mass distribution over the globe: a sea-level rise of order of decimetres in equatorial and mid-latitude regions and a sea-level fall of up to 2 metres in polar regions, mostly around Antarctica.

**Keywords:** Earth system model, greenhouse gas scenario, global warming, sea-level change

## 1 Introduction

Multiple studies of greenhouse-gas effects on the Earth system, considered as an integral system of abiotic and biotic components, have recently been undertaken (e.g., Brovkin et al. 1997; Gallee et al. 1992; Joos et al. 1999; Prinn et al. 1999; among many others). It is due to the widely acknowledged assertion that the Earth's atmospheric temperature is increasing (e.g.. Houghton et al. 2001), thus causing a number of after-effects, including melting of the previously ice-covered areas (i.e., deglaciation). The latter, together with an increased ocean water temperature, is alleged to cause the phenomenon commonly termed "sea-level rise".

However, sea-level, which largely mirrors the geoid, will not necessarily rise in all parts of the globe, and may even fall in some. This was demonstrated in historical changes in sea-level dating back to the last Ice Age (e.g., Farrell and Clark 1976; Kaufmann 2002; Donato et al. 2000; Feng and Hager 1999), including studies around Australia (e.g., Chappell 1983; Lambeck and Nakada 1990; Milne and Mitrovica 1998), and future climatological projections (e.g., Weaver et al. 2001).

The present study deals with future projections of sea-level by way of climate change on geological time scales. This work is distinctive from other contributions, notably works by Lambeck and Nakada (1990) and Peltier (1999), because it extends climatological studies into the future, whereas others look back at the last glacial period.

Two different greenhouse-warming scenarios are explored here: 2.5% and 1% increase of the initial concentration of Carbon Dioxide ($CO_2$) per annum. A brief description of the global ocean circulation model (GOCM) used in this study is given in Section 2. Section 3 presents the scenarios considered. The results obtained and following discussion are given in Section 4. Section 5 contains some concluding remarks and prospects for future research.

## 2 The Global Climate Model

The global climate model used in this study was the Earth System Climate Model from the University of Victoria, Canada, (UVic ESCM), which is of intermediate complexity (Weaver et al. 2001) and couples the atmosphere-ocean-ice processes. The models of intermediate complexity fill the gap between inductive (conceptual) and quasi-deductive (comprehensive) global climate models (Claussen et al. 2001), thus allowing less detailed - though still highly realistic - multiple projections over long time intervals. The UVic ESCM numerically represents the primary thermodynamic and hydrological components of the Earth's climate system in a spherical grid with latitudinal resolution of 3.6° and longitudinal resolution of 1.8° (i.e., 100 by 100 nodes).

In UVic ESCM, the surface balances and feedbacks of heat, freshwater and momentum of motion are represented using Fanning and Weaver's (1996) diffusive energy-moisture balance atmospheric model. The Modular Ocean Model (Pacanowski 1996) simulates general ocean circulation using primitive equations of motion. The thermodynamics and elastic-viscous plastic dynamics of sea-ice are represented by models with different levels of complexity (e.g., Semtner 1976; Bitz et al. 2001; Holland et al. 2001), which allow estimation of the properties of sea-ice and variations in ice-cover distribution by employing some parameterizations and the momentum balance equations.

Importantly, the rigid-lid approximation was used here as an ocean surface condition. This approximation assumes the static ocean surface elevation thus excluding the high-frequency external gravity waves from consideration. In the model, information on the absolute sea-level variability is provided in the form of water surface pressure onto the surface lid. This variable is regarded as the implicit ocean free surface in terms of pressure.

The UVic ESCM provides a reliable basis for our long-term (~2,000 years) experiments to simulate present-day and last glacial maximum climates (Weaver et al. 2001) with outcomes for a number of oceanographic and atmospheric parameters. This model was previously used in climatological simulations up to year 2500. Thus the novelty of our study consists in a significant extension of the projection period to geological time scales.

In our case, the parameters of interest include surface air temperature (SAT; measured/modelled at ~1.5m above the ground surface), sea-ice and snow volumes averaged globally, as well as global patterns of precipitation, seawater salinity and potential temperature (temperature of a parcel of water at the sea surface after it has been raised adiabatically - in an insulated container so it does not exchange heat with its surroundings - from some depth in the ocean), and surface pressure (Ps) in a column of water (Weaver et al. 2001; Pacanowski 1996) due to the rigid-lid approximation. These parameters are of primary importance in the estimation of sea-level changes.

## 3 Scenarios Used

In the first series of experiments, the ESCM was run until annually averaged surface fluxes became close to zero, thus indicating stabilisation of the system's conditions. This typically takes 2000 years of integration. After such a run, the model output was verified versus present-day climatology. It is claimed (Weaver et al. 2001) that the output is in good agreement with present-day observations. The model's equilibrium state then served a starting point for further numerical experiments under different scenarios of $CO_2$ increase.

In order to estimate the time-scale of sea-ice and snow/glacier melting, the linear option for $CO_2$ increase was used. The rate of the concentration change was 9.00 and 3.65 ppm/year (2.5% and 1% of the initial concentration, respectively) (Fig. 1). The latter linear rate of increase, which is the present day value, was adopted in the Intergovernmental Panel on Climate Change studies (IPCC 2001). Two and three times higher concentrations, which are also probable scenarios, were studied by, e.g., Hoyme and Zielke (2001). The linear increase of the greenhouse gas concentration in this study resulted in a non-linear change of the SAT (Fig. 2), which is attributed to the radiative properties of the Earth's atmosphere. First, the atmospheric radiance budget is significantly imbalanced with the fast long-wave irradiance accumulation coming from the Earth surface. Then later on, the balance tends to be restored with the increase of the irradiance from the atmosphere to space.

The projection under the 2.5% rate of $CO_2$ in crease will be presented and discussed in this paper as an illustrative example because the results of the both mentioned above experiments are similar in terms of SAT and Ps, with the only significant difference present in the time-scales of the projected changes (Fig. 3). It is clear that the experiment with the 1% $CO_2$ increase rate takes much longer (about 40%) to reach the same degree of sea-ice and snow melting than the model run with 2.5% increase.

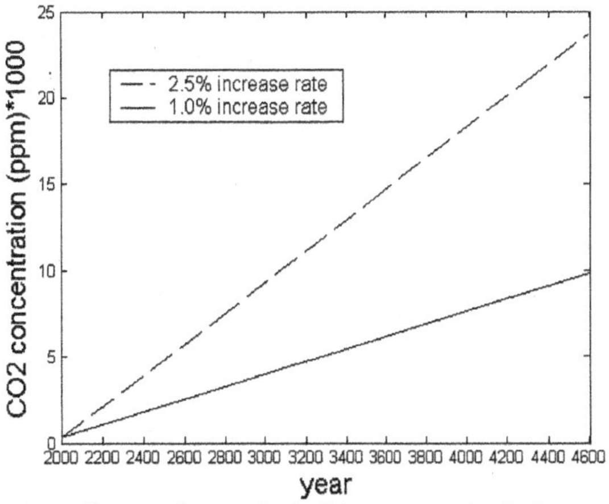


**Fig. 1.** Change of atmospheric $CO_2$ concentration in the two long-term greenhouse experiments tested.

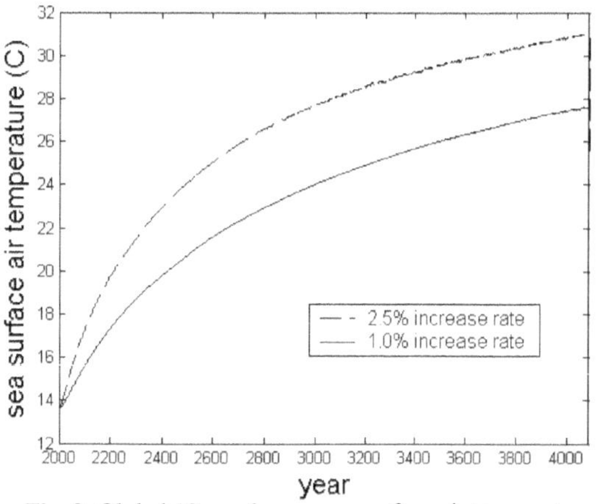

**Fig. 2.** Global 10-yearly average surface air temperature

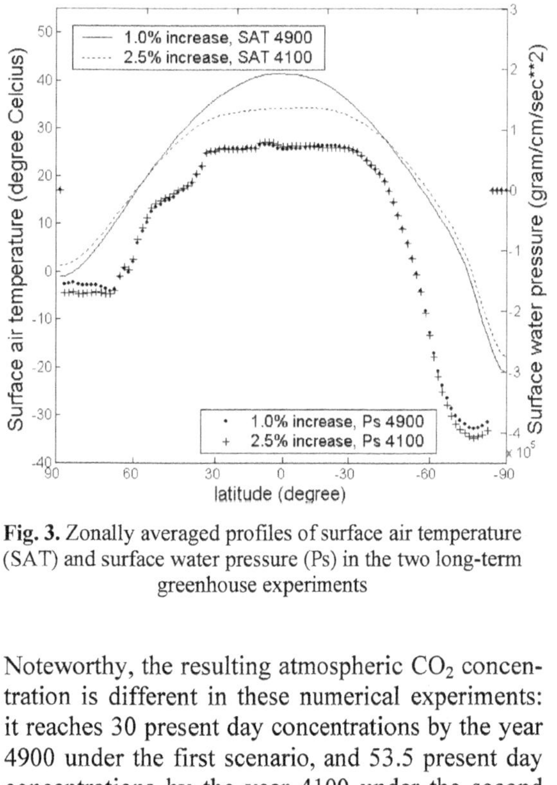

**Fig. 3.** Zonally averaged profiles of surface air temperature (SAT) and surface water pressure (Ps) in the two long-term greenhouse experiments

Noteworthy, the resulting atmospheric $CO_2$ concentration is different in these numerical experiments: it reaches 30 present day concentrations by the year 4900 under the first scenario, and 53.5 present day concentrations by the year 4100 under the second scenario. Such a difference is attributed to the large thermal inertia of the ocean equilibrating the inner conditions to the prescribed surface forcing. This process lasts from hundreds to a thousand years. Therefore, a relatively fast $CO_2$ accumulation in the atmosphere under the 2.5% scenario have a very similar to the 1.0% scenario "footprint" in the ocean, as well as in SAT, directly depending on the ocean's thermal state.

## 4 Results Obtained and Discussion

The experiment under the 2.5% scenario was carried out for a period of 2100 years, from year 2000 to year 4100. By 4100, more than half of the initial snow volume has permanently melted (Fig. 4). After that, the value fluctuates around this level insignificantly. In the ESCM model, snow forms when precipitation falls in the areas with the SAT is $\leq -5^{o}$C. Such conditions persist over Antarctic regions (Fig. 3) where snow still accumulates. Meanwhile, the sea-ice has melted almost completely (Fig. 4), with some insignificant masses persisting to form in polar regions during the winter periods.

With the general increase in atmospheric temperature, precipitation increases, leading to overall positive differences (Fig. 5). Following the pattern of precipitation (Fig. 5) and due to the melting of sea-ice in polar regions, both providing fresh-water influx into the oceans, the differences in water salinity between the years 4100 and 2000 are generally negative (i.e., lower salinity) with the lowest values in polar regions (Fig. 6). The processes of snow and sea-ice melting also define the redistribution of water temperature in the oceans (Fig. 7), with absolute minima around the Antarctic exhibiting a lower rate of surface water temperature growth compared to subtropical and equatorial regions. All these factors lead to negative differences in the values of surface water pressure, which solely reflects the change of water mass, thus meaning a decrease of sea-level in these areas. A rough estimate of the sea-level change (Fig. 8) was derived from the surface water pressure change as

$$\Delta H = \Delta Ps/\rho_w g \qquad (1)$$

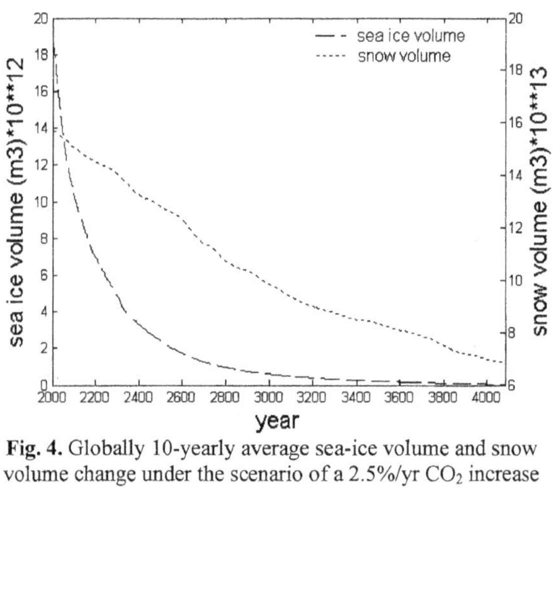

**Fig. 4.** Globally 10-yearly average sea-ice volume and snow volume change under the scenario of a 2.5%/yr $CO_2$ increase

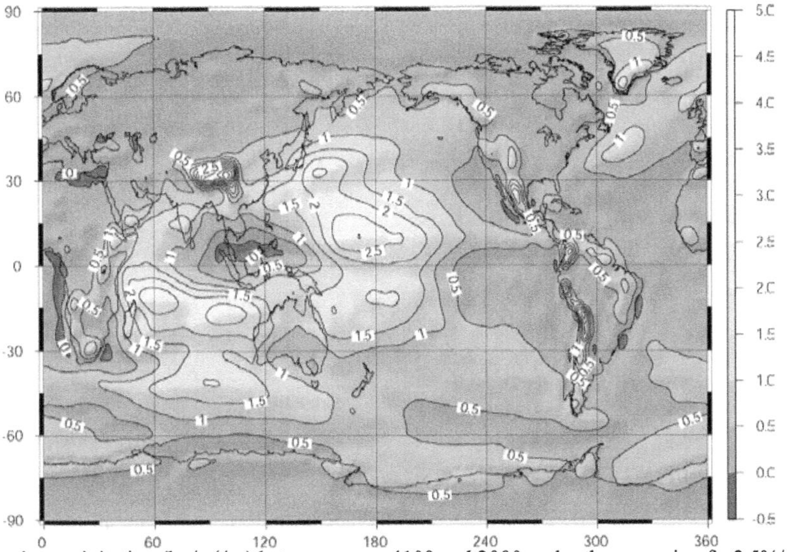

**Fig. 5.** Differences in precipitation (kg/m²/yr) between years 4100 and 2000 under the scenario of a 2.5%/yr $CO_2$ increase

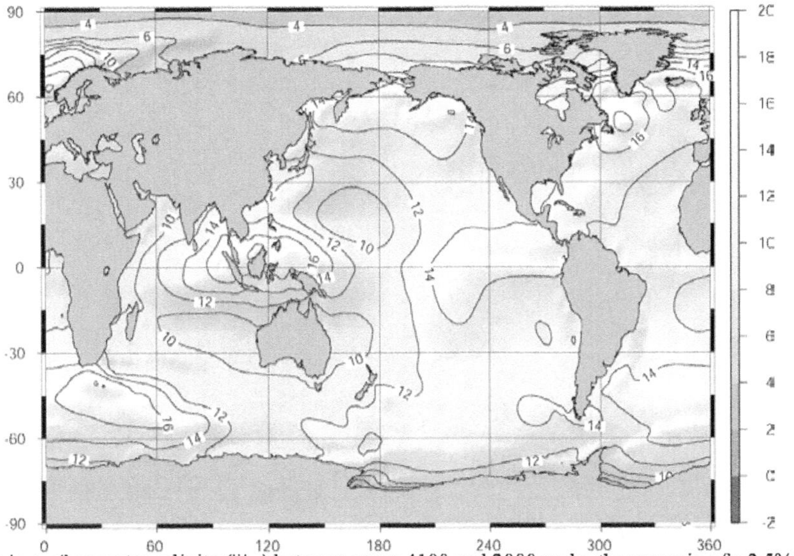

**Fig. 6.** Differences in surface water salinity (‰) between years 4100 and 2000 under the scenario of a 2.5%/yr $CO_2$ increase

where Δ is the difference of corresponding parameters, $\rho_w$ is the mean sea water density, $g$ is the gravitational acceleration. Equation (1) demonstrates the same tendencies as the Ps variations (not shown). It is clear that the melt-water from the polar ice shields does not distribute uniformly over the oceans. For instance, Australia undergoes an increase in sea-level to its north and a decrease to its south under this scenario. Meanwhile, the spatial distribution of positive differences in water salinity and temperature between years 4100 and 2000 generally coincide with the distribution of positive difference in sea level near southern Africa in the Indian Ocean and between Australia and South America in the Pacific.

## 5 Summary and Future Work

In this study, the coupled Earth System Climate Model from University of Victoria, Canada (UVic ESCM) of intermediate complexity was implemented to obtain long-term (2000-year) projections of global climate change and ocean mass redistribution. Two continuous numerical experiments were conducted, but only the greenhouse-warming

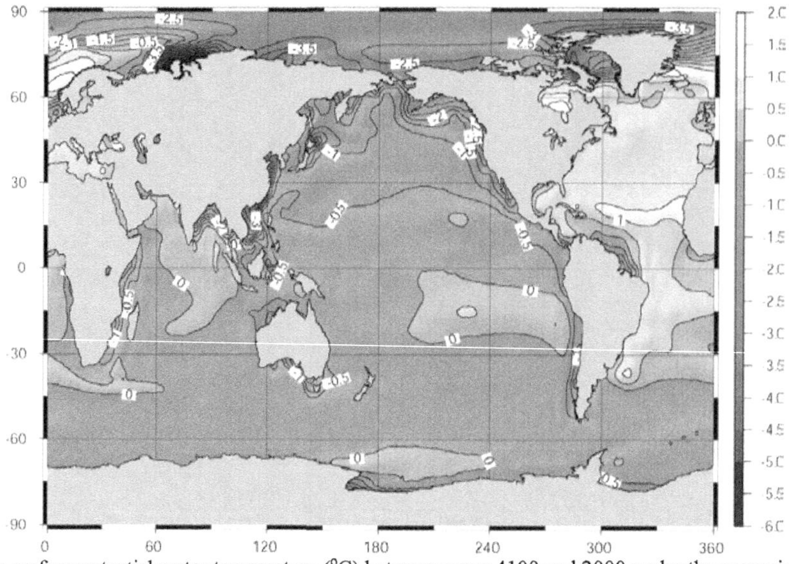


**Fig. 7.** Differences in surface potential water temperature (°C) between years 4100 and 2000 under the scenario of a 2.5%/yr $CO_2$ increase

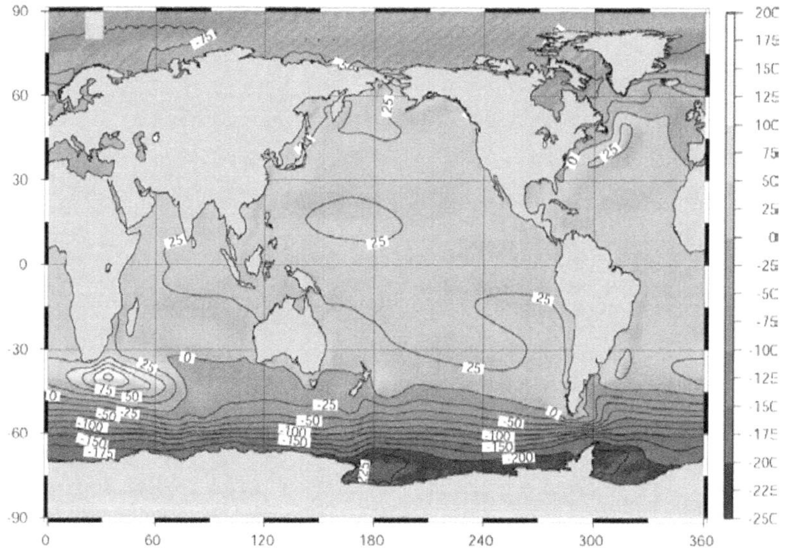


**Fig. 8.** Differences in sea-level (cm) between years 4100 and 2000 under the scenario of a 2.5%/yr $CO_2$ increase

scenario with a 2.5% per year rate of $CO_2$ increase was presented here. The results demonstrated that the UVic ESCM reflects the general physical tendencies in the change of oceanic parameters, thus providing a basis for the later application to the Synthetic Earth Gravity Model under refinement at Curtin University of Technology (Kuhn and Featherstone 2004).

Importantly, projected non-homogeneous sea-level redistribution will affect the physical Earth (e.g., the gravity field, location of the geocentre, and the Earth's spin axis). The assessment of these effects can only be carried out after the separation of the steric effect from mass changes influencing the general sea-level change. In our further research, the available vertical profiles of water salinity, potential temperature and pressure will be used to perform this separation. Meanwhile, more detailed analysis and coarse quantification of possible sea-level changes form the scope of the present contribution.

The inclusion of a continental ice model coupled with the UVic ESCM, which will take place in the near future, will allow a deeper insight into the future climate changes, and a better-based analysis of such projections. These will allow for a fuller evaluation of the expected changes to the geocentre, the Earth's rotation vector, and thence the geoid.

*Acknowledgment.* This study was funded by Australian Research Council Discovery-Project grant DP0345583. The authors thank the reviewers for their constructive comments.

## References

Bitz CM, Holland MM, Weaver AJ, Eby M (2001) Simulating the ice-thickness distribution in a coupled climate model. J Geophys Res 106: 2441–2464

Brovkin V, Ganopolski A, Svirezhev Y (1997) A continuous climate-vegetation classification for use in climate-biosphere studies. Ecological Modelling 101: 251-261

Chappell J (1983) Evidence for smoothly falling sea-level relative to north Queensland, Australia, during the past 6,000 yr, Nature 302: 406-408

Claussen M, Mysak LA, Weaver AJ, Crucifix M, Fichefet T, Loutre M-F, Weber SL, Alcamo J, Alexeev VA, Berger A, Calov R, Ganopolski A, Goosse H, Lohman G, Lunkeit F, Mokhov II, Petoukhov V, Stone P, Wang Zh (2001) Earth system models of intermediate complexity: closing the gap in the spectrum of climate system models, Climate Dynamics 18: 579-586

Donato GD, Vermeersen LLA, Sabadini R (2000) Sea-level changes, geoid and gravity anomalies due to Pleistocene deglaciation by means of multi-layered analytical Earth models, Tectonophysics 320: 409-418

Fang M, Hager BH (1999) Postglacial sea-level: energy method, Global and Planetary Change 20: 125-156

Fanning AG, Weaver AJ (1996) An atmospheric energy-moisture model: Climatology, interpentadal climate change and coupling to an ocean general circulation model. J Geophys Res 101: 15111–15128

Farrell WE, Clark JA (1976) Postgacial sea-level, Geophys J of the Royal Astronomical Society 46 (3): 647-667

Gallee H, Van Ypersele JP, Fichefet T, Tricot C, Berger AL (1992) Simulation of the last glacial cycle by a coupled 2-D climate-ice sheet model. Part 2: Response to insolation and CO2. J Geophys Res 97: 15713-15740

Gasperini P, Sabadini R, Yuen DA (1986) Excitation of the Earth's rotational axis by recent glacial discharges. Geophys Res Lett 13: 533-536

Holland MM, Bitz CM, Eby M, Weaver AJ (2001) The role of ice ocean interactions in the variability of the North Atlantic thermo-haline circulation. J Clim 14: 656–675

Houghton JT, Ding Y, Griggs DJ, Noguer M, van der Linden PJ, Dai X, Maskell K, Johnson CA (eds) (2001) Climate change 2001: the scientific basis. Contribution of working group 1 to the third assessment report of the intergovernmental panel on climate change. Cambridge Univ Press, Cambridge

Hoyme H, Zielke W (2001) Impact of climate changes on wind behaviour and water levels at the German North Sea coast. Estuarine, Coastal & Shelf Sci 53(4): 451-458

James TS, Ivins ER (1997) Global geodetic signatures of the Antarctic Ice Sheet. J Geophys Res 102: 605-633

Joos F, Plattner G-K, Stocker TF, Marchal O, Schmittner A (1999) Global warming and marine carbon cycle feedbacks on future atmospheric CO2. Science 284: 464-467

Kaufmann G (2002) Predictions of secular geoid changes from Late Pleistocene and Holocene Antarctic ice-ocean mass balance, Geophys J Int 148: 340-347

Kuhn M, Featherstone WE (2004) Construction of a synthetic Earth gravity model by forward gravity modelling, in: Sansò F (ed) A Window on the Future of Geodesy, Springer, Berlin, pp 350-355

Lambeck K, Nakada M (1990) Late Pleistocene and Holocene sea-level change along the Australian coast. Palaeogeog Palaeoclimat Palaeoecol 89: 143-176

Milne GA, Mitrovica JX (1998) The influence of a time-dependent ocean-continent geometry on predictions of post-glacial sea-level change in Australia and New Zealand, Geophys Res Lett 25: 793-796

Mitrovica JX, Davies JL, Shapiro II (2001) Recent mass balance of the polar ice sheets inferred from patterns of global sea-level change, Nature 409: 1026-1029

Pacanowski RC (1996) MOM 2 Version 2 Documentation User's Guide and Reference Manual. GFDL Ocean Technical Report 3.2. Princeton

Peltier WR (1999) Global sea level rise and isostatic adjustment, Global and Planetary Change 20: 93-123

Prinn R, Jacoby H, Sokolov A, Wang C, Xiao X, Yang Z, Eckaus R, Stone P, Ellerman D, Melillo J, Fitzmaurice J, Kicklighter D, Holian G, Liu Y (1999) Integrated global system model for climate policy assessment: feedbacks and sensitivity studies, Climatic Change 41: 469-546

Semtner AJ (1976) A model for the thermodynamic growth of sea ice in numerical investigations of climate, J Physical Oceanography 6: 379-389

Tamisiea ME, Mitrovica JX, Milne GA, Davis JL (2001) Global geoid and sea-level changes due to present-day ice mass fluctuations. J Geophys Res 106: 30849-30865

Weaver AJ, Eby M, Wiebe EC, Bitz CM, Duffy PB, Ewen TL, Fanning AF, Holland MM, MacFadyen A, Damon Matthews H, Meissner KJ, Saenko O, Schmittner A, Wang H, Yoshimori M (2001) The UVic Earth System Climate Model: model description, climatology, and applications to past, present and future climates. Atmosphere-Ocean 39(4): 361-428

# Crossover Adjustment of New Zealand Marine Gravity Data, and Comparisons with Satellite Altimetry and Global Geopotential Models

M.J. Amos[1,2] W.E. Featherstone[2] J. Brett[3]

1. Land Information New Zealand, Private Box 5501, Wellington, New Zealand; Fax: +64-4-460-0112; E-mail: mamos@linz.govt.nz
2. Western Australian Centre for Geodesy, Curtin University of Technology, GPO Box U1987, Perth, WA 6845, Australia; Fax: +61-8-9266-2703; E-mail: W.Featherstone@curtin.edu.au
3. Intrepid Geophysics, 138 Grand Prom, Doubleview, WA 6018, Australia; Fax: +61-8-9244-9313; E-mail: john@intrepid-geophysics.com

**Abstract.** This paper summarises the crossover adjustment of approximately 90,000-line-km of ship-track gravity observations around New Zealand. The adjustment reduced the standard deviation of the ~$10^6$ crossovers from ~2.0 mgal to ~0.3 mgal. These data were then used to assess four different grids of satellite-altimeter-derived gravity anomalies. The KMS02 altimeter grid was selected for use around New Zealand as it gave a better fit to the coastal ship-track data. Least-squares collocation was then used to 'drape' the altimetry onto the crossover-adjusted ship-tracks to counter the well-known problems with satellite altimeter data near the coast. The precision of this merged ship-altimeter gravity dataset is estimated to be 3.5 mgal.

**Keywords.** Gravity, marine gravimetry, satellite altimetry, crossover adjustment, geopotential model

## 1 Introduction

Marine gravity observations in the vicinity of New Zealand (NZ) have been collected over the past 45 years by various agencies at different times for different purposes (Fig. 1). Until recently, these observations were stored in different formats, in terms of different (horizontal and gravity) datums, and no attempt had been made to ensure consistency among individual cruises, let alone the datasets. The problems with offsets and tilts in marine gravimetry are well known (e.g., Wessel and Watts 1988). Therefore, a crossover adjustment of all the observations surrounding NZ has now been carried out by Intrepid Geophysics under contract to Land Information New Zealand (LINZ) to bring them into a single, coherent, internally consistent dataset.

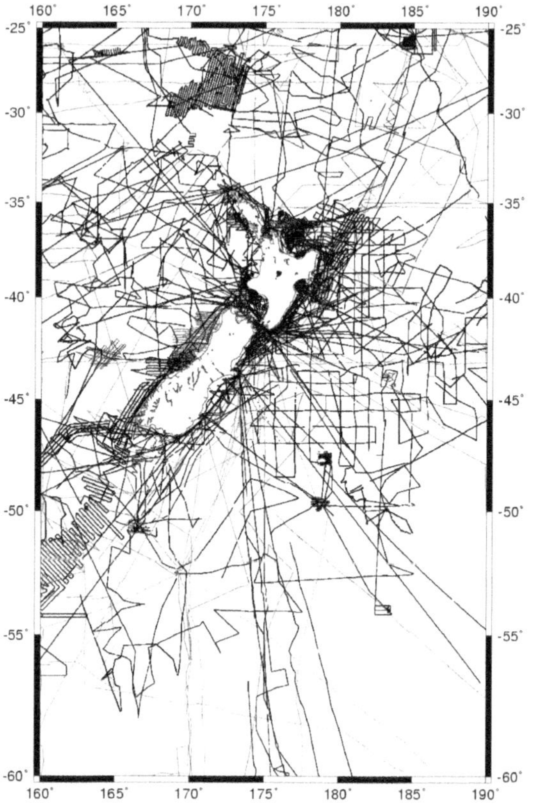


**Fig 1.** Coverage of ship-track gravity observations around New Zealand (Mercator projection)

This paper briefly describes the crossover adjustment of the ship-track gravity data (summarised from Brett 2004). The unadjusted and adjusted ship-track gravity anomalies are then compared with various satellite-altimeter-derived gravity anomaly grids (Table 1), and the EGM96 (Lemoine et al. 1998) and GGM01S (Tapley et al. 2004) global geopotential models (cf. Featherstone 2003; Denker and Roland 2004). The crossover-adjusted ship-track gravity anomalies are then used to select the most appropriate altimeter-derived gravity anomalies for the computation of a new NZ gravimetric geoid model. Finally, we describe the merging of the altimeter-derived and crossover-adjusted ship-track anomalies using least-squares collocation (LSC) (cf. Strykowski and Forsberg 1998).

| *Data* | *Resolution* | *Reference* |
|---|---|---|
| KMS02 | 2' | Andersen et al (2004) |
| SSv11.2 | 2' | Sandwell and Smith (1997) |
| GMGA02 | 2' | Hwang et al (2002) |
| GSFC00 | 2' | Wang (2001) |
| *Data* | *URL* | |
| KMS02 | ftp://ftp.kms.dk/pub/GRAVITY | |
| SSv11.2 | http://topex.ucsd.edu/marine_grav/mar_grav.html | |
| GMGA02 | ftp://gps.cv.nctu.edu.tw/pub/data/marine_gravity/ | |
| GSFC00 | http://magus.stx.com/mssh/mssh.html | |

**Table 1.** Summary of some recent public domain satellite-altimeter-derived marine gravity anomalies

## 2 The Crossover Adjustment

The crossover adjustment of the ship-track gravity observations surrounding NZ (Fig. 1) was carried out by Intrepid Geophysics under contract to LINZ (Brett 2004). This section summarises the main points; copies of Brett (2004) are available from the first-named author.

A total of 3,119,289 ship-track gravity observations were collated from recent surveys conducted for NZ's UNCLOS (United Nations Convention on the Law of the Sea) continental shelf claim, the NZ Institute of Geological and Nuclear Sciences (GNS), the US National Oceanographic and Atmospheric Administration (NOAA) and Geoscience Australia (GA). The area was restricted to $160°E \leq \lambda \leq 190°E$ and $25°S \leq \phi \leq 60°S$ (2,401,932 points) since this is the region over NZ which the gravimetric geoid model is required.

Where necessary, the gravity datum was transformed from the 1959 NZ (Potsdam) datum to IGSN71 by subtracting 15.21 mgal. All horizontal positions were assumed to be on a geocentric horizontal geodetic datum because the survey methods had not been stored in the respective databases (except the UNCLOS data, which is on WGS84). Also where necessary, the free-air gravity anomalies were converted to GRS80 (Moritz 1980); otherwise they were recomputed on GRS80 using a second-order free-air correction and Somigliana normal gravity.

The ship-track data were then checked for spikes using a fourth-difference examination of each profile. In addition, statistical reporting (min, max, mean, and standard deviation) was also performed on all data. Any outlier values were then examined more closely with an interactive data viewer and editor, and a judgement was made as to whether the spike or feature should be removed or retained.

When initially imported, the data for each ship cruise were stored as a single, long 'line' of data. A necessary step prior to crossover adjustment was to split the cruise data into shorter, approximately straight-line segments. The advantage of this is that two straight lines either do not cross, or if they do cross, then there is a single crossover. Given this precondition, it is possible to design a highly efficient algorithm to locate all crossovers. This process discarded no data. The only outcome was that all points were grouped into line segments for the purpose of identifying crossovers. The estimated crossover correction was applied to the cruise as a whole, taking no account of the breakdown into 'lines'.

The computer software used by Intrepid Geophysics also allows for the horizontal positions of the gravity observations to be adjusted. However, experiments indicated that this made little difference to the results (i.e., crossover statistics), so the horizontal positions were left unchanged.

The datasets were ranked according to their perceived reliability. This ranking determined the preferred processing order. Starting with the UNCLOS datasets, internal and external crossovers were computed. On the basis of this and the area of coverage the 'res00-11' cruise was ranked highest. Systematic offsets (i.e., biases) were applied to each of the other UNCLOS datasets to reduce the misclosure statistics for the UNCLOS cruises as a whole. The next ranked dataset was the GNS data, followed by the NOAA data, and then the GA data. This order was determined on the basis of internal crossover statistics, data coverage and visual inspection of the raw data.

After applying the offsets to the individual UNCLOS datasets they were concatenated into a single dataset. This was then adjusted using "loop closure levelling". This procedure is a single process that consists of several steps. Firstly, the crossovers of a dataset are identified (as described above). Each crossover was then quantified (bias only), where two crossovers were within ~1 km only one bias was evaluated. The misclosure errors around closed loops were then distributed using least-squares procedures for network adjustment to produce a correction function. The final levelling adjustment, at every observation point, was then interpolated from the correction function using an Akima spline.

The loop closure levelling was then applied to the GNS dataset. The levelled GNS data was gridded and the Intrepid GridMerge process used to determine an offset of 4.35 mgal to align it with the UNCLOS data. The GNS data was then appended to the UNCLOS data at the loop closure levelling repeated. The same process was followed to progressively include the NOAA (5.94 mgal offset) and GA (8.16 mgal offset) datasets.

Comparing the misclosures at the crossover points in Tables 2 and 3 shows a 714% improvement in the standard deviation (STD) of the crossovers, as well as a significant reduction in the mean differences. Table 4 gives the statistics of the ship-track gravity anomalies before and after the crossover adjustment.

| *Data* | *X-overs* | *Max* | *Min* | *Mean* | *STD* |
|---|---|---|---|---|---|
| UNCLOS | 345 | 79.7 | 0.0 | 7.6 | 12.9 |
| GNS | 57512 | 271.3 | 0.0 | 2.5 | 7.6 |
| NOAA | 971988 | 236.1 | 0.0 | 0.7 | 0.7 |
| GA | 36271 | 52.6 | 0.0 | 1.6 | 2.7 |
| All data | 1069289 | 271.3 | 0.0 | 0.8 | 2.0 |

**Table 2.** Misclosure statistics for the original ship-track observations (mgal)

| *Data* | *X-overs* | *Max* | *Min* | *Mean* | *STD* |
|---|---|---|---|---|---|
| UNCLOS | 345 | 12.1 | 0.0 | 0.45 | 1.39 |
| GNS | 57512 | 68.9 | 0.0 | 0.19 | 1.50 |
| NOAA | 971988 | 1.9 | 0.0 | 0.09 | 0.08 |
| GA | 36271 | 14.9 | 0.0 | 0.04 | 0.11 |
| All data | 1069244 | 93.4 | 0.0 | 0.05 | 0.28 |

**Table 3.** Misclosure statistics for the adjusted ship-track observations (mgal)

| *Adjustment* | *Max* | *Min* | *Mean* | *STD* |
|---|---|---|---|---|
| Original | 477.0 | -813.6 | 4.1 | 43.2 |
| Adjusted | 455.6 | -807.7 | 6.9 | 42.6 |

**Table 4.** Statistics of the original and adjusted ship-track observations (mgal; 2,401,932 points)

## 3 Comparisons with Altimeter- and GGM-derived Gravity Anomalies

Firstly, it is well known that satellite-altimeter-derived gravity anomalies are less accurate close to the coast. This is due to factors such as poorly tracked altimetry close to the coast (Deng et al., 2002), poor shallow-water tidal models, and poor wet delay corrections (e.g., Andersen and Knudsen 2000). In addition, there are significant differences close to the Australian coast among altimeter-derived anomalies derived by different groups (Featherstone 2003). This is also the case in NZ, albeit to a lesser extent than near Australia (Fig. 2 and Table 5). Note that the range in Fig. 2 has been truncated for display purposes (cf. Table 5).

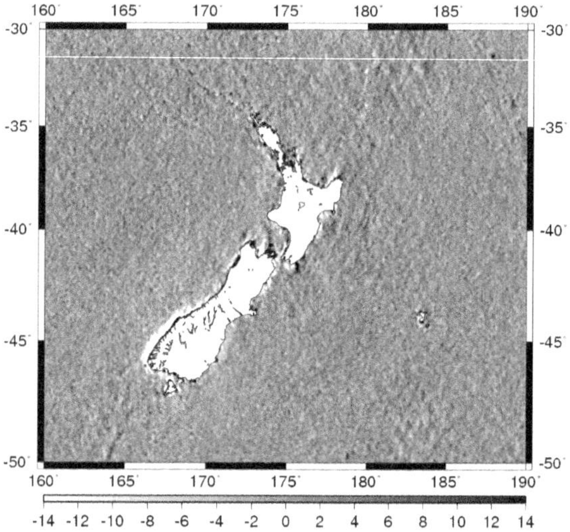


**Fig 2.** Difference between KMS02 and SSv11.2 gravity anomalies around NZ (mgal; Mercator projection)

| *Data* | *Max* | *Min* | *Mean* | *STD* |
|---|---|---|---|---|
| KMS02-SSv11.2 | 139.4 | -79.4 | -0.1 | 3.0 |
| KMS02-GMGA02 | 371.8 | -337.3 | -0.0 | 4.2 |
| KMS02-GSFC00 | 117.2 | -103.2 | 0.1 | 2.9 |
| SSv11.2-GMGA02 | 380.8 | -334.7 | 0.0 | 4.0 |
| SSv11.2-GSFC00 | 123.9 | -129.3 | 0.2 | 3.2 |
| GMGA02-GSFC00 | 334.7 | -366.2 | 0.1 | 4.2 |

**Table 5.** Statistics of the differences between altimeter-derived gravity anomalies around NZ (920,918 pts mgal)

The largest differences among the altimeter grids occur along the western coast of NZ's South Island (centred at: ~45°S, ~167°E); see the example in Fig. 2. This is due to a combination of the problems with coastal satellite altimetry, coupled with the very steep gravity gradients at the boundary of the Australian and Pacific plates. The latter will give a large Gibbs's phenomenon when transforming the sea surface heights/gradients to gravity anomalies because there is no gravity data on land. From Table 5, the comparisons that involve the GMGA02 grid give the largest maximum and minimum differences. These differences are concentrated as several 'spikes' located close to the NZ and Chatham

Island (183°E, 44°S) coasts. This shows that it is the least consistent with the other grids, which are reasonably self-consistent (Table 5).

Next, the various altimeter-derived anomalies (Table 1) were compared with the crossover-adjusted anomalies in order to select the best grid for future NZ geoid computations. The altimeter-derived gravity anomalies (assumed to also be on a geocentric horizontal datum) were bi-cubically interpolated to the locations of the ship-track data. The statistics in Table 6 only use the dense ship-track data in a 50-400-km band around NZ and the Chatham Islands (Fig. 3). This is because the altimeter data are less reliable within ~50 km from the NZ coast (Fig. 2). The altimeter data are probably more reliable than the ship-track data in the open oceans, especially in areas with sparse data coverage where the crossover adjustment is poorly constrained (e.g., south of 55°S; see Fig. 1).

| *Data* | *Max* | *Min* | *Mean* | *STD* |
|---|---|---|---|---|
| KMS02 | 194.5 | -108.7 | 1.7 | 8.2 |
| SSv11.2 | 196.0 | -109.1 | 1.6 | 8.2 |
| GMGA02 | 197.3 | -107.7 | 1.7 | 8.2 |
| GSFC00 | 193.9 | -107.4 | 2.2 | 8.0 |

**Table 6.** Statistics of the differences between the altimeter-derived altimetry grids and the 890,290 crossover-adjusted NZ ship-track observations within 50–400 km of the coast (mgal)

From the results in Table 6, no single altimeter grid is significantly better than another in the 50-400-km region around NZ. However, an analysis of the comparison between the altimetry grids and all 2,401,932 adjusted ship-track data reveals that the KMS02 altimetry grid gives an overall better fit. As such, this grid was selected for use because it will reduce the amount of 'draping' required to fit it to the ship-track data (see section 4).

Finally, the 2,401,932 original and crossover-adjusted anomalies were compared with gravity anomalies implied by the EGM96 (Lemoine et al. 1998) and GGM01S (Tapley et al. 2004) (Table 7). Because of increased noise in the high-degree coefficients, GGM01S was truncated at spherical harmonic degree and order 90; EGM96 was used to degree and order 360. Acknowledging that the ship-track data has different frequency content to the long-wavelength GGMs, Table 7 indicates that the adjusted ship-track data give a better agreement with the GGMs, thus providing further verification of the success of the crossover adjustment.

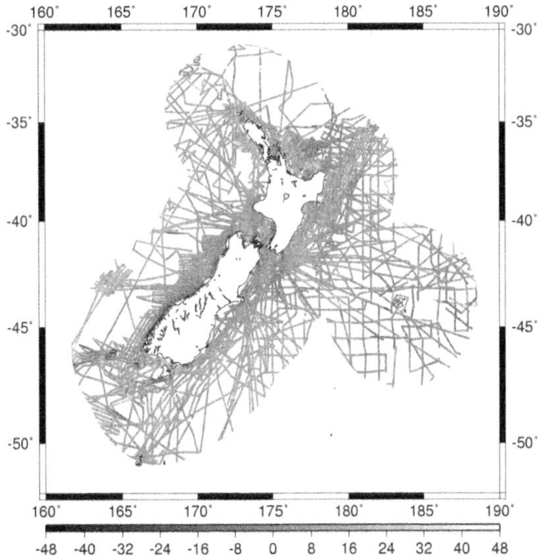


**Fig 3.** Difference between the adjusted NZ ship-track observations and KMS02 altimeter-derived gravity anomalies up to 400 km from coast (mgal; Mercator projection)

| | *Max* | *Min* | *Mean* | *STD* |
|---|---|---|---|---|
| *Original* | | | | |
| EGM96 | 500.5 | -784.3 | -0.4 | 22.7 |
| GGM01S | 494.5 | -795.7 | -1.8 | 37.4 |
| *Adjusted* | | | | |
| EGM96 | 479.1 | -778.4 | 2.3 | 22.6 |
| GGM01S | 473.2 | -789.8 | 0.9 | 37.1 |

**Table 7.** Statistics of the differences between the 2,401,932 original and adjusted NZ ship-track observations and gravity anomalies implied by global geopotential models (mgal)

## 4 Operational Merging using LSC

Now that it has been proven that the crossover adjustment has been successful and, in turn, allowed the identification of KMS02 as the 'best' grid of altimeter-derived gravity anomalies around NZ, we now aim to use the crossover-adjusted ship-tracks to correct for the poorer altimeter data near the coast (cf. Strykowski and Forsberg 1998; also see Figs. 2 and 3). This was achieved using the least-squares collocation (LSC) interpolation routines in the GRAVSOFT suite (Tscherning et al. 1992).

The LSC 'draping' procedure broadly followed the procedures of Strykowski and Forsberg (1998). Firstly, the crossover-adjusted ship-track data was supplemented with land gravity information. Next, the differences between the ship-track/land data and the KMS02 altimeter data within the study area were determined. These differences were then 'softly' gridded (predicted with LSC with a relatively large standard deviation) to a 2 arc-minute correction grid over the computation area. A second-order Markov covariance model was used with a correlation length of 20 km and 3 mgal RMS noise of the gravity data. These parameters were optimised by testing them over a range of 5 – 100 km and 1 – 5 mgal, respectively (see next paragraph). The correction grid was then added to the pre-gridded altimetry data. This yields an altimetry data set that is consistent with the ship-track data, thus correcting the well-known coastal errors in the altimetry data.

Importantly, this LSC combination procedure only used the dense ship-track data within 400 km of the NZ coast and Chatham Island (Fig. 3). This is because the sparse oceanic ship-track data are considered less reliable than the altimeter data because the crossover adjustment is poorly constrained beyond this distance. The LSC data combination was [partly] independently tested by extracting 2,328 observations (~0.2%) from the adjusted ship-track data within 400 km of the coast. These observations were selected by removing every 2,328th record from the ship-track data file. These data were not used in the LSC combination, but used later to test the results; it also allowed empirical optimisation of the choice of RMS noise and correlation length. Coincidently, this optimisation resulted in the same values for the noise and correlation length as adopted by Strykowski and Forsberg (1998). The comparison between the 2,328 extracted marine observations and the KMS02 altimetry anomalies before and after draping revealed a significant improvement in the fit (Table 8). An additional comparison was made between all of the ship-track anomalies (Table 9). This also demonstrates an improved fit between the datasets after the LSC draping has been performed.

From this comparison, we cautiously estimate the precision/accuracy of the LSC combined gravity anomalies to be ~3.5 mgal, which is a GLOPOV (General Law of Propagation of Variance) combination of the estimated error in the ship-tracks from the crossover adjustment (0.3 mgal; Table 3) and the STD of the differences with the independent points (3.2 mgal; Table 8). The final marine gravity grid (LSC combined using all the ship-track data) is shown in Fig. 4.

| *Grid* | *Max* | *Min* | *Mean* | *STD* |
|---|---|---|---|---|
| Original | 61.2 | -89.7 | 0.7 | 9.9 |
| Draped | 32.0 | -32.1 | 0.0 | 3.2 |

**Table 8.** Statistics of the differences between 2328 ship-track observations (within 400 km of the coast) and the KMS02 anomalies before and after draping (mgal)

| *Grid* | *Max* | *Min* | *Mean* | *STD* |
|---|---|---|---|---|
| Original | 486.3 | -789.5 | 1.4 | 11.2 |
| Draped | 486.2 | -789.5 | 0.9 | 9.1 |

**Table 9.** Statistics of the differences between 2,401,932 ship-track observations and the KMS02 anomalies before and after draping (mgal)

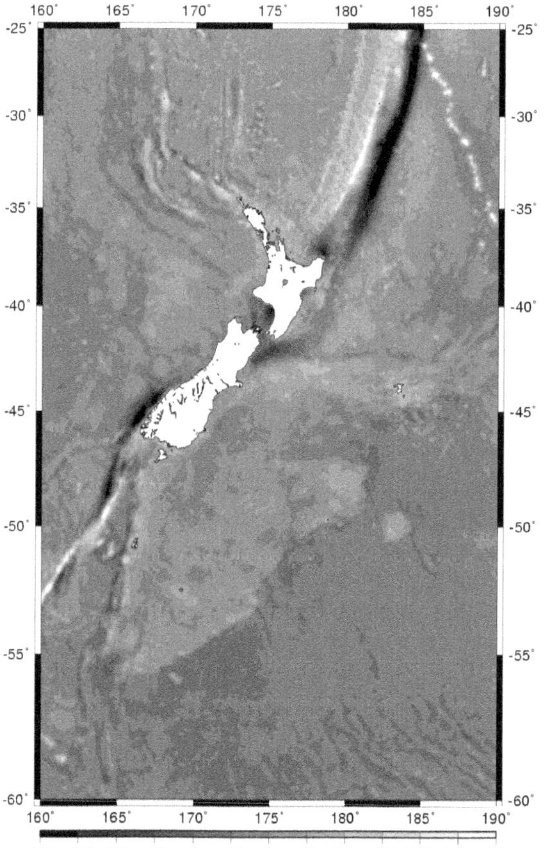


**Fig 4.** The final NZ 2' marine gravity anomaly grid combining adjusted ship-track observations and KMS02 altimeter-derived gravity anomalies (mgal; Mercator projection)

## 5 Conclusion

This paper has summarised the crossover adjustment of approximately 90,000-line-km of ship-track gravity observations around New Zealand. The standard deviation of the ~$10^6$ crossovers was reduced from ~2.0 mgal to ~0.3 mgal. These data were then used to assess four different grids of satellite-altimeter-derived gravity anomalies. After excluding known problematic data areas, this showed no single altimeter grid as being significantly better than another. The KMS02 grid was selected for use around New Zealand as it had a better fit than the others when the coastal and offshore areas were included in the comparison. This grid was then 'draped' onto the crossover adjusted ship-tracks using least-squares collocation to counter the well-known problems with satellite altimeter data near the coast. The precision of this merged dataset is estimated to be 3.5 mgal.

*Acknowledgements*: This research is funded by Land Information New Zealand and a Curtin University Postgraduate Scholarship. We are greatly indebted to the individuals and organisations that supplied data, without which this study could not have been started. Finally, we would like to thank the reviewers for their time taken to consider this manuscript.

## References

Andersen OB, Knudsen P, Trimmer R (2004) Improving high-resolution altimetric gravity field mapping (KMS02). Proc IUGG General Assembly, Sapporo, Japan 2003, Springer, Berlin, in press

Andersen OB, Knudsen P (2000) The role of satellite altimetry in gravity field modelling in coastal areas, Phys Chem. Earth 25(1): 17-24

Brett J (2004) Marine gravity crossover adjustment for New Zealand, Report to Land Information New Zealand, Intrepid Geophysics, Melbourne, Australia

Deng XL, Featherstone WE, Hwang C (2002) Estimation of contamination of ERS-2 and Poseidon satellite radar altimetry close to the coasts of Australia, Marine Geodesy 25(4): 249-271

Denker H, Roland M (2004) Compilation and evaluation of a consistent marine gravity data set surrounding Europe., Proc IUGG General Assembly, Sapporo, Japan 2003, Springer, Berlin, in press

Featherstone WE (2003) Comparison of different satellite altimeter-derived gravity anomaly grids with ship-borne gravity data around Australia, in: Tziavos IN (ed) Gravity and Geoid 2002, Dept of Surv & Geodesy, Aristotle Univ of Thessaloniki, pp.326-331

Hwang C, Hsu H-Y, Jang R-J (2002) Global mean sea surface and marine gravity anomaly from multi-satellite altimetry: applications of deflection-geoid and inverse Vening Meinesz formulae, J Geod 76(8): 407-418

Lemoine FG, Kenyon SC, Factor RG, Trimmer RG, Pavlis NK, Chinn DS, Cox CM, Klosko SM, Luthcke SB, Torrence MH, Wang YM, Williamson RG, Pavlis EC, Rapp RH, Olson TR (1998) The development of the joint NASA GSFC and the National Imagery and Mapping Agency (NIMA) geopotential model EGM96, NASA Tech rep NASA/TP-1998-206861, Goddard Space Flight Center, Greenbelt, Maryland

Moritz H (1980) Geodetic reference system 1980, Bull Géod 54(4): 395-405

Sandwell DT, Smith WHF (1997) Marine gravity anomaly from GEOSAT and ERS 1 satellite altimetry, J Geophys Res 102(B5): 10039-10054

Strykowski G, Forsberg R (1998) Operational merging of satellite airborne and surface gravity data by draping techniques. in: Forsberg R, Feissl M, Dietrich R (eds) Geodesy on the Move, Springer, Berlin, 207-212

Tapley BD, Bettadpur S, Watkins M, Reigber Ch (2004) the gravity recovery and climate experiment: mission overview and early results. Geophys Res Lett 31, L09607: doi10.1029/2004GL019920

Tscherning CC, Forsberg R, Knudsen P (1992) The GRAVSOFT package for geoid determination. in: Holota P, Vermeer M (eds.) Proc 1st Continental Workshop on the Geoid in Europe, Prague, May, ISBN 80-901319-2-1, pp. 327-334

Wang YM (2001) GSFC00 mean sea surface, gravity anomaly, and vertical gravity gradient from satellite altimeter data. J Geophys Res 106(C12): 31167-31174

Wessel P, Watts AB (1988) On the accuracy of marine gravity measurements, J Geophys Res 94(B4): 7685-7729

# Results of the International Comparison of Absolute Gravimeters in Walferdange (Luxembourg) of November 2003

Olivier Francis
University of Luxembourg, 6, rue Coudenhove-Kalergi, L-1359 Luxembourg
and European Center for Geodynamics and Seismology, 19, rue Josy Welter, L-7256 Walferdange

Tonie van Dam
European Center for Geodynamics and Seismology, 19 rue Josy Welter, L-7256 Walferdange

M. Amalvict, M. Andrade de Sousa, M. Bilker, R. Billson, G. D'Agostino, S. Desogus, R. Falk, A. Germak, O. Gitlein, D. Jonhson, F. Klopping, J. Kostelecky, B. Luck, J. Mäkinen, D. McLaughlin, E. Nunez, C. Origlia, V. Palinkas, P. Richard, E. Rodriguez, D. Ruess, D. Schmerge, S. Thies, L. Timmen, M. Van Camp, D. van Westrum, H. Wilmes

**Abstract.** The results of an international comparison of absolute gravimeters held in Walferdange Luxembourg in November 2003 are presented. The absolute meters agreed with one another with a standard deviation of less than 2 μGal (1 Gal = 1 cm/s$^2$) (if we exclude one prototype instrument from the analysis). For the first time, the ability of the operators was put to the test. The comparison indicates that the errors due to the operator are less than 1 μGal, i.e. within the observational errors.

## 1 Introduction

On November 3rd to November 7th 2003, Luxembourg's European Center for Geodynamics and Seismology (ECGS) hosted an international comparison of absolute gravimeters in the Underground Laboratory for Geodynamics in Walferdange (WULG). This is the first time in the history of geophysics and metrology that 15 absolute gravimeters were brought together in the same location for simultaneous observations. Teams from all over the world including the United States and Brasil, as well as teams from Europe participated, in the comparison (Table 1).

Table 1. Participants in the International Comparison of Absolute Gravimeters.

| Country | Institution | Absolute gravimeter(s) | Relative gravimeter(s) |
|---|---|---|---|
| Austria | Bundesamt für Eich- und Vermessungswesen (BEV), Vienna | JILAg-6 | |
| Belgium | Observatoire Royal de Belgique (ORB), Brussels | FG5-202 | Scintrex CG3M-256 |
| Brazil | Observatorio Nacional , Rio de Janeiro | FG5-223 | |
| Finland | Finnish Geodetic Institute, (FGI), Masala | FG5-221 | |
| France | École et Observatoire des Sciences de la Terre (EOST), Strasbourg | FG5-206 | |
| Germany | Bundesamt für Kartographie und Geodäsie (BKG), Frankfurt | FG5-301 | |
| Germany | Institut für Erdmessung (IfE), Universität Hannover, Hannover | FG5-220 | |
| Italy | Istituto di Metrologia "G. Colonnetti" (IMGC), Turin | IMGC-02 | |
| Luxembourg | European Center for Geodynamics and Seismology, ECGS | FG5-216 | Scintrex-CG5 021210008 |
| Czech Republic | RIGTC, Geodetic Observatory Pecny | FG5-215 | |
| Spain | Instituto Geográfico Nacional (IGN), Madrid | A10-006 | |
| Spain | Instituto Geográfico Nacional (IGN), Madrid | FG5-211 | |
| Switzerland | Swiss Federal Office of Metrology and Accreditation (METAS), Bern-Wabern | FG5-209 | Scintrex CG3M-494 |
| UK | Proudman Oceanographic Laboratory (POL), Bidston | FG5-103 | |
| USA | United States Geological Survey | A10-008 | |

In 1999, a laboratory (Figure 1) dedicated to the intercomparison of absolute gravimeters was built within the WULG. The laboratory lies 100 meters below the surface at a distance of 300 m from the entrance of the mine. To transport the 350 kilograms of equipment (the typical weight of an absolute gravimeter and its peripherals) over the 300 meters to the lab, electric golf carts were used. The cart travels on a smooth newly installed concrete surface.

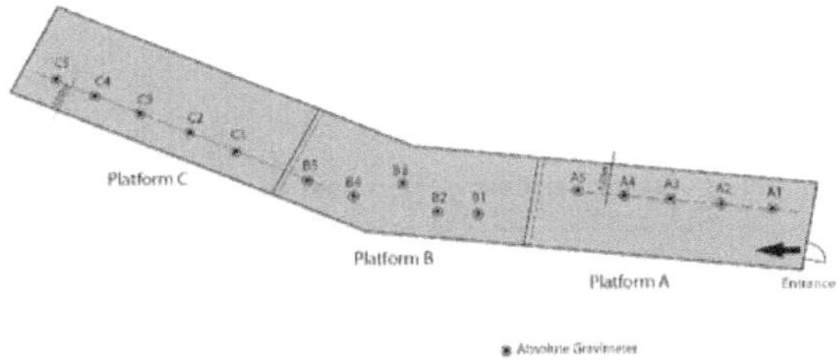


**Fig. 1** Underground laboratory where 15 gravimeters can be setup at the same time (40 m length and 3.6 wide)

The WULG is environmentally stable (i.e. constant temperature and humidity within the lab), and is extremely well isolated from anthropogenic noise. It has the power and space requirements to be able to accommodate up 15 instruments operating simultaneously (Figure 2).

**Fig. 2** Picture taken during the comparsion of absolute gravimeters in the Underground Laboratoty for Geodynmaics in Walferdange

Absolute gravimeters are used in geophysics for monitoring gravity variations due to mass changes within the Earth (i.e. the motion of magma underneath volcanoes), mass changes within the Earth's upper layers (i.e. the seasonal variations of continental water storage that might be related to global warming), density changes and vertical displacement caused by deformations of the Earth's crust (i.e. tectonic deformations associated with the build up and release of strain during an earthquake). In metrology, absolute gravimeters are used in the determination of standards derived from the kilogram (ampere, pressure, force). However, because these instruments are 'absolute', to verify that the instruments are operating properly, they must be regularly compared to other instruments of the same accuracy. Being absolute instruments, these gravimeters cannot really be calibrated. Only some of their components (such as the atomic clock or the laser) can be calibrated by comparison with known standards. The only way one currently has to verify their good working order is via a simultaneous intercomparison with other absolute gravimeters of the same and/or if possible even of a different model, to put in evidence systematic errors..

During a comparison, we cannot estimate how accurate the meters are: in fact, as we have no way to know the true value of g, we can only investigate the relative offset between instruments. This means that all instruments can suffer from the same unknown and undetectable systematic error. In addition, differences larger than the uncertainty of the measurements, is an indication of possible systematic error.

Intercomparisons of this type are difficult to arrange which is why they have only officially been organized every 4 years by the Bureau International des Poids et Mesures (BIPM) in Sèvres, France. This time scale is not sufficient for most users as most also regularly deploy their instruments for field observations.

For the first intercomparison in Walferdange, 15 meters from 13 countries including 5 types of absolute gravimeters were present: 1 Jilag, 11 FG5s with bulk and fiber interferometer, 2 A10s, and 1 prototype from IMGC. For the first time, simultaneous observations were taken. In addition an original experiment was conducted to estimate the error due to the operators.

## 2 Protocol

Ideally to compare gravimeters, they should measure at the same site at the same time. Obviously, this is practically impossible. The comparison was spread over three days. The first day, each instrument was installed at one of the 15 sites. The second day, as the WULG is composed of three different platforms, all instruments moved to another site on a different platform and again on the third day. Overall, each instrument occupied at least 3 sites one on each platform. We also planned the observations in such a way, that two different instruments which occupied the same site, did not measure at another common site again. This allows us to compare each instrument to as many other instruments possible.

Some teams arrived a few days before the comparison and others teams did stay longer afterward. Those extra measurements were also included in the final adjustment.

## 3 Data processing

Raw data of the absolute gravimeters consist of vectors of time intervals between successive positions of the falling object during the drops. To obtain the gravity value, a linear equation representing the equation of motion is fit to the raw data including the vertical gravity gradient which has been measured with relative meters. The procedures followed are the same as at the comparisons in Sèvres (Francis and van Dam, 2003). Geophysical corrections are applied to the raw gravity data: earth tides using observed tidal parameters from the superconducting gravimeter installed in a gallery next to the laboratory, atmospheric pressure using a constant admittance and polar motion effect using pole positions from IERS.

The vertical gravity gradients were measured by three different operators (O. Francis, M. Van Camp and P. Richard) with two Scintrex CG3-M and one Scintrex CG5. Comparisons between the rubidium clocks and the barometers were carried out by M. Van Camp and R. Falk. The results of these comparisons were used in data processing. We did not have any laser calibration as we are not equipped for this.

Most of the data were processed with the "g-soft" version 4.0 from Micro-g Solutions which runs on Microsoft Windows®. However, the Jilag gravimeter operating with old electronics is not compatible and the program, "Replay", from "Olivia" was used. This early version of the software contains the same coded algorithms for computing the g-values and the geophysical corrections as in "g-soft". The only difference is in the data input format.

## 4 Errors due to the operators

An original experiment to estimate the operators' error has been performed with the agreement of all the participants. After the third day, all the operators of the FG5s and 1 A-10 left their instruments in the hands of experts from Micro-g Solutions, manufacturer of the FG5. The instruments remained at the same site but were run by Micro-g engineers. The results (Figure 3) show that the measurements agree within the error bar of the observations. There are two exceptions: the FG5#211 due to a bad collimation of the laser corrected by Micro-g and the FG5#216 which was operated during the comparison by Micro-g for which we do not have an explanation yet.

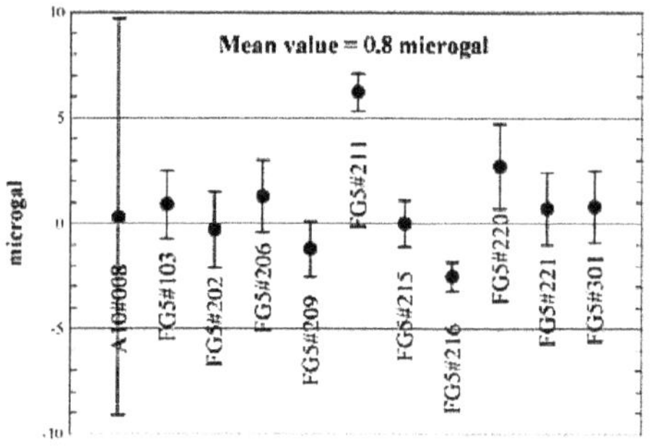


**Fig. 3** Difference in the gravity values as measured by the usual operators and the expert operators from Micro-g.

## 5 Adjustment of the data

Data from one instrument (A10#006) were discarded due to problem with the power supply. Data at site A1 were not included in the final adjustment as only one instrument occupied the site. The data of the prototype gravimeter IMGC-2 were not included in the adjustment because an offset of –46.7 μGal was detected and would have biased the adjustment. The data from the FG5#211 were corrected for an offset of –2.7 μGal 1 due to the collimation error (see previous section).

All the gravimeters could not occupy all the sites. To compare their measurements, the following least-square adjustment has been performed:

$$g_{ik} = g_k + e_i$$

where $g_{ik}$ is the gravity value at the site k given by the instrument i, $g_k$ is the adjusted value at the site k and $e_i$ is the uncertainty containing a systematic component (the offset) and a stochastic component. We assume a systematic error of 2 μGal for the FG5s and Jilag and 5 μGal for the A10 as specified by the manufacturer. For the stochastic component, we took the average value at each site and calculate the difference for each instrument with the average value. Then we computed the standard deviation of this difference, which was used to estimate the precision (the stochastic part) of each instrument.

The results of the adjustment using the complete set of data are displayed in Table 2 and Figures 4.

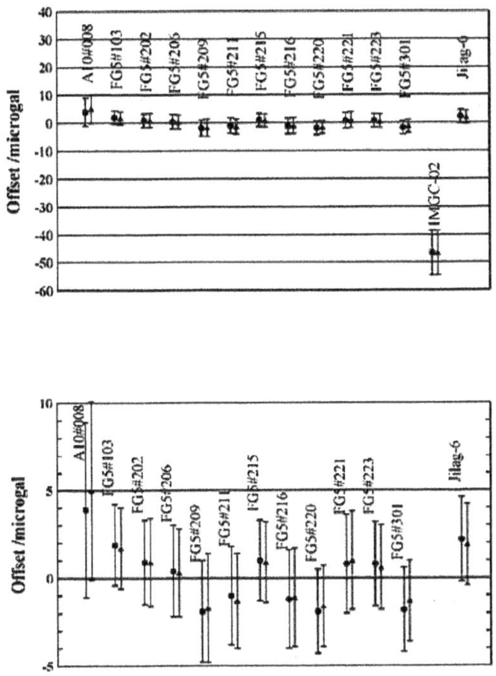


**Fig. 4** Relative offsets between the gravimeters for the unweighted (black dots) and weighted (red triangles) adjustments.

The standard deviation of the relative offset between the different instruments varies from 1.8 for the unweighted solution to 1.9 μGal for the weighted solution if we exclude the prototype instrument IMGC-02 which has an offset of –46.7 μGal. It is worth noting that all the error bars cross the zero line. The A-10 shows the largest offset and uncertainty as we could expect from the specifications of the instruments which should have a precision and accuracy of 10 μGal.

## Conclusions

The international comparison of absolute gravimeters in Walferdange shows an agreement between the participating gravimeters at 1.9 μGal (1 standard deviation) if we exclude one prototype instrument. This the best agreement ever achieved during an intercomparison. This excellent result is due to the coincidence of a few favorable factors: a very good site with stability in temperature and low microseismic noise, excellent operators, short duration of the experiment (3 days), interaction between the participants working all together in the same lab, and the last but not the least, Micro-g Solutions experts support during the experiment.

This historic experiment marks the recognition of the WULG as an international absolute gravimeter intercomparison. It is expected, that these intercomparisons will occur regularly as a complement to the comparisons at the BIPM.

## References

Francis O., van Dam T.M., Processing of the Absolute data of the ICAG01, Cahiers du Centre Européen de Géodynamique et de Séismologie, vol.22, 45-48, 2003.

**Table 2.** Relative offsets between the gravimeters for the unweighted and weighted adjustments

| Instrument | Unweighted offset Average /μGal | Error /μGal | Weighted offset Average /μGal | Error /μGal | Difference /μGal |
|---|---|---|---|---|---|
| A10#008 | 3,9 | 5,0 | 5 | 5,1 | -1,1 |
| FG5#103 | 1,9 | 2,3 | 1,7 | 2,3 | 0,2 |
| FG5#202 | 0,9 | 2,4 | 0,9 | 2,5 | 0 |
| FG5#206 | 0,4 | 2,6 | 0,3 | 2,5 | 0,1 |
| FG5#209 | -1,9 | 2,9 | -1,7 | 3,1 | -0,2 |
| FG5#211 | -1 | 2,8 | -1,3 | 2,7 | 0,3 |
| FG5#215 | 1 | 2,3 | 0,9 | 2,3 | 0,1 |
| FG5#216 | -1,2 | 2,8 | -1,1 | 2,8 | -0,1 |
| FG5#220 | -1,9 | 2,4 | -1,6 | 2,3 | -0,3 |
| FG5#221 | 0,8 | 2,8 | 1 | 2,8 | -0,2 |
| FG5#223 | 0,8 | 2,4 | 0,6 | 2,4 | 0,2 |
| FG5#301 | -1,8 | 2,4 | -1,3 | 2,3 | -0,5 |
| IMGC-02 | -46,7 | 8,1 | -46,7 | 7,9 | 0 |
| Jilag-6 | 2,2 | 2,4 | 1,9 | 2,3 | 0,3 |
| Std | 1,8 | | 1,9 | | |
| Mean | | | | | -0.09 |

# A New Small Cam-Driven Absolute Gravimeter

J.E. Faller, A.L. Vitouchkine
JILA, National Institute of Standards and Technology and University of Colorado, 440 UCB, Boulder, Colorado, USA 80309-0440

**Abstract.** We have developed a small, portable absolute gravimeter that employs a cam-based dropping mechanism. The resulting high data rate (100 drops in 30 s) serves to compensate for the short (3.0 cm) measurement distance. We think of this instrument as being both a "portable FG5" and an "absolute LaCoste & Romberg" (or Scintrex). Its realizable measurement accuracy appears to be 3 μGal. This compact (easily portable) and absolute instrument will be useful as a field gravimeter. It will also be useful in standards and calibration applications since it produces a g determination of high accuracy.

**Keywords.** Absolute gravity, gravimeter, relative gravimeter, gravity instrumentation

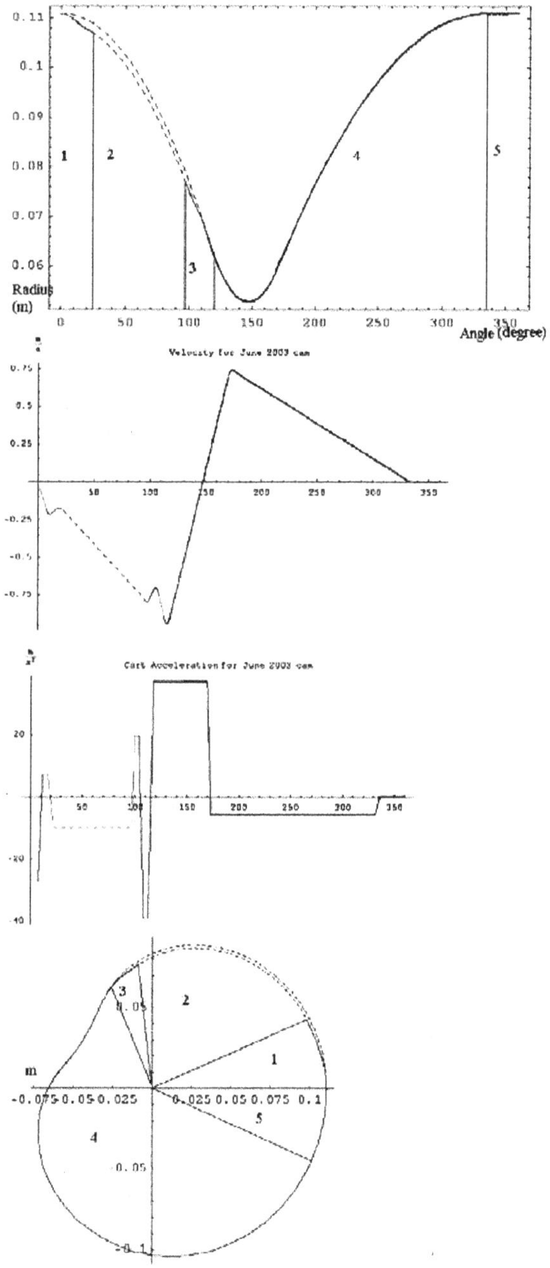


Fig. 1

## Introduction

Today's absolute instruments, though capable of achieving an accuracy of 2–3 μGal, $2\text{–}3\times10^{8}$ m/s$^2$, are still fairly bulky in comparison with relative instruments (Faller 2002). Relative instruments can be used to study various geophysical phenomena, but their use requires multiple repeat-measurements at a minimum of two sites—one of which is presumed to be stable—to determine gravity, or gravity changes, at a location of interest. A smaller and simpler absolute gravimeter, provided it has the requisite accuracy, should therefore be of considerable interest for various field studies since, being absolute, it avoids the difficult transfer problems associated with otherwise required relative measurements at a variety of sites.

One approach to a small field-usable absolute instrument is to measure g by dropping atoms and measuring their rate of fall using atom interferometry (Peters et al. 2001). Furthermore, since atoms are small, they have the advantage of producing negligible instrumental recoil when they're dropped. However, up to this point, the bulk of the attendant lasers and electronics has resulted in instruments that are not yet of an easily field-usable size. In the meantime, we have developed a smaller instrument, using more traditional technologies, that will meet both today's and tomorrow's gravity-measurement needs.

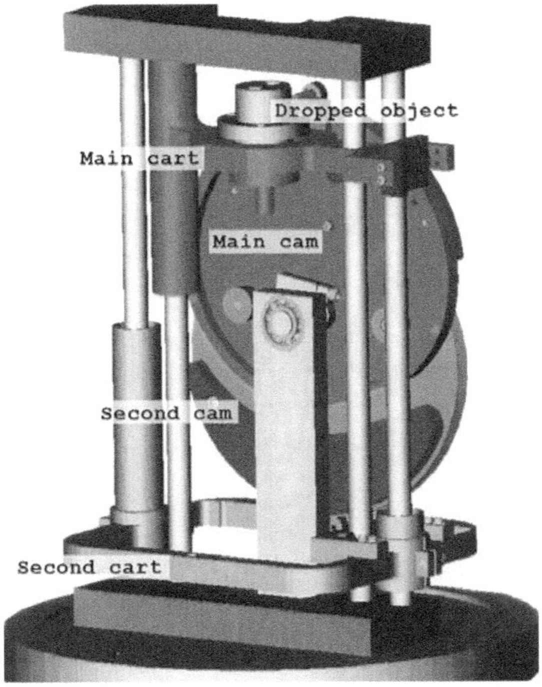


Fig. 2

## The New Gravimeter

The new instrument uses a simple cam-driven dropping mechanism (Figure 1) to create a free-fall drop every 0.3 s. Through the use of an auxiliary mass that is driven by a second co-rotating cam, the instrument is "inertially compensated," as shown in Figure 2. That is to say, the instrument—through the second-cam-driven motions of this auxiliary mass—keeps the weight of the instrument supported by the floor constant throughout the measurement cycle; thus there are no systematic (i.e., tied to the measurement process) recoil effects. This is our instrument's answer to the fact that its dropped object weighs a lot more than do the 100,000 or so atoms that would be "dropped" in an atom-interferometer-based instrument.

At the start of each drop, the (corner-cube-containing) dropped object rests kinematically on the upper cart. When this cart accelerates downwards faster than g, the dropped object, having lost its base of support, goes into free fall. The shape of this carefully designed cam is such that after release, the upper cart then maintains a (nearly) constant "lift-off" (separation between it and the dropped object) until the end of the free-fall period. At the end of the drop, this object is gently caught, slowed down, and returned to the top position, at which point the entire process repeats itself again. The release-drop-catch-lift cycle occurs in one revolution of the cam.

Both cams are machined out of titanium using a numerically controlled milling machine. Once made (requiring two days of machine time), they have proven to be very robust; over a million "drops" have been carried out with our latest cam design without any evidence of wear. Two important changes have been incorporated into this most

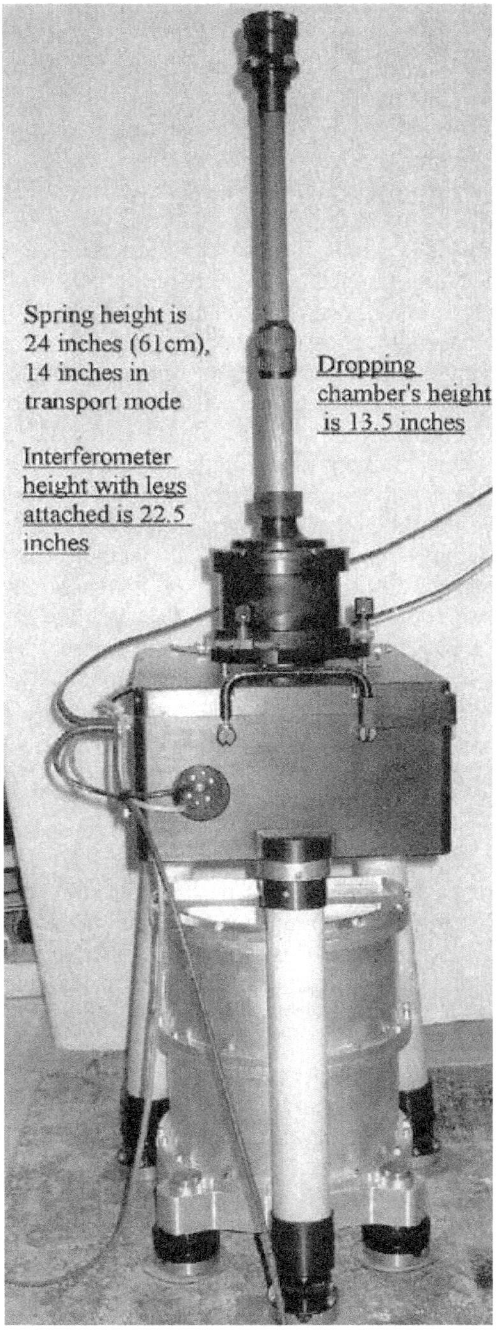


Fig. 3

recent cam design (see Figure 1). First, the cam (now) has a constant radius for the last 20+ degrees, with the result that the cart is momentarily at rest before the measurement starts; this results in considerably less rotation of the dropped object on its release. Second, the return to the top following the catch at the bottom is done more efficiently, with the result that more time is available for the actual measurement. The measurement is made during 3.0 cm of undisturbed free-fall during the 5.85 cm total drop length. One consequence of this short drop is that 200 measurements can be made every minute. Another is that the dropping chamber's height is only 34 cm and its weight is just 14 kg.

A photograph of the instrument is given in Figure 3. The vacuum chamber containing the dropped test mass is positioned under the interferometer. The vacuum is maintained at a pressure of $10^{-4}$ Pa by a miniature 4 l/sec ion pump. A passive, and purposefully chosen, 1.2-second period (simple) spring that supports the reference arm's corner cube sits directly on top of the interferometer. The spring assemblage in its working state is 61 cm high (35.5 cm in its transport mode) and its weight is 3.4 kg.

Because the instrument makes four measurements during each period of the spring, the scatter caused by the large—contributing accelerations of the order of 150 μGals—motion of the spring in its fundamental mode cancels out. [The average of any four sequential measurements is the "correct value" in the case were the spring amplitude does not change—which is nearly true as a result of its high Q (low damping) for the up-and-down motion of the supported mass.] Thus, though this spring design is very simple, it is very effective in reducing the set-to-set (typically 120 drops per set) scatter as the coherence of the fundamental motion causes this source of scatter to essentially cancel out, leaving the spring free to isolate the reference cube against other (higher-frequency) random noise sources. Needless to say, the other corner cube is completely isolated from ground vibrations while it is freely falling.

The interferometer base weighs 24.7 kg with the three legs (which are detachable for transport) contributing 4.2 kg to that total. Its height, with legs attached, is 57 cm. Contained within this base is a modified Michelson interferometer and a polarization (intensity) stabilized laser. [The laser is stabilized by keeping the intensity of the two orthogonally polarized modes equal in intensity. Lasers of this type have a long-term drift rate of the order of 2 parts in $10^9$/year in the laser frequency.] As the test mass falls, fringes occur at the output of

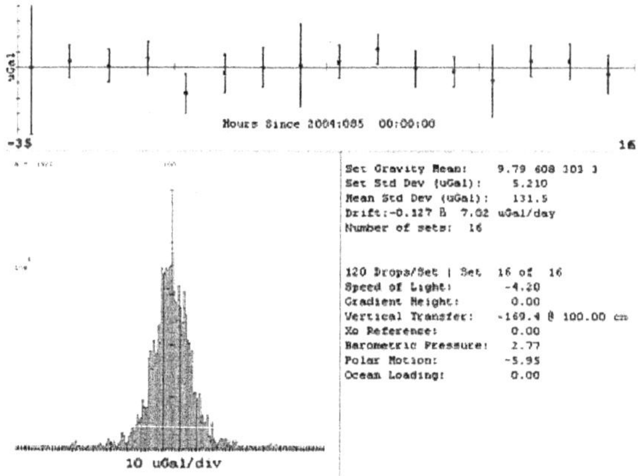


**Fig. 4**

the interferometer. The maximum fringe frequency of this short drop instrument is 2.5 MHz. An avalanche photodiode converts the interferometer's optical output to an electrical signal. A high-speed comparator generates the zero-crossing signals. The TTL (square wave) signal is then sent to the time interval counter. The internal clock is slaved to a 10 MHz rubidium frequency standard.

The signal is processed in a laptop computer that uses FG5 software compiled to suit our system. The data are least-squares fits to a function that uses the measured vertical gravity gradient. The software also makes the required corrections for the solid Earth tides, polar motion, barometric pressure, speed of light, and ocean loading as well as allowing for the transfer of the measured value to any preferred datum height.

## Results

Measurements, mostly taken in JILA's sub-basement "spec-lab" location (a good but not great site; it is 3–5 times noisier than is the Table Mountain Gravity Observatory) range from 60–300 μGals for 120 drop sets. For ten sets, the set-to-set standard deviation varies (from night to night) between 3 and 8 μGals; the number "agrees" within the measurement uncertainties and the gravity-gradient-transfer uncertainty with the value determined by an FG5 instrument. Figure 4 shows a typical overnight run in which one data set (36 seconds of data taking) is taken every hour. The large error bar measurement at the start occurred because the spring had no chance (a few minutes being required) to settle down before the run was started. The fact that its mean agrees well with subsequent sets is the proof of the pudding regarding the accuracy gain that results from our four measurements per spring period rate of sampling! The most recent—and for this location best ever—once-an-hour–through-the-night run at this site had a set-to-set variation of 3 μGals. Furthermore, "cut-off tests" show only one or two μGal sensitivity to the actual starting fringe.

**Conclusion**

At this point, this new instrument appears to be working as well as we could have hoped for. As a descendant of the JILA g (Niebauer et al. 1987) and FG5 (Niebauer et al. 1995) heritages, its design has benefited from what we learned from designing these apparatuses as well as from some new thinking. Our new instrument (perhaps it should be called the FC1) is compact and robust, allowing it to be used both as a field and a laboratory instrument of high precision and accuracy. We think of it as constituting both a portable FG5 and an absolute LaCoste & Romberg/Scintrex.

**Acknowledgements.** This work has been supported in part by NIST as a part of its precision measurement program. The authors would also like to acknowledge the superb fabrication skills of Kim Hagen, a member of JILA's instrument shop.

## References

Faller, J. E. (2002). Thirty Years of Progress in Absolute Gravimetry: A Scientific Capability Implemented by Technological Advances. *Metrologia*, 39:425-428.

Niebauer, T. M., M. P. McHugh and J. E. Faller (1987). Galilean Test for the Fifth Force. *Physical Review Letters,* 59:609-612.

Niebauer, T. M., G. S. Sasagawa, J. E. Faller, R. Hilt and F. Klopping (1995). A New Generation of Absolute Gravimeters. *Metrologia,* 32:159-180.

Peters, S., K. Y. Chung and S. Chu (2001). High-precision Gravity Measurements Using Atom Interferometry. *Metrologia,* 38:25-61.

# Absolute Gravity Measurements in Australia and Syowa Station, Antarctica

Y. Fukuda, T. Higashi, S. Takemoto, S. Iwano
Graduate school of Science, Kyoto University, Sakyo-ku, Kyoto 606-8502, Japan.

K. Doi, K. Shibuya
National Institute of Polar Research, Kaga 1-chome, Itabashi-ku, Tokyo 173-8515, Japan

Y. Hiraoka, I. Kimura
Geographical Survey Institute, 1, Kitasato, 305-0811, Japan

H. McQueen
Research School of Earth Sciences, Australian National University, Canberra ACT 0200, Australia

R. Govind
Geoscience Australia, Department of Industry, Tourism and Resources,
GPO Box 378 CANBERRA ACT 2601, Australia

**Abstract.** Absolute gravity measurements have been carried out in Perth and Canberra, Australia, and at Syowa Station, Antarctica, as a part of activities of the 45th Japanese Antarctic Research Expedition during 2003–2004.

Using a FG-5 (#210), the measurements in Perth were conducted at the Perth Observatory on November 25 and 26, 2003, and those in Canberra were conducted at Mt Stromlo Observatory and at Tidbinbilla Deep Space Tracking Station from March 26 to April 16, 2004. The gravity values obtained at the Perth observatory, Mt Stromlo and Tidbinbilla are 979403619.9 μGal, 979549870.7 μGal and 979576114.3 μGal, respectively. Using a FG-5 (#111), Geoscience Australia and Micro-g Solutions occupied the same points at Perth and Tidbinbilla in June 2003. The gravity values at Tidbinbilla agreed very well at 0.1 μGal, while the values at Perth differed by 13.4 μGal due to unknown causes.

The measurements at Syowa Station, Antarctica were carried out using two FG-5s (#210 and #203) from December 30th, 2003 to February 1st, 2004. There is a category "A" point of the International Absolute Gravity Base Station Network (IAGBN) in the gravity observation hut at Syowa Station. The gravity values obtained by #203 and #210 are 982524322.8 μGal and 982524324.5 μGal, respectively such that both agreed at 1.7 μGal. There are two previous gravity values obtained by a FG-5 in 1995 and 2001. The secular rate of the gravity changes estimated from the values is –0.27±0.42 μGal/yr. This value is consistent with an estimated rate from postglacial rebound models.

**Keywords.** absolute gravity measurements, IAGSN, GGP, superconducting gravimeter, Antarctica, postglacial rebound

## 1. Introduction

An international collaboration for precise gravity obervation, the Global Geodynamics Project (GGP) was started in 1997 (Crossley et al., 1999), and phase II of the project has been undertaken since July, 2003. One of the main goals of the GGP is to detect very weak gravity signals from the Earth's deep interior using a global network of superconducting gravimeters (SG), which extends from Ny-Alesund in the Arctic region to Syowa Station in Antarctica. An SG possesses high sensitivity and long-term stability, both of which are essential for the studies of gravity changes. However, because it is a relative gravimeter, calibration by means of an absolute gravimeter (AG) is required. Moreover, due to instrumental drift even if it is very small, SGs are unable to monitor secular gravity changes. Therefore repeated AG measurements are not only very important to control the calibration factor, but also to remove the drift of the SG and consequently, to study secular gravity changes or slow gravity variations. From these point of views, we have been conducting AG measurements in Japan (Ogasawara et al., 2001; Imanishi et al., 2002; Tamura et al., 2004), in Indonesia (Fukuda et al., 2004) and in Antarctica (Iwano et al., 2003) as well.

On the other hand, AG measurements provide nationwide fundamental basis for local and regional gravity surveys and consequently a reference for the height system of the nation as well. From this point of view, Geoscience Australia commissioned Micro-g Solutions to conduct an absolute gravity survey over Australia. The survey was undertaken in June, 2003 using FG5-#111, and measurements at 12 stations including Perth and Tidbinbilla were completed (Micro-g Solutions, 2003).

Under these circumstances, to enhance the absolute gravity station network in the southern hemisphere for the studies of sea level changes, post-glacial rebound (PGR) as well as secular gravity changes due to other reasons, and calibrate SGs at the GGP sites in Canbera and in Syowa, we have carried out AG measurements in Australia and Antarctica, as a part of activities of the 45th Japanese Antarctic Research Expedition (JARE-45) during 2003–2004. There is no doubt that we benefit from extending the AG network to the Antarctica.

In this paper, we report the AG measurements and their results, and briefly discuss the gravity changes at Syowa station from the viewpoint of Post Glacial Rebound (PGR).

## 2 Absolute Gravity Measurements

### 2.1 Absolute gravity measurements in Perth

The measurements in Perth were conducted on November 25 and 26, 2003, using the FG5-#210 of Kyoto University at the Perth observatory located approximately 25 km from the center of Perth. We obtained more than 5200 effective drops.

The data were reprocessed by means of the "g4.0, Micro-g Solutions" software including ocean-loads effects and other standard corrections. The results are summarized in Table 1. The final gravity value on the mark is 979403619.9 μGal . This value is 13.4 μGal larger than the value of 979403606.51 μGal which was obtained by Micro-G Solutions us-

Table 1. Results of the absolute gravity measurements.

| Station Name | Site Code | Lat | Lon | Height | Gravity* | SD** | Drops | dg/dz | Date | FG-5 |
|---|---|---|---|---|---|---|---|---|---|---|
| | | deg (S) | deg (E) | m | μGal | μGal | | n μGal/cm | | |
| Perth Observatory | 7999.0117 | 32.0075 | 116.1350 | 380 | 979403619.9 | 11.5 | 5298 | 3.396 | 03/11/25 -03/11/26 | #210 |
| Syowa Station | IAGBN(A) | 69.0067 | 39.5856 | 21.49 | 982524322.8 | 31.0 | 67208 | 3.34 | 03/12/28 -04/01/17 | #203 |
| Syowa Station | IAGBN(A) | 69.0067 | 39.5856 | 21.49 | 982524324.5 | 14.2 | 54775 | 3.34 | 04/01/17 -04/02/01 | #210 |
| Syowa Station | spare pier | 69.0067 | 39.5856 | 21.49 | 982524323.6 | 31.0 | 43075 | 3.34 | 04/01/17 -04/01/31 | #203 |
| Syowa Station | spare pier | 69.0067 | 39.5856 | 21.49 | 982524327.0 | 12.5 | 69226 | 3.34 | 03/12/28 -04/01/17 | #210 |
| Mt Stromlo | Seismic vault | 35.3188 | 148.9963 | 685 | 979569856.6 | 9.2 | 12146 | 2.650 | 04/03/23 -04/03/26 | #210 |
| Mt Stromlo | SG room | 35.3206 | 149.0075 | 762.51 | 979549870.7 | 12.8 | 61459 | 2.784 | 04/03/29 -04/04/15 | #210 |
| Tidbinbilla CDSCC | 9699.9404 | 35.4010 | 148.9821 | 640.81 | 979576114.3 | 12.5 | 9155 | 2.604 | 04/04/14 -04/04/16 | #210 |

* Gravity value on the marks
** Single drop standard deviations

ing FG5-#111, in June 2003. The estimated uncertainties of the measurements were 2 to 3 μGal. Thus the discrepancy is too large to be considered as measurement errors. A part of the discrepancy may be seasonal gravity changes due to groundwater table or other reasons, but until we have more information we cannot draw any firm conclusions.

## 2.2 Absolute gravity measurements at Syowa station, Antarctica

The status of AG measurements, as well as other geodetic observations at Syowa Station is well summarized in Shibuya (1993) and Shibuya et al. (2003). There is a category "A" point of the International Absolute Gravity Base Station Network (IAGBN (A); Boedecker and Fritzer, 1986) in the gravity observation hut, which was established in 1991. Since then, AG measurements have been conducted 4 times at the IAGBN (A) point. The first measurements were conducted by means of GA 60 gravimeter during JARE-33 (1991-1992) (Fujiwara et al., 1994), the second measurements by means of the second type of National Astronomical Observatory of Mizusawa gravimeter (NAOM2, Tsubokawa and Hanada, 1986) and the Absolute Gravimeter with Rotating Vacuum Pipe (AGRVP) apparatus (Hanada et al., 1987) were conducted during JARE-34 (1992-1993). After that time, FG5-#104 was employed for the measurement during JARE-36 (1994-1995) (Yamamoto, 1996) and FG-5-#203 during JARE-42 (2000-2001) (Kimura, 2002).

In parallel with this, continuous SG observation using TT70-#016 started from March 22, 1993 in the same gravity observation hut. The SG observation by TT70-#016 continued until November 2004. Then, TT70 was replaced with a new compact tidal SG, CT-#043, which has a cryocooler capable of obtaining temperatures below the vaporization point of liquid helium. This removes the need for a separate liquifier. Even though about 7 months parallel observations were conducted with the old (TT70-#016) and the new (CT-#043) SGs before the replacement, calibration of the new SG with an absolute meter. was also necessary Thus, with the aim of calibrating the SG, as well as detecting secular gravity changes by comparing the previous AG values, we carried out AG measurements during JARE-45 summer season (from December, 2003 to February. 2004).

Access to Syowa Station is only possible on board Icebreaker Shirase once a year. Shirase departs from Japan in the middle of November, calls at Fremantle, Australia, and arrives at Syowa Station in the middle of December. During Shirase's call at Fremantle, we arrived with FG5-#210 which was employed for the measurements at Perth observatory. At the same time, Shirase also loaded another FG5-#203 belonging to the Geographical Survey Institute of Japan.

We made AG measurements at Syowa Station using these two FG-5s; one occupied the IAGBN (A) point on one isolated pier and the other occupied a spare pier in the same gravity observation hut. Exchanging the occupied points to check the instrumental offset between two meters, parallel observations with CT-#043 were conducted from December 30th, 2003 to February 1st, 2004.

The results are summarized in Table 1. Because of unspecified instrumental troubles, the scatter of #203 was 2 to 3 times larger than that of #210. However the mean values agreed well to 1.7 μGal on the IAGBN (A) point and 3.4 μGal on the spare pier.

## 2.3 Absolute gravity measurements in Canberra

Shirase departed from Syowa Station in the middle of February and called at Sydney on March 20, 2004. FG5-#210 was then unloaded to carry out AG measurements in Canberra.

A variety of geodetic observations were conducted at Mt Stromlo Observatory, which is located approximately 15 km from Canberra. A superconducting gravimeter (CT-#031) observation has been hosted there since January 1997 as one of the GGP observation sites (Sato et al., 1998). There were several gravity benchmarks which had been ocupied a number of times for AG measurements (Murakami et al., 1997; Amalvict et al., 2001). However, in a very sad affair, a major bushfire attacked the site and much of the Mt Stromlo Observatory was destroyed in the fire in January 2003. The library above the gravity benchmarks was burnt out completely, while the SG (CT-#031) barely survived. Rebuilding of the SG site is planned in the near future and a change of the absolute gravity values is expected then. Thus we decided to occupy two gravity marks at Mt Stromlo observatory; one in the same building as the SG aiming a good calibration, and the other in a seismic vault about 1 km away from the SG site aiming a future reference for absolute gravity changes.

We first conducted the AG measurements in the seismic vault from March 24 to 26, and then conducted parallel measurements with CT-#031 at the SG site from March 26 to April 14. We obtained more than 60,000 drops with a single drop standard deviation of 12.8 μGal as summarized in Table 1.

There is another absolute gravity benchmark near Canberra at Tidbinbilla Deep Space Tracking Station located approximately 13km south of Mt Stromlo. Because AG measurements were conducted at that benchmark by Micro-g Solutions using FG5-#111 in June 2003, as previously described, we decided to occupy the same benchmark to confirm the instrumental reliability after its long voyage. The result is summarized in Table 1. The gravity value obtained by Micro-G Solutions was 979576114.4 μGal which agreed with our result very well at the 0.1 μGal level.

## 3 Discussion

### 3.1 Comparison of the absolute gravity values

Discrepancies sometimes occur between the values measured by different absolute gravimeters (Vitushkin et al., 2002). Thus, we must assess the reliability of the absolute measurements by comparing the obtained gravity values.

The first comparison is at the Perth observatory. There is a discrepancy of about 13 μGal between the value obtained by Micro-G Solutions and ours. However, since the measurements were conducted in different seasons, i.e., those of Micro-G Solutions in June while ours in November, the values could not be compared directly. There might be seasonal gravity changes due to groundwater variations (e.g., van Dam et al., 2001), and we need reoccupation of the point to clarify this question.

The comparison of the values in Syowa Station is more straightforward. We conducted the measurements by FG5-#203 and #210 side by side, and the occupied points were exchanged to make doubly sure. The values on the IAGBN (A) point agreed at 1.7 μGal, a value which is within the specification of FG-5 (±2 μGal). The discrepancy on the spare pier was 3.4 μGal. This value is slightly larger but it is still within the specification WRT the mean value.

The gravity values obtained by FG5-#210 (this study) and FG5-#111 (Micro-G Solutions) at Tidbinbilla agreed very well at 0.1 μGal. There might be seasonal gravity changes at the site because it is not on bedrock, but we consider that such seasonal gravity changes would be small because the measurements were conducted almost in the same season, i.e., in June and in April.

Just after FG5-#210 went back to Japan, it was sent to Micro-G Solutions for a regular overhaul in June, 2004. We made a special request to conduct a pre- and post-service run of it to verify whether the instrument gave proper gravity values or not. FG5-#210 was taken to Table Mountain Gravity Observatory (TMGO), and it was compared with FG5-#111 on the same pier. The result is summarized in Table 2 and shows that FG5-#210 was determined to be operating within the specification prior to service.

### 3.2 Gravity changes at Syowa Station

So far, AG measurements have been conducted 4 times at Syowa Station since 1991. The gravity values obtained by the first and the second measurements might have a few tens μGals of uncertainties, and they were less reliable than the recent FG-5

Table 2. Comparison results of pre-and post- service verification run of FG5-#210 performed at TMGO by Micro-G solutions.

| Instrument | g-value | Uncertainty | Precision | Sets |
|---|---|---|---|---|
| | μGal | μGal | μGal | |
| FG5-#111 | 979622836.1 | 2.10 | 0.40 | 24 |
| FG5-#210 preservice | 979622835.2 | 2.02 | 0.23 | 24 |
| FG5-#210 postservice | 979622836.9 | 2.07 | 0.20 | 24 |

Table 3. Absolute gravity values at Syowa Station.

| JARE | g-value | date | nstrument |
|---|---|---|---|
| | μGal | | FG5- |
| JARE-36 | 982524326.9 | 1995/01/20 -1995/02/11 | #104 |
| JARE-42 | 982524328.2 | 2000/12/29 -2001/01/25 | #203 |
| JARE-45 | 982524323.7 | 2003/12/28 -2004/02/01 | #203 #210 |

measurements. Therefore we will discuss the secular gravity changes using the gravity values after JARE-36.

The gravity values are summarized in Table 3, and Fig. 1 shows the gravity changes. The estimated rate of the gravity change is –0.27±0.42 μGal /yr.

In Antarctica, several model calculations were undertaken to estimate crustal motion and secular gravity changes produced by PGR (James and Ivins, 1998; Nakada *et al.*, 2000). Typical model estimation of the crustal uplift at Syowa Station is less than or about 2mm/yr, and the estimated gravity decrease is 0.2–0.5 μGal/yr. While recent results of VLBI experiments (Fukuzaki et al, 2004) give a uplift velocity of 1.94 ± 1.59 mm/yr at Syowa Station, the rate of gravity decrease observed by this study does not contradict the values provided by the models, though the observed value still has large uncertainty and is not yet significant.

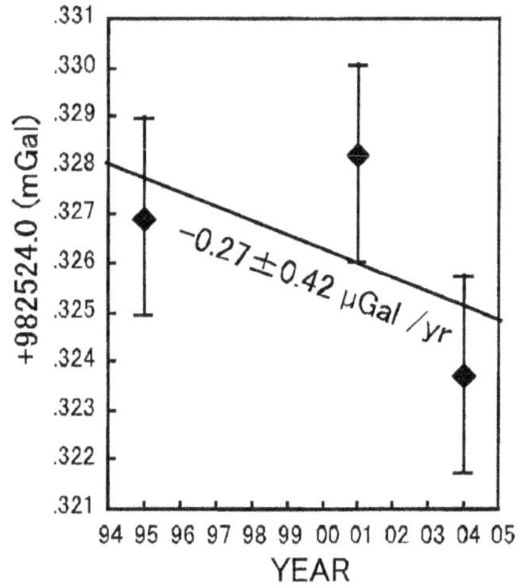

Fig. 1. Gravity change rate at Syowa Station.

## 4. Concluding remarks

We successfully conducted AG measurements in Perth and Canberra, Australia, and at Syowa Station, Antarctica. The result in Syowa Station suggests a gravity decrease due to PGR. While the uplift rate estimated by VLBI measurements may be affected by the baseline change rate between the VLBI stations, AG measurements can literally detect absolute gravity changes or absolute vertical crustal movements by isolated measurements at the site. Thus absolute gravimeter measurements are indispensable for these studies. The estimated uplift and/or gravity decrease at Syowa Station are not so large compared with those in West Antarctica. But both are detectable by AG measurements as well as present space geodetic techniques. Moreover, combining AG measurements and other geodetic techniques provides information on mass redistribution. Therefore, repeated measurements in future will continue to provide valuable information on rates of deformation in response to ice retreat.


**Acknowledgments**

We are deeply indebted to the Perth observatory, the Mt Stromlo Observatory and Tidbinbilla Deep Space Tracking Station for the AG measurements. We thank to all the JARE-45 members led by Prof. H. Kanda for their kind support. We also thank to Dr. H. Ikeda of JARE-44 for his assistance during the absolute measurements at Syowa Station. This work was partially supported by the Grant-in-Aid for Scientific Research from the Ministry of Education, Culture, Sports, Science and Technology (No. 14340132)


## References


Amalvict, M., McQueen, H. and Govind, R. (2001): Absolute gravity measurements and calibration of SG-CT031 at Canberra, 1999-2000, J. Geod. Soc. Jpn., **47**, 410-416.

Boedecker, G. and Fritzer, T. (1986): International Absolute Gravity Basestation Network, Status Report 1986. International Association of Geodesy Special Study Group 3.87, Veroffentlichungen der Bayerischen Kommission fur die Internationale Erdmessung der Bayerischen Akademie der Wissenschaften, Heft Nr. **47**, 68pp.

Crossley, D., Hinderer J., Casula, G., Francis, O., Hsu, H.T., Imanishi, Y., Jentzsch, G., Kaarianen, J., Merriam, J., Meurers, B, Neumeyer, J., Richter, B., Shibuya, K., Sato, T. and van Dam, T. (1999): Network of Superconducting Gravimeters, Benefits a Number of Disciplines, Eos Trans. AGU, 80, no. 11, 121, 125-126.

Fujiwara, S., Watanabe, K. and Fukuda, Y. (1994): Measurement of absolute gravity at Syowa Sataion, Antarctica, Bull. Geogr. Surv. Inst., **40**, 1-6.

Fukuda, Y., Higashi, T., Takemoto, S., Abe, M., Dwipa, S., Kusuma, D. S., Andan, A., Doi, K., Imanish, Y. and Arduino, G. (2004): First absolute gravity measurements in Indonesia, J. Geodynamics, in press.

Fukuzaki, Y., Shibuya, K. Doi and Jike, T. (2004): The latest result of Antarctic VLBI experiments carried out in

Syowa Station, 102nd Meeting of the Geodetic Society of Japan, Abstract, 15-16 (in Japanese)

Imanishi Y., Higashi, T. and Fukuda Y. (2002): Calibration of the superconducting gravimeter T011 by parallel observation with the absolute gravimeter FG5 #210 - a Baysican approch, Geophys, J. Int., **151**, 867-878.

Iwano, S., Kimura, I. and Fukuda, Y. (2003): Calibration of the superconducting gravimeter TT70 #016 at Syowa Station by parallel observation with the absolute gravimeter FG5 #203. Polar Geoscil, **16**, 22-28.29-52.

James, T.S. and Ivins, E.R. (1998): Predictions of Antarctic crustal motions driven by present-day ice sheet evolution and by isostatic memory of the Last Glacial Maximum, J. Geophys. Res., **103**, 4993-5017.

Kimura, I. (1996): Gravity measurements with the portable absolute gravimeter FG5 at Antarctica (II), Bull. Geogr. Surv. Inst., **97**, 17-23 (in Japanese).

Micro-g Solutions (2003): Report for Australian Regional Absolute Gravity Survey 2003, Unpublished Report for Geosience Australia, 8p.

Murakami, Msk., Murakami, Mkt., Nitta, K., Yamaguchi, H., Yamamoto H., Karasawa, M., Nakahori, Y., Doi, K., Murphy., B., Govind, R., Morse, M. and Gladwin, M. (1997): Absolute determination of gravity for the purpose of establishment of precise reference frame for mean sea level change monitoring in the Southwestern Pacific, IAG Symposia, **117**, 32-39.

Nakada, M., Kimura R, Okuno J, Moriwaki K, Miura H and Maemoku H (2000): Late Pleistocene and Holocene melting history of the Antarctic ice sheet derived from sea-level variations Marine Geology **167**, 85-103

Ogasawara, S., Higashi, T., Fukuda, Y. and Takemoto, S. (2001): Calibration of a Superconducting Gravimeter with an Absolute Gravimeter FG-5 in Kyoto, J. Geod. Soc. Japan, **47**, 404-409.

Sato, T., McQueen, H., Murphy, B., Lambeck, K., Hamano, Y., Asari, K., Tamura, Y. and Ooe, M. (1998): On the observations of gravity tides and Earth's free oscillations with a superconducting gravimeter CT#031 at Mt. Stromlo, Australia., Proc. 13th Int. Symp. on Earth Tides, Brussels, 583.

Shibuya, K. (1993): Syowa Station; observatory for global geodesy in Antarctica (a review), Proc. NIPR Symp. Antarct. Geosci., **6**, 26-36.

Shibuya, K., Doi, K. and Aoki, S. (2003): Ten year's progress of Syowa Station, Antarctica, as a global geodesy netwark site. Polar Geoscil, **16**, 29-52.

Tamura, Y., Sato, T., Fukuda, Y. and Higashi, T. (2004): Calibration of the scale factor of a superconducting gravimeter at Esashi Station, Japan by means of absolute measurements, Submitted to J. Geodesy.

Tsubowaka, T. and Hanada, H. (1986): The ILOM transportable absoute gravimeter, a description of the instrument and results of the measurements in Tohoku District. Proc. Int. Lat. Obs. Mizusawa, **25**, 17-64 (in Japanese)

van Dam, T. M., Wahr, J.M., Milly, P. C.D. and Francis, O. (2001): Gravity changes due to continental water storage, J. Geod. Soc. Japan, **47**, 249-254.

Vitushkin L., Becker M., Jiang Z., Francis O., van Dam T.M., Faller J., Chartier J.-M., Amalvict M., Bonvalot S., Debeglia N., Desogus S., Diament M., Dupont F., Falk R., Gabalda G., Gagnon C. G. L., Gattacceca T., Germak A., Hinderer J., Jamet O., Jeffries G., Ka"ker R., Kopaev A., Miard J., Lindau A., Longuevergne L., Luck B., Maderal E.N., Ma"kinen J., Meurers B., Mizushima S., Mrlina J., Newell D., Origlia C., Pujol E.R., Reinhold A., Richard Ph., Robinson I.A., Ruess D., Thies S., Van Camp M., Van Ruymbeke M., de Villalta Compagni M.F., and Williams S. (2002) : Results of the Sixth International Comparison of Absolute Gravimeters, ICAG-2001, Metrologia, **39**, 407-424.

Yamamoto, H. (1996): Gravity measurements with the portable absolute gravimeter FG5 at Antarctica, Bull. Geogr. Surv. Inst., **85**, 18-22 (in Japanese).

# Unified European Gravity Reference Network 2002 (UEGN02) – Status 2004

G. Boedecker
Bavarian Academy of Sciences and Humanities / BEK, München, Germany. Fax +49 8923031100
boe@bek.badw-muenchen.de

O. Francis .
European Centre for Geodynamics and Seismology / University of Luxembourg,Luxembourg, Fax +352 33 14 87 88, olivier@ecgs.lu

A. Kenyeres
FOMI Satellite Geodetic Observatory, Penc, Hungary; Fax +36 27 374982, kenyeres@sgo.fomi.hu

**Abstract.** The last previous realisation of the Unified European Gravity Reference Network was completed in 1994 (UEGN94) covering 11 West European countries. Since that time significant work has been done e.g. in the frame of the UNIGRACE project that focused on the establishment of absolute gravity stations in Central and East European countries.

The past European Subcommission of the IGGC recommended to continue with the unification of the gravity reference networks at the continental level. During the IAG2001 Scientific Assembly in Budapest, it was decided to establish a new continental gravimetric reference network.

The total UEGN2002 network area covers 25 countries, the number of stations expected was around 1000, but has already exeeded this number. The data processing starts from raw absolute and relative gravity observations. The tidal corrections are computed by the ECGS in Luxembourg in a uniform manner and based on state-of-the-art models.

At the current status (2004) the work is focused on 12 countries with 404 absolute and 33000 relative gravity observations at some 1500 stations. Main problems are data bugs. Hence, appropriate tools had to be developed.

The paper presents details of the adjustment, earth tide corrections as well as analyses of the data and adjusted gravity values. After this partial UEGN02 adjustment, the remaining countries that also showed their interest to participate are again invited to submit data.

**Keywords.** Reference network, gravity reference

## 1 Introduction

### 1.1 Role of Gravity Reference Networks

Because bulk field gravimetry still is carried out by relative spring gravimeters, it is necessary to link these to an absolute reference. Decades ago, only a few absolute observations worldwide provided the basis for extended relative gravity reference networks. Today, as absolute gravity meters become more easily available, they become more fieldworthy and portable, also their number is increasing. Consequently, the number of absolute stations is increasing and the contribution of relative meters to reference networks is decreasing. In each case where a new national reference or an extended field campaign is envisaged, the benefit-cost-relation will lead to a new optimum design appropriate to the specific situation, see e.g. Boedecker (2002). Relative observations are not only important for reference networks, their use in combined absolute / relative networks will also benefit in their calibration.

### 1.2 European cooperation / Preceding works

After the old global gravity reference network IGSN71 (Morelli et al. 1971) proved to be not a sufficiently accurate basis for new gravity observations any more, many countries designed and observed national networks. In 1994, an attempt was made on behalf of the IGC Subcommission Western Europe to unify a number of European national networks (Boedecker, Marson, Wenzel 1994) and named UEGN94 ('Unified European Gravity Networks 1994'). The network covered 11 countries comprising 499 stations with 123 absolute and 14532 relative gravity observations; see fig. 1.

From 1998 to 2002 the project UNIGRACE ('Unification of Gravity Systems in Central and Eastern Europe') was carried out, coordinated by the German BKG ('Bundesamt für Kartographie und Geodäsie') with contractors from all participating countries, see fig. 2. See Reinhart, Richter, Wilmes 1998. In the framework of that project, absolute observations at 19 stations in 12 countries in central and eastern Europe have been observed and were made available to the national survey agencies and to the UEGN project.

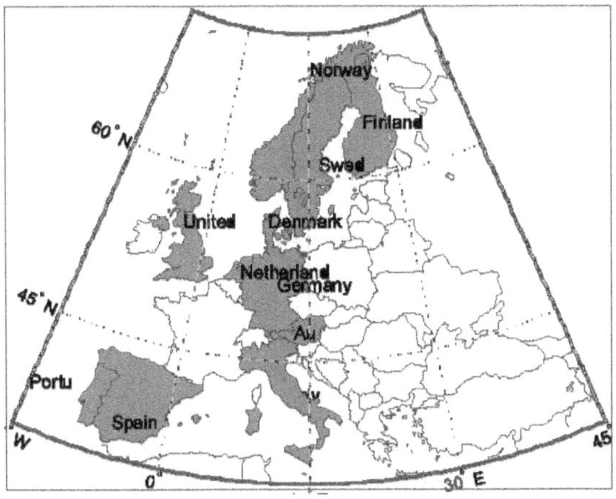


**Figure 1: UEGN94**

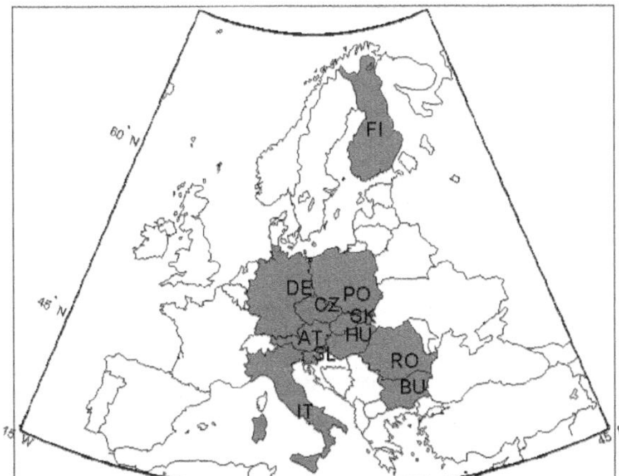


**Figure 2: UNIGRACE**

These works for a unification of European gravity reference networks were continued by the UEGN02. A first meeting to agree on the guidelines took place on May 13-14, 2001, in Vienna.

The work was delayed because of problems with details of the ties between absolute and relative observation – a key problem for the correct gravity level of the network.

Some further delay is caused by incorrect data: As a matter of fact, this is a bigger problem than in other much larger data sets, because in this case we are dealing with (currently) 33000 observations each of which has its own evolution and is not the result of one unique process as is the case in other type of geodetc observations like GPS observation series or similar. For this reason, some software tools had to be developed in order to make the observation series and procedures more transparent.

The current status of the UEGN02 activities is illustrated in figure 3:

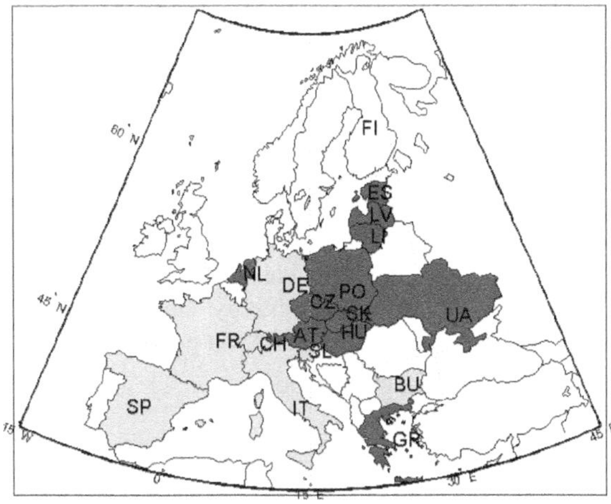


**Figure 3: UEGN02:** Activity status 08/04
dark: adjustment computation, data cleaning
medium: some data available
light: data envisaged

## 2 Data collection

The data were collected in four types of formatted files: station data, absolute observations, relative observations. Some national agencies do not agree to have their data published. For this reason, no original data set will be transferred without permission of the original owner.

## 3 Tidal reduction

All the raw observations of the relative gravity meters have been corrected for the body Earth tides and ocean loading and attraction effect using stateofthe art models. The tidal parameters for the body Earth tides come from the Dehant-Defraigne-Wahr (1999) model for an inelastic non-hydrostatic Earth (Dehant et al., 1999) including a delta factor of 1.16 for the long periods tides. The body tides prediction is computed with the etgtab software written by Wenzel (http://www-geod.kugi.kyoto-u.ac.jp/iagetc/etcdat/etgtab/etgtab.txt

On the other hand, the ocean loading and attraction parameters were calculated using the ocean tides model CSR3.0 and the Green's function of the PREM with the Load89 software (Francis O. and P. Mazzega, 1993). The oceanic loading tidal prediction is then calculated using the ocean loading parameters in a separate program. The tidal correction is the sum of the contributions: body Earth tides and oceanic loading and attraction gravity effect.

The data from the absolute gravimeters were provided already corrected for tides. It means that the tides correction for these data could be slightly different. It was outside the scope of this work to reprocess all the absolute gravity data from the raw observations. ).

## 4 Adjustment model

The adjustment model is identical to the adjustment of UEGN94. Absolute gravity observations are immediate observations of station gravity.

Absolute gravity observations :

$$v_i = g_j - \tilde{g}_i$$

$v_i$ residual of observation i

$g_j$ unknown station gravity

$\tilde{g}_i$ observed gravity incl. tidal reduction

Relative gravity meter obervations are taken as a series of consecutive observations within one day by one gravity meter. A series is modeled by station gravity values, an offset (per series), linear drift of the readings with time (per series) and a linear calibration factor for the entire set of observations of the meter. Relative observations are the raw observations – corrected for e.g.periodic errors, if necessary or available, converted by the manufacturers calibration table to care for nonlinear components, and reduced by tidal effects (see below).

Relative gravity observation :

$$v_i = g_j + o_l + z_i f_g + t_i d_l - r_i$$

$v_i$ residual of observation i

$g_j$ unknown gravity of station j

$o_l$ orientation of series l

$z_i$ raw gravimeter reading i

$f_g$ linear calibration factor of gravimeter g

$t_i$ time of observation i

$d_l$ linear drift for series l

$r_i$ gravimeter reading i incl. tides & pre - calib.

The intention was to use the original observations at definition height of the absolute observation and to connect to all other stations including the immediate ground marker by relative observations. This would have avoided introducing prior 'gradient corrections' etc. However, frequently the original relative gravity observations were not carried out at the height of the absolute observations; hence, the ties between absolute observations and the network is a mix of immediate relative observations and intermediate use of gradients.

To account for the variations in accuracy of the absolute observations, the internal error variances estimations were introduced, as communicated by the agencies providing the data.

The weights of the relative observations were assumed constant within one series and the series weights were estimated from an iterative variance component estimation.

Initial weights :

$$p_l = \frac{\sigma_0^2}{\sigma_l^2}$$

$\sigma_0^2$ zero variance

$\sigma_l^2$ variance of series l

Weight iteration :

$$p_l = \frac{\sigma_0^2}{\frac{[vv]_l}{n_l - 2} + \varepsilon}$$

$\sigma_0^2$ zero variance

$[vv]_l$ squared sum of residuals of series l

$n_l$ number of obsns. of series l

$\varepsilon$ small number

## 5 Execution of adjustment

The adjustment programme was developed in MATLAB. The challenges of this work are in the data organization etc. E.g., relative observation series have to be identified from the contiguous observation set, station data checked; series of only two observations do not provide network information and are removed.

The data were processed to a certain point, then metadatafiles were transferred to ECS Luxembourg (O. Francis) for tidal reduction. After this, the processing continued. Variants of the adjustment are based on the same data set, hence the tidal

reduction computation was necessary only once so far.
Because of the restrictions of this paper format, it is not possible to provide a full and detailed image of the screen for the practical work on the network. Rather, a sequence of cutout figures tries to provide at least a rough impression, see appendix.
The network figures (see appendix) show e.g. normal stations, absolute stations, stations with observations by parties from other countries, station names, station gravity standard deviation estimates, absolute and relative residuals above some threshold (arrows), observation series with residuals above some threshold (red dotted line).
The graphical user interface is not just a nice toy, but it is essential because otherwise a clarification of bugs in the large data set of individual observations would not be possible.
A standard programme run including all original data file input and output, graphics, parameter estimation for about 10.000 parameters, iterative covariance component estimates in six iterations, but without tidal reduction and inversion of the normals takes less than 1 minute on a normal PC under Windows.
The inversion for computing the standard deviations of the parameters takes between half a minute to several minutes depending on the quality of the normal matrix.

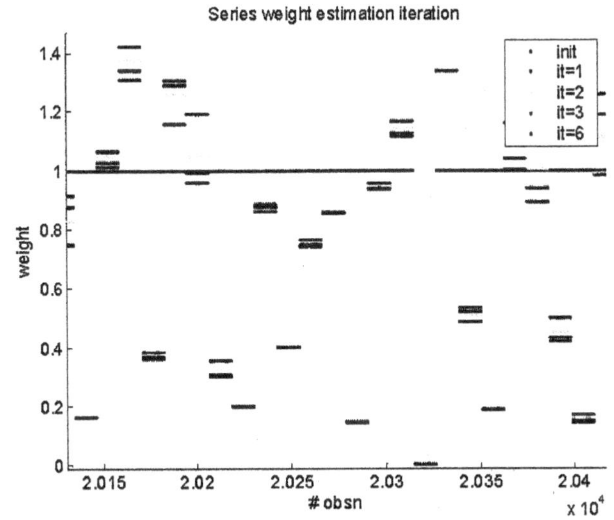


**Figure 4: Iterative weight estimation**
Further explanation see text

The weight estimation for relative observations starts with unit weight (black in the sample figure 4), except the series is downweighted to zero as a result of a preceding bug analysis (see figure). The iteration subsequently yields the weights depicted above.
As can be recognized, the convergence is quite stable and fast.
The key numbers of the sample run are given below. The histogram of residuals shows that 50% of the residuals are less than 0.007 mGal.

| | |
|---|---|
| Absolute obsns: | 405 |
| Relative obsns: | 33092 |
| Rel. gravimeters | 44 |
| Series | 3989 |
| Stations: | 1446 |
| Parameters: | 9445 |

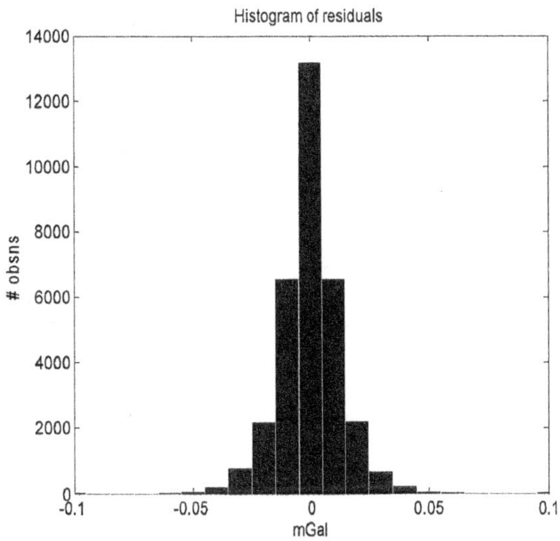


**Figure 5: Histogram of residuals**

## 6 Conclusions and outlook

- A subnet of the UEGN02 has successfully been processed
- The national network structures are very diverse
- Observation data bugs are a major problem
- This project completes the UNIGRACE objectives
- The accuracy is satisfactory
- The benefits are very manifold
- All European countries are called to participate
- Some of the countries contributing already are called to deliberate to submit more complete data in order to have greater benefit

- It is envisaged to complete the project in 2005
- The further procedure, e.g. publication (also stations ?) has to be agreed by the participating countries

## 7 Acknowledgements

This work has been made possible through the contributions of many colleagues in institutes and agencies in the participating countries, Austria, Czech Republic, Croatia, Estonia, Finland, France, Germany, Greece, Hungary, Italy, Latvia, Lithuania, Netherlands, Poland, Slovakia, Slovenia, Spain, Switzerland, Ukraine. Others offered there active participation in computations such as Völgyesi and Csapo from Hungary, Simek from Czech Republic and Klobusiak from Slovakia; these offers are appreciated.

## References

Boedecker, G.: World gravity standards – present status and future challenges. Metrologia 2002, 39 no.5, 429-434

Boedecker, G.; Marson, I.; Wenzel, H.-G. : The Adjustment of the Unified European Gravity Network 1994 (UEGN 94). In: Sünkel, H. / Marson, I. (eds.): Gravity and Geoid. Joint IGC/ICG Symposium Graz 1994. IAG Symposium No. 113, Springer 1995, 82-91

Dehant, V., P. Defraigne and J. M. Wahr, Tides for a convective Erath, Journal of Geophysical Research, 104, 1035-1058, 1999.

Francis O. and Mazzega P., Global charts of ocean tide loading effects, Journal of Geophysical Research, 95(C7), 11411-11424, 1990.

Morelli, C., Gantor, C., Honkasalo, T., McConnell, R.K., Tanner, J.G., Szabo, B., Votila, V. & Whalen, C.T.-1971- The International Gravity Standardisation Net. Pub. Spec. no. 4 du Bulletin Géodésique., 1974

Reinhart, E.; Richter, B.; Wilmes, H.: "UNIGRACE – Ein Projekt zur Vereinheitlichung der Schwere-Referenzsysteme in Zentral-Europa" Mitteilungen des Bundesamtes für Kartographie und Geodäsie, Band1, Frankfurt a. M. 1998;
see also:
INCO: International Scientific Cooperation Projects (1998-2002) Contract Number: ERBIC15CT970805 FINAL REPORT

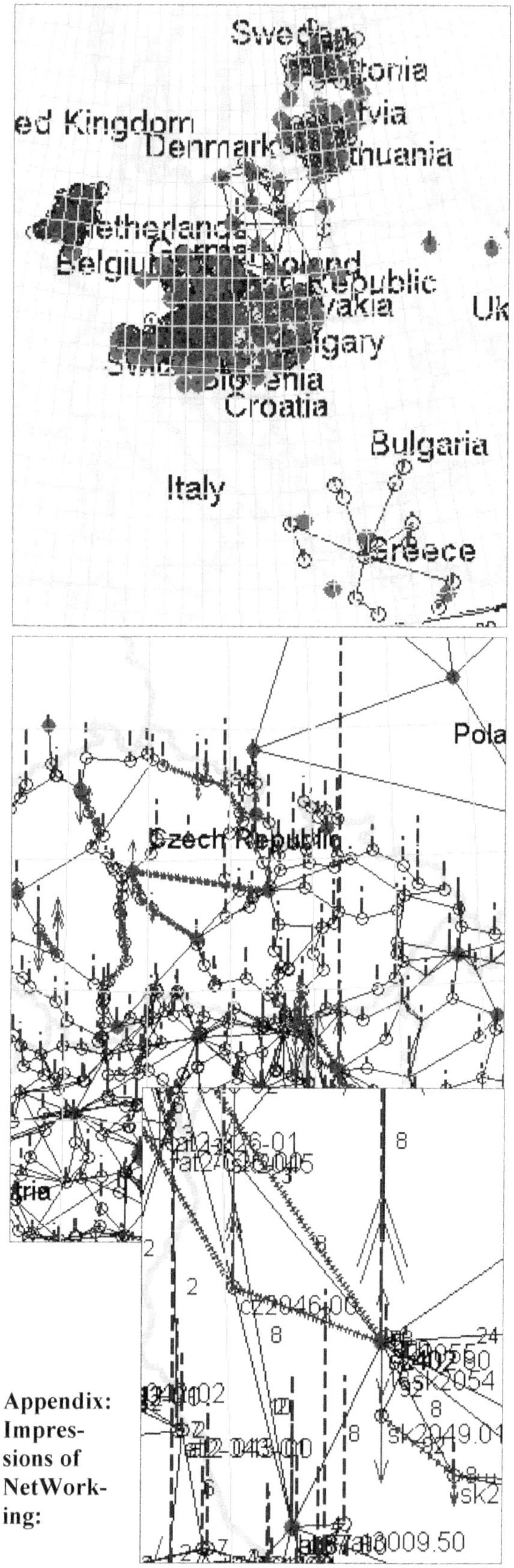


**Appendix: Impressions of NetWorking:**

**Appendix:**
**Sample output**

```
UEGN results
2004  8  19  19  12
 Stations files read:
D:\UEGN02\at\stations_at_03.txt4
D:\UEGN02\cz\stations_cz_02.txt4
...
Absolute observations files read:
D:\UEGN02\at\absobs_at_03.txt  5
D:\UEGN02\cz\absobs_cz_02.txt  5

Relative observations files read:
D:\UEGN02\at\relobs_at_4b.txt  5
D:\UEGN02\cz\relobs_cz_2a.txt  5

Stations with identical coordinates 3D :
at0-059-20 atA-059-20
cz107.10   hu107.10
cz107.10   sk107.10
cz108.00   hu108
...
 Parameters:
names (seq: grav, lincal, orient, drift), adjusted values, mse:
at0-021-00     980867.380     0.002
at0-021-01     980868.928     0.003
at0-050-00     980741.267     0.002
at0-058-20     980842.055     0.003
  180.041        -0.089      0.000
  191.001         6.083      0.001
  192.001        -3.727      0.001
   192.05         2.749     0.006
  193.001       -23.301      0.001

...
```

```
 Parameters not solved for:
 Station Gravity:
794 gr299RAFI
797 gr309KYMH
{
 Observations per
instrument: # instr, # obsns
   1  -900.002      1008
   2  -625.001       110
   3   -90.002       178
   4     3.004      302
   5     4.001      26
   6    51.002      1934
...
Observations per stations: / # rel obsns, # abs obsns
at0-021-00  65  0
at0-021-01  17  0
hu87        7  0
hu88        0  2
hu88.0      54  2
hu89        0  1
...
Absolute outliers:
hu88     930826  980765.513000   0.048 AXIS-FG5-107 0.002 Gschwind
hu88     971213  980765.610000  -0.049 JILAG-6 0.002 Ruess

 Bad series:
  Bad series:
 mse, maxr:  0.057230 0.060223
 (StatName, date, time, instr, raw, obs(inc cal/tid), tides, resid, status)
at1I-SCHLb 010929 1837  -900.002    119.817    134.397  -0.056  +0.031 0
at2-171-20 010929 1946  -900.002    17.798     19.951  -0.032  -0.013 0
at2-171-21 010929 2012  -900.002    42.309     47.471  -0.023  -0.060 0
at1I-SCHLb 010929 2114  -900.002    119.777    134.402  -0.006  +0.042 0
```

# Determination of gravity anomalies from torsion balance measurements

L. Völgyesi, G. Tóth
Department of Geodesy and Surveying, Budapest University of Technology and Economics, H-1521 Budapest, Hungary, Müegyetem rkp. 3.
G. Csapó
Eötvös Loránd Geophysical Institute of Hungary, H-1145 Budapest, Hungary, Kolumbusz u. 17-23.

**Abstract.** There is a dense network of torsion balance stations in Hungary, covering an area of about $40000\,km^2$. These measurements are a very useful source to study the short wavelength features of the local gravity field, especially below 30 km wavelength. Our aim is thus to use these existing torsion balance data in combination with gravity anomalies. Therefore a method was developed, based on integration of horizontal gravity gradients over finite elements, to predict gravity anomaly differences at all points of the torsion balance network. Test computations were performed in a Hungarian area extending over about $800\,km^2$. There were 248 torsion balance stations and 30 points among them where Δg values were known from measurements in this test area.

**Keywords.** Gravity anomalies, torsion balance measurements.

## 1 The proposed method

Let's start from the fundamental equation of physical geodesy:

$$\Delta g = g - \gamma = -\frac{\partial T}{\partial r} - \frac{2T}{R},$$

where $T$ is the potential disturbance and $R$ is the mean radius of the Earth (Heiskanen, Moritz 1967). Changing of gravity anomaly $\Delta g$ between two arbitrary points $P_1$ and $P_2$ is:

$$(\Delta g_2 - \Delta g_1) = -\left[\left(\frac{\partial T}{\partial r}\right)_2 - \left(\frac{\partial T}{\partial r}\right)_1\right] - \frac{2}{R}(T_2 - T_1)\,.$$

In a special coordinate system ($x$ points to North, $y$ to East and $z$ to Down) the changing of gravity anomaly:

$$(\Delta g_2 - \Delta g_1) = \left[\left(\frac{\partial T}{\partial z}\right)_2 - \left(\frac{\partial T}{\partial z}\right)_1\right] - \frac{2}{R}(T_2 - T_1)\,.$$

Let's estimate the order of magnitude of term $(2/R)(T_2 - T_1)$ which is:

$$\frac{2}{R}(T_2 - T_1) = \frac{2\gamma}{R}\Delta N_{12}\,, \qquad (1)$$

where $\Delta N_{12}$ is the changing of geoid undulation between the two points. If the changing of geoid undulation between two points is 1 m, than the value of (1) is 0.3 mGal (1 mGal = $10^{-5}$m/s$^2$). Taking into account the average distance between the torsion balance stations and supposing not more than dm order of geoid undulation's changing, the value of (1) can be negligible.

Applying the notation $T_z = \partial T / \partial z$ for the partial derivatives, the changing of gravity anomalies between the two points $P_1$ and $P_2$ is:

$$(\Delta g_2 - \Delta g_1) = (T_z)_2 - (T_z)_1\,.$$

So in the case of displacement vector $d\mathbf{r}$ the elementary change of gravity anomaly $\Delta g$ will be:

$$d\Delta g = \nabla(\Delta g)\cdot d\mathbf{r} = \frac{\partial \Delta g}{\partial x}dx + \frac{\partial \Delta g}{\partial y}dy + \frac{\partial \Delta g}{\partial z}dz = T_{zx}dx + T_{zy}dy + T_{zz}dz\,.$$

Integrating this equation between points $P_1$ and $P_2$ we get the changing of gravity anomaly:

$$(\Delta g_2 - \Delta g_1) = \int_1^2 d\Delta g = \int_1^2 T_{zx}dx + \int_1^2 T_{zy}dy + \int_1^2 T_{zz}dz\,, \qquad (2)$$

where $T_{zx} = W_{zx} - U_{zx}$ , $T_{zy} = W_{zy} - U_{zy}$ and $T_{zz} = W_{zz} - U_{zz}$ ; $W_{zx}$ and $W_{zy}$ are horizontal gradients of gravity measured by torsion balance, $W_{zz}$ is the measured vertical gradient, $U_{zx}$ and $U_{zy}$ are the normal value of horizontal gravity gradients, and $U_{zz}$ is the normal value of vertical gradient. According to Torge (1989):

$$U_{zx}=\frac{\partial\gamma}{\partial x}=\frac{\gamma_e\beta}{M}\sin 2\varphi\ ,\quad U_{zy}=0\ ,$$
$$U_{zz}=\gamma\left(\frac{1}{M}+\frac{1}{N}\right)+2\omega^2$$

where $M$ and $N$ is the curvature radius in the meridian and in the prime vertical, $\gamma=\gamma_e(1+\beta\sin^2\varphi)$ is the normal gravity on the ellipsoid. With the values of the Geodetic Reference System 1980, the following holds at the surface of the ellipsoid:

$$U_{zx}=8.1\sin 2\varphi\ ns^{-2}$$
$$U_{zz}=3086\ ns^{-2}\ .$$

Let's compute the first integral on the right side of equation (2) between the points $P_1$ and $P_2$. Before the integration a relocation to a new coordinate system is necessary; the connection between the coordinate systems (x,y) and the new one (u,v) can be seen on Figure 1. Denote the direction between the points $P_1$ and $P_2$ with $u$ and be the coordinate axis $v$ perpendicular to $u$. Denote the azimuth of $u$ with $\alpha_{12}$ and point the $z$ axis to down, perpendicularly to the plane of (xy) and (uv)!

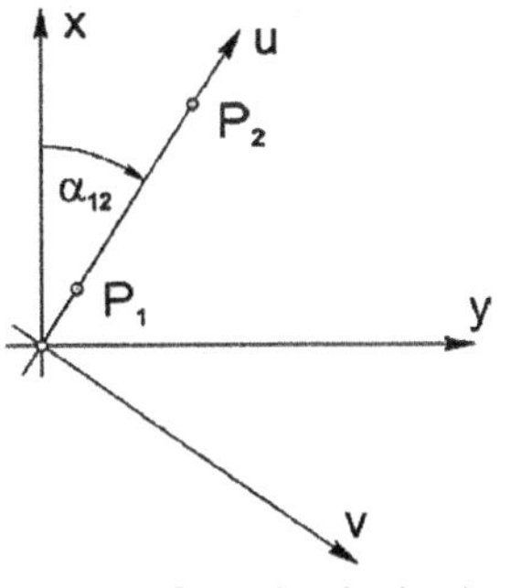


**Fig. 1** Coordinate transformation (x,y)→(u,v)

The transformation between the two systems is:

$$\left.\begin{aligned}x&=u\cos\alpha_{12}-v\sin\alpha_{12}\\ y&=u\sin\alpha_{12}+v\cos\alpha_{12}\end{aligned}\right\}\ .$$

Using these equations, the first derivatives of any function $W$ are:

$$\frac{\partial W}{\partial u}=\frac{\partial W}{\partial x}\frac{\partial x}{\partial u}+\frac{\partial W}{\partial y}\frac{\partial y}{\partial u}=\frac{\partial W}{\partial x}\cos\alpha_{12}+\frac{\partial W}{\partial y}\sin\alpha_{12}$$

$$\frac{\partial W}{\partial v}=\frac{\partial W}{\partial x}\frac{\partial x}{\partial v}+\frac{\partial W}{\partial y}\frac{\partial y}{\partial v}=-\frac{\partial W}{\partial x}\sin\alpha_{12}+\frac{\partial W}{\partial y}\cos\alpha_{12}$$

From this first equation, if $W=T_z$ than

$$T_{zu}du=\left(T_{zx}\cos\alpha_{12}+T_{zy}\sin\alpha_{12}\right)du=T_{zx}dx+T_{zy}dy\ ,$$

because

$$\begin{pmatrix}dx\\ dy\end{pmatrix}=\begin{pmatrix}\cos\alpha_{12}\\ \sin\alpha_{12}\end{pmatrix}du\ .$$

If points $P_1$ and $P_2$ are close to each other as required, integrals on the right side of equation (2) can be computed by the following trapezoid integral approximation formula:

$$\int_1^2\left(T_{zx}dx+T_{zy}dy\right)=\int_1^2 T_{zu}du\approx\frac{s_{12}}{2}\left[(T_{zu})_1+(T_{zu})_2\right]\quad(3)$$

$$\int_1^2 T_{zz}dz\approx\frac{\Delta h_{12}}{2}\left[(T_{zz})_1+(T_{zz})_2\right]\quad(4)$$

where $s_{12}$ is the horizontal distance between points $P_1$ and $P_2$, and $\Delta h_{12}$ is the height difference between these two points.

The value of integral (4) depends on the vertical gradient disturbance $T_{zz}$ and the height difference between the points. If points are at the same height (on a flat area) and in case of small vertical gradient disturbances the third integral in (2) can be neglected. (E.g. the value of (4) is 0.25 mGal in case of $\Delta h_{12}=50m$ and $[(T_{zz})_1+(T_{zz})_2]/2=50E$ ).

So, discarding the effect of (4) the differences of gravity anomalies between two points can be computed by the approximate equation:

$$\begin{aligned}(\Delta g_2-\Delta g_1)\approx\frac{s_{12}}{2}\{\ &[(T_{zx})_1+(T_{zx})_2]\cos\alpha_{12}\\ &+[(T_{zy})_1+(T_{zy})_2]\sin\alpha_{12}\ \}\end{aligned}\ .\quad(5)$$

## 2 Practical solutions

If we have a large number of torsion balance measurements, it is possible to form an interpolation net (a simple example can be seen in Figure 2) for determining gravity anomalies at each torsion balance points (Völgyesi, 1993, 1995, 2001). On the basis of Eq. (5)

$$(\Delta g_k-\Delta g_i)=C_{ik}\quad(6)$$

can be written between any adjacent points, where

$$C_{ik} = s_{ik}\left\{ \frac{(W_{zx} - U_{zx})_i + (W_{zx} - U_{zx})_k}{2} \cos\alpha_{ik} + \frac{(W_{zy})_i + (W_{zy})_k}{2} \sin\alpha_{ik} \right\} \quad (7)$$

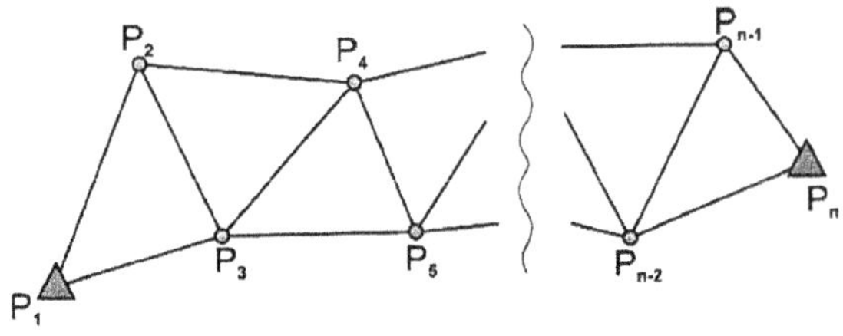


**Fig. 2** Interpolation net connecting torsion balance points

For an unambiguous interpolation it is necessary to know the real gravity anomaly at a few points of the network (triangles in Figure 2). Let us see now, how to solve interpolation for an arbitrary network with more points than needed for an unambiguous solution, where gravity anomalies are known. In this case the $\Delta g$ values can be determined by adjustment.

The question arises what data are to be considered as measurement results for adjustment: the real torsion balance measurements $W_{zx}$ and $W_{zy}$, or $C_{ik}$ values from Eq. (7). Since no simple functional relationship (observation equation) with a measurement result on one side and unknowns on the other side of an equation can be written, computation ought to be made under conditions of adjustment of direct measurements, rather than with measured unknowns – this is, however, excessively demanding in terms of storage capacity. Hence concerning measurements, two approximations will be applied: on the one hand, gravity anomalies from measurements at the fixed points are left uncorrected – thus, they are input to adjustment as constraints – on the other hand, $C_{ij}$ on the left hand side of fundamental equation (6) are considered as fictitious measurements and corrected. Thereby observation equation (6) becomes:

$$C_{ik} + v_{ik} = \Delta g_k - \Delta g_i \quad (8)$$

permitting computation under conditions given by adjusting indirect measurements between unknowns (Detrekői, 1991).

The first approximation is possible since reliability of the gravity anomalies determined from measurements exceeds that of the interpolated values considerably. Validity of the second approximation will be reconsidered in connection with the problem of weighting.

For every triangle side of the interpolation net, observation equation (8):

$$v_{ik} = \Delta g_k - \Delta g_i - C_{ik} \quad (9)$$

may be written. In matrix form:

$$\underset{(m,1)}{\mathbf{v}} = \underset{(m,2n)}{\mathbf{A}} \underset{(2n,1)}{\mathbf{x}} + \underset{(m,1)}{\mathbf{l}}$$

where $\mathbf{A}$ is the coefficient matrix of observation equations, $\mathbf{x}$ is the vector containing unknowns $\Delta g$, $\mathbf{l}$ is the vector of constant terms, $m$ is the number of triangle sides in the interpolation net and $n$ is the number of points. The non-zero terms in an arbitrary row $i$ of matrix $\mathbf{A}$ are:

$$[\ldots \; 0 \; +1 \; -1 \; 0 \; \ldots]$$

while vector elements of constant term $\mathbf{l}$ are the $C_{ik}$ values.

Gravity anomalies fixed at given points modify the structure of observation equations. If, for instance, $\Delta g_k = \Delta g_{k0}$ is given in (8), then the corresponding row of matrix $\mathbf{A}$ is:

$$[\ldots \; 0 \; 0 \; -1 \; 0 \; \ldots]$$

the changed constant term being: $C_{ij} - \Delta g_{k0}$, that is $\Delta g_k$, and of coefficients of $\Delta g_k$ are missing from vector $\mathbf{x}$, and matrix $\mathbf{A}$, respectively, while corresponding terms of constant term vector $\mathbf{l}$ are changed by a value $\Delta g_{k0}$.

Adjustment raises also the problem of weighting. Fictive measurements may only be applied, however, if certain conditions are met. The most important condition is the deducibility of covariance matrix of fictive measurements from the law of error propagation, requiring, however, a relation yielding fictive measurement results, – in the actual case, Eq. (7). Among quantities on the right-hand side of (7), torsion balance measurements $W_{zx}$ and $W_{zy}$ may be considered as wrong. They are about equally reliable $\pm 1E$ ($1E = 1\,Eötvös\ Unit = 10^{-9}s^{-2}$), furthermore, they may be considered as mutually independent quantities, thus, their weighting coefficient matrix $\mathbf{Q}_{WW}$ will be a unit matrix. With the knowledge of $\mathbf{Q}_{WW}$, the weighting coefficient matrix $\mathbf{Q}_{CC}$ of fictive measurements $C_{ik}$ after Detrekői (1991) is:

$$\mathbf{Q}_{CC} = \mathbf{F}^*\mathbf{Q}_{WW}\mathbf{F} = \mathbf{F}^*\mathbf{F}$$

$\mathbf{Q}_{WW} = \mathbf{E}$ being a unit matrix. Elements of an arbitrary row $i$ of matrix $\mathbf{F}^*$ are:

$$\left[\left(\frac{\partial C_{ik}}{\partial W_{zx}}\right)_1 \left(\frac{\partial C_{ik}}{\partial W_{zx}}\right)_2 ,\ldots, \left(\frac{\partial C_{ik}}{\partial W_{zx}}\right)_n \right.$$
$$\left. \left(\frac{\partial C_{ik}}{\partial W_{zy}}\right)_1 \left(\frac{\partial C_{ik}}{\partial W_{zy}}\right)_2 ,\ldots, \left(\frac{\partial C_{ik}}{\partial W_{zy}}\right)_n \right]$$

For the following considerations let us produce rows $\mathbf{f}_1^*$ and $\mathbf{f}_2^*$ of matrix $\mathbf{F}^*$ (referring to sides between points $P_1 - P_2$ and $P_1 - P_3$ respectively):

$$\mathbf{f}_1^* = [\ \frac{s_{12}\sin\alpha_{12}}{2}, \frac{s_{12}\sin\alpha_{12}}{2}, 0, 0, \ldots, 0,$$
$$\frac{s_{12}\cos\alpha_{12}}{2}, \frac{s_{12}\cos\alpha_{12}}{2}, 0, 0, \ldots, 0\ ]$$

and

$$\mathbf{f}_1^* = [\ \frac{s_{13}\sin\alpha_{13}}{2},\ 0, \frac{s_{13}\sin\alpha_{13}}{2}, 0, 0, \ldots, 0,$$
$$\frac{s_{13}\cos\alpha_{13}}{2},\ 0, \frac{s_{13}\cos\alpha_{13}}{2}, 0, 0, \ldots, 0\ ].$$

Using $\mathbf{f}_1^*$, variance of $C_{ik}$ value referring to side $P_1 - P_2$ is:

$$m^2 = \frac{s_{12}^2}{4}\left(2\sin^2\alpha_{12} + 2\cos^2\alpha_{12}\right) = \frac{s_{12}^2}{2}$$

while $\mathbf{f}_1^*$ and $\mathbf{f}_2^*$ yield covariance of $C_{ik}$ values for sides $P_1 - P_2$ and $P_1 - P_3$:

$$\mathrm{cov} = \frac{s_{12}s_{13}}{4}\left(\sin\alpha_{12}\sin\alpha_{13} + \cos\alpha_{12}\cos\alpha_{13}\right).$$

Thus, fictive measurements may be stated to be correlated, and the weighting coefficient matrix contains covariance elements at the junction point of the two sides. If needed, the weighting matrix may be produced by inverting this weighting coefficient matrix. Practically, however, two approximations are possible: either fictive measurements $C_{ij}$ are considered to be mutually independent, so weighting matrix is a diagonal matrix; or fictive measurements are weighted in inverted quadratic relation to the distance.

By assuming independent measurements, the second approximation results also from inversion, since terms in the main diagonal of the weighting coefficient matrix are proportional to the square of the side lengths. The neglection is, however, justified, in addition to the simplification of computation, also by the fact that contradictions are due less to measurement errors than to functional errors of the computational model (Völgyesi, 1993).

## 3 Test computations

Test computations were performed in a Hungarian area extending over about 800 $km^2$. In the last century approximately 60000 torsion balance measurements were made mainly on the flat territories of Hungary, at present 22408 torsion balance measurements are available. Location of these 22408 torsion balance observational points and the site of the test area can be seen on Figure 3.

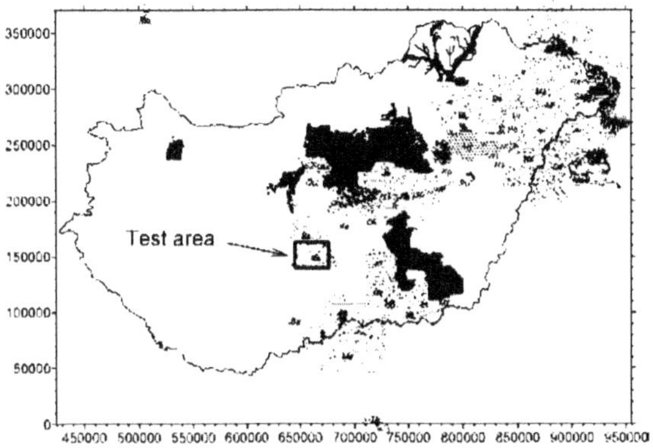


**Fig. 3** Location of torsion balance measurements being stored in computer database, and the site of the test area

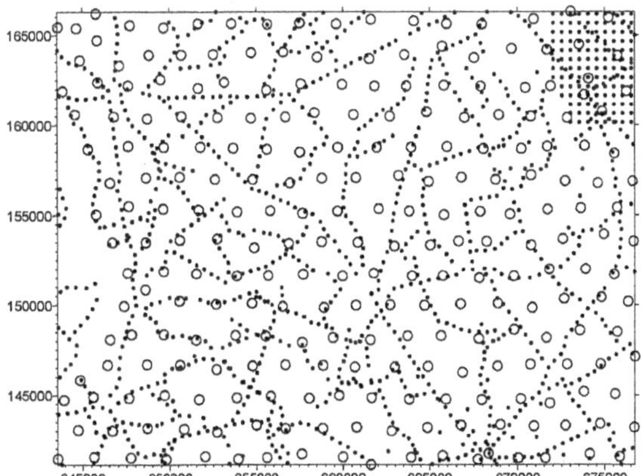


**Fig. 4** Gravity measurements (marked by dots) and torsion balance points (marked by circles) on the test area

The nearly flat test area can be found in the middle of the country, the height difference between the lowest and highest points is less than 20 m. There were 248 torsion balance stations and 1197 gravity measurements on this area. 30 points from these

248 torsion balance stations were chosen as fixed points where gravity anomalies $\Delta g$ are known from gravity measurements and the unknown gravity anomalies were interpolated on the remaining 218 points. Location of torsion balance stations (marked by circles) and the gravity measurements (marked by dots) can be seen on Figure 4.

The isoline map of gravity anomalies $\Delta g = g - \gamma$ ($\gamma$ is the normal gravity) constructed from 1197 $g$ measurements can be seen on Figure 5. Small dots indicate the locations of measured gravity values. Measurements were made by Worden gravimeters, by accuracy of ±20-30 μGal. At the same time the isoline map of gravity anomalies constructed from the interpolated values from 248 torsion balance measurements can be seen on Figure 6. Small circles indicate the locations of torsion balance points.

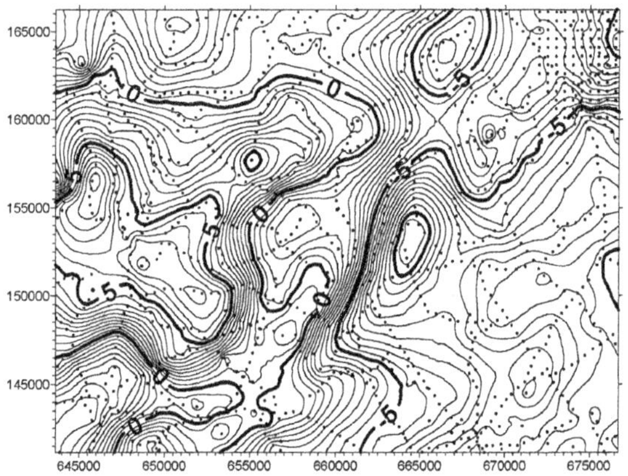


**Fig. 5** Gravity anomalies from $g$ measurements on the test area

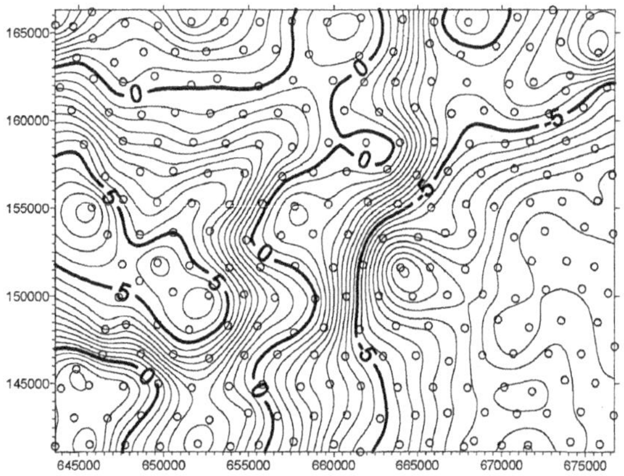


**Fig. 6** Interpolated gravity anomalies from $W_{zx}$ and $W_{zy}$ gradients measured by torsion balance on the test area

More or less a good agreement can be seen between these two isoline maps. In order to control the applicability and accuracy of interpolation, we compared the given and the interpolated gravity anomalies. $\Delta g$ values were determined for each torsion balance points from gravity measurements by linear interpolation on the one hand and gravity anomalies for the same points from gravity gradients measured by torsion balance on the other hand. Isoline and surface maps of differences between the two types of $\Delta g$ values can be seen on Figures 7 and 8. The differences are about ±1–2 mGal the maximum difference is 4 mGal.

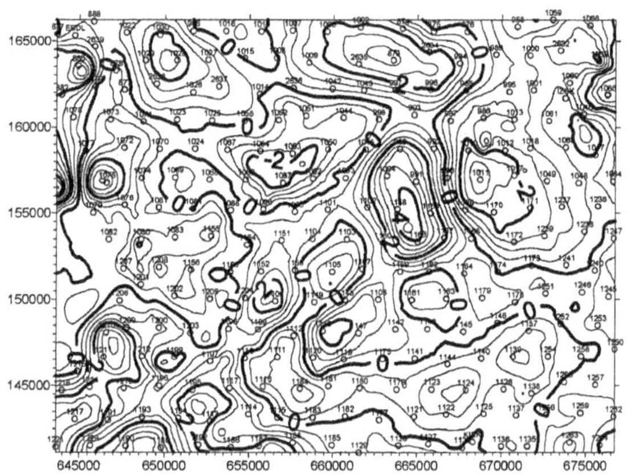


**Fig. 7** Isoline map of differences between the measured and the interpolated gravity anomalies on the test area

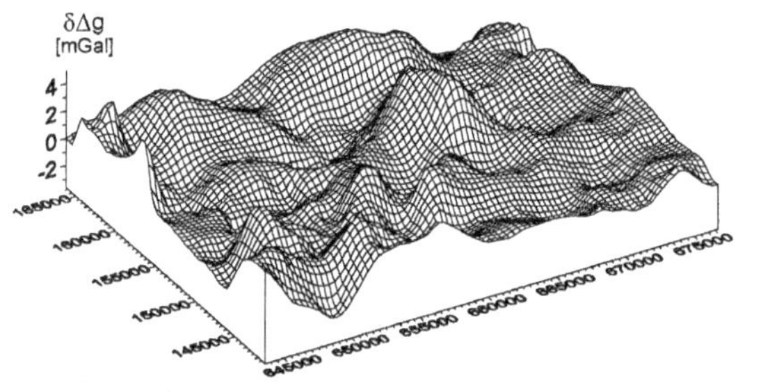


**Fig. 8** Surface map of differences between the measured and the interpolated gravity anomalies on the test area

Finally the standard error characteristic to interpolation, determined by

$$m_{\Delta g} = \pm\sqrt{\frac{1}{n}\sum_{i=1}^{n}(\Delta g_i^{mes.} - \Delta g_i^{\text{int.}})^2}\,,$$

was computed (where $\Delta g_i^{mes.}$ is the gravity anomaly from gravity measurements, $\Delta g_i^{\text{int.}}$ is the interpolated value from torsion balance measurements and $n = 248$ is the number of torsion balance stations).

Standard error $m_{\Delta g} = \pm 1.281\,\text{mGal}$ indicates that horizontal gradients of gravity give a possibility to determine gravity anomalies from torsion balance measurements by *mGal* accuracy on flat areas.

In case of a not quite flat area (like our test area) accuracy of interpolation would probably be increased by taking into consideration the effect of vertical gradients by integral (4), but unfortunately we haven't got the real vertical gradient values of torsion balance points on our test area yet. It would be important to investigate the effect of vertical gradient for the interpolation in the future.

## Summary

A method was developed, based on integration of horizontal gradients of gravity $W_{zx}$ and $W_{zy}$, to predict gravity anomalies at all points of the torsion balance network. Test computations were performed in a characteristic flat area in Hungary where both torsion balance and gravimetric measurements are available. Comparison of the measured and the interpolated gravity anomalies indicates that horizontal gradients of gravity give a possibility to determine gravity anomalies from torsion balance measurements by *mGal* accuracy on flat areas. Accuracy of interpolation would probably be increased by taking into consideration the effect of vertical gradients.

## Acknowledgements

We should thank for the funding of the above investigations to the National Scientific Research Fund (OTKA T-037929 and T-37880), and for the assistance provided by the Physical Geodesy and Geodynamic Research Group of the Hungarian Academy of Sciences.

## References

Detrekői Á. (1991) Adjustment calculations. Tankönyvkiadó, Budapest. (in Hungarian)

Heiskanen W, Moritz H. (1967) Physical Geodesy. W.H. Freeman and Company, San Francisco and London.

Torge W. (1989) Gravimetry. Walter de Gruyter, Berlin – New York.

Völgyesi L. (1993) *Interpolation of Deflection of the Vertical Based on Gravity Gradients.* Periodica Polytechnica Civ.Eng., Vol. 37. Nr. 2, pp. 137-166.

Völgyesi L. (1995) *Test Interpolation of Deflection of the Vertical in Hungary Based on Gravity Gradients.* Periodica Polytechnica Civ.Eng., Vol. 39, Nr. 1, pp. 37-75.

Völgyesi L. (2001) *Geodetic applications of torsion balance measurements in Hungary.* Reports on Geodesy, Warsaw University of Technology, Vol. 57, Nr. 2, pp. 203-212.

# Decadal Ocean Bottom Pressure Variability and its Associated Gravitational Effects in a Coupled Ocean-Atmosphere Model

R.J. Bingham, K. Haines
Environmental Systems Science Centre, University of Reading, 3 Earley Gate, Reading, RG6 6AL, UK

**Abstract.** The launch of the Gravity Recovery and Climate Experiment (GRACE) satellite mission in March 2002 has made timely the study of geophysical processes that redistribute the Earth's mass. This study uses the Hadley Centre coupled ocean-atmosphere model HadCM3 to examine the ocean's role in mass redistribution on inter-annual to decadal timescales. The leading empirical mode of inter-annual bottom pressure variability is a striking, basin-wide, oscillation between the Atlantic and Pacific Oceans. Our analysis suggests that this mode is primarily a wind driven phenomenon. We find some evidence for such a mode in a re-analysis of the global ocean, although the indirect nature of this evidence means no certain conclusions can yet be drawn. Thus, we consider the gravitational effects of this mode and the potential of current geodetic missions to detect it. A surprising result is that oceanic mass redistribution can lead to decadal trends in the zonal harmonic $J_2$, with a slope of approximately one-third that observed in geodetic measurements of $J_2$, all of which is normally attributed to post glacial rebound.

**Keywords.** ocean bottom pressure, inter-annual variability, time-dependent gravity, coupled model

## 1 Introduction

Prompted by recent geodetic observations (*Cox and Chao*, 2002) a number of studies have raised the possibility that mass redistribution in the ocean may lead to significant inter-annual changes in the Earth's gravitational field (e.g. *Chao et al.*, 2003; *Chen et al.*, 2003). Using a coupled ocean-atmosphere model *Leuliette et al.* (2002) have also shown that ocean mass redistribution related to climate change may be detectable by GRACE. Yet, the true nature of oceanic mass redistribution, or, equivalently, ocean bottom pressure ($p_b$) variability, and its relationship with the wider dynamics of the ocean, is still poorly understood. This is due, primarily, to the difficulty in obtaining measurements of $p_b$ at spatial scales relevant to the large-scale ocean circulation.

Until recently there have been two main approaches, occupying opposite ends of the spatial-scale spectrum, used to observe mass redistribution in the oceans. A limited number of in-situ measurements have been made with bottom pressure recorders. These are difficult to deploy and recover, and can only survive the extreme conditions of the ocean floor for relatively short periods. Moreover, since the $p_b$ variability at any particular location may be strongly influenced by local topographic features, bottom pressure recorders generally do not reflect the large-scale features of the ocean circulation. Alternatively, space geodetic techniques, such as Satellite Laser Ranging (SLR), have provided a global measure of mass redistribution within the whole of the Earth system for the past two decades. However, SLR derived changes in the Earth's gravitational field are limited to long spatial scales (>8000km), thus making it difficult to attribute gravitational signals to particular geophysical processes.

The launch of the GRACE satellite mission in March 2002 - with the objective of measuring temporal changes in the Earth's gravitational field to spatial scales of 200km - promises to revolutionise our understanding mass redistribution within the Earth system. To utilise these data to their fullest potential it is important that we understand how changes in the ocean circulation are related to $p_b$ variability, and how this in-turn manifests itself as changes in the Earth's gravitational field.

The first attempt to quantify large-scale changes in $p_b$ was made by *Gill and Niiler* (1973). They calculated seasonal averages of $p_b$ variability for the North Atlantic and North Pacific, and concluded that the dominant factor influencing $p_b$ variation on seasonal time scales was the wind, via Ekman pumping, with baroclinic effects becoming significant within $15^o$ of the equator. Maximum amplitudes of 0.8cm and 1.5cm of equivalent water thickness were found in the western North Atlantic and North Pacific basins, respectively. A study into the seasonal cycle of global $p_b$ in a primitive-equation model was conducted by *Ponte* (1999) who found that over most of the deep ocean large-scale $p_b$ signals did not vary by more than 1cm.

This paper examines the nature and causes of inter-annual $p_b$ variability, and the gravitational effects of this variability. In the next section we briefly outline the HadCM3 model used in this study and how $p_b$ is calculated from its output. In section 3 we consider the nature and causes of $p_b$ variability in HadCM3, whilst in section 4 we consider the gravitational consequences of this variability. The last section provides a summary of our findings.

## 2 Model and Data

### 2.1 Description of HadCM3

The Hadley Centre coupled atmosphere-ocean model HadCM3 provides the $p_b$ dataset used in this study. The atmospheric component of HadCM3 has a resolution of 2.5° in latitude and 3.75° in longitude, and has 19 vertical levels. The oceanic component has a horizontal resolution of 1.25° in both latitude and longitude, and has 20 depth levels, which vary in thickness from 10m near the surface to 500m at the bottom. The model was initialised from the *Levitus* (1994) climatology, and simulates a realistic and stable present day mean climate without the need of unphysical flux adjustments required in many coupled models in order to prevent climate drift (*Gordon et al.*, 2000).

HadCM3 successfully represents many of the important climate phenomena. The model has a realistic thermohaline circulation with a maximum Atlantic meridional over-turning strength of approximately 20Sv occurring at 45N. The NINO3 index, defined as the anomalous sea-surface temperature averaged over the region 150W-90W, 5S-5N is used to characterise the El Nino Southern Oscillation. The HadCM3 NINO3 index has a standard deviation that matches the observed index to within observational uncertainty, and a power spectrum that shows, in close agreement with observations, a clear spectral peak at 3-4 years (*Collins et al.*, 2001).

### 2.2 Calculating bottom pressure

Assuming the ocean is in hydrostatic balance, and given that, at the frequencies of interest in this study, the ocean surface responds as an inverted barometer to changes in the over-lying atmospheric pressure (*Wunsch and Stammer*, 1997), $p_b$ is approximated by

$$p_b = g \int_{-H}^{0} \rho dz + g\rho_0 \eta + \overline{p}_a, \qquad (1)$$

where $\rho$ is density, $g$ is acceleration due to gravity, $z = -H$ is bathymetric depth, $z = \eta$ is the height of the sea surface, corrected for the inverse barometer response, relative to a reference level $z = 0$, and $\overline{p}_a$ is the spatial mean of atmospheric surface pressure over the global ocean. As a further approximation, valid because variations in the density of seawater are small compared to the mean density, $\rho_0$, replaces $\rho$ in the second term on the right of equation (1).

A 100 year monthly mean $p_b$ timeseries was constructed from the model output of HadCM3. The formulation of the ocean component of HadCM3 - like many ocean models - is Boussinesq. Consequently, volume rather than mass is conserved. Changes in $\eta$ are replaced by an equivalent pressure on the, so-called, rigid lid at $z = 0$. As a consequence of the way this rigid-lid pressure is diagnosed from the model output, a net change of density, and therefore mass, of the fixed volume does not lead to a compensating change in global mean sea-level. Following *Greatbatch* (1994) this was accounted for by adjusting the $p_b$ by a spatial constant at each time-step to close the oceanic mass budget. The validity of this correction depends on certain assumptions discussed more fully elsewhere (e.g *Ponte*, 1999). In this study we did not consider the small inter-annual changes in $p_b$ due to $\overline{p}_a$ or freshwater fluxes in and out of the ocean. Throughout this paper $p_b$ will be given in equivalent centimetres of water thickness found by dividing $p_b$ by $g\rho_0$.

## 3 Bottom Pressure Analysis

The area weighted mean power spectrum (Figure 1) for all model grid points shows that the annual period dominates $p_b$ variability in HadCM3. The clear peaks in the spectrum at the semi-annual and third-annual periods reflects the fact that the seasonal cycle is not purely sinusoidal. Previous ocean model studies (e.g. *Ponte*, 1999; *Condi and Wunsch*, 2004) have found that over much of the open ocean the amplitude of the seasonal $p_b$ cycle is less than 1cm, while coastal regions, the North Pacific, and parts of the Southern Ocean are regions with significant seasonal $p_b$ variability. Similarly, the seasonal $p_b$ cycle in HadCM3 is weak over much of the open ocean, while the RMS variance of the seasonal $p_b$ cycle in the North Pacific and Southern Ocean is up to 3cm. In the Indian Ocean, however, the RMS variance of the seasonal $p_b$ signal in HadCM3 is up to 4cm, somewhat greater than previous studies have found, although the phase of the signal is similar.

Although Figure 1 shows that HadCM3 does not have any significant periodic modes of $p_b$ variability below the annual frequency there is still significant power in the $p_b$ signal at inter-annual to decadal frequencies. If this power is to translate into a significant gravitational signal then it must correspond to large scale patterns of $p_b$ variability, rather than just being uncorrelated red noise.

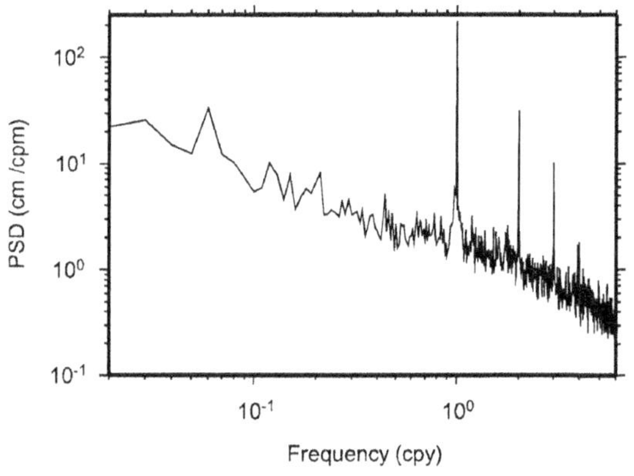


Figure 1: An area weighted mean power density spectrum of $p_b$ in HadCM3 (no smoothing).

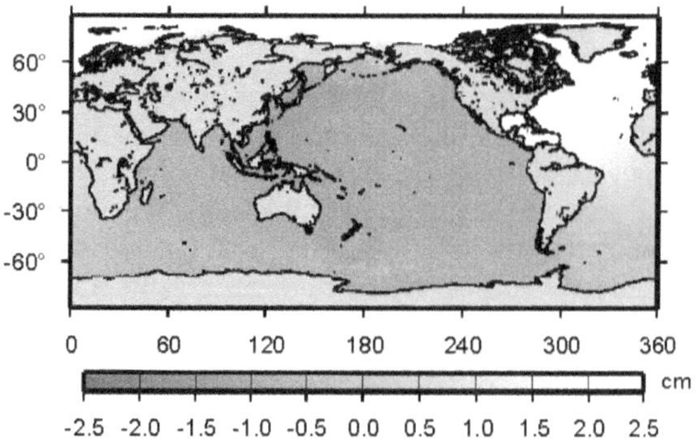


Figure 2: The 1st EOF of $p_b$ scaled by the RMS value of the associated PC.

To look for large-scale modes of inter-annual $p_b$ variability the mean annual cycle was removed from the detrended $p_b$ timeseries and an application of a 13 month box-car filter removed the remaining intra-annual variability. The detrending was necessary because HadCM3 does exhibit some long-term drift, particularly in the deep-water properties, that may affect $p_b$. The method of empirical orthogonal function (EOF) analysis (*Emery and Thomson*, 2001) was then used to partition the $p_b$ timeseries into orthogonal modes, each consisting of a spatial function, or EOF, and a temporal function, or principal component (PC). Often the first few of these empirical modes will correspond to dynamical modes of the analysed system, although this is not necessarily the case. The Greatbatch correction ensures that the spatial mean of each EOF will be zero, so for each mode a $p_b$ change at one model grid point must offset by an opposite $p_b$ change elsewhere.

The first EOF (EOF1) of inter-annual $p_b$ variability is shown in Figure 2, scaled by the RMS value of the associated PC (PC1), and accounts for 38% of the total inter-annual variance. The basin-scale nature of this leading $p_b$ mode (henceforth BP1) is striking; the North Atlantic and Arctic Oceans oscillate out of phase with the Pacific, where the largest amplitude changes occur in the western Pacific, and to some extent with the Indian Ocean. BP1 accounts for 60-90% of the inter-annual $p_b$ variability in the North Atlantic, and for 30-60% in the Pacific. This difference reflects the fact that in the Pacific other processes have a significant influence on inter-annual $p_b$ variability in HadCM3. PC1 (a scaled version of which is shown by the thick line in Figure 6) shows significant variability occurring on decadal time-scales, and in Section 4 we consider the gravitational effect of this mode. The pattern in Figure 2 would represent a volume deficit in the Pacific of $1.5 \times 10^{12}\mathrm{m}^3$ . The second $p_b$ mode accounts for only 10% of the total inter-annual $p_b$ variance and will not be considered further here.

## 3.1 Composites of other fields

High-low composites, based on PC1, were used as a straightforward way of establishing the relationship between BP1 and other fields in HadCM3. A threshold value was set, equivalent to an spatial mean $p_b$ signal in the North Atlantic of approximately $\pm$2cm, such that approximately 25% of the timeseries frames were included in each of the high and low composites. The low composite is subtracted from the high composite to give the high-low composite.

From the sea-surface height (SSH) high-low composite (Figure 3a) it is clear that the greatest changes in SSH occur where the gradient of the mean dynamic topography is steepest. This is particularly evident in the North Pacific where large changes occur along the axis of the Kuroshio extension, and in the North Atlantic where the largest changes occur along the Gulf Stream and North Atlantic Current. Both of these currents are fronts separating the sub-tropical and sub-polar gyres. A similar composite of atmospheric surface pressure (Figure 3b) shows a significant low over the North Pacific and a significant high over the North Atlantic. Together Figures 3a and 3b suggest that BP1 may be due to changes in the wind driven gyre circulation, particularly the sub-polar and sub-tropical gyres in the North Pacific and North Atlantic.

## 3.2 Gyre Circulation

Many papers have considered the nature and causes of decadal variability in the oceans. Using a simple linear model *Frankignoul et al.* (1997) concluded that stochastic wind stress forcing may explain a substantial part of the decadal variability of the oceanic gyres. *Sturges et al.* (1998) showed that in the central North Atlantic large decadal-scale fluctuations

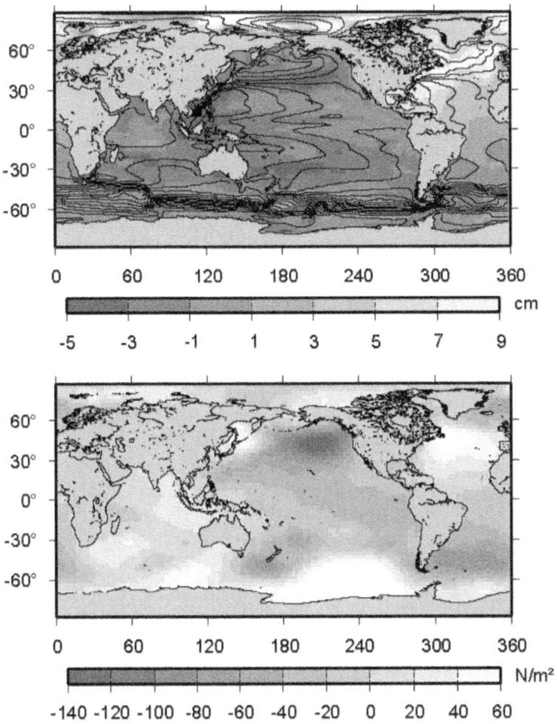

Figure 3: High-low composites, based on PC1, of: (a) Sea-surface height, with contours representing the mean dynamic topography; (b) Atmospheric surface pressure.

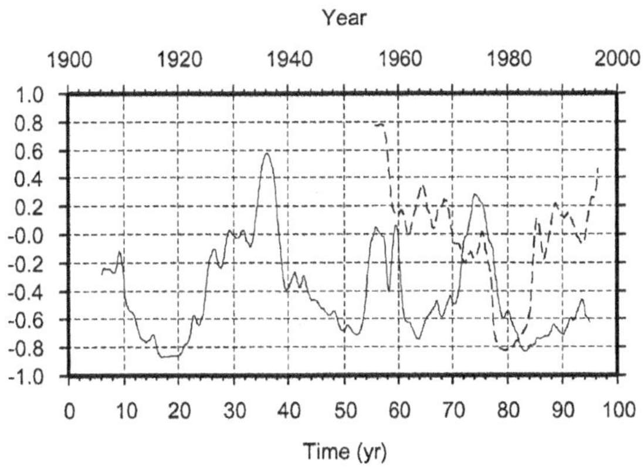

Figure 4: The 10 year running correlation between the NA and NP timeseries in HadCM3 (solid line and bottom axis) and from a re-analysis of the global ocean (dashed line and top axis). Horizontal axes refer to the centre point of the correlation window. (The horizontal position of one curve relative to the other is arbitrary.)

in sea-level could be explained by low-frequency Rossby waves forced by wind. Similarly, *Qiu* (2002) showed that persistent trends in SSH in the North Pacific between 1992 and 1998, as observed by Topex/Poseidon, were the result of surface wind-stress curl forcing accumulating westward along the first-mode baroclinic Rossby wave characteristics.

Given these results, and the patterns of anomalous SSH and atmospheric surface pressure associated with BP1 (Figures 3a and 3b), we hypothesise the following mechanism to explain this mode of $p_b$ variability: In the North Pacific, the overlying anomalous low pressure has associated winds that tend to spin down the sub-tropical gyre, whilst at the same time spinning up the sub-polar gyre. This results in a basin wide decrease in both SSH and $p_b$. In a similar fashion, the correlated anomalous high pressure overlying the North Atlantic gives rise to a basin wide increase in SSH and $p_b$. For the opposite surface pressure patterns the situation is reversed.

### 3.3 Oceanographic evidence

As Figure 3a demonstrates, in HadCM3 the dynamics that give rise to BP1 also produce a similar oscillation in SSH. Let NP and NA represent the spatial means of the SSH in the North Pacific and North Atlantic (filtered in the same way as the $p_b$). Then an anti-correlation between the observed NP and NA timeseries would indicate the presence, in the physical ocean, of a mode similar to the HadCM3 leading $p_b$ mode. Yet, during the 1990s the altimetric NP and NA timeseries are in fact positively correlated at 0.4. However, Figure 4 (solid line) shows that over any 10 year period the correlation between the HadCM3 NP and NA timeseries varies considerably. In fact, the degree of anti-correlation between the NA and NP timeseries in any 10 year period can be taken as an indirect indication the strength of BP1 in that period.

A longer "observational" SSH timeseries (1950-1999) comes from a re-analysis of the global ocean (*Carton et al.*, 2000). A 10 year running correlation between the NP and NA timeseries from this dataset also shows considerable variability (Figure 4 dashed line). From 1985 onward they have become increasingly positively correlated, in agreement with the altimetric data, but at other times, most notable the 10 year period centred on 1979, they have been strongly anti-correlated indicating that maybe a $p_b$ mode similar to the HadCM3 mode was occurring from the mid 70's to mid 80's.

## 4 Time-dependent Gravity Analysis

In the previous section we saw that in HadCM3 there is a dominant basin-scale mode of inter-annual $p_b$ variability, namely BP1, that indirect oceanographic evidence suggests may also occur in the ocean. More direct evidence may come from space geodetic techniques such as SLR and GRACE, but as we saw above, because the statistical properties of this mode are not stationary it may be difficult to detect it in the restricted data windows that are a part of such techniques. In this section we consider how BP1 would influence the Earth's gravity field if it were to occur in the ocean.

Stokes coefficients, to degree and order 50 (where a coefficient of degree $l$ corresponds to a spatial-scale of 20000km/$l$), were calculated according to *Gegout and Cazenave* (1993) equation (5) for the BP1 signal and, for comparison, the HadCM3 seasonal $p_b$ cycle. From these, space/time degree amplitude spectra were computed according to *Wahr et al.* (1998) equation (24). Figure 5 shows that BP1 (thick line) has less power than the seasonal signal (thin line) at all wavelengths, and indeed it is expected that the seasonal signal will make the largest oceanic contribution to the time-dependent gravity field at the time-scales resolvable by GRACE. In fact, a greater oceanic contribution to the time-dependent gravity field will be made at higher frequencies which creates an aliasing problem for the GRACE mission. Also, BP1 has a smaller percentage of its power at shorter wavelengths compared to the seasonal signal.

In Figure 5 we also assess the ability of GRACE to detect BP1 by comparing the spectra of BP1 with the power of the GRACE errors (dashed line) after 10 years of GRACE data - where 10 years is chosen as an optimistic estimate of the GRACE mission lifetime - inferred from the calibrated errors for the March 2004 solution as computed by the Center for Space Research, Texas. To take account of the non-stationarity of BP1, Figure 5 also shows the spectra of BP1 for its strongest and weakest 10 year periods (dotted lines). This shows that at the present accuracy GRACE could resolve the BP1 signal to spatial scales of approximately 1600km, and only in the worst case would BP1 remain below the level of the GRACE errors at all wavelengths. It is worth noting that the errors inferred from the March 2004 solution are conservative, and, with refinements in reprocessing the GRACE data, errors are likely to be much smaller.

## 4.1 The effect of BP1 on $J_2$

Although GRACE will determine the Earth's gravity field with unprecedented spatial resolution, SLR to geodetic satellites will remain the best way of determining the longest wavelength features of the Earth's gravitational field. From its inception in the late 1970s until recently, SLR observations showed a secular downward trend in the Earth's dynamical oblateness ($J_2 = -\sqrt{5}\Delta C_{20}$) of approximately $-3.0 \times 10^{-11}\mathrm{yr}^{-1}$, thought to be due, primarily, to post glacial rebound. *Cox and Chao* (2002) reported a sudden reversal of this trend occurring in 1997/98 that implied a rapid equatorward redistribution of mass in the Earth system. Given the abrupt nature of the change, and its coincidence with the strong 1997/98 El Nino, they suggested oceanic mass re-

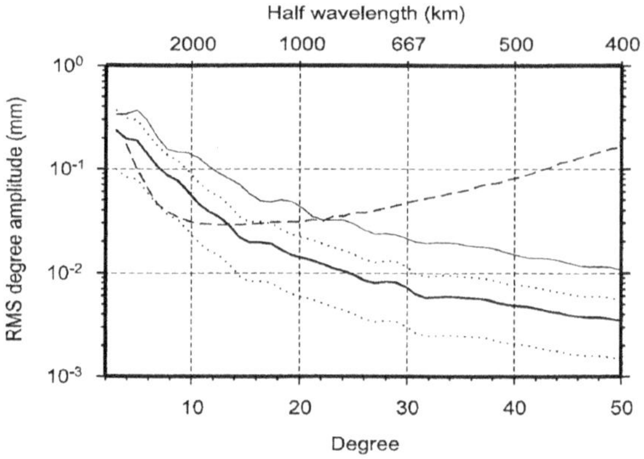


Figure 5: Degree amplitude spectra for: the BP1 signal (thick line); strongest and weakest 10 year BP1 signal (dotted lines); the seasonal $p_b$ cycle (thin line); GRACE errors for 10 years of data inferred from calibrated GRACE errors for March 2004 (dashed line).

distribution as a possible cause. *Dickey et al.* (2002) concluded sub-polar glacial melting may have made a large contribution to this observed $J_2$ anomaly, while *Chao et al.* (2003) suggested it was also related the Pacific Decadal Oscillation.

Although BP1 most obviously represents a zonal redistribution of mass, it also redistributes mass meridionally and therefore contributes significantly to the $J_2$ signal in HadCM3 (Figure 6), accounting for 58% of the total inter-annual $J_2$ variance. Figure 6 shows that BP1 is responsible for significant decadal-scale trends in $J_2$: The 20 year slope magnitude has a mean value of $0.4 \times 10^{-11}\mathrm{yr}^{-1}$ and a standard deviation of $0.3 \times 10^{-11}\mathrm{yr}^{-1}$. The maximum 20 year slope has a magnitude of $1.0 \times 10^{-11}\mathrm{yr}^{-1}$, which is about one-third of that observed in the SLR timeseries of $J_2$. For comparison, Figure 6 also shows the amplitudes of the annual harmonic of $J_2$ in HadCM3, and that of the observed $J_2$ signal, which, as it includes annual mass redistribution within the complete Earth system, is greater than the amplitude of HadCM3 ocean-only annual harmonic. Even though no claim is made here that a mode such as BP1 was responsible for the observed change in $J_2$ (in fact, as we saw in the previous section, this is unlikely), the HadCM3 $J_2$ timeseries is noteworthy because is shows oceanic mass redistribution can, in theory, give rise to significant persistent trends in $J_2$.

# 5 Conclusions

In this paper we have investigated the potential of the ocean to produce significant long-term variability in the Earth's gravity field using a 100 year $p_b$ dataset from the coupled ocean-atmosphere model HadCM3. In the model the time-integrated response of the ocean to the overlying wind field leads to a

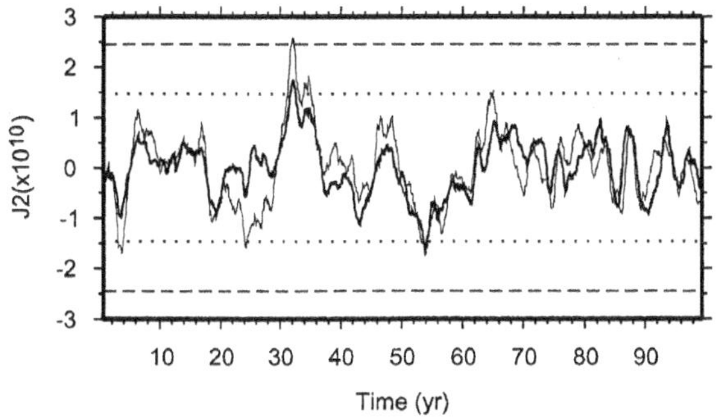


Figure 6: Variability of zonal harmonic $J_2$ that would result from HadCM3 inter-annual bottom pressure variability (thin line) and due solely to BP1 (thick line). Also shown are the amplitude of the annual harmonic for the HadCM3 $J_2$ signal (dotted lines), and for the observed $J_2$ signal (dashed lines).

dominant basin-scale mode of inter-annual $p_b$ variability (BP1) consisting mainly of an Atlantic-Pacific basin dipole. The mechanism of gyre spin-up and spin-down that is accompanied in HadCM3 by significant $p_b$ changes, is a realistic mechanism which could be produced in the real ocean under similar conditions of decadal timescale atmospheric wind variations. This is therefore a feasible mechanism by which the ocean-atmosphere system could lead to decadal trends in the Earth's gravity field.

Although BP1 has large spatial scales and dominates the 100 years of model output when considered in its entirety, the stochastic nature of the wind forcing means that its power in a restricted data window, of any few decades, varies considerably. Therefore, existing geodetic data may be insufficient to distinguish such variability from the post glacial rebound signal. The non-stationarity of BP1 also explains the absence of the characteristic sea-surface height signature of BP1 from the altimetry data. Nevertheless, a re-analysis of the global ocean provides some indirect evidence to suggest that variability similar to that associated with BP1 may have occurred in the past, although unfortunately before the commencement of reliable observations of the Earth's gravity field. It may therefore be possible in the future to use geodetic techniques to observe decadal changes in the ocean circulation, and also to study long-term climate variability.

**Acknowledgements.** This work was supported by the UK Natural Environment Research Council grant NER/A/S/2000/01001.

## References

Carton, J., G. Chepurin, X. Cao, and B. Giese (2000), A Simple Ocean Data Assimilation analysis of the global upper ocean 1950-95. Part I: Methodology, *J. Phys. Oceanogr.*, *30*, 294–309.

Chao, B., A. Au, J. Boy, and C. Cox (2003), Time-variable gravity signal of an anomalous redistribution of water mass in the extratropic Pacifi c during 1998-2002, *Geochemistry Geophysics Geosystems*, *4*(11), 1096, doi: 10.1029/2003GC000589.

Chen, J., C. Wilson, X. Hu, and B. Tapley (2003), Large-scale mass redistribution in the oceans, 1993-2001, *Geophys. Res. Lett.*, *30*(20), 2024, doi:10.1029/2003GL018048.

Collins, M., S. Tett, and C. Cooper (2001), The internal climate variability of HadCM3, a version of the Hadley Centre coupled model without flux adjustments, *Clim. Dyn.*, *17*, 61–81.

Condi, F., and C. Wunsch (2004), Measuring gravity fi eld variability, the geoid, ocean bottom pressure fluctuations, and their dynamical implications, *JGR*, *109*, C02013, doi:10.1029/2002JC001727.

Cox, C., and B. Chao (2002), Detection of a large-scale mass redistribution in the terrestrial system since 1998, *Science*, *297*, 832–833.

Dickey, J., S. Marcus, O. de Viron, and I. Fukumori (2002), Recent Earth oblateness variations: unravelling climate and postglacial rebound effects, *Science*, *298*, 1975–1977.

Emery, W., and R. Thomson (2001), *Data analysis methods in physical oceanography*, 638 pp., Elsevier.

Frankignoul, C., P. Muller, and E. Zorita (1997), A simple model of the decadal response of the ocean to stochastic wind forcing, *J. Phys. Oceanogr.*, *27*, 1533–1546.

Gegout, P., and A. Cazenave (1993), Temporal variations of the Earth gravity fi eld for 1985-1989 derived from LAGEOS, *Geophys. J. Int.*, *114*(2), 347–359.

Gill, A., and P. Niiler (1973), The theory of seasonal variability in the ocean, *Deep-Sea Research*, *20*, 141–177.

Gordon, C., C. Cooper, C. Senior, H.Banks, J. Gregory, T. Johns, J. Mitchell, and R. Wood (2000), The simulation of SST, sea ice extents and ocean heat transports in a version of the Hadley Centre coupled model without flux adjustments, *Clim. Dyn.*, *16*, 147–168.

Greatbatch, R. (1994), A note on the representation of steric sea-level in models that conserve volume rather than mass, *J. Geophys. Res.*, *99*, 12,767–12,771.

Leuliette, E., R. Nerem, and G. Russell (2002), Detecting time variations in gravity associated with climate change, *J. Geophys. Res.*, *107*(B6), 2112, doi:10.1029/2001JB000404.

Levitus, S. (1994), World Ocean Atlas, Volume 4: Temperature, NOAA/NESDIS E/OC21, US Department of Commerce, Washington DC.

Ponte, R. (1999), A preliminary model study of the large-scale seasonal cycle in the bottom pressure over the global ocean, *J. Geophys. Res.*, *104*(C1), 1289–1300.

Qiu, B. (2002), Large-scale variability in the midlatitude subtropical and subpolar North Pacifi c Ocean: Observations and causes, *J. Phys. Oceanogr.*, *32*, 353–375.

Sturges, W., B. Hong, and A. Clarke (1998), Decadal wind forcing of the North Atlantic subtropical gyre, *J. Phys. Oceanogr.*, *28*, 659–668.

Wahr, J., M. Molenaar, and F. Bryan (1998), Time variability of the Earth's gravity fi eld: Hydrological and oceanic effects and their possible detection using GRACE, *J. Geophys. Res.*, *103*, 30,205–30,229.

Wunsch, C., and D. Stammer (1997), Atmospheric loading and the oceanic "inverted barometer" effect, *Rev. Geophys.*, *35*, 79–107.

# Gravity Changes in the Fennoscandian Uplift Area to be Observed by GRACE and Absolute Gravimetry

J. Müller, L. Timmen, O. Gitlein, H. Denker
Institut für Erdmessung, University of Hannover, Schneiderberg 50, 30167 Hannover, Germany
mueller@ife.uni-hannover.de

**Abstract.** Gravity changes due to the glacial isostatic adjustment in Fennoscandia entails a temporal geoid variation of about 3 mm and a gravity change of about 10 $\mu$Gal over five years (linear trend) in the centre of the land uplift area. This gravity change can be observed by absolute gravimetry with an accuracy of $\pm 1 - 2$ $\mu$Gal and by GRACE with a geoid accuracy at the few mm level for long wavelengths. Since 2003, annual absolute gravity measurements have been intensified to improve the knowledge about the land uplift, about its accuracy and its spatial structure. Independently, relative gravimetry, tide gauge observations and GPS measurements are carried out by various institutions in dedicated campaigns or during their routine work. All Nordic countries contribute in different ways to the project (preparation of the sites, campaigns, collection of auxiliary data, computations and analyses).

The main result of the project will be reduced point-wise gravity observations, which are combined with GPS height observations to derive an independent, spatially and temporally smoothed model of the geoid change. It can be compared with the land uplift determined from the GRACE data, i.e., gound-truth for the satellite measurements.

**Keywords.** Postglacial rebound, absolute gravimetry, temporal gravity changes, GRACE

## 1 Geodetic Determination of the Fennoscandian Land Uplift

In Fennoscandia, the Earth's crust has been rising continuously as a result of deglaciation since the last glacial maximum. The region is dominated by the Precambrian basement rocks of the Baltic Shield and comprises mainland Norway, Sweden, Finland, the Kola Peninsula, and Russian Karelia. The spatial extension is about 2000 km in diameter (e.g. Ekman 1996); see also Fig. 1 for the approximate shape and location. To monitor and investigate the postglacial rebound, various geodetic data have been collected since 1892 (mareographs, levellings, relative gravity measurements since 1966). According to Ekman (1996), these observations reveal a maximum orthometric height change of 10.2 mm/a over the Bothnian Bay and show symmetry around the maximum, closely correlated to the former Late Pleistocene Fennoscandian Ice Sheet. The height change in the centre is associated with a maximum gravity variation of -2 $\mu$Gal per year. Based on these numbers, a geoid change of 0.6 mm/a has been derived from observations for the central area (e.g. Ekman and Mäkinen 1996). Correspondingly, this conversion factor from gravity change to geoid change is typical for this region. During their investigations an eustatic sea-level rise of 1.2 mm/a has been assumed from tide gauge observations (Nakiboglu and Lambeck 1991, Lambeck et al. 1990).

Since 1993 permanent GPS stations have been established in Scandinavia to implement a further geodetic method with several advantages compared to the classical techniques (permanent data acquisition, homogeneous point distribution, large extension of the measurement area, low cost). In this respect, the project BIFROST (Baseline Inferences for Rebound Observations, Sea Level, and Tectonics) was based on GPS technique and geophysical modelling and has delivered a maximum height variation (with respect to an ellipsoid) of more than 11 mm/a (see Fig. 1, cf. Milne et al. 2001, Johansson et al. 2002, Scherneck et al. 2003). The location of the centre and the geometrical structure of the uplift process differ from the previous model, with no clear zero line and more regional structures being visible. Another uplift measurement method is presented in Kuo et al. (2004). They combine satellite altimetry and tide gauge data to obtain vertical crustal motion at tide gauges in the Baltic Sea region. This independent solution can be used well for the comparison with the vertical motion as observed by GPS or other techniques in nearby sites.

During the mission duration of GRACE (about five years), a temporal geoid change of 3 mm can be expected in the centre of the Fennoscandian land uplift area, corresponding to a gravity change of about 10 $\mu$Gal. As the geoid derived from GRACE data

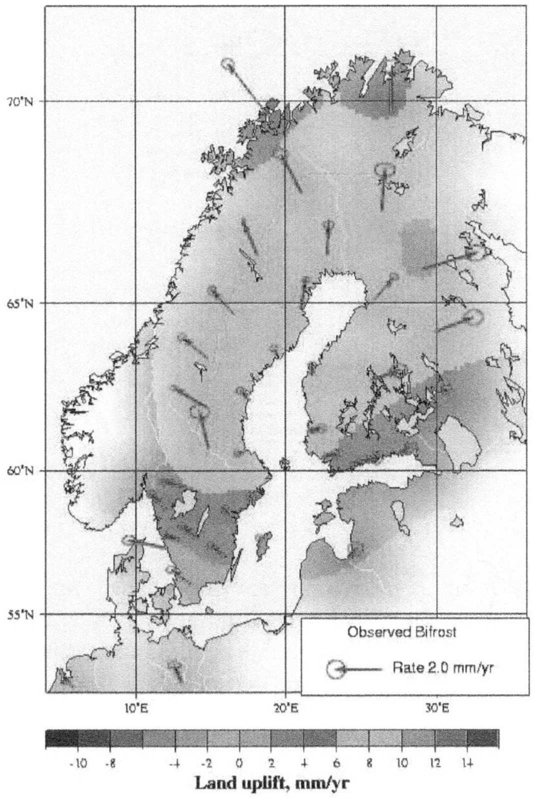


**Fig. 1.** The land uplift as determined by BIFROST, 1993-2000, in mm/a (see Scherneck et al. 2003).

can be determined monthly with an accuracy at the few mm level at a spatial resolution of 600 km (a little bit worse for the older 2002 solutions, cf. Tapley et al. 2004), the land uplift causes a measurable signal in the observations (NRC 1997, Wahr and Velicogna 2002). The problem of separating this effect from other time-variable gravity changes is a big challenge for the satellite as well as for the terrestrial approaches. These changes are caused by a wide range of mass movements, like oceanic and continental water mass variations, by atmospheric mass changes as well as by solid-Earth processes which show up at different time scales. Furthermore, the impact of the various effects in the GRACE data is different from those in the terrestrial gravity data, where also local and more regional phenomena (e.g. ground water table variations) affect the measurements. In contrast, the GRACE observations contain more long-wavelength temporal effects, such as an annual atmospheric contribution, which is removed in post-processing using appropriate models, so that only residual signals remain. The temporal high-frequency signals (below one month), e.g. tidal effects, are reduced during the GRACE data pre-processing, the spatial short-wavelength signals cancel out because of physical reasons, i.e. smoothing through upward-continuation in satellite altitude. In section 4 below, some examples of GRACE-derived temporal gravity variations are shown. For absolute gravimetry, the strategy is to reduce known temporal effects (e.g. tidal and atmospheric components) using available models, to minimize seasonal variations (e.g. from hydrology or atmosphere) by always carrying out the measurements in the same period of each year, to consider local changes by corresponding auxiliary measurements (e.g. of the ground water table). Finally, remaining periodic (and episodic) signals may smooth out by averaging over five years, further signals may be averaged out by the combination of oberservations in co-located sites ('spatial smoothing'); there are, e.g., several sites close to the center of the uplift area. A further idea is to apply smoothing filters like spherical caps to reduce the errors.

Therefore, absolute gravimetry offers a good chance for GRACE validation. For obvious reasons, Scandinavia is one of the preferable areas to apply terrestrial geodetic methods (absolute gravimetry, GPS) with the objective to obtain reliable accurate ground-truth information. Due to the excellent infrastructure, the measurement stations are easily accessible. Scandinavian research institutes have investigated the rebound effect over many decades (e.g. Scherneck et al. 2003 or Mäkinen et al. in these proceedings). They proved the capability of terrestrial point measurement techniques (levelling, relative gravimetry) to determine the land uplift along east-west profiles (e.g. Ekman and Mäkinen 1996). With the availability of faster and more precise geodetic methods (absolute gravimetry combined with GPS), the geophysical phenomenon can be observed more effectively now than in the past.

## 2 Absolute Gravimetry

Besides geometrical methods like GPS, absolute gravimetry is a further terrestrial geodetic technique to study the vertical motion. In addition and complementary to the other geodetic measurements, absolute gravimetry has the following positive characteristics:

- accuracy ± 10 to 30 nm/s$^2$ (10 nm/s$^2$ = 1 $\mu$Gal) for one station determination,
- north-south profiles are possible (no calibration problems due to large gravity differences),

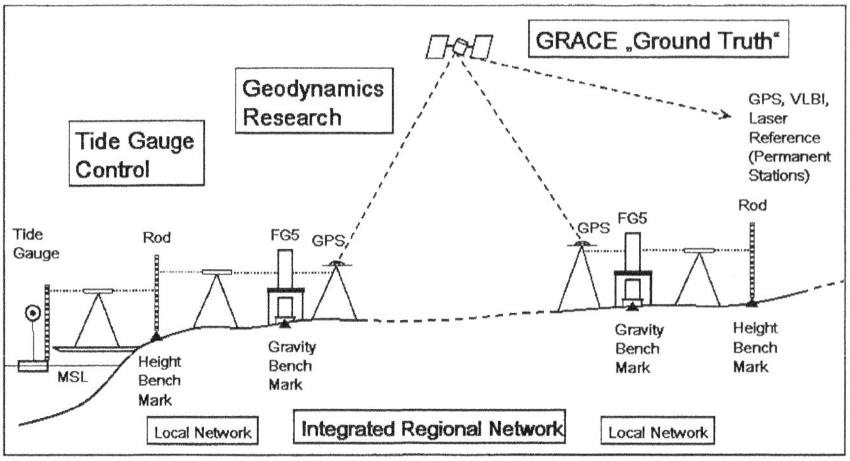


**Fig. 2.** Integration of different geodetic techniques to survey gravity and geoid variations with time in the Fennoscandian land uplift area.

- independent validation method for direct comparison with GPS, VLBI, SLR, and superconducting gravimetry,
- combined with geometrical methods, vertical surface deformations and subterranean mass movements (e.g. from tectonics, hydology or mantle dynamics ) can be separated.

The benefit of absolute gravimetry has already been shown in different research projects covering areas of global or regional extensions and monitoring variations caused by mass movements (Wilmes et al. 2004, Lambert et al. 2001). The International Absolute Gravity Basestation Network (IAGBN) serves, among others, to determine large scale tectonic plate movements (Boedecker and Fritzer 1986). Applying absolute gravimetry to the determination of crustal deformations, secular gravity changes should be measured with a precision better than $\pm$ 10 nm/s$^2$ per year. This can be achieved for GRACE by annual measurements over five years. Therefore absolute gravimetry can be used to validate GRACE results, i.e., it can provide point-wise ground-truth for the satellite measurements (gravity anomaly changes with time).

In addition, the terrestrial results can be used to calculate a model of the temporal geoid changes in Fennoscandia. This calculation is based upon an approach for gravity disturbances proposed by Hotine (1969), which allows the conversion from temporal gravity and height changes into geoid changes

$$\dot{N} = \frac{R}{4\pi\gamma} \int\int_\sigma H(\psi) \left(\dot{g} + \frac{2\gamma}{r}\dot{h}\right) d\sigma,$$

with $\dot{N}, \dot{h}, \dot{g}$ are the temporal changes of geoid height, ellipsoidal height and gravity; $\gamma$ is normal gravity, $R$ mean Earth radius, $r$ radius of computation point P, $\psi$ spherical distance and

$$H(\psi) = \mathrm{cosec}\frac{\psi}{2} - \ln\left(1 + \mathrm{cosec}\frac{\psi}{2}\right).$$

As the integration has to be performed over the whole Earth in principle, appropriate mathematical procedures must be applied. But this equation shows obviously that measured gravity changes from absolute gravimetry and ellipsoidal height changes from GPS have to be combined to obtain geoid height changes, which can also be used for the validation of the corresponding GRACE gravity field quantities. The independent comparison with reduced absolute gravity observations and/or derived quantities supports the reliability of the GRACE-based gravity field which is difficult to verify at all, as no better global gravity field model does exist.

## 3 Project Realisation

A joint project for the survey of the land uplift in Fennoscandia has been set up. The Working Group for Geodynamics of the Nordic Geodetic Commission (NKG) serves as a platform to organize the project. Besides the Institut für Erdmessung (IfE) from Hannover, the following institutions are joining the project: National Survey and Cadastre (KMS, Copenhagen/Denmark), Finnish Geodetic Institute (FGI, Masala/Finland), Bundesamt für Kartographie und Geodäsie (BKG/Germany), Institute of Mapping Sciences, Agricultural University of Norway (Ås), Statens Kartverk (SK, Hønefoss/Norway), Onsala

Space Observatory, Chalmers University of Technology (Onsala/Sweden), National Land Survey of Sweden (Gävle). In 2002 the IfE of the University of Hannover has received a new FG5 absolute gravity meter from Micro-g Solutions, Inc. (Erie, Colorado), which is a "state-of-the-art" instrument. In 2004, also the Institute of Mapping Sciences, Agricultural University of Norway has received a new FG5, so that togehter with those of the FGI and the BKG four absolute gravimeters were used in this project (see also Wilmes et al. 2005).

The project strategy and objectives may be summarized as follows:

- absolute gravity determinations at up to 30 sites in 2003 and 2004,
- repetition measurements in the years 2005, 2006 and 2007,
- simultaneous GPS measurements (gravity sites are located close to permanent GPS sites),
- ensuring geometrical connections between gravity sites and tide gauges (mainly by GPS),
- auxiliary levelling measurements to excentres (control of local variations, ties between gravity and GPS stations),
- elaboration of reduction models (air mass movements, ocean tides, etc.),
- integration of other geodetic data sources (e.g., already existing FG5 measurements),
- providing the products of the project (gravity anomaly changes, two-dimensional model describing the geoid change),
- validation of GRACE results.

Fig. 2 demonstrates the principle to combine different geodetic methods. All absolute gravity sites are connected to permanent GPS stations, which again are connected to tide gauges and further control points. Fig. 3 shows the project plan which has been developed by the NKG working group for geodynamics for 2004. The absolute gravity sites occupied by IfE in 2004 are indicated by the blue circles. The use of more than one absolute gravimeter allows parallel and control measurements on identical sites and helps to identify possible offsets or systematic discrepancies of the instruments. Such a strategy is necessary to improve the absolute accuray of the observed gravity network.

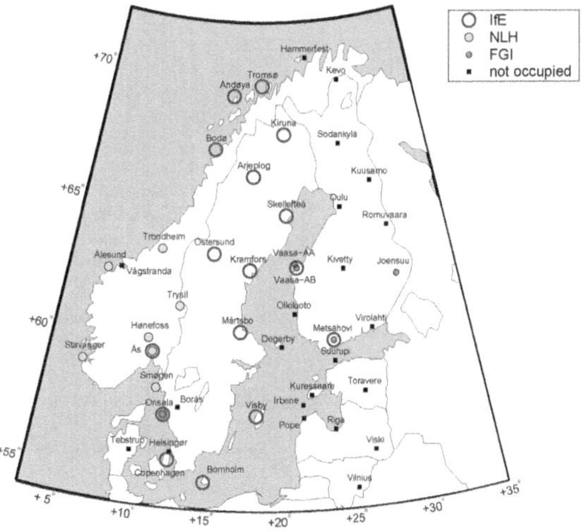


**Fig. 3.** Observation plan developed by the NKG working group for geodynamics. The sites occupied by IfE in 2004 are indicated by big blue circles.

### Accuracy of the observations

First absolute gravity measurements of the IfE were carried out in June, July and August 2003. At several locations (e.g. in Onsala, Metsähovi) reference and parallel measurements between different FG5s took place. Also re-occupations of one site by the same instrument or other FG5s, but at different epochs were performed. From these checks, the accuracy of observations with the FG5 of the IfE were found in the order of about $\pm 2 - 3$ $\mu$Gal. An rms discrepancy in the same order of magnitude did exist between the IfE and the FGI absolute gravimeters. No significant difference was visible between the BKG and the IfE instruments. Six of the sites were re-occupied in May/June 2004, some new sites were added and observed for the first time in 2004. The actual observation plan is shown in Fig. 3, where Vaasa and Metsähovi again have been observed by two FG5s (of FGI and IfE, respectively) and Onsala by three instruments (FGI, NLH and IfE) at the same time. There, the preliminary results of the 2004 measurements agreed within about $\pm 2$ $\mu$Gal with each other, but both seem to show an increase of about $2-3$ $\mu$Gal with respect to the 2003 values, whereas a decrease due to the land uplift has been expected. The reasons for this behaviour is manifold: First, there is a non-neglible impact from ground water changes (very dry summer in 2003), which has still to be considered. The records of local ground water in Metsähovi, e.g., indicate a change of $+2$ $\mu$Gal between the absolute gravity measurement epochs in 2003 and 2004. Then the measurement campaigns took place at different

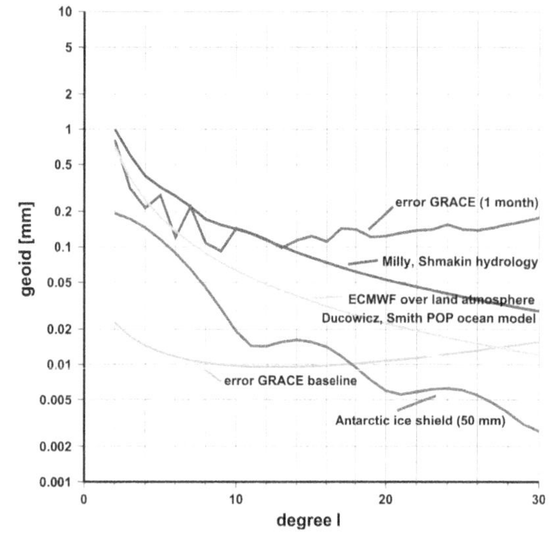


**Fig. 4.** Spectral GRACE error curves (that of the recent monthly solutions and the originally expected one) in terms of geoid heights together with corresponding curves of those effects causing temporal gravity variations like atmosphere, hydrology and ocean. The curves have been provided by Dr. Petrovic (GFZ Potsdam).

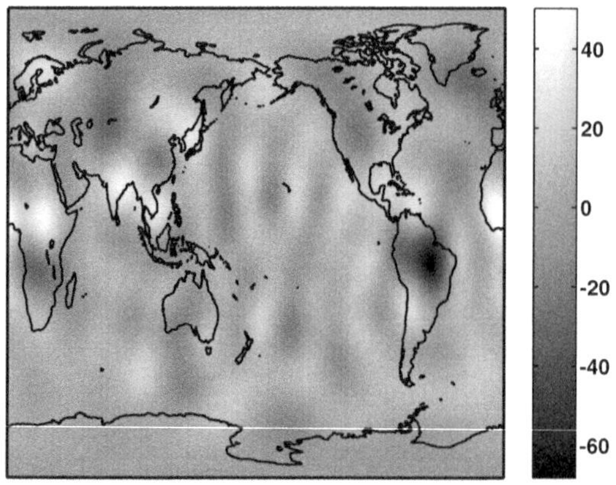


**Fig. 5.** Difference of two monthly GRACE solutions (September - July 2003) represented in terms of gravity anomalies (units: $nm/s^2$) where the oceanic and atmospheric contributions have already been subtracted; only spherical harmonics up to degree and order 13 have been considered.

seasons of the year. Thirdly, the wind driven sea-level changes ('surges') in the Gulf of Bothnia connected with varying loading and attraction effects led to significant gravity changes which have to be accounted for (Ekman 1996b). These preliminary results show, that the reduction of the point-wise terrestrial gravity observations is not trivial. To get a better feeling for how far local and regional effects may affect absolute gravity measurements, the collection of data from rain water tubes, ground water tables, nearby lakes and other has been extended this year. Also the establishment of specific measurement facilities at some sites by the contributing agencies was very helpful in this respect.

## 4 GRACE Contribution

The main result of the project will be reduced point-wise gravity observations and a spatially and temporally smoothed model of the land uplift (gravity and geoid changes), which both can be used for the validation of the GRACE data, i.e. ground-truth for the satellite measurements. Terrestrial gravimetry determines temporal gravity changes point-wise and allows - after reducing the pure vertical motion using GPS measurements - a direct comparison with gravity changes as derived from GRACE data. In this context, the spatial resolution of the satellite results has to be considered. Up to now only the very long wavelengths can significantly be determined from the differences of the monthly GRACE solutions. Fig. 4 shows the GRACE error curves (the actual one and the expected error) together with the contribution of the main temporal gravity changes which have to be removed from the GRACE gravity field model. The atmospheric and oceanic impact as well as the tidal effect is reduced during the standard GRACE processing by applying corresponding models, so that the residual monthly GRACE solutions are mainly affected by the hydrological part (see Fig. 5). As mentioned above, the idea for comparison with the terrestrial gravity data is, that this contribution, which shows a seasonal behaviour, is reduced when measurements are taken at almost the same period of the year. Remaining residual parts will be averaged out to a certain extent by collecting data over several years. Fig. 6 gives a first impression of the annual difference between 2003 and 2004 deduced from GRACE. Here, the difference of two 3-month means (February, March and April) has been computed from the monthly GRACE solutions of 2004 and 2003 considering spherical harmonics up to degree and order 20. The order of magnitude might be correct (a little bit too high), it corresponds to the preliminary absolute gravity measurements. Certainly, also non-PGR signals are still present in this difference and have to be further invesigated.

## 5 Summary and Outlook

Our first absolute gravity observations indicate that the expected accuracy for the determination of the

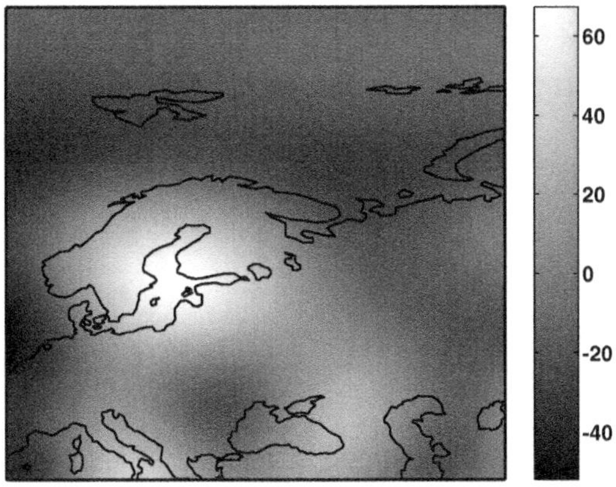


**Fig. 6.** Difference of two 3-month means (Feb., March, April) from monthly GRACE solutions of 2004 and 2003 over Northern Europe represented in terms of gravity anomalies (units: $nm/s^2$) where the oceanic and atmospheric contributions have already been subtracted. The seasonal hydrological signal should have been cancelled out to some extent. Here, spherical harmonics up to degree and order 20 have been considered.

gravity (and GPS height) variations can be achieved. We obtained an accuracy of about $\pm 2 - 3$ $\mu$Gal for a single absolute gravity determination. The reductions of the various effects, however, have to be carried out very carefully applying models as well as auxiliary measurements. Here the reduction of local signals is most challenging. We expect to overcome this problem by appropriate temporal (over five years) and spatial (neighbouring sites) smoothing. The application of different FG5 absolute gravimeters offers good possibilities for inter-comparisons and instrumental checks which helps to improve the absolute accuracy of the network. Measured changes of absolute gravity and GPS heights can be transformed to a temporal geoid variation of Fennoscandia and used as ground-truth for GRACE validation.

Also the processing of the GRACE derived monthly gravity field solutions has to be further improved to get rid of all non-PGR signals in the GRACE-derived gravity field quantities. The set up of a model for the geoid variation in Fennoscandia by the combination of GRACE and terrestrial data has to be investigated in more detail in the future.

**Acknowledgment.** The research was financially supported by the Geotechnologien-Projekt of the German Federal Ministry of Education and Research (Bundesminsterium für Bildung und Forschung BMBF) and the German Research Foundation (Deutsche Forschungsgemeinschaft DFG). This is publication no. GEOTECH-90 of the programme GEOTECHNOLOGIEN.

## References

Boedecker, G. and Fritzer, Th. (1986): International Absolute Gravity Basestation Network. Veröff. Bayer. Komm. für die Internat. Erdmessung der Bayer. Akad. d. Wissensch., Astron.-Geod. Arb. 47, München.

Ekman, M. (1996): A consistent map of the postglacial uplift of Fennoscandia. Terra Nova 8, 158-165.

Ekman, M. (1996b): Extreme annual means in the Baltic Sea level during 200 years. Small Publications in Historical Geophysics, No. 1-12, Summer Institute for Historical Geophysics, Åland Islands, 1995-2003.

Ekman, M. and Mäkinen, J. (1996): Recent postglacial rebound, gravity change and mantle flow in Fennoscandia. Geophyical Journal International 126, 229-234.

GRACE web pages: http:\\www.csr.utexas.edu\grace\ or http:\\op.gfz-potsdam.de\grace\index_GRACE.html.

Hotine, M. (1969): Mathematical Geodesy. ESSA Monograph 2, U.S. Dept. of Commerce, Washington, D.C.

Johansson, J.M., Davis, J.L., Scherneck, H.-G., Milne, G.A., Vermeer, M., Mitrovica, J.X., Bennett, R.A., Jonsson, B., Elgered, G., Elsegui, P., Koivula, H., Poutanen, M., Rönnäng, B.O., Shapiro, I.I. (2002): Continuous GPS measurements of postglacial adjustment in Fennoscandia, 1. geodetic results. J. Geophys. Res. 107, B8, ETG 3, 1-27.

Kuo, C. Y., C. K. Shum, A. Braun, and J. X. Mitrovica (2004), Vertical crustal motion determined by satellite altimetry and tide gauge data in Fennoscandia, Geophys. Res. Lett., 31, L01608, doi:10.1029/ 2003GL019106.

Lambeck, K., Johnston, P., Nakada, M. (1990): Holocene glacial rebound and sea-level change in NW Europe. Geophysical Journal International 103, 451-468.

Lambert, A., Courtier, N., Sasagawa, G.S., Klopping, F., Winester, D., James, T.S., Liard, J.O. (2001) New constraints on Laurentide postglacial rebound from absolute graviy measurements. Geophysical Research Letters 28, 10, 2109-2112.

Mäkinen, J., Engfeldt, A., Harsson, B.G., Ruotsalainen, H., Strykowski, G., Oja, T., Wolf, D. (2004): The Nordic land uplift gravity lines 1966-2003. This volume.

Milne, G.A., Davis, J.L., Mitrovica, J.X., Scherneck, H.-G., Johannson, J.M., Vermeer, M., Koivula, H. (2001): Space-geodetic constraints on glacial isostatic adjustment in Fennoscandia. Science 291, 23, 2381-2385.

Nakiboglu, S.M. and Lambeck, K. (1991): Secular sea-level change. In: Glacial isostacy, sea level and mantle rheology. eds. Sabadini et al., Kluwer Academic Publishers, 237-258.

NRC (1997): Satellite gravity and the geosphere. National Research Council, National Academic Press, Washington.

Scherneck, H.-G., Johannson, J.M., Koivula, H., van Dam, T., Davis, J.L. (2003): Vertical crustal motion observed in the BIFROST project. Journal of Geodynamics 35, 425-441.

Tapley, B., Bettadpur, S., Ries, J., Thompson, P., Watkins, M. (2004): GRACE measurements of mass variability in the Earth System. Science 305, 503-505.

Wahr, J. and Velicogna, I. (2002): What might GRACE contribute to studies of postglacial rebound? Space Sciences Series of ISSI 18, Kluwer Academic Publishers, Dordrecht.

Wilmes, H., Falk, R., Roland, E., Lothhammer, A., Reinhold, A., Richter, B., Plag, H.-P., Mäkinen, J. (2005): Long-term gravity variations in Scandinavia from repeated absolute gravity measurements in the period 1991 to 2003. Proceedings (CD publ.) of IAG Symposium GGSM2004, Porto.

# Recovery of global time-variations of surface water mass by GRACE geoid inversion

G. Ramillien, A. Cazenave
LEGOS CNRS UMR5566 – CNES, Observatoire Midi-Pyrénées,
14, Avenue Edouard Belin, 31400 Toulouse, France

Ch. Reigber, R. Schmidt, P. Schwintzer
GeoForschungsZentrum (GFZ) Potsdam,
Telegrafenberg, 14473 Potsdam, Germany

**Abstract.** Successfully launched in mid-March 2002, the GRACE (*Gravity Recovery & Climate Experiment*) mission currently provides monthly maps of the variations of the geo-potential, at the unprecedented resolution of 200-300 km (max. degree ~100-120). Tiny time variations of the Earth's gravity field (a few of mm in terms of geoid height) measured by GRACE are due to global redistributions of water and air masses inside the fluid envelops at the surface of the planet. As GRACE measures the effects of all the fluid and solid mass variations integrated vertically over time and space, it cannot distinguish the different sources of anomaly. We present here a new inverse approach, based on the least-squares criteria, for separating efficiently the contributions of the main surface fluid reservoirs (atmosphere, oceans, land waters including soil wetness, groundwater and snow cover) from the very first GRACE geoids made recently available (2002-2003). Geoid coefficients estimated for each hydrological contribution are then converted into harmonic coefficients of the water mass anomaly at the surface of the Earth, and expressed in terms of mm of equivalent-water thickness. The proposed inversion method also provides the *a posteriori* uncertainties on the estimated harmonic coefficients. In spite of the increasing noise on the observed GRACE coefficients, the "land water" solutions, developed until degree 30 (~600 km of spatial resolution) reveal important seasonal variations of the water mass (up to 200-300 mm of equivalent-water height) inside large drainage basins of the globe, like Amazon and Congo rivers. The monthly solutions are also used to compute time-series of the total mass of the oceans (i.e. non-steric component of the sea level) that is consistent with the one derived from radar altimeter data, as well as time-series of regional water storage in studied basins. In particular, from the comparison between the evolution of both oceanic and continental waters, it is now possible to estimate the water mass balance between these two global reservoirs.

**Keywords.** GRACE time-variable gravimetry, L2 inverse method, global hydrology.

## 1 Introduction

After a successful launch on the 2002 March 17, the *Gravity Recovery & Climate Experiment* (GRACE) mission developed by National Aeronautics and Space Administration in the USA and Deutsches Zentrum fur Luft and Raumfahrt in Germany is currently mapping the gravity field of the Earth over its 5-year lifetime. For the very first time, this satellite technique will provide an efficient way to measure the changes of gravity with an accuracy better than ~1 cm rms in terms of geoid height and an unprecedented resolution (200-400 km spacially and 1 month temporally). The gravity variations that GRACE can detect include: changes as a result of surface and deep currents in the oceans; changes in soil and ground water storage on lands; mass changes of the ince sheets and glaciers; air and water vapor mass change within the atmosphere; and variations of mass inside the solid Earth. However, the main disadvantage is that GRACE products are values of integrated mass over a vertical column of matter, thus it is not possible to distinguish between the different sources of geoid anomaly.

Several authors have already demonstrated the ability of GRACE to monitor water storage variability on continental areas of several hundred of km or larger, anywhere in the world (Dickey *et al.*, 1997; ESA, 2000; Wahr *et al.*, 1998; Rodell & Famiglietti, 1999, 2001, 2002; Velicogna *et al.*, 2001; Swenson & Wahr, 2002). In particular, comparing variations of soil moisture and groundwater contents from a network of stations in Illinois (~145 800 km$^2$), Rodell & Famiglietti (2001) concluded that these signals would be detectable by GRACE for basin sizes greater than 200 000 km$^2$ with a precision of a few mm in water thickness.

We developed a new method based on an inverse approach using least-square criteria to unravel the different contributions of the main surface fluid reservoirs: atmosphere, oceans and total continental water storage. To constrain the inversion, we included independent information derived from outputs of the global atmospheric, hydrological and oceanic models. We applied this to the time-series of monthly GRACE geoid recently made available for 05/2002 – 03/2004.

## 2 Method

### 2.1 The forward problem

The monthly time-variable geoid $\delta G(t)$ corresponds to the difference between the monthly geoid $G(t)$ measured by GRACE, and the static mean component $G_0$. This latter term represent the main contribution to the gravity field (nearly 99 per cent):

$$\delta G(t) = G(t) - G_0 \tag{1}$$

$\delta C_{nm}(t)$ and $\delta S_{nm}(t)$ are the normalized Stokes coefficients (from the dimensionless ones provided by GRACE data pre-processing) expressed in terms of millimetres of geoid, which depend upon time $t$. They are provided by the GFZ and the CSR to the GRACE users. $n$ and $m$ are the harmonic degree and order respectively. $\delta G(t)$ can be developed as a series of these harmonic coefficients:

$$\delta G(t)=\sum_{n=1}^{N}\sum_{m=0}^{n}(\delta C_{nm}(t)\cos(m\lambda)+\delta S_{nm}(t)\sin(m\lambda))P_{nm}(\cos\theta) \tag{2}$$

where $\theta$ is the co-latitude, $\lambda$ is longitude and $P_{nm}$ is the normalized associated Legendre function.
Neglecting the mass variations inside the solid Earth induced by tectonics such as earthquakes, postglacial rebound and mantle convection, let us assume that $\delta G$ is merely the sum of $K$ surface fluid contributions and these contributions are not correlated in time and space:

$$\delta G(t)=\sum_{k=1}^{K}\delta G_k(t)=A\,\delta G_k \tag{3}$$

where $A$ is the "separating" matrix formed by a column of identity blocks.
For a given surface density of mass $\delta h(t)$ at the surface of the Earth, the corresponding geoid anomaly coefficients can be predicted from the surface density ones $\delta C^h{}_{nm}(t)$ and $\delta S^h{}_{nm}(t)$ using a fast linear filtering:

$$\begin{Bmatrix}\delta C_{nm}(t)\\ \delta S_{nm}(t)\end{Bmatrix}=W_n^0\begin{Bmatrix}\delta C^h_{nm}(t)\\ \delta S^h_{nm}(t)\end{Bmatrix} \tag{4}$$

where $W^0{}_n$ is an isotropic and stationary function that weights the harmonic coefficients (e.g. Ramillien, 2002):

$$W_n^0=\frac{4\pi G R\rho_W}{(2n+1)\gamma(\theta)}(1+z_n) \tag{5}$$

$z_n$ represent the Love numbers for the elastic response of a surface-loaded Earth and $\gamma(\theta)$ is the normal gravity on the reference ellipsoid, respectively. $G$ ($\approx$ 6,67.10$^{-11}$ m$^3$kg$^{-1}$s$^{-2}$) is the gravitational constant and R ($\approx$6378 km) is the mean radius of the Earth. $\rho_W$ ($\approx$ 1000 kgm$^{-3}$) and $\rho_E$ ($\approx$5517 kgm$^{-3}$) are the mean water and Earth densities respectively.
Considering the elastic deformation of the solid Earth under the variable load via the Love numbers $z_n$, the Stokes coefficients become:

$$\begin{Bmatrix}\delta C_{nm}(t)\\ \delta S_{nm}(t)\end{Bmatrix}=\frac{(1+z_n)R^2}{(2n+1)M}\iint_S \delta q(\theta,\lambda,t)\begin{Bmatrix}\cos\\ \sin\end{Bmatrix}(m\lambda)P_{nm}(\cos\theta) \tag{6}$$

where $\delta q(\theta, \lambda, t)$ is the surface load. $S$ and $M$ is the total surface and mass of the Earth, respectively.
Atmospheric mass redistribution is classically deduced from atmospheric surface pressure data. The variable load $\delta q(\theta, \lambda, t)$ is related to the pressure anomaly $\delta p(\theta, \lambda, t)$ through:

$$\delta q(\theta,\lambda,t)=\frac{\delta p(\theta,\lambda,t)}{\rho_W\gamma(\theta)} \tag{7}$$

This latter formula is also used to compute the oceanic surface load from the ocean bottom pressure data provided by ocean general circulation models (OGCMs).
Land waters considered here include soil water, groundwater and snow depth. These three contributions provided by global hydrology models are expressed in terms of equivalent water height $\delta h(\theta, \lambda, t)$ and the corresponding surface load is given by:

$$\delta q(\theta,\lambda,t)=\rho_{W,S}\,\delta h(\theta,\lambda,t) \tag{8}$$

where $\rho_{W,S}$ can be either liquid water or snow density.

### 2.2 The inverse problem

According to Eq.3 there are $K$ times more Stokes coefficients to adjust than observed ones. Therefore the problem is highly underdetermined. Because of non-uniqueness of this inverse problem, new *a priori* information need to be included as new constraints for recovering the coefficients of all water contributions. We propose to use a generalized least-square inversion developed by Tarantola (1987) for estimating the Stokes coefficients of the $K$ fluid contributions. The estimates are built as linear combination of optimal fitted parameters:

$$\Gamma_k(t)=\Gamma_k^0(t)+C_k\xi \tag{9}$$

where the vector $\xi$ is the solution of the expression:

$$(C_D+C_M+AC_kA^T)\xi=\Gamma^{obs}(t)-A\Gamma_k^0(t) \quad (10)$$

where $\Gamma^{obs}(t)$ is the vector formed by the list of all the observed geoid coefficients and $\Gamma^0_k(t)$ is the initial guess of *k-th* contribution. $C_D$ and $C_M$ are the covariance matrices of the *a priori* error for GRACE observations and the chosen model corresponding to the fluid contribution number *k*, respectively. $C_k$ is the covariance matrix that describes the statistics of the water mass variations in the reservoir *k*, which are determined by the analysis of available model (see sec. 2.3). The *a posteriori* covariance matrix is computed using:

$$C_k^{prime}=C_k-C_kA^T(AC_kA^T+C_D+C_M)^{-1}AC_k \quad (11)$$

The uncertainties associated to the fitted Stokes coefficients are given by the root-mean square of the diagonal elements of the latter matrix:

$$\sigma_k(t)=\sqrt{C_k^{prime}(diag)} \quad (12)$$

The surface density coefficients, expressed in terms of surface load variations, are estimated using an inverse filtering:

$$\begin{Bmatrix}\delta C^h_{nm}(t)\\ \delta S^h_{nm}(t)\end{Bmatrix}=W_n^{-1}\begin{Bmatrix}\delta C_{nm}(t)\\ \delta S_{nm}(t)\end{Bmatrix} \quad (13)$$

where $W_n^{-1}$ is the predicting filter from Eq.5, that is weighted by a stabilizing function that tapers the amplitudes of the filter at high degrees, for avoiding the development of short-wavelength instabilities in the prediction.

### 2.3 The covariance matrix construction

We assume here that the surface fluid fields are stationary processes. If $D_k(\Delta t)$ is the matrix formed by the list of all the Stokes coefficients up to the maximum degree $N=100$, related to the global fluid model number *k*, for a period of analysis of $\Delta t$ months, then an estimation of $C_k$ for this period of time and this *k*-th contribution is given by:

$$C_k=\left[D_k(\Delta t)-\overline{D_k}\right]^T\left[D_k(\Delta t)-\overline{D_k}\right] \quad (14)$$

where $\overline{D_k}$ is the time-mean values of the model coefficients computed during $\Delta t$. Several tests have shown that should be ~2-3 months around the chosen month *t* of GRACE observations to be inverted. Greater values of this parameter lead to numerical smoothing and so provide less precise geoid solution.

## 3 Statistical constraints from models

Model outputs are used as first guess $\Gamma^0_k$ and for evaluating the $C_D$ and $C_M$ matrices, for a set of twelve months of a so-called "standard year" by statistical analysis, and before solving the linear system (Eq.10).

### 3.1 Atmosphere

Global maps of surface pressure from 6-hr re-analysis of ground observations coupled with equations of general circulation are currently provided by National Center for Environment Predictions (NCEP) and European Centre for Medium Weather Forecast (ECMWF). Errors are around ~1 mm bar on the average for a monthly gridded value, but this uncertainty can be regionally and locally more important and correlated with the station distribution over the globe (poor coverage and quality for Antarctica). Monthly mean surface pressure fields are available from 1979 to 1996 for NCEP and from 1979 to 1993 for ECMWF.

### 3.2 Oceans

Maps of the ocean bottom pressure are routinely provided by several OGCM. In the present study, we used : (i) POCM_4C is described in Semtner & Chervin (1992), the model is run from 1979 to 1998 and forced by daily ECMWF re-analysis of heat, freshwater flux (evaporation minus precipitation and river runoff) and wind stress, its resolution is approximately 0.25° latitude and longitude, permitting detection of the western boundary currents; (ii) OPA from Laboratoire d'Océanographie Dynamique et de Climatologie (LODYC) (Madec & Imbard, 1996; Madec *et al.*, 1999): a global model forced by the European Remote Sensing (ERS) satellite-derived surface winds and ECMWF heat and freshwater fluxes, The modelled period is 1992-2000 inclusive and the correction to bottom pressure as a result of the Boussinesq approximation is applied (Greatbatch, 1994); (iii) ECCO assimilation (Stammer *et al.* 2000) of *in situ* data (vertical temperature and salinity profiles) and mean sea surface measurements by Topex/Poseidon satellite altimetry from 1993 to present, which provides global maps of ocean bottom pressure every 12 hours (http://www.ecco-group.org).

### 3.3 Land waters

Land water mass variations are deduced from outputs of global hydrological models. These

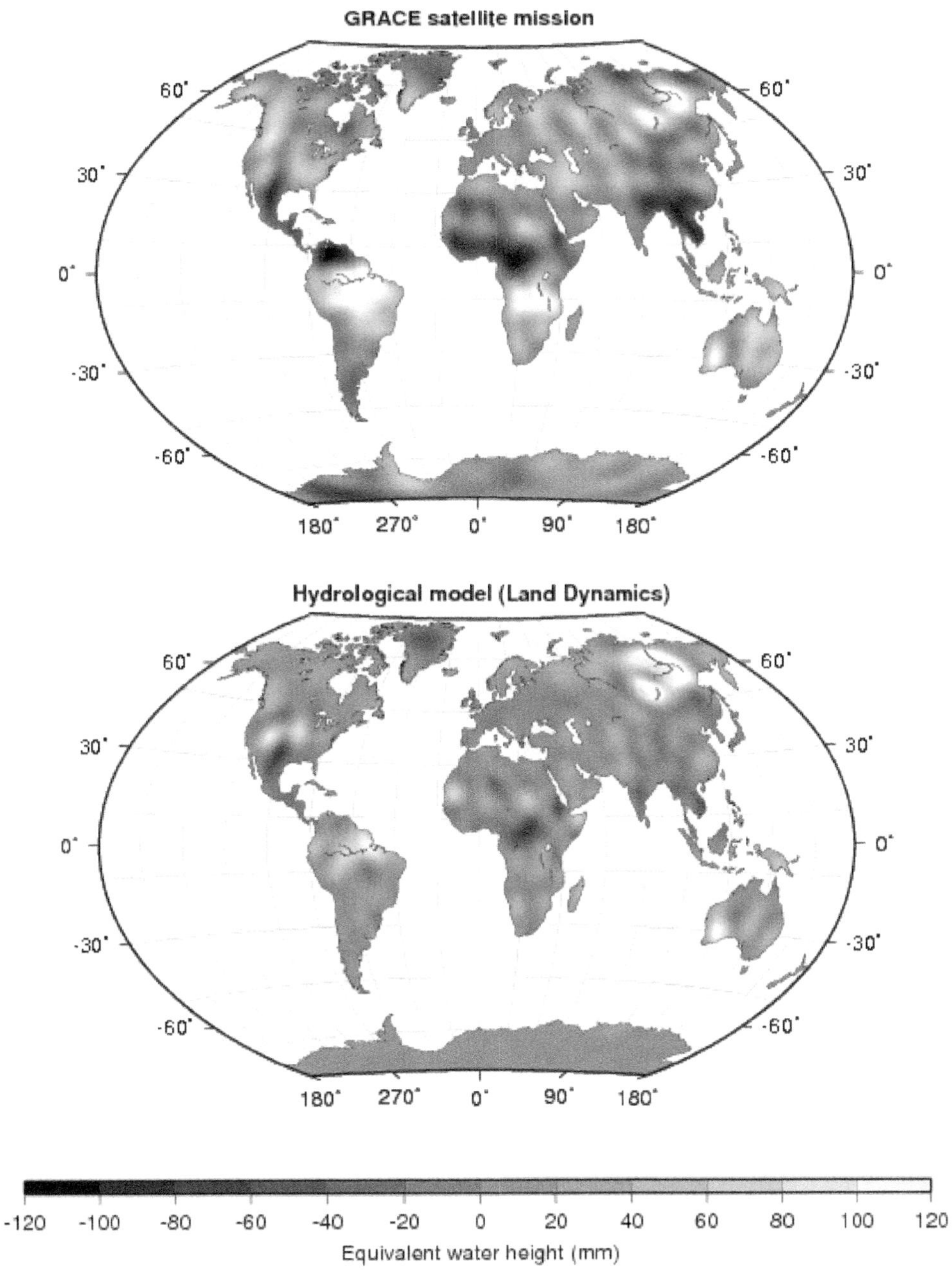


**Figure 1:** Global seasonal variations of the land waters: soil moisture + grounwater + snow depth (April-May minus November 2002) estimated from inversion of GRACE geoids (top), and according to the LaD model (Milly & Shmakin, 2002) (bottom). Spatial resolution here is 660 km (cut-off degree ~30). Note the differences of amplitude with no data in the model over Antarctica.

predictions correspond to water stored in the root-zone (i.e., soil wetness), water in the shallow aquifers (underground water) and snow. We used the prognostics from two models : (i) The Land Dynamics (LaD) model developed by Milly & Shmakin (2002), this model provides global (exclusive of Antarctica and Greenland) monthly 1° x 1° gridded time series of snow depth, soil moisture and groundwater for 1981-1998; (ii) Global Soil Wetness Project (GSWP) developed by ~8 groups including the hydrological model from Douville *et al.* (1999) based on the Interactions between Soil Biosphere and Atmosphere (ISBA) land surface scheme (Noilhan & Planton, 1989), this model provides monthly 1° x 1° gridded time series of soil wetness and snow depth for only two years (1987-1988), once again excluding Greenland and Antarctica.

## 4 Application: Inversion of time-series of monthly GRACE geoids

The adjustment according to Eq. 9 and 10 was performed for the monthly GRACE geoids recently made available by GFZ and CSR from April/May 2002 to June 2004, considering the spherical harmonic coefficients up to degree/order 30 (noise in the GRACE data becomes critical for higher degrees) and excluding the $C_{20}$ term. $C_{00}$ and the one-degree coefficients are automatically set to zero, thus any change in the total mass over continents is balanced by opposite change in the oceans.

These provided monthly gravity fields should represent the hydrology on the continents plus the non-tidal variations of the atmosphere and the oceans, so we used statistical constraints in Eq.14 from the LaD model and "residual" signals for atmosphere and ocean computed from ECMWF and NCEP, and POCM and ECCO assimilation outputs respectively. Therefore, the three fluid contributions ($K=3$) to recover for each month are the total continental water storage (soil wetness, groundwater and snow depth), and non-tidal atmospheric and oceanic residuals.

Because of numerical instabilities of high-degree coefficients in the matrices estimated from the models, the linear system become ill-conditioned. In practice, the matrix of Eq.10 is always symmetric by construction. If it is definite positive, so one can use the Cholesky factorisation to solve the linear system. In any case, LU and Singular Value Decomposition (SVD) would also be used but reveal to be time-consuming for fitting ~4000 at degree 30.

From the resulting vectors $\Gamma_k(t)$, the coefficients estimated for the continental water storage were used for computing the difference fields over the seasonal period (April/May 02 minus November 02) as the main cycle, and compared to the LaD model as presented on Fig.1. The principal features visible on the solutions exceed in amplitude the structures predicted by the hydrological models. The L2 adjustment also provides the *a posteriori* uncertainties (from Eq. 11-12) associated with the fitted Stokes coefficients, they are typically of ~15 mm of equivalent water thickness for the total water storage over the continents.

Considering the GRACE solutions, the global temporal trend of the oceanic mass (i.e., "eustatic" contribution expressed in terms of sea level rise) is ~1,8 mm/y at the spatial resolution of 2000 km (deg. 10) and for the recent period 2002-2004, this estimate is comparable to the value deduced from satellite altimetry (~2,8 mm/y using ten years of T/P data). We found a positive and small trend of +0,2 mm/y of equivalent water height for all the continental areas (gain of mass), whereas global hydrological model such as LaD tend to predict an important negative trend. GRACE solutions also reveal an important loss of mass over the ice sheets for the studied period, around –2,0 mm/y, and in particular about –1,6 mm/y for Antarctica. Unfortunately, the presence of noise in the GRACE geoids for degree higher than 10-15 remains an important limiting factor for improving the estimation of net balance estimates for global water mass reservoirs.

## 5 Conclusion

We successfully applied a new L2 method for solving a very unconstrained gravity inverse problem of "de-correlation" of fluid contributions by introducing independent information from different numerical models. Instrument errors and other a priori uncertainties on GRACE data and the physical model were included in this least-square approach. Using the monthly predictions and the signals variance-covariance matrices derived from a hydrological model as *a priori* information, the inversion of the GRACE monthly fields yields mass redistribution patterns comparable to models but of greater amplitude. GRACE is demonstrating its unique capability to monitor large scale temporal variations in particular of the water mass on the continents. This new approach offers the advantage of not relying on model forecasts. More information on the development of the method of separation with models are provided in Ramillien *et al.*, 2004.

## References

Dickey J. and National Research Council Commission, (NRC), 1997, **Satellite Gravity & The Geosphere**, *National Academy Press*, Washington D.C., pp 112.

Douville H., E. Basile, P. Caille, D. Giard, J. Noilhan, L. Peirone and Taillefer, 1999,

"Global Soil Wetness Project: forecast an assimilation experiments performed at Météo-France", J. Meteorological Soc. Of Japan, 77(1B), 305-316

European Space Agency, (ESA), 2000, **From Eötvös to Milligal**, Final Report, No. 13392/98/NL/GD.

Greatbatch R., 1994, "A note on the representation of steric sea level in models that conserve volume rather than mass", *J. Geophys. Res.*, **99**, 12767-12771.

Madec G. and M. Imbard, 1996, "A global mesh to overcome the North pole singularity", *Clim. Dyn.*, **12**, 381-388.

Madec G., P. Delecluse, M. Imbard and C. Lévy, 1999, OPA 8.1 "Ocean General Circulation Reference Model", Tech. Rep. 10, Institut Pierre Simon de Laplace, Paris.

Milly P. C. D. and A. B. Shmakin, 2002, "Global modeling of land water and energy balances: 1.The Land Dynamics (LaD) model", *J. of Hydrometeorology*, **3**, 283-299.

Noilhan J. and S. Planton, 1989, "A simple parametrization of the land surface processes for meteorological models", *Monthly Weather Review*, **117**, 536-549.

Ramillien G., 2002, "Gravity/magnetic potential of uneven shell topography", *J. of Geodesy*, **76**, 139-149.

Ramillien G., A. Cazenave and O. Brunau, 2004, "Global time variations of hydrological signals from GRACE satellite gravimetry", *Geophys. J. Int.*, **158**, 813-826.

Rodell M. and J. S. Famiglietti, 1999, "Detectability of variations in continental water storage from satellite observations of the time dependent gravity field", *Water Resources Res.*, **35**, 9, 2715-2723.

Rodell M. and J. S. Famiglietti, 2001, "An analysis of terrestrial water storage variations in Illinois with implications for the Gravity Recovery and Climate Experiment (GRACE)", *Water Resources Res.*, **37**, 5, 1327-1339.

Rodell M. and J. S. Famiglietti, 2002, "The potential for satellite-based monitoring of groundwater storage changes using GRACE: the High Plains Aquifer, central U.S.", *J. of Hydrology*, **263**, 245-256.

Semtner A. J. and R. M. Chervin, 1992, "Ocean general circulation from a global eddy-resolving model", *J. Geophys. Res.*, **97**, 5493-5550.

Stammer D., C. Wunsch, R. Giering, C.Eckert, P. Heimbach, J. Marotzke, A. Adcroft, C. N. Hill and J. Marshall, 2000, "The global ocean circulation and transport during 1992 – 1997 estimated from ocean observations and a general circulation model", Parts 1, 2, 3, ECCO Report Series.

Swenson S. and J. Wahr, 2002, "Methods for inferring regional surface-mass anomalies from Gravity Recovery and Climate Experiment (GRACE) measurements of time variable gravity", *J. Geophys. Res.*, **107** (B9), ETG, 1-13.

Tarantola A., 1987, **Inverse Problem Theory**, Elsevier, Amsterdam, pp. 613.

Velicogna I., J. Wahr, H. Van den Dool, 2001, "Can surface pressure be used to remove atmospheric contributions from GRACE data with sufficient accuracy to recover hydrological signals ?", *J. Geophys. Res.*, **106** (B8), 16415-16434.

Wahr J., M. Molenaar and F. Bryan, 1998, "Time variability of the Earth's gravity field: hydrological and oceanic effects and their possible detection using GRACE", *J. Geophys. Res.*, **103** (B12), 30205-20229.

# Seasonal Gravity Field Variations from GRACE and Hydrological Models

Ole B. Andersen, Danish Space Center, Copenhagen, Denmark, oa@spacecenter.dk.

Jacques Hinderer, Institut de Physique du Globe de Strasbourg (CNRS/ULP UMR 7516), Strasbourg, France.

Frank G. Lemoine, NASA GSFC, Space Geodesy, Code 926, Greenbelt, MD, 20771, USA.

**Abstract.**
This study present an investigation of the newly released 18 monthly gravity field solutions from the GRACE twin space-crafts with emphasis on the global scale annual gravity field variations observed from GRACE and modeled from hydrological models as annual changes in terrestrial water storage.

Four global hydrological models covering the same period in 2002-2003 as the GRACE observations were investigated to for their mutual consistency in estimates of annual variation in terrestrial water storage and related temporal changes in gravity field. The hydrological models differ by a maximum of 2 μGal or nearly 5 cm equivalent water storage in selected regions. Integrated over all land masses the standard deviation among the annual signal from the four hydrological models are 0.6 μGal equivalent to around 1.4 cm in equivalent water layer thickness.

The estimated accuracy of the annual variation in gravity from GRACE is around 0.4 μGal (0.9 cm water layer thickness) on 2000 km length scales. This makes the GRACE observations of terrestrial water storage on global annual scales more accurate than present-day hydrological models.

**Keywords, GRACE, gravity, annual signal, hydrology.**

## 1 Introduction

GRACE were successfully launched on March 17th, 2002 through the joint partnership between the US National Aeronautics and Space Administration (NASA) and the German Deutsche Forschungsanstalt fur Luft und Raumfart (DLR). The twin identical GRACE satellites co-orbit the Earth at 480 km altitude with an inclination of 89°. Through highly accurate ranging of the 200-km distance between the twin GRACE satellites monthly gravity fields are produced over the 5-year mission. Spatial and temporal variations in the Earth's gravity field are manifested through changes in the distance between the spacecrafts as they orbit the Earth (Tapley and Reigber, 2000; Bettadpur et al., 1999). Such a mission has been advocated for many years (i.e. Kaula, 1970, Rummel, 1979, Colombo and Chao, 1992) and GRACE is now bringing a new era to satellite geodesy. Because of the low orbiting altitude GRACE will be able to map both spatial and temporal changes in the Earth's gravity field with more detail and accuracy than any other existing satellite.

Changes in the gravity field is an integral indicator of mass re-distribution within the Earth system, and GRACE gravity data can be used to study a number of such processes: terrestrial water storage; changes due to surface and deep currents in the ocean, exchange of water between land and ocean like ice-melting and river runoff; variations of mass within the Earth; and changes in the atmosphere (Wahr, 1998).

Temporal gravity field changes from GRACE have been presented by Andersen and Hinderer (2005), and Wahr et al., (2004). Wahr et al visually compared GRACE observations with one hydrological model on annual scales, whereas Andersen and Hinderer studied inter-annual gravity changes from GRACE.

## 2 The GRACE Monthly Gravity Fields

The monthly Level-2 GRACE gravity fields provided by CSR (Center of Space Research, The University of Texas at Austin) were used for this investigation. Presently 18 monthly solutions have been released. These are irregularly spaced over nearly two years covering the following months of 2002 (April/May, August through November), the following months of 2003 (February, March, April, May, July through December), and January through March for 2004.

A number of different processes in the Earth's system cause gravity field variations on the time scales investigated in this paper. In the GRACE level-2 monthly estimates temporal gravitational accelerations due to the solid Earth and ocean Tides; Atmosphere; Barotropic ocean variability; Pole tide and N-body perturbations are accounted for using various models (Bettadpur, 2003a 2003b). Assuming these corrections to be accurate (i.e., Knudsen, 2002; Ray et al., 2003) leaves the hydrology/terrestrial water storage as the largest contributor to the temporal gravity changes on annual scales.

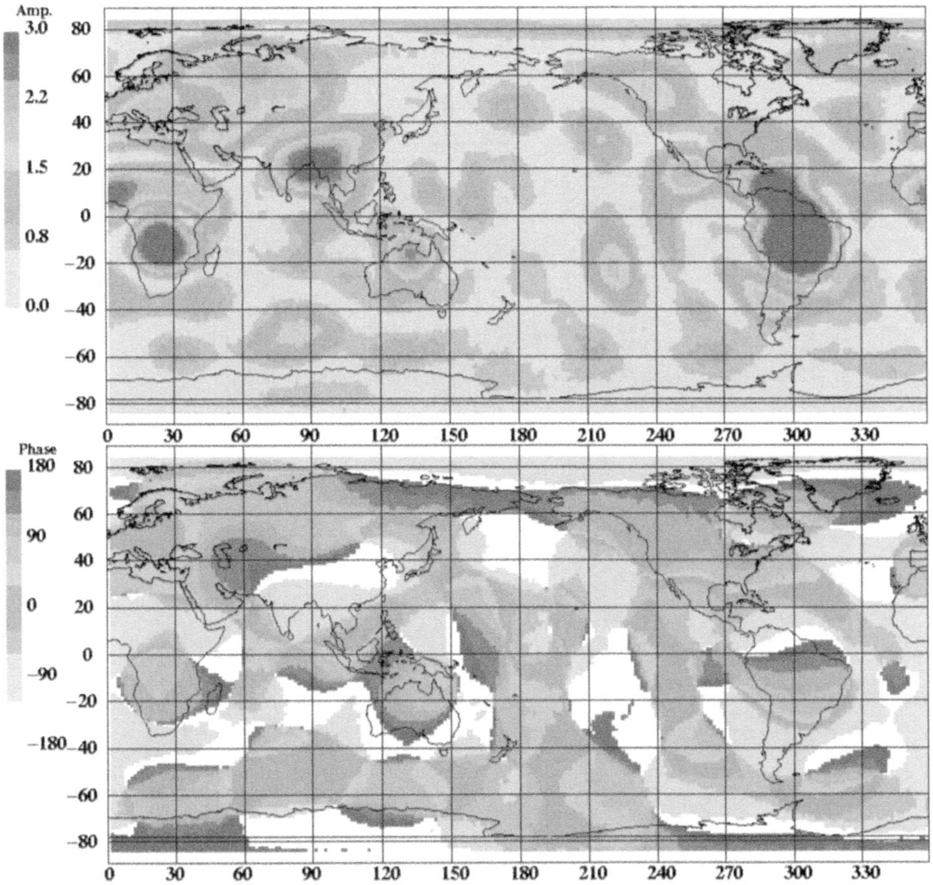


**Figure 1**. Annual Gravity Signal from GRACE. Upper panel shows the amplitude in μGal, and the lower panel shows the phase in degrees. The phase correspond to the day number that the amplitude peaks.

## 3 Annual gravity field variation from GRACE

The 18 monthly gravity-fields were initially differentiated with the 111 days mean GRACE gravity field model GGM01S to obtain the monthly deviations from the mean gravity field. These monthly residuals were then expanded from degree 2 to 10. Calibrated standard deviations associated with the geo-potential coefficients for each monthly solution are the current best error estimate

(Bettadpur, 2003). Degree zero and one terms were not included in the GRACE solutions, and the $C_{20}$ term was omitted as it shows large un-physical variability during 2002.
These were expanded to the same degrees and all data were submitted to a stochastic least squares inversion (Jackson, 1979). In this fit annual varying cosine and sine components were fitted centered at Jan 1st 2003. The estimated annual varying gravity signal is shown in Figure 1.

The upper panel in Figure 1 shows the amplitude in μGal of the annual gravity signal observed from GRACE. The lower panel shows the phase with zero degree corresponding to Jan 1st.

Amplitudes larger than 4 μGal are found in an extended region over South America going from 20°S to 20°N with the maximum amplitudes reaching 7.2 μGal in the Amazon Basin. Over the central southern Africa amplitudes peaks at 4.2 μGal and over Bangladesh at 3.4 μGal. Amplitudes larger than 2 μGal are found in the following regions: Central Africa; Northern Territory (Australia); western part of Russia and Mexico. Over the eastern Canada, Europe, and in the Arctic amplitude exceeds 1 μGal. The central Pacific Ocean around 20°S, and southern Indian Ocean, exhibit annual variations exceeding one μGal.

The phases are shown in the lower panel. The southern Hemisphere anomalies (Amazon, Southern Africa and Australia) all peaks near the Spring (March) Equinox or a little later (Amazon), whereas the anomalies over the northern Hemisphere (Sahara and Bangladesh, Canada) peaks 180 days later near the fall Equinox (September). In Europe the signal peaks in January and for Russia it is around April/May. The two peaks on the African continent are in opposite phase, and the large anomaly over South America also changes phase by 180° as it crosses the Equator.

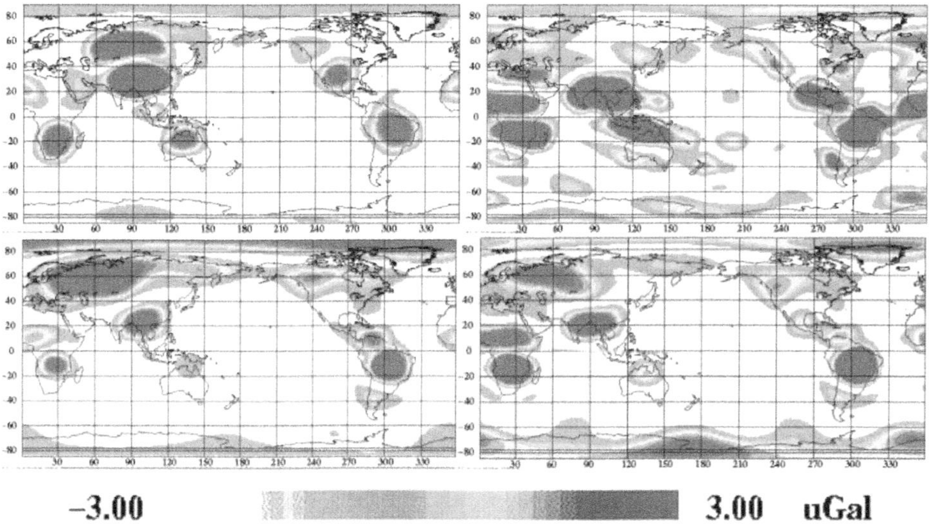


**Figure 2.** Annual Gravity signal (amplitude) from the four global hydrological models investigated: Au & Chao (upper left) and CPC upper right, LADWorld (lower left) and GLDAS (lower right)

## 4 Annual gravity field variation from hydrology

The dominant contributor to the annual gravity field variation on land can be subscribed to terrestrial water storage (Wahr et al., 1998). Four global hydrological models were investigated for their annual variations in soil moisture to investigate their mutual consistency and to verify the annual signal observed by GRACE. These four hydrological models are believed to be the most accurate global hydrological models and are: The Climate Prediction Center (CPC) model by Fan & Van den Dool. (2004); the LADWorld model by Milly and Schmekin (Milly et al., 2003); the Global Land Data Assimilation System GLDAS by Rodel et al., (2004); and the Au & Chao model from NASA/GSFC (Au et al., 2003). It should be noticed that the LADWorld model only covers the first 18 months of 2002 and 2003, and that Polar Regions including Greenland have been eliminated from the analysis.

Each model was expanded into spherical harmonic expansion to degree and order 10 similar to the expansion used for the GRACE data. The water stored in the soil can be approximated by a thin layer of water of thickness $h(t)$. The gravity change due to the mass of this water can be expressed as a free air anomaly change using the Bouguer plate approximation (Telford et al., 1976; Ray, 2002; Knudsen and Andersen, 2001) like $\Delta g(t) = 41.89\ h(t)$ assuming the water density to be 1 g/cm3. $h(t)$ is the height of the thin layer of water in meters, and the resulting gravity change $\Delta g(t)$ is in micro Gal (μGal). The Bouguer plate approximation might introduce errors of up to 10 percent (Crossly et al., 2003).

Figure 2 shows the annual variation in gravity field from the 4 hydrological models. The amplitudes are generally in good agreement with the amplitude obtained by GRACE. The major differences are found over Russia, which can be subscribed to how the different models treat snow-coverage (Andersen and Hinderer, 2004, Wahr et al. 2004). The phases are in generally in fair agreement over most regions except for North America. Over the ocean the spherical harmonic expansion of the terrestrial hydrological signal generates artificial signal.

## 5 Annual gravity field error

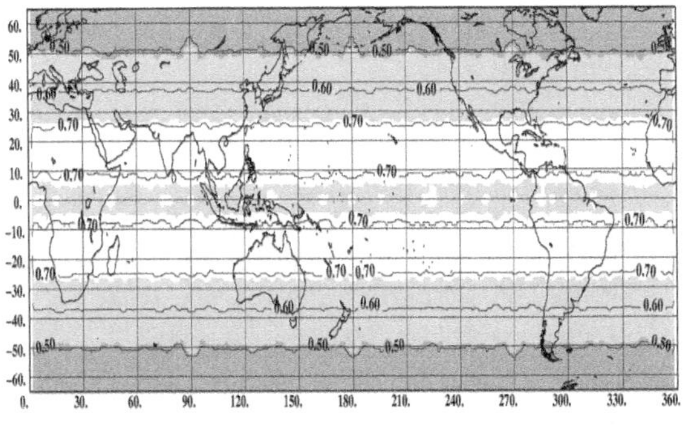

**Figure 3** Estimated amplitude error for the annual gravity field variation from the applied stochastic inversion. Units are in μGal. The largely zonal structure reflects the largely zonal structure of the error in the GRACE monthly gravity field solutions

The stochastic inversion yields an error estimate based on the calibrated standard deviations associated with the errors provided on each of the 18 monthly input data. The estimated error is turned into an error-field for the amplitude of the annual variation is shown in Figure 3. The error has a zonal structure with a mean value of 0.55 μGal with a maximum of 0.75 μGal. In terms of equivalent water level this corresponds to a mean value of 1.2 cm and a maximum of 1.8 cm in the zonal. The $C_{20}$ term was re-introduced for the error computation, as it accounts for most of the signal, as the a-priori errors for the monthly solutions are nearly zonal because of the increasing number of observations with latitude

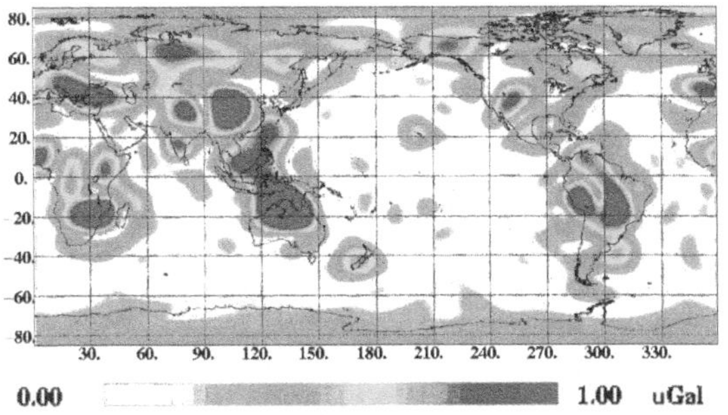

**Figure 4** Estimated standard deviation of the annual gravity field variation from the four global hydrological models (CPC, LADWorld, GLDAS and Au&Chao). Units are in μGal.

An error estimate of the ability to model global variation in the gravity fields from variations in terrestrial water storage can be obtained by computing the consistency or standard deviation among the four "best" global hydrological models described above.

The standard deviation of annual amplitude from the four hydrological models is shown in Figure 4. The maximum values reaches 2.3 μGal over Australia, which is subscribed to the CPC model being different to the other models in this region. The average standard deviation computed over the land regions among the four models is 0.65 μGal, which is equivalent to 1.5 cm of water. The similar analysis carried out over only the three more well-known models (CPC, LADWorld and GLDAS) increases the averaged standard deviation to 0.72 μGal over the continents.

## 6 Conclusions

The investigation clearly demonstrates the existence of an annual gravity field signal in the GRACE monthly gravity field data. Amplitudes larger than 4 μGal are found in an extended region over South America going from 20°S to 20°N with the maximum amplitudes reaches 7.1 μGal in the Amazon Basin. Over the central southern Africa amplitudes peaks at 4.2 uμGal and over Bangladesh at 3.4 μGal. Overall the annual variation explains roughly 60% of the variance in the 18 monthly fields.

The predicted accuracy of the amplitude of the annual gravity variation from GRACE is around 0.4 μGal which equivalents to roughly 0.9 cm of water on 2000 km length scales. This was compared with an error estimate for the annual variation in gravity field based on hydrological models. The error estimate was computed from the standard deviation among the four "best" global hydrological models available. The analysis shows that on global scales GRACE is able to provide important new insight into annual gravity field variations, as the GRACE a-posteriori error is roughly 1.5 times smaller than the standard deviation among the present-day hydrological models. This makes the GRACE observations more accurate than present-day hydrological models for the study of changes in terrestial water storage on global annual scales.

This paper focused on the annual changes in gravity as caused by hydrology assuming the corrections for atmospheric and oceanographic to be "perfect". GRACE observes the integrated gravity change and changes in atmospheric or oceanographic not accounted will map into the accuracy estimates presented. Possible errors in atmospheric pressure on these scales could be large at high latitude and consequently the error can be even lower that the presented results.

**Acknowledgements.** This analysis is partly sponsored by the Danish Research Agency and the NASA GEST office. The authors are thankful to the reviewers as well as B. F. Chao, and colleagues at NASA/GSFC for constructive comments, to CSR for providing the GRACE monthly gravity fields and to colleagues in hydrology for providing their models.

## References

Andersen, O, and J. Hinderer, Global Inter-Annual Gravity Changes from GRACE: Early Results, Geophys Res Lett. Geophys. Res. Lett., Vol. 32, No. 1, L01402 10.1029/2004GL020948, 2005.

Au, A.Y., B.F. Chao, M. Rodell and T.J. Johnson Global Soil Moisture Field and the Effects on the Earth's Gravity. (abstract) Workshop on Hydrology from Space, http://gos.legos.free.fr/, Toulouse, France, 2003

Bettadpur, S., Gruber, Th., Watkins, M., Grace Science Data System Design, *Supp. EOS Transaction EGU* 80 (46), 1999.

Bettadpur, S., Level-2 Gravity Field Product user Handbook, *GRACE 327-734, CSR publ. GR-03-01*, 17pp., 2003

Bettadpur, S., UTCSR Level-2 Processing Standards Document for Level-2 products Release 0001, *GRACE 327-742, CSR publ. GR-03-03*, 16 pp., 2003

Colombo, O., and B. F. Chao, Global Gravity Change in 2001, *Proc, IAG Symposium 112 Geodesy and Physics of the Earth*, (Ed) C. Reigber, 71-75, Springer Verlag, Heidelberg, 1992.

Crossley, D., Hinderer, J., Llubes, M., & Florsch, N., The potential of ground gravity measurements to validate GRACE data, *Advances in Geosciences, 1*, 1-7, (2003)

Fan, Y., and H. van den Dool, The Climate Prediction Center global monthly soil moisture data set at 0.5_ resolution for 1948 to present, *J. Geophys. Res.*, 109, D10102, doi:10.1029/2003JD004345, 2004

Jackson, D. D: The Use of A Priori Data to Resolve Non-Uniqueness in Linear Inversion", *Geophys. J. Royal Astron. Soc.*, v.57, 137-158, 1979

Kaula. W. M. The Terrestial Environment Solid Earth and Ocean Physics, *NASA Report CR 159*, 1970

Knudsen, P. Ocean Tides in GRACE monthly averaged gravity fields, *Space Sciences Reviews, 108*, 261-270, 2003

Knudsen P. and O. Andersen, Correcting GRACE gravity fields for ocean tide effects *Geophys Res. Lett, 29(8)*, 19-1 19-4, 2002

Milly, P. C. D., and A. B. Shmakin, Global modeling of land water and energy balances. Part I: The land dynamics (LaD) model. *Journal of Hydrometeorology*, 3(3), 283-299, 2002

Rodell, M., P. R. Houser, U. Jambor, J. Gottschalck, K. Mitchell, C.-J. Meng, K. Arsenault, B. Cosgrove, J. Radakovich, M. Bosilovich, J. K. Entin, J. P. Walker, D. Lohmann, and D. Toll, The Global Land Data Assimilation System, Bull. Amer. Meteor. Soc., 85 (3), 381–394, 2004.

Ray, R. D., D. D. Rowlands and G. Egbert, Tidal models in a new era of satellite gravimetry, *Space Science Reviews, 108*, 271-282, 2002

Rummel. R., Determination of the short wavelength components of the gravity field from satellite to satellite tracking or satellite gradiometry: An attempt to an identification of problem areas, *Manus. Geodetica 4*, 107-148, 1979

Tapley, B D. and Reigber, C. The GRACE mission status and future plans, *EOS Trans, 81, F311*, 2000

Telford, W. M., Geldart, L. P. Sheriff, R. E. Keys, D. A., 1976 *Applied Geophysics*, Cambridge University Press, 1976

Wahr, J., Molenaar, M., Bryan, F. Time variability of the Earth's gravity field: Hydrological and oceanic effects and their possible detections using GRACE, *J. Geophys. Res., 103(B12),* 30205-30229, 1998

Wahr, John; Swenson, S.; Zlotnicki, V. & Velicogna, I., Time-variable gravity from GRACE: First results, *Geophys. Res. Lett.*, Vol. 31, No. 11, L1150110 (2004)

# Mass redistribution from global GPS timeseries and GRACE gravity fields: inversion issues

J. Kusche, E.J.O. Schrama
DEOS,
TU Delft, Kluyverweg 1, PO box 5058, 2600 GB Delft, The Netherlands

**Abstract.** Monitoring hydrological redistributions through their integrated gravitational effect is one of the primary aims of the GRACE mission. Yet it has been proposed that at larger scales this may be achieved independently by measuring and inverting the elastic loading associated with redistributing masses, e. g. with the IGS network. This is particularly interesting as long as GRACE monthly gravity solutions not (yet) match the targeted baseline accuracies at the lower degrees. In this contribution (1) we describe a joint inversion technique, (2) we introduce a physically motivated regularization that guarantees stable inversion results if only GPS data is used, (3) we apply this technique to GPS data provided by the IGS service covering recent years, and (4) finally we compute what the relative contribution of GRACE and GPS would be in a joint inversion.

**Keywords.** Mass redistribution, GPS, GRACE, inversion

## 1 Introduction

It is one of the primary objectives of the GRACE mission to monitor hydrological mass redistributions through their integrated gravitational effect. In August 2004, the first set of monthly gravity fields has been released to the public by the GRACE project, covering April 2002 - April 2004. Yet it has been proposed by Blewitt et al., (2001), Blewitt and Clarke (2003), that this may be achieved independently at larger scales by geometrically measuring the elastic response of the Earth to the loading associated with redistributing masses, e. g. with the IGS network. Since GRACE has not (yet) matched the targeted baseline, with respect to the accuracy of low–degree monthly gravity fields, this prospect appears particularly interesting.

In any case, the analyst is forced to constrain solved-for mass configurations either by low-degree truncation of the spherical harmonic (SH) expansion, by spatial averaging (Swenson and Wahr 2002), or by regularization operations employing mathematically or physically motivated constraints. Combining satellite gravity and geometrical displacements in a joint inversion may help in releasing these constraints and improving the reliability of estimates, but in particular it offers prospects for cross-validation and for a more realistic quality assessment. We describe a concept for such a joint inversion.

We have, in this study, analyzed global weekly GPS station coordinates and covariance matrices made available through the IGS service, for the time span June 26, 1999 – June 20, 2004. One can show that parametrization issues such as truncation degree and station distribution affect the estimation of low-degree mass redistributions like load center of mass (see Wu et al., 2002) and seasonal changes of low–degree SH coefficients from these data. We show how a regularized estimation technique that applies only a physically motivated constraint over oceanic areas renders the truncation problem in a less serious one. Then, linear combinations of SH coefficients beyond degree 4 contain significant information. Finally, we quantify what the effect of adding GRACE monthly solutions in a joint inversion would be.

## 2 Inverse Methodology

As usual, we assume that 1) all relevant processes occur in a thin layer at the Earth's surface, 2) gravitational and geometrical response of the Earth can be described by an elastic load Love number (LLN) theory (Farrell 1972). Load thickness $T = \frac{\Delta\sigma}{\rho_w}$ will be expressed as an equivalent water column, where $\Delta\sigma$ and $\rho_w$ are the surface density anomaly (departure from multi–year mean) and the mean density of seawater. A spherical harmonic expansion for $\Delta\sigma$ reads

$$\Delta\sigma(\lambda, \theta) = \quad (1)$$

$$a\rho_w \sum_{l,m} (\Delta C^{\sigma}_{lm} \cos m\lambda + \Delta S^{\sigma}_{lm} \sin m\lambda) P_{lm}(\cos\theta)$$

with $a$, $\Delta C^{\sigma}_{lm}$, $\Delta S^{\sigma}_{lm}$, $P_{lm}$ being a reference radius, the dimensionless SH coefficients of the density anomaly, and the fully normalized Legendre polynomials. One then relates the load SH coefficients to those of observable quantities, such as geoid change and crustal deformation. The choice of degree–1 LLN's implicitly defines the coordinate system to which subsequent inversions refer (Blewitt and Clarke 2003): in this study we work with $l_1 = 0.134, h_1 = -0.269$ corresponding to the center of figure (CF) frame. On adopting Farrells's loading theory, with $g_l = \frac{3\rho_w}{\rho_e}\frac{1}{2l+1}$ the spectral mapping for geoid, height, lateral displacement SH coefficients is simply

$$\Delta C^{g}_{lm} = g_l(1 + k_l)\Delta C^{\sigma}_{lm} \quad (2)$$

$$\Delta C^{h}_{lm} = g_l h_l \Delta C^{\sigma}_{lm} \quad (3)$$

$$\Delta C^{\psi}_{lm} = g_l l_l \Delta C^{\sigma}_{lm} \quad (4)$$

$\rho_e$ is the mean density of Earth, and $k_l, h_l, l_l$ are load Love numbers. When relating these coefficients to observations, one can add nuisance parameters, unknown datum offsets, etc. For a time-averaged geoid coefficient (like 4 weeks), i.e.

$$\Delta C^{g}_{lm} = \frac{1}{\Delta t}\int_t^{t+\Delta t} g_l(1 + k_l)\Delta C^{\sigma}_{lm}(\tau)d\tau \quad (5)$$

or for GPS displacements in height, east, north

$$\Delta h = \quad (6)$$
$$a\sum_{l,m} \left(\Delta C^{h}_{lm} \cos m\lambda + \Delta S^{h}_{lm} \sin m\lambda\right) P_{lm}(\cos\theta)$$
$$+\mathbf{e}_h \cdot \mathbf{\Delta x} - a\Delta s$$

$$\Delta x = \quad (7)$$
$$\frac{a}{\sin\theta}\sum_{l,m} m(-\Delta C^{\psi}_{lm} \sin m\lambda + \Delta S^{\psi}_{lm} \cos m\lambda)$$
$$\times P_{lm}(\cos\theta) + \mathbf{e}_x \cdot \mathbf{\Delta x} + \mathbf{e}_y \cdot \epsilon$$

$$\Delta y = \quad (8)$$
$$-a\sum_{l,m} (\Delta C^{\psi}_{lm} \cos m\lambda + \Delta S^{\psi}_{lm} \sin m\lambda)$$
$$\times \frac{\partial}{\partial\theta} P_{lm}(\cos\theta) + \mathbf{e}_y \cdot \mathbf{\Delta x} - \mathbf{e}_x \cdot \epsilon$$

Here $\mathbf{\Delta x}, \epsilon, \Delta s$ would account for an unknown (or subject to validation) set of Helmert parameters, and $\mathbf{e}_h, \mathbf{e}_x, \mathbf{e}_y$ means the unit vectors corresponding to the local frame. On assuming the degree of expansion is sufficiently chosen, eqs. (2–8) constitutes a Gauss–Markov model

$$\mathbf{y} + \mathbf{e} = \mathbf{A}\mathbf{x} \qquad E\{\mathbf{e}\} = \mathbf{0}, \qquad E\{\mathbf{e}\mathbf{e}'\} = \mathbf{C} \quad (9)$$

where the unknown $\Delta C^{\sigma}_{lm}, \Delta S^{\sigma}_{lm}$ together with nuisance parameters are collected in $\mathbf{x}$. In connection with GPS data, this inverse problem has been posed by Blewitt and Clarke (2003). Here we consider additionally the combination with GRACE monthly gravity field models. It is obvious that estimability (or correlation) of unknowns, the degree of truncation, and the role of other constraining operations have to be discussed in detail. In this contribution we propose a physically motivated regularization technique, if GPS data only is to be inverted. With combining GRACE and GPS, one would expect these problems less serious, but possibly additional 'calibration' factors would have to be introduced. An account on the relative contribution of both data sources in a combined solution is given in section 5.

## 3 Estimability, truncation, and regularization

It is clear that technically one can estimate only a limited number of SH coefficients from a limited number of observations, however any truncation of the SH expansion leads to a potential aliasing of unmodelled higher–frequency mass redistribution signals into the solution. Ways to circumvent the problem include the choice of tailored base functions, extraction of stable linear combinations of SH coefficients, and regularization. When the 'neglected' part of the model (9) is denoted by $\tilde{\mathbf{A}}\tilde{\mathbf{x}}$, that is $\mathbf{y} + \mathbf{e} - \tilde{\mathbf{A}}\tilde{\mathbf{x}} = \mathbf{A}\mathbf{x}$, this will create a bias $\mathbf{b} = E\{\hat{\mathbf{x}} - \mathbf{x}\}$

$$\mathbf{b} = (\mathbf{A}'\mathbf{C}^{-1}\mathbf{A})^{-1}\mathbf{A}'\mathbf{C}^{-1}\tilde{\mathbf{A}}\tilde{\mathbf{x}} \quad (10)$$

for a least squares estimate. For degree–1 inversion, Wu et al. (2002) have investigated this effect on the basis of different load models $\tilde{\mathbf{x}}$. The biases becomes

$$\mathbf{b} = (\mathbf{A}'\mathbf{C}^{-1}\mathbf{A} + \alpha\Xi)^{-1}(\mathbf{A}'\mathbf{C}^{-1}\tilde{\mathbf{A}}\tilde{\mathbf{x}} - \alpha\Xi\mathbf{x}) \quad (11)$$

in an $\alpha\Xi$–regularized estimate. From (11) we see that by choosing an appropriate regularization operator one might in principle seek to minimize the total bias. The design of regularization operators and choice of the correct degree of smoothness is however a challenging problem, see Xu and Rummel (1994) for a general discussion; a recent account in satellite gravity modelling is Ditmar et al. (2003), and for geomagnetic modelling from discrete heterogeneous data see Korte and Holme (2003). In the present case, the land–ocean inbalance in the station distribution causes an inherent problem. On the other hand, with oceanic tides being already removed in the IGS GPS data analysis, and assuming the inverse barometric (IB) hypothesis as valid, remaining oceanic mass variations are much smaller than those over continental area. Therefore we introduce an optimization principle that adds a constraint on the oceanic variability of $\Delta\sigma$ to the least squares sum,

$$\mathcal{J}_\alpha = ||\mathbf{y} - \mathbf{Ax}||^2_{\mathbf{C}^{-1}} + \alpha \int_{\mathcal{O}} (\Delta\sigma)^2 d\omega \,. \tag{12}$$

By $\mathcal{O} \subset \Omega$ we mean the ocean area. This comes down to solving the common type of normal equation system $(\mathbf{A}'\mathbf{C}^{-1}\mathbf{A}' + \alpha\Xi_{\mathcal{O}})\hat{\mathrm{x}} = \mathbf{A}'\mathbf{C}^{-1}\mathbf{y}$, where building the regularization matrix requires either spherical harmonic analysis of $\bar{Y}_{lm}$ ocean grids (O is the ocean function),

$$\Xi_{\mathcal{O};lm,l'm'} = \int_\omega O\, \bar{Y}_{lm}\, \bar{Y}_{l'm'}\, d\omega \tag{13}$$

$$= \frac{1}{4\pi} \sum_{l,m} \bar{O}_{l''m''} \int_\Omega \bar{Y}_{lm} \bar{Y}_{l'm'} \bar{Y}_{l''m''} d\omega$$

or recursive algorithms involving Clebsch–Gordon coefficients. For such recursions see Blewitt and Clarke (2003) and the literature cited there. $\alpha$ comes from a numerical ocean model, and/or in combination with tradeoff investigations. We have, in a similar way, implemented regularization operators that constrain surface gradients $(\nabla^*\Delta\sigma)^2$ and surface Laplacian $(\Delta^*\Delta\sigma)^2$ in (12) instead.

## 4 Analysis of IGS GPS coordinates

We have applied our analysis to weekly IGS GPS combination coordinate sets spanning 1999.5–2004.5 (June 26, 1999 – June 20, 2004), obtained from the IGS ftp site as SINEX files (`igsyyPwwww.snx`) and equipped with full variance–covariance matrices. In total 158 sites which contribute for more than 2 years in the mentioned time span were selected. Time series with step functions due to earthquakes or antenna changes have been excluded. Linear station movements have been removed, and all time series are centered about the mean. Note further that from summer 2001 on there is a significant increase in the number of sites.

Within the inversion, we solved for a maximum resolution of $l = 7$ for each week. Regularization with $\Xi_{\mathcal{O}}$, and $\alpha = 0.9 \cdot 10^5$ corresponding to an (average) a–posteriori cumulative variability of $\hat{\mathrm{x}}'\Xi_{\mathcal{O}}\hat{\mathrm{x}}$ of $(0.014\mathrm{m})^2$ (an average value taken from an ocean bottom pressure model) was applied. We exclude degree–1 deformation from the regularization. Furthermore, we removed some 1% of observations where either height, east or north displacements did not pass Pope's $\tau$ test at 95% level of significance (Koch 1999).

Fig. 1 shows the estimated load moment evolution, expressed as a variation of the CF frame with respect to the center of mass of Earth plus load (CM) system. Here $X = \frac{1}{M}\mathbf{m}_x = \frac{1}{M}\frac{4\pi a^4 \rho_w}{3}\Delta C^\sigma_{11}$ etc. The period 1999.5–2001.0, after scaling displacements by Earth's mass $M = 6.0 \cdot 10^{24}$km, fits very close to Fig. 1 from Blewitt et al. (2001), who solved for degree–1 only. One should add that degree–1 deformation can be separated from a (nuisance) Helmert translation, when vertical and horizontal displacements are analyzed. Statistical correlations vary between -0.1 and -0.6 in our regularized inversion, dependent on site distribution and coordinate direction.

In Fig. 2 we show some estimated low–degree spherical harmonic coefficients. Their amplitudes are indeed sensitive to the strenght of the regularization. Unconstrained coefficients ($\circ$), however, are probably too large. For comparison, two models from SLR analysis have been plotted (Nerem et al., 2000, Cheng et al., 2002). For higher degrees ($l > 2$) there is typically less power in our regularized estimates than in those obtained from SLR analyses. Moreover, the figures suggest that energy is transferred from the $Z$ annual oscillation to $C_{20}$. This pattern reappears when we repeat our computation with different station subsets, so it is probably not related to the increasing number of IGS stations. However, care has to be exercized. The amplitude of $C_{20}$ appears sensitive towards the chosen optimization principle.

We show two plots of spatial mass distribution in Fig. 3, at resolution of $l = 2 \ldots 6$, and for the months August 2002 and August 2003 obtained as a four–week average from our analysis. In this case we have removed the NCEP atmospheric pressure contribution in a post–analysis step. This means, these plots should basically be comparable to what GRACE delivers at the large spatial scales. Cumulative a–posteriori error estimates reach 0.02m…0.05m over land areas, dependent on site distribution. Finally, Fig. 4 gives the corresponding surface mass change ($l = 2 \ldots 6$), as computed for the same months from the CPC/LDAS land hydrology model (Fan et al.,

2003), and Fig. 5 shows these changes as derived from the GRACE level 2 products. We have used the corresponding GFZ fields and referred them to the GFZ mean field. Interpretation of the misfits will be a complex task that we postpone here, but at a first glance there is a remarkable correspondance between our inversion, the model, and the GRACE–derived solution.

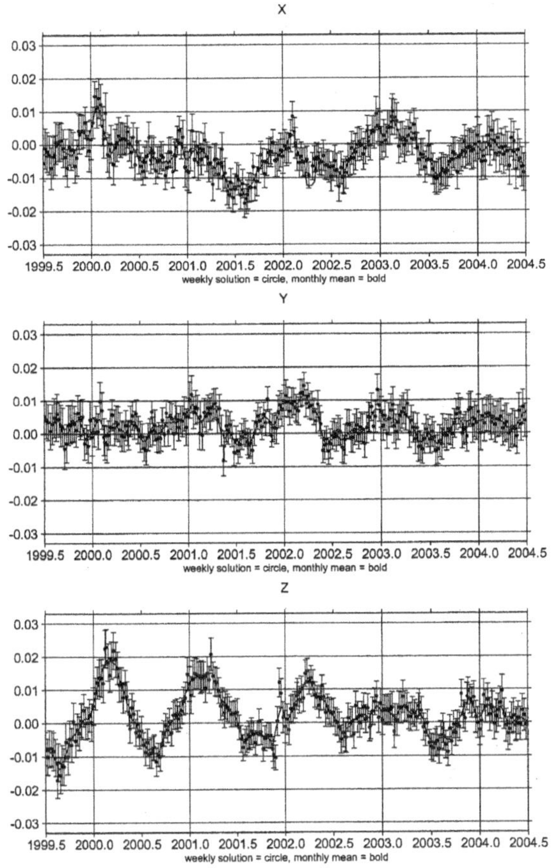


**Fig.1.** Load moment expressed as CF frame variation [m] w.r.t. CM from GPS inversion, in X-, Y-, Z-direction. Weekly estimates (dots), and running 30day mean (line).

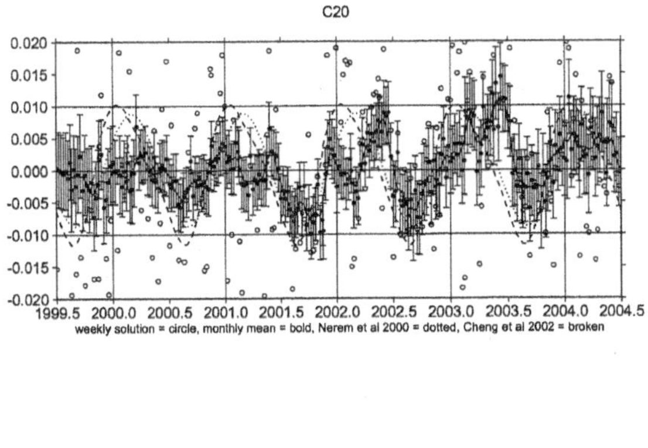


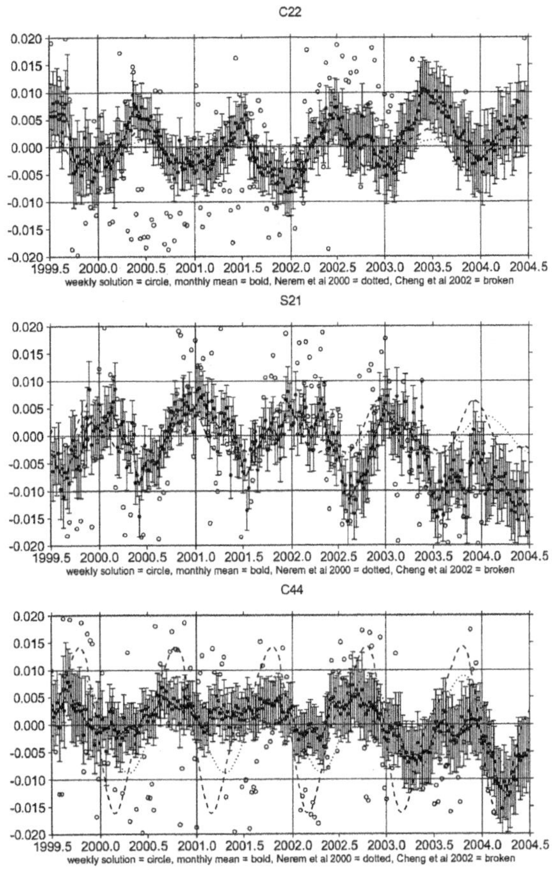


**Fig.2.** Low–degree load harmonic coefficients [m] from GPS inversion. Weekly estimates (dots), running 30day mean (line), and two models from SLR analyses. (○) marks weekly estimates without regularization.

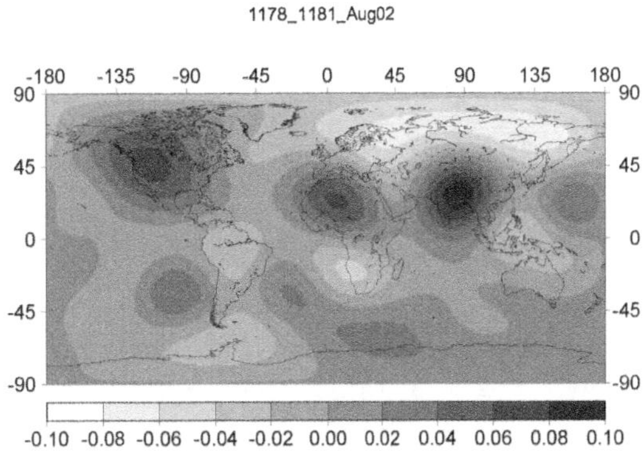

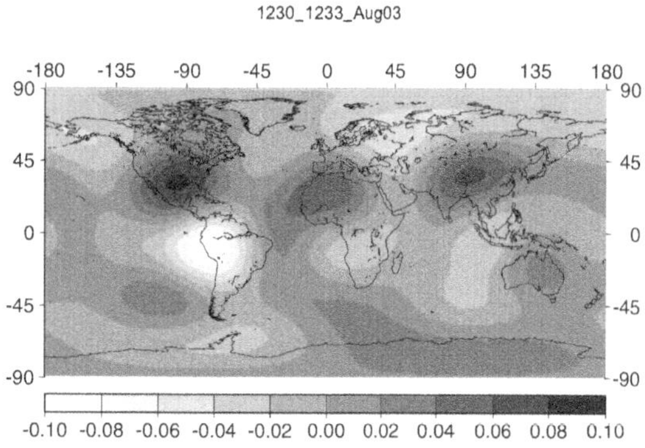


**Fig.3.** Monthly surface density variation in equivalent water height [m], from GPS inversion, $l = 2 \ldots 6$. Atmospheric effect removed (from NCEP). Upper figure: August 2002, lower figure: August 2003.

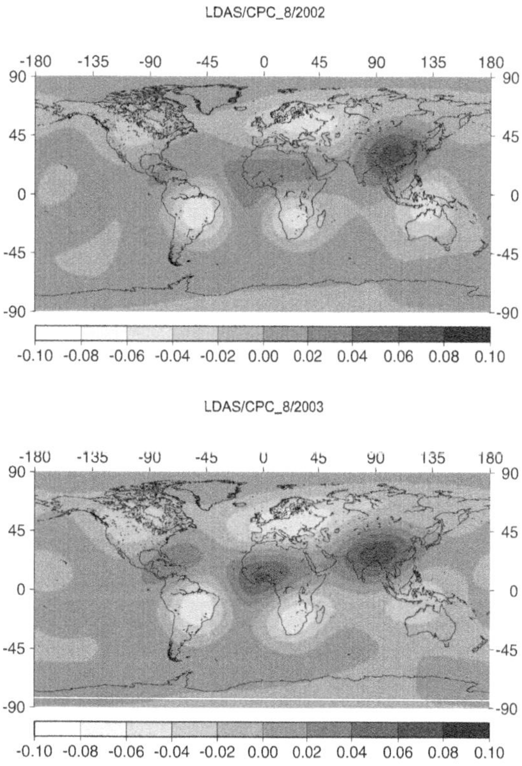


**Fig.4.** Monthly surface density variation in equivalent water height [m], from CPC/LDAS land hydrology, $l = 2 \ldots 6$. Upper figure: August 2002, lower figure: August 2003.

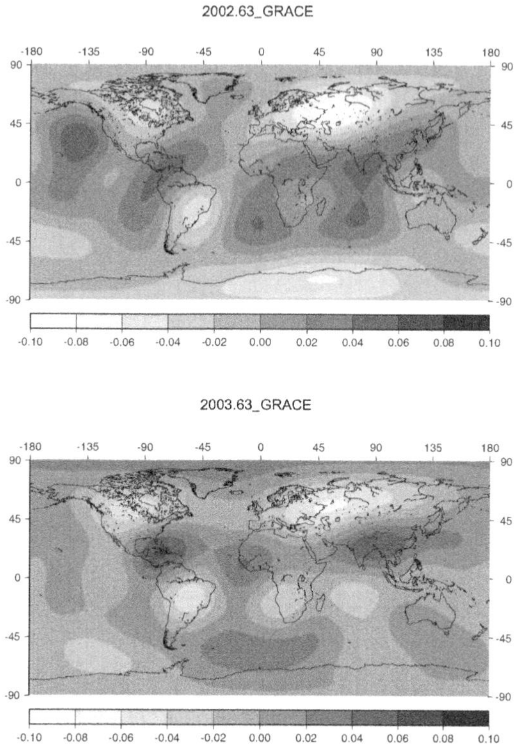


**Fig.5.** Monthly surface density variation in equivalent water height [m], from the GRACE project, $l = 2 \ldots 6$. Upper figure: August 2002, lower figure: August 2003.

## 5 Simulation of GRACE full covariance and relative contribution

What would be the contribution of GPS in a joint inversion, that is, when one combines GRACE gravity coefficients (5) and GPS station displacements (6,7,8)? In order to assess this question, we have used the variance–covariance matrix $\mathbf{C}^{h,\psi}$ of 120 IGS stations that we extracted from the GPS week 1239 (2003.75) SINEX file for the computation in section 4 together with the variance–covariance information of the GFZ EIGEN–GRACE01S model (Reigber et al., in prep.). EIGEN–GRACE01S has been computed from some 39 days and its accuracy is comparable to the first set of monthly gravity models (P. Schwintzer, pers. comm.). However, because EIGEN–GRACE01S is not given with a full covariance matrix, we have build up such a matrix $\mathbf{C}^{g}$ from simulation with the energy balance approach and 30 days of simulated orbits (Ilk et al., 2003) and scaled subsequently degree–wise to match the actual EIGEN–GRACE01S error degrees. The contribution of GRACE can then be judged by evaluating the partial redundancy for a particular gravity coefficient $\Delta C^g_{lm}$ (Koch 1999), and the contribution of the GPS site distribution to $\Delta C^\sigma_{lm}$ is $1 - q_{lm}$, where

$$q_{lm} = \mathbf{a}^{g\,\prime}_{lm}(\mathbf{A}^{g\prime}\mathbf{C}^{g^{-1}}\mathbf{A}^{g} + \mathbf{A}^{h,\psi\prime}\mathbf{C}^{h,\psi^{-1}}\mathbf{A}^{h,\psi})^{-1}\mathbf{A}^{g\prime}\mathbf{C}^{g^{-1}}\mathbf{e}_{lm}\,. \quad (14)$$

Here $\mathbf{a}^g_{lm}$ is the corresponding column of the design matrix, and $\mathbf{e}_{lm}$ a vector with 1 in the position corresponding to $l, m$ and 0 elsewhere.

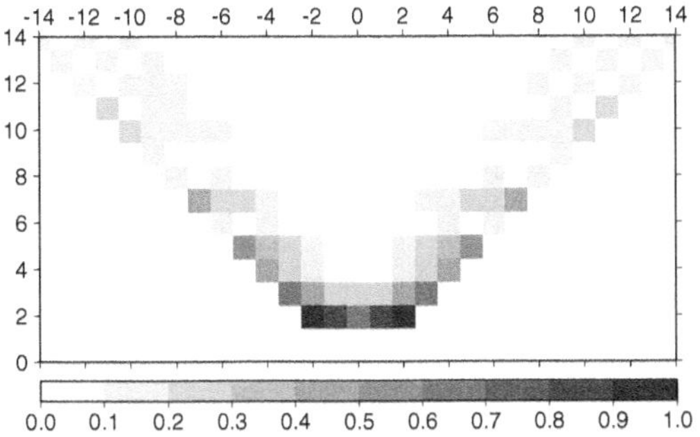


**Fig.6.** Projection of the relative contribution of GPS (120 IGS sites, week 1239) in a joint GRACE–GPS inversion. 1 means coefficient is completely determined from GPS, 0 from GRACE.

The projection shows clearly that the gain through adding GPS lies in the very low degrees, and to some extend in tesseral harmonics. There is a 50% contribution up to degree 5. However, with increasing GPS site density and increasing GRACE accuracy such projection will have to be reconsidered.

## 6 Conclusions

We have analyzed publicly available weekly IGS GPS time series from the years 1999.5-2004.5 for mass redistribution through loading inversion. Truncation and regularization issues affect the estimation of low–degree mass changes; due to nonhomogeneous site distribution and the amplification of errors. Our results show that a regularized inversion technique, which takes only the physically motivated constraint of limited oceanic variability into account, renders this parametrization problem into a less significant one. SH coefficients well beyond degree 4 appear to contain significant information. GRACE should change this picture, but dependent on the quality of its monthly solutions, there is a potential for improvement by data combination.

**Acknowledgements**

We acknowledge the use of IGS GPS station coordinate SINEX files made publicly available through ftp. Thanks go also to the GRACE project for using their level 2 products.

## References

Blewitt G, Lavallee D, Clarke P, Nurutdinov K (2001). *A new global mode of Earth deformation: seasonal cycle detected.* Science 294:2342–2345

Blewitt G, Clarke P (2003). *Inversion of Earth's changing shape to weigh sea level in static equilibrium with surface mass redistribution.* JGR 108(B6), doi:10.1029/2002JB002290

Cheng MK, Gunter B, Ries JC, Chambers DP, Tapley BD *Temporal variation in the Earth's gravity field from SLR and CHAMP GPS data.* Gravity and Geoid 2002 (GG2002), I.N. Tziavos (Ed.), submitted

Ditmar P, Kusche J, Klees R (2003). *Computation of spherical harmonic coefficients from gravity gradiometry data to be acquired by the GOCE satellite: regularization issues.* J Geodesy, 77(7–8):465-477

Fan Y, Van del Dool H, Mitchell K, Lohmann D (2003) *A 51-year reanalysis of the U.S. land-surface hydrology.* GEWEX News, 13(2)

Farrell W (1972). *Deformation of the Earth by surface loads* Rev Geophys Space Phys 10(3):761–797

Ilk KH, Visser P, Kusche J (2003). *Satellite Gravity Field Missions, Final Report Special Commission 7.* Travaux IAG, Vol. 32, General and technical reports 1999-2003, Sapporo

Koch KR (1999). *Parameter estimation and hypothesis testing in linear models.* Springer, Berlin

Korte M, Holme R (2003) *Regularization of spherical cap harmonics.* Geophys J Int 153:253–262

Nerem RS, Eanes RJ, Thompson PF, Chen JL (2000) *Observations of annual variations of the Earth's gravitational field using satellite laser ranging and geophysical models.* GRL 27(12):1783–1786

Reigber C, Schmidt R, Flechtner F, König R, Meyer U, Neumayer KH, Schwintzer P, Zhu S (in prep.) *First EIGEN gravity field model based on GRACE mission data only.*

Swenson S, Wahr J (2002). *Methods for inferring regional surface-mass anomalies from Gravity Recovery and Climate Experiment (GRACE) measurements of time-variable gravity.* JGR 107(B9), doi:10.1029/2001JB000576

Wu X, Argus DF, Heflin MB, Ivins ER, Webb FH (2002). *Site distribution and aliasing effects in the inversion for load coefficients and geocenter motion from GPS data.* GRL 29(24), doi:10.1029/2002GL016324

Xu P, Rummel R (1994). *Generalized ridge regression with applications in determination of potential fields.* Man Geo, 20:8–20

# The Fennoscandian Land Uplift Gravity Lines 1966–2003

J Mäkinen
Finnish Geodetic Institute, Geodeetinrinne 2, FIN-02430 Masala, Finland

A. Engfeldt
Lantmäteriet, Lantmäterigatan 2, SE-801 82 Gävle, Sweden

B.G. Harsson
Norwegian Mapping Authority, Kartverksveien 21, N-3507 Hønefoss, Norway

H. Ruotsalainen
Finnish Geodetic Institute, Geodeetinrinne 2, FIN-02430 Masala, Finland

G. Strykowski
National Survey and Cadastre, Rentemestervej 8, DK-2400 København NV, Denmark

T. Oja
Estonian Land Board, Mustamäe tee 51, EE10602 Tallinn, Estonia

D. Wolf
GeoForschungsZentrum Potsdam, Telegrafenberg, D-14473 Potsdam, Germany

**Abstract.** The Fennoscandian Land Uplift Gravity Lines (sometimes called Nordic Land Uplift Gravity Lines) consist of four east-west profiles across the Fennoscandian postglacial rebound area, along the approximate latitudes 65°, 63°, 61°, and 56°N. Repeated relative gravity measurements have been so far performed 1975–2000 (65°N), 1966–2003 (63°N), 1976–1983 (61°N), and 1977–2003 (56°N). The line 63°N has most observations. From the measurements along it up to 1993, Ekman and Mäkinen (1996) deduced the ratio –0.20 μgal/mm between surface gravity change and uplift relative to the Earth's center of mass. Since that time, more gravity measurements have been taken. On the eastern part of the line 63°N, they result in slightly smaller estimates for the rate of gravity. New estimates of uplift and model predictions are also available. The updated gravity change combined with various estimates of uplift gives ratios between –0.16 and –0.20 μgal/mm. On the western part of the line 63°N an apparently anomalous change in gravity difference requires further study. In the future, the measurements will be performed using absolute gravity techniques.

**Keywords.** Postglacial rebound, gravity change, Fennoscandia, glacial isostatic adjustment

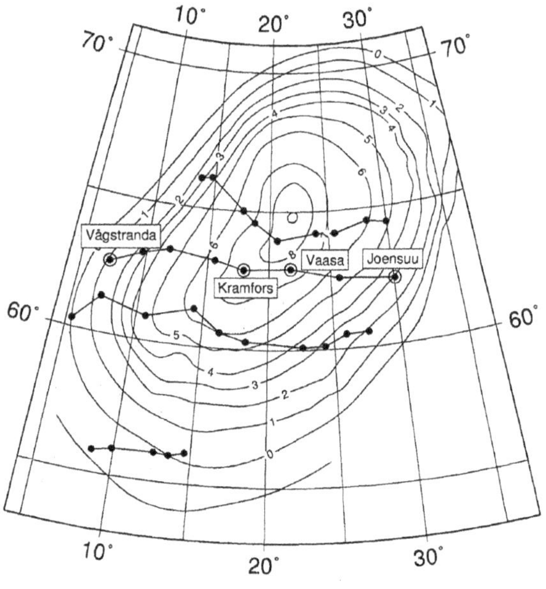


**Fig. 1.** The Fennoscandian Land Uplift Gravity Lines. The named stations on the line 63°N are used in this paper. The isolines show postglacial rebound rates (mm/yr) relative to mean sea level according to Ekman (1996).

# 1 Introduction

## 1.1 Background

The Fennoscandian Land Uplift Gravity Lines are high-precision relative gravity profiles across the Fennoscandian postglacial rebound (PGR) area (Fig. 1). They run approximately east-west, along the latitudes 65°, 63°, 61°, and 56°N. The concept and methodology largely go back to the recently deceased A. Kiviniemi (1930–2004), who started measurements on the Finnish section of the 63°N line in 1966 (Kiviniemi, 1974). Other sections (Pettersson, 1974) and other lines followed (Table 1). The work has been performed in Nordic cooperation (with guests from institutions around the world) and coordinated through the Nordic Geodetic Commission (http://www.nkg.fi), in particular its Working Group for Geodynamics. Kiviniemi (1974) and Mäkinen et al. (1986) described the measurement methods. Mäkinen et al. (1986) gave the station descriptions, an account of the computation methods, and a detailed catalogue of results until 1986.

**Table 1.** Relative gravity campaigns on the Fernnoscandian Land Uplift Gravity Lines 1966–2003. FI=Finnish section.

| Line | Full campaigns | Partial campaigns |
|---|---|---|
| 65°N | 1975, 1980, 1981 | 1999, 2000 (FI) |
| 63°N | Every 5 years 1966–2003 | Every 1–3 years 1966–2003 (FI) |
| 61°N | 1976, 1983 | |
| 56°N | 1977, 1984, 2003 | |

When the work started, the general features of the vertical motion associated with the Fennoscandian PGR were rather well known. The research agenda was to determine $\dot{g}$, the rate of change in surface gravity, and compare it with $\dot{h}$, the rate of crustal uplift. The ratio $\dot{g}/\dot{h}$ would then allow conclusions on the physics of the uplift (Kiviniemi, 1974).

## 1.2 Uplift and gravity change

Two simple geometric images of the uplift process have often been used to construct bounds for $\dot{g}/\dot{h}$. Assume first that the uplift is due to decompression with no additional mass. This leads to the "free air model" $\dot{g}/\dot{h}$ = -0.31 µgal/mm. If, on the other hand, there is no decompression, but additional mass in the upper mantle (density 3300 kg/m$^3$), the Bouguer approximation leads to the ratio $\dot{g}/\dot{h}$ = –0.17 µgal/mm, the "Bouguer model". This latter figure is close to the relationship (–1 µgal)/(6.5 mm) = –0.15 µgal/mm, obtained by Wahr et al. (1995) using a number of models of the glacial isostatic adjustment (GIA) with Maxwell rheology of the mantle. See also Fang and Hager (2001).

In the discussion above, $\dot{h}$ refers to the Earth's center of mass ("absolute uplift"). GPS observations give absolute uplift rates. Uplift rates from tide gauges refer to the mean sea level ("apparent uplift"), and rates from repeated precise levelling to the geoid. We thus have

$$\dot{h} = \dot{H} + \dot{N} \quad (1a)$$

$$\dot{H} = \dot{H}_a + \dot{H}_e \quad (1b)$$

Here $\dot{H}_a$ is apparent uplift, $\dot{H}$ is uplift relative to the geoid, $\dot{N}$ is the uplift of the geoid, and $\dot{H}_e$ (neglecting changes in sea surface topography) the eustatic rise in mean sea level.

In this paper we are concerned with uplift differences between stations. From them, $\dot{H}_e$ is eliminated, i.e. $\dot{H}_a$ differences and $\dot{H}$ differences are identical. However, $\dot{H}$ differences need to be corrected for $\dot{N}$ to obtain $\dot{h}$. Ekman and Mäkinen (1996) estimated (using their observed $\dot{g}/\dot{H}$) that $\dot{N}$ differences are roughly 6% of $\dot{H}$ differences. The GIA model by Milne et al. (2001) gives approximately the same percentage.

To avoid a clogged notation, we use $\dot{g}$, $\dot{h}$, $\dot{H}$, $\dot{N}$ to refer both to values at sites and to differences between sites. It will be clear from the context, which one is intended.

## 1.3 Previous estimates

The purpose of this paper is to examine the new data on the line 63°N. Therefore we give a summary of the numerical results of Ekman and Mäkinen (1996). They divided the line into two parts (Fig. 1): the western part Vågstranda–Kramfors (W), and the eastern part Vaasa–Joensuu (E). Writing symbolically with conventional one-sigma standard errors

$$(\dot{g}/\dot{h})_{obs} = \frac{-1.52 \pm 0.20 \text{ µgal/yr}}{6.9 \pm 0.5 \text{ mm/yr}} \quad \text{(W)} \quad (2a)$$

$$(\dot{g}/\dot{h})_{obs} = \frac{-1.00 \pm 0.14 \text{ µgal/yr}}{-4.7 \pm 0.5 \text{ mm/yr}} \quad \text{(E)} \quad (2b)$$

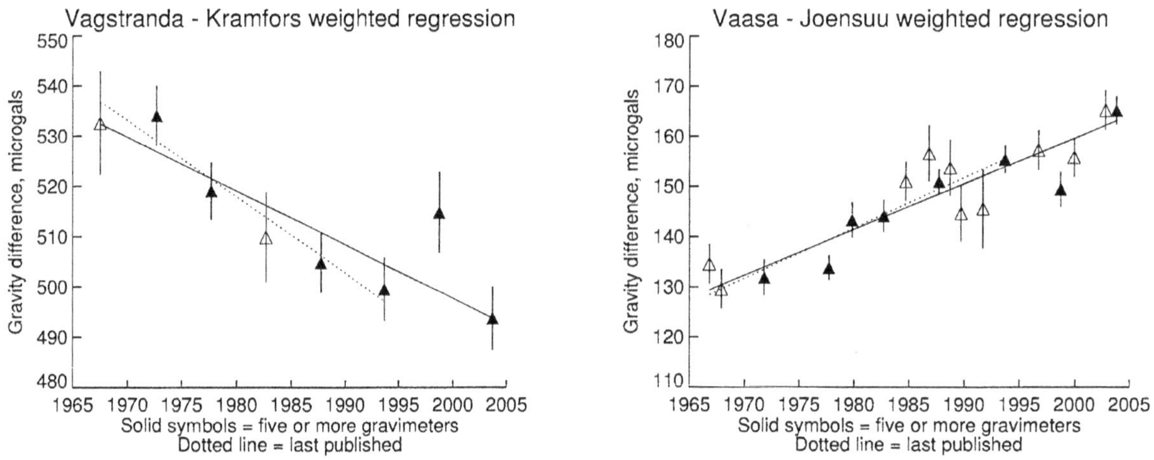


**Fig. 2.** Results of the relative gravity measurements on the western part (left) and on the eastern part (right) of the line 63°N. Each point is the unweighted mean of all gravimeters in the particular year. Regression lines are calculated with the number of gravimeters as weights. The solid line uses all data, the dotted line uses data till 1993. Error bars (one-sigma) are explained in the text.

With the relevant Student's t-distributions the ratios correspond to the 95% confidence intervals

$$(\dot{g}/\dot{H})_{obs} = -0.220 \pm 0.086 \text{ µgal/mm (W)} \quad (3a)$$

$$(\dot{g}/\dot{H})_{obs} = -0.213 \pm 0.080 \text{ µgal/mm (E)} \quad (3b)$$

After an iterative calculation of $\dot{N}$ to obtain $\dot{h} = \dot{H} + \dot{N}$, Ekman and Mäkinen (1996) found the corresponding intervals for $\dot{g}/\dot{h}$

$$(\dot{g}/\dot{h})_{obs} = -0.208 \pm 0.086 \text{ µgal/mm (W)} \quad (4a)$$

$$(\dot{g}/\dot{h})_{obs} = -0.200 \pm 0.080 \text{ µgal/mm (E)} \quad (4b)$$

The combined 95% confidence interval is

$$(\dot{g}/\dot{h})_{obs} = -0.204 \pm 0.058 \text{ µgal/mm} \quad (5)$$

whence they concluded that the free air model is ruled out by the observations. In the following we will see to which extent new results since 1993 modify the numbers in equations (2)...(5).

## 2 New gravity results

Since 1993, there have been two campaigns on the eastern part of the line 63°N, in 1998 and 2003. Their results do not agree with the earlier ones (Fig. 2, left). Weighted regression now gives

$$\dot{g} = -1.07 \pm 0.24 \text{ µgal/yr (W)} \quad (6)$$

where the error estimate is the conventional one-sigma. This is a large change from the previous estimate −1.52 µgal/yr (numerator of equation 2a).

There is no comparable anomaly on the eastern part of the line, with five new campaigns (Fig. 3). Weighted regression gives

$$\dot{g} = +0.91 \pm 0.09 \text{ µgal/yr (E)} \quad (7)$$

again with the conventional one-sigma error estimate. This $\dot{g}$ is slightly smaller than the previous estimate +1.00 µgal/yr (numerator of equation 2b).

We are currently screening the results in more detail. Fig. 3 (left) shows the results on the western part for single gravimeters. So far, we have for instance noticed that the same instrument is responsible for the largest value both in 1998 and 2003. Moreover, there are additional measurements in 1998, which will shortly become available to us. Further analysis of the ostensible "jump" in the time series must therefore wait. Fig. 4 (right) shows the single-gravimeter results for the eastern part.

The estimate of the standard deviation of a single gravimeter, calculated by pooling all the sums of squares around the annual means, is on the western part of the line $m_0 = 17.8$ µgal. On the eastern part $m_0 = 7.8$ µgal. Error bars in Fig. 2 are calculated by dividing the appropriate value of $m_0$ by the square root of the number of gravimeters used in the particular year.

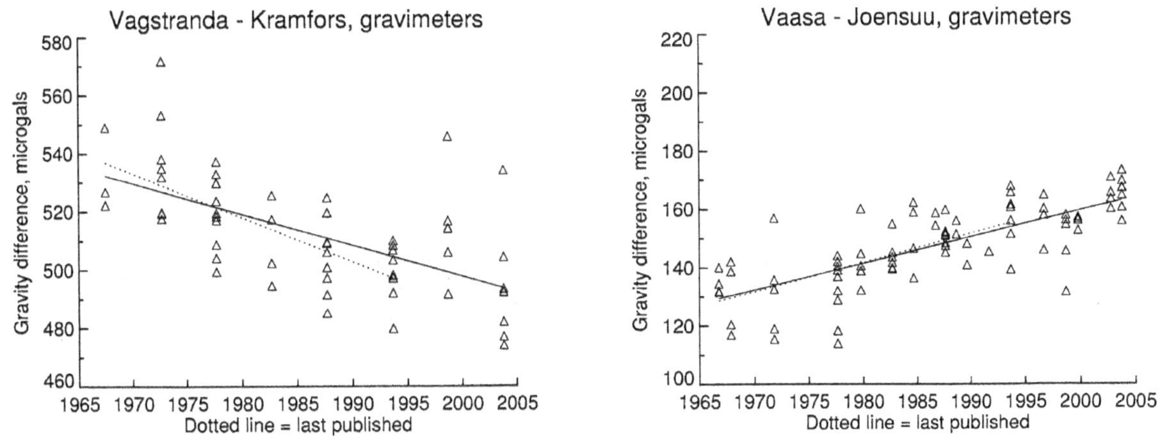


**Fig. 4.** Results of the relative gravity measurements on the western part (left) and on the eastern part (right) of the line 63°N. Each point represents a single gravimeter. Simple regression then produces the same slope as the weighted regression of Fig.2. The solid line uses all data, the dotted line uses data till 1993.

## 3 New uplift results and ratios

On the eastern part of the line, the value of $\dot{H}$ in the denominator of equation (2b), i.e.,

$$\dot{H} = -4.7 \pm 0.5 \text{ mm/yr} \qquad (6)$$

results from two Finnish precise levellings, the same that underly Fig. 1. Now $\dot{H}$ values from three levellings are available (Mäkinen and Saaranen, 1998). Furthermore, the land uplift stations at Joensuu and Vaasa are within 0.1 km and 2 km, respectively, of continuous GPS stations in the FinnRef® network. For these stations, $\dot{h}$ values are available both from FinnRef® (Mäkinen et al., 2003) and BIFROST (Johansson et al., 2002) processing of five years of GPS observations. We use the winter-edited results (op. cit.)

In addition, we employ predicted $\dot{H}_a$ values from the GIA model by Lambeck et al. (1998). The main observational constraints for this GIA model are relative sea level histories, although it has been fine tuned by a fit to the tide gauge rates computed by Ekman (1996). The GIA model by Milne et al. (2001) uses the same ice history, but the rheology is fitted to 3-D velocities from the BIFROST project (Johansson et al., 2002).

The uplift results and corresponding ratios $\dot{g}/\dot{h}$ are collected in Table 2. The new estimate of $\dot{g}$ and most of the new uplift estimates bring the ratio closer to the Bouguer model. On the western part of the line the gravity measurements and the resulting $\dot{g}$ value are still under scrutiny. Therefore we have not computed new $\dot{g}/\dot{h}$ ratios.

**Table 2.** The ratio $\dot{g}/\dot{h}$ on the eastern part of the line 63°N, computed using different values of $\dot{g}$, and $\dot{h}$ or $\dot{H}$. The old $\dot{g}$ is from equation (2b) and the new $\dot{g}$ from equation (5). The sources of $\dot{h}$ or $\dot{H}$ are described in the text. The $\dot{H}$ values are marked by an asterisk. They are multiplied by 1.06 to obtain $\dot{h}$. The previous estimate of $\dot{g}/\dot{h}$ is shown in boldface.

| Source of $\dot{h}$ or $\dot{H}$ | $\dot{h}$ or $\dot{H}$ * mm/yr | $\dot{g}/\dot{h}$ old $\dot{g}$ μgal/mm | $\dot{g}/\dot{h}$ new $\dot{g}$ μgal/mm |
|---|---|---|---|
| Two levellings 1892–1955 | –4.7* | **–0.21** | –0.18 |
| Three levellings 1892–2004 | –5.5* | –0.17 | –0.16 |
| Continuous GPS, FinnRef® | –5.7 | –0.18 | –0.16 |
| Continuous GPS, BIFROST | –4.8 | –0.20 | –0.18 |
| Lambeck et al. (1998) GIA model | –4.2* | –0.22 | –0.20 |
| Milne et al. (2001) GIA model | –5.3 | –0.19 | –0.17 |

## 4. Summary and discussion

We have used new gravity observations and new uplift results on the eastern part of the 63°N line to compute new estimates of the ratio between gravity change and vertical motion connected with post-glacial rebound. On this part, the previous estimate is –0.21 μgal/mm. The new estimates are –0.16...–0.20 μgal/mm. They are thus closer to the ratio for the Bouguer model, which is –0.17 μgal/mm. On the western part of the line an apparently anomalous gravity change must be further studied before any conclusions can be drawn.

For this kind of study, relative gravity measurements are less accurate and more laborious than absolute measurements. Therefore the time series will be continued with absolute gravimeters. The end points of the 63°N line, and the Danish section of the 56°N line have already been equipped with absolute stations. For the other lines, we are studying the possibility of using portable absolute meters directly on the outdoor sites.

## Acknowledgments

The measurements since 1993 were made (in addition to the authors) by L. Engman, L.-Å. Haller, E. Roland, P. Rouhiainen, H. Skatt, H. Virtanen, and K. Wieczerkowski. H.G. Scherneck provided predictions from the GIA model by Milne et al. (2001). Digitized isobases of the map by Ekman (1996) were supplied by B.-G. Reit. Discussions with M. Ekman were instructive and inspiring.

## References

Ekman M. (1996). A consistent map of the postglacial uplift in Fennoscandia. *Terra Nova*, 8, pp. 158–165.

Ekman, M. and J. Mäkinen (1996): Mean sea surface topography in the Baltic Sea and its transition area to the North Sea: A geodetic solution and comparisons with oceanographic models. *J. Geophys. Res.*, 101 (C5), pp. 11993–11999.

Fang, M. and B.H. Hager (2001): Vertical deformation and absolute gravity. *Geophys. J. Int.*, 146, pp. 539–548.

Johansson, J.M., Davis, J.L., Scherneck, H.-G., Milne, G.A., Vermeer, M., Mitrovica J.X., Bennett, R.A., Ekman, M., Elgered, G., Elósegui, P., Koivula, H., Poutanen, M., Rönnäng, B.O., Shapiro, I.I. (2002): Continuous GPS measurements of postglacial adjustment in Fennoscandia, 1. Geodetic results. *J. Geophys. Res.*, 107 (B8), DOI 10.1029/2001JB000400.

Kiviniemi A. .(1974): High precision measurements for studying the secular variation in gravity in Finland. *Publ. Finn. Geod. Inst.*, 78, 64 p.

Lambeck K., C. Smither and J. Ekman (1998): Tests of glacial rebound models for Fennoscandia based on instrumented sea- and lake-level records. *Geophys. J. Int.*, 135, pp. 375–387.

Mäkinen, J., M. Ekman, Å. Midtsundstad and O. Remmer (1986): The Fennoscandian land uplift gravity lines 1966–1984. *Rep. Finn. Geod. Inst.*, 85:4, 238 p.

Mäkinen J. and V. Saaranen (1998): Determination of postglacial land uplift from the three precise levellings in Finland. *J. Geodesy*, 72, pp. 516–529.

Mäkinen, J., H. Koivula, M. Poutanen and V. Saaranen (2003): Vertical velocities in Finland from permanent GPS networks and from repeated precise levelling. *J. Geodynamics*, 35, pp. 443–456.

Milne G.A., Davis J.L., Mitrovica J.X., Scherneck H.-G., Johansson J.M., Vermeer M.,. Koivula H. (2001): Space-geodetic constraints on glacial isostatic adjustment in Fennoscandia. *Science*, 291 (23 March 2001), pp. 2381–2385.

Pettersson, L. (1975): Studium av sekulär ändring i tyngdkraften utefter latitud 63° mellan Atlanten och Bottenhavet. In: *The 7th meeting of the Nordic Geodetic Commission*, Copenhagen, May 6–10, 1974. Referat, Bind I. Danish Geodetic Institute, pp. 103–125.

Wahr J, Han D, Trupin A (1995): Predictions of vertical uplift caused by changing polar ice volumes on a viscoelastic earth. *Geophys. Res. Letters*, 22, pp. 977–980.

# Temporal Gravity Variations in GOCE Gradiometric Data

F. Jarecki, J. Müller
Institut für Erdmessung, Universität Hannover, Schneiderberg 50, D-30167 Hannover, Germany

S. Petrovic, P. Schwintzer
GeoForschungsZentrum Potsdam, Department 1: "Geodesy and Remote Sensing",
Telegraphenberg A17, D-14473 Potsdam, Germany

**Abstract.** The accuracy and spatial resolution expected from GOCE gravity gradiometry might deteriorate due to temporal gravity variations, which are mainly induced by mass redistributions in the System Earth. These mass redistributions occur in the atmosphere at various time scales, in the oceans as ocean tides and currents and on the continents as solid Earth tides, loading and hydrological effects. Opposite to GRACE, GOCE is developed to measure the static gravity field. Therefore, any time variable effects have to be removed from the measured data prior to further processing in a dealiasing step.
In this report the effects mentioned above are analysed by means of available models in order to investigate if they deliver significant contributions to the GOCE gradiometer measurements. Simulations of the different system parts (Atmosphere, Hydrosphere, Cryosphere and Solid Earth) are run. The gravitational effect of the simulated mass changes is expanded into spherical harmonics from which gravitational gradients are computed along a simulated GOCE orbit. The resulting gradients are compared with the specifications of the GOCE gradiometer. Furthermore, the residual (i.e. atmospheric and oceanic parts removed) time variable part of the gravity field as detected by GRACE is discussed with respect to its impact on GOCE gradiometry.

**Keywords.** gravity satellite mission, GOCE, GRACE, gravity gradiometry, global hydrological model, temporal gravity field variations

## 1 Temporal variability of the Earth's gravity field

The main cause of the variability of the Earth's gravity field beside tides are redistributions of masses in the atmosphere, oceans, cryosphere and in the continental water storage. In order to achieve the goal of the GOCE mission (a high-resolution static gravity field) it is necessary to eliminate both the high-frequency and the seasonal effects using available physical models, like:

- atmosphere: ECMWF, NCEP
- ocean tides: FES2004 or CSR 4.0
- solid Earth tides: IERS CONVENTIONS
- ocean models: barotropic (e.g. JPL), baroclinic
- continental water storage: global hydrological models like WGHM (Döll et al. 2003), H96 (Huang et al. 1996, Fan and Van den Dool 2004), LaD (Milly and Shmakin 2002) or NCEP
- models for ice shields (e.g. Sasgen 2004).

Both the high-frequency and the seasonal impact of the atmosphere and the ocean can be successfully dealiased using the above models. However, the differences arising when using different atmospheric or ocean models cannot be neglected either. Hence, the remaining modeling errors can produce time-variable aliasing effects, see Han et al. (2004) and Thompson et al. (2004).

The order of magnitude of the seasonal changes induced by atmosphere, oceans, hydrology and Antarctic ice is presented in Fig. 1. A direct comparison of the presented power curves is not quite fair, since the data coverage of different physical quantities extends over different parts of the Earth surface. The atmospheric and oceanic influences seem at first to be considerably stronger than the hydrological ones, at least up to degree and order $n, m = 10$ in the spherical harmonic expansion. However, the combined influence of the atmosphere and oceans is less than the influence of the continental water storage due to the inverse barometer effect.

Adding the influence of the Antarctic ice shield to the hydrological signal brings no significant change in the degree variances. The only considerable difference can be seen in degree two, which is in any case uncertain due to different datum definitions in various physical models.

In order to check the reliability of the existing hydrological models, the gravity variations induced by them were compared both with the effects recovered from GFZ's monthly GRACE gravity field solutions

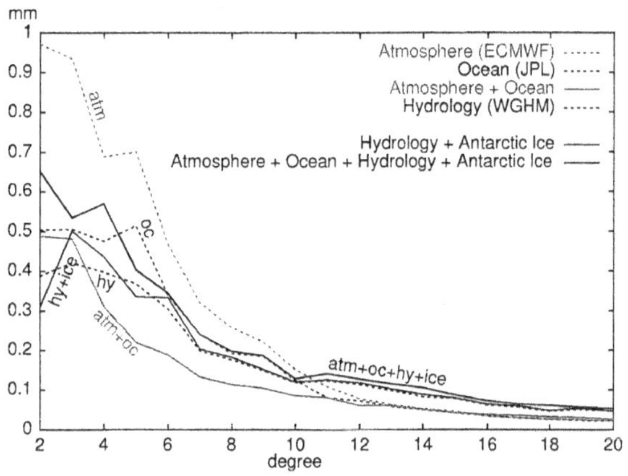


**Fig. 1.** Influence of different physical components on the geoid variations (degree variances for the year 2002 computed using monthly mean values)

(which are already dealiased for the atmosphere and oceans, see Flechtner 2003), and with each other.

Variations of the Earth's gravity field deduced from GRACE monthly solutions and from different hydrological models have been compared in space domain. (A graphical presentation of this study can be found as Fig. 2 in Jarecki et al. (2005) on the CD-ROM providing the full proceedings of this meeting. A reasonable reproduction of this figure in B/W is not possible).

In some regions (e.g. on the South American continent) the coincidence between the variabilities deduced from GRACE and from WGHM is better than between those from GRACE and the other considered hydrological model, in some regions (for instance in North Asia) the opposite holds. The latter is probably due to the fact that LaD includes some modeling of snow which is not contained in WGHM. The well pronounced features, especially in large tropical river basins, are clearly visible in all three representations. However, the GRACE observed variations are significantly larger than both hydrological predictions. Table 1 quantifies the global comparison of the gravity field variations deduced from GRACE and WGHM. The differences are considerable and lie in the same order of magnitude as the variations of the hydrological part of the geoid. The correlation coefficients are not especially high. Substituting H96 or LaD for WGHM yields very similar results.

Comparison of the variations of geoid undulations deduced from two different hydrological models leads to similar conclusion, see Table 2. The differences lie almost in the same order of magnitude as the variations themselves, although the correlation coefficients are rather high. More on these topics can be found in Schmidt et al. (2004).

The hydrological models in their present form have to be regarded as rather uncertain. Therefore, the use of the series of GRACE monthly gravity field solutions in order to characterize sub-seasonal, seasonal and interannual hydrological variations and the remaining effects which are not contained in the atmospheric and oceanic models can be recommended. The GOCE results should be combined with a GRACE derived static field to stabilize the long-wavelength part of the gravity model solution.

## 2 Temporal Variations of GOCE Gradients

This section concentrates on investigations, whether hydrological models and/or monthly GRACE gravity fields are useful to reduce corresponding variations in the gravity gradients $V_{ij}$ measured by the upcoming GOCE satellite. The indices $i$ and $j$ denote the direction of differentiation, six different gradients build the symmetric gravity gradient tensor $\mathbf{V}$.

Before addressing the hydrological models, one should take a look on the amplitudes of the investigated signals, i.e. gravity variations between monthly GRACE solutions, dealiased for the atmospheric and oceanic signals, at GOCE altitude. As the variations are computed from gravity field differences between two months (e.g. $M2$ and $M1$), the appropriate signal degree variances (for radial gravity gradients $V_{rr}$, see next paragraph resp. eq. (3)) are given by

$$\sigma_n^2(\Delta V_{rr}) = \sum_{m=0}^{n} \lambda_{Vrr}^2 \cdot \left((c_{nm,M2} - c_{nm,M1})^2 + (s_{nm,M2} - s_{nm,M1})^2\right) \quad (1)$$

with the scaling factor for radial gravity gradients

$$\lambda_{Vrr}^2 = \left(\frac{GM}{r}\right)^2 \left(\frac{(n+1)(n+2)}{r^2}\right)^2 \left(\frac{a}{r}\right)^{2n}. \quad (2)$$

These signal degree variances are shown in Fig. 2. They reach only small values, remaining after con-

**Table 1.** Comparison of geoid changes in mm deduced from GRACE and WGHM

| months | GRACE | | | WGHM | | | GRACE-WGHM | | | correlation coefficient |
|---|---|---|---|---|---|---|---|---|---|---|
| | min | max | WRMS | min | max | WRMS | min | max | WRMS | |
| 05/02–08/02 | -11.28 | 9.62 | 3.62 | -7.19 | 8.37 | 1.70 | -6.57 | 7.46 | 2.80 | 0.67 |
| 08/02–11/02 | -8.23 | 9.68 | 2.11 | -3.68 | 5.47 | 1.05 | -7.31 | 5.44 | 1.99 | 0.36 |
| 11/02–03/03 | -12.61 | 8.90 | 2.90 | -13.17 | 4.89 | 1.74 | -5.78 | 6.76 | 1.96 | 0.75 |
| 03/03–05/03 | -9.46 | 4.12 | 2.05 | -3.89 | 3.41 | 0.73 | -7.37 | 3.82 | 1.82 | 0.48 |

**Table 2.** Comparison of geoid changes in mm deduced from WGHM and LaD

| months | WGHM | | | LaD | | | WGHM-LaD | | | correlation coefficient |
|---|---|---|---|---|---|---|---|---|---|---|
| | min | max | WRMS | min | max | WRMS | min | max | WRMS | |
| 05/02–08/02 | -7.19 | 8.37 | 1.70 | -8.92 | 11.16 | 2.30 | -5.57 | 2.50 | 0.97 | 0.93 |
| 08/02–11/02 | -3.68 | 5.47 | 1.05 | -4.84 | 6.42 | 1.12 | -2.25 | 4.33 | 0.66 | 0.82 |
| 11/02–03/03 | -13.17 | 4.89 | 1.74 | -9.95 | 5.73 | 1.95 | -3.42 | 4.57 | 0.96 | 0.87 |
| 03/03–05/03 | -3.89 | 3.41 | 0.73 | -4.63 | 5.60 | 1.05 | -2.85 | 1.90 | 0.59 | 0.84 |

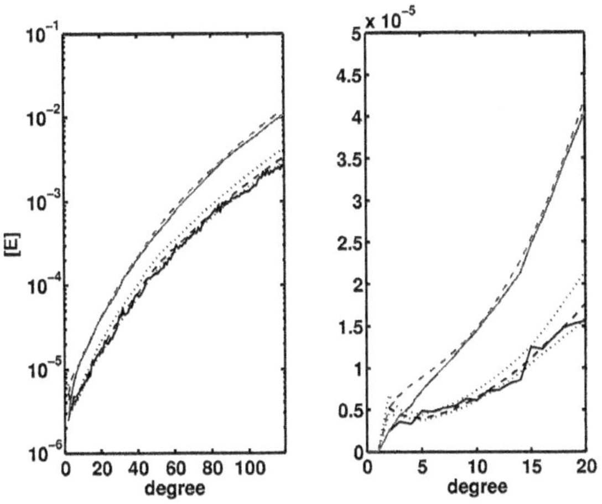


**Fig. 2.** Signal degree variances of GOCE-like gradient differences in 260 km altitude from averaged monthly GRACE fields (solid line), compared with the monthly GRACE performance (error degree variances extrema between consecutive months (dotted) and mean of the used months per degree (dashed); both from calibrated error models). The mean values are indicated as cummulative degree variances as well (grey lines). The figure on the right focusses on the low degree part.

secutive subtraction of the monthly models and upward continuation. The resulting signal amplitudes are in the range of their calibrated error degree variances, i.e. they are not significant. Nevertheless, the results from section 1 indicate at least a small hydrological contribution in the low-degree part of the monthly GRACE gravity fields. Comparing the cummulative mean degree variances, the signal does not exceed the error curve at the lower degrees due to the reduced accuracy of the very low coefficients. In contrast, the degree-wise signal exceeds the error curve between $n = 4$ and 11 and has some peaks at higher degrees. But especially because of missing temporal equidistancy of the GRACE models (the time span between two consecutive models varies between one and four months), this averaged view is not really satisfying. A closer look onto the single differences (e.g. shown in Table 1 and 4) shows much better results for longer time spans. So the extrema of the error degree variances for each degree are given, too. This shows, that the mean signal slightly exceeds the error of the best monthly difference up to $n = 22$. Considered over the whole spectrum, it even drops below this best-case scenario several times and it is always below its worst error estimate. At the level of $n = 12 \ldots 15$ the signal curve clearly approaches and even intersects the best-case error curve (lower dotted line in the right plot of Fig. 2), encouraging us to search for the maximum degree where the hydrology signal comes up in GOCE gradients.

The mentioned hydrological models, especially the combination of WGHM and the ice-model (GFZ, see section 1), here denoted with WGHM+ice, which contains at least a simple consideration of antarctic ice mass balance, are used to derive spherical harmonics of the corresponding gravity effect. Due to the resolution of the models, their monthly representations are truncated at a maximum degree of $n = 100$. From these spherical harmonics it is easy to compute gravity gradients in an Earth orientated frame, whose radial component according to eq. (3) is quite comparable with the radial component of the latter GOCE gradient, either in the local orbit reference frame (see e.g. Ditmar and Klees 2002) or in the gradiometer reference frame (see e.g. Pail 2004). Because of the good global comparability even in different reference frames (Müller 2003), this radial component is used here. Furthermore, one could compute exact GOCE gradients from the spherical harmonic expansion to accomplish exact comparisons when applying simulated GOCE orbits and orientations. The Earth-pointing radial component $V_{rr}$ is computed by

$$V_{rr} = \frac{GM}{R^3} \sum_{n=0}^{n_{max}} (n+1)(n+2) \left(\frac{R}{h_1}\right)^{n+3} \cdot \sum_{m=0}^{l} (\bar{c}_{lm} \cos m\lambda + \bar{s}_{lm} \sin m\lambda) P_{lm}(\cos\vartheta). \quad (3)$$

Continuing the previous section, the hydrology and ice models are the first to be investigated at gradient level. Figure 3 shows the minimum and maximum monthly changes and their RMS for the whole WGHM+ice period (1992 to 2003) on a global 1°grid of $V_{rr}$ in 260 km altitude, which is representative for the GOCE orbit (see ESA 1999). The spherical harmonics are used completely up to $n = 100$. The amplitude reached by these gravity changes from modelled hydrology and ice is not critical for GOCE gradiometry at all, as it does not reach the mE-level (see ESA 1999) even in extreme constellations. Furthermore, the time scale represented by these models is not fitting the gradiometer's measurement bandwidth (MBW).

Although it does not seem to affect the gradiometric measurements itself, studies were carried out to find a threshold for the maximum degree for application of GRACE temporal solutions in the gradients to reduce possible systematic influences. This might improve the error estimates as well as the whole analysis of the gradiometric data.

Table 3 shows the global RMS of the consecutive $V_{rr}$ changes of the eleven available monthly GRACE solutions produced by the University of Texas' Center for Space Research (see Flechtner 2004) in mE. The last column indicates the mean correlation coefficient between each monthly difference model derived from GRACE on one hand and the WGHM+ice-model on the other hand. Depending on the treatment of ice, WGHM is able to explain half the power of the GRACE-derived gradients up to $n_{max} = 10$ or 13, with a mean correlation of about 40%. These results are comparable with those for the geoid height changes presented in section 1. So, neglecting the rather unrealistic ice model, it is better to use GRACE data only up to degree $n = 10$ for hydrology dealiasing of GOCE gradients. Taking the ice into account, degree $n = 13$ looks more reasonable. Furthermore, one has to note the much better agreement of seasonal differences in $V_{rr}$ from GRACE and WGHM than of monthly ones, see Table 4.

This implies that the model data might be rather unuseful for the reduction of gradients, especially with respect to the MBW, and shows that GRACE might considerably contribute to the improvement of hydrological models.

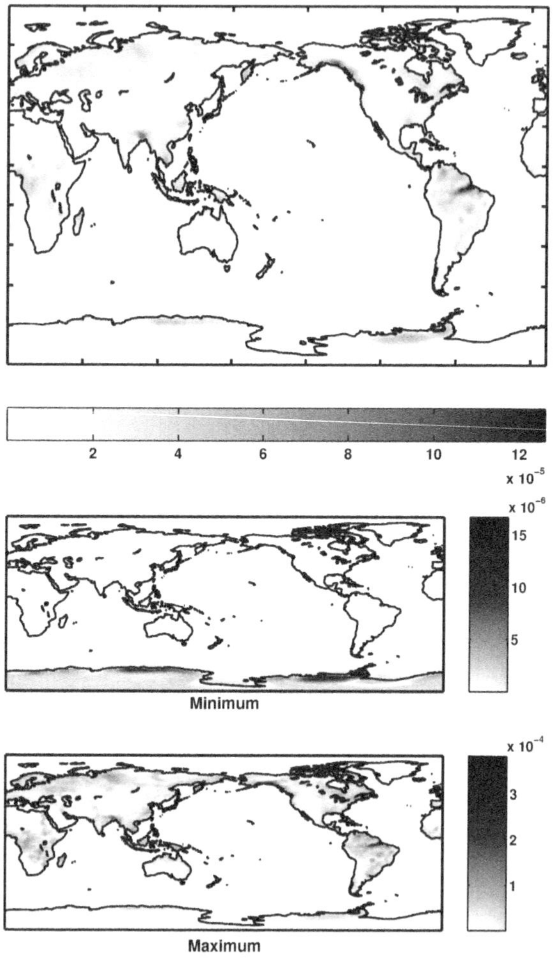


**Fig. 3.** RMS, min and max of magnitude of monthly $V_{rr}$ changes from hydrology and antarctic ice in altitude 260 km in [E]. Note the small but regular influence of the ice.

Referring to Fig. 2 again, there is no difference in the significance of the GRACE derived gradient signal, but the signal strength increases several orders of magnitude at the higher degrees. Therefore, just cutting of the GRACE models at low degrees, like suggested above, produces a large formal omission error (about 10 mE for $n_{max} = 13$, i.e. 500 times the used signal). To include this signal, which is contaminated by a raising noise level too, the GRACE models were evaluated completely up to $n_{max} = 120$ (the WGHM+ice-models up to the considered maximal expansion of $n_{max} = 100$) and filtered by a moving average filter in the spatial domain to remove the noise.

The results are shown in Table 5, which presents the approximated maximum degree of spherical harmonic expansion compared with the averaging window applied and global RMS and correlations for the derived gradient grids. Each filtered high resolution GRACE/WGHM+ice model combination shows

**Table 3.** Global RMS of the gradient changes in $V_{rr}$ in units [mE] w.r.t. the maximum spherical degree utilised

| $n_{max}$ | GRACE | WGHM | WGHM+ice | GRACE −WGHM | GRACE −(WGHM+ice) | correlation coefficient GRACE−(WGHM+ice) |
|---|---|---|---|---|---|---|
| 10 | 0.016 | 0.010 | 0.011 | 0.011 | 0.014 | 0.384 |
| 13 | 0.021 | 0.011 | 0.014 | 0.017 | 0.016 | 0.416 |
| 20 | 0.037 | 0.016 | 0.017 | 0.034 | 0.035 | 0.213 |
| 27 | 0.062 | 0.017 | 0.019 | 0.060 | 0.060 | 0.125 |
| 120/100 | 8.942 | 0.022 | 0.023 | 8.942 | 8.942 | 0.001 |

**Table 4.** Gradient changes in $V_{rr}$ deduced from GRACE and WGHM+ice models, $n_{max}$=13, in units [mE]

| | GRACE | | | WGHM+ice | | | GRACE−(WGHM+ice) | | | correlation |
|---|---|---|---|---|---|---|---|---|---|---|
| months | min | max | RMS | min | max | RMS | min | max | RMS | coefficient |
| 08/02 - 11/02 | -0.152 | 0.089 | 0.022 | -0.057 | 0.045 | 0.012 | -0.106 | 0.075 | 0.020 | 0.403 |
| 11/02 - 03/03 | -0.097 | 0.221 | 0.029 | -0.066 | 0.153 | 0.015 | -0.074 | 0.091 | 0.021 | 0.722 |
| 03/03 - 04/03 | -0.039 | 0.061 | 0.011 | -0.018 | 0.015 | 0.004 | -0.041 | 0.055 | 0.011 | 0.163 |
| 04/03 - 05/03 | -0.052 | 0.058 | 0.012 | -0.032 | 0.035 | 0.006 | -0.042 | 0.040 | 0.012 | 0.288 |

large differences. Correlation is below the 1% level again and the global RMS differences seem to be unaffected by the hydrological model. Filtering does not enhance the (limited) usefulness of a complete $n_{max} = 120/100$ model. Obviously, spatial filtering does not seem to be an adequate approach for the highly location-sensitive gradient data. Therefore, one should apply an appropriate filtering approach in the spherical harmonics, for which the simple cut-off procedure performed in this study is just an example. A combination of methods might be another way to go: The medium resolution cut-off model with $n_{max} = 27$ looks slightly enhanced (correlation grows from 12.5% to 16.7% resp. 19.9%) by the filtering. Nevertheless most of the monthly GRACE signal beyond $n_{max} = 13$ does not seem to be explained by the hydrological models. Therefore, these residuals represent a significant non-hydrological signal, a hydrological signal, which is for some reason not contained in the models or just measurement noise. If hydrology dealiasing should be applied to GOCE gradients, although it turns out to play a minor role here, but reduces systematic parts, it seems reasonable to rely only on those parts cross-validated by the GRACE hydrology signal.

Nevertheless, hydrology seems to be the main systematic signal in the monthly GRACE solutions. Hence, the GRACE dealiasing products for ocean and atmospheric mass changes prove to be good. Therefore they have been used to calculate gradients along simulated GOCE tracks, which led to signals in the mE/$\sqrt{\text{Hz}}$-range. Considered spectrally, they turn out to be below the actual gradiometer performance. Figure 4 shows combined atmosphere and ocean influence on the test orbit as PSD plot. The lower plot presents the superposition of these signals and the influences of ocean and solid Earth tides (note the 1 cpr peak), which turn out to be the principal influence for gradiometry. Besides third body potential (neglected in the plots), which is at the same order of magnitude and can easily be modelled, the curves show the time variable gravity components proposed for dealiasing of GOCE gradients (see Abrikosov and Schwintzer 2004). Same as in the case of hydrology, they do not reach the actual GOCE performance, but the dealiasing with validated models should improve

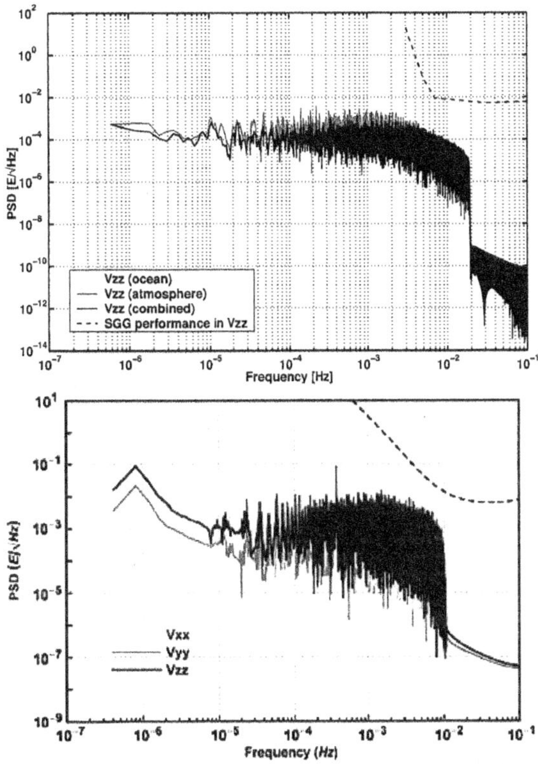


**Fig. 4.** Signal PSD of dealiasing gradients along GOCE test orbits, based on GRACE dealiasing for ocean and atmosphere on the top and beneath the complete time varying gravity including solid Earth and ocean tides.

**Table 5.** Global RMS of the $V_{rr}$ changes in [mE] from filtered high-resolution models

| $n_{max}$ | averaging window | approx. $n_{max}$ | GRACE | WGHM+ice | GRACE −(WGHM+ice) | correlation coefficient |
|---|---|---|---|---|---|---|
| 27 | 37°x 37° | 10 | 0.018 | 0.007 | 0.018 | 0.167 |
| 27 | 27°x 27° | 13 | 0.018 | 0.009 | 0.018 | 0.199 |
| 120/100 | 37°x 37° | 10 | 0.177 | 0.007 | 0.177 | 0.007 |
| 120/100 | 27°x 27° | 13 | 0.276 | 0.009 | 0.276 | 0.004 |
| 120/100 | 13°x 13° | 27 | 0.772 | 0.014 | 0.772 | 0.006 |

the stochastic behaviour of the gradiometric data.

## 3 Conclusion and Outlook

The investigated long-term geophysical models do not indicate that a reduction of geophysical effects for GOCE gradiometric measurements should be necessary. Neither the well-known and relatively "fast sampled" oceanographic and atmospheric mass redistribution models nor the much "slower" hydrology simulations, which are distorted additionally by internal inconsistency and inconsistency with GRACE, deliver simulated gradient influences in the amplitude of the GOCE gradiometer specifications. Nevertheless it might be reasonable to consider the dealiasing for stochastic reasons. Furthermore, it has to be pointed out, that this study focusses on the gradiometric part of the GOCE mission. To get a complete dealiasing strategy, it is essential to have a close look on the satellite-to-satellite-tracking parts too, i.e. the GOCE orbit determination. These observations will be used for the detection of the spatial and temporal longer wavelength part of the gravity field. It is obvious that in this part the large scale physical effects, e.g. the tides of the ocean and the solid Earth, will play an important role, which - in contradiction to the effect on the gradients - have to be considered.

**Acknowledgments.** The German Ministry of Education and Research (BMBF) supports the GOCE-GRAND project, in whose context this study was carried out, within the GEOTECHNOLOGIEN geoscientific programme under grant 03F0329D. This is publication no. GEOTECH-88 of that programme.
The authors thank Chris Milly, Yun Fan and Huug van den Dool, as well as Petra Döll and Andreas Güntner for providing the LaD, H96 and WGHM data, respectively.

## References

Abrikosov, O., Schwintzer, P., 2004. Recovery of the Earth's Gravity Field from GOCE Satellite Gravity Gradiometry: a Case Study. GOCE, The Geoid and Oceanography. ESA-SP569, ESA Publications Division, Nordwijk.

Ditmar, P., Klees, R., 2002. A Method to Compute the Earth's Gravity Field from SGG/SST Data to be Acquired by the GOCE Satellite. DUP Science, Delft.

Döll, P., Kaspar, F., Lehner, B., 2003. A global hydrological model for deriving water availability indicators: model tuning and validation. J. Hydrol. 270, 105-134.

ESA, 1999. Gravity Field and Steady-State Ocean Circulation Explorer, Reports for Assessment: The Four Candidate Earth Explorer Core Missions, ESA-SP1233(1), ESA Publications Division, Nordwijk.

Fan, Y., Van den Dool, H., 2004. The CPC global monthly soil moisture data set at 1/2 degree resolution for 1948-present. J. Geophys. Res. 109, D10102, doi:1029/2003JD004345.

Flechtner, F., 2003. AOD1B Product Description Document. GRACE Project Documentation, JPL 327-750, Rev. 1.0, JPL Pasadena, Ca.

Flechtner, F., 2004. GRACE Science Data System Monthly Report January 2004. GRACE Information System and Data Center, GFZ Potsdam.

Han, S.-C., Jekeli, C., Shum, C. K., 2004. Time-variable aliasing effects of ocean tides, atmosphere, and continental water mass on monthly mean GRACE gravity field. J. Geophys. Res. 109, B 04403, doi:10.1029/2003/JB002501.

Huang, J., Van den Dool, H. M., Georgakakos, K. P., 1996. Analysis of model-calculated soil moisture over the United States (1931-1993) and applications to long-range temperature forecasts. J. of Climate 9, 1350-1362.

Jarecki, F., Müller, J., Petrovic, S., Schwintzer, P., 2005. Temporal Gravity Variations in GOCE Gradiometric Data. In: Proceedings of the Gravity, Geoid and Space Missions GGSM2004 IAG International Symposium, Porto, Portugal, August 30th to September 3rd, 2004. CD-ROM.

Milly, P. C. D., Shmakin, A. B., 2002. Global modeling of land water and energy balances. Part I: The Land Dynamics (LaD) Model, J. of Hydrometeorology 3(3), 283-299.

Müller, J., 2003. GOCE gradients in various reference frames and their accuracies. Adv. in Geosciences 1, 33-38.

Pail, R., 2004. GOCE Quick-Look Gravity Field Analysis: Treatment of Gravity Gradients Defined in the Gradiometer Reference Frame. GOCE, The Geoid and Oceanography. ESA-SP569, ESA Publications Division, Nordwijk.

Sasgen, I., 2004. Geodetic signatures of glacial changes in Antarctica: Rates of geoid-height change and radial displacement due to present and past ice-mass variations. Diploma thesis, GFZ Potsdam.

Schmidt, R., Schwintzer, P., Flechtner, F., Reigber, Ch., Güntner, A., Döll, P., Ramillien, G., Cazenave, A., Petrovic, S., Jochmann, H., Wünsch, J., 2004. GRACE Observations of Changes in Continental Water Storage. Submitted to Global and Planetary Change.

Thompson, P. F., Bettadpur, S. V., Tapley, B. D., 2004. Impact of short period, non-tidal, temporal mass variability on GRACE gravity anomalies. Geophy. Res. Lett. 31, L06619, doi:10.1029/2003GL019285.

# Estimating GRACE Aliasing Errors

Ki-Weon Seo
Department of Geological Sciences
University of Texas, Austin TX 78712, kiweon@geo.utexas.edu

Clark R. Wilson
Department of Geological Sciences
University of Texas, Austin TX 78712

**Abstract.** Synthetic GRACE aliasing errors are examined for land, ocean and atmospheric signals. Simulated GRACE data are generated using least square fits of spherical harmonics to potential differences. Mass variations are produced by GLDAS (Global Land Data Assimilation Scheme), ECCO (Estimating the Circulation and Climate of Ocean), and the difference between NCEP and ECMWF (as an estimate of dealiasing error). Additional studies include single spherical harmonics and isolated (Gaussian) loads. Aliasing errors are measured by their degree-order spectrum, and arise from inadequate space-time sampling. Because atmospheric variations are rapid, they cause large aliasing errors at low degree, below 15. To better understand these, we examine aliasing due to sinusoidally oscillating single harmonics, and use Kaula's resonance solution to interpret the aliases. Single temporal frequency Gaussian mass load variations illustrate aliasing variations associated with geographical location.

**Keywords.** GRACE, Aliasing, Kaula's resonance solution.

## 1 Introduction

GRACE (Gravity Recovery And Climate Experiment) satellite mission launched March 2002 is providing monthly Stokes coefficients up to degree and order 120. With time varying Stokes coefficients from month to month, water mass load changes over oceans and land can be estimated.

Two main sources of errors that corrupt GRACE spatial resolution are measurement and aliasing errors. The current GRACE measurement error level is about 40-50 times greater than the pre-launch error estimate (Seo and Wilson 2004(b) and Wahr et. al. 2004), but preliminary research with GRACE Stokes coefficients showed that temporal water mass variations over entire oceans and large scale river basins (for example, the Amazon basin) are detectable (Andersen 2004, Chambers et. al. 2004, Seo and Wilson 2004(a) and Wahr et. al. 2004). The reprocessed GRACE Stokes coefficients will have less measurement error (Tapley and Reigber 2004), so spatial resolution should improve.

Inadequate space-time sampling of temporal gravity field changes causes aliasing errors. Thompson et. al. (2004) and Han et. al. (2004) showed that GRACE aliasing errors are lower than current measurement error levels using synthetic data. However, measurement errors will decrease in reprocessing of GRACE measurements so aliasing error will remain the most serious error source. In this study, we estimate GRACE aliasing error with synthetic climate data, and explain features of aliasing errors from sinusoidally varying single spherical harmonics (SH) and Gaussian-shaped isolated mass load changes. The understanding of aliasing errors will be useful for future research to reduce these aliasing errors and improve spatial resolution.

## 2 Aliasing errors from climate data

To estimate GRACE aliasing error, we use potential differences between two satellites. Han et. al. (2004) showed that GRACE measurements can be transformed to potential differences. We assume the altitudes of satellites are fixed, and Stokes coefficients are estimated using a least square fit of spherical harmonics to potential differences. This simplified approach should simulate general features of GRACE aliasing errors.

**FIG. 1**

12hr GLDAS (Global Land Data Assimilation Scheme), 12hr ECCO (Estimating the Circulation and Climate of Ocean), and the difference between 6hr NCEP and ECMWF at September, 2003 are used as time varying air and water mass models. The atmospheric model difference estimates the de-aliasing error of atmospheric surface pressure. The aliasing errors from GLDAS and ECCO are defined to be the difference between estimated Stokes coefficients for a month and the true monthly mean coefficients. The aliasing error from atmospheric surface pressure is the Stokes coefficients estimate of the difference NCEP minus ECMWF because the an atmospheric model is already used in the processing as a dealias model.

Figure (1) shows the degree amplitudes of true monthly signals (solid lines) and aliasing errors (dotted lines). In the figure aliasing errors from land and oceans rapidly increase up to degree 15, and slowly increase after degree 15. The error amplitude of oceans is one order of magnitude greater than that of land. The atmospheric aliasing error, on the other hand, is significant at low degree under 15, and shows similar amplitude to oceanic effects over degree 15.

**FIG 2.**

Spatial-temporal spectra of time varying gravity signal are useful to explain general aliasing features. Figure (2) shows different spatial-temporal spectra between GLDAS and NCEP. GLDAS variations are mainly at low temporal frequencies, thus the aliasing from GLDAS is not severe. NCEP shows that high temporal frequencies are significant, especially at 1 and 2 day periods. These high frequency signals could cause low degree (below 15) aliasing error in figure (1).

**FIG 3.**

Figure (3) illustrates the map views of land and atmospheric aliasing errors in figure (1). In figure 3(a), the Amazon basin shows the biggest aliasing error because the signal amplitude is the largest. The atmospheric aliasing error (the aliasing error from imperfect atmospheric model for dealiasing) in figure 3(b) shows the error is greater at low latitude than high latitude. The model difference (dealiasing error) would be larger over high latitudes such as Antarctica than low latitudes. This unexpected aliasing error feature can be explained by the fact that GRACE samples more densely and frequently over the poles than near equator. Therefore, imperfect atmospheric models at low latitudes generate higher aliasing errors. We assume an inverted barometer (IB) response in the oceans. Thus the atmospheric error over the oceans is exactly zero.

## 3 Aliasing errors from single SH

From the aliasing error estimates in figure 1 and 3, it is complicated to explain the features of aliasing in the previous section because broad ranges of temporal and spatial frequencies contribute to the aliasing errors together. Therefore, we try to estimate the aliasing error for very simple cases, single spatial and temporal frequency signals. We also consider separately zonal and sectoral SH. The directional properties of these different harmonic signals cause different aliasing error within the same degree due to GRACE range rate measurements along a north-south direction.

### 3.1 Degree 5 Order 0 SH

Figure 4 presents aliasing error from a (5,0) SH oscillating sinusoidally from 0.5 to 128 days in periods. In figure 4(a), red circles are degree amplitudes of true 30 days mean, and blue solid lines are those of estimated Stokes coefficients from 30 days range rate measurement. Figure 4(b) is the aliasing error spectra versus degree and order. At sinusoidal periods of 0.5-1 days, alias error at degrees (2-15) is significant. At longer periods, aliasing error dominates from degree 15 and above. For a 0-order harmonic, temporal variations at periods less than 1 day are very undersampled. This explains why rapid atmospheric variations cause aliasing errors at low degree. Errors from degree 15 can be predicted by Kaula's resonance formula, (Lambeck, 1988)

$$N = m\dot{\theta} \,/\, (l-2p+q) \qquad (1)$$

N=15 or 16 rev/day (for GRACE), $\dot{\theta}$=1 cycle/day, l and m are degree and order, (p and q are the coefficients of inclination and eccentricity functions used to represent global gravity potential by Kaula (1966). Aliasing error is sensitive to the wavelength of gravity potential that approximately equals the length between successive ground tracks of a satellite in longitude (Lambeck, 1988)). For GRACE, eccentricity is almost zero so q=0. The conditions for resonance are (assuming N=15), m=15 when l=15,17,19,21,..., m=30 when l=30,32,32,34,... and more for m=45,60,... Figure 4 shows these resonance degrees and orders very well.

**FIG.4**

### 3.2 Degree 5 Order 5 SH

Figure 5 presents aliasing error from a (5,5) SH as in figure 4. As in the (5,0) case, the (5,5) case shows different aliasing features at short and long periods. At long periods, the pattern of aliasing error has a more complex relation with Kaula's resonance solution. Since the signal has non-zero order terms, the resonance order should be a linear combination of signal order and Kaula-resonant order. The resonance will happen at m=15-5 when l=10,12,14,.., m=15+5 when l=20,22,24, and more for m=30-5,30+5,45-5,....

FIG. 5

## 4 Aliasing errors from isolated mass load

A more realistic case than a single SH, allowing a study of the effect of geographic location, is a Gaussian distributed load at a single temporal frequency. The radius of the distribution (half-amplitude) is 2,000km.

### 6.1 Gaussian distribution at north pole

For a Gaussian distribution centered at the North Pole, the spatial spectrum includes zonal SH only. Figure 6 shows the aliasing errors, and it is very similar to (5,0) case.

FIG 6

The aliasing errors peak at degree 15, and then amplitudes decrease. The aliasing error contribution from different degree zonal SH may reduce the amplitudes over degree 15 by canceling out errors. The primary aliasing errors are near order 15, 30 and 45.

### 6.2 Gaussian distribution at equator

Figure 7 shows aliasing errors from Gaussian distribution centered at Equator. The aliasing error features are different from the North Pole case. Now the spatial spectrum includes zonal, sectoral and tesseral terms so aliasing error must come from various spatial spectra. The errors peak at degrees 15, 30 and 45. Alias variance appears to have a Gaussian distribution centered at main resonance order 15, 30 and 45. The primary aliasing errors are along tesseral SH.

FIG. 7

The different aliasing error features between figure 6 and 7 associated with geographic locations imply that an anti-aliasing filter might be constructed for specific river basins, depending on their latitude, to recover water mass loads.

## 7 Summary

Synthetic GRACE aliasing errors from land and ocean models rapidly increase up to degree and order 15. Atmospheric aliasing error is significant below degree 15. The errors are dependent on mass load signal amplitude, temporal-spatial frequency and geographic location.

To understand aliasing error, single spatial and temporal frequencies are generated, and those errors are estimated. Kaula's resonance solution helps explain the aliasing error in this simple case.

Different aliasing errors from different geographic location imply that latitude specific anti-aliasing processing or filtering might be possible. The simulated study of aliasing errors at river basins may be useful to construct anti-aliasing filter.

## Acknowledgement

We thank Shin-Chan Han for discussion on aliasing error estimation. This research was funded by NASA grants NAG5-8798, NGT5-30448 and NNG04GF22G. Additional support was provided by the Geology Foundation of the University of Texas.

## References

Andersen OB (2004) Annual gravity field variation from GRACE – Initial analysis (abstract). Eos Trans. AGU. 85(17) Spring Meet. Suppl., U31A-06.

Chamber DP, Wahr J and Nerem RS (2004) Preliminary observations of global ocean mass variations with GRACE, Geophys. Res. Lett., 31, L13310, doi:10.1029/2004GL020461.

Han SC, Jekeli C and Shum CK (2004) Time-variable aliasing effects of ocean tides, atmosphere and continental water mass on monthly mean GRACE gravity field, J. Geophys. Res. 109 B04403 doi:10.1029/2003JB002501.

Kaula, WM (1966) Theory of satellite geodesy. Blaisdell, Waltham, Mass.

Lambeck K (1988) Geophysical Geodesy-The Slow Deformations of the Earth, 719, Oxford

Seo KW and Wilson CR (2004a) Water storage variations and associated errors estimates from GRACE (abstract), EOS Trans. AGU, 85(17) Spring Meet. Suppl., U31A-05.

Seo KW and Wilson CR (2004b) Simulated estimation of hydrological loads from GRACE, J Geodesy, accepted.

Tapley BD and Reigber C (2004) Status and early results from the GRACE mission (abstract), EOS Trans. AGU, 85(17) Spring Meet. Suppl., U31A-01

Thomson PF, Bettadpur SV and Tapley BD (2004) Impact of short period, non-tidal, temporal mass variability on GRACE gravity estimates, Geophys. Res. Lett., 31, L06619 doi:10.1029/2003GL019285.

Wahr J, Swenson S, Zlotnicki V and Velicogna I (2004) Time-variable gravity from GRACE: First results, Geophys. Res. Lett., 31, L11501, doi:10.1029/2004GL019779.

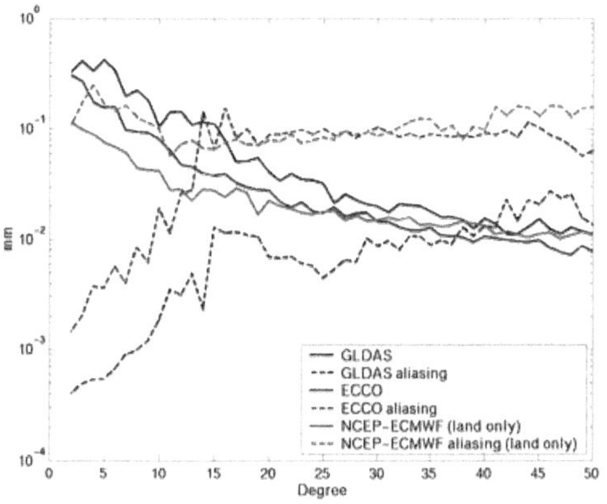


Figure 1: Degree amplitudes of true monthly means and aliasing errors in units of geoid.

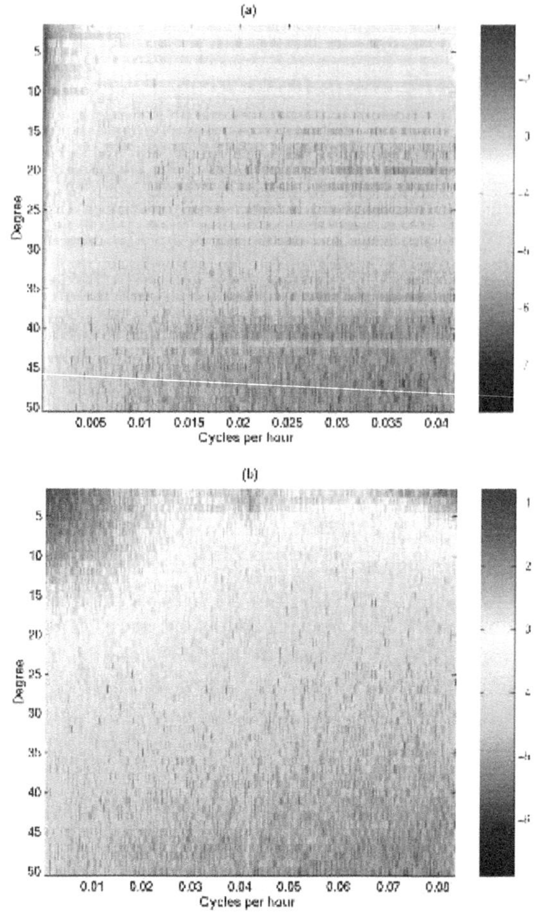


Figure 2: SH degree and temporal frequency spectra of (a) GLDAS and (b) NCEP in geoid units (log scale).

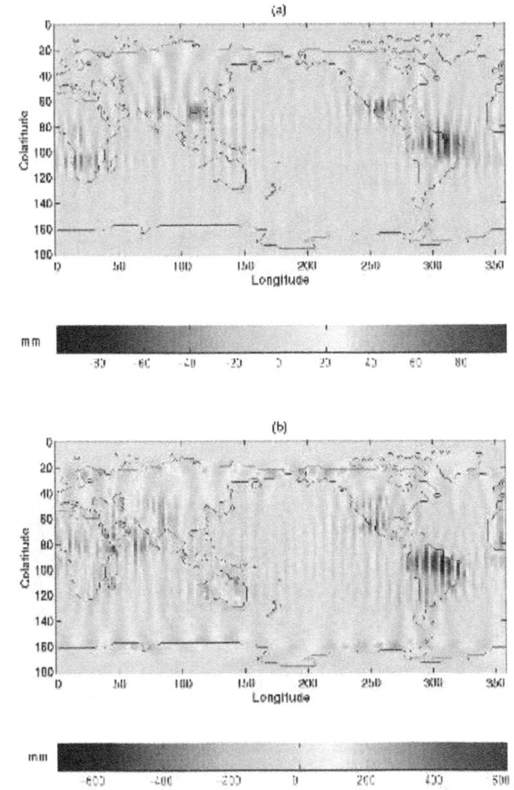

Figure 3: Map view of aliasing errors for GLDAS (a) and the difference between ECMWF and NCEP (b).

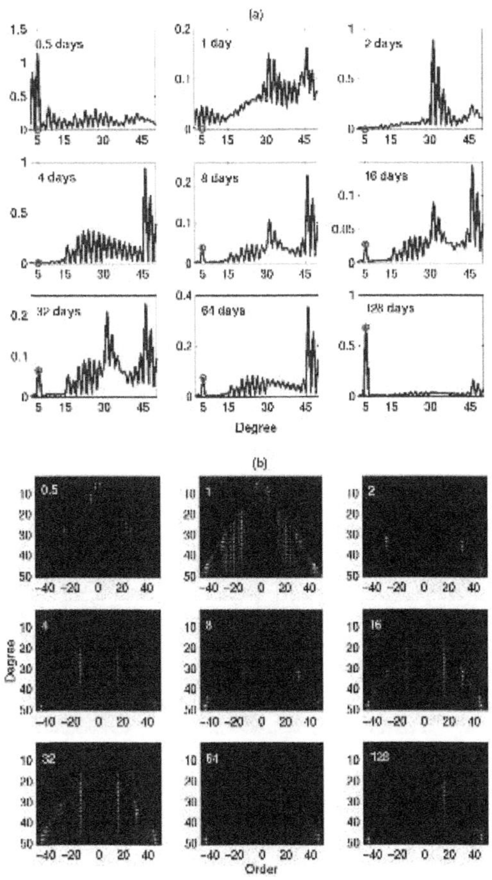

Figure 4: (a) Degree amplitude of Stokes coefficients from 30 days of data (blue lines) and true 30 day mean (red circles). The (5,0) harmonics are sinusoidally oscillating at periods from 0.5 to 128 days. (b) Aliasing error spectra vs. degree and order

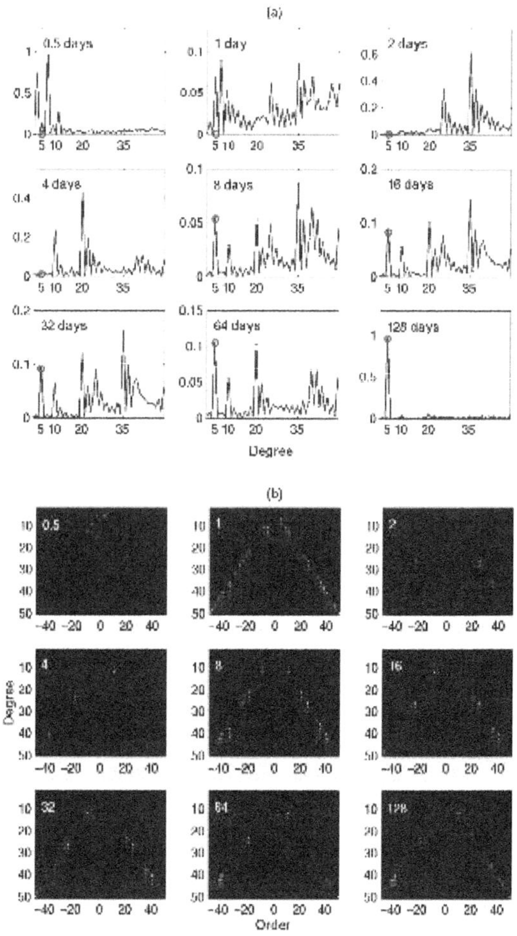


Figure 5: (a) Degree amplitude of Stokes coefficients from 30 days of data (blue lines) and true 30 day mean (red circles). The (5,5) harmonics are sinusoidally oscillating at periods from 0.5 to 128 days. (b) Aliasing error spectra vs. degree and

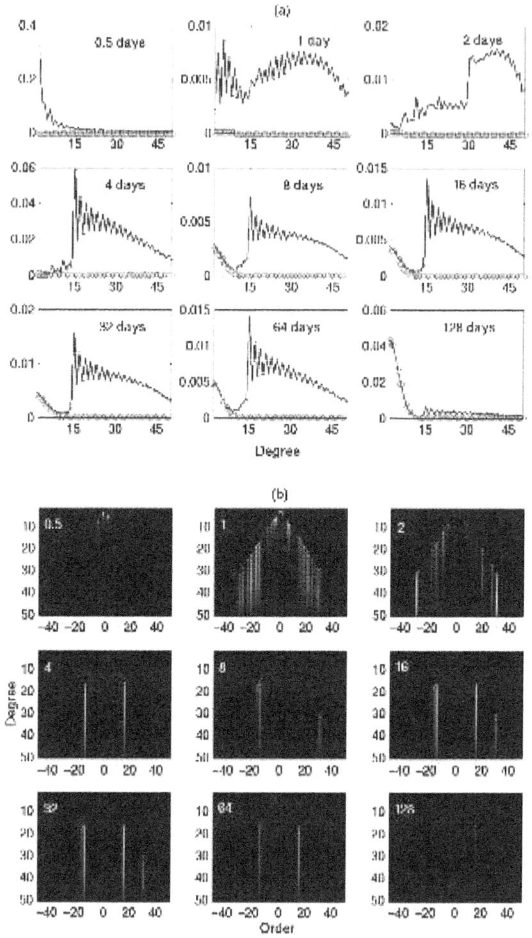


Figure 6: (a) Degree amplitude of Stokes coefficients from 30 days of data (blue lines) and true 30 day mean (red circles) for Gaussian loads at north pole. The Gaussian distributions are sinusoidally oscillating at periods from 0.5 to 128 days. (b) Aliasing error spectra vs. degree and order

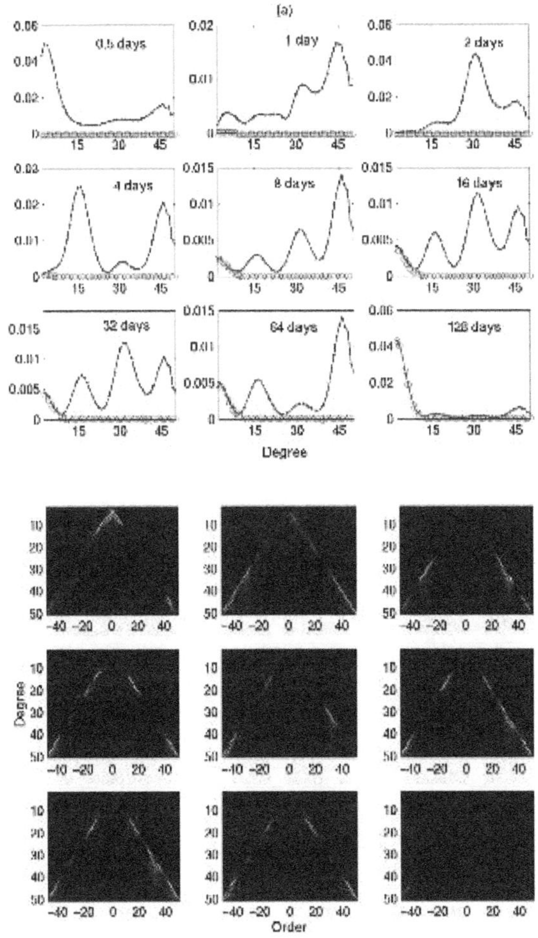


Figure 7: (a) Degree amplitude of Stokes coefficients from 30 days of data (blue lines) and true 30 day mean (red circles) for Gaussian loads at equator. The Gaussian distributions are sinusoidally oscillating at periods from 0.5 to 128 days. (b) Aliasing error spectra vs. degree and order

# Methods to Study Co-seismic Deformations Detectable by Satellite Gravity Mission GRACE

Wenke SUN and Shuhei OKUBO
Earthquake Research Institute, University of Tokyo, Tokyo, Japan
sunw@eri.u-tokyo.ac.jp

**Abstract.** Co-seismic deformations cannot be compared directly with those observed by gravity satellite missions because of the spatial resolution limit of the missions and the signal attenuation of the gravity field. For this purpose, the conventional dislocation theory for a spherical earth model can be used because it is expressed in the form of a spherical harmonic. In this study, analytical expressions of degree variances of the co-seismic geoid and gravity changes for shear and tensile sources are derived and calculated for three real earthquakes. Those results are compared with expected errors of GRACE to elucidate whether or not co-seismic geoid and gravity changes are detectable by gravity satellite missions. Behaviors of the degree variances for four independent seismic sources are investigated. Results indicate that both the gravity and geoid changes are near two orders of magnitude larger than the precision of the gravity missions in low harmonic degrees. Based on these results, we derived the minimum magnitudes of earthquakes detectable by GRACE. We concluded that co-seismic deformations for an earthquake with a seismic magnitude above m =7.5 are expected to be detected by GRACE.

**Key Words:** Co-seismic Deformation, Geoid, Gravity, Gravity Mission, Dislocation, Earthquake.

## 1. Introduction

Satellite gravity missions, such as the Gravity Recovery and Climate Experiment (GRACE) [NRC, 1997], are available for gravity field determination from space. Use of GRACE allows measurement of temporal gravity variations caused by various geophysical processes. The primary objective of GRACE is to provide an unprecedented accurate, global, and high-resolution estimate of constant and time-variable components of the earth's gravity field every 30 days over a 5-year period [Wahr et al., 1998]. It is anticipated that the gravity missions will yield extremely wide geophysical applications in geosciences.

Temporal gravity variations of global nature result from atmospheric mass redistribution, ocean circulation, polar ice melting or aggregation, the visco-elastic response of the Earth's lithosphere to past and present loads, etc. [*Chao et al.* 2000; *Chao* 2003]. In addition to these processes, earthquakes are expected to produce significant global gravity perturbations that are detectable through analysis of gravity missions. To study quantitatively whether co-seismic deformations are detectable from space, *Gross and Chao* [2001] investigated this problem using normal mode technique based on *Chao and Gross* [1987]. They computed the degree variance

$$c_n^2 = \sum_{m=0}^{n} \left( C_{nm}^2 + S_{nm}^2 \right) \tag{1}$$

of the co-seismic gravitational potential anomaly, which gives the contribution of the degree $n$ terms to the total variance (Heiskanen and Moritz, 1967),

$$V = \frac{GM}{a} \sum_{n=0}^{\infty} \left( \frac{a}{r} \right)^{n+1} \sum_{m=-n}^{n} (C_{nm} \cos m\varphi + S_{nm} \sin m\varphi) P_n^m (\sin\theta) \tag{2}$$

Comparing the degree amplitude spectra of some earthquakes with expected GRACE sensitivity, they concluded that co-seismic effects of great earthquakes such as the 1960 Chilean or 1964 Alaska events can cause global gravitational field changes that are sufficiently large as to be detectable by GRACE. Mass redistribution in the Earth caused by an earthquake changes not only the gravity field, but also global rotation or polar motion. A related study can be found in *Chao et al.* [1996]. More applications of GRACE can be found in *Chen et al.* [2004] and *Tapley et al.* [2004].

In this study, we investigate the same problem by two parallel methods: 1. truncated co-seismic deformation, and 2. degree variance of co-seismic deformation using dislocation Love numbers. The former presents a way to observe spatial distribution of individual or group spherical harmonic degrees of co-seismic deformations. The latter is based on the degree variance of co-seismic deformations. We investigate co-seismic geoid and gravity changes by observing the distribution of their degree variances in comparison to the expected sensitivity of satellite gravity missions. Results for co-seismic deformations for large earthquakes are discussed with respect to their detectability. Note that this study leads to an identical conclusion to that of *Gross and Chao* [2001] using the normal mode scheme.

## 2. Co-seismic Geoid and Gravity Changes and Their Degree Amplitude Spectra

Assume that an inclined point dislocation be located on the polar axis in a compressible and self-gravitating spherical earth. Furthermore, and assume the fault line is in the direction of $\varphi = 0$. According to the quasi-static dislocation theory, co-seismic geoid and gravity changes at an observation point $(a, \theta, \varphi)$ can be expressed as [*Sun and Okubo*, 1993]

$$\zeta^{ij} = \sum_{n,m} k_{nm}^{ij} Y_n^m(\theta, \varphi) \cdot \nu_i n_j f_\varsigma \tag{3}$$

$$\delta g^{ij} = \sum_{n,m} (n+1) k_{nm}^{ij} \cdot Y_n^m(\theta, \varphi) \nu_i n_j f_g \tag{4}$$

where $k_{nm}^{ij}$ (related to the gravitational potential change) are the dislocation Love numbers defined by *Sun and Okubo* [1993], functions of the spherical harmonic degree, order, source depth, and source type. Components of the slip vector and its normal on the infinitesimal fault area *dS* are $\nu_i$ and $n_j$, with total dislocation *U*. Gravity on the earth surface is $g_0$; $a$ is the radius of the Earth and $Y_n^m(\theta, \varphi)$ is the spherical harmonic function of degree $n$ and order $m$. The dislocation factors, $f_\varsigma = UdS / a^2$ and $f_g = g_0 UdS / a^3$, define the earthquake magnitude and give the unit of geoid and gravity changes.

There are nine total solutions for all possible sources. However, only four independent solutions exist if the earth model is spherically symmetric and isotropic. In this study, we choose the following four independent solutions: $ij$ = 12, 32, 22, and 33. They represent strike-slip, dip-slip, horizontal tensile and vertical tensile, respectively. Note that components of $ij$ = 22 include two parts: $m$ = 0 and 2; in this study, we calculate and discuss only the deformations of $m$ = 0: the computation of $m$ = 2 can be derived easily from the component of $ij$ = 12. Details can be found in *Sun and Okubo* [1993].

Dislocation Love numbers $k_{nm}^{ij}$ can be obtained numerically for a spherically symmetric earth model [*Sun and Okubo*, 1993] such as the 1066A [*Gilbert and Dziewonski*, 1975] Subsequently, gravity and geoid changes can be calculated by the above summations in (3) and (4). Figure 1 gives numerical results of the dislocation Love numbers $k_{nm}^{ij}$ of the four types of seismic sources at a depth of 32 km with the 1066A model as a function of the spherical harmonic degree $n$ up to 2000. Once a dislocation source or earthquake parameter is provided, co-seismic deformations can be calculated easily using these Love numbers.

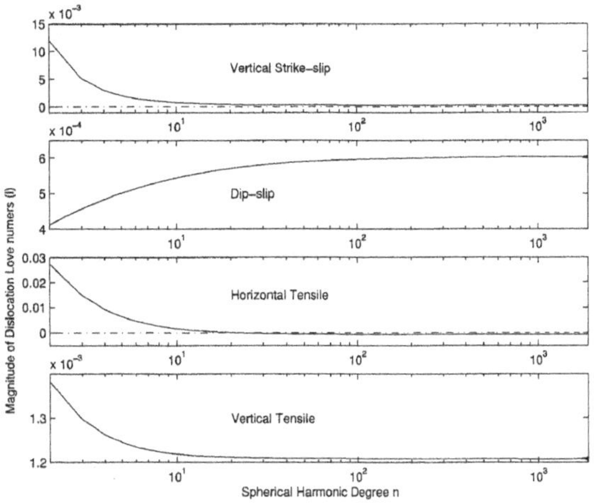


Fig. 1. Normalized dislocation Love numbers $k_{nm}^{ij}$ of four types of seismic sources at a depth of 32 km.

On the other hand, a dislocation vector $\mathbf{v}$ and its normal $\mathbf{n}$ can be expressed in terms of dip-angle $\delta$ and slip-angle $\lambda$ of the fault as

$$\mathbf{n} = \mathbf{e}_3 \cos\delta - \mathbf{e}_2 \sin\delta \tag{5}$$

$$\mathbf{v} = \mathbf{e}_3 \sin\delta \sin\lambda + \mathbf{e}_1 \cos\lambda + \mathbf{e}_2 \cos\delta \sin\lambda \tag{6}$$

It is a shear dislocation problem if the dislocation vector $\mathbf{v}$ runs parallel to the fault plan. For a tensile opening, $\mathbf{v} = \mathbf{n}$. Then for an arbitrary shear and tensile faults on the polar axis, according to Eq. (3) the co-seismic geoid changes can be written as

$$\zeta^{Shear} = \sum_{n=2}^{\infty} \Big\{ \cos\lambda \big[ -k_{n2}^{12} \sin\delta Y_n^2(\theta, \varphi) + k_{n1}^{13} \cos\delta \cdot Y_n^1(\theta, \varphi) \big] + \sin\lambda \Big[ \frac{1}{2} \big( k_{n0}^{33} - k_{n0}^{22} \big) \sin 2\delta \cdot Y_n^0(\theta, \varphi) + k_{n1}^{32} \cos 2\delta Y_n^1(\theta, \varphi) \Big] \Big\} \cdot f_\varsigma \tag{7}$$

$$\zeta^{Tensile} = \sum_{n=2}^{\infty} \Big[ \big( k_{n0}^{33} \cos^2\delta + k_{n0}^{22} \sin^2\delta \big) Y_n^0(\theta, \varphi) - k_{n1}^{32} \sin 2\delta Y_n^1(\theta, \varphi) \Big] \cdot f_\varsigma \tag{8}$$

There are the similar expressions for gravity changes, omitted here for shortening.

Based on (7) and (8), it is straightforward to investigate whether co-seismic deformations are detectable by the satellite gravity missions, e.g., GRACE. For this purpose, amplitude spectra of the above co-seismic geoid and gravity changes will be computed for degrees $n$ = 2–100. From (7)–(8) it can be seen that the terms of degrees $n$=0 and $n$=1 vanish because the total mass of the earth is constant and the origin of the reference frame is located at the center of mass of the earth model.

Comparing Eqs. (7) and (8) to (2) indicates that the coefficients (the dislocation love numbers and the geographical parameters of the fault) of $Y_n^m(\theta,\varphi)$ in (7) and (8) are nothing but the Stokes coefficients. The angular order $m$ vanishes except $m$ = 0, 1 and 2 because the source is chosen at the polar axis and also because of the symmetric property of the source functions. Therefore the computation of the degree variance is slightly easier than the full spectrum-distributed coefficients. The degree variances for shear and tensile sources can be written straightforwardly as the following equations.

$$\left(c_n^{Shear}\right)_\varsigma^2 = \Big[\left(k_{n2}^{12}\sin\delta\cos\lambda\right)^2 + \left(k_{n1}^{13}\cos\delta\cos\lambda\right)^2 + \left(\frac{k_{n0}^{22}}{2}\sin 2\delta\sin\lambda\right)^2 + \left(\frac{k_{n0}^{33}}{2}\sin 2\delta\sin\lambda\right)^2 + \left(k_{n1}^{32}\cos 2\delta\sin\lambda\right)^2\Big]\cdot\left(f_\varsigma\right)^2 \quad (9)$$

$$\left(c_n^{Tensile}\right)_\varsigma^2 = \Big[\left(k_{n0}^{33}\cos^2\delta\right)^2 + \left(k_{n0}^{22}\sin^2\delta\right)^2 + \left(k_{n1}^{32}\sin 2\delta\right)^2\Big]f_\varsigma \quad (10)$$

It is seen that the degree variances involve not only the dislocation Love numbers, but also the geometrical position of the fault described by the dip-angle $\delta$ and slip-angle $\lambda$, and the dislocation factors. It is also found that the degree variances for a shear fault movement include both the shear and tensile components; whereas the degree variances for a tensile fault also include the two components, but are unconnected with the dip-angle.

In practical calculation, parameters such as source type $ij$ and harmonic order $m$ are determined according to the selected source types. Then for the four independent sources, the degree variances can be simplified as the following formulas.

$$\left(c_n^{12}\right)_\varsigma = \left|k_{n2}^{12}\right|\frac{UdS}{a^2},\quad \left(c_n^{32}\right)_\varsigma = \left|k_{n1}^{32}\right|\frac{UdS}{a^2}$$

$$\left(c_n^{22}\right)_\varsigma = \left|k_{n0}^{22}\right|\frac{UdS}{a^2},\quad \left(c_n^{33}\right)_\varsigma = \left|k_{n0}^{33}\right|\frac{UdS}{a^2}$$

Those formulas show that the degree variances are proportional to the absolute values of the dislocation Love numbers because they are functions of one of the three angular orders 0, 1, or 2. For each harmonic degree $n$, the only variable is the dislocation Love number, so that the root-square of the dislocation number is its absolute value. Therefore, the dislocation Love numbers themselves give their degree variances, multiplied by the dislocation factors that are determined by a selected earthquake. These degree variances can be obtained easily from the above dislocation Love numbers in Fig. 1.

## 3. Degree Variance of the Four Independent Solutions

A great fault is considered equivalent to that of the Alaska earthquake (1964, $m_w$=9.2) with parameters of [*Savage and Hastie*, 1966] 600 km length, 200 km width, 20 km depth, and 10 m dislocation. The degree variances are computed for the above four independent types of sources. Thereby, we can observe behaviors of respective sources. The dislocation factors yield $f_\varsigma$=29.56 mm for geoid change. Consequently, the degree variances of the four source types can be written as.

$$\left(c_n^{ij}\right)_\varsigma = 29.56\left|k_{nm}^{ij}\right| \text{ (mm)} \quad (11)$$

Correspondingly, degree variances of the geoid for the four sources are calculated and plotted in Fig. 2. This figure also show the expected instrument errors of the GRACE measurements [*Gross and Chao*, 2001].

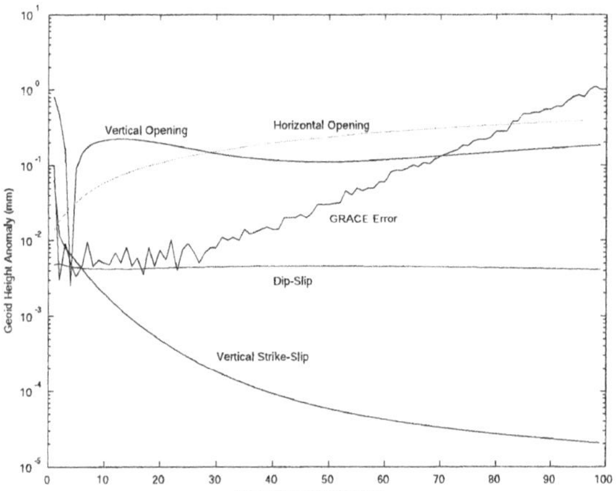

Fig. 2. Degree amplitude spectra for the co-seismic geoid height changes caused by four types of sources at 20 km depth in the 1066A earth model.

The relative contribution to each component of the geoid change can be seen clearly from Fig. 2 by observing the magnitude distribution of these degree variances. An important property shown in Fig. 2 is that the two tensile sources have stronger power than the two shear sources. This property means that co-seismic deformations of the tensile sources are expected to be larger than those of the shear sources. It implies that even for the same geographical fault size (or moment) of an earthquake, their contributions to geoid and gravity changes may be different. Comparing the degree variances with the GRACE error indicates that GRACE can detect the co-seismic geoid changes for the two tensile sources because they are almost two orders larger in magnitude until the first 70 degrees. The geoid change for the two shear sources is all equal to or less than the GRACE error; also, they are difficult to detect. Figure 2 also show that the degree strength has a different contribution at different harmonic degree *n*. The degree variance of vertical strike-slip source decreases rapidly as degree *n* increases.

Table 1 shows that we may easily derive the minimum magnitudes of earthquakes which are expected to be detected by the gravity missions according to the above results and discussions. It indicates that if an earthquake is as large as the magnitude of m = 9.0 (for source types 12 and 32) or m = 7.5 (for source types 22 and 33), the corresponding deformations are expected to be detected by GRACE. Note that the co-seismic geoid and gravity changes are expected to be easily detected by the future GRACE follow-on. If it offers about two orders better accuracy than GRACE [*Watkins et al.*, 2000; *NRC*, 1997], the minimum magnitude of earthquakes expected to be detected by GRACE follow-on is m = 7.5 for source types 12 and 32 or m = 6.0 for source types 22 and 33.

Table 1. Minimum magnitude of earthquakes expected to be detected by GRACE.

| Source type *ij* | Seismic magnitude |
|---|---|
| 12 | $m = 9.0$ |
| 32 | $m = 9.0$ |
| 22 | $m = 7.5$ |
| 33 | $m = 7.5$ |

## 4. Case Study – the Alaska (1964, 2002) and Hokkaido (2003) Earthquakes

In this section, as a case study, let us consider three actual earthquakes: the Alaska earthquakes occurred in 1964 ($m_W$=9.2) and in 2002 (m=7.9), and the Hokkaido (Japan) earthquake happened in 2003 (m=8.0). We can learn whether the co-seismic geoid and gravity can be detected by observing their degree variances in comparison to the GRACE error. Case 1: the 1964 Alaska earthquake. According to the parameters, it is known that $\delta = 9°$ and $\lambda = 90°$, $f_\varsigma = 29.56$mm and $f_g = 4.556\mu$gal. Case 2: the 2002 Alaska earthquake. According to the fault parameters, we have $\delta = 90°$ and $\lambda = 0°$, $f_\varsigma = 0.6356$mm and $f_g = 0.0980\mu$gal. Case 3: the 2003 Hokkaido earthquake. In this case, $\delta = 20°$ and $\lambda = 90°$, $f_\varsigma = 0.4100$mm and $f_g = 0.0630\mu$gal. For the three cases, the degree variances of geoid and gravity changes can be calculated using the dislocation Love numbers given in Fig. 1 and plotted in Fig. 3 for the geoid change. The GRACE errors are also plotted in the figures for comparison.

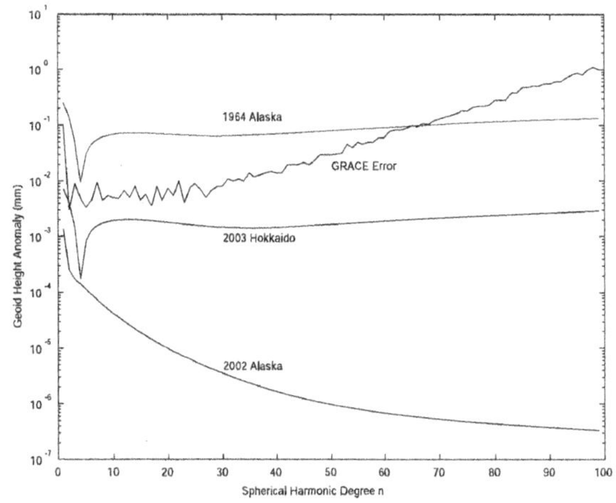


Fig. 3. Degree amplitude spectra for the co-seismic geoid height changes caused by three earthquakes: the 1964 Alaska (9.2), the 2002 Alaska (7.9), and the 2003 Hokkaido (8.0).

Figure 3 show that the 1964 Alaska earthquake causes the co-seismic geoid changes in low degree part with near two orders in magnitude larger than the GRACE error. This fact indicates that the global geoid changes are sufficiently large to be detected by GRACE. The figure also show that the geoid changes caused by the 2002 Alaska earthquake and the 2003 Hokkaido earthquake are smaller than the GRACE error: apparently, they are too small to be detected by GRACE. Although the geoid effects at degree 3 show the same level with the GRACE error, considering other physical effects and the rather long measurement period, they are less likely to be detected in actual practice. Note that results of the 1964 Alaska earthquake show the same property as that given by *Gross and Chao* [2001].

Comparison of the degree variances of the 2002 Alaska and the 2003 Hokkaido earthquakes showed that geoid effects for the 2003 Hokkaido earthquake are much larger than those of the 2002 Alaska earthquake at almost all the harmonic degrees, but that their fault sizes are almost identical. This fact further confirms that the geographical position, especially the dip angle, plays an important role in these geophysical effects. This importance is understandable: the dip slip source involves tensile components, which are shown to be much larger than the shear sources.

Note that the above results represent co-seismic deformations of a point source. In practice, if the fault size is extremely large or compatible with the distance between the source and satellite, the geometrical shape of the fault should be considered. A point source is sufficient if the fault size is sufficiently small in comparison to the distance from satellite to the earth surface. The source depth is another factor affecting the magnitude of deformations. Nevertheless, the effect of source depth is considered to be relatively small compared to the fault size because co-seismic deformation, especially the geoid change, is relatively less sensitive to depth [*Sun and Okubo*, 1998]. Note that the finite size of the fault might be important for the GRACE follow-on mission. To reduce contamination from hydrology, oceanography, and other factors, we address extremely small regions because relatively small follow-on errors will permit us to examine such small regions. Those regions should be sufficiently small that the fault size plane extent for a large event could be important. An integration of a point source over the fault plane is required to compute accurate co-seismic deformations by a limited fault size. However, this study is intended to observe the magnitude of co-seismic deformations. For that reason, a rough approximation is acceptable.

We emphasize that co-seismic gravity changes are difficult to distinguish in practice because of complications of the gravity field. Ideally, to distinguish co-seismic geoid and gravity changes, the gravity field should be observed just before and after the seismic event. In this case, all other long temporal effects on gravity change should be relative small and can therefore be neglected. In practice, however, GRACE provides us with a complete gravity observation in a one month time interval. During that one month, the earth undergoes many geophysical changes that engender temporal gravity changes. In other words, the temporal gravity variations are expected to comprise many physical effects such as tidal changes, atmospheric changes, rain or snow fall, and so on. On the other hand, in some time scale, other temporal gravity variations are equal in size, or even larger than, the co-seismic deformations. Therefore all of these gravity changes should be well modeled or observed before the co-seismic gravity changes can be detected.

The above results and discussions imply that co-seismic geoid and gravity changes are almost impossible to detect by GRACE. However, for an earthquake with a magnitude greater than m = 7.9, such as m > 8.0, as shown in Table 1, co-seismic geoid and gravity changes are anticipated to be detectable by GRACE. On a hopeful note, geoid and gravity changes caused by the equivalent of the 2002 Alaska earthquake may be detectable from space by projects like the GRACE follow-on mission because the forthcoming GRACE follow-on gravity mission is expected to have better accuracy, by more than two orders, than GRACE.

## 5. Truncated Co-seismic Deformations in the Domain of Spherical Harmonic Degree

Except the above normal mode technique [*Gross and Chao*, 2001] and the degree variance method [*Sun and Okubo*, 2004a], we may investigate the same problem from another way, i.e., the truncated co-seismic deformations [Sun and Okubo, 2004b]. The dislocation theory for a spherical earth [*Sun and Okubo*, 1993] can be used to calculate a co-seismic geoid or gravity change for each individual or group, since sometimes we may need to consider the cumulative co-seismic deformations because the gravity satellite missions cannot detect whole components of degree $n$ as a result of its spatial resolution limit. For this purpose, we should calculate co-seismic deformation with a truncation, so that truncated co-seismic deformations can be calculated and be compared to the gravity missions. Therefore, we define here a concept of truncated co-seismic geoid and gravity changes, i.e., co-seismic deformations are calculated only for an interested degree band, i.e., the summations in Eqs. (3) and (4) are performed from $N_1$ to $N_2$ to yield

$$\zeta^{ij} = \sum_{n=N_1}^{n=N_2} \sum_m k_{nm}^{ij} Y_n^m(\theta,\varphi) \cdot \nu_i n_j f_\varsigma \tag{12}$$

$$\delta g^{ij} = \sum_{n=N_1}^{n=N_2} \sum_m (n+1) k_{nm}^{ij} \cdot Y_n^m(\theta,\varphi) \nu_i n_j f_g \tag{13}$$

Notice that when $N_1 = N_2$, it is nothing but the spatial distribution of the co-seismic deformations for each harmonic degree. Comparing the numerical results calculated by (12) and (13) with the expected error of GRACE leads to the same conclusion as listed in Table 1. Refer Sun and Okubo (2004b) for detail.

## 6. Summary

Co-seismic deformations in the spectral domain are considered so that the deformations can be investigated by individual spherical harmonic degrees. Degree variances of the co-seismic geoid and gravity changes for shear and tensile sources are derived analytically. Numerical investigation was carried out to observe whether co-seismic geoid and gravity changes are detectable by satellite gravity missions. Numerical results of the degree variances were used for comparison with the expected sensitivity of GRACE. Then, we undertook case studies of Alaska (1964, 2002) and Hokkaido earthquakes (2003). The corresponding modeled co-seismic deformations indicated that both gravity and geoid changes are near two orders larger than the GRACE precession. This study derived minimum earthquakes to be detected by GRACE mission. The results engender the conclusion that co-seismic deformations for an earthquake with a seismic magnitude of m = 7.5 are expected to be detectable by GRACE. Finally, it is pointed out that the above discussions are based on the expected GRACE error obtained by simulations. The real GRACE observation data show that its real accuracy is somehow lower than the expected one. However, it is believed that the basic conclusion obtained in this research keeps unchanged.

**Acknowledgments.** This research was supported financially by JSPS research grants.

## References

Chao, B.F. and R.S. Gross, Changes in the Earth's rotation and low-degree gravitational field induced by earthquakes, Geophys. J. R. astr. Soc., 91, 569–596, 1987.

Chao, B.F., R.S. Gross, and Y. Han, Seismic excitation of the polar motion, 1977–1993, PAGEOPH, 146, 407–419, 1996.

Chao, B.F., V. Dehant, R.S. Gross, R.D. Ray, D.A. Salstein, M.M. Watkins, and C.R. Wilson, Space geodesy monitors mass transports in global geophysical fluids, Eos, Transactions, American Geophysical Union, 81, 247–250, 2000.

Chao, B.F., Geodesy is not just for static measurements any more, Eos, Transactions, American Geophysical Union, 84, 145–156, 2003.

Chen J. L., C. R. Wilson, B. D. Tapley, J. C. Ries (2004), Low degree gravitational changes from GRACE: Validation and interpretation, Geophys. Res. Lett., 31, L22607, doi:10.1029/2004GL021670.

Gilbert F. and A.M. Dziewonski, An application of normal mode theory to the retrieval of structural parameters and source mechanisms from seismic spectra, *Phil. Trans. R. Soc. London A, 278*, 187–269, 1975.

Gross, R.S. and B.F. Chao, The gravitational signature of earthquakes, in Gravity, Geoid, and Geodynamics 2000, edited by M.G. Sideris, pp. 205–210, IAG Symposia Vol. 123, Springer-Verlag, New York, 2001.

Heiskanen, W.H., and H. Moritz, 1967, Physical Geodesy, W.H. Freeman and Co., San Francisco.

National Research Council, NAS, Satellite Gravity and the Geosphere, ed. J. O. Dickey, Washington, D.C., 1997.

Savage J.C. and L.M., Hastie, Surface deformation associated with dip-slip faulting, *J. Geophys. Res., 71*, 4897–4904, 1966.

Sun W. and S. Okubo, Surface potential and gravity changes due to internal dislocations in a spherical earth – I. Theory for a point dislocation, *Geophys. J. Int., 114*, 569–592, 1993.

Sun, W. and S. Okubo, 1998. Surface potential and gravity changes due to internal dislocations in a spherical earth - II. Application to a finite fault, Geophys. J. Int., 132, 79-88

Sun, W. and S. Okubo, 2004a. Co-seismic Deformations Detectable by Satellite Gravity Missions – a Case Study of Alaska (1964, 2002) and Hokkaido (2003) Earthquakes in the Spectral Domain, J. Geophys. Res., Vol. 109, No. B4, B04405, doi:10.1029/2003JB002554.

Sun, W. and S. Okubo, 2004b. Truncated Co-seismic Geoid and Gravity Changes in the Domain of Spherical Harmonic Degree, EPS, 56, 881-892.

Tapley B. D., S. Bettadpur, M. Watkins, C. Reigber (2004), The gravity recovery and climate experiment: Mission overview and early results, Geophys. Res. Lett., 31, L09607, doi:10.1029/2004GL019920.

Wahr, J., M. Molenaar, and F. Bryan, Time variability of the Earth's gravity field: Hydrological and oceanic effects and their possible detection using GRACE, J. Geophys. Res., 103, 30205–30230, 1998.

Watkins, M. M., W. M. Folkner, B. F. Chao, and B. D. Tapley, The NASA EX-5 Mission: A laser interferometer follow-on to GRACE, IAG Symp. GGG2000, Banff, July, 2000.

# The gradiometric-geodynamic boundary value problem

Gy. Tóth

[1] *Budapest University of Technology and Economics, Department of Geodesy and Surveying and Physical Geodesy and Geodynamics Research Group of BUTE-HAS, H-1521 Budapest, Hungary, Müegyetem rkp. 3, Fax: +36 14633192, E-mail:* gtoth@sci.fgt.bme.hu.

**Abstract** The role of gravity gradients is investigated in the framework of geodetic-geodynamic boundary value problems. The time variation of the Eötvös tensor can be separated into three parts. The first part is a surface movement term, the second is the time variation of the gravity field in the original point, and the third is a coupling term. The first and third terms can be computed by a third order derivative tensor of the gravity potential. These terms can be formulated in spherical and planar approximation.

Gravity gradients have the advantage over gravity measurements that certain gravity gradient combinations are insensitive to site movements, thereby allowing the time variation of the gravity field to be determined without repeated height measurements. The solution of the corresponding gradiometric-geodynamic boundary value problem is shown with repeated gravity gradient measurements of the torsion balance.

Several test computations with a simple mass density model were performed in order to analyze the time variation of gravity gradients and give estimates of their respective magnitudes.

**Keywords.** time variation, gravity gradient tensor, boundary value problem

## 1 Introduction

The dynamic processes of the Earth produce mass redistributions in various parts of the Earth, which in turn cause surface deformations and gravity field variations to take place as functions of time and position. Thus repeated measurements related to position and gravity potential or its first and second order derivatives give important information on our dynamic planet, the Earth.

Repeated position and gravity measurements give values related to two functions describing position $\mathbf{r}(\mathbf{r}, t)$ and gravity potential $W(\mathbf{r}, t)$ as a function of position $\mathbf{r}$ and time $t$. By differencing these functions with respect to a certain initial epoch $t_0$ we get the following time dependent surface deformation $\delta\mathbf{r}$ and gravity potential variation $\delta W$

$$\begin{aligned}\delta\mathbf{r} &= \mathbf{r}(t) - \mathbf{r}(t_0) \\ \delta W &= W(t) - W(t_0)\,.\end{aligned} \tag{1}$$

The differences can be computed either at the undeformed position $\mathbf{r}$, which is the Eulerian approach, or at the deformed position $\mathbf{r} + \delta\mathbf{r}$ in the Lagrangean approach. The time-dependent difference of gravity potential or any related gravity field paremeter will be called disturbance (denoted by the prefix '$\delta$') in the Eulerian approach or anomaly (denoted by the prefix '$\Delta$') in the Lagrangean approach (Heck, 1986). If the time variable $t$ is considered to be discrete, we have two discrete set of functions, considered as unknowns of the problem, characterizing geodynamical processes of the Earth. To further simplify the matter it will be assumed that there are only two epochs, therefore there are only two unknown functions to be solved for.

If we take repeated measurements on a deforming boundary surface and seek for the deformation and potential anomaly/disturbance we have a geodetic-geodynamic boundary value problem (GGBVP). Several authors discussed such problems, see e.g. Biró, (1983) and Heck, (1986). The measurements considered were repeated leveling, gravity, vertical gravity gradient and position (GPS) measurements, for example. In the present paper we will consider repeated gravity gradient measurements, i.e. repeated determination of the Eötvös tensor $\mathbf{E}$ of second order potential derivatives ($\mathbf{E} = V_{ij}=\partial^2 V/\partial i\partial j$, $i,j\in\{x,y,z\}$).

In the first part of the paper we discuss the relation between time-dependent gravity gradient disturbance $\delta\mathbf{E}$, surface deformation $\delta\mathbf{r}$ and time-dependent gravity gradient anomaly $\Delta\mathbf{E}$, and linearization of the problem. Next linearization and approximation errors will be investigated through a simple numerical example. Finally it will be shown how to treat the problem of time-dependent gravity disturbance determination from time-dependent gravity gradient disturbances.

## 2 Surface deformation and time-dependent gravity gradient anomalies

Let us suppose that at initial epoch $t_0$ the position of an arbitrary point P of the surface S is described by the vector $\mathbf{r}(t_0)$. At a later epoch $t$ the point P has moved to a new position P'($\mathbf{r}'$) on S' (Fig. 1). The Eötvös tensor $\mathbf{E}(t_0)$ at P has changed to $\mathbf{E}'(t)$ at the new point P' as well. The two gravity gradient tensors at epochs $t_0$ and $t$ will be considered in the following as known from repeated gravity gradient measurements.

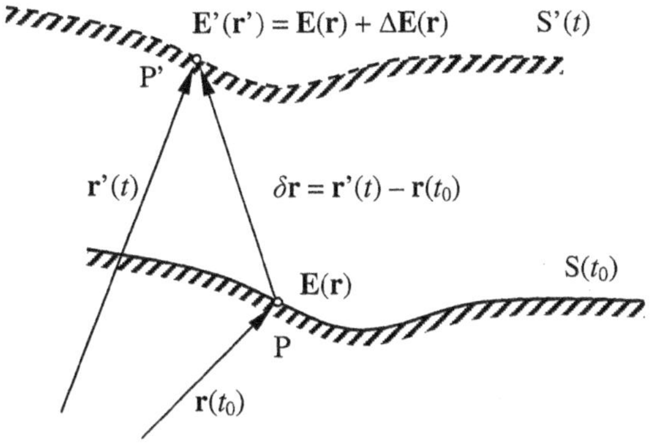


**Fig. 1** Definitions of surface deformation and gravity gradient anomaly

The time-dependent gravity gradient anomaly $\Delta\mathbf{E}(\mathbf{r})$ is therefore known from repeated measurements and it is by definition

$$\Delta\mathbf{E}(\mathbf{r}) = \mathbf{E}'(\mathbf{r}') - \mathbf{E}(\mathbf{r}) . \tag{1}$$

The gravity gradient anomaly $\Delta\mathbf{E}(\mathbf{r})$ contains both the effects of gravity field variation and surface deformation. Hence to separate these two effects, the relation between $\Delta\mathbf{E}(\mathbf{r})$ and time-dependent gravity gradient disturbance $\delta\mathbf{E}(\mathbf{r})$, which is independent of the deformation $\delta\mathbf{r}$, have to be found. We have to consider that by definition at P'

$$\delta\mathbf{E}(\mathbf{r}') = \mathbf{E}'(\mathbf{r}') - \mathbf{E}(\mathbf{r}')$$

holds. It follows immediately, that

$$\Delta\mathbf{E}(\mathbf{r}) = \delta\mathbf{E}(\mathbf{r}') + \mathbf{E}(\mathbf{r}') - \mathbf{E}(\mathbf{r})$$

is valid. Finally if we introduce time-dependent gravity gradient disturbance at P instead of P' we have the following equation:

$$\Delta\mathbf{E}(\mathbf{r}) = \delta\mathbf{E}(\mathbf{r}) + (\delta\mathbf{E}(\mathbf{r}') - \delta\mathbf{E}(\mathbf{r})) + \mathbf{E}(\mathbf{r}') - \mathbf{E}(\mathbf{r}) \tag{2}$$

It can be seen clearly, that in Eq. (2) two functions have to be differenced between points P and P': $\delta\mathbf{E}(\mathbf{r})$ and $\mathbf{E}(\mathbf{r})$. These two tensor differences can be expressed mathematically by expanding the *tensor* functions $\delta\mathbf{E}(\mathbf{r})$ and $\mathbf{E}(\mathbf{r})$ into Taylor series at point P. If all but the first terms are neglected, a *linear approximation* is achieved. This means that $\delta\mathbf{E}(\mathbf{r}') - \delta\mathbf{E}(\mathbf{r})$ is approximated by the product of the deformation vector $\delta\mathbf{r}$ and the derivative of the tensor $\delta\mathbf{E}(\mathbf{r})$ computed at P, which will be denoted by grad $\delta\mathbf{E}(\mathbf{r})$. Mathematically this is a third order tensor composed of 27 elements, which expresses the rate of change of the tensor at P. That is, if we move by $d\mathbf{r}$ to a point P' very close to P, a second-order tensor $\mathbf{T}$ will change as grad$\mathbf{T}$ d$\mathbf{r}$. Similarly, $\mathbf{E}(\mathbf{r}') - \mathbf{E}(\mathbf{r})$ is the product of the deformation vector $\delta\mathbf{r}$ and the derivative of the tensor, grad $\mathbf{E}(\mathbf{r})$. Hence the following approximations are introduced

$$\begin{aligned} \delta\mathbf{E}(\mathbf{r}') - \delta\mathbf{E}(\mathbf{r}) &= \mathrm{grad}\delta\mathbf{E}(\mathbf{r})\delta\mathbf{r} \\ \mathbf{E}(\mathbf{r}') - \mathbf{E}(\mathbf{r}) &= \mathrm{grad}\mathbf{E}(\mathbf{r})\delta\mathbf{r} \end{aligned} ,$$

and finally we get the following expression

$$\Delta\mathbf{E}(\mathbf{r}) = \delta\mathbf{E}(\mathbf{r}) + \mathrm{grad}\delta\mathbf{E}(\mathbf{r})\delta\mathbf{r} + \mathrm{grad}\mathbf{E}(\mathbf{r})\delta\mathbf{r} . \tag{3}$$

The second term on the right side of Eq. (3) describes the spatial change of time-dependent gravity gradient disturbance due to surface deformation, whereas the third term is the spatial change of the gravity gradient tensor itself caused by the deformation $\delta\mathbf{r}$. This is the observation equation of repeated determinations of the Eötvös tensor. It contains two unknown parameters, namely functions $\delta\mathbf{r}$ and $\delta\mathbf{E}$, and is valid at points P($\mathbf{r}$) of the undeformed boundary S.

The gravity gradient tensor $\mathbf{E}$ can be additively decomposed into a normal part $\mathbf{E}_U$, and a disturbing part $\mathbf{E}_T$, similarly to the corresponding decomposition of the gravity potential $W$ into a normal potential $U$ and disturbing potential $T$. This is a usual process in gravity field modeling since it is far more convenient to use residual gravity field quantities rather than original ones. This decomposition leads to a slightly modified version of Eq. (3),

$$\begin{aligned} \Delta\mathbf{E}(\mathbf{r}) &= \delta\mathbf{E}(\mathbf{r}) + \mathrm{grad}\delta\mathbf{E}(\mathbf{r})\delta\mathbf{r} \\ &\quad + \mathrm{grad}(\mathbf{E}_U + \mathbf{E}_T)(\mathbf{r})\delta\mathbf{r} \end{aligned} . \tag{4}$$

For comparison and later reference, the time-dependent gravity anomaly $\Delta\mathbf{g}$ can be written in linearized form in the following form:

$$\begin{aligned} \Delta\mathbf{g}(\mathbf{r}) &= \delta\mathbf{g}(\mathbf{r}) + \mathrm{grad}\delta\mathbf{g}(\mathbf{r})\delta\mathbf{r} \\ &\quad + \mathrm{grad}(\boldsymbol{\gamma} + \mathbf{g}_T)(\mathbf{r})\delta\mathbf{r} \end{aligned} . \tag{5}$$

In spherical approximation the normal part of the grad $\mathbf{E}$ $\delta\mathbf{r}$ term is written as

$$\mathrm{grad}\mathbf{E}_U\delta\mathbf{r} = \begin{bmatrix} U_{xx} \\ U_{xy} \\ U_{xz} \\ U_{yy} \\ U_{yz} \\ U_{zz} \end{bmatrix} \cong \frac{3GM}{r^4} \begin{bmatrix} 0 & 0 & 1 \\ 0 & 0 & 0 \\ 1 & 0 & 0 \\ 0 & 0 & 1 \\ 0 & 1 & 0 \\ 0 & 0 & -2 \end{bmatrix} \begin{bmatrix} \delta x \\ \delta y \\ \delta z \end{bmatrix} \tag{6}$$

in the local $(x, y, z)$ = (East, North, Up) coordinate system. Equation (6) can easily be derived from the third derivatives of the Eötvös tensor in terms of spherical coordinates $(r, \varphi, \lambda)$. Since the expressions (7a-j) of *all* of the third derivatives seem nonexistent in the literature, for completeness we include them here. To obtain our result, the $U = GM/r$ potential function has to be inserted into expressions (3a-j). Since this potential function depends solely on the radial variable $r$, only those derivatives are nonzero ($U_{xxz}$, $U_{yyz}$, $U_{zzz}$) which contain a pure radial derivative term. From the symmetry of the third derivative tensor (e.g. $U_{xxz} = U_{xzz} = U_{zxx}$) the nonzero pattern of the matrix on the right hand side of (6) now follows.

$$
\begin{aligned}
V_{xxx} &= \frac{1}{r^3\cos^3\varphi}V_{\lambda\lambda\lambda} - \frac{3\tan\varphi}{r^3\cos\varphi}V_{\varphi\lambda} + \frac{3}{r^2\cos\varphi}V_{r\lambda} - \frac{2}{r^3\cos^3\varphi}V_{\lambda}\\
V_{xxy} &= \frac{1}{r^3\cos^2\varphi}V_{\varphi\lambda\lambda} + \frac{2\tan\varphi}{r^3\cos^2\varphi}V_{\lambda\lambda} - \frac{\tan\varphi}{r^3}V_{\varphi\varphi} + \frac{1}{r^2}V_{r\varphi} - \frac{1}{r^3\cos^2\varphi}V_{\varphi}\\
V_{xxz} &= \frac{1}{r^2\cos^2\varphi}V_{r\lambda\lambda} - \frac{2}{r^3\cos^2\varphi}V_{\lambda\lambda} - \frac{\tan\varphi}{r^3}V_{r\varphi} + \frac{1}{r}V_{rr} + \frac{2\tan\varphi}{r^3}V_{\varphi} - \frac{1}{r^2}V_{r}\\
V_{xyy} &= \frac{1}{r^3\cos\varphi}V_{\varphi\varphi\lambda} + \frac{2\tan\varphi}{r^3\cos\varphi}V_{\varphi\lambda} + \frac{1}{r^2\cos\varphi}V_{r\lambda} + \frac{2\tan^2\varphi}{r^3\cos\varphi}V_{\lambda}\\
V_{xyz} &= \frac{1}{r^2\cos\varphi}V_{r\varphi\lambda} - \frac{2}{r^3\cos\varphi}V_{\varphi\lambda} + \frac{\tan\varphi}{r^2\cos\varphi}V_{r\lambda} - \frac{2\tan\varphi}{r^3\cos\varphi}V_{\lambda}\\
V_{yyy} &= \frac{1}{r^3}V_{\varphi\varphi\varphi} + \frac{3}{r^2}V_{r\varphi} - \frac{2}{r^3}V_{\varphi}\\
V_{yyz} &= \frac{1}{r^2}V_{r\varphi\varphi} - \frac{2}{r^3}V_{\varphi\varphi} + \frac{1}{r}V_{rr} - \frac{1}{r^2}V_{r}\\
V_{yzz} &= \frac{1}{r}V_{rr\varphi} - \frac{2}{r^2}V_{r\varphi} + \frac{2}{r^3}V_{\varphi}\\
V_{zzx} &= \frac{1}{r\cos\varphi}V_{rr\lambda} - \frac{2}{r^2\cos\varphi}V_{r\lambda} + \frac{2}{r^3\cos\varphi}V_{\lambda}\\
V_{zzz} &= V_{rrr}
\end{aligned}
\qquad (7a\text{-}j)
$$

On the surface of the Earth the $3GM/r^4$ term is approximately equal to 0.73 mE/m ($1\text{E} = 10^{-9}\ \text{s}^{-2}$), hence in contrast to the normal part of gravity variation with respect to height ($\delta z$), which is $2GM/r^3 = 308$ μGal/m, the normal part of the time variation of the gravity gradient tensor in practice can be neglected for surface deformation of reasonable size, as it will be seen in the following from a simple numerical example.

## 3 Estimation of linearization and approximation errors

One can introduce several approximations into Eq. (2):

- linearization at P, yielding Eq. (3)
- neglection of second and third terms of the linearized equation Eq. (3)

To make an estimate of these errors, a simple 2D model of gravity field variation and surface deformation will be considered.

Let us define a density anomaly of a homogeneous sphere of mass $m$ buried at depth $z = 0$ producing anomalous gravity field on the surface $z = r$ (Fig. 2). The time variation of mass $\delta m$ is placed at $z = r'$, i.e. at depth $d' = r - r'$. The surface deformation is an upward movement defined by the function $\delta\mathbf{r} = \delta z(x)$. For this simple geometry it is easy to calculate the time-dependent gravity and vertical gravity gradient anomaly $\Delta g$ and $\Delta V_{zz}$, respectively (Table 1). The normal gravity is approximated in a spherical geometry.

**Table 1.** Time-dependent gravity and vertical gravity gradient anomaly for the simple model in Fig. 1. $M$ and $R$ denote the Earth's mass and mean radius, respectively.

| Time-dependent anomaly | $\delta V_{zz}$<br>$\delta g$ | $\text{grad}V_{zz}\,\delta z$<br>$\text{grad}g\,\delta z$ | $\text{grad}\,\delta V_{zz}\,\delta z$<br>$\text{grad}\,\delta g\,\delta z$ |
|---|---|---|---|
| $\Delta V_{zz}$ | $2G\dfrac{\delta m}{d'^3}$ | $-6G\left(\dfrac{m}{r^4}+\dfrac{M}{R^4}\right)\delta z$ | $-6G\dfrac{\delta m}{d'^4}\delta z$ |
| $\Delta g$ | $-G\dfrac{\delta m}{d'^2}$ | $2G\left(\dfrac{m}{r^3}+\dfrac{M}{R^3}\right)\delta z$ | $2G\dfrac{\delta m}{d'^3}$ |

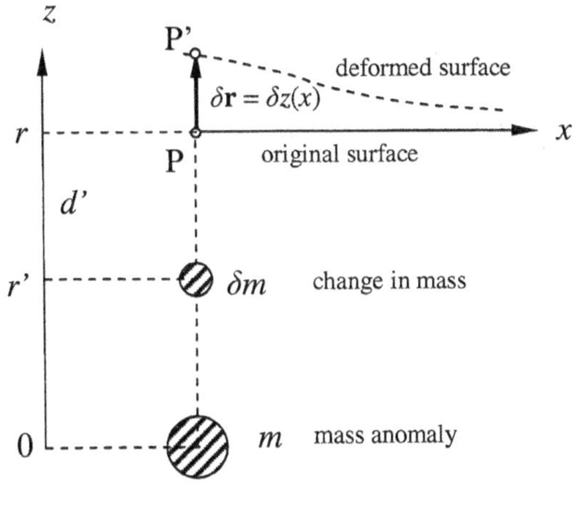


**Fig. 2** Simple model of time variation of gravity and height

In order to have numerical values of the corresponding terms of Table 1, we placed a mass anomaly $m = 10^{13}$ kg at depth $r = 5000$ m. The change in mass $\delta m$ occurs at depths $d = 20$ and 100 m and has the values $\delta m = 8\cdot10^5$ kg and $10^8$ kg, respectively. For surface deformation $\delta\mathbf{r} = \delta z(x)$ the simple model $\delta z(x) = z_0\exp(-(x/x_0)^2)$ was considered with the parameters $z_0 = 1$ m and $x_0 = 200$ m.

It is interesting to see the results for the time-dependent vertical gravity gradient and gravity anomalies in Figs. 3 and 4, respectively. First the big difference in the normal gravity field terms for gravity anomalies and vertical gravity gradients should be noticed. Whereas for gravity gradient anomalies the normal gravity field term is about 4 order of magnitudes *smaller* than the time-dependent vertical gravity gradient disturbance $\delta\mathbf{E} = \delta V_{zz}$ and thus it can be neglected, it is the dominant term for time-dependent gravity anomaly. Exactly, it is almost by an order of magnitude *larger* than the time-dependent gravity disturbance itself! This means that the indirect gravity change induced by the surface deformation (vertical movement of the point) should always be determined in repeated gravity measurements by precise GPS positioning, for example, whereas no information on height variation is necessary for repeated vertical gravity gradient measurements.

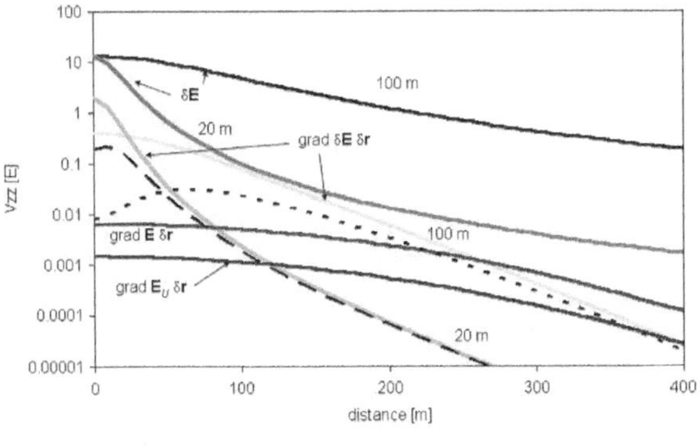


**Fig. 3** Magnitudes of vertical gravity gradient anomaly terms in Eq. (4) for the simple model of Fig. 2. Dashed lines show linearization errors of the grad$\delta$**E**$\delta$**r** term. Linearization errors of the other two terms (grad$\mathbf{E}_T\,\delta\mathbf{r}$ and grad$\mathbf{E}_U\,\delta\mathbf{r}$) are below $10^{-5}$ E. The unit of the logarithmical vertical axis is Eötvös.

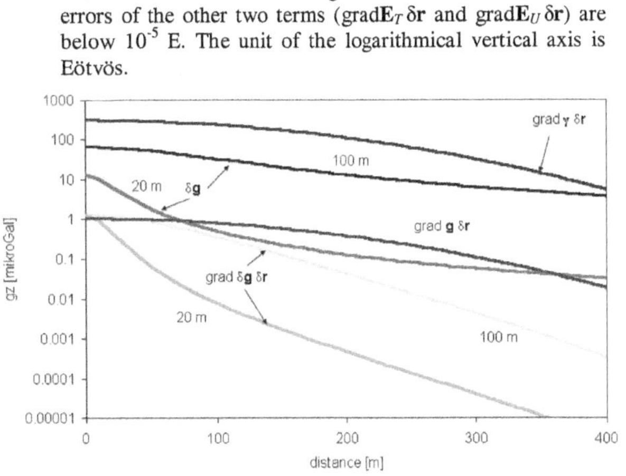


**Fig. 4** Magnitudes of gravity anomaly terms in Eq. (5) for the simple model of Fig. 2. The unit of the logarithmical vertical axis is µGal.

Another difference one should notice in these figures is the relative magnitude of time-dependent disturbances for shallow (20 m) and more deep (100 m) change in masss. These two timely mass variations give exactly the same maximum magnitude of the signal at $x = 0$ in case of vertical gravity gradients, but the shallow mass gives about an order of magnitude smaller signal relative to the more deep mass in gravity disturbances. This means that as expected, gravity gradients are more sensitive to changes of shallow masses, giving larger signal.

Finally, if one considers the coupling term of vertical gravity gradient and deformation (grad δ**E** δ**r** in Eq. 4), it is the largest one, although yet at least about an order of magnitude smaller than the time-dependent gravity gradient disturbance itself, for a relatively large vertical surface deformation of maximum 1 m. It is the largest for shallow mass variations and decreases (relative to the time variation of gravity gradient itself) with increasing source depth. For practical cases, however, if the surface deformation is smaller, and the source depth is larger, this term can be neglected as well.

Linearization errors of time-dependent vertical gravity gradient anomalies are also shown in Fig. 3. These errors are largest for the gradδ**E** δ**r** term, but the maximum value is 0.2 E. Linearization errors of the other two terms (grad$\mathbf{E}_T\,\delta\mathbf{r}$ and grad$\mathbf{E}_U\,\delta\mathbf{r}$) have a maxima of $2.6\cdot10^{-6}$ E and $4.6\cdot10^{-10}$ E, respectively.

To summarize, repeated *gravity gradient* measurements have the relative advantage over repeated measurements of *gravity* that no precise repeated height determination is needed, and in general the simplification

$$\Delta\mathbf{E}(\mathbf{r}) \cong \delta\mathbf{E}(\mathbf{r}) \qquad (8)$$

to the measurement equation Eq. (4) is perfectly acceptable.

If it is suspected that unwanted shallow mass variations "contaminate" the measured gravity gradient signal, one can try to place a second measurement site close to the first one and compare the two observations to eliminate any unnecessary short wavelength signal.

## 3 A geodynamic boundary value problem of repeated torsion balance measurements

In this final part of the paper it will be shown how to derive time-dependent vertical gravity gradient anomalies from repeated observations of the Eötvös torsion balance by solving the corresponding geodetic-geodynamic boundary value problem.

Let us assume that repeated torsion balance $W_{xz}$, $W_{yz}$ observations are available on the surfaces S and S', respectively (Fig. 1). Since the disturbing potential $T = W - U$ is a harmonic function, time-dependent potential disturbance $\delta T$ is also a harmonic function. Following what was said in the preceeding Section, repeated torsion balance observations can be regarded as measurements of time-dependent gravity gradient disturbances $\delta W_{xz}$, $\delta W_{yz}$., since terms containing the surface deformation $\delta\mathbf{r}$ can be neglected. Therefore the only unknown function of this gradiometric-geodynamic boundary value problem (GGBVP) is the time-dependent vertical gravity gradient disturbance $\delta W_{zz}$, if the horizontal positions of the measurements are known with sufficient accuracy.

The solution of this GGBVP is similar to the solution of the corresponding gradiometric boundary value problem, discussed in Tóth (2003). First a vector-valued time-dependent gravity gradient disturbance combination

$$\boldsymbol{\delta}\mathbf{W}^{(1)} = \delta W_{xz}\mathbf{e}_x + \delta W_{yz}\mathbf{e}_y \qquad (9)$$

has to be formed and expanded into vector spherical harmonics. Next the connection between this measurement combination and the unknown function has to be established in the spherical harmonic spectral domain and finally this spectral relation is transformed back to the spatial domain. The result is the following integral of the measurements over the unit sphere σ:

$$\delta W_{zz} = \iint_{\sigma} K_{zz}^{(1)}(r,\psi)(\delta W_{xz}\cos\alpha^* + \delta W_{yz}\sin\alpha^*)d\sigma \qquad (10)$$

Here $\alpha^*$ denotes the azimuth of the source point measured at the evaluation point, $\psi$ is the spherical distance of source and evaluation points and $K_{zz}^{(1)}(r,\psi)$ is the kernel function of the solution. The spectral form of this kernel is

$$K_{zz}^{(1)}(r,\psi) = \sum_{\ell=2}^{\infty}\left(\frac{R}{r}\right)^{\ell+3}\frac{2\ell+1}{\ell}P_\ell^1(\cos\psi) \qquad (11)$$

where $P_\ell^1(\cos\psi)$ denotes the Legendre function of degree $\ell$ and order 1 and $R$ is the mean Earth radius. The infinite sum in Eq. (11) can be computed analytically, and the closed form of the kernel function is

$$K_{zz}^{(1)}(s,x) = s^4\left(\frac{2}{L^3} - 3\right)\sqrt{1-x^2} + \frac{s^3}{\sqrt{1-x^2}}\left(\frac{s-x}{L} + x\right) \qquad (12)$$

where $s = R/r$, $x = \cos\psi$ and $L = \sqrt{1 - 2sx + s^2}$. This kernel function is shown in Fig. 5 when the computation height $h = r - R$ is 1 m.

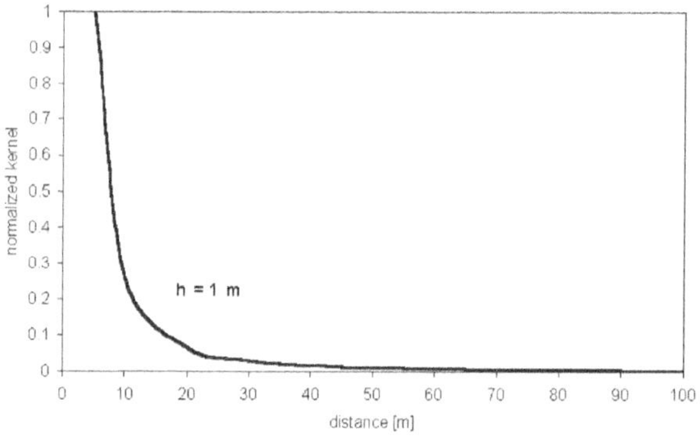


boundary value problem of repeated torsion balance measurements at computation height $h$ = 1 m as a function of distance $s$.

The kernel function in Fig. 5 indicates that the computation of time-dependent vertical gravity gradient anomalies from repeated torsion balance measurements seems feasible, since the observations may be restricted to a limited computation area due to the rapid decrease of weights assigned by the kernel function to the measurements.

## 4 Conclusions and outlook

We have formulated and examined the measurement equations of repeated determinations of the Eötvös tensor. The magnitudes of the different terms in this equation have been estimated from a simple model of gravity field variation and surface deformation. By examining this simple model it was found that in practice the problem can often be set up and solved without repeated height determinations, which is an advantage of repeated gravity gradient measurements over repeated gravity observations. This is a feature that comes from weak coupling of gravity gradients and differential changes of observation positions. Of course these conclusions should be confirmed by further model computations.

Investigations in this paper can be regarded as complementary to similar investigations for other repeated gravity field measurements found in the literature (Heck, 1986).

Finally, it was shown that it is feasible to compute time variations of other gravity field parameters in the frame of a geodetic-geodynamic boundary value problem approach. Here the example of computing time-dependent vertical gravity gradient anomalies was examined in more detail. We conclude and argue therefore, that repeated gravity gradient determinations may contribute usefully to the solution of local geodynamic problems.

## Acknowledgements

The above investigations were funded by the Bolyai Scholarship of Hungarian Academy of Sciences and OTKA projects T037929, T046418. The helpful comments of D. Tsoulis are also gratefully acknowledged.

## References

Biró P (1983). Time Variation of Height and Gravity. Herbert Wichmann Verlag, Karlsruhe.

Heck, B (1986). Time-Dependent Geodetic Boundary Value Problems. Proc. Int. Symp. Figure and Dynamics of the Earth, Moon, and Planets. Prague, Czechoslovakia, September 15-20, 1986. Edited by P. Holota, pp 195-225.

Tóth Gy (2003) The Eötvös spherical horizontal boundary value problem – gravity anomalies from gravity gradients of the torsion balance. In: Gravity and Geoid 2002. 3rd Meeting of the IGGC. Tziavos (ed.), Gravity and Geoid 2002, Ziti Editions, pp 102-107.

# Relation between the geological conditions and vertical surface movements in the Pannonian basin

L. Völgyesi
Department of Geodesy and Surveying, Budapest University of Technology and Economics;
Research Group of Physical Geodesy and Geodynamics of the Hungarian Academy of Sciences,
H-1521 Budapest, Hungary, Müegyetem rkp. 3.

G. Csapó, Z. Szabó
Eötvös Loránd Geophysical Institute of Hungary, H-1145 Budapest, Hungary, Kolumbusz u. 17/23.

**Abstract.** Vertical surface movements can be attributable to two effects in young sedimentary basins (like the Pannonian Basin): compaction of sediments, on the one hand, and structural movements, on the other. The depths of the Quaternary, Lower and Upper Pannonian base were determined from different geological data along the main levelling lines and compared to the vertical surface movements. In spite of uncertainties in geological data, in most of the deep sedimentary regions positive correlation can be found between sediment thickness and subsidence. As a consequence of subsidence, gravity changes should be detected at surface points because the points change their position in the Earths' gravity field. Calculating the effects of the vertical surface movements, the variation of gravity in general should be about 0.3 μGal/year (1 μGal = $10^{-8}$m/s$^2$), but at special places it may reach more than 2 μGal/year in the Pannonian basin.

**Keywords:** vertical surface movements, sediment thickness, compaction of sediments, variation of gravity

## 1 Geological conditions

Apart from occasional foundation problems of levelling base points, in the Pannonian basin – where the surface is covered by young, unconsolidated sediments – vertical surface movements may be attributed to two factors: compaction of sediments and structural movements.

As 85 % of the area of Hungary belongs to the Pannonian basin, compaction of sediments (depending on their composition and thickness) influences significantly the levelling data. Therefore it is expedient to investigate the relations between sediments (their extension and thickness) and surface movement data. If compaction is determined and corrected for, crustal movements of geodynamic origin can be separated. Before dealing with compaction, a short review of basin evolution is given as revealed by geological and geophysical data.

The Pannonian basin is the central part of the Carpathian basin, surrounded by the Alps, the Carpathians and the Dinarides. The Pannonian basin itself is divided into smaller sub-basins of different types by inselbergs, representing the basement on the surface. The largest of them is on its western part the Danube-Rába basin, shared by Hungary and Slovakia, while in the East the Great Hungarian Plain contains the Békés basin and the two deepest depressions, the Makó-Hódmezővásárhely and the Derecske troughs.

Evolutions of the basins are in close connection with the emergence of the Alp-Carpathian system, consequently with the collision of the African and European plates. In the collision zone, microplates of different origin and in different stages of tectonization took up their present location during the Upper Cretaceous–Tertiary phase of the Alpian orogeny. The mantle diapir induced by the subducting plates resulted in thinning and subsidence of the crust. Apart from normal growth faults strike-slip faults played an important role in the evolution of the sub-basins, especially in the late phase of subsidence, leaving their mark even in the Pliocene sediments.

Both the time of initiation and the rate of subsidence of the different sub-basins may differ significantly. Palaeogene sediments are rare, mostly in the northern part of the Pannonian basin (in the so-called North-Hungarian Palaeogene basin). The subsidence of the whole basin started in the Neogene. At the beginning, the shallow and subsequently deeper marine sediments overlay the basement. The filling-up process started with the enormous prograding delta system, originating from the emerging Carpathians and the separation of the Pannonian-sea from the Mediterranean, resulting in its becoming gradually a fresh-water lake. In the last phase of filling-up, the delta system was replaced by a more or less uniform lacustrine sedimentation. This tripartition of sedimentation agrees with the ratio of subsidence and sediment accumulation: in the beginning the rate of sedimentation was less than that of subsidence, in the period of delta formation, especially in its earlier stages, subsidence and sediment accumulation were balanced, keeping the sea at a constant level, while in the last phase sedimentation rate exceeded subsidence, finishing the filling-up process.

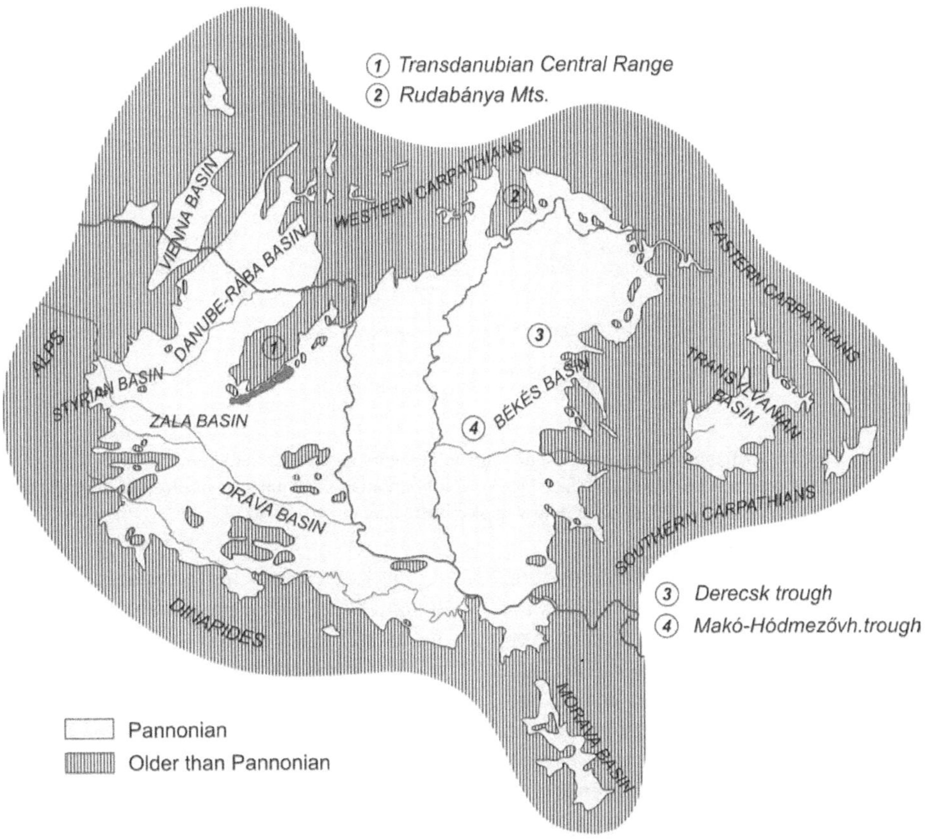


**Fig. 1** Distribution of Pannonian sediments in the Carpathian Basin, according to *Carte Geologique Internationale dé l'Europe* 1:1500 000 (1969).

Lithology of the sedimentary fill depends on the sedimentation environment and on the source rocks, the rocks of the surrounding mountain chain. Deep-water sediments are fine-grained, undisturbed, locally of high carbonate content. Dipping layers of delta sediments represent transition from near-shore coarse sand toward fine-grained shale when receding from the actual coastline. This topic is dealt with the extensive literature on seismic stratigraphy; just mentioning a few (Pogácsás 1980, 1984, Mattick et al. 1988).

Comparing the above data of geophysical methods, based on physical parameters, and those of geological methods, studying lithological and palaeontological conditions, insoluble contradictions arise. A short review of the geological approach is given according to (Jámbor 1985):

The extension of the Pannonian sea changed in time, sometimes flooding even the Vienna- and Styrian basins in the West and the Transylvanian basin in the East (see Figure 1). Formations of the Pannonian sea, representing the 12–2.4 Ma interval – classified as Pannonian (s.l.), subdivided into Lower Pannonian, or Pannonian (s.str.) and Upper Pannonian, or Pontian, Dacian, Romanian, representing the Pliocene (5.4–2.4 Ma) – cover the whole basin area, except the mountains (see Figure 2). The Pannonian–Sarmatian contact can be traced easily along the mountain rims by lithology and palaeontology, but in the deep basins the identification of this contact is ambiguous, due to similar facies (fine-grained shale, marl). Accordingly, depths to the Pannonian base data are burdened with significant errors.

**Fig. 2** Distribution of Pannonian sediments in Hungary according to (Jámbor 1979)
1 – Older than Pannonian formations on the surface (15%), 2 – Pannonian sediments on or close to the surface (28%), 3 – Pannonian below thick Pleistocene cover (57%).

There are similar problems in the determination of the Pannonian–Pleistocene boundary. Near the coastlines Pleistocene deposits overlay Upper Pannonian by sharp discordance and they differ from Pannonian sediments both in grain-size and colour: Pleistocene sediments are represented by gravel and wind-blown sand. In the deep basins, however, sedimentation was continuous with similar facies: shale, aleurite, fine-grained sand. Determining the Upper Pannonian–Pleistocene contact, geologists have to use well-logging data, too. The generally accepted 2.4 Ma age was determined by palaeomagnetic investigations (Rónai 1985).

Thickness of overburden plays an important role in the character of the Pannonian sediments. In deep basins, Pannonian shales, marls are hard rocks; they keep their consistency if soaked, while the coarse-grained layers are sandstones and conglomerates. In near-surface conditions layers of similar composition and age appear as plastic clay and loose sand, gravel, respectively.

Depth to the Pannonian base varies significantly. During the 12 million years a difference of 5200 m developed between the Rudabánya Mts. and the Makó–Hódmezővásárhely trough, an equivalent of 0.4 mm/year average subsidence. At the end of the Pannonian, the Transdanubian Central Range tilted towards the south-east, accordingly its north-western side had risen about 150–200 m, while its foreland, the Danube–Rába basin subsided significantly.

The Lower and Upper Pannonian base, as well as thickness of Quaternary maps were constructed by the Geological Institute of Hungary (MÁFI), originally in the scale of 1:200,000 – in areas of detailed investigation in the scale of 1:25,000 – and published in the scale of 1:500,000 (Csiky et al. 1986, Csiky 1987, Síkhegyi 1995). About the reliability of these maps (Jámbor 1985) has written the following: "All available borehole data were used, although in the published map only a fraction of these holes are plotted, because of technical reasons. Constructing the isolines, for interpolating between boreholes the filtered gravity map of the National Petroleum Company (OKGT) and, at some places, their seismic data were utilized."

The majority of deep boreholes are oil- and gas-exploration holes, with rare coring in the Pannonian sequence. For marking out geological horizons, well logging provided information. Naturally, in well logging curves, based on different physical parameters, those horizons can be identified unambiguously which represent sharp contrast between density, resistivity or velocity of under- and overlying rocks; the main such horizon being the basement. If Lower Pannonian sediments were deposited directly on the basement, that horizon is reliable, however, to decide whether older than Pannonian sediments occur or not, is very difficult, even if the borehole has penetrated deep enough. The Lower and Upper Pannonian, as well as the Upper

Pannonian and Quaternary contacts carry even more ambiguity.

The distribution of oil-exploration deep boreholes is extremely disproportionate, comparing the in-field and between fields borehole spacing. Interpolation by gravity data is only justified in the case of the Pannonian base horizon, and within areas of similar geological build-up. Seismic reflections can be correlated with geological horizons but their reliability depends on their age, i.e. on the applied technique.

## 2 Geological conditions and vertical surface movements

Concluding the above-written aspects, it is obvious that geological data are burdened by heavy ambiguities. On the other hand, the average least square error of vertical surface movement is ±0.97 mm/year (Joó 1996). In our investigation, to decrease ambiguities as much as possible, the following method was used: Starting out from the surface movement map (Joó 1995), data were sampled at a 5 km rate along the levelling lines. This method was chosen to avoid the interpolation errors introduced by using geological data in map construction. Thus our data set can be regarded reliable and independent from any geological considerations. Location map of levelling lines is given in Figure 3.

Next, the depths data of the following geological horizons were determined in the sampling points: the Pannonian base, Lower and Upper Pannonian contact and base Quaternary. This latter was deduced from the Quaternary thickness map, as the few metres thickness of the Holocene can be neglected, compared to other possible errors. These data were plotted for 24 levelling lines (in lines 23 and 25 no useful sediment data were found), and a few interesting examples are presented in Figure 4. The numbering of these graphs corresponds to that of the levelling lines in Figure 3, where even the directions of the profiles are also given. On the horizontal axis the distance from the starting point is given in km, on the vertical axes, on the left side the depths data of the geological horizons in m, while on the right side the vertical surface movement data are given in mm/year.

## 3 Time variation of gravity

Accuracy of gravimeters has reached such a high level nowadays that formerly neglected side effects in the μGal range have to be taken into consideration, even if geodynamic processes are the subjects of our investigation. Vertical surface movements are generally treated as vertical crustal movements although movements of different origin cannot be separated by geodetic measurements.

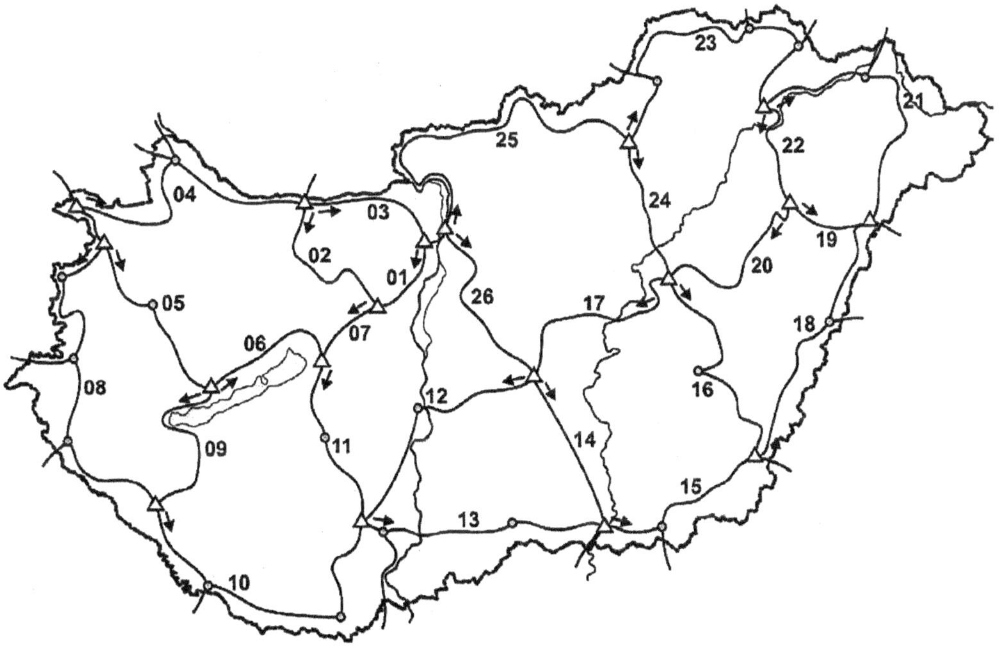


**Fig. 3** Location map of levelling profiles with direction of sampling

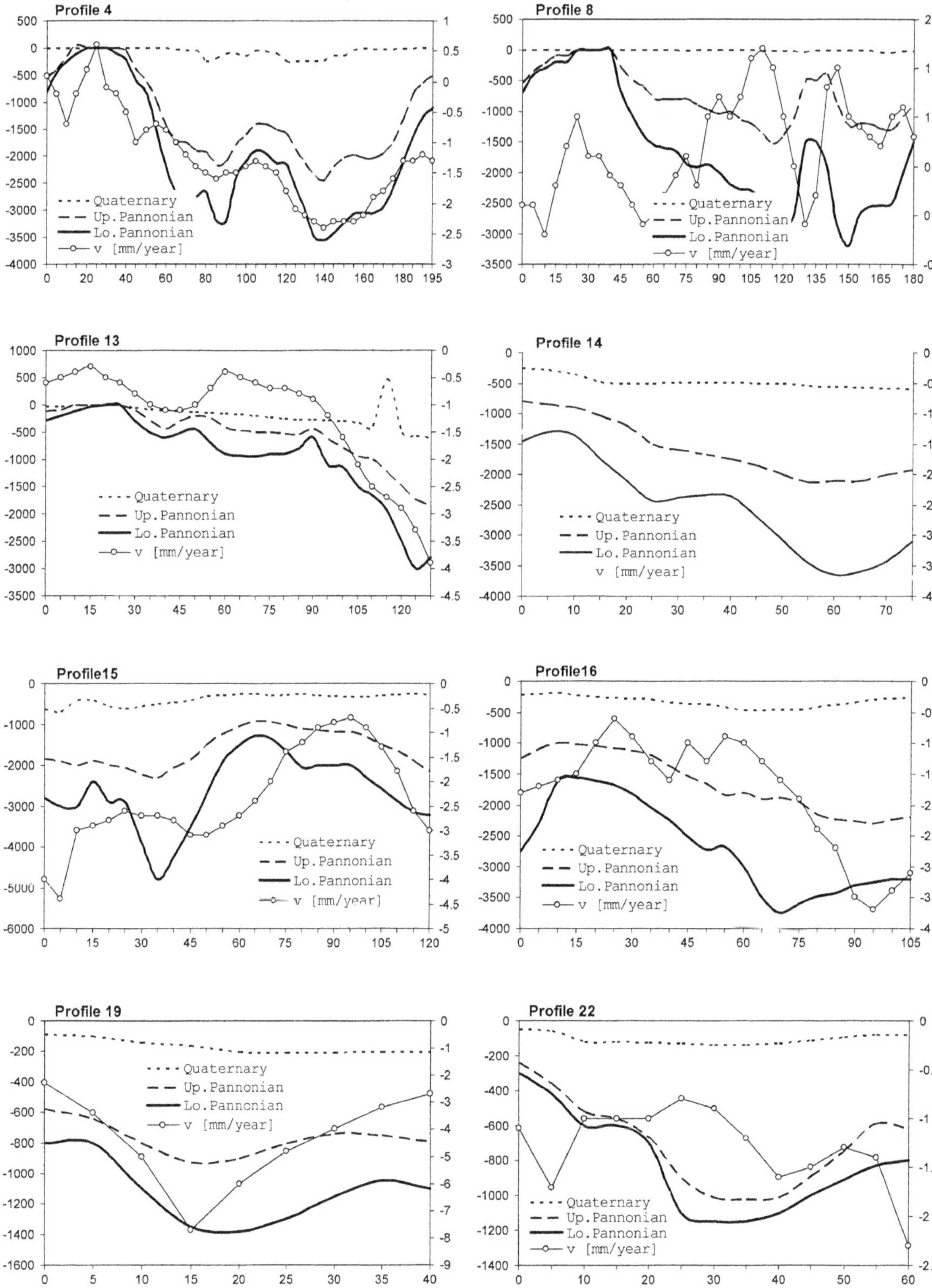


**Fig. 4** Sediment thickness and velocity of vertical surface movement along levelling lines 4, 8, 13, 14, 15, 16, 19, 22. Depths to 1 –Qaternary base, 2 –Upper Pannonian base, 3 –Lower Pannonian base. 4 – velocity of vertical surface movement

In the course of vertical surface movement, surface points change their position in the Earths' gravity field. With the present accuracy of gravimeters, this change ($\delta g$) – depending on the vertical gradient (*VG*) – can be determined as follows:

$$\delta g = \partial g/\partial h \cdot \Delta h$$

The well-known normal value of the vertical gradient is: $\partial g/\partial h$ = 0.3086 mGal/m, but the actual value of *VG* may differ significantly (Csapó–Völgyesi 2002). Accordingly, a given $\Delta h$ surface movement results in different $\delta g$ gravity change.

In Hungary, the average value of vertical surface movements is 1 mm/year, but locally this value may reach as much as 8 mm/year (see profile 19 in Fig. 4.). So the average gravity change is about 0.3 μGal/year, while the maximum is more than 2 μGal/year computed from the vertical surface movements.

## 4 Conclusions

It can be seen that in spite of all uncertainty, positive correlation exists between sediment thickness and surface movement. Preliminary correlation computations show significant differences in the degree of correlation between sub-basins. The highest deviation was found in the Zala basin, with negative correlation (see profile 8 in Fig. 4.). Accordingly, no uniform correlation can be expected for the whole area of the Pannonian basin between sediment thickness and surface movement, or between surface movement and gravity variation, respectively.

With regard to the type and tectonic conditions of each sub-basin, those sections of the levelling lines should be collected for further correlation computations where these parameters are similar. We are continuing our investigation in this direction.

## Acknowledgement

We should thank for the funding of the above investigations to the National Scientific Research Fund (OTKA T-037929).

## References

Csapó G., Völgyesi L. 2002: Determination and reliability of vertical gradients (*VG*) based on test measurements in Hungary (in Hungarian with English abstract). Magyar Geofizika 43, 4. pp. 151–160

Csiky G. et al. 1986: Pannonian formations of Hungary. Map of base Transdanubian Super-group (Upper Pannonian). M=1:500,000. Geol. Inst. of Hungary

Csiky G. ed. 1987: Map of base Peremarton Super-group (Lower Pannonian). M=1:500,000. Geol. Inst. of Hungary

Jámbor Á. 1985: Explanatory notes to the geological maps of Pannonian Formations of Hungary, M=1:500,000. Geol. Inst. of Hungary

Joó I. ed. 1995: The National Map of vertical movements of Hungary (M=1:1,000,000)

Joó I. 1996: Vertical movements of the Earth's surface in Hungary. Geodézia és Kartográfia XLVIII. 4. pp. 6–12

Mattick R. E., Rumpler J., Phillips R. L. 1985: Seismic stratigraphy and depositional framework of sedimentary rocks in the Pannonian Basin in southeastern Hungary. AAPG Memoir 45. pp. 117-146

Pogácsás Gy. 1980: Evolution of Hungary's Neogene depressions in the light of geophysical surface measurements (in Hungarian with English abstract). Földtani Közlöny 110. 3–4. pp.485–497

Pogácsás Gy. 1984: Seismic stratigraphic features of Neogene sediments in the Pannonian Basin. Geophysical Transactions 30. 4. pp. 373–410

Rónai A. 1985: The Quaternary of the Great Hungarian Plain. Geol. Hung. Ser. Geol. Vol. 21. 446 p.

Síkhegyi F. 1995 after Franyó 1992: The thickness of the Quaternary sediments in Hungary, M=1:500,000. Geol. Inst. of Hungary

# Modelling gravity gradient variation due to water mass fluctuations

L. Völgyesi, G. Tóth
Department of Geodesy and Surveying, Budapest University of Technology and Economics;
Research Group of Physical Geodesy and Geodynamics of the Hungarian Academy of Sciences,
H-1521 Budapest, Hungary, Müegyetem rkp. 3.

**Abstract.** Two case studies were considered where gravity field changes were detected resulting from water mass variations. The first case is an urban water reservoir where the maximum daily change of water is 40000 $m^3$. The 3D model of the water mass allowed us to build an accurate polyhedral model of the variation of mass changes of the water. This mass density variation model made it possible to compute and compare variations of various gravitational field functionals. Gravity and full gravity gradient tensor changes were computed on a regular grid for the model area. Gravity changes were also compared with actual gravity field measurements made with two LaCoste & Romberg (LCR) gravimeters. The measured gravity change was nearly 30 μGal relative to 5000 $m^3$ water mass. A good agreement was found between the computed and measured changes. The second case is the water level fluctuations of the Danube River in Budapest during the great flood in 2002. In this case also modelling and measurements were compared. We found the gravitational change to be very sensitive to the actual distance of the point from the river bank.

**Keywords.** Water mass variations, changing of gravity anomalies, changing of full gravity gradient tensor

## 1 The gravitational effect of water level fluctuations of a water reservoir

The gravity effect of the daily water level fluctuation of the Gellért Hill reservoir of Budapest, having 80000 $m^3$ capacities, was studied by relative gravity measurements, and model calculations were carried out to verify the measurements. Modelling sketch of this water reservoir can be seen in Figure 1. The underground water reservoir consists of two basins, the concrete ceiling is covered by 3–4 m of gravel and soil. One metre of water level change in either of the two basins corresponds to 5000 $m^3$ of water mass.

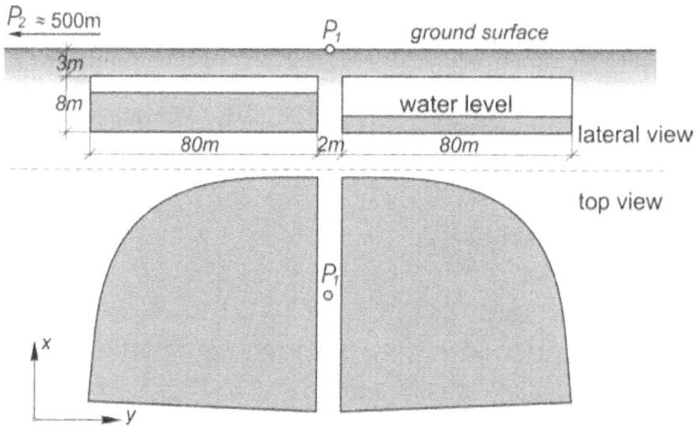


**Fig. 1** Modelling sketch of the Gellért Hill water reservoir

For the gravity measurements one of the gravity stations was located above the reservoir at point $P_1$, the other station $P_2$ was at a distance of 500 m, where the water level change of the reservoir had already no gravitational effect. The observations were repeated several times with two gravimeters in the morning and in the afternoon. The volumes of water at the time of measurements were computed using the dimensions of the reservoir and the water level registration of the Metropolitan Waterworks.

**Table 1.** Gravimeter measurements above the Gellért Hill water reservoir, measured by G.Csapó (Csapó et al, 2003), (x = δΔg /ΔH)

| day/ | | Gravimeter LCR 1919G | | | | | Gravimeter LCR 963G | | | | |
|---|---|---|---|---|---|---|---|---|---|---|---|
| part | | Δg | δΔg | H | ΔH | x | Δg | δΔg | H | ΔH | x |
| 1. | a.m. | 477 | 126 | 7.35 | 4.82 | 26.1 | 475 | 116 | 7.11 | 4.59 | 25.5 |
| | p.m. | 351 | | 2.53 | | | 359 | | 2.52 | | |
| 2. | a.m. | 468 | 92 | 7.29 | 3.71 | 24.8 | 475 | 122 | 6.94 | 3.55 | 34.4 |
| | p.m. | 376 | | 3.58 | | | 353 | | 3.39 | | |
| 3. | a.m. | 484 | 102 | 7.42 | 3.60 | 28.3 | 477 | 93 | 7.22 | 3.91 | 23.8 |
| | p.m. | 382 | | 3.82 | | | 384 | | 3.31 | | |
| mean | | | | | | 26.4 | | | | | 27.9 |
| average | | | | | | 27.2 μGal/m | | | | | |

The observed $\Delta g$ values are presented in Table 1, where $\Delta g$ means the observed relative gravity in μGal (1 μGal = $10^{-8}$m/s$^2$) for the given gravimeter and $H$ means the height (in m) of water in the reservoir. The change per metre was calculated from the actual $\delta\Delta g$ and $\Delta H$ data, where $\delta\Delta g$ means the gravity variation due to the daily water level change. The result of these measurements provided a value of 27.2 μGal/m, which means that e.g. 5 m water level change (corresponds to 25000 m$^3$ of water mass) causes 0.136 mGal change of gravity at point $P_1$.

These measurements were controlled by theoretical computations by the software Mod3D written by I. Cerovský (Cerovský et al, 2004). Mod3D is an interactive 3D geophysical gravity and magnetic modelling software; it has been developed to create 3D geophysical models in a user friendly interactive environment. Mod3D computes anomalous gravitational field components $\mathbf{g}(g_x, g_y, g_z)$ and the full gravitational tensor elements $W_{xx}$, $W_{yy}$, $W_{xy}$, $W_{zx}$, $W_{zy}$, $W_{zz}$. Anomalous gravitational fields are computed using formulae for polyhedral bodies. The application of polyhedron volume element can provide a more realistic geometrical description of boundary surface of a geological body (e.g. topographic surface, without height jumps) than the description made by rectangular prism (parallelepiped) models. Using the polyhedron density model the generated second derivatives of the gravitational potential are more smooth and realistic function than the ones provided by prism model (Benedek, 2002). If the calculation level is near to the surface of the gravitational source, the accuracy of related quantities of gravitation can be increased by a detailed description of this surface in the vicinity of calculation.

Model computations were carried out for the right side basin in Figure 1 in case of thickness 5m (25000 m$^3$) water mass. The result of this computation provided a value of 0.15 mGal at point $P_1$. The difference between measured (0.136 mGal) and computed (0.15 mGal) values may be due to the uncertainty in the relative position of the observation point and the reservoir.

Computations were performed not only for the point $P_1$, but for a few hundred meters surroundings of the water reservoir. In Figure 2 changes of gravity in the surroundings of water reservoir can be seen. Maximum gravity changes are exactly above the edge of water mass, and the change decreases strongly in the function of distance from the water reservoir.

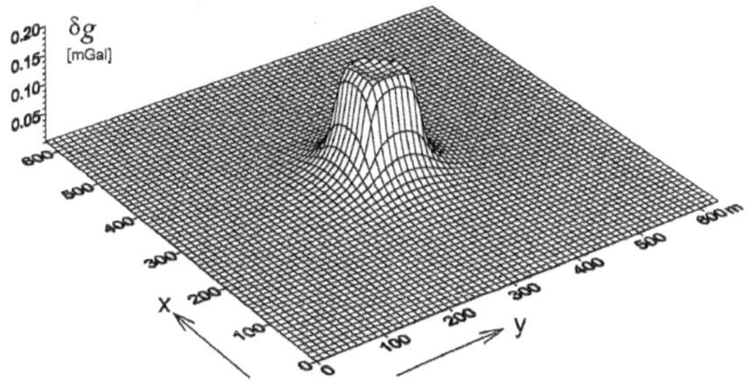


**Fig. 2** Changes of gravity in the surroundings of water reservoir for the case of 50000 m$^3$ water mass.

At the same time the elements of full gravitational tensor were computed too for this water mass. The variations of Eötvös tensor elements $W_{xx}$, $W_{yy}$, $W_{\Delta}$, $W_{xy}$, $W_{zx}$, $W_{zy}$, $W_{zz}$ can be seen in Figures 3, 4, 5, 6, 7, 8 and 9 respectively in Eötvös unit (1 Eötvös Unit = 1E = $10^{-9}s^{-2}$).

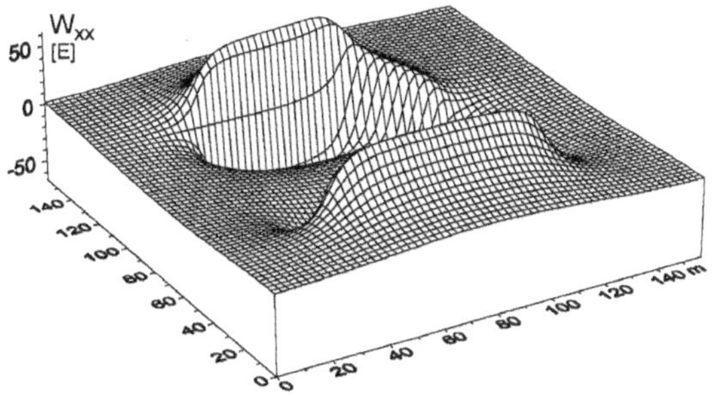


**Fig. 3** Changes of gravity gradient $W_{xx}$

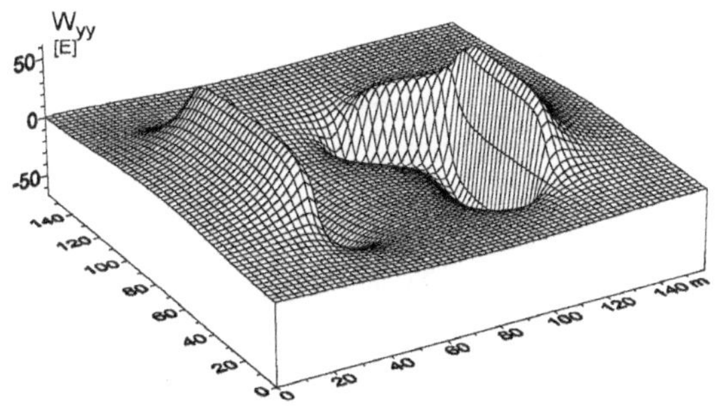


**Fig. 4** Changes of gravity gradient $W_{yy}$

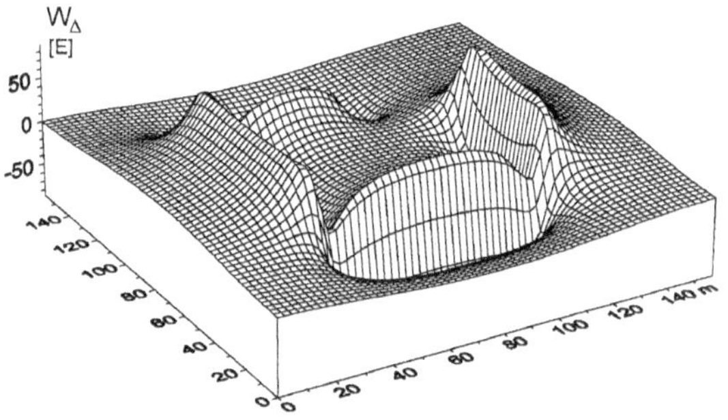


**Fig. 5** Changes of curvature data $W_\Delta = W_{yy} - W_{xx}$

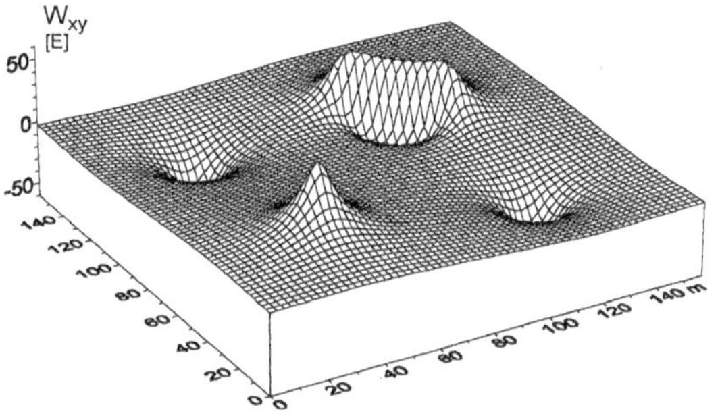


**Fig. 6** Changes of curvature data $W_{xy}$

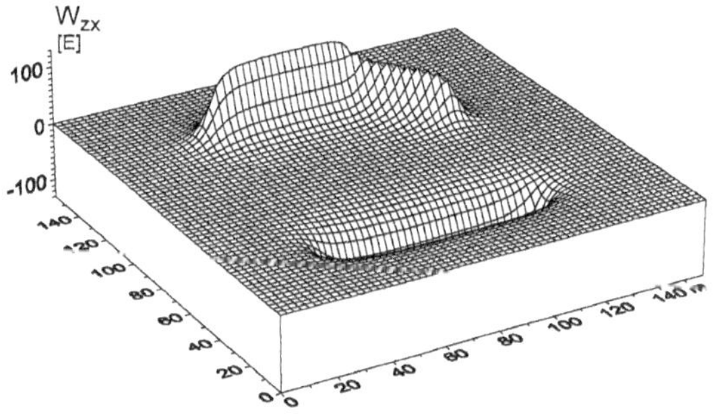


**Fig. 7** Changes of horizontal gravity gradient $W_{zx}$

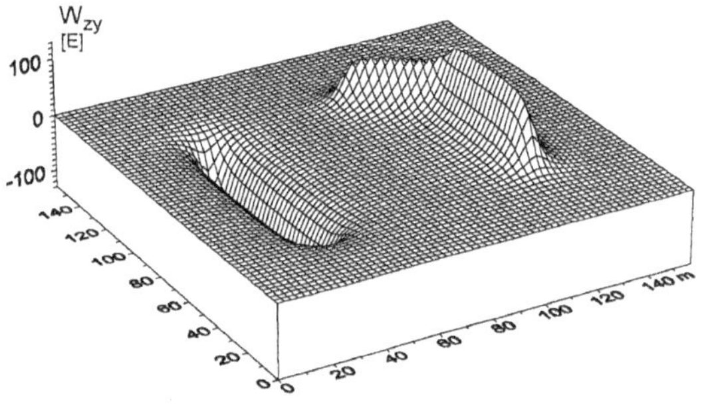


**Fig. 8** Changes of horizontal gravity gradient $W_{zy}$

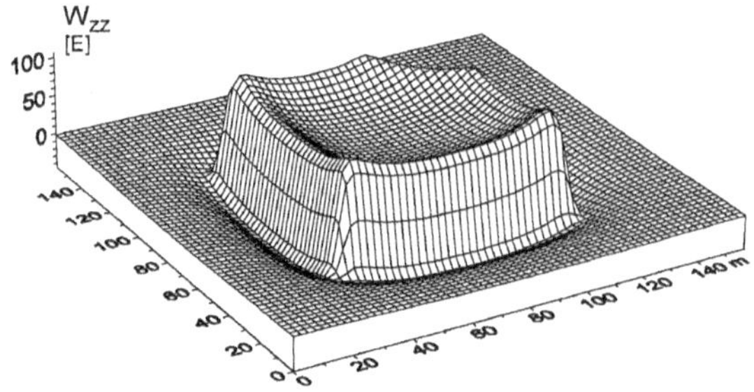


**Fig. 9** Changes of vertical gravity gradient $W_{zz}$

As it can be seen in Figures 3, 4, 5 and 6 the changes of curvature data are between -50 and +50 E at the height of 4 m above the water mass. The biggest changes of $W_{xx}$, $W_{yy}$, $W_{xy}$ are exactly above the edge of water mass, while in case of $W_\Delta$ at the corner points. The changes decrease strongly in the function of distance from the water mass. The changes of horizontal gradients of gravity $W_{zx}$ and $W_{zy}$ are two times bigger than the changes of curvature data. As it can be seen in Figures 7 and 8 the maximum changes are between -100 and +100 E above the edge of water mass. The changing of vertical gradient is more than 100 E above the water mass, which is about 3% of the normal value.

## 2 The gravitational effect of water level fluctuations of the Danube River

In August of 2002 there was a great flood of Danube River and gravity measurements were carried out to study the effect of water level fluctuations of the river. One station was located on the embankment at Szabadság Bridge in Budapest, the other one about 500 m to the west, at the foot of Gellért Hill where the water of Danube had already no significant gravitational effect (see Figure 10). It is important to remark, that the gravity stations are influenced not only by the level of the Danube, but by ground-water fluctuations as well.

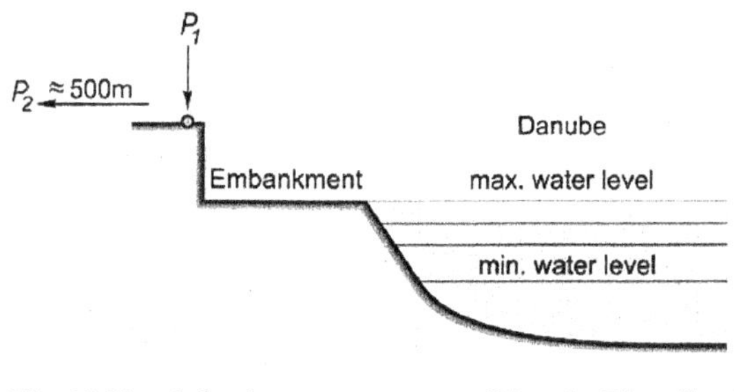


**Fig. 10** Sketch for the measurements of Danube River flood

Measurements were carried out by two LCR gravimeters on four subsequent days. Daily observations were carried out in a $P_1$–$P_2$–$P_1$–$P_2$–$P_1$–$P_2$ sequence. The maximum subsidence of the level of the Danube was 4.16 m and the respective gravity change 41±8 μGal (Csapó, Szabó, Völgyesi 2003).

To verify the measurements, model calculations were carried out applying the software Mod3D written by I. Cerovský (Cerovský et al, 2004). We have calculated the gravitational effect of the water mass of Danube flood using a three-dimensional model. The 3D model of the Danube flood can be seen in Figure 11. The heights of topography and the position of water mass were given a 10×10 m grid spacing. Dimension of the model is 640×640 m. Anomalous gravitational field components and the gravitational tensor elements were computed along the central cross sectional part (*M section*) of this model.

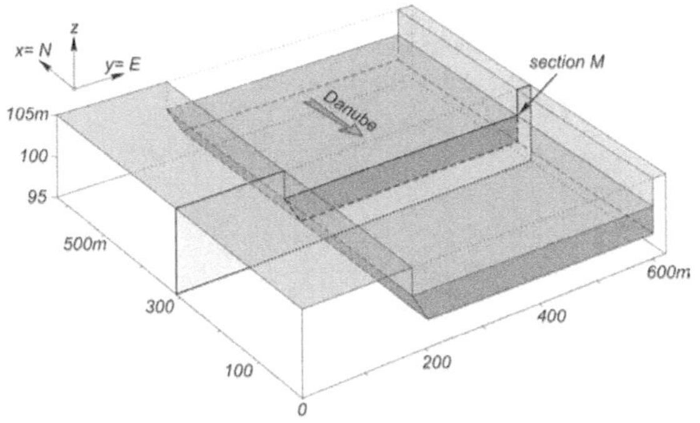


**Fig. 11** The 3D model of Danube River's water mass

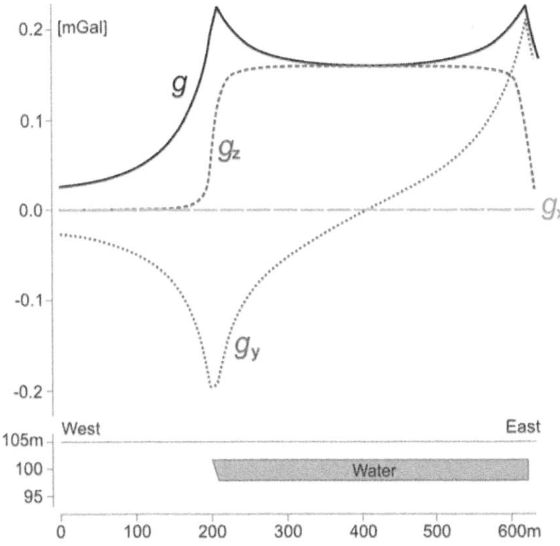


**Fig. 12** The gravitational field components $\mathbf{g}(g_x, g_y, g_z)$, computed along the central cross sectional part (*M section*) of the model.

In Figure 12, the gravity effects of the water masses versus distance are plotted in a section perpendicular to the river bank. In the lower part of the figure the section of water mass model is presented, while in the upper part the respective gravity effects. As it can be seen, in the immediate neighbourhood of the river bank the gravity effect can reach a value as high as 150 μGal, and decreases quickly moving away from the river bank. In case of 4 m high flood only a few μGal variation can be registered at a distance of 40-50 m. In point $P_1$, the gravity effect of 4 m high water mass was found to be about 40 μGal, which is in good agreement with the measured 41 μGal.

Elements of full gravitational tensor were computed too for the water mass of Danube flood. The variations of Eötvös tensor elements $W_{xx}$, $W_{yy}$, $W_{xy}$, $W_{zx}$, $W_{zy}$, $W_{zz}$ were computed along the central cross sectional part of the model. Results can be seen in Figure 13. The changes of gradients are very similar to the results referring to the water reservoir; the maximum values can be seen exactly along the river banks.

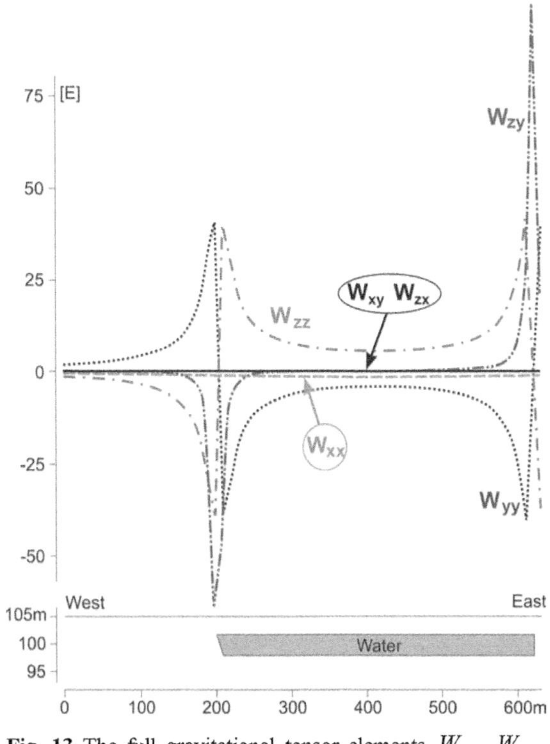


**Fig. 13** The full gravitational tensor elements $W_{xx}$, $W_{yy}$, $W_{zz}$, $W_{xy}$, $W_{zx}$, $W_{zy}$ in a section perpendicular to the river bank.

## 3 Conclusion

Investigation of time variation of gravity is important in gravimetry. Two case studies were considered where gravity field changes were detected resulting from water mass variations. One case is an urban water reservoir where the maximum daily change of water is 40000 $m^3$; the other is the water level fluctuations of the Danube River in Budapest during the great flood in 2002. In the case of water reservoir because of the rapid water level change, the gravity change can reach a value as high as 0.1-0.2 mGal, and gravity gradient changes may be 50-100 E in a few hours. This is the same case for the Danube River; the water level change may cause the same magnitude of variations of gravity and gravity gradients in a few days.

The other important conclusion of our results is: gravity stations should not be located at places where the movement of large volumes of water may be present. The measurements and model calculations prove that in the case of gravity base networks, locating stations near river banks or any other places where the changes of water mass may occur, one has to consider the gravity effect of water level fluctuations.

## Acknowledgements

We should thank for the funding of the above investigations to the National Scientific Research Fund (OTKA T-037929 and T-37880). In addition we should thank to I. Cerovský for the MOD3D gravity and magnetic modelling software.

## References

Benedek J. (2002): The application of polyhedron volume element in the calculation of gravity related quantities. Geomatika Közlemények V, MTA FKK Geodéziai és Geofizikai Kutatóintézet, Sopron, pp. 191-206. (In Hungarian)

Cerovský I., Meurers B., Pohánka V., Frisch W., Goga B. (2004): Gravity and magnetic 3D modeling software - Mod3D, in Meurers, B. and Pail, R. (eds): Proc. 1st Workshop on Int. Gravity Field Research, Österr. Beitr. Met. Geoph., 163-168.

Csapó G., Szabó Z., Völgyesi L. (2003): Changes of gravity influenced by water level fluctuations based on measurements, and model computations. Reports on Geodesy, Warsaw University of Technology, Vol. 64, Nr. 1, pp. 143-153.

GPSR Compliance

*The European Union's (EU) General Product Safety Regulation (GPSR) is a set of rules that requires consumer products to be safe and our obligations to ensure this.*

*If you have any concerns about our products, you can contact us on ProductSafety@springernature.com*

In case Publisher is established outside the EU, the EU authorized representative is:

Springer Nature Customer Service Center GmbH
Europaplatz 3
69115 Heidelberg, Germany

**Batch number: 10755939**

Printed by Printforce, the Netherlands